P9-DNQ-757

MANUAL OF THE GRASSES OF THE UNITED STATES

A. S. Hitchcock

Second Edition
Revised by AGNES CHASE

IN TWO VOLUMES

VOLUME TWO

DOVER PUBLICATIONS, INC.
NEW YORK

Published in Canada by General Publishing
Company, Ltd., 30 Lesmill Road, Don Mills,
Toronto, Ontario.
Published in the United Kingdom by Constable
and Company, Ltd., 10 Orange Street, London WC 2.

This Dover edition, first published in 1971, is an
unabridged republication of the second revised
edition, as published by the United States Government Printing Office in 1950 as U. S. Department
of Agriculture Miscellaneous Publication No. 200.
The first edition of the work was published in 1935.
For convenience in handling, the text is published
in two volumes in this paperback edition.

International Standard Book Number: 0-486-22718-9
Library of Congress Catalog Card Number: 70-142876

Manufactured in the United States of America
Dover Publications, Inc.
180 Varick Street
New York, N. Y. 10014

MANUAL OF THE GRASSES OF THE UNITED STATES

VOLUME I

MANUAL OF
THE GRASSES OF THE
UNITED STATES

TRIBE 12. PANICEAE

127. ANTHAENÁNTIA Beauv.

Spikelets obovoid; first glume wanting; second glume and sterile lemma about equal, 5-nerved, the broad internerves infolded, densely villous, the sterile lemma with a small palea and sometimes with a staminate flower; fertile lemma cartilaginous, brown, with narrow pale hyaline margins, boat-shaped, 3-nerved, subacute. Erect perennials with short creeping rhizomes, narrow, firm, flat blades, the uppermost much reduced, and narrow panicles, the slender branches ascending or appressed. Type species, *Anthaenantia villosa*. Name from Greek *anthos*, flower, and *enantios* contrary. (Beauvois misinterpreted the structure of the spikelet.)

In pine barrens *A. rufa* may be an important element in the natural pasture.

Blades erect or spreading, rather blunt or rounded at the apex, linear, folded at base; panicle usually purple.. 1. A. RUFA.
Blades ascending or spreading (on the average shorter and broader than in *A. rufa*), tapering to the apex, rounded at base; panicle usually pale............................... 2. A. VILLOSA.

1. Anthaenantia rúfa (Ell.) Schult. (Fig. 821.) Culms slender, 60 to 120 cm. tall; blades elongate, 3 to 5 mm. wide, often scabrous; panicle 8 to 15 cm. long, usually purple; spikelets 3 to 4 mm. long. ♃ —Moist pine barrens, Coastal Plain, North Carolina to Florida and eastern Texas.

2. Anthaenantia villósa (Michx.) Beauv. (Fig. 822.) Differing from *A. rufa* in the wider, mostly shorter, spreading blades and in the usually pale panicles. ♃ —Dry pine barrens, Coastal Plain, North Carolina to Florida and Texas.

FIGURE 821.—*Anthaenantia rufa*, × 1. (Amer. Gr. Natl. Herb. 290, N. C.)

128. TRICHÁCHNE Nees

(*Valota* Adans., inadequately published)

Spikelets lanceolate, in pairs, short-pediceled, in 2 rows along one side of a slender rachis; first glume minute, glabrous; second glume and sterile lemma about as long as the fruit, 3- to 5-nerved, copiously silky; fertile lemma cartilaginous, lanceolate, acuminate, usually brown, the flat white hyaline margins broad. Perennials with slender erect or ascending racemes, approximate to rather distant along a slender main axis, forming a white to brownish silky panicle. Type species, *Trichachne insularis*. Name from Greek *thrix* (*trich-*), hair, and *achne*, chaff, alluding to the silky spikelets.

Trichachne insularis is not relished by cattle, hence the name sourgrass by which it is called in the West Indies; *T. californica* is a constituent of the ranges of the Southwest, and furnishes fair forage.

Fruit 4 mm. long; spikelets tawny-villous.. 1. T. INSULARIS.
Fruit 3 mm. or less long (rarely 3.5 mm.); spikelets white-villous.
    Spikelets long-silky, the hairs exceeding the spikelet; fruit 3 to 3.5 mm. long.
        Panicle branches stiffly ascending or spreading, comparatively few-flowered; fruit
            oblong-lanceolate, gradually pointed... 3. T. PATENS.
        Panicle branches appressed, densely flowered; fruit obovate, abruptly pointed, the point
            scarcely indurate... 2. T. CALIFORNICA.
    Spikelets short-silky, the hairs not exceeding the spikelet; fruit 2.4 mm. long.
                                    4. T. HITCHCOCKII.

1. Trichachne insuláris (L.) Nees.

SOURGRASS. (Fig. 823.) Culms sub-erect from a hard scaly hairy swollen base, 1 to 1.5 m. tall; leaves numerous; the sheaths sparsely hirsute;

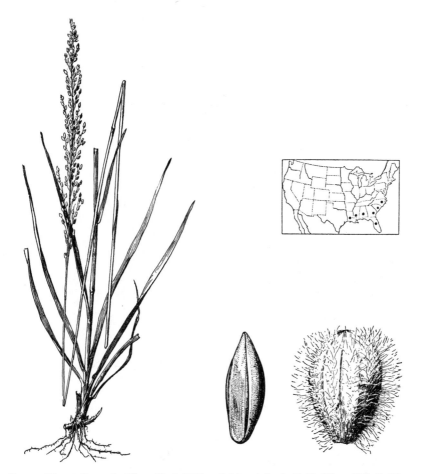

FIGURE 822.—*Anthaenantia villosa*. Plant, × ½; spikelet and floret, × 10. (Chase 4605, N. C.)

blades elongate, 8 to 15 mm. wide; panicle 15 to 30 cm. long, the slender racemes mostly 10 to 15 cm. long, somewhat nodding; spikelets approximate, excluding the hairs about 4 mm. long, the tawny hairs much exceeding them. ♃ —Low open ground and waste places, Florida, Alabama (Mobile), southern Texas, and southern Arizona; Mexico; West Indies to Argentina.

FIGURE 823.—*Trichachne insularis*. Plant, × ½; spikelet and floret, × 10. (Baker and Wilson 602, Cuba)

2. Trichachne califórnica (Benth.) Chase. COTTONTOP. (Fig. 824.) Culms erect from a knotty swollen felty-pubescent base, 40 to 100 cm. tall; leaves numerous, the sheaths glabrous to sparsely pilose; blades mostly less than 12 cm. long, 3 to 5 mm. wide, from nearly glabrous to densely puberulent; panicle mostly 5 to 10 cm. long, the few racemes usually 3 to 5 cm. long, occasionally longer, erect or nearly so; spikelets approximate, excluding the hairs 3 to 4 mm. long, the white to purplish hairs much exceeding them, often spreading, the middle internerves of the sterile lemma glabrous. ♃ (*T. saccharata* Nash.)—Plains and dry open ground, Texas and Oklahoma to Colorado, Arizona, and Mexico; South America.

3. Trichachne pátens Swallen. (Fig. 825.) Culms tufted, erect, 40 to 90 cm. tall; sheaths more or less papillose-pilose, the lowermost densely felty-pubescent; blades 5 to 15 cm. long, 1 to 4 mm. wide, scabrous; panicle 10 to 18 cm. long, the racemes stiffly ascending or spreading; spikelets remote, 4 mm. long, densely silky, the hairs exceeding the spikelet; fruit 3 mm. long, acute. ♃ —Dry fields, prairies, and roadsides, Texas.

4. Trichachne hitchcóckii (Chase) Chase. (Fig. 826.) Culms tufted and branching at base, leafy below, slender, 30 to 50 cm. tall; sheaths and

FIGURE 824.—*Trichachne californica*, × 1. (Hitchcock 13608, Tex.)

blades nearly glabrous to puberulent, sometimes densely so toward base, the blades 2 to 5 cm. long, 2 to 3 mm. wide; panicle long-exserted, 6 to 10 cm. long, the few racemes 3 to 4 cm. long, mostly rather remote and erect; spikelets 2.5 to 3 mm. long, densely silky-villous, the prominent nerves not hidden, the grayish hairs not exceeding the spikelet. ♃ —Dry plains, Texas; northern Mexico.

129. DIGITÁRIA Heister. CRABGRASS
(*Syntherisma* Walt.)

Spikelets in twos or threes, rarely solitary, subsessile or short-pediceled, alternate in 2 rows on one side of a 3-angled winged or wingless rachis; spikelets lanceolate or elliptic, nearly planoconvex; first glume minute or wanting; second glume equaling the sterile lemma or shorter; fertile lemma cartilaginous, the hyaline margins pale. Annual or perennial, erect to prostrate, often weedy grasses, the slender racemes digitate or approximate on a short axis. Type species, *Digitaria sanguinalis*. Name from Latin *digitus*, finger, alluding to the digitate inflorescence of the type species.

The species are in the main good forage grasses. *Digitaria sanguinalis*, the common crabgrass, is a weed in cultivated soil. In the Southern States, where it produces an abundant growth in late summer on fields from which crops have been gathered, it is utilized for forage and is sometimes cut for hay. This species and *D. ischaemum* are common weeds in lawns. They form a fine green growth at first but start late and die in the fall.

FIGURE 825.—*Trichachne patens.* Plant, ✕ 1; spikelet and floret, ✕ 10. (Reed 11, Tex.)

1a. Rachis winged or flat-margined, the margin as wide as the central rib; plants annual, creeping at least at base.
Rachis bearing scattered long fine hairs (these rarely wanting); spikelets narrow, acuminate, nearly glabrous.. 2. D. HORIZONTALIS.
Rachis not bearing hairs; spikelets elliptic, acute, pubescent.
Plants perennial, stoloniferous... 7. D. LONGIFLORA.
Plants annual. Culms erect or decumbent spreading.
Sheaths glabrous; fertile lemma brown.
Spikelets 2 mm. long, 1 mm. wide, the hairs or most of them capitellate.
3. D. ISCHAEMUM.
Spikelets 1.5 to 1.7 mm. long, about 0.6 mm. wide, the hairs not capitellate.
Sterile lemma with 5 distinct nerves; spikelets sparingly pubescent, 1.7 mm. long; fertile lemma light brown; racemes, if more than 2, not digitate.
4. D. FLORIDANA.
Sterile lemma with 3 distinct nerves; spikelets distinctly pubescent, 1.5 mm. long, fertile lemma dark brown, racemes usually all digitate.
5. D. VIOLASCENS.
Sheaths pilose or villous; fertile lemma pale.
Spikelets 1.5 to 1.7 mm. long; pedicels terete, glabrous............... 6. D. SEROTINA.
Spikelets 2.5 to 3.5 mm. long; pedicels angled, scabrous...... 1. D. SANGUINALIS.
1b. Rachis wingless or with a very narrow margin (see also *D. horizontalis*), triangular; plants not creeping (except in *D. texana*), annual or perennial.
2a. Fertile lemma pale or gray.
Plants annual, decumbent and rooting at base. Spikelets 3 mm. long, glabrous or nearly so... 8. D. SIMPSONI.
Plants perennial.
Spikelets densely or sparsely villous; racemes 5 to 10.
Spikelets 2.8 to 3.5 mm. long, sparsely to densely villous........... 14. D. RUNYONI.
Spikelets 2 to 2.5 mm. long, rather sparsely villous................ 13. D. TEXANA.
Spikelets glabrous to obscurely appressed-pubescent on the internerves; racemes 2 to 5, some of them naked at base for 1 to 1.5 cm.
First glume broad, hyaline, minute but obvious; spikelets 3.2 mm. long, glabrous.
15. D. PAUCIFLORA.
First glume obsolete or nearly so; spikelets 2.5 to 2.8 mm. long, obscurely to obviously appressed-pubescent.
Racemes 2 to 4; culms ascending from a curved base; sheaths papillose-pilose.
16. D. SUBCALVA.
Racemes 5 to 10; culms erect; sheaths conspicuously villous.
17. D. ALBICOMA.
2b. Fertile lemma dark brown. Plants erect or at least not rooting at the decumbent base; annual or sometimes apparently perennial.
Second glume and sterile lemma glabrous (see also *D. laeviglumis* under *D. filiformis*).
12. D. GRACILLIMA.
Second glume and sterile lemma capitellate-pubescent.
Spikelets 2 to 2.5 mm. long.. 10. D. VILLOSA.
Spikelets 1.5 to 1.7 mm. long.
Blades folded or involute, flexuous.................................. 11. D. DOLICHOPHYLLA.
Blades flat.. 9. D. FILIFORMIS.

1. Digitaria sanguinális (L.) Scop.

CRABGRASS. (Fig. 827.) Plant branching and spreading, often purplish, rooting at the decumbent base, the culms sometimes as much as 1 m. long, the flowering shoots ascending; sheaths, at least the lower, papillose-pilose; blades 5 to 10 mm. wide, pubescent to scaberulous; racemes few to several, 5 to 15 cm. long, rarely longer, digitate, with usually 1 or 2 whorls a short distance below; spikelets about 3 mm. long; first glume minute but evident; second glume about half as long as the spikelet, narrow, ciliate; sterile lemma strongly nerved, the lateral internerves appressed-pubescent, the hairs sometimes spreading at maturity (*D. fimbriata* Link); fertile lemma pale. ⊙ —Fields, gardens, and waste places, a troublesome weed in lawns and cultivated ground throughout the United States at low and medium altitudes, more common in the East and South; temperate and tropical regions of the world. Native of Europe. A specimen with nearly glabrous sheaths and inflorescences of 2 racemes collected by

FIGURE 826.—*Trichachne hitchcockii.* Plant, X 1; spikelet and floret, X 10. (Type.)

Tracy in Mississippi, said to be introduced, has been erroneously referred to *Syntherisma barbatum* (Willd.) Nash (*Digitaria barbata* Willd.).

DIGITARIA SANGUINALIS var. CILI-ÁRIS (Retz.) Parl. Sterile lemma, pectinate-ciliate, the stiff cilia 1.5 mm. long. Along railroad, Berks County, Pa. Waif from Asia.

2. Digitaria horizontális Willd. (Fig. 828.) Resembling *D. sanguinalis,* the culms more slender, the racemes mostly subracemose, very slender, lax, the rachis scarcely winged, bearing scattered long fine spreading hairs (these rarely wanting); spikelets narrow, about 2 mm. long; first glume minute or obsolete; second glume half as long as the spikelet. ⊙ (*Syntherisma setosum* Nash; *S. digitatum* Hitchc.)—Waste places, southern and central Florida; ballast, Mobile, Ala.; tropical regions of North America and South America.

3. Digitaria ischaémum (Schreb.) Schreb. ex Muhl. SMOOTH CRABGRASS.

(Fig. 829.) Erect or usually soon decumbent-spreading, resembling *D. sanguinalis* but not so coarse or tall; foliage glabrous, bluish or purplish; racemes mostly 2 to 6, 4 to 10 cm. long, the rachis with thin wings wider than the midrib; spikelets about 2 mm. long; first glume hyaline, obscure; second glume and sterile lemma as long as the dark fertile lemma, pubescent with capitellate hairs. ⊙ (*Syntherisma humifusum* Rydb.)— Waste places, often a troublesome weed in lawns. Quebec to Georgia, west to Washington and California; introduced from Eurasia. The first glume is so thin as to be apparently wanting. DIGITARIA ISCHAEMUM var. MISSISSIPPIÉNSIS (Gattinger) Fernald. Taller, the racemes mostly 5 to 7, often 10 or even 15 cm. long; first glume often more easily seen. ⊙ —Maryland, Indiana, Illinois, Virginia, Tennessee, South Carolina, and Georgia.

4. Digitaria floridána Hitchc. (Fig. 830.) Culms tufted, decumbent at base, 20 to 30 cm. tall; foliage glabrous except for a few long hairs around the mouth of the sheath; blades 4 to 7 cm. long, 3 to 6 mm. wide; racemes 3 or 4, rather distant on the axis, 3 to 6 cm. long, the rachis wings wider than the midrib; spikelets 1.5 to 1.7 mm. long, rather sparingly pubescent; first glume wanting; second glume and sterile lemma about as long as the light-brown fertile lemma. ⊙ —Sandy pine woods, Florida (Hernando County). The inflorescence resembles that of *D. filiformis,* but the rachis is winged; the spikelets are smaller than those of *D. ischaemum.*

5. Digitaria violáscens Link. (Fig. 831.) Annual or apparently perennial; culms numerous in a tuft, spreading at base, slender, 10 to 40 cm. tall; leaves mostly clustered near the base, the sheaths glabrous; blades flat, mostly less than 5 cm. long, 3 to 6 mm. wide, the upper culm blade distant, reduced; racemes slender, 2 to 5, usually 2 or 3, digitate or some-

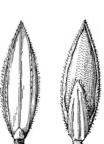

times approximate on a short axis 3 to 6 cm. long, at maturity spreading or curved, the rachis flat, winged, about 0.7 mm. wide; spikelets closely set, elliptic, acutish, minutely pubescent, about 1.5 mm. long; first glume wanting; second glume about three fourths as long as the spikelet; sterile lemma as long as the spikelet, with

FIGURE 827.—*Digitaria sanguinalis.* Plant, × ½; two views of spikelet, and floret, × 10. (Norton 566, Kans.)

FIGURE 828.—*Digitaria horizontalis.* Plant, × 1; spikelet and floret, × 10. (Nash 996, Fla.)

Figure 830.—*Digitaria floridana.* Plant, × 1; spikelet and fertile floret, × 10. (Type.)

Figure 829.—*Digitaria ischaemum.* Plant, × 1; spikelet and floret, × 10. (Jones 1761, Vt.)

three distinct nerves and 1 or 2 obscure pairs; fertile lemma acute, dark brown at maturity. ☉ ♃ —Open pineland in sandy soil, Indiana and Kentucky; Georgia and Florida to Arkansas and Texas; tropical America; tropical Asia.

6. Digitaria serótina (Walt.) Michx. (Fig. 832.) Creeping, sometimes forming extensive mats; flowering culms ascending or erect, 10 to 30 cm. tall; leaves crowded on the creeping culms, the blades short; sheaths villous; blades 2 to 8 cm. long, 3 to 7 mm. wide; racemes usually 3 to 5, slender, often arcuate, 3 to 10 cm. long, the rachis with thin wings wider than the midrib; spikelets pale, about 1.7 mm. long; first glume wanting; second glume about one-third as long as the sterile lemma, both finely pubescent; fertile lemma pale. ☉ —Pastures and waste places, Coastal Plain, Pennsylvania to Florida and Louisiana; Philadelphia (ballast); Cuba.

7. Digitaria longiflóra (Retz.) Pers. (Fig. 833.) Stoloniferous; culms ascending, 20 to 40 cm. tall, glabrous; sheaths glabrous; ligule membranaceous, 1 mm. long; blades 1 to 4 cm. long, 3 to 5 mm. wide, flat, glabrous; racemes 2 to 4, 3 to 8 cm. long, usually curved, the rachis flat, 0.5 to 0.8 mm. wide; spikelets 1.5 mm. long, elliptic, minutely pubescent. ♃ —Ditches and sandy ground, southern Florida; tropical regions of the Old World; introduced in the American Tropics.

8. Digitaria simpsóni (Vasey) Fernald. (Fig. 834.) Resembling *D. sanguinalis* in habit; sheaths papillose-pilose, those of the innovations compressed-keeled; blades not more than 6 mm. wide, softly pilose; racemes 4 to 8, ascending, pale, 8 to 12 cm. long, the triangular rachis narrowly margined; spikelets about 3 mm. long; first glume hyaline, obsolete or nearly so; second glume and sterile lemma finely 7- to 9-nerved, glabrous

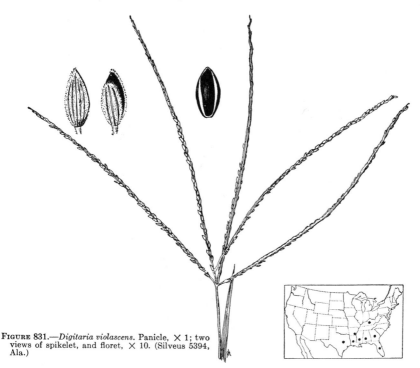

FIGURE 831.—*Digitaria violascens.* Panicle, × 1; two views of spikelet, and floret, × 10. (Silveus 5394, Ala.)

or very obscurely pubescent, barely exceeding the pale, slightly apiculate fertile lemma. ⊙ —Sandy fields, Florida, rare; Isla de Pinos, Cuba.

9. Digitaria filifórmis (L.) Koel. (Fig. 835, *A.*) Culms in small tufts, slender, usually erect, 10 to 60 cm. tall, rarely taller, those of a tuft very unequal; lower sheaths pilose, the upper mostly glabrous; blades erect, usually 5 to 15 cm. long (longer in more robust plants), 1 to 4 mm. wide; racemes mostly 1 to 5, unequal, erect or ascending, mostly less than 10 cm. long, somewhat distant, not fascicled; spikelets 1.5 to 1.7 mm. long; first glume wanting; second glume and sterile lemma pubescent with short capitellate hairs, sometimes nearly glabrous, the glume shorter than the spikelet; fertile lemma dark brown, slightly apiculate. ⊙ —Sandy fields and sterile open ground, New Hampshire to Iowa and Oklahoma, south to Florida, Texas, and Mexico. A form with

FIGURE 832.—*Digitaria serotina.* Plant, × 1; two views of spikelet, and floret, × 10. (Tracy 4653, Miss.)

FIGURE 833.—*Digitaria longiflora*, Plant, × ½. Stolon and panicle, × 1; spikelet and floret, × 10. (Silveus 4405, Fla.)

glabrous spikelets from Manchester, N. H., has been described as *D. laeviglumis* Fernald (835, *B.*).

10. Digitaria villósa (Walt.) Pers. (Fig. 836.) Perennial at least in the Southern States, in large tufts, purplish at base; culms 0.75 to 1.5 m. tall, rarely branching; sheaths, at least the lower, grayish villous, sometimes sparsely so; blades elongate, 3 to 6 mm. wide, often flexuous, from softly pilose to nearly glabrous; racemes 2 to 7, narrowly ascending, rarely somewhat spreading, very slender, usually 15 to 25 cm. long, rather distant, often naked at base, sometimes interrupted; spikelets 2 to 2.5 mm. long, usually densely pubescent with soft capitellate hairs, the hairs longer than in *D. filiformis*, and some-times only obscurely capitellate, the spikelets otherwise very like those of *D. filiformis*. ♀ —Sandy fields and woods, Maryland to Missouri, south to Florida and Texas; Cuba, Mexico. This species and *D. filiformis* seem to intergrade to some extent. Plants from peninsular Florida with less strongly pubescent sheaths, 2 to 4 elongate racemes, and spikelets with longer hairs have been distinguished as *D. leucocoma* (Nash) Urban.

11. Digitaria dolichophýlla Henr. (Fig. 837.) Slender wiry perennial, 50 to 115 cm. tall; blades elongate, folded or involute, flexuous, about 1 mm. wide; racemes mostly 1 to 3, erect, 5 to 20 cm. long, usually 10 to 20 cm., very slender, loosely flowered; spikelets about 1.5 mm.

long, the capitellate hairs rather stiff and appressed; fruit dark brown. ♃ (Has been confused with *D. panicea* (Swartz) Urban.)—Moist pine barrens and open ground, southern Florida; Cuba, Puerto Rico.

12. Digitaria gracíllima (Scribn.) Fernald. (Fig. 838.) Perennial in dense tufts; culms 60 to 100 cm. tall, erect; lower sheaths appressed-villous; blades elongate, 1 to 2 mm. wide,

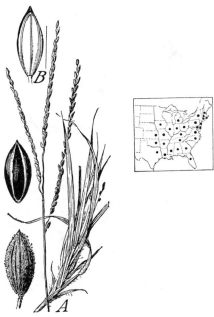

FIGURE 835.—*A, Digitaria filiformis*. Plant, × 1; spikelet and floret, × 10. (Bissell, Conn.) *B, D. laeviglumis*. Spikelet, × 10. (Type coll.)

often involute, more or less flexuous; racemes mostly 2 or 3, distant (rarely as many as 5 and fairly approximate), very slender; spikelets rather remote, relatively long pediceled, about 2.3 mm. long, glabrous; first glume ob-

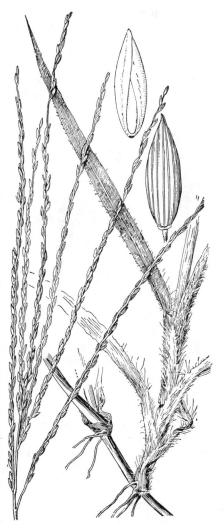

FIGURE 834.—*Digitaria simpsoni*. Plant, × 1; spikelet and floret, × 10. (Curtiss 6422, Fla.)

FIGURE 836.—*Digitaria villosa*. Plant, × 1; spikelet and floret, × 10. (Curtiss 5300, Fla.)

FIGURE 837.—*Digitaria dolichophylla.* Plant, × 1; spikelet and floret, × 10. (Tracy 9058, Fla.)

solete, the second one-fourth to half as long as the dark-brown fertile lemma; sterile lemma scarcely equaling the fruit. ♃ —Sandy soil, high pineland, peninsular Florida, rare. A tall plant from Grasmere with 3 to 5 racemes, the spikelets having second glumes about two-thirds as long as the fertile lemma, has been differentiated as *D. bakeri* (Nash) Fernald.

13. Digitaria texána Hitchc. (Fig. 839.) Perennial, erect or somewhat decumbent and branching at base; culms 30 to 60 cm. tall; lower sheaths, rarely all the sheaths, villous or velvety-pubescent, the uppermost glabrous; ligule prominent; blades flat, the lower villous, the upper glabrate, 10 to 15 cm. long, 3 to 5 mm. wide; racemes mostly 5 to 10, slender, pale, ascending or erect, 5 to 12 cm. long, the axis 1 to 4 cm. long; rachis angled, the scabrous margins much narrower than the whitish center; spikelets mostly rather distant, 2 to 2.5 mm. long, from short-villous to nearly glabrous, the silky hairs not at all capitellate; first glume obsolete; second glume and sterile lemma as long as the pale acute fertile lemma.

♃ —Sandy oak woods or sandy prairie, southern Texas.

14. Digitaria runyóni Hitchc. (Fig. 840.) Perennial; culms ascending, 40 to 70 cm. tall, the base often long-creeping and rooting, many-noded; sheaths densely villous or the upper glabrate; blades flat, the lower densely velvety-villous, the upper sparingly pilose or glabrous, mostly less than 10 cm. long, 3 to 6 mm. wide; racemes 5 to 10, on an axis 1 to 4 cm. long, mostly suberect, 7 to 12 cm. long, pale, sometimes naked at base, the rachis flat-triangular; spikelets narrowly lanceolate, acute, 2.8 to 3.5 mm. long; first glume minute or obsolete; second glume and sterile lemma equal, sparsely to densely villous on the internerves, the lemma

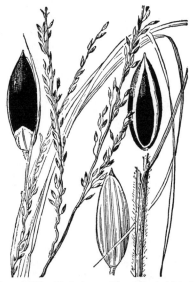

FIGURE 838.—*Digitaria gracillima.* Plant, × 1; two views of spikelet, and floret, × 10. (Type.)

glabrous on the middle internerves; fertile lemma acuminate, usually a little shorter than the spikelet, pale at maturity. ♃ —Sand dunes and sandy prairies along the coast, southern Texas.

15. Digitaria pauciflóra Hitchc. (Fig. 841.) Perennial; culms erect or somewhat decumbent at base, 0.5 to

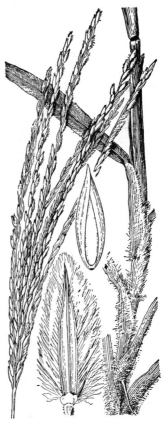

FIGURE 839.—*Digitaria texana.* Plant, X 1; spikelet and floret, X 10. (Type.)

FIGURE 840.—*Digitaria runyoni.* Plant, X 1; spikelet and floret, X 10. (Type.)

1 m. tall, very slender, sparingly branching; foliage grayish-villous, the blades 6 to 12 cm. long, about 2 mm. wide; racemes 2 or 3, ascending or erect, 5 to 11 cm. long, the filiform rachis naked for 1 to 1.5 cm. at base, or with distant abortive spikelets; spikelets rather distant, elliptic, about 3.2 mm. long, glabrous; first glume minute with a hyaline erose margin; second glume and sterile lemma finely nerved, as long as the grayish fertile lemma. ♃ —Pinelands, southern Florida.

16. Digitaria subcálva Hitchc. (Fig. 842.) Perennial; culms tufted, slender, ascending from a curved base, 40 to 100 cm. tall; sheaths papillose-pilose;

blades flat, scabrous, the lower pilose, 3 to 15 cm. long, 1 to 3 mm. wide; racemes 2 to 4, narrowly ascending, 5 to 12 cm. long, approximate, the rachis slender, triangular, mostly naked at base for 1 to 1.5 cm.; spikelets 2.5 to 2.8 mm. long, acute; first glume obsolete; second glume and sterile lemma slightly shorter than the acute pale or drab fruit, the internerves from obscurely to distinctly appressed silky-pubescent. ♃ — Known only from Plant City, Fla.

17. Digitaria albicóma Swallen. (Fig. 843.) Culms 65 to 75 cm. tall, erect, simple or branched at the base; lower sheaths densely villous, the upper elongate, glabrous or papillose-

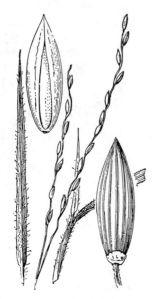

FIGURE 841.—*Digitaria pauciflora.* Plant, X 1; spikelet and floret, X 10. (Type.)

pilose toward the base; blades 10 to 30 cm. long, 3 to 5 mm. wide, pilose, the margins scabrous; racemes 5 to 9, 8 to 12 cm. long, ascending or spreading, naked at base; spikelets solitary or paired, 2.5 to 2.8 mm. long, glabrous, one subsessile, the other pedicellate; first glume obsolete; the second narrow, 3-nerved; sterile lemma as long as the fruit, 5- to 7-nerved; fruit 2.5 to 2.8 mm. long, dark brown. ⨅ —Open sandy woods. Known only from Chinsegut Hill Sanctuary, Brooksville, Fla.

DIGITARIA PÉNTZII Stent. Culms densely tufted, erect, stoloniferous, with conspicuously hairy sheaths; racemes few to several, ascending to spreading, approximate on a short axis; spikelets about 3 mm. long, villous, the first glume well developed. ⨅ —Introduced from South Africa. On trial as a pasture grass in the Southern States.

DIGITARIA DECUMBENS Stent. Similar to *D. pentzii,* extensively stoloniferous or creeping, the culms less densely tufted and more leafy; sheaths nearly glabrous; racemes spreading at maturity; spikelets 2.7 to 3 mm. long, glabrous or sparingly silky on the internerves. ⨅ —Introduced from South Africa, and grown as a pasture grass in Florida and southern California. This and the preceding are not known to set seed and are planted by cuttings.

FIGURE 842.—*Digitaria subcalva.* Plant, X 1; spikelet and floret, X 10. (Type.)

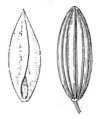

FIGURE 843.—*Digitaria albicoma.* Spikelet and floret, X 10. (Type.)

DIGITARIA SWAZILANDENSIS Stent. Culms tufted, compressed, leafy, 25 to 50 cm. tall, erect to spreading, and with slender wiry stolons, hairy at the nodes; blades flat, rather soft; racemes 2 or 3, digitate, pale, 5 to 8 cm. long; spikelets 2.3 mm. long; first glume minute, the second half as long as the spikelet; sterile lemma strongly nerved, obscurely pilose on the margin; fruit drab at maturity. ⨅ —Introduced from South Africa. Grown at experiment stations, Tifton, Ga., and Gainesville, Fla.

130. LEPTOLÓMA Chase

Spikelets on slender pedicels; first glume minute or obsolete; second glume 3- to 5-nerved, nearly as long as the 5- to 7-nerved sterile lemma, a more or less prominent stripe of appressed silky hairs down the internerves and margins of each, the sterile lemma empty or enclosing a minute nerveless rudimentary palea; fertile lemma cartilaginous, elliptic, acute, brown, the delicate hyaline margins enclosing the palea. Branching perennials with brittle culms, felty-pubescent at base, flat blades, and open or diffuse panicles, these breaking away at maturity, becoming tumbleweeds. Type species, *Leptoloma cognatum.* Name from Greek *leptos,* thin, and *loma,* border, alluding to the thin margins of the lemma.

Spikelets 2.5 to 3 mm. long; culms spreading from a knotty, often densely hairy, base.
1. L. COGNATUM.
Spikelets 4 mm. long; plants branching at base, producing long slender rhizomes.
2. L. ARENICOLA.

1. Leptoloma cognátum (Schult.) Chase. FALL WITCHGRASS. (Fig. 844.) Ascending from a decumbent knotty often densely hairy base, often forming large bunches, pale green, leafy; culms 30 to 70 cm. long; blades mostly less than 10 cm. long, 2 to 6 mm. wide, rather rigid; panicle one-third to half the entire height of the plant, purplish and short-exserted at maturity, very diffuse, the capillary branches soon widely spreading, pilose in the axils, the spikelets solitary on long capillary pedicels, narrowly elliptic, 2.5 to 3 mm. long, abruptly acuminate. ♃ (*Panicum cognatum* Schult., *Panicum autumnale* Bosc.)— Dry soil and sandy fields, New Hampshire to Minnesota, south to Florida, Texas, and Arizona; northern Mexico. A fairly palatable grass.

2. Leptoloma arenícola Swallen. (Fig. 844A.) Culms 30 to 40 cm. long, branching at base, with slender rhizomes as much as 50 cm. long, sometimes branching, the scales thin, softly pubescent; lower sheaths and blades softly pubescent, the upper glabrous; blades flat, 4 to 13 cm. long, 2 to 4 mm. wide; panicle nearly half the entire height of the plant, at maturity wider than long, few-flowered, the branches stiffly spreading, scabrous, bearing 2 to 5 spikelets near the ends and a few long stiff capillary 1-flowered branchlets, the lower bearing in addition 1 to few sterile branchlets; spikelets narrowly elliptic, acu-

minate, 4 mm. long, with 5 to 7 pale nerves, the internerves densely silky with appressed dark-purple hairs; fertile lemma 3.4 mm. long, dark brown with pale hyaline margins. ♃ —Sand hills, Kennedy County, Tex.

131. STENOTÁPHRUM Trin.

Spikelets embedded in one side of an enlarged and flattened corky rachis tardily disarticulating toward the tip at maturity, the spikelets remaining attached to the joints; first glume small; second glume and sterile lemma about equal, the latter with a palea or staminate flower; fertile lemma chartaceous. Creeping stoloniferous perennials, with short flowering culms, rather broad and short obtuse blades, and terminal and axillary racemes. Type species, *Stenotaphrum glabrum* Trin. Name from Greek, *stenos,* narrow, and *taphros,* trench, referring to the cavities in the rachis.

1. Stenotaphrum secundátum (Walt.) Kuntze. ST. AUGUSTINE GRASS. (Fig. 845.) Culms branching, compressed, the flowering shoots 10 to 30 cm. tall; blades mostly less than 15 cm. long, longer on the innovations, in rich soil 4 to 10 mm. wide; racemes 5 to 10 cm. long; spikelets solitary or in pairs, rarely threes, 4 to 5 mm. long. ♃ —Moist, especially mucky soil, mostly near the seashore,

South Carolina to Florida and Texas; American Tropics. Cultivated as a lawn grass in the coastal cities; also in Marin County, Calif., and escaping. The lawns have a coarse texture but are otherwise satisfactory. Propagated by cuttings of the stolons. A variegated form with leaves striped with white is used as a basket plant. Called by gardeners var. *variegatum*.

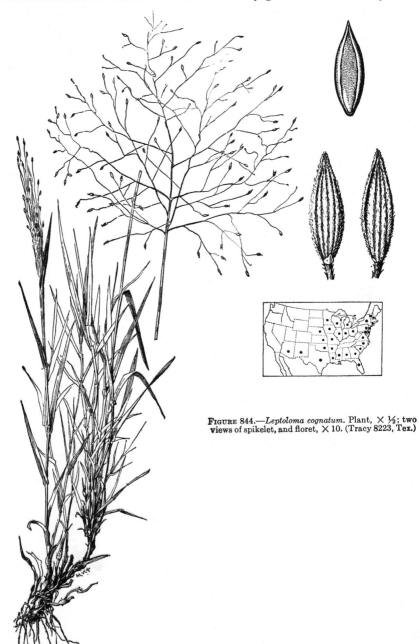

FIGURE 844.—*Leptoloma cognatum*. Plant, × ½; two views of spikelet, and floret, × 10. (Tracy 8223, Tex.)

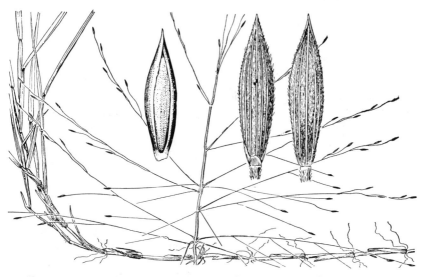

FIGURE 844A.—*Leptoloma arenicola.* Base and panicle, X ½; spikelet and floret, X 10. (Type.)

132. ERIÓCHLOA H. B. K.   CUPGRASS

Spikelets more or less pubescent, solitary or sometimes in pairs, short-pediceled or subsessile, in two rows on one side of a narrow rachis, the back of the fertile lemma turned from the rachis; lower rachilla joint thickened, forming a more or less ringlike, usually dark-colored callus below the second glume, the first glume reduced to a minute sheath about this and adnate to it; second glume and sterile lemma about equal, the lemma usually enclosing a hyaline palea or sometimes a staminate flower; fertile lemma indurate, minutely papillose-rugose, mucronate or awned, the awn often readily deciduous, the margins slightly inrolled. Annual or perennial, often branching grasses, with terminal panicles of several to many spreading or appressed racemes, usually approximate along a common axis. The species are called cupgrasses because of the tiny cup made by the first glume at the base of the spikelet. Type species, *Eriochloa distachya* H. B. K. Name from Greek *erion*, wool, and *chloa*, grass, alluding to the pubescent spikelets and pedicels.

A West Indian species, *E. polystachya* H. B. K. (*E. subglabra* (Nash) Hitchc.), called malojilla in Puerto Rico, is used for forage. This has been tried along the Gulf Coast from Florida to southern Texas and has given excellent results in southern Florida and at Biloxi, Miss. It is similar in habit to Para grass, producing runners but less extensively, is suited to grazing, and will furnish a good quality of hay. It will not withstand either cold or drought. The name carib grass has been proposed for it. In Arizona *E. gracilis* has some value for forage in the national forests.

Spikelets, including slender awns, 7 to 10 mm. long................................ 1. E. ARISTATA.
Spikelets not more than 6 mm., awnless or awn-tipped.
    Pedicels with erect hairs at least half as long as the spikelet, racemes dense, erect or appressed; spikelets relatively blunt (see also *E. gracilis*).
        Blades 2 to 3 mm. wide, elongate.. 2. E. SERICEA.
        Blades 5 to 15 mm. wide, not more than 15 cm. long........................ 3. E. LEMMONI.
    Pedicels scabrous or short-pubescent; spikelets acuminate or acute.
        Plants perennial.

Rachis velvety to villous; spikelets narrowly ovate......................... 8. E. MICHAUXII.
Rachis scabrous only; spikelets lanceolate...................................... 7. E. PUNCTATA.
Plants annual.
  Rachis scabrous only; racemes slender. Introduced..................... 4. E. PROCERA.
  Rachis pubescent; racemes stouter.
    Blades glabrous; fruit apiculate....................................... 5. E. GRACILIS.
    Blades pubescent; fruit with an awn about 1 mm. long........... 6. E. CONTRACTA.

1. Eriochloa aristáta Vasey. (Fig. 846.) Annual; culms erect or spreading at base, 50 to 80 cm. tall; blades flat, mostly 10 to 12 mm. wide, gla-

brous or scabrous; racemes several, ascending, overlapping, 3 to 4 cm. long, the rachis pilose, the pedicels bearing several long stiff hairs; spikelets about 5 mm. long, the glume and sterile lemma tapering into awns (awn of the glume about as long as the spikelet), appressed-villous on the lower half or two-thirds, the upper part scaberulous only; fruit 3.5 mm. long, apiculate. ☉ —Open ground, Arizona and California (near Yuma); northern Mexico.

2. Eriochloa serícea (Scheele) Munro. (Fig. 847.) Perennial, in dense tufts; culms simple, erect, 50 to 100 cm. tall, the lowermost sheaths felty-pubescent; blades elongate, 2 to 3 mm. wide, flat or mostly involute,

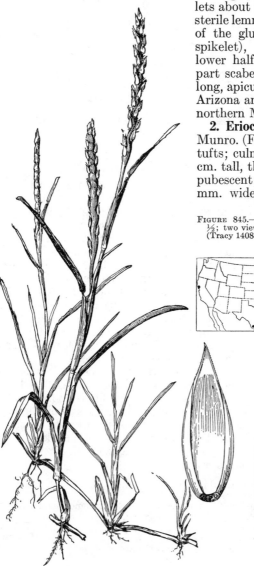

FIGURE 845.—*Stenotaphrum secundatum*. Plant, ×
½; two views of spikelet, and fertile floret, × 10.
(Tracy 1408, Miss.)

densely puberulent at the junction with the sheath; racemes several, appressed, somewhat distant, usually not overlapping, mostly 1.5 to 3 cm. long, the rachis hirsute, the pedicels with copious stiff hairs half as long as the spikelet; spikelets 4 mm. long, rather turgid, short-villous, the glume and sterile lemma acutish; fruit 3 mm. long, apiculate. ♃ —Prairies and hills, Texas and Oklahoma.

3. Eriochloa lemmóni Vasey and Scribn. (Fig. 848.) Annual; culms decumbent at base, 30 to 60 cm. tall; blades flat, only the larger as much as 15 cm. long, 5 to 15 mm. wide, velvety-pubescent on both surfaces; racemes erect, the upper overlapping, 1.5 to 3 cm. long, the axis and rachis densely villous, the pedicels with several long hairs; spikelets 4 mm. long, rather turgid, villous except the apex, abruptly narrowed to a short obtuse point; fruit 3 mm. long, slightly apiculate. ⊙ —Canyons, southern Arizona and northern Mexico.

4. Eriochloa prócera (Retz.) C. E. Hubb. (Fig. 849.) Annual; culms spreading at base, 40 to 60 cm. tall; blades flat, 2 to 4 mm. wide; racemes loose, slender, ascending, 3 to 5 cm.

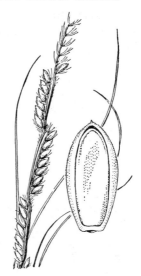

FIGURE 847.—*Eriochloa sericea*. Plant, × 1; floret, × 10. (Reverchon 1170, Tex.)

FIGURE 848.—*Eriochloa lemmoni*. Plant, × 1; floret, × 10. (Peebles and Harrison 4703, Ariz.)

long, the rachis scabrous only; spikelets 3 to 3.5 mm. long, appressed-pubescent, except toward the tip, the glume and sterile lemma acuminate; fruit 2 mm. long, the slender awn about 0.5 mm. long. ⊙ (*E. ramosa*

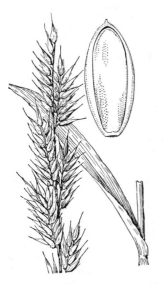

FIGURE 846.—*Eriochloa aristata*. Plant, × 1; floret, × 10. (Thornber 98, Ariz.)

Kuntze.)—Introduced on the university campus at Tucson, Ariz.; Cuba; tropical Asia.

FIGURE 849.—*Eriochloa procera*, ✕ 10. (Griffiths 1516, Ariz.)

5. Eriochloa grácilis (Fourn.) Hitchc. (Fig. 850.) Annual; culms erect or decumbent at base, 40 to 100 cm. tall; blades flat, glabrous, mostly 5 to 10 mm. wide; racemes several to numerous, approximate, ascending to slightly spreading, 2 to 4 cm. long, the axis and rachis softly pubescent, the pedicels short-pilose; spikelets 4 to 5 mm. long, rather sparsely appressed-pubescent, acuminate, or the glume sometimes tapering into an awn-point as much as 1 mm. long; sterile lemma empty; fruit about 3 mm. long, apiculate. ☉ —Open ground, often a weed in fields, Oklahoma and western Texas to southern California, south through the highlands of Mexico. (This species has been referred to *E. acuminata* (Presl) Kunth, an unidentified species of Mexico.)

ERIOCHLOA GRACILIS var. MÍNOR (Vasey) Hitchc. Mostly smaller, with more crowded, less acuminate spikelets, the pedicels with a few long hairs at the summit; fertile lemma about as long as the glume and sterile lemma (excluding the short points), obtuse or slightly apiculate. ☉ —Open ground, Texas, New Mexico, and Arizona; Mexico.

6. Eriochloa contrácta Hitchc. PRAIRIE CUPGRASS. (Fig. 851.) Annual; culms erect or sometimes decumbent at base, pubescent at least about the nodes, 30 to 70 cm. tall; blades pubescent, usually not more than 5 mm. wide; panicle usually less than 15 cm. long, contracted, cylindric, the racemes appressed, closely overlapping, 1 to 2 cm. long, the axis and rachises villous; spikelets 3.5 to 4 mm. long, excluding the awn-tip, appressed-villous; glume awn-tipped; sterile lemma slightly shorter, acuminate, empty; fruit 2 to 2.5 mm. long, with an awn nearly 1 mm. long. ☉ —Open ground, ditches, low fields, and wet places, Nebraska to Colorado, Louisiana and Arizona; introduced in Missouri and Virginia. Differing from *E. gracilis* in the pubescent foliage, subcylindric panicle, and the awned fruit.

7. Eriochloa punctáta (L.) Desv. (Fig. 852.) Perennial; culms in tufts, usually 50 to 100 cm. tall; blades flat, mostly 5 to 10 mm. wide, glabrous; racemes several, ascending, overlapping, 3 to 5 cm. long, the axis, rachises, and pedicels scabrous only; spikelets 4 to 5 mm. long, lanceolate, rather sparsely appressed-pilose; glume tapering to an awn-point about 1 mm. long; sterile lemma a little shorter than the glume, empty; fruit about half as long as the glume, with an awn 1 mm. long or more. ♃ —Marshes, river banks, and moist ground, southwestern Louisiana and southern Texas; American Tropics.

8. Eriochloa michaúxii (Poir.) Hitchc. (Fig. 853.) Perennial; culms erect, rather stout, 60 to 120 cm. tall; blades flat or, on the innovations, sometimes involute, elongate, 2 to 14 mm. wide, usually less than 1 cm., glabrous; racemes ascending or spreading, usually numerous, 3 to 5 or even to 15 cm. long, the axis 15 to 30 cm. long, this and the rachises densely velvety-pubescent; spikelets narrowly ovate, 4 to 5 mm. long, appressed-villous, acute; sterile floret usually with a well-developed palea and stamens; fruit 3 to 4 mm. long, hirsutulous at apex, apiculate or with an awn not more than 0.3 mm. long. ♃ (*E. mollis* Kunth.)—Brackish or fresh meadows and marshes and sandy

FIGURE 850.—*Eriochloa gracilis*, Plant, × ½; two views of spikelet, and floret, × 10. (McDougal, Ariz.)

FIGURE 851.—*Eriochloa contracta*. Panicle, × 1; floret, × 10. (Hitchcock 13420, Tex.)

prairies, southeastern Georgia and Florida. A form with narrow blades

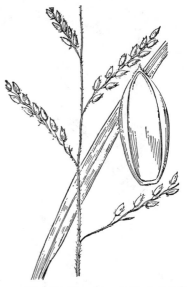

FIGURE 853.—*Eriochloa michauxii*. Plant, × 1; floret, × 10. (Amer. Gr. Natl. Herb. 297, Fla.)

and relatively few racemes, the axis and rachis puberulent, has been described as *E. mollis* var. *longifolia* Vasey. It grades into the typical form with broader blades and more numerous racemes; the sterile floret contains a staminate flower.

ERIOCHLOA MICHAUXII var. SIMPSÓNI Hitchc. Resembling the narrow-leaved form of the species; racemes few, appressed; sterile lemma empty. ♃ —Moist places, Fort Myers to Cape Sable, Fla.

ERIOCHLOA VILLÓSA (Thunb.) Kunth. Tall annual with few to several racemes, the rachis and pedicels very woolly, the rather blunt, turgid pubescent spikelets about 5 mm. long. ☉ —Ballast, near Portland, Oreg., occasionally cultivated; adventive in Colorado. Eastern Asia. (Had been confused with *E. nelsoni* Scribn. and Smith of Mexico.)

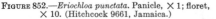

FIGURE 852.—*Eriochloa punctata*. Panicle, × 1; floret, × 10. (Hitchcock 9661, Jamaica.)

133. BRACHIÁRIA (Trin.) Griseb.

Spikelets solitary, rarely in pairs, subsessile, in 2 rows on one side of a 3-angled, sometimes narrowly winged rachis, the first glume turned toward the rachis; first glume short to nearly as long as the spikelet; second glume and sterile lemma about equal, 5- to 7-nerved, the lemma enclosing a hyaline palea and sometimes a staminate flower; fertile lemma indurate, usually papillose-rugose, the margins inrolled, the apex rarely mucronate or bearing a short awn. Branching and spreading annuals or perennials, with linear blades and several spreading or appressed racemes approximate along a common axis.

Type species, *Brachiaria erucaeformis*. Name from Latin *brachium*, arm, alluding to the armlike racemes.

Spikelets densely silky-pubescent; plants perennial............................ 1. B. CILIATISSIMA.
Spikelets glabrous; plants annual.
   Spikelet flat-beaked beyond the fruit.................................... 2. B. PLATYPHYLLA.
   Spikelet not beaked beyond the fruit.................................... 3. B. PLANTAGINEA.

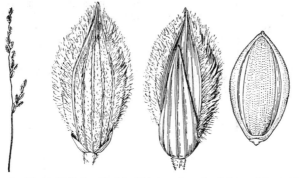

FIGURE 854.—*Brachiaria ciliatissima*. Panicle, × 1; two views of spikelet, and floret, × 10. (Type.)

1. Brachiaria ciliatíssima (Buckl.) Chase. (Fig. 854.) Perennial, producing long leafy stolons with short internodes, rooting at the swollen nodes, the blades short, firm, divaricately spreading; flowering culms erect or ascending, 15 to 40 cm. tall, the nodes bearded; sheaths sparsely to densely pilose; blades 3 to 7 cm. long, 3 to 5 mm. wide, tapering to a sharp point, usually ciliate along the lower part of the thick white margin; panicle finally long-exserted, 3 to 6 cm. long, the few branches erect or ascending, 1 to 2 cm. long; spikelets 4 mm. long; first glume three-fourths the length of the spikelet, glabrous; second glume and sterile lemma about equal, 5-nerved, the marginal part densely white-silky; fruit 3 mm. long. ♃ —Open sandy ground, Texas, Oklahoma, and Arkansas (Benton County).

2. Brachiaria platyphýlla (Griseb.) Nash. (Fig. 855.) Annual; culms decumbent, rooting at the lower nodes; blades rather thick, 4 to 12 cm. long, 6 to 12 mm. wide; panicle short-exserted or included at base; racemes 2 to 6, distant, 3 to 8 cm. long, ascending or spreading, the rachis

winged, 2 mm. wide; spikelets ovate, 4 to 4.5 mm. long, about 2 mm. wide; first glume scarcely one-third the length of the spikelet, blunt; second glume and sterile lemma equal, exceeding the fruit and forming a flat beak beyond it, 3- to 5-nerved, with transverse veinlets toward the summit; fruit 3 mm. long, elliptic, papillose-roughened. ☉ (*B. extensa* Chase.)—Low, sandy, open ground, Georgia, Florida; Missouri; Arkansas, southern Louisiana, Texas, and Oklahoma; Cuba.

3. Brachiaria plantagínea (Link) Hitchc. (Fig. 856.) Resembling *B. platyphylla*, more widely creeping, usually taller, blades commonly wider; rachis 1 to 1.5 mm. wide, the margins infolded; first glume strongly clasping; transverse veinlets wanting or obscure on the second glume and sterile lemma, these not pointed beyond the fruit. ☉ —Open, mostly moist, ground, Metcalf, Ga.; ballast, Philadelphia, Pa., and Camden, N. J.; Mexico to Bolivia and Brazil.

Brachiaria erucaefórmis (J. E. Smith) Griseb. (Fig. 857.) Spreading annual with rather delicate erect racemes and pubescent spikelets 2.5

mm. long. ⊙ —Has been cultivated in grass gardens, occasionally
escaped. Old World.

Brachiaria subquadripára (Trin.)
Hitchc. Creeping leafy perennial;
culms 25 to 60 cm. long; blades
flat, 5 to 10 cm. long, 4 to 8 mm.
wide; racemes mostly 3 to 5, spreading, rather distant; spikelets 3.5 to
4 mm. long, elliptic, glabrous. ♃
—Occasionally planted in southern
Florida, thriving in dry weather and
showing some promise as a forage
grass. Asia.

Figure 855.—*Brachiaria platyphylla*. Plant, × ½; two views of spikelet, and floret, × 10. (Nealley, Tex.)

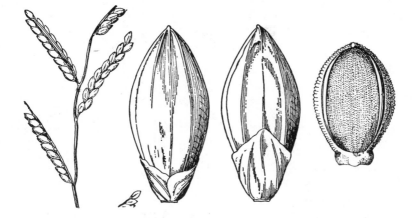

FIGURE 856.—*Brachiaria plantaginea.* Panicle, × 1; two views of spikelet, and floret, × 10. (Pringle 3904, Mex.)

FIGURE 857.—*Brachiaria erucaeformis.* Panicle, × 1; two views of spikelet, and floret, × 10. (Cult.)

134. AXÓNOPUS Beauv.

Spikelets depressed-biconvex, not turgid, oblong, usually obtuse, solitary, subsessile, and alternate, in 2 rows on one side of a 3-angled rachis, the back of the fertile lemma turned from the rachis; first glume wanting; second glume and sterile lemma equal, the lemma without a palea; fertile lemma and palea indurate, the lemma oblong-elliptic, usually obtuse, the margins slightly inrolled. Stoloniferous or tufted perennials, rarely annuals, with usually flat or folded, abruptly rounded or somewhat pointed blades, and few or numerous, slender spikelike racemes, digitate or racemose along the main axis. Type species, *Axonopus compressus.* Name from Greek *axon,* axis, and *pous,* foot.

One of the species, *A. affinis,* is a predominant pasture grass in the alluvial or mucky soil of the southern Coastal Plain. It is of little importance on sandy soil and does not thrive on the uplands. *Axonopus compressus* is used as a lawn grass, for which purpose it is propagated by setting out joints of the stolons.

Spikelets 4 to 5 mm. long, glabrous; midnerve of glume and sterile lemma evident.
1. A. FURCATUS.
Spikelets 2 to 3 mm. long, sparsely appressed-silky; midnerve of glume and sterile lemma suppressed.
Second glume and sterile lemma scarcely, if at all, pointed beyond the fruit; blades 2 to 4 mm., rarely to 6 mm., wide; nodes glabrous .... 3. A. AFFINIS.
Second glume and sterile lemma distinctly pointed beyond the fruit; blades mostly 8 to 10 mm. wide; nodes often bearded .... 2. A. COMPRESSUS.

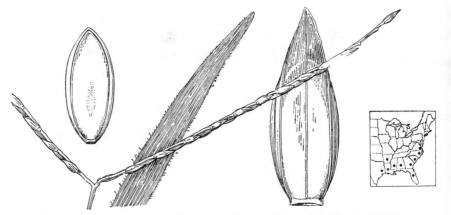

FIGURE 858.—*Axonopus furcatus*. Plant, X 1; spikelet and floret, X 10. (Combs 1205, Fla.)

1. Axonopus furcátus (Flügge) Hitchc. (Fig. 858.) Plants stoloniferous; culms compressed, tufted, erect, or decumbent at base, 40 to 100 cm. tall; blades flat, mostly 5 to 10 mm. wide, glabrous, ciliate, or even hirsute; racemes 2, digitate, rarely a third below, spreading, 5 to 10 cm. long; spikelets 4 to 5 mm. long (rarely less), glabrous, acute, glume and sterile lemma 5-nerved; fruit about two-thirds as long as the spikelet. ♃ —Marshes, river banks, and moist pine barrens, on the Coastal Plain, southeastern Virginia to Florida, Texas, and Arkansas. (The name *Anastrophus paspaloides* has been misapplied to this species. *Digitaria paspalodes* Michx., upon which it is based, is *Paspalum distichum* L.)

2. Axonopus compréssus (Swartz) Beauv. (Fig. 859.) Stoloniferous; culms 15 to 50 cm. tall, relatively stout, compressed, the nodes usually densely pubescent; stolons elongate with short internodes and short, broad, obtuse blades; culm blades 8 to 25 cm. long, mostly 8 to 12 mm. wide, the uppermost greatly reduced, the margins ciliate; racemes 2 to 5, mostly 4 to 8 cm. long, ascending, the upper two conjugate, the others remote on the axis; spikelets 2.2 to 2.5, occasionally to 2.8, mm. long, sparsely pilose, the second glume and

sterile lemma distinctly pointed beyond the fruit. ♃ —Moist ground, roadsides, and waste places, southern Florida and Louisiana; Mexico and the West Indies to Bolivia and Brazil.

3. Axonopus affínis Chase. (Fig. 860.) Tufted or stoloniferous; culms slender, glabrous, 25 to 35 cm. tall, rarely as much as 75 cm., sometimes forming dense mats; sheaths compressed, keeled; blades as much as 28 cm. long, usually less than 15 cm., 2 to 6 mm. wide, flat or folded; racemes 2 to 4, 2 to 10 cm. long, ascending; spikelets 2 mm. long, oblong-elliptic, subacute, the second glume and sterile lemma covering the fruit or slightly pointed beyond it, sparsely silky-pilose. ♃ —Moist mucky or sandy meadows, open woods and waste places, North Carolina to Florida and west to Oklahoma and Texas; Cuba and southern Mexico; Venezuela and Colombia to Argentina. Naturalized and common in Australia.

135. REIMARÓCHLOA Hitchc.

Spikelets strongly dorsally compressed, lanceolate, acuminate, rather distant, subsessile, and alternate in 2 rows along one side of a narrow, flattened rachis, the back of the fertile lemma turned toward it; both glumes wanting, or the second sometimes present in the terminal spike-

FIGURE 860.—*Axonopus affinis.* Two views of spikelet, and floret, × 10. (Type.)

FIGURE 859.—*Axonopus compressus.* Plant, × ½; two views of spikelet, and floret, × 10. (Combs 413, Fla.)

free nearly half its length. Spreading or stoloniferous perennials, with flat blades and slender racemes, these subdigitate or racemose along a short axis, stiffly spreading or reflexed at maturity. Type species, *Reimaria acuta* Flügge (*Reimarochloa acuta* Hitchc.). Named for J. A. H. Reimarus, and Greek *chloa*, grass.

1. Reimarochloa oligostáchya (Munro) Hitchc. (Fig. 861.) Glabrous; culms compressed, often long-decumbent and rooting at the lower nodes, the flowering shoots, 20 to 40 cm. tall; sheaths loose; blades 2 to 4 mm. wide; racemes 1 to 4, mostly 2 or 3, 5 to 8 cm. long; spike-

let; sterile lemma about equaling the fruit, the sterile palea obsolete; fertile lemma scarcely indurate, faintly nerved, acuminate, the margins inrolled at the base only, the palea lets about 5 mm. long. ♃ (*Reimaria oligostachya* Munro.)—In water or wet soil, Florida; Cuba. In general aspect resembles *Paspalum vaginatum* Swartz.

FIGURE 861.—*Reimarochloa oligostachya.* Plant, X 1; two views of spikelet, and floret, X 10. (Curtiss 3596A, Fla.)

136. PÁSPALUM L.

Spikelets planoconvex, usually obtuse, subsessile, solitary or in pairs, in 2 rows on one side of a narrow or dilated rachis, the back of the fertile lemma toward it; first glume usually wanting; second glume and sterile lemma commonly about equal, the former rarely wanting; fertile lemma usually obtuse, chartaceous-indurate, the margins inrolled. Perennials in the United States (except *P. boscianum* and *P. convexum*), with one to many spikelike racemes, solitary, paired, or several to many on a common axis. Type species, *Paspalum dissectum*. Name from Greek *paspalos*, a kind of millet.

Several species inhabiting meadows and savannas furnish considerable forage. *Paspalum dilatatum* is valuable for pasture, especially for dairy cattle in the Southern States, where it has been cultivated under the name water grass and recently Dallis grass. In the Hawaiian Islands, Australia, and some other countries, where it is called paspalum or paspalum grass, it is valuable as a pasture grass. *P. pubiflorum* var. *glabrum* is rather abundant in some regions and is considered a good forage grass. Vasey grass, *P. urvillei*, is used to a limited extent for hay and, when young, for pasture; the panicles, after the spikelets have fallen, also make excellent whisk brooms for brushing lint. In the Southern States (Virginia to Florida and even to California) *P. distichum*, because of its extensively creeping stolons, is useful for holding banks of streams and ditches.

1a. Rachis foliaceous, broad and winged.
 Racemes falling from the axis, rachis extending beyond the uppermost spikelet.
 　　　　　　　　　　　　　　　　　　　　　　　　　　　　3. P. FLUITANS.
 Racemes persistent on the axis; rachis with a spikelet at the apex.
 　Spikelets 2 mm. long, obovate-oval............... 1. P. DISSECTUM.
 　Spikelets more than 3 mm. long, pointed.................. 2. P. ACUMINATUM.
1b. Rachis not foliaceous nor winged (slightly winged in *P. boscianum*).
 2a. Racemes 2, conjugate or nearly so at the summit of the culm, rarely a third below.
 　Spikelets elliptic to narrowly ovate.
 　　Plants with creeping rhizomes or stolons.
 　　　Second glume and sterile lemma glabrous; spikelets flattened 4. P. VAGINATUM.
 　　　Second glume pubescent; spikelets relatively turgid.................. 5. P. DISTICHUM.
 　　Plants in dense tufts, without creeping rhizomes.............................. 11. P. ALMUM.
 　Spikelets suborbicular, broadly ovate or obovate.
 　　Spikelets concavo-convex, sparsely long-silky around the margin; plant stoloniferous.
 　　　　　　　　　　　　　　　　　　　　　　　　31. P. CONJUGATUM.
 　　Spikelets plano-convex, not silky-margined; plants not stoloniferous.
 　　　Spikelets 3 to 3.5 mm. long.. 9. P. NOTATUM.
 　　　Spikelets 2 to 2.5 mm. long.. 10. P. MINUS.
 2b. Racemes 1 to many, racemose on the axis, not conjugate.
 　3a. First glume developed on at least one of the pair of spikelets (often obsolete in some pairs in Nos. 22 and 23).
 　　Spikelets turgidly biconvex... 48. P. BIFIDUM.
 　　Spikelets plano-convex.
 　　　Plants without rhizomes; culms tufted; spikelets pubescent.......... 24. P. LANGEI.
 　　　Plants with stout scaly rhizomes; culms mostly solitary; spikelets glabrous.
 　　　　Blades flat, 8 to 15 mm. wide............................... 22. P. UNISPICATUM.
 　　　　Blades folded at base, terete above, not more than 2 mm. wide.
 　　　　　　　　　　　　　　　　　　　　　　　　23. P. MONOSTACHYUM.
 　3b. First glume normally wanting (occasionally developed on 1 to few spikelets in a raceme).
 　　4a. Racemes terminal and axillary, the axillary sometimes hidden in the sheaths and perfecting grains cleistogamously, terminal inflorescence of 1 to 3, rarely to 6 racemes (see also *P. unispicatum* and *P. monostachyum*).
 　　　5a. Spikelets not more than 1.8 mm. long (or sometimes 1.9 in *P. debile* and *P. propinquum*), usually 1.5 to 1.7 mm. (see also exceptional *P. ciliatifolium*).
 　　　Blades conspicuously ciliate, otherwise nearly glabrous.

Blades relatively short, rounded at base and recurved-ascending; foliage aggregate toward the base, the upper culm relatively naked; spikelets glabrous, mostly 1.5 to 1.6 mm. long...... 12.   P. LONGEPEDUNCULATUM.

Blades mostly elongate, suberect, not aggregate toward the base; spikelets pubescent, 1.7 to 1.9 mm. long................................. 20.   P. PROPINQUUM.

Blades and sheaths conspicuously pubescent throughout.

Culms slender, erect or suberect; foliage not aggregate at base; blades suberect, usually not more than 5 mm. wide..................... 13.   P. SETACEUM.

Culms stouter, mostly spreading; foliage more or less aggregate at base; blades spreading, usually more than 5 mm. wide..................... 14.   P. DEBILE.

5b. Spikelets 2 to 2.5 mm. long (or 1.8 to 1.9 mm. in *P. ciliatifolium* and *P. propinquum*).

Foliage, except margins, glabrous as a whole or nearly so (sparsely pubescent in exceptional *P. ciliatifolium* and lower sheaths usually pubescent in *P. rigidifolium*).

Blades stiff, usually not more than 6 mm. wide; spikelets mostly 2.2 to 2.4 mm. long.. 21.   P. RIGIDIFOLIUM.

Blades from lax to rather firm, if firm more than 6 mm. wide; spikelets not more than 2.1 mm. long.

Spikelets mostly 2 mm. long, rounded at summit; blades mostly more than 8 mm. wide.. 19.   P. CILIATIFOLIUM.

Spikelets 1.8 to 1.9 mm. long, slightly pointed; blades not more than 8 mm. wide.. 20.   P. PROPINQUUM.

Foliage conspicuously pubescent (or sparsely so in exceptional specimens of *P. pubescens*).

Culms erect or nearly so.

Blades from sparsely to rather densely pilose, rather thin.
18.   P. PUBESCENS.

Blades puberulent on both surfaces with long hairs intermixed or the lower surface nearly or quite glabrous except for a few long hairs along midrib and margin, usually rather firm........................... 17.   P. STRAMINEUM.

Culms widely spreading or prostrate.

Foliage coarsely hirsute; plants commonly relatively stout.
15.   P. SUPINUM.

Foliage finely puberulent; plants usually grayish olivaceous.
16.   P. PSAMMOPHILUM.

4b. Racemes terminal on the primary culm or leafy branches, no truly axillary racemes.

6a. Spikelets conspicuously silky-ciliate around the margin, the hairs as long as the spikelet or longer.

Racemes commonly 3 to 5; culms geniculate at base........... 32.   P. DILATATUM.

Racemes commonly 12 to 18; culms erect................................. 33.   P. URVILLEI.

6b. Spikelets not ciliate.

7a. Fruit dark brown and shining.

Plants perennial; spikelets 2.2 to 2.8 mm. long, elliptic or obovate-oval.

Spikelets obovate, turgid, the sterile lemma wrinkled; culms erect, densely cespitose.. 43.   P. PLICATULUM.

Spikelets elliptic, depressed, not turgid; culms decumbent or floating at the base.

Plants terrestrial, culms decumbent at base................. 44.   P. TEXANUM.

Plants aquatic, lower part of culms floating........ 45.   P. HYDROPHILUM.

Plants annual; spikelets 2 to 3 mm. long, suborbicular or broadly obovate.

Spikelets suborbicular, 2 to 2.2 mm. long, glabrous........ 46.   P. BOSCIANUM.

Spikelets broadly obovate, 2.2 to 3 mm. long, pubescent.
47.   P. CONVEXUM.

7b. Fruit pale to stramineous (brown but not shining in *P. virgatum*).

8a. Plants robust, 1 to 2 m. tall.

Spikelets pubescent at least toward the summit; fruit brown at maturity.
42.   P. VIRGATUM.

Spikelets glabrous; fruit pale.

Culms ascending; leaves crowded toward the base.... 39.   P. DIFFORME.

Culms erect or suberect, leafy throughout.

Glume and sterile lemma slightly inflated and wrinkled, green.
40.   P. FLORIDANUM.

Glume and sterile lemma not inflated and wrinkled, rusty-tinged.
41.   P. GIGANTEUM.

8b. Plants not robust, if more than 1 m. tall, culms relatively slender.

9a. Spikelets suborbicular or broadly obovate or broadly oval.
    Spikelets turgidly plano-convex, 3.5 to 4 mm. long.... 39. P. DIFFORME.
    Spikelets depressed plano-convex or lenticular, 2.2 to 3.4 mm. long.
        Spikelets solitary; glume and sterile lemma firm.
            Spikelets orbicular, 3 to 3.2 mm. long, scarcely one-third as thick;
            blades usually equaling the base of the panicle or overtopping it.
                                                            36. P. CIRCULARE.
            Spikelets longer than broad, more than one-third as thick; panicle
            usually much exceeding the blades.
                Sheaths and blades pilose, mostly conspicuously so.
                                                            35. P. LONGIPILUM.
                Sheaths and blades from glabrous to sparsely pilose.
                                                            34. P. LAEVE.
        Spikelets paired and solitary in the same raceme (rarely all solitary or
        all paired).
            Spikelets 2.2 to 2.5 mm. (rarely to 2.8 mm.) long; foliage not con-
            spicuously villous.................................... 37. P. PRAECOX.
            Spikelets 2.7 to 3.4 mm. long; lower sheaths and blades mostly con-
            spicuously villous at least at base............ 38. P. LENTIFERUM.
9b. Spikelets elliptic to oval or obovate.
    Culms decumbent at base, rooting at the lower nodes (occasional plants
    in dry situations erect), branching.
        Spikelets turgidly plano-convex, 3 to 3.2 mm. long; culms rather stout.
                                                            6. P. PUBIFLORUM.
        Spikelets depressed plano-convex; culms rather slender.
            Spikelets glabrous... 7. P. LIVIDUM.
            Spikelets pubescent............................. 8. P. HARTWEGIANUM.
    Culms erect to spreading, not rooting at the nodes.
        Racemes solitary, rarely paired; spikelets usually solitary, 1.3 to 1.6
        mm. long.. 30. P. SAUGETII.
        Racemes 2 or more, commonly 3 to 8.
            Spikelets about 1.3 mm. long, obovate, glandular-pubescent.
                                                            25. P. BLODGETTII.
            Spikelets 1.5 mm. or more long, elliptic or elliptic-obovate, the ob-
            scure pubescence not glandular.
                Nodes or some of them appressed-pilose; spikelets green or purplish.
                                                            26. P. CAESPITOSUM.
                Nodes glabrous; spikelets pale or brownish.
                    Spikelets 1.7 to 2 mm. long; racemes slender, lax.
                        Foliage glabrous or nearly so; spikelets elliptic-obovate.
                                                            27. P. LAXUM.
                        Foliage softly pilose; spikelets broadly ovate.
                                                            29. P. VIRLETII.
                    Spikelets 2.2 to 2.5 mm. long; racemes rigid.
                                                            28. P. PLEOSTACHYUM.

1. Dissécta.—Blades flat; rachis foliaceous. Aquatics, subaquatics,
or plants of wet ground.
    1. Paspalum disséctum (L.) L.
(Fig. 862.) Glabrous, olive green,
creeping, freely branching, the flowering branches ascending, 20 to 60 cm.
long; blades thin, 3 to 6 cm. long,
4 to 5 mm. wide; panicles terminal
and axillary, the racemes 2 to 4,
usually erect, 2 to 3 cm. long; rachis
2 to 3 mm. wide; spikelets solitary,
obovate, subacute, 2 mm. long. ♃
—On muddy and sandy banks of
ponds and ditches or in shallow water,
New Jersey; Illinois to Oklahoma,
south to Florida and Texas; Cuba.

2. Paspalum acuminátum Raddi.
(Fig. 863.) Culms decumbent at base,
sometimes extensively creeping, 30
to 100 cm. long; blades 4 to 12 cm.
long; 5 to 12 mm. wide; racemes 3
to 5, erect or ascending, 3.5 to 7
cm. long; rachis 3 to 3.5 mm. wide;
spikelets solitary, 3.5 mm. long,
abruptly pointed. ♃ —In shallow
water or wet open ground, from
southern Louisiana and Texas to
Argentina.
    3. Paspalum flúitans (Ell.) Kunth.
(Fig. 864.) Annual aquatic; culms
mostly submerged, rooting at the
nodes, 30 to 100 cm. long; sheaths

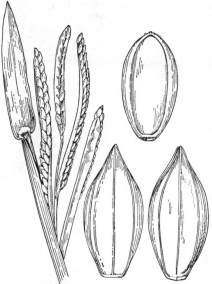

FIGURE 862.—*Paspalum dissectum.* Panicle, × 1; two views of spikelet, and floret, × 10. (Commons 85, Del.)

FIGURE 863.—*Paspalum acuminatum.* Panicle, × 1; two views of spikelet, and floret, × 10. (Arsène 3132, Mex.)

glabrous or pilose, with an erect auricle 1 to 5 mm. long on each side, the sheaths of the floating branches inflated, commonly long-hirsute and purple-spotted; blades usually 10 to 20 cm. long, 10 to 15 mm. wide (sometimes 25 cm. long and 2.5 cm. wide); panicles mostly 10 to 15 cm. long, of numerous ascending, spreading or recurved racemes, 3 to 8 cm. long, falling

entire, the rachis 1.3 to 2 mm. wide; spikelets solitary, elliptic, 1.3 to 1.8 mm. long, acute or acuminate, pilose with delicate hairs, sometimes obscurely so, the sterile lemma with a V-shaped pink marking at base. ⊙ —(*P. mucronatum* Muhl.; included in *P. repens* Bergius in Manual, ed. 1.) Floating in sluggish streams or standing water or creeping in wet places, Virginia to Illinois, Kansas

FIGURE 864.—*Paspalum fluitans.* Panicle, × 1; two views of spikelet, and floret, × 10. (Combs 912, Fla.)

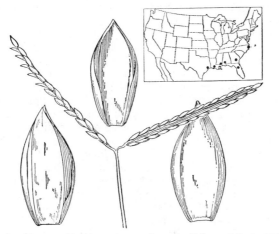

FIGURE 865.—*Paspalum vaginatum.* Panicle, × 1; two views of spikelet, and floret, × 10. (Hitchcock 9866, Jamaica.)

and Oklahoma, south to Florida and Texas; Venezuela.

PASPALUM RACEMÓSUM Lam. Branching annual; blades 5 to 12 cm. long, 1 to 2 cm. wide; panicles tawny to purple; racemes numerous, 1 to 2 cm. long; spikelets about 2.7 mm. long, pointed; sterile lemma transversely fluted each side of the midnerve. ⊙ —Sometimes cultivated for ornament. Peru.

2. Dísticha.—Creeping, with wiry compressed culms and stolons or rhizomes; racemes mostly 2, paired or approximate.

4. Paspalum vaginátum Swartz. (Fig. 865.) Flowering culms 8 to 60 cm. tall; sheaths usually overlapping; blades 2.5 to 15 cm. long, 3 to 8 mm. wide, tapering to an involute apex; racemes at first erect, usually spreading or reflexed at maturity, 2 to 5 cm. long; rachis 1 to 2 mm. wide; spikelets solitary, 3.5 to 4 mm. long, ovate-lanceolate, acute, pale-stramineous; first glume rarely developed; midnerve of the second glume and sterile lemma usually suppressed. ♃ — Seacoasts and brackish sands, often forming extensive colonies, North Carolina to Florida and Texas, south to Argentina; tropics of Eastern Hemisphere.

5. Paspalum dístichum L. KNOT-GRASS. (Fig. 866.) Resembling *P. vaginatum,* sometimes with exten-

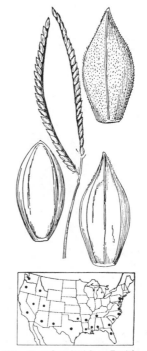

FIGURE 866.—*Paspalum distichum.* Panicle, × 1; two views of spikelet, and floret, × 10. (Hitchcock 9394, Jamaica.)

sively creeping stolons with pubescent nodes; racemes 2 to 7 cm. long, commonly incurved; spikelets 2.5 to 3.5 mm. long, elliptic, abruptly acute, pale green; first glume frequently developed; second glume appressed-pubescent, the midnerve in glume and sterile lemma developed. ♃ — Ditches and wet, rarely brackish places, New Jersey to Florida and Texas, Tennessee, and Arkansas, west to California and north along the coast to Washington; Idaho; south to Argentina; warm coasts of the Eastern Hemisphere.

PASPALUM PAUCISPICÁTUM Vasey. Resembling vigorous specimens of *P. distichum*, but with 3 to 5 racemes with mostly paired spikelets. ♃ — A specimen collected by Palmer in 1888, said to be from "Southern California," is in the United States National Herbarium. The locality is doubtful, the species ranging from Sonora to Oaxaca.

3. Lívida.—Culms compressed; racemes few to several, mostly plants of alkaline soil.

6. Paspalum pubiflórum Rupr. ex Fourn. (Fig. 867.) Culms decumbent at base, 40 to 100 cm. tall; sheaths, at least the lower, sparsely papillose-pilose; blades flat, usually 10 to 15 cm. long, 6 to 14 mm. wide, usually with a few stiff hairs at the rounded base; racemes mostly 3 to 5, 2 to 10 cm. long, rather thick, erect to spreading, the rachis 1.2 to 2 mm. wide; spikelets obovate, pubescent, about 3 mm. long. ♃ (*P. hallii* Vasey and Scribn.)—Moist open ground, banks, low woods, along streams and irrigation ditches, especially in alkaline clay soil, Louisiana and Texas; Mexico and western Cuba.

PASPALUM PUBIFLORUM var. GLÁ-BRUM Vasey ex Scribn. Somewhat more robust, the sheaths less pilose, the racemes commonly longer and

FIGURE 867.—*Paspalum pubiflorum.* Panicle, × 1; two views of spikelet, and floret, × 10. (Hitchcock 5555, Mex.)

FIGURE 868.—*Paspalum lividum.* Panicle, × 1; two views of spikelet, and floret, × 10. (Arséne 3176, Mex.)

often more than 5; spikelets glabrous. ♃ (*P. geminum* Nash; *P. laeviglume* Scribn.)—Moist low open ground, woods, and ditch banks, North Carolina, Ohio, and Indiana to Florida, west to Kansas and Texas; adventive, Chester, Pa.

7. Paspalum lívidum Trin. LONG-TOM. (Fig. 868.) Glabrous; culms solitary or few in a tuft, from a decumbent or creeping base, 50 to 100 cm. tall; blades 15 to 25 cm. long, 3 to 6 mm. wide; racemes usually 4 to 7, ascending, flexuous; rachis 1.5 to 2 mm. wide, dark livid purple; spikelets 2 to 2.5 mm. long, obovate, subacute. ♃ —Low ground, wet savannas, and swamps, and along streams and ditches, Alabama to Texas and Mexico, south to Argentina; Cuba.

to 9 cm. long; rachis 1 to 1.5 mm. wide; spikelets imbricate, about 3 mm. long, elliptic, apiculate, softly pubescent. ♃ (*P. buckleyanum* Vasey.)—Wet prairies, alkaline meadows, and along irrigation ditches, sometimes growing in the water, southern Texas and throughout Mexico.

4. Notáta.—Culms in dense tufts, compressed, leafy at base; sheaths keeled; racemes 2, rarely 3, paired or nearly so; spikelets solitary, glabrous.

9. Paspalum notátum Flügge. BAHIA GRASS. (Fig. 870.) Culms 15 to 50 cm. tall from a short, stout, woody, horizontal rhizome; blades flat or folded; racemes recurved-ascending, usually 4 to 7 cm. long; spikelets

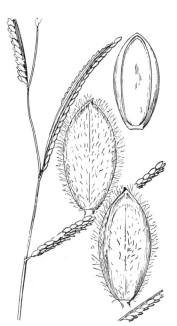

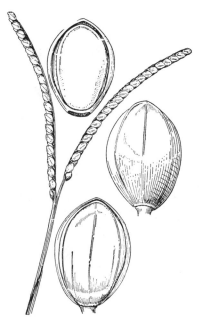

FIGURE 869.—*Paspalum hartwegianum*. Panicle, × 1 two views of spikelet, and floret, × 10. (Buckley, Tex.)

FIGURE 870.—*Paspalum notatum*. Panicle, × 1; two views of spikelet, and floret, × 10. (Chase 6639, P. R.)

8. Paspalum hartwegiánum Fourn. (Fig. 869.) Culms ascending from a decumbent base, 50 to 150 cm. tall; blades 10 to 35 cm. long, 2 to 6 mm. wide, the margins very scabrous; racemes usually 4 to 7, ascending, 2

ovate to obovate, 3 to 3.5 mm. long, smooth and shining. ♃ —Introduced sparingly in New Jersey, North Carolina, Florida, Louisiana, and Texas; Mexico and the West Indies to South America.

PASPALUM NOTATUM var. SAÚRAE Parodi. A more hardy form, 40 to 70 cm. tall, with blades to 35 cm. long, the racemes 2 or 3, rarely to 5, sub-erect, the spikelets 2.8 to 3 mm. long, is showing promise of becoming an important forage and erosion-control grass in the Southern States. This has been found in lawns at Wilmington, N. C., Pensacola, Fla., and in several localities in Texas. It has been called the "Pensacola strain." An introduction from Paraguay belongs to this form and has come to be known as the "Paraguay strain." It has been confused with *Paspalum minus* Fourn., a distinct species that occurs in a few localities in Texas. ♃ —Paraguay and Argentina.

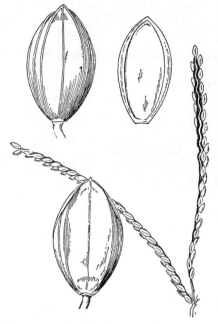

FIGURE 872.—*Paspalum almum.* Panicle, × 1; two views of spikelet, and floret, × 10. (Type.)

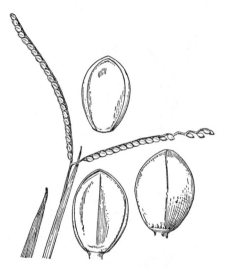

FIGURE 871.—*Paspalum minus.* Panicle, × 1; two views of spikelet, and floret, × 10. (Type coll.)

10. Paspalum mínus Fourn. (Fig. 871.) Resembling *P. notatum*, commonly in denser mats; culms rarely more than 30 cm. tall; racemes more slender; spikelets 2 to 2.5 mm. long, less shining than those of *P. notatum.* ♃ —Open slopes and savannas, eastern Texas; Mexico to West Indies and Paraguay.

11. Paspalum álmum Chase. COMBS PASPALUM. (Fig. 872.) Culms in very dense tufts; blades flat, 2 to 3 mm. wide, long-hirsute on the upper surface at base, papillose-hirsute on the lower surface toward the ends, the margins stiffly ciliate toward base; racemes slender, approximate, scarcely paired, occasionally 3, ascending, 5 to 9 cm. long; rachis 1 mm. wide, minutely wing-margined; spikelets 3 mm. long, 1.8 to 2 mm. wide, obovate-elliptic; sterile lemma slightly concave. ♃ —Sandy or silty clay loam, Jefferson County, Tex.; Brazil, Paraguay, and Argentina. An excellent forage grass.

5. Setácea.—Culms compressed from a knotted base or very short rhizome; blades mostly flat; inflorescence terminal and axillary, the axillary sometimes hidden in the sheaths; racemes 1 to few, slender, subcylindric; spikelets in pairs, crowded. Species closely related with frequent intergrades.

12. Paspalum longepedunculátum LeConte. (Fig. 873.) Culms slender, ascending or suberect, 25 to 80 cm. tall; leaves mostly aggregate at the base, the sheaths ciliate on the margin; blades usually folded at base, 4 to 10 cm. long, rarely longer, 3 to 8 mm. wide, stiffly papillose-ciliate on the margin, the hairs 1.5 to 3 mm. long; racemes on very slender finally elongate peduncles, 1 or 2, rarely 3, on the primary, 1 on the axillary peduncles; racemes arching, 3 to 8 cm. long; spikelets about 1.5 mm. long, elliptic-obovate, glabrous. ♃ —Sandy soil, mostly in low pine land or flat woods, Virginia and Kentucky to Florida and Mississippi.

13. Paspalum setáceum Michx. (Fig. 874.) Culms slender, erect, usually 30 to 50 cm. tall; sheaths pilose;

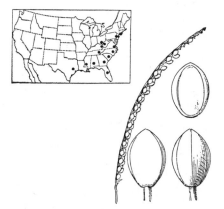

Figure 874.—*Paspalum setaceum.* Raceme, × 1; two views of spikelet, and floret, × 10. (Hitchcock 300, S. C.)

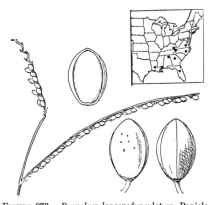

Figure 873.—*Paspalum longepedunculatum.* Panicle, × 1; two views of spikelet, and floret, × 10. (Nash 2074, Fla.)

blades rather firm, erect or nearly so, linear, about 10 to 12 cm. long, 2 to 6 mm. wide, densely pilose on both surfaces and papillose-ciliate on the margin; racemes on slender peduncles, solitary or sometimes 2, arching, 5 to 7 cm. long; spikelets elliptic-obovate, about 1.5 mm. long, glabrous or minutely pubescent. ♃ —Sandy soil, usually open woods, mostly on or near the Coastal Plain, Long Island to Florida and Texas; Ohio and West Virginia to Tennessee; Mexico.

14. Paspalum débile Michx. (Fig. 875.) Differing from *P. setaceum* in the stouter, more spreading culms, the foliage more crowded at base,

densely grayish villous, the blades on the average wider; racemes more commonly 2; spikelets 1.8 to 1.9 mm. long, pubescent. ♃ —Sandy, mostly dry soil, barrens and flatwoods, Long Island to Florida and Texas; Mexico and Cuba.

15. Paspalum supínum Bosc ex Poir. (Fig. 876.) Culms relatively stout, widely spreading, 30 to 90 cm. tall; sheaths usually hirsute; blades 15 to 25 cm. long, 8 to 15 mm. wide, hirsute; racemes usually 2 to 4, rarely to 6, 4 to 10 cm. long; spikelets elliptic-obovate, 2 mm. long, gla-

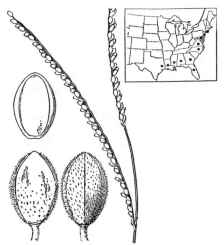

Figure 875.—*Paspalum debile.* Panicle, × 1; two views of spikelet, and floret, × 10. (Nash 946, Fla.)

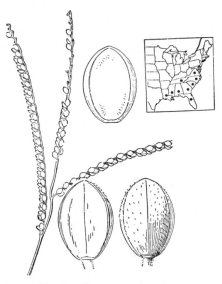

FIGURE 876.—*Paspalum supinum.* Panicle, × 1; two views of spikelet, and floret, × 10. (Chase 4572, N. C.)

17. Paspalum stramíneum Nash. (Fig. 878.) Yellowish green, the culms erect, 40 to 100 cm. tall; blades 6 to 25 cm. long, rarely longer, 6 to 15 mm. wide, puberulent on both surfaces and sparsely pilose as well, or the lower surface nearly glabrous; racemes 2 or 3, rarely 4, 6 to 14 cm. long, the axillary often wholly or partly included in the sheaths, short racemes commonly borne in basal sheaths; spikelets suborbicular, 2.1 to 2.2 mm. long, pale, from densely pubescent to glabrous. ⚥ (*P. bushii* Nash.)—Sandy soil, in open ground or open woods, Indiana to Minnesota, Texas, Arizona, and northwestern Mexico.

18. Paspalum pubéscens Muhl. (Fig. 879.) Culms ascending, 45 to 90 cm. tall, often pilose at the summit; sheaths usually pilose toward the summit; blades 8 to 20 cm. long,

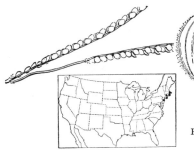

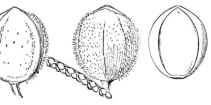

FIGURE 877.—*Paspalum psammophilum.* Panicle, × 1; two views of spikelet, and floret, × 10. (Graves, N. Y.)

brous, or the glume minutely pubescent. ⚥ —Dry, sandy, open ground and old fields, Virginia to Florida and west to Louisiana.

16. Paspalum psammóphilum Nash. (Fig. 877.) Forming dense grayish-olivaceous mats, the culms usually prostrate, 25 to 100 cm. long; sheaths appressed-pubescent; blades 4 to 16 cm. long, 4 to 11 mm. wide, densely appressed-pubescent; racemes 1 to 3, commonly 2, 4 to 9 cm. long, the axillary ones wholly or partly included in the sheaths; spikelets suborbicular, 2 mm. long, the glume densely pubescent. ⚥ — Dry sandy soil, mostly near the coast, Massachusetts to New Jersey.

2 to 10 mm. wide (rarely larger), pilose on both surfaces; racemes 1 to 3, 4 to 17 cm. long; spikelets about 2 mm. long, suborbicular, usually glabrous. ⚥ (*P. muhlenbergii* Nash.)—Open ground or open woods, common in old fields and pastures, especially in sandy regions, Vermont to Florida, west to Michigan, Kansas, and Texas.

19. Paspalum ciliatifólium Michx. (Fig. 880.) Culms erect to spreading, 35 to 90 cm. tall; sheaths glabrous or the lower puberulent; blades 10 to 35 cm. long, 7 to 20 mm. wide (rarely larger), usually strongly ciliate along the margin and glabrous otherwise; racemes 1 to 3, usually 7 to 10 cm.

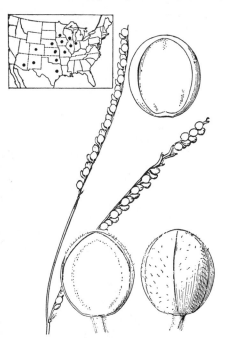

FIGURE 878.—*Paspalum stramineum.* Panicle, × 1; two views of spikelet, and floret, × 10. (Type.)

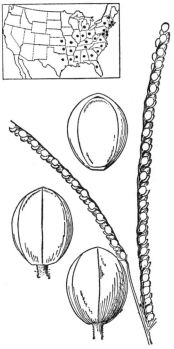

FIGURE 879.—*Paspalum pubescens.* Panicle, × 1; two views of spikelet, and floret, × 10. (Hitchcock 298, Ga.)

long; spikelets about 2 mm. long, suborbicular, the glumes often minutely pubescent. ♃ (*P. chapmani* Nash; *P. eggertii* Nash; *P. blepharophyllum* Nash; *P. epile* Nash.)—Open ground or open woods, mostly sandy, New Jersey to Florida, Minnesota, Kansas, and Texas; Honduras and the West Indies. This species is exceedingly variable. Pubescence on foliage and spikelets varies in a single plant. Rather stout, somewhat paler, seacoast plants, with firmer blades scarcely ciliate, are the form described as *P. epile.* Plants with softly pubescent lower sheaths, and blades but slightly ciliate, are the form described as *P. eggertii.* The shape of the spikelet varies in a single raceme from elliptic-obovate to suborbicular. The spikelets tend to become rounder at maturity, but both mature and immature are found of both shapes.

20. Paspalum propínquum Nash. (Fig. 881.) Resembling *P. ciliati-*

folium, the blades firmer and narrower, the spikelets slightly smaller,

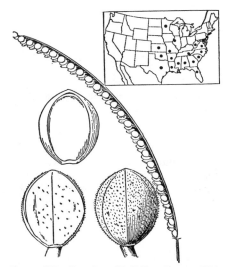

FIGURE 880.—*Paspalum ciliatifolium.* Raceme, × 1; two views of spikelet, and floret, × 10. (Nash 1426, Fla.)

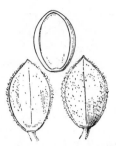

FIGURE 881.—*Paspalum propinquum.* Two views of spikelet, and floret, × 10. (Type.)

puberulent; racemes 1 or 2, 7 to 14 cm. long; spikelets usually 2.2 to 2.4 mm. long, obovate-elliptic, glabrous or nearly so, rarely pubescent. ♃ —Sand barrens and high pineland, North Carolina and peninsular Florida to Texas.

6. Dimorphóstachys.—Inflorescence terminal and axillary; racemes one to few, slender; spikelets in pairs, the first glume usually developed on one of the pair, often on both, or sometimes obsolete on both.

22. Paspalum unispicátum (Scribn. and Merr.) Nash. (Fig. 883.) Culms 1 to few in a tuft from horizontal scaly rhizomes, erect or ascending, 50 to 80 cm. tall, simple or with a

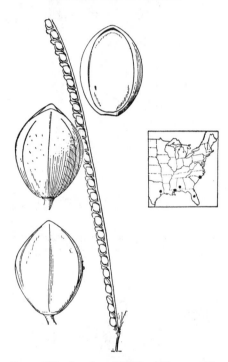

FIGURE 882.—*Paspalum rigidifolium.* Raceme, × 1; two views of spikelet, and floret, × 10. (Type.)

subacute. ♃ —Sandy savannas and sand barrens overlying limestone, peninsular Florida; West Indies; Veracruz, Mexico, to Panama.

21. Paspalum rigidifólium Nash. (Fig. 882.) Culms erect, rather stiff, purplish, 25 to 75 cm. tall; sheaths glabrous or the lower grayish-pubescent; blades firm, linear, mostly 10 to 15 cm. long, 2 to 5 mm. wide, usually not wider than the summit of the sheath, glabrous or minutely

FIGURE 883.—*Paspalum unispicatum.* Raceme, × 1; two views of spikelet, and floret, × 10. (Type.)

single erect leafy branch; blades flat,
rather stiff, 10 to 30 cm. long, 8 to
15 mm. wide, stiffly papillose-ciliate
on the margin, sparsely papillose-
hirsute on both surfaces, or scaberulous only; racemes usually solitary, 1 terminal and 1 from the axil
of the uppermost sheath, 6 to 20
cm. long; spikelets about 3.2 mm.
long, elliptic; first glume on the
primary spikelet minute, sometimes
obsolete, on secondary spikelet mostly half to three-fourths as long as
the spikelet. ♃ —Meadows, savannas, open slopes, and banks,
southern Texas to Venezuela and
Argentina; Cuba.

23. Paspalum monostáchyum

Vasey. (Fig. 884.) Culms 1 to few
from horizontal scaly rhizomes, erect,
50 to 120 cm. tall; blades elongate,
slender, terete, firm; racemes 1 or 2,
10 to 30 cm. long; spikelets 3 to 3.5

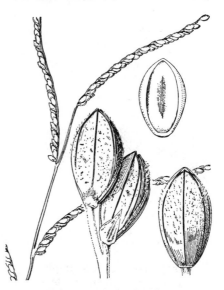

FIGURE 885.—*Paspalum langei.* Panicle, × 1; two
views of spikelet, and floret, × 10. (Pringle 3991,
Mexico.)

mm. long, subovate-elliptic, the pedicels of the pair nearly equal; first
glume often developed in few to
several of the primary spikelets, commonly wanting or rudimentary. ♃
(*P. solitarium* Nash.)—Moist places
in flatwoods or coastal dunes, southern
Florida and Texas.

24. Paspalum lángei (Fourn.)

Nash. (Fig. 885.) Culms ascending,
30 to 100 cm. tall; blades flat, rather
thin, 10 to 40 cm. long, 6 to 15 mm.
wide, glabrous to sparsely pubescent,
the lower tapering to a narrow base;
peduncles 1 to 3 from the upper
sheath, often also from middle
sheaths; racemes 2 to 5, 4 to 10 cm.
long; spikelets 2.2 to 2.6 mm. long,
elliptic-obovate, pubescent and glan-
dular-speckled; first glume minute or
obsolete on the primary spikelet,
one-fourth to one-third as long as
the spikelet on the secondary. ♃
(*Dimorphostachys ciliifera* Nash; *Paspalum ciliiferum* Hitchc.)—Moist
woods and shaded slopes and banks,
occasionally in open ground, mostly
at low altitudes, Florida, Louisiana,
Texas; Greater Antilles to Venezuela.

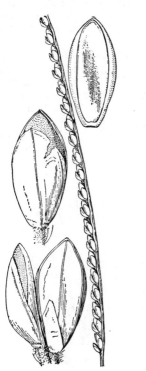

FIGURE 884.—*Paspalum monostachyum.* Raceme, × 1;
two views of spikelet, and floret, × 10. (Type.)

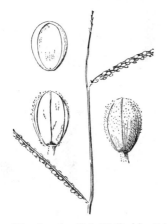

FIGURE 886.—*Paspalum blodgettii.* Panicle, × 1; two views of spikelet, and floret, × 10. (Simpson, Fla.)

7. Caespitósa.—Culms simple or with a single branch, its leaf sometimes hidden in the parent sheath, the inflorescence appearing to be axillary; racemes few to several.

25. Paspalum blodgéttii Chapm. (Fig. 886.) Cespitose, with tough, commonly somewhat swollen and bulblike base, the scales densely pubescent; culms erect, slender, 40 to 100 cm. tall; lower leaves crowded; blades flat, 5 to 25 cm. long, mostly 5 to 10 mm. wide; racemes usually 3 to 8, slender, remote, 2 to 8 cm. long; spikelets about 1.3 mm. long, obovate, the glume glandular-pubescent. ♃ (*P. simpsoni* Nash; *P. gracillimum* Nash.)—Open or brushy calcareous soil, southern Florida; Yucatan, Honduras, British Honduras, Bahamas, and the Greater Antilles.

26. Paspalum caespitósum Flügge. (Fig. 887.) Cespitose, bluish green; culms erect, rather wiry, 30 to 60 cm. tall; blades flat, folded or involute, 5 to 20 cm. long, rarely longer, 4 to 10 mm. wide; racemes usually 3 to 5, relatively thick, remote, ascending, 1.5 to 6 cm. long; spikelets 1.5 to 1.8 mm. long, elliptic, sparsely appressed-pubescent to nearly glabrous. ♃ —Mostly in partly shaded humus in limestone soil or

rock, sometimes in sandy pinelands; southern Florida, Mexico, Central America, and the West Indies.

27. Paspalum láxum Lam. (Fig. 888.) Culms mostly 50 to 75 cm. tall, compressed, rigid, ascending; blades more or less involute, mostly 20 to 30 cm. long, 3 to 8 mm. wide, usually glabrous; racemes usually 3 to 5, mostly remote, 3 to 10 cm. long; spikelets about 2 mm. long, elliptic-obovate, the glume pubescent. ♃ (*P. glabrum* Poir.)—Sandy and limestone soils, characteristic of coconut groves, Key West, Fla.; West Indies.

28. Paspalum pleostáchyum Doell. (Fig. 889.) Culms 40 to 100 cm. tall, in rather large tough clumps, glabrous, or scabrous below the panicle, leafy; sheaths densely ciliate on the margins, villous across the collar, otherwise glabrous or sometimes papillose-hispid; blades as much as 55 cm. long, 4 to 8 mm. wide, flat or becoming folded, stiffly ascending, more or less pubescent above, the margins scabrous; racemes 3 to 15, ascending or stiffly spreading, 7 to 14 cm. long; spikelets 2.2 to 2.5 mm. long, elliptic-obovate, glabrous. ♃ —On rocks or in sand or clay near the seacoast, Marathon Key, Fla.; Cuba to Brazil.

FIGURE 887.—*Paspalum caespitosum.* Panicle, × 1; two views of spikelet, and floret, × 10. (Poiteau, Dominican Republic.)

29. Paspalum virlétii Fourn. (Fig. 890.) Culms 40 to 75 cm. tall; nodes, sheaths and blades softly pilose; blades flat, lax, 8 to 15 cm. long, 5 to 10 mm. wide, slightly narrowed to the base; racemes 4 or 5, slender, spreading, 2 to 7 cm. long, the margin of the slender rachis sometimes with a few long hairs; spikelets 2 mm. long, 1.5 mm. wide, broadly ovate; glume obscurely pubescent to glabrous. ♃ —Sandy soil, bottom

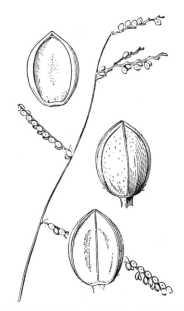

FIGURE 890.—*Paspalum virletii.* Panicle, × 1; two views of spikelet, and floret, × 10. (Type.)

of Sycamore Canyon, near Ruby, Santa Cruz County, Ariz.; northern Mexico. Rare.

8. Rupéstria.—Tufted perennials with slender culms and narrow blades; racemes slender, usually solitary; spikelets minute.

30. Paspalum saugétii Chase. (Fig. 891.) Culms 15 to 40 cm. tall, slender, densely tufted, glabrous, the nodes appressed-pubescent; blades 3 to 15 cm. long, 3 to 7 mm. wide, flat, or involute in drying, rather thick, glabrous or sometimes sparsely pilose; racemes solitary, sometimes 2, 2 to 4

FIGURE 888.—*Paspalum laxum.* Panicle, × 1; two views of spikelet, and floret, × 10. (Richard's specimen in Paris Herbarium.)

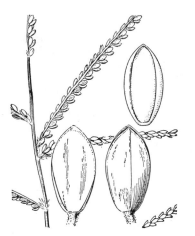

FIGURE 889.—*Paspalum pleostachyum.* Panicle, × 1; two views of spikelet, and floret, × 10. (Ekman 15756, Cuba.)

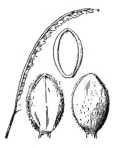

FIGURE 891.—*Paspalum saugetii.* Raceme, × 1; two views of spikelet, and floret, × 10. (Type.)

FIGURE 892.—*Paspalum conjugatum*. Plant, × ½; two views of spikelet, and floret, × 10. (Baker 90, Cuba.)

cm. long, erect or falcate; spikelets solitary or paired, 1.3 to 1.6 mm. long, oval, blunt, appressed-pubescent. ♃ —Rocky, mostly limestone soil, Florida (south of Royal Palm State Park) and the Greater Antilles.

9. Conjugáta.—Stoloniferous; blades flat; racemes 2, paired, rarely a third below, slender; spikelets flattened concavo-convex, solitary, silky-fringed.

31. Paspalum conjugátum Bergius. (Fig. 892.) Extensively creeping, with long leafy stolons and ascending suberect flowering branches, 20 to 50 cm. tall; nodes of stolons usually conspicuously pilose; blades rather thin, 8 to 12 cm. long, 5 to 15 mm. wide, usually glabrous; racemes widely divaricate, 8 to 12 cm. long; spikelets 1.4 to 1.8 mm. long, ovate, light yellow, the margin conspicuously ciliate-fringed. ♃ —A common weed in cultivated and waste ground, southern Florida to Texas, south to Argentina; West Indies; tropics of Old World.

10. Dilatáta.—Rather stout, in leafy clumps; blades flat; racemes few to numerous, spikelets in pairs, flat, silky-fringed.

32. Paspalum dilatátum Poir. DALLIS GRASS. (Fig. 893.) Culms tufted, leafy at base, mostly 50 to 150 cm. tall, ascending or erect from a decumbent base; blades 10 to 25 cm. long, 3 to 12 mm. wide; racemes usually 3 to 5, spreading, 6 to 8 cm. long; spikelets ovate-pointed, 3 to 3.5 mm. long, fringed with long white silky hairs and sparsely silky on the surface. ♃ —In low ground, from rather dry prairie to marshy meadows, New Jersey to Tennessee and Florida, west to Oklahoma and Texas; adventive in Oregon, Colorado, Arizona, and California; native of South America. Widely known as paspalumgrass, water-paspalum, water grass, or more commonly, simply paspalum. Introduced into the southern United States from Uruguay or Argentina

about the middle of the last century, now common throughout the Gulf States. Valuable pasture grass. Dallis grass was named for A. T. Dallis of La Grange, Ga., who grew it extensively.

33. Paspalum urvíllei Steud. VASEY GRASS. (Fig. 894.) Culms in large clumps, erect, mostly 1 to 2 m. tall; lower sheaths coarsely hirsute or occasionally glabrous; blades mostly elongate, 3 to 15 mm. wide, pilose at base; panicle erect, 10 to 40 cm. long, of about 12 to 20 rather crowded, ascending racemes, 7 to 14 cm. long; spikelets 2.2 to 2.7 mm. long, ovate, pointed, fringed with long white silky hairs, the glume appressed-silky. ♃ (*P. larranagai* Arech.; *P. vaseyanum* Scribn.)—Along ditches and roadsides and in waste ground, mostly in rather moist soil; Virginia to Florida and west to Texas; southern California, south to Argentina. Introduced from South America.

11. Laévia.—Rather tall, simple or occasionally with reduced flowering branches; blades mostly flat; racemes few to several; spikelets broadly oval to orbicular, depressed planoconvex, glabrous.

34. Paspalum laéve Michx. (Fig. 895.) Culms erect or ascending, leafy at base, 40 to 100 cm. tall; sheaths keeled, glabrous or nearly so; blades usually folded at base, flat or folded above, 5 to 30 cm. long, 3 to 10 mm. wide, glabrous to ciliate or sparsely pilose on the upper surface or sometimes toward the base beneath; racemes usually 3 or 4, spreading, 3 to 10 cm. long; spikelets broadly oval, 2.5 to 3 mm. long. ♃ (*P. angustifolium* LeConte; *P. australe* Nash.)—Meadows, open woods, old fields, and waste ground, New Jersey to Ohio, Florida, Arkansas, and eastern Texas.

35. Paspalum longípilum Nash. (Fig. 896.) Similar to *P. laeve*, usually less leafy at base, sheaths and blades pilose; racemes somewhat more lax than in *P. laeve*. ♃ (*P. plenipilum* Nash.)—Damp, mostly sandy soil, savannas, open woods, and wet pine

FIGURE 893.—*Paspalum dilatatum.* Plant, × ½; two views of spikelet, and floret, × 10. (Hitchcock 297, La.)

FIGURE 894.—*Paspalum urvillei.* Plant, × ½; two views of spikelet, and floret, × 10. (Chase 4388, La.)

Figure 895.—*Paspalum laeve*. Plant, × ½; two views of spikelet, and floret, × 10. (Chase 2600, D. C.)

barrens, New York to Tennessee, Florida, Arkansas, and Texas.

36. Paspalum circuláre Nash. (Fig. 897.) Culms in dense leafy clumps, 30 to 80 cm. tall; sheaths pilose to nearly glabrous; blades mostly erect, commonly about equaling the inflorescence, 15 to 30 cm. long, 5 to 10 mm. wide, usually pilose on the upper surface; racemes 2 to 7, mostly

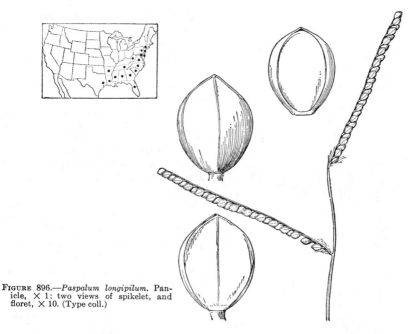

FIGURE 896.—*Paspalum longipilum.* Panicle, X 1; two views of spikelet, and floret, X 10. (Type coll.)

suberect, 5 to 12 cm. long; spikelets nearly orbicular, about 3 mm. long. ♃ (*P. praelongum* Nash.)—Fields, meadows, and open waste ground, Massachusetts to Georgia and Mississippi, west to Kansas and Texas.

37. Paspalum praécox Walt. (Fig. 898.) Culms erect from short scaly rhizomes, 50 to 100 cm. tall; sheaths keeled, glabrous, or the lower villous; blades 15 to 25 cm. long, 3 to 7 mm. wide, glabrous or nearly so; racemes usually 4 to 6, ascending to arcuate-spreading, 2 to 7 cm. long, the common axis very slender; rachis about 1.5 mm. wide, purplish; spikelets usually solitary and paired in each raceme, strongly flattened, suborbicular, 2.2 to 2.8 mm. long, the glume and sterile lemma thin and fragile. ♃ —Wet pine barrens, borders of cypress swamps, moist places in flatwoods, and wet savannas, in the Coastal Plain, North Carolina to central Florida and along the Gulf to Texas.

38. Paspalum lentíferum Lam. (Fig. 899.) Similar to *P. praecox;*

culms more robust, sometimes as much as 150 cm. tall; sheaths less strongly keeled; blades usually more or less pilose; racemes usually 4 or 5; spikelets 2.7 to 3.4 mm. long, broadly oval. ♃ (*P. glaberrimum* Nash; *P. tardum* Nash; *P. kearneyi* Nash; *P. amplum* Nash.)—Moist pine barrens, borders of flatwoods, and cypress swamps, and in savannas on the Coastal Plain, from Virginia to southern Florida and along the Gulf to Texas.

12. Floridána.—Mostly robust, culms simple; blades mostly flat; racemes few; spikelets large, rather turgid, glabrous.

39. Paspalum diffórme LeConte. (Fig. 900.) Culms solitary or few from a short knotty rhizome, rather stout, 35 to 75 cm. tall; leaves commonly crowded at the base; blades 10 to 15 cm. long, 5 to 10 mm. wide, usually pilose on the upper surface toward base; racemes 2 to 4, ascending to suberect, 3.5 to 8 cm. long; spikelets 3.5 to 4 mm. long, oval to obovate. ♃ —Moist sandy soil in

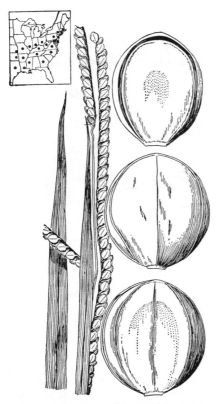

FIGURE 897.—*Paspalum circulare.* Panicle, × 1; two views of spikelet, and floret, × 10. (Chase 3836, Md.)

bust, taller; foliage glabrous or nearly so; racemes longer, more spreading. ♃ —Brackish marshes and low, sandy, mostly open ground, southern New Jersey to central Florida, west to Kentucky, Illinois, southeastern Kansas, and Texas.

41. Paspalum gigantéum Baldw. ex Vasey. (Fig. 902.) Culms mostly solitary from short scaly rhizomes, erect, 1.5 to 2 m. tall; leaves numerous at base; blades elongate, 10 to 20 mm. wide, glabrous or nearly so; racemes commonly 3 or 4, 10 to 20 cm. long; spikelets oval, about 3.5 mm. long, usually russet-tinged. ♃ (*P. longicilium* Nash.)—Moist sandy soil, open ground, stream banks, flatwoods, and hammocks, on the Coastal Plain from Georgia to southern Florida; Mississippi (Biloxi).

open ground and in flatwoods, in the Coastal Plain, South Carolina, to Orange County, Fla., and west near the Gulf to Louisiana.

40. Paspalum floridánum Michx. (Fig. 901.) Culms solitary or few from short stout scaly rhizomes, 1 to 2 m. tall; sheaths villous to nearly glabrous; blades firm, flat or folded, 15 to 50 cm. long, 4 to 10 mm. wide, usually villous at least on the upper surface toward base; racemes usually 2 to 5, 4 to 12 cm. long; spikelets crowded, oval, about 4 mm. long. ♃ —Low moist sandy soil, pine woods, flatwoods, savannas, and low prairies, in the Coastal Plain from Maryland to central Florida and along the Gulf to Texas, north in the valleys to Missouri and Oklahoma. PASPALUM FLORIDANUM var. GLABRÁTUM Engelm. ex Vasey. More ro-

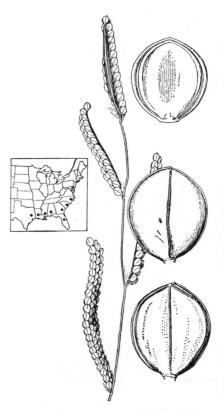

FIGURE 898.—*Paspalum praecox.* Panicle, × 1; two views of spikelet, and floret, × 10. (Stone 377, S.C.)

13. Virgáta.—Robust; blades firm with sharp-cutting edges; racemes several to numerous. Mostly tropical species.

42. Paspalum virgátum L. (Fig. 903.) Culms in large dense clumps, erect, 1 to 2 m. tall; sheaths papillose-hirsute at margin and summit; blades elongate, flat, 1 to 2.5 cm. wide; panicle slightly nodding, 15 to 25 cm. long; racemes usually 10 to 16, ascending or drooping, 5 to 15 cm. long; spikelets crowded, obovate, about 2.2 to 2.5 mm. long, brownish, pubescent along the margin at least toward the summit. ♃ —Open, mostly moist or swampy ground, southern Texas (Brownsville) to South America; throughout the West Indies.

Paspalum intermédium Munro ex Morong. Coarse, densely tufted perennial; sheaths compressed, keeled, the lower rather soft and papery; blades folded toward the base, the margins sharply hispid-serrate; panicle dense, the numerous racemes narrowly ascending or somewhat spreading; rachis rather prominently papillose-hispid-ciliate; spikelets about 2 mm. long, acute, glabrous, conspicuously purple-tinged. ♃ —Introduced from South America. Escaped along roadsides near Tifton, Ga.

14. Plicátula.—Perennials and annuals with compressed purplish culms; blades flat or folded; racemes few to several; spikelets rather turgid, drab, turning brown or dark olivaceous; fruit dark brown, shining.

43. Paspalum plicátulum Michx. BROWNSEED PASPALUM. (Fig. 904.) Culms in small tufts with numerous leafy shoots, suberect, 50 to 100 cm. tall; blades folded at base, usually flat above, rather firm, elongate, 3 to 10 mm. wide, usually pilose near base; racemes mostly 3 to 10, arcuate-spreading, 3 to 10 cm. long; spikelets usually 2.5 to 2.8 mm. long, obovate-oval, brown at maturity, glabrous or the glume appressed-pubescent, the sterile lemma with short transverse wrinkles just inside the slightly raised

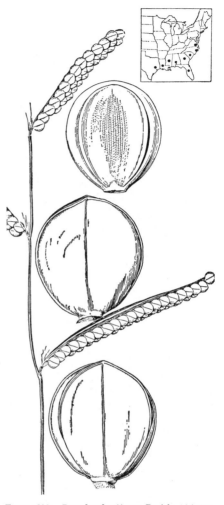

FIGURE 899.—*Paspalum lentiferum.* Panicle, × 1; two views of spikelet, and floret, × 10. (Harper 1629, Ga.)

margin. ♃ —Open ground or wet wood borders, Georgia and Florida to Texas, south to Argentina; throughout the West Indies.

PASPALUM NICORAE Parodi. Widely creeping, branching rhizomes; culms slender, erect or ascending; sheaths and blades, at least the lower, sparsely pilose, the blades sometimes minutely pubescent on the upper surface; racemes 3 or 4, appressed or ascending, the axis and rachis slender; spikelets about 3 mm. long, similar to those of *P. plicatulum* but slightly narrower and the sterile lemma less wrinkled. ♃ —Grown at the experiment station, Gainesville, Fla., the seed from southern Brazil.

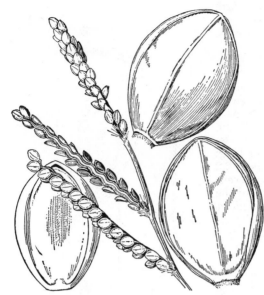

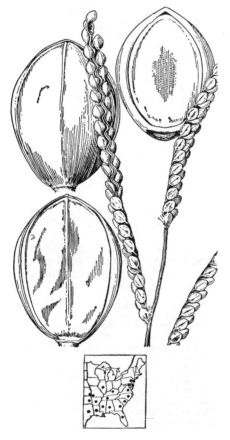

FIGURE 900.—*Paspalum difforme.* Panicle, × 1; two views of spike-let, and floret, × 10. (Type.)

44. Paspalum texánum Swallen. (Fig. 905.) Culms 70 to 110 cm. tall, erect or ascending from long rhizomes; sheaths much longer than the internodes, glabrous or papillose-hirsute toward the keeled summit, the uppermost bladeless; blades elongate, 2 to 6 mm. wide, flat, papillose-hirsute or papillose only to glabrous on both surfaces; racemes 4 to 6, ascending to suberect, 6 to 9 cm. long, the axis 6 to 13 cm. long; spikelets usually paired, 2.4 to 2.7 mm. long, 1.4 to 1.6 mm. wide, glabrous, the pedicels often to 2 mm. long; glume and sterile lemma thin, brownish, covering the fruit or slightly pointed beyond it, the lemma usually cross-wrinkled inside the margin; fruit 2.3 to 2.4 mm. long, chestnut brown at maturity. ♃ —Moist ground, southeastern Texas.

45. Paspalum hydróphilum Henr. (Fig. 906.) Aquatic; culms compressed, the submerged part lush, 1 to 2.5 m. long, with tufts of long roots at the nodes; sheaths and blades glabrous, the blades flat, lax, 7 to 15 cm. long, 3 to 7 mm. wide, glabrous; racemes 2 or 3, ascending, 5 to 10

FIGURE 901.—*Paspalum floridanum.* Panicle, × 1; two views of spikelet, and floret, × 10. (Chase 4221, Fla.)

cm. long; spikelets mostly paired, 2.8 to 3 mm. long, 1.3 to 1.4 mm. wide, elliptic, glabrous; glume and sterile lemma thin, olive brown, covering the fruit or minutely pointed beyond it; fruit light brown at maturity. ♃ —Irrigation ditches, Louisiana; southern Brazil and Paraguay.

46. Paspalum bosciánum Flügge. BULL PASPALUM. (Fig. 907.) Rather succulent annual, branching at base and commonly from the middle nodes, usually conspicuously brownish purple, glabrous as a whole; culms 40 to 60 cm. long, ascending or widely spreading; sheaths broad, loose; blades 10 to 40 cm. long, 8 to 15 mm. wide, papillose-pilose on upper surface near base; racemes 4 to 12, usually 4 to 7 cm. long; rachis 2 to 2.5 mm. wide; spikelets crowded, obovate-orbicular, 2 to 2.2 mm. long, glabrous, rust brown at maturity. ⊙ (Depauperate specimens have been described as *P. scrobiculatum* L.)—Moist or wet open ground, along ditches and ponds, sometimes a weed in cultivated fields, Pennsylvania (ballast), Virginia to Florida, Louisiana, Arkansas, and Texas, south to Brazil.

47. Paspalum convéxum Humb. and Bonpl. ex Willd. (Fig. 908.) Culms

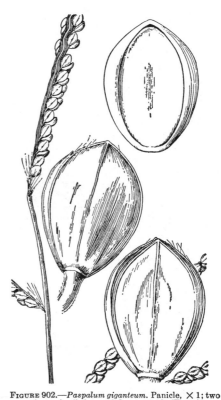

FIGURE 902.—*Paspalum giganteum.* Panicle, × 1; two views of spikelet, and floret, × 10. (Type.)

mostly 20 to 40 cm. tall, geniculate-ascending or widely spreading, leafy,

FIGURE 903.—*Paspalum virgatum.* Panicle, × 1; two views of spikelet, and floret, × 10. (Hitchcock 9555, Jamaica.)

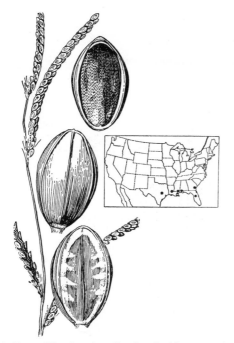

FIGURE 904.—*Paspalum plicatulum.* Panicle, × 1; two views of spikelet, and floret, × 10. (Chase 7061, Ga.)

branching from the lower and middle nodes; leaves conspicuously papillose-pilose or sometimes nearly glabrous; racemes 1 to 4, ascending or spreading, 2 to 4 cm. long; spikelets 2.2 to 3 mm. long, obovate, pubescent to glabrous; fruit dark brown, shining. ⊙ —Roadsides, Texas (Jasper County); northern Mexico to Brazil; Cuba, Trinidad.

Paspalum scrobiculátum L. Stouter and with larger spikelets, unequally biconvex, the sterile lemma loose and wrinkled. ⊙ —Ballast, Camden, N. J.; Abilene, Tex.; Asia; cultivated in India.

15. Bífida.—A single species approaching *Panicum;* spikelets turgid; a minute first glume usually developed.

48. Paspalum bífidum (Bertol.) Nash. (Fig. 909.) Culms erect from short rhizomes, 50 to 120 cm. tall; blades flat, 10 to 50 cm. long, 3 to 14 mm. wide, villous to nearly glabrous;

racemes usually 3 or 4, at first erect, 4 to 16 cm. long; rachis slender, subflexuous; spikelets distant to irregularly approximate, elliptic-obovate, 3.3 to 4 mm. (rarely to 4.2 mm.) long; second glume and sterile lemma conspicuously nerved. ♃ —Sandy pine and oak woods, occasionally in hammocks, nowhere common, on the Coastal Plain from Virginia to Florida, Tennessee, Texas, and Oklahoma. A specimen from Virginia with villous foliage, long-exserted panicle, and spikelets 4.2 mm. long has been named *P. bifidum* var. *projectum* Fernald. The various differences, pubescence, long-exserted panicles, diverging lower raceme, spikelets 4.2 mm. long, and longer first glume are found, not coordinated, in occasional specimens throughout the range.

16. Malacóphylla.—A single species in North America; both glumes of spikelet suppressed (the second half as long as the spikelet in one species), the fertile

FIGURE 905.—*Paspalum texanum.* Panicle, × 1; spikelet, and floret, × 10. (Type.)

FIGURE 906.—*Paspalum hydrophilum.* Node with rootlets, and panicle, × 1; two views of spikelet, and floret, × 10. (Silveus 4199, La.)

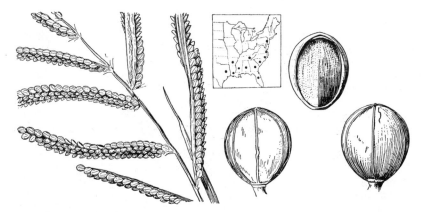

FIGURE 907.—*Paspalum boscianum.* Panicle, × 1; two views of spikelet, and floret, × 10. (Kearney 152, Fla.)

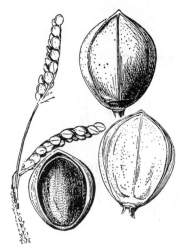

FIGURE 908.—*Paspalum convexum.* (Palmer 592 in 1886; Mexico.)

lemma strongly longitudinally ridged.

Paspalum malacóphyllum Trin. RIBBED PASPALUM. Culms rather coarse, 1 to 2 m. tall; blades flat, 8 to 35 mm. wide, the lower narrowed to a slender base; panicles nodding, the usually numerous racemes approximate; spikelets 1.8 to 2 mm. long, glabrous; second glume wanting; fertile lemma strongly ridged. ♃ — Mexico to Bolivia and Argentina. Introduced in the Southern States. Occasionally grown for hay and sometimes used in soil conservation work.

FIGURE 909.—*Paspalum bifidum.* Panicle, × 1; two views of spikelet, and floret, × 10. (Curtiss 5590, Fla.)

137. PÁNICUM L. PANICUM

Spikelets more or less compressed dorsiventrally, in open or compact panicles, rarely racemes; glumes 2, herbaceous, nerved, usually very unequal, the first often minute, the second typically equaling the sterile lemma, the latter of the same texture and simulating a third glume, bearing in its axil a membranaceous or hyaline palea and sometimes a staminate flower, the palea rarely wanting; fertile lemma chartaceous-indurate, typically obtuse, the nerves obsolete, the margins inrolled over an enclosed palea of the same texture. Annuals or perennials of various habit. Type species, *Panicum miliaceum.* *Panicum*, an old Latin name for the common millet (*Setaria italica*).

Panicum miliaceum, proso millet, is cultivated to a limited extent in this country for forage. In Europe it is sometimes cultivated for the seed which is used for food. Two species are commonly cultivated in the lowland tropics for forage, *P. maximum,* Guinea grass, an African species, said to have been introduced into Jamaica in 1774, and *P. purpurascens,* Para grass, introduced into Brazil from Africa. Certain native species are constituents of wild hay or of the range. *P. virgatum,* switch grass, and *P. stipitatum,* of the eastern half of the United States, *P. bulbosum* and *P. obtusum,* of the Southwest, and *Panicum texanum* in Texas furnish hay or forage. The seeds of *P. sonorum* Beal are used for food by the Cocopa Indians.

Axis of branchlets extending beyond the base of the uppermost spikelet as a point or bristle 1 to 6 mm. long.. SUBGENUS 1. PAUROCHAETIUM.
Axis of branchlets not extending into a bristle. (In *P. geminatum* and *P. paludivagum* the somewhat flattened axis is pointed but not bristle-form.)
Basal leaves usually distinctly different from those of the culm, forming a winter rosette; culms at first simple, the spikelets of the primary panicle not perfecting seed, later usually becoming much branched, the small secondary panicles with cleistogamous fruitful spikelets.. SUBGENUS 2. DICHANTHELIUM.
Basal leaves similar to the culm leaves, not forming a winter rosette; spikelets all fertile.
SUBGENUS 3. EUPANICUM.

Subgenus 1. Paurochaetium

Blades elongate, usually more than 15 cm. long, narrowed toward the base.
Spikelets about 3.5 mm. long... 3. P. REVERCHONI.
Spikelets about 2 mm. long, or less... 1. P. CHAPMANI.
Blades usually less than 10 cm. long, not narrowed toward the base; spikelets 2.5 to 3 mm. long.
Blades of midculm long-acuminate, usually 2 to 4 mm. wide............. 2. P. RAMISETUM.
Blades of midculm abruptly acute, usually 4 to 7 mm. wide................ 4. P. FIRMULUM.

Subgenus 2. Dichanthelium

Blades elongate, not more than 5 mm. wide, 20 times as long as wide; autumnal phase branching from the base only (from the lower nodes in *P. werneri*).
1. DEPAUPERATA.
Blades not elongate (or if so, more than 5 mm. wide and autumnal phase not branching from base).
Plants branching from the base, finally forming rosettes or cushions, the foliage soft, lax.
Blades prominently ciliate except in *P. laxiflorum*............................. 2. LAXIFLORA.
Plants branching from the culm nodes or rarely remaining simple.
Blades long, stiff; autumnal phase bushy-branched above.
Spikelets turgid, attenuate at base; mostly pustulose-pubescent; blades conspicuously striate, tapering from base to apex........................... 3. ANGUSTIFOLIA.
Spikelets scarcely turgid, not attenuate at base; blades tapering to both ends.
4. BICKNELLIANA.
Blades not long and stiff (somewhat so in *P. oligosanthes, P. malacon, P. commonsianum,* and *P. equilaterale*); not bushy-branched.
Plants not forming a distinct winter rosette; spikelets attenuate at base, papillose.
14. PEDICELLATA.
Plants forming a distinct winter rosette; spikelets not attenuate at base.
Spikelets turgid, blunt, strongly nerved (not strongly turgid in *P. oligosanthes*); blades rarely as much as 1.5 cm. wide (sometimes 2 cm. in *P. ravenelii* and *P. xanthophysum*).
Sheaths or some of them, papillose-hispid (sometimes all glabrous in *P. helleri*); spikelets 3 to 4 mm. long (2.7 to 3 mm. in *P. wilcoxianum*).
13. OLIGOSANTHIA.
Sheaths glabrous or minutely puberulent; spikelets 1.5 to 2.5 mm. long, asymmetrically pyriform; culms wiry... 12. LANCEARIA.
Spikelets not turgid, blunt, nor strongly nerved (somewhat so in *P. roanokense* and *P. caerulescens*).
Ligule of conspicuous hairs, usually 3 to 5 mm. long.
Sheaths glabrous or only the lowermost somewhat pubescent..... 7. SPRETA.
Sheaths strongly pubescent... 8. LANUGINOSA.
Ligule obsolete or nearly so (manifest in *P. oricola, P. tsugetorum,* and *P. curtifolium*).
Spikelets nearly spherical at maturity; blades glabrous, firm, cordate. Plants usually sparingly branching.. 10. SPHAEROCARPA.
Spikelets usually obovate or elliptic.
Blades of midculm elongate, less than 1.5 cm. wide. Culms usually tall; spikelets pointed, abruptly so in the velvety *P. scoparium*.
15. SCOPARIA.
Blades of midculm not elongate (somewhat so in *P. equilaterale*).

Blades cordate, 1 to 3 cm. wide (5 to 12 mm. in *P. ashei*). Spikelets
pubescent.
Spikelets 2.5 to 3 mm. long. Sheaths glabrous or minutely puberulent.
16. COMMUTATA.
Spikelets 3 to 5 mm. long (sometimes but 2.7 mm. long in the hispid-
sheathed *P. clandestinum*)................................... 17. LATIFOLIA.
Blades not cordate, less than 1 cm. wide.
Sheaths crisp- or appressed-pubescent. Blades firm; spikelets pubescent.
9. COLUMBIANA.
Sheaths glabrous or ciliate only in autumnal phase (sparsely pilose in
P. curtifolium, the lower velvety in *P. mattamuskeetense*, rarely
pilose in *P. roanokense* and *P. caerulescens*).
Vernal culms delicate (sometimes scarcely so in *P. albomarginatum*
and *P. tenue*); spikelets 1.5 mm. or less long (1.6 to 1.7 mm. in
P. tenue).. 11. ENSIFOLIA.
Vernal culms slender but not delicate, rarely less than 40 cm. tall;
spikelets 2 to 2.9 mm. long (1.5 mm. in *P. microcarpon* and *P.
caerulescens*).
Lower internodes short, upper elongate, producing a nearly naked
culm, leafy at base; spikelets narrowly ovate, 2.7 to 2.9 mm.
long... 5. NUDICAULIA.
Lower internodes not shorter, the vernal culms about evenly leafy
throughout; spikelets elliptic or obovate, not more than 2.5
mm. long... 6. DICHOTOMA.

1. Depauperata

Spikelets about 3.5 mm. long, beaked..................... 5. P. DEPAUPERATUM.
Spikelets 3 mm. long or less (sometimes 3.2 mm. long in *P. perlongum*), not beaked.
Culms single or few in a tuft; spikelets turgid, blunt, 2.7 to 3.2 mm. long, prairie plants.
6. P. PERLONGUM.
Culms in large tufts; spikelets not turgid, 2.2 to 2.7 mm. long; plants of woods.
Sheaths pilose; spikelets 2.2 to 2.7 mm. long, pilose.............. 7. P. LINEARIFOLIUM.
Sheaths glabrous; spikelets 2.2 to 2.3 mm. long, glabrous or sparingly pilose.
8. P. WERNERI.

2. Laxiflora

Sheaths retrorsely pilose; spikelets papillose-pilose.
Blades ciliate and more or less pilose on the surface; spikelets 2 mm. long.
10. P. XALAPENSE.
Blades glabrous or nearly so on the surface and margin; spikelets 2.2 mm. long.
9. P. LAXIFLORUM.
Sheaths not retrorsely pilose; spikelets pubescent or glabrous.
Spikelets pubescent, about 2 mm. long.................... 11. P. CILIATUM.
Spikelets glabrous.
Blades glabrous on the surface................... 12. P. POLYCAULON.
Blades pilose on the surface................... 13. P. STRIGOSUM.

3. Angustifolia

Nodes bearded; plants grayish-villous; autumnal blades flat.
Spikelets 2 mm. long................................ 16. P. CHRYSOPSIDIFOLIUM.
Spikelets 2.5 to 2.8 mm. long................................ 17. P. CONSANGUINEUM.
Nodes not bearded; plants villous only at base, or nearly glabrous; autumnal blades involute
or flat.
Autumnal blades flat; lower panicle branches spreading or reflexed, or loosely ascending.
Spikelets 2 mm. long; panicle branches loosely ascending.......... 15. P. BENNETTENSE.
Spikelets 2.5 to 2.8 mm. long; panicle branches widely spreading at anthesis.
18. P. ANGUSTIFOLIUM.
Autumnal blades involute; lower panicle branches more or less ascending.
Spikelets pointed beyond the fruit, fusiform.
Spikelets 3.3 to 3.5 mm. long..................... 20. P. FUSIFORME.
Spikelets mostly 2.3 to 2.5 mm. long (or to 3 mm. before maturity).
19. P. PINETORUM.
Spikelets not pointed beyond the fruit; obovate.
Plants glabrous or nearly so. Autumnal culms erect.
Spikelets subsecund along the suberect panicle branches.... 23. P. NEURANTHUM.

Spikelets not subsecund, the panicle loose and open.................... 22. P. OVINUM.
Plants pubescent, at least on the lower half.
Spikelets about 2.4 mm., rarely only 2.1 mm., or as much as 2.8 mm., long; vernal blades 7 to 12 cm. long; autumnal blades not falcate.
21. P. ARENICOLOIDES.
Spikelets not more than 2 mm. long; vernal blades 4 to 6 cm. long; autumnal blades much crowded, falcate.. 14. P. ACICULARE.

4. Bicknelliana

Spikelets 2.5 to 2.8 mm. long; blades not more than 9 mm. wide.......... 24. P. BICKNELLII.
Spikelets 3 mm. long; blades as much as 12 mm. wide.................... 25. P. CALLIPHYLLUM.

5. Nudicaulia

A single species... 26. P. NUDICAULE.

6. Dichotoma

1a. Nodes, at least the lower, bearded.
Spikelets 1.5 to 1.6 mm. long, glabrous (occasional individuals with pubescent spikelets).
27. P. MICROCARPON.
Spikelets 2 mm. long or more.
Spikelets glabrous, 2 mm. long.
Autumnal phase erect, branched like a little tree; primary blades rarely more than 5 mm. wide.................... 33. P. DICHOTOMUM.
Autumnal phase top-heavy-reclining; primary blades 6 to 10 mm. wide.
34. P. BARBULATUM.
Spikelets pubescent.
Blades all velvety; autumnal phase branching from upper nodes.
29. P. ANNULATUM.
Blades glabrous or nearly so, or only the lowermost velvety.
Primary blades mostly erect; autumnal phase sparingly branching, the branches erect; blades and panicles not much reduced.................... 32. P. BOREALE.
Primary blades spreading; blades and panicles of autumnal phase much reduced.
Spikelets 2 mm. long; autumnal phase profusely branching.... 28. P. NITIDUM.
Spikelets 2.2 mm. long or more; autumnal phase less profusely branching.
Sheaths and upper nodes glabrous.................... 31. P. CLUTEI.
Lower sheaths and all nodes pubescent.............. 30. P. MATTAMUSKEETENSE.
1b. Nodes not bearded.
2a. Spikelets pubescent.
Culms erect, never becoming vinelike.
Primary blades spreading; panicles purplish; fruit exposed at summit.
31. P. CLUTEI.
Primary blades erect; panicles green; fruit covered (woods forms with spreading blades may be distinguished from P. dichotomum by pubescent spikelets, 2.2 mm. long)................................ 32. P. BOREALE.
Culms soon prostrate, vinelike, the branches divaricate.
Plants bright green, culms lax; spikelets not more than 2.1 mm. long.
38. P. LUCIDUM.
Plants grayish green, culms stiff; spikelets 2.5 mm. long........ 39. P. SPHAGNICOLA.
2b. Spikelets glabrous.
Culms soon prostrate.
Plants bright green, culms lax; spikelets not more than 2.1 mm. long.
38. P. LUCIDUM.
Plants grayish green, culms stiff; spikelets 2.5 mm. long........ 39. P. SPHAGNICOLA.
Culms erect, or the autumnal phase topheavy, never prostrate.
Spikelets not more than 1.6 mm. long; panicles narrow; plants glaucous bluish green.
37. P. CAERULESCENS.
Spikelets 2 mm. long or more; panicles open.
Blades erect, firm; spikelets turgid, strongly nerved; plants grayish olive green.
36. P. ROANOKENSE.
Blades spreading; spikelets not turgid.
Spikelets 2.2 mm. long or more, pointed; sheaths bearing pale glandular spots.
35. P. YADKINENSE.
Spikelets not more than 2 mm. long, not pointed.
Autumnal phase erect, branched like a little tree; primary blades rarely more than 5 mm. wide; second glume shorter than fruit and sterile lemma.
33. P. DICHOTOMUM.

Autumnal phase topheavy-reclining; primary blades 6 to 10 mm. wide; second
glume equaling fruit and sterile lemma.................. 34. P. BARBULATUM.

7. Spreta

Panicle narrow, one-fourth to one-third as wide as long............................ 40. P. SPRETUM.
Panicle open, two-thirds as wide as long, or more.
  Spikelets 1.5 mm. long.. 41. P. LINDHEIMERI.
  Spikelets 1.3 mm. long or less.
    Culms and sheaths glabrous.. 43. P. LONGILIGULATUM.
    Culms and sheaths appressed-pubescent.
      Spikelets 1.2 to 1.3 mm. long.............. 42. P. LEUCOTHRIX.
      Spikelets not more than 1 mm. long.. 44. P. WRIGHTIANUM.

8. Lanuginosa

1a. Spikelets not more than 2 mm. long.
  2a. Plants grayish, velvety-pubescent.
    Spikelets 1.3 to 1.5 mm. long; autumnal blades involute-pointed (see also *P. albe-
marlense*)... 51. P. AUBURNE.
    Spikelets 1.8 to 2 mm. long; autumnal blades flat.
      Plants dark or olive green when dry: spikelets 1.9 to 2 mm. long.... 52. P. THUROWII.
      Plants light or yellow green when dry.
        Autumnal phase prostrate, branching from base and lower nodes, forming close
mats; blades not ciliate. Around hot springs...................... 57. P. THERMALE.
        Autumnal phase ascending or spreading, branching from middle and upper nodes,
the reduced, fascicled blades strongly ciliate................ 50. P. LANUGINOSUM.
  2b. Plants pubescent, often villous, but not velvety.
    3a. Culms conspicuously pilose with long, horizontally spreading hairs. Culms branch-
ing before expansion of primary panicles........................ 53. P. PRAECOCIUS.
    3b. Culms variously pubescent, if pilose the hairs not long and horizontally spreading.
      4a. Vernal blades glabrous or nearly so on the upper surface, firm in texture.
        Autumnal culms branching from the lower nodes, forming a spreading bunch 10
to 15 cm. high; Pacific slope..................... 55. P. OCCIDENTALE.
        Autumnal culms branching from the middle nodes, forming widely spreading mats;
Atlantic slope (see also form of *P. huachucae* var. *fasciculatum*).
            49. P. TENNESSEENSE.
      4b. Vernal blades pubescent on upper surface, sometimes pilose near base and mar-
gins only.
        5a. Spikelets 1.3 to 1.5 mm. long; vernal blades long-pilose on upper surface.
          Autumnal phase widely decumbent-spreading, forming a mat; vernal culms soon
geniculate-spreading; plants olivaceous.................. 46. P. ALBEMARLENSE.
          Autumnal phase erect or leaning, never forming a mat; plants yellowish green.
            Axis of panicle pilose, panicle branches tangled, the lower drooping.
              47. P. IMPLICATUM.
            Axis of panicle puberulent only, panicle branches not tangled, the lower
ascending................................ 45. P. MERIDIONALE.
        5b. Spikelets 1.6 to 2 mm. long; vernal blades pilose or pubescent.
          Upper surface of blades pilose; spikelets 1.8 to 2 mm. long; autumnal phase
decumbent-spreading.
            Spikelets pointed; culms weak and lax................. 58. P. LANGUIDUM.
            Spikelets obtuse; culms not weak and lax.
              Culms leafy below, branching from base and lower nodes; Maine to Minne-
sota.. 54. P. SUBVILLOSUM.
              Culms evenly leafy, branching from upper nodes; Pacific slope.
                56. P. PACIFICUM.
          Upper surface of blades appressed-pubescent or pilose toward the base only;
spikelets 1.6 to 1.8 mm. long; autumnal phase not decumbent-spreading.
            48. P. HUACHUCAE.
1b. Spikelets 2.2 mm. long or more.
  Spikelets 2.2 to 2.5 mm. long.
    Pubescence on culms horizontally spreading; autumnal phase freely branching.
      59. P. VILLOSISSIMUM.
    Pubescence on culms appressed or ascending; autumnal phase rather sparingly branch-
ing.
      Upper internodes shortened, the leaves approximate, the blades often nearly equaling
the panicle.
        Blades glabrous or nearly so on the upper surface; spikelets 2.2 to 2.5 mm. long;
first glume glabrous.................. 60. P. BENNERI.

Blades sparsely hispid on the upper surface; spikelets 2.2 to 2.3 mm. long; first glume pubescent.. 63. P. SCOPARIOIDES.
Upper internodes not shortened, the copious pubescence silky.
61. P. PSEUDOPUBESCENS.
Spikelets 2.7 to 2.9 mm. long.
Culms stiff; blades conspicuously ciliate; southern Atlantic coast........ 62. P. OVALE.
Culms weak; blades not ciliate; Pacific coast.................................... 64. P. SHASTENSE.

9. Columbiana

1a. Spikelets 2 to 3.2 mm. long, mostly elliptic.
Winter blades 5 to 10 cm. long; spikelets 2 mm. long; plants blue-green.
69. P. WILMINGTONENSE.
Winter blades 1 to 3 cm. long.
Spikelets 3.2 mm. long; first glume conspicuously distant................. 65. P. MALACON.
Spikelets not more than 2.9 mm. long; first glume not distant.
Spikelets 2.8 to 2.9 mm. long; vernal blades 8 to 15 cm. long.......... 66. P. DEAMII.
Spikelets not more than 2.4 mm. long; vernal blades not more than 8 cm. long.
Spikelets about 2.4 mm. (2.2 to 2.4 mm.) long; panicle open, the branches stiffly spreading.. 67. P. COMMONSIANUM.
Spikelets 2 to 2.1 mm. long; panicle rather dense, the branches ascending.
68. P. ADDISONII.
1b. Spikelets not more than 1.9 mm. long, obovate, turgid.
Culms crisp-puberulent or appressed-pubescent with crimped hairs; plants bluish or grayish green; panicle about 3 to 7 cm. long.
Spikelets 1.8 to 1.9 mm. long... 70. P. TSUGETORUM.
Spikelets 1.5 to 1.6 mm. long... 71. P. COLUMBIANUM.
Culms appressed or ascending-pilose; plants olivaceous; panicle rarely more than 3 cm. long. Spikelets not more than 1.5 mm. long, rounded and turgid.
Spikelets 1.5 mm. long; culms rather stout; autumnal phase branching from all the nodes... 72. P. ORICOLA.
Spikelets 1.3 to 1.4 mm. long; culms very slender; autumnal phase with branches mostly aggregate toward the summit............. 71. P. COLUMBIANUM var. THINIUM.

10. Sphaerocarpa

Culms spreading; blades obscurely nerved; panicle nearly as broad as long.
73. P. SPHAEROCARPON.
Culms erect or ascending; blades rather strongly nerved; panicle néver more than two-thirds as broad as long, usually less.
Spikelets 1.5 to 1.6 mm. long; blades lanceolate, the upper not reduced.
74. P. POLYANTHES.
Spikelets 1 to 1.2 mm. long; blades tapering from base to apex, the upper much smaller than the lower.. 75. P. ERECTIFOLIUM.

11. Ensifolia

Ligules about 1 mm. long; sheaths or some of them sparsely spreading-pilose.
83. P. CURTIFOLIUM.
Ligules obsolete or nearly so; pubescence if present not spreading.
Blades prominently white-margined, firm; spikelets densely puberulent.
Blades puberulent beneath, often above; sheaths and sometimes lower internodes ascending-pubescent.. 76. P. TENUE.
Blades glabrous; sheaths glabrous or minutely ciliate only.
Uppermost culm blades much reduced; culms branching from lower nodes only, the branches repeatedly branching........................... 77. P. ALBOMARGINATUM.
Uppermost culm blades about as long as the others; culms bearing short branches from the upper and middle nodes................................. 78. P. TRIFOLIUM.
Blades not white-margined or very obscurely so (or if white margin evident spikelets only 1.1 mm. long); spikelets glabrous or puberulent.
Culms branching only at base; plants soft, light green.................. 82. P. VERNALE.
Culms branching at the nodes; plants firm or at least not soft.
Spikelets glabrous.
Spikelets 1.1 to 1.2 mm. long; blades rarely as much as 5 cm. long.
84. P. CHAMAELONCHE.
Spikelets 1.2 to 1.5 mm. long.
Blades elongate, at least some of them 8 to 10 cm. long.
85. P. GLABRIFOLIUM.
Blades not more than 3 cm. long...................... 81. P. ENSIFOLIUM.

Spikelets puberulent.
  Spikelets 1.1 mm. long. Winter blades bluish green, not glossy.
                                                 80. P. CONCINNIUS.
  Spikelets 1.3 to 1.5 mm. long.
    Blades involute, falcate, with long stiff hairs on margin near base. Plants stiff
      and wiry... 86. P. BREVE.
    Blades not involute, or at tip only, not falcate.
      Plants bright green; winter blades conspicuous, glossy green.
                                      79. P. FLAVOVIRENS.
      Plants olive; winter blades not conspicuous nor glossy.... 81. P. ENSIFOLIUM.

12. Lancearia

Spikelets, 1.5 to 1.6 mm. long... 87. P. PORTORICENSE.
Spikelets 2 mm. long or more.
  Blades, or some of them, at least 8 mm. wide, glabrous on the upper surface; fruit papil-
    lose-roughened.. 90. P. WEBBERIANUM.
  Blades not more than 6 mm. wide (or if wider, puberulent on the upper surface); fruit
    smooth and shining.
    Spikelets 2.4 to 2.6 mm. long. Blades narrowed toward the base.
                                        91. P. PATENTIFOLIUM.
    Spikelets not more than 2.1 mm. long.
      Blades firm, glabrous above; culms stiffly ascending............... 88. P. LANCEARIUM.
      Blades lax, softly puberulent on both surfaces; culms decumbent.... 89. P. PATULUM.

13. Oligosanthia

Nodes bearded; blades velvety-pubescent beneath.
  Plants lax, soft-velvety throughout; spikelets not more than 3 mm. long.
                                        93. P. MALACOPHYLLUM.
  Plants stiff, pubescence harsh; spikelets about 4 mm. long.............. 97. P. RAVENELII.
Nodes not bearded (or but obscurely so in *P. wilcoxianum*); blades not velvety.
  Panicle narrow, branches erect, or spreading only at anthesis. Blades erect.
    Spikelets not more than 3 mm. long; blades not more than 6 mm. wide.
                                        92. P. WILCOXIANUM.
    Spikelets 3.7 to 4 mm. long; blades 8 to 20 mm. wide.
      Blades papillose-hispid.. 98. P. LEIBERGII.
      Blades glabrous on both surfaces..................................... 99. P. XANTHOPHYSUM.
  Panicle about as wide as long.
    Spikelets narrowly obovate, subacute; plants olivaceous; appressed-pubescent.
                                        96. P. OLIGOSANTHES.
    Spikelets broadly obovate, turgid, blunt; plants green, the pubescence, if present, not
      appressed.
      Blades erect, not more than 6 mm. wide; plants copiously hirsute throughout.
                                      92. P. WILCOXIANUM.
      Blades ascending or spreading, rarely less than 8 mm. wide, usually wider; plants not
        hirsute throughout.
        Spikelets 3.2 to 3.3 mm. long; blades firm; sheaths or some of them more or less
          hispid.. 95. P. SCRIBNERIANUM.
        Spikelets not more than 3 mm. long; blades rather thin; sheaths, or some of them,
          glabrous or sparsely hispid... 94. P. HELLERI.

14. Pedicellata

Culms erect or leaning; blades thin, 5 to 9 cm. long, narrowed toward the base.
                                        100. P. PEDICELLATUM.
Culms decumbent; blades thick, not more than 5 cm. long, not narrowed toward the base.
                                        101. P. NODATUM.

15. Scoparia

Pubescence soft-villous or velvety. Spikelets abruptly pointed............ 102. P. SCOPARIUM.
Pubescence when present not velvety.
  Spikelets elliptic.
    Spikelets 3 mm. long; second glume and sterile lemma pointed beyond the fruit.
                                      103. P. ACULEATUM.
    Spikelets not more than 2.8 mm. long; second glume and sterile lemma scarcely, if at
      all, pointed beyond the fruit.

Culms glabrous; sheaths not viscid-spotted; spikelets 2.2 to 2.9 mm. long, sparsely
   pubescent_____ 104. P. RECOGNITUM.
Culms pilose with ascending hairs, the nodes densely pubescent; sheaths conspicu-
   ously viscid-spotted; spikelets 1.8 to 2.2 mm. long, densely pubescent.
                                                         105. P. MUNDUM.
Spikelets ovate, that is, broadest below the middle.
Sheaths or some of them hispid, rarely glabrous; autumnal phase with crowded branch-
   lets_____ 106. P. SCABRIUSCULUM.
Sheaths glabrous; autumnal phase sparingly branching_____ 107. P. CRYPTANTHUM.

16. Commutata

Plants glaucous, glabrous; basal blades conspicuously ciliate; vernal culms usually solitary.
                                                         110. P. MUTABILE.
Plants not glaucous.
Blades nearly linear, that is, with parallel margins; first glume about half as long as the
   spikelet_____ 112. P. EQUILATERALE.
Blades lanceolate.
Culms crisp-puberulent; blades usually rigid, symmetrical, rarely more than 10 mm.
   wide; spikelets about 2.5 mm. long_____ 108. P. ASHEI.
Culms glabrous or softly puberulent; blades firm or lax; spikelets 2.7 to 3.2 mm. long.
Culms erect, or autumnal phase leaning; blades symmetrical, broadly cordate.
                                                         109. P. COMMUTATUM.
Culms decumbent; blades usually asymmetrical and falcate, narrowed to the scarcely
   cordate base_____ 111. P. JOORII.

17. Latifolia

Sheaths strongly papillose-hispid, at least the lower and those of the branches.
                                                         113. P. CLANDESTINUM.
Sheaths glabrous or softly villous.
Nodes glabrous; spikelets 3.4 to 3.7 mm. long_____ 114. P. LATIFOLIUM.
Nodes bearded; spikelets 4 to 4.5 mm. long_____ 115. P. BOSCII.

Subgenus 3. Eupanicum

1a. Plants annual.
Inflorescence consisting of several more or less secund spikelike racemes; fruit transversely
   rugose_____ 3. FASCICULATA.
Inflorescence a more or less diffuse panicle.
Spikelets tuberculate_____ 13. VERRUCOSA.
Spikelets not tuberculate.
First glume not more than one-fourth the length of the spikelet, truncate or triangu-
   lar-tipped_____ 4. DICHOTOMIFLORA.
First glume usually as much as half the length of the spikelet, acute or acuminate.
Blades linear; spikelets more than 1.7 mm. long, the second glume and sterile lemma
   pointed beyond the fruit_____ 5. CAPILLARIA.
Blades ovate-lanceolate; spikelets about 1.3 mm. long, the second glume and sterile
   lemma not pointed beyond the fruit_____ 7. TRICHOIDEA.
1b. Plants perennial.
2a. Spikelets short-pediceled along one side of the rachises, forming spikelike racemes
   (compare Agrostoidia with 1-sided but not spikelike panicle branches).
First glume nearly equaling the sterile lemma.
Racemes spreading; fruit not more than one-third the length of the spikelet.
                                                         17. GYMNOCARPA.
Racemes appressed; fruit nearly as long as the spikelet_____ 15. OBTUSA.
First glume much shorter than the sterile lemma.
Fruit transversely rugose.
Nodes glabrous_____ 1. GEMINATA.
Nodes bearded_____ 2. PURPURASCENTIA.
Fruit not rugose_____ 16. HEMITOMA.
2b. Spikelets in open or sometimes contracted or congested panicles (somewhat 1-sided
   in Agrostoidia).
Fruit transversely rugose (obscurely so in P. plenum)_____ 8. MAXIMA.
Fruit not transversely rugose.
Spikelets villous_____ 14. URVILLEANA.
Spikelets glabrous.
Sterile palea enlarged and indurate at maturity, expanding the spikelet. Blades

scarcely wider than their sheaths; spikelets about 2.3 mm. long, borne toward
the ends of the few slender branches.. 12. Laxa.
Sterile palea, if present, not enlarged.
Plants with conspicuous creeping scaly rhizomes.
Spikelets long-pediceled, not secund, arranged in an open or contracted
panicle.. 9. Virgata.
Spikelets short-pediceled, more or less secund along the nearly simple panicle
branches... 11. Agrostoidia.
Plants without creeping scaly rhizomes.
Panicles narrow and few-flowered; culms erect and wiry; blades drying in-
volute.. 10. Tenera.
Panicles open or contracted, many-flowered.
Spikelets short-pediceled along the nearly simple panicle branches.
11. Agrostoidia.
Spikelets long-pediceled, the panicle open............................... 6. Diffusa.

1. Geminata

Spikelets 3 mm. long; glumes and sterile lemma papery.................. 117. P. paludivagum.
Spikelets not more than 2.4 mm. long; glumes and sterile lemma not papery.
116. P. geminatum.

2. Purpurascentia

A single species.. 118. P. purpurascens.

3. Fasciculata

Spikelets 5 to 6 mm. long.. 124. P. texanum.
Spikelets 2 to 4 mm. long.
Spikelets strongly reticulate-veined, 2 to 3 mm. long, glabrous.... 120. P. fasciculatum.
Spikelets scarcely reticulate-veined or only near apex.
Spikelets not more than 2 mm. long, glabrous..................... 119. P. reptans.
Spikelets more than 3 mm. long, pubescent.
Rachis scabrous but not bristly; spikelets acuminate-pointed, 4 to 4.5 mm. long.
121. P. adspersum.
Rachis and/or pedicels bristly-hirsute; spikelets acute, pubescent, or sometimes
glabrous, 3 to 4 mm. long.
Rachis and pedicels bristly-hirsute; blades lanceolate, rarely more than 7 mm. wide.
123. P. arizonicum.
Rachis scabrous, only the pedicels bristly-hirsute; blades ovate-lanceolate, as much
as 2 cm. wide.. 122. P. ramosum.

4. Dichotomiflora

Plants perennial; blades elongate, 2 to 3 mm. wide.................... 127. P. lacustre.
Plants annual; blades mostly 5 to 15 mm. wide.
Sheaths glabrous.. 125. P. dichotomiflorum.
Sheaths papillose... 126. P. bartowense.

5. Capillaria

Panicles drooping; spikelets 4.5 to 5 mm. long....................... 138. P. miliaceum.
Panicles erect; spikelets not more than 4 mm. long.
Panicles more than half the length of the entire plant.
Panicles narrow, usually less than half as broad as long................. 128. P. flexile.
Panicles as broad as long.
Fruit without scar at base... 133. P. capillare.
Fruit with a lunate scar at base..................................... 134. P. hillmani.
Panicles not more than one-third the entire height of the plant.
Spikelets not more than 2.2 mm. long, acute but not long-acuminate (see also P. hir-
sutum).
Culms relatively stout; blades about 1 cm. wide; spikelets turgid.
129. P. gattingeri.
Culms slender; blades not more than 6 mm. wide; spikelets not turgid.
Spikelets 1.7 mm. long; foliage green.
Axillary pulvini pilose... 130. P. philadelphicum.
Axillary pulvini glabrous... 132. P. tuckermani.
Spikelets 2.1 to 2.2 mm. long; foliage conspicuously tinged with purple, the blades
erect.. 131. P. lithophilum.

Spikletes 2.7 to 4 mm. long, acuminate.
First glume about one-third the length of the spikelet, subacute or blunt.
137. P. STRAMINEUM.
First glume usually more than half the length of the spikelet, acuminate.
First glume more than three-fourths the length of the spikelet; spikelets 4 mm. long.
136. P. PAMPINOSUM.
First glume half to two-thirds the length of the spikelet; spikelets not more than
3.3 mm. long.. 135. P. HIRTICAULE.

6. Diffusa

Second glume and sterile lemma elongate, at least three times as long as the fruit.
139. P. CAPILLARIOIDES.
Second glume and sterile lemma not elongate.
Culms as much as 1 cm. thick; blades 2 cm. wide or more .... 144. P. HIRSUTUM.
Culms slender; blades not more than 1 cm. wide.
Spikelets 4 to 4.2 mm. long; midnerves of glumes and sterile lemma scabrous toward
the apex.. 142. P. LEPIDULUM.
Spikelets usually less than 3.5 mm. long.
Blades hirsute on both surfaces (sometimes glabrescent), not at all glaucous.
143. P. GHIESBREGHTII.
Blades glabrous on both surfaces or with a few hairs on either surface, glaucous above.
Panicle much exceeding the leaves; spikelets 3 to 3.5 mm. long (rarely 3.7 mm.).
141. P. HALLII.
Panicle usually equaled or exceeded by the uppermost blades; spikelets 2 to 2.6 mm.
long.. 140. P. FILIPES.

7. Trichoidea

A single species.. 145. P. TRICHOIDES.

8. Maxima

Culms with a cormlike base.. 148. P. BULBOSUM.
Culms not cormlike at base.
Nodes hirsute; ligules 4 to 6 mm. long; fruit strongly rugose............. 146. P. MAXIMUM.
Nodes glabrous; ligules 2 mm. long; fruit obscurely rugose................. 147. P. PLENUM.

9. Virgata

Spikelets not more than 2.5 mm. long, first glume less than half the length of the spikelet.
Panicle loosely flowered; first glume truncate, about one-fifth the length of the spikelet.
149. P. REPENS.
Panicle rather densely flowered; first glume triangular, about one-third the length of the
spikelet.. 150. P. GOUINI.
Spikelets 3 to 7 mm. long (sometimes less than 3 mm. in P. virgatum var. cubense); first
glume more than half the length of the spikelet.
Panicle elongate, strongly contracted; seacoast plants.
Culms rarely 1 m. tall, solitary from the nodes of the horizontal rhizome.
153. P. AMARUM.
Culms 1 to 2 m. tall, in dense tufts................................... 154. P. AMARULUM.
Panicle diffuse, or only slightly contracted; plants sometimes of salt marshes but not
littoral.
Spikelets 6 to 8 mm. long; culms solitary, with a creeping base.... 152. P. HAVARDII.
Spikelets less than 5 mm. long (in exceptional specimens 6 mm. long); culms erect, pro-
ducing numerous scaly rhizomes................................... 151. P. VIRGATUM.

10. Tenera

A single species.. 155. P. TENERUM.

11. Agrostoidea

Rhizomes present; culms but little compressed; spikelets set obliquely on the appressed
pedicels.
Panicles open; spikelets 3.4 to 3.8 mm. long (shorter in exceptional specimens).
162. P. ANCEPS.
Panicles more or less contracted; spikelets not more than 2.8 mm. long.
163. P. RHIZOMATUM.
Rhizomes wanting; culms strongly compressed with keeled sheaths; spikelets not obliquely
disposed (except in P. abscissum).

Ligule ciliate; basal leaves half as long as the culm or more; panicle much exceeding the upper leaves.
Spikelets not more than 2.7 mm., usually 2.5 mm. long, the first glume less than half that length; ligule 2 to 3 mm. long.. 160. P. LONGIFOLIUM.
Spikelets 3 to 3.5 mm. long; first glume two-thirds to three-fourths that length; ligule less than 1 mm. long... 161. P. COMBSII.
Ligule erose or lacerate, not ciliate; basal leaves in short tufts, the upper usually nearly equaling the terminal panicle.
Fruit stipitate; spikelets 2.5 to 2.8 mm. long, conspicuously secund.
159. P. STIPITATUM.
Fruit not stipitate; spikelets not conspicuously secund.
Sheaths much broader at the summit than the base of the blades, truncate or auriculate... 156. P. ABSCISSUM.
Sheaths about as wide at the summit as the base of the blades.
Spikelets 1.8 to 2 mm., in occasional specimens 2.2 mm. long; panicle branches ascending or spreading.. 157. P. AGROSTOIDES.
Spikelets about 2.5 mm. long; panicle branches erect or nearly so.
158. P. CONDENSUM.

12. Laxa

A single species.. 164. P. HIANS.

13. Verrucosa

Spikelets about 2 mm. long, glabrous, warty.................. 165. P. VERRUCOSUM.
Spikelets more than 3 mm. long, hispid.................. 166. P. BRACHYANTHUM.

14. Urvilleana

A single species.. 167. P. URVILLEANUM.

15. Obtusa

A single species.. 168. P. OBTUSUM.

16. Hemitoma

A single species.. 159. P. HEMITOMON.

17. Gymnocarpa

A single species.. 170. P. GYMNOCARPON.

SUBGENUS 1. PAUROCHAÉTIUM Hitchc. and Chase

Perennials; culms tufted, erect, blades not more than 7 mm. wide; panicle slender, the branches short, appressed, the ultimate branchlets bearing 1 to several spikelets, produced beyond the uppermost spikelet as a bristle 1 to 6 mm. long; spikelets much swollen on the face, glabrous, strongly nerved; fruit transversely rugose, apiculate.

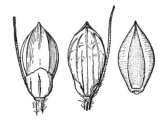

FIGURE 910.—*Panicum chapmani.* Panicle, × 1; two views of spikelet, and floret, × 10. (Type.)

1. Panicum chapmáni Vasey. (Fig. 910.) Culms ascending or spreading, slender, wiry, 40 to 100 cm. tall; blades erect, rather firm, 15 to 40 cm. long, 2 to 5 mm. wide, more or less involute when dry; panicle mostly 20 to 30 cm. long; bristle 3 to 6 mm. long; spikelets 2 to 2.2 mm. long, obovate; first glume about one-third as long as the spikelet, obtuse or truncate. ♃ —Coral sand and shell mounds, southern Florida; Bahamas; Yucatan.

2. Panicum ramisétum Scribn. (Fig. 911.) Culms erect or ascending from short horizontal rhizomes, 25 to

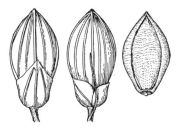

FIGURE 911.—*Panicum ramisetum*. Two views of spikelet, and floret, × 10. (Type.)

60 cm. tall; blades 5 to 12 cm. long, 2 to 4 mm. wide; panicle 5 to 20 cm.

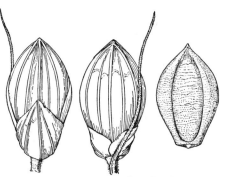

FIGURE 913.—*Panicum firmulum*. Two views of spikelet, and floret, × 10. (Type.)

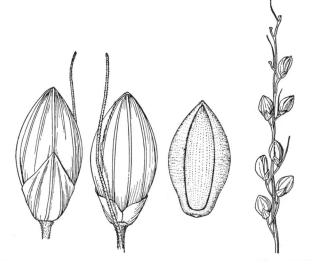

FIGURE 912.—*Panicum reverchoni*. Panicle, × 2; two views of spikelet, and floret, × 10. (Type.)

long; bristle not exceeding the spikelet; spikelets about 2.5 mm. long, obovate; first glume about half as long as the spikelet. ♃ —Sandy plains and prairies, southern Texas and northern Mexico.

3. **Panicum reverchóni** Vasey. (Fig. 912.) Culms stiffly erect, from short rhizomes, 30 to 70 cm. tall; blades erect, stiff, 5 to 20 cm. long, 2 to 3 mm. wide; panicle 5 to 20 cm. long; spikelets 1 to 4 to a branchlet, the bristle equaling or exceeding the spikelet; spikelets 3.5 to 3.8 mm. long, elliptic; first glume about half as long

as the spikelet. ♃ —Rocky or sandy prairies and limestone hills, Texas.

4. **Panicum fírmulum** Hitchc. and Chase. (Fig. 913.) Culms ascending or decumbent at base, 30 to 40 cm. tall, rather loosely tufted from creeping knotted rhizomes as much as 5 cm. long; blades ascending or spreading, firm, 4 to 10 cm. long, 4 to 7 mm. wide; bristle 1 to 2 times as long as the spikelet; spikelets 3 to 3.2 mm. long, obovate; first glume half as long as the spikelet. ♃ — Sandy prairies, southern Texas.

SUBGENUS 2. DICHANTHÉLIUM Hitchc. and Chase

Perennial, from a crown, rarely from short matted rhizomes, surrounded by a more or less well-marked rosette of usually short winter leaves, in spring producing simple culms with mostly narrowly lanceolate blades and terminal panicles with numerous spikelets, these rarely perfecting seed (or occasionally in Lanuginosa group and in *P. clandestinum*); early culms branching at some or all of the nodes (in a few species from the base only) after the maturity of the primary panicles or sometimes before; branches often repeatedly branching, the short branchlets more or less fascicled and bearing usually much reduced leaves; the terminal one or two joints of the primary culm often finally falling, the whole producing an autumnal phase usually strikingly different from the vernal phase; secondary panicles reduced, the latest more or less included in the sheaths, the spikelets cleistogamous and perfecting their grains. The species of this subgenus are usually known as dichotomous panicums because they are related to *Panicum dichotomum*.

Key to the species of subgenus 2 irrespective of the groups

1a. Spikelets glabrous.
  2a. Spikelets 3 mm. long or more, strongly nerved.
    Spikelets pointed; blades elongate.. 5. P. DEPAUPERATUM.
    Spikelets blunt; blades not elongate.
      Spikelets 3.2 to 3.3 mm. long; blades firm; sheaths, or some of them, hispid.
                                       95. P. SCRIBNERIANUM.
      Spikelets not more than 3 mm. long; blades rather thin; sheaths glabrous or sparsely hispid.. 94. P. HELLERI.
  2b. Spikelets less than 3 mm. long.
    3a. Second glume and sterile lemma exceeding the fruit and pointed beyond it. Spikelets 2.2 to 2.9 mm. long.
      Blades clustered toward the base.. 26. P. NUDICAULE.
      Blades not clustered toward the base.
        Sheaths, at least the secondary, hispid................................ 106. P. SCABRIUSCULUM.
        Sheaths glabrous.
          Blades firm; fruit 1.5 mm. long................................ 107. P. CRYPTANTHUM.
          Blades thin; fruit nearly 2 mm. long................................ 35. P. YADKINENSE.
    3b. Second glume and sterile lemma not pointed beyond the fruit.
      4a. Ligule manifest, 1 to 3 mm. long.
        Culms rather stout; ligule 2 to 3 mm. long; sheaths glabrous........ 40. P. SPRETUM.
        Culms slender; ligule 1 mm. long; sheaths sparsely pilose...... 83. P. CURTIFOLIUM.
      4b. Ligule obsolete.
        5a. Spikelets 1.5 mm. or less long.
          Nodes bearded.. 27. P. MICROCARPON.
          Nodes not bearded.
            Culms and blades pilose.. 13. P. STRIGOSUM.
            Culms glabrous.
              Blades conspicuously ciliate; plants branching at base only.
                              12. P. POLYCAULON.
              Blades not ciliate; plants branching from middle or upper nodes.
                Vernal culms 50 cm. tall or more; spikelets turgid, strongly nerved; autumnal phase erect, with fascicled branches shorter than the primary internodes.. 37. P. CAERULESCENS.
                Vernal culms usually much less than 50 cm. tall; autumnal phase spreading or reclining.
                  Spikelets 1.1 to 1.2 mm. long; blades rarely as much as 5 cm. long.
                              84. P. CHAMAELONCHE.
                Spikelets 1.2 to 1.4 mm. long.
                  Blades elongate, at least some of them 8 to 10 cm. long.
                            85. P. GLABRIFOLIUM.
                  Blades not more than 3 cm. long...................... 81. P. ENSIFOLIUM.
      5b. Spikelets 2 mm. long or more.
        Blades elongate, some of them 20 times as long as wide; spikelets 2.2 to 2.8 mm. long.

Blades erect; branches, when present, from the lower nodes only.
8. P. WERNERI.
Blades spreading; branches from upper nodes................... 24. P. BICKNELLII.
Blades not elongate, about 10 times as long as wide.
Culms soon prostrate, vinelike; branches divaricate.
Plants bright green; culms lax; spikelets not more than 2.1 mm. long.
38. P. LUCIDUM.
Plants grayish green; culms stiff; spikelets 2.5 mm. long.
39. P. SPHAGNICOLA.
Culms not vinelike; branches not divaricate.
Spikelets 2.3 to 2.6 mm. long.
Blades, or some of them, at least 8 mm. wide; fruit papillose-roughened.
90. P. WEBBERIANUM.
Blades not more than 6 mm. wide; fruit smooth and shining.
91. P. PATENTIFOLIUM.
Spikelets 2 mm. long.
Culms wiry, crisp-puberulent; blades ciliate at base.
88. P. LANCEARIUM.
Culms glabrous; blades not ciliate.
Blades erect, firm; spikelets turgid, strongly nerved; plants grayish
olive.. 36. P. ROANOKENSE.
Blades spreading; spikelets not turgid.
Autumnal phase branched like a little tree; nodes glabrous or some
sparsely pilose... 33. P. DICHOTOMUM.
Autumnal phase topheavy-reclining; nodes, at least the lower,
bearded, rarely glabrous............................ 34. P. BARBULATUM.
1b. Spikelets pubescent.
6a. Spikelets 3 mm. or more long.
7a. Blades elongate, those of the midculm at least 15 times as long as wide.
Secondary panicles from basal sheaths only.
Spikelets pointed, about 3.5 mm. long.................................. 5. P. DEPAUPERATUM.
Spikelets blunt, about 3 to 3.2 mm. long........................ 6. P. PERLONGUM.
Secondary panicles from upper branches.
Spikelets attenuate at base, pustulose-pubescent; lowermost sheaths softly villous.
20. P. FUSIFORME.
Spikelets not attenuate at base, not pustulose; lowermost sheaths glabrous or
hispid.
Upper leaves approximate, sheaths glabrous.................. 112. P. EQUILATERALE.
Upper leaves distant; at least the lower sheaths hispid.... 103. P. ACULEATUM.
7b. Blades not elongate, usually less than 10 times as long as wide.
8a. Blades velvety-pubescent beneath.
Spikelets 3 mm. long; plants velvety-villous throughout.
93. P. MALACOPHYLLUM.
Spikelets 4 mm. long or more.
Sheaths ascending-hirsute; ligule 3 to 4 mm. long.............. 97. P. RAVENELII.
Sheaths downy-pubescent; ligule obsolete.............. 115. P. BOSCII var. MOLLE.
8b. Blades not velvety-pubescent beneath.
9a. Sheaths glabrous or minutely puberulent only.
Nodes bearded; spikelets 4 mm. long or more............................ 115. P. BOSCII.
Nodes not bearded; spikelets not more than 3.8 mm. long.
Spikelets 3.5 to 3.8 mm. long; blades 2 cm. wide or more.
114. P. LATIFOLIUM.
Spikelets scarcely more than 3 mm. long.
Spikelets turgid, blunt; blades mostly less than 1 cm. wide.
94. P. HELLERI.
Spikelets not turgid; blades more than 1 cm. wide.
Panicle narrow, the branches ascending; spikelets on long stiff pedicels.
25. P. CALLIPHYLLUM.
Panicle as broad as long, the branches spreading.
Plants glaucous; basal blades conspicuously ciliate.
110. P. MUTABILE.
Plants not glaucous; basal blades not ciliate, or at the base only.
Culms erect, or autumnal phase leaning; blades symmetrical, broadly
cordate... 109. P. COMMUTATUM.
Culms decumbent; blades usually unsymmetrical and falcate, nar-
rowed to the scarcely cordate base................... 111. P. JOORII.

9b. Sheaths pubescent.
  Pubescence ascending or appressed.
    Spikelets 3 to 3.2 mm. long; first glume conspicuously remote.
                                          65. P. MALACON.
    Spikelets 3.5 to 4 mm. long; first glume not remote.... 96. P. OLIGOSANTHES.
  Pubescence spreading, sometimes sparse.
    Plants robust, about 1 m. tall; blades usually 2 cm. or more wide.
                                   113. P. CLANDESTINUM.
    Plants rarely more than 50 cm. tall; blades rarely more than 1.5 cm. wide.
      Panicle about as wide as long; blades ascending or spreading.
        Spikelets attentuate at base, 3.5 to 4 mm. long.... See 14. PEDICELLATA.
        Spikelets not attentuate at base, not more than 3.3 mm. long.
          Spikelets 3.2 to 3.3 mm. long; blades firm; sheaths, or some of them,
            more or less hispid_____ 95. P. SCRIBNERIANUM.
          Spikelets not more than 3 mm. long; blades rather thin; sheaths, or
            some of them, glabrous or sparsely hispid_____ 94. P. HELLERI.
      Panicle narrow, the branches erect (sometimes ascending in *P. wilcoxianum*),
        or spreading at anthesis only; blades erect.
        Spikelets not more than 3 mm. long; blades not more than 6 mm. wide.
                                   92. P. WILCOXIANUM.
        Spikelets 3.7 to 4 mm. long; blades 8 to 20 mm. wide.
          Blades papillose-hispid_____ 98. P. LEIBERGII.
          Blades glabrous on both surfaces_____ 99. P. XANTHOPHYSUM.
6b. Spikelets less than 3 mm. long.
  10a. Blades elongate, not more than 5 mm. wide; secondary panicles at the base only
    or wanting.
    Culms single or few in a tuft; spikelets turgid, 2.7 to 3 mm. long.
                                     6. P. PERLONGUM.
    Culms in large tufts; spikelets not turgid, not more than 2.7 mm. long.
      Sheaths pilose_____ 7. P. LINEARIFOLIUM.
      Sheaths glabrous_____ 8. P. WERNERI.
  10b. Blades usually not elongate; secondary panicles not at the base.
    11a. Spikelets attentuate at base, mostly prominently pustulose. Blades narrow,
      stiff, strongly nerved, tapering from base to apex.
      Nodes bearded; plants grayish-villous; autumnal blades flat.
        Spikelets 2 mm. long_____ 16. P. CHRYSOPSIDIFOLIUM.
        Spikelets 2.5 to 2.8 mm. long_____ 17. P. CONSANGUINEUM.
      Nodes not bearded; plants villous only at the base, or nearly glabrous.
        Autumnal blades flat.
        Spikelets 2 mm. long; panicle branches loosely ascending.
                               15. P. BENNETTENSE.
        Spikelets 2.5 to 2.8 mm. long; panicle branches widely spreading at anthesis.
                               18. P. ANGUSTIFOLIUM.
      Autumnal blades involute; lower panicle branches more or less ascending.
        Spikelets pointed beyond the fruit, fusiform_____ 19. P. PINETORUM.
        Spikelets blunt, obovate.
          Plants glabrous or nearly so.
            Spikelets subsecund along the suberect panicle branches.
                                   23. P. NEURANTHUM.
            Spikelets not subsecund; panicle loose and open_____ 22. P. OVINUM.
          Plants pubescent, at least on the lower half.
            Spikelets about 2.4 mm. long; vernal blades 7 to 12 cm. long; autumnal
              blades not falcate_____ 21. P. ARENICOLOIDES.
            Spikelets not more than 2 mm. long; vernal blades 4 to 6 cm. long; au-
              tumnal blades falcate_____ 14. P. ACICULARE.
  11b. Spikelets not attentuate at base.
    12a. Sheaths retrorsely pilose. Blades soft and lax.
      Blades ciliate and more or less pilose on the surface; spikelets 2 mm. long.
                                   10. P. XALAPENSE.
      Blades glabrous or nearly so on the surface and margin; spikelets 2.2 mm. long.
                                   9. P. LAXIFLORUM.
    12b. Sheaths not retrorsely pilose.
      13a. Ligule manifest, mostly 2 to 5 mm. long, at least 1 mm. long.
        Sheaths, or all but the lowest, glabrous; spikelets not more than 1.6 mm. long.
          Panicle narrow, one-fourth to one-third as wide as long.... 40. P. SPRETUM.
          Panicle open, nearly as wide as long.

Spikelets 1.5 mm. long.. 41. P. LINDHEIMERI.
Spikelets 1.1 mm. long............................. 43. P. LONGILIGULATUM.
Sheaths pubescent.
Ligule 1 mm. long; sheaths sparsely pilose; spikelets 1.4 mm. long.
83. P. CURTIFOLIUM.
Ligule usually more than 1 mm. long.
Ligule 1 to 1.5 mm. long. Culms and sheaths appressed-pubescent;
spikelets 1.5 mm. long or more.
Spikelets 2.8 to 2.9 mm. long................................. 66. P. DEAMII.
Spikelets less than 2 mm. long.
Spikelets 1.8 to 1.9 mm. long; plants bluish green.
70. P. TSUGETORUM.
Spikelets 1.5 mm. long, nearly globular; plants olivaceous.
72. P. ORICOLA.
Ligule 2 to 5 mm. long.
Spikelets 1 to 1.3 mm. long; culms and sheaths softly appressed-pubescent.
Spikelets 1.2 to 1.3 mm. long................................ 42. P. LEUCOTHRIX.
Spikelets not more than 1 mm. long.............. 44. P. WRIGHTIANUM.
Spikelets mostly more than 1.5 mm. long, if less, pubescence spreading.
See 8. LANUGINOSA.
13b. Ligule obsolete or less than 1 mm. long.
14a. Nodes bearded (*P. scoparium* may appear to be bearded).
Spikelets nearly 3 mm. long; plants velvety-villous throughout.
93. P. MALACOPHYLLUM.
Spikelets rarely as much as 2.5 mm. long; plants not pubescent throughout.
Spikelets 1.5 to 1.6 mm. long................................ 27. P. MICROCARPON.
Spikelets 2 mm. long or more (sometimes 1.8 mm. in *P. mundum*).
Blades all velvety; autumnal phase usually sparingly branching.
29. P. ANNULUM.
Blades glabrous, or only the lower pubescent or velvety.
Autumnal phase profusely branching, the branchlets forming large
clusters at the nodes of the primary culms. Upper sheaths usually glandular spotted... 28. P. NITIDUM.
Autumnal phase sparingly branching.
Lower sheaths or blades velvety-pilose.
Sheaths and upper nodes glabrous.................... 31. P. CLUTEI.
Lower sheaths and all the nodes pubescent.
30. P. MATTAMUSKEETENSE.
Lower sheaths ascending-pilose or glabrate, not velvety, the blades
glabrous or papillose-ciliate toward the base.
105. P. MUNDUM.
14b. Nodes not bearded.
15a. Plants densely gray-velvety throughout, a viscid, glabrous ring below
the nodes... 102. P. SCOPARIUM.
15b. Plants not gray-velvety.
16a. Sheaths or some of them pilose or hispid.
Pubescence papillose-hispid, papillose-pilose, or sometimes glabrate.
Spikelets glabrous, ovate, pointed beyond the fruit, the first glume
short, broadly acute........................... 106. P. SCABRIUSCULUM.
Spikelets pubescent or pilose, obovate or elliptic, the first glume
longer, acute.
Blades about 2 cm. wide, often as much as 3 cm.; fruit distinctly
shorter than the spikelet.................. 113. P. CLANDESTINUM.
Blades less than 15 mm. wide; fruit nearly as long as the spikelet.
Blades flat, 8 to 15 mm. wide.............. 104. P. RECOGNITUM.
Blades involute-acuminate, not more than 6 mm. wide.
92. P. WILCOXIANUM.
Pubescence ascending-pilose.
Spikelets 2.8 to 2.9 mm. long................................ 66. P. DEAMII.
Spikelets not more than 2.5 mm. long.
Spikelets 2 to 2.5 mm. long.
Winter blades elongate, 5 to 10 cm. long; plants bluish green;
spikelets 2 mm. long................. 69. P. WILMINGTONENSE.
Winter blades 1 to 3 cm. long; plants olivaceous.
Spikelets about 2.4 mm. long; panicle open, branches stiffly
spreading................................... 67. P. COMMONSIANUM.

Spikelets 2 to 2.1 mm. long; panicle rather dense, branches
ascending_____ 68. P. ADDISONII.
Spikelets not more than 1.7 mm. long.
Blades white-margined; spikelets 1.6 to 1.7 mm. long, elliptic.
76. P. TENUE.
Blades not white-margined; spikelets 1.3 to 1.4 mm. long, nearly
globular_____ 71. P. COLUMBIANUM var. THINIUM.
16b. Sheaths glabrous or puberulent only.
  17a. Spikelets spherical, not more than 1.8 mm. long. Blades cordate,
    ciliate at base_____ See 10. SPHAEROCARPA.
  17b. Spikelets not spherical.
    18a. Culms soon prostrate, vinelike; branches divaricate.
    Plants bright green; culms lax; spikelets not more than 2.1 mm.
    long_____ 38. P. LUCIDUM.
    Plants grayish green; culms stiff; spikelets 2.5 mm. long.
    39. P. SPHAGNICOLA.
    18b. Culms not vinelike; branches not divaricate.
      19a. Spikelets asymmetrically pyriform, strongly nerved; culms
      wiry_____ See 12. LANCEARIA.
      19b. Spikelets not pyriform.
        20a. Blades elongate, especially the upper, about 20 times as
        long as wide. Spikelets about 2.5 mm. long, on long pedicels.
        24. P. BICKNELLII.
        20b. Blades not elongate. (See continuation.)
(Continuation.)
21a. Spikelets 2 mm. long or more.
Spikelets 2.5 to 3 mm. long; blades cordate, usually 1 cm. or more wide.
Plants glaucous; basal blades conspicuously ciliate_____ 110. P. MUTABILE.
Plants not glaucous; basal blades ciliate at base only.
Culms crisp-puberulent; blades rarely more than 1 cm. wide; spikelets about 2.5 mm.
long_____ 103. P. ASHEI.
Culms glabrous or obscurely puberulent; blades usually 1.5 cm. wide or more; spike-
lets 2.7 to 3 mm. long_____ 109. P. COMMUTATUM.
Spikelets not more than 2.3 mm. long; blades not cordate, usually less than 1 cm. wide.
Blades conspicuously ciliate, soft, lax, crowded at the base_____ 11. P. CILIATUM.
Blades not ciliate or at base only, not crowded at the base.
Blades not more than 6 mm. wide; plants not branching or rarely branching from
near the base_____ 8. P. WERNERI.
Blades 7 mm. wide or more; plants branching from middle and upper nodes.
Primary blades spreading; panicle purplish; fruit exposed at summit.
31. P. CLUTEI.
Primary blades erect; panicle green; fruit covered_____ 32. P. BOREALE.
21b. Spikelets not more than 1.7 mm. long.
Culms crisp-puberulent; spikelets turgid_____ 71. P. COLUMBIANUM.
Culms glabrous.
Blades white-margined, firm.
Blades puberulent beneath, often above_____ 76. P. TENUE.
Blades glabrous.
Uppermost blades much reduced; culms branching from lower nodes only, the
branches repeatedly branching_____ 77. P. ALBOMARGINATUM.
Uppermost blades about as long as the others; culms bearing short branches from
middle and upper nodes_____ 78. P. TRIFOLIUM.
Blades not white-margined or very obscurely so (or if white margin is evident, spikelets
only 1.1 mm. long).
Culms branching only at base; plants soft, light green_____ 82. P. VERNALE.
Culms branching at the nodes.
Spikelets 1.1 mm. long; winter blades bluish green, not glossy.
80. P. CONCINNIUS.
Spikelets 1.3 to 1.5 mm. long.
Blades involute, falcate, with long stiff hairs on margin near base; plants stiff
and wiry_____ 86. P. BREVE.
Blades not involute or at tip only, not falcate.
Plants bright green; winter blades conspicuous, glossy green.
79. P. FLAVOVIRENS.
Plants olive; winter blades not conspicuous nor glossy____ 81. P. ENSIFOLIUM.

1. Depauperáta.—Ligule less than 1 mm. long; blades elongate, the basal ones not forming a distinct rosette in autumn; spikelets strongly 7- to 9-nerved. Autumnal phase with short branches from lower nodes.

5. Panicum depauperátum Muhl. (Fig. 914.) Vernal phase with culms several to many in a tuft, slender but rather stiff, erect or nearly so; sheaths glabrous or papillose-pilose; blades 6 to 15 cm. long, 2 to 5 mm. wide, often involute in drying; panicle exserted, usually not much exceeding the leaves, 4 to 8 cm. long, few-flowered; spikelets 3.2 to 3.8 mm. long, elliptic, pointed, glabrous or sparsely pubescent; second glume and sterile lemma extending beyond the fruit, forming a beak. Autumnal phase similar, the reduced panicles partly concealed in the basal leaves. ♃ —Open sterile woods, Quebec and Nova Scotia to Minnesota, south to Georgia and Texas.

6. Panicum perlóngum Nash. (Fig. 915.) Vernal phase similar to that of *P. depauperatum;* the tufts smaller, usually pilose, the panicle narrower; spikelets 2.7 to 3.2 mm. long, oval, blunt, sparingly pilose, the glume and sterile lemma not extending beyond the fruit. Autumnal phase similar, the reduced panicles numerous. ♃ —Prairies and dry soil, Indiana to Manitoba and North Dakota, south to Colorado and Texas.

7. Panicum linearifólium Scribn. (Fig. 916.) Vernal phase in dense tufts; culms slender, erect, 20 to 45 cm. tall; sheaths papillose-pilose; blades erect, usually overtopping the panicles, 2 to 4 mm. wide; panicle long-exserted, 5 to 10 cm. long, the flexuous branches ascending; spikelets 2.2 to 2.7 mm. long, oblong-elliptic, obtuse, sparsely pilose. Autumnal phase similar, the reduced panicles hidden among the basal leaves. ♃ —Dry woods, Quebec and Maine to Wisconsin, south to Georgia and Texas.

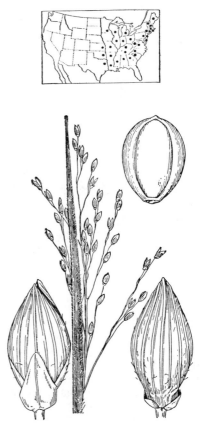

FIGURE 914.—*Panicum depauperatum.* Panicle, × 1; two views of spikelet, and floret, × 10. (Amer. Gr. Natl. Herb. 78, D. C.)

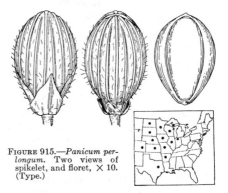

FIGURE 915.—*Panicum perlongum.* Two views of spikelet, and floret, × 10. (Type.)

8. Panicum wernéri Scribn. (Fig. 917.) Vernal phase similar to that of *P. linearifolium,* the culms usually stiffer, blades firmer, shorter and wider (15 cm. long or less); nodes

FIGURE 916.—*Panicum linearifolium.* Two views of spikelet, and floret, × 10. (Type.)

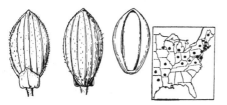

FIGURE 917.—*Panicum werneri.* Two views of spikelet, and floret, × 10. (Type.)

usually sparingly pilose; sheaths glabrous; spikelets 2.1 to 2.4 mm. long, nearly or quite glabrous. Autumnal phase similar to the vernal, sometimes late in the season bearing simple branches from the lower nodes. ♃ —Sterile woods and knolls, Quebec and Maine to Minnesota, Tennessee, Virginia, Kentucky, and Texas. Intergrades with *P. linearifolium.*

2. **Laxiflóra.**—Tufted, erect to spreading; foliage aggregate toward base, light green, soft, the basal blades not in distinct rosettes in autumn; ligule nearly obsolete; primary panicles long-exserted; spikelets obovate, obtuse, turgid, 5- to 7-nerved. Autumnal phase branching near base, forming close flat tuft, with reduced panicles.

9. **Panicum laxiflórum** Lam. (Fig. 918.) Vernal culms 20 to 60 cm. tall, erect or geniculate below; nodes bearded with reflexed hairs; sheaths retrorsely pilose; blades 10 to 20 cm. long, 7 to 12 mm. wide, glabrous or sparsely ciliate; panicle 8 to 12 cm. long, lax, few-flowered, the lower branches often reflexed; spikelets 2.2 to 2.3 mm. long, papillose-pilose. Autumnal blades scarcely reduced, much exceeding the secondary panicles. ♃ —Rich or damp woods, Virginia to Florida and Alabama.

10. **Panicum xalapénse** H. B. K. (Fig. 919.) Vernal culms and blades on the average shorter than in *P. laxiflorum,* the blades pilose on one or both surfaces or nearly glabrous, usually short-ciliate; spikelets 1.9 to 2 mm. long, pilose. Autumnal phase with usually denser tufts and shorter blades. ♃ —Woods, Maryland to Illinois and Missouri, south to Florida and Texas; Mexico; Guatemala; Dominican Republic. Originally described from Xalapa (Jalapa), Mexico. PANICUM XALAPENSE var. STRICTIRÁMEUM Hitchc. and Chase. Vernal

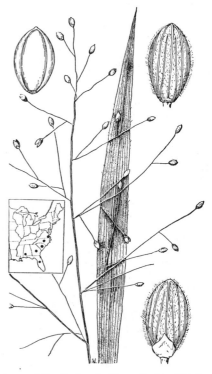

FIGURE 918.—*Panicum laxiflorum.* Panicle, × 1; two views of spikelet, and floret, × 10. (Curtiss 6635, Fla.)

panicles more compact, branches ascending, spikelets 1.7 mm. long; blades shorter, narrower. ♃ — Dry woods, Coastal Plain, South Carolina to Texas; Tennessee.

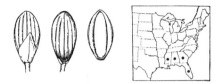

FIGURE 921.—*Panicum polycaulon.* Two views of spikelet, and floret, × 10. (Type.)

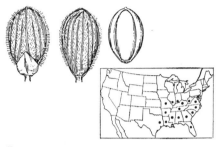

FIGURE 919.—*Panicum xalapense.* Two views of spikelet, and floret, × 10. (Type.)

11. Panicum ciliátum Ell. (Fig. 920.) Vernal culms 5 to 30 cm. tall; sheaths ciliate on the margin; blades 3 to 6 cm. long, 3 to 8 mm. wide, the uppermost often much smaller, ciliate with stiff hairs 2 to 3 mm. long; panicle 3 to 4 cm. long, the axis pilose, branches spreading; spikelets 1.8 to 2 mm. long, pilose. Autumnal mats with slightly smaller blades. ♃ —Low pinelands and hammocks, Coastal Plain, North Carolina to Florida and Texas; Mexico.

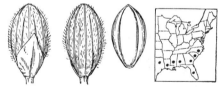

FIGURE 920.—*Panicum ciliatum.* Two views of spikelet, and floret, × 10. (Type.)

12. Panicum polycaúlon Nash. (Fig. 921.) Vernal culms 10 to 20 cm. tall; blades mostly narrower than in *P. ciliatum,* panicle similar; spikelets 1.5 to 1.6 mm. long (rarely as much as 2 mm.), glabrous. Autumnal mats very dense. ♃ —Low pine woods, Coastal Plain, Georgia, Florida, Alabama, and Mississippi; West Indies; British Honduras.

13. Panicum strigósum Muhl. (Fig. 922.) Vernal culms 15 to 30 cm. tall, the culms and sheaths sparsely pilose; nodes bearded; blades mostly 5 to 7 mm. wide, pilose on both surfaces, stiffly ciliate; panicle 4 to 6 cm. long, axis and branches pilose; spikelets 1.3 to 1.5 mm. long, glabrous. Autumnal phase a dense mat. ♃ — Sandy woods, Virginia and Tennessee to Florida and Texas; Mexico and Cuba to Colombia.

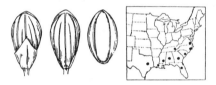

FIGURE 922.—*Panicum strigosum.* Two views of spikelet, and floret, × 10. (Type.)

3. Angustifólia.—Densely tufted, grayish green; ligules not more than 1 mm. long; blades narrow, usually stiff, with prominent nerves, sometimes longitudinally wrinkled, often ciliate at base; spikelets attenuate at base, rather strongly 7-nerved, papillose-pubescent; first glume narrow and sheathing at base. Autumnal culms repeatedly branching, forming bushy crowns; blades greatly reduced.

14. Panicum aciculáre Desv. ex Poir. (Fig. 923.) Vernal culms ascending from a spreading base, 20 to 50 cm. tall, appressed-pubescent below; lower sheaths villous; blades spreading or ascending, narrowed to an involute point, glabrous or the lower sparsely pilose, the middle culm blades 4 to 6 cm. long, 2 to 5

FIGURE 923.—*Panicum aciculare.* Plant, × 1; two views of spikelet, and floret, × 10. (Vernal phase, Chase 7148, N. C.; autumnal phase, Hitchcock 317, N. C.)

mm. wide; panicle 3 to 7 cm. long, the flexuous branches spreading at maturity; spikelets 1.9 to 2 mm. long, obovate. Autumnal phase bushy branching, the culms 10 to 30 cm. long, spreading, forming dense cushions, the blades involute, sharp-pointed, usually arcuate, mostly 1 to 3 cm. long. ♃ —Sandy pine woods Coastal Plain, New Jersey; Virginia to northern Florida, Arkansas, Oklahoma, and Texas; West Indies, northern South America.

ascending, 8 to 15 cm. long (the lower shorter), 4 to 7 mm. wide, acuminate; panicle short-exserted, 5 to 7 cm. long, the flexuous branches loosely ascending; spikelets 2 mm. long, obovate-ellipsoid, papillose-villous. Autumnal phase stiffly ascending, sparingly branching at the middle and upper nodes, the branches and numerous flat reduced blades narrowly ascending, the blades mostly 4 to 5 cm. long. ♃ —Known only from dry sandy savannalike park

FIGURE 924.—*Panicum bennettense.* Two views of spikelet, and floret, × 10. (Duplicate type.)

FIGURE 925.—*Panicum chrysopsidifolium.* Two views of spikelet, and floret, × 10. (Type.)

15. Panicum bennetténse M. V. Brown. (Fig. 924.) Vernal culms erect, 30 to 70 cm. tall, obscurely appressed-puberulent; lower sheaths sparsely papillose-pubescent; blades

surrounding Bennett Civil War Memorial, near Durham, N. C.

16. Panicum chrysopsidifólium Nash. (Fig. 925.) Vernal culms ascending or spreading, 30 to 45 cm.

tall, grayish-villous, especially below, the nodes bearded; sheaths villous; blades 5 to 10 cm. long, 3 to 5 mm. wide, villous on both surfaces; panicle 4 to 6 cm. long; spikelets 2 mm. long, obovate, villous. Autumnal phase spreading, forming mats; blades flat, becoming papery with age. ♃ — Sandy pine woods, Coastal Plain, Virginia to Florida, Arkansas and Texas; West Indies.

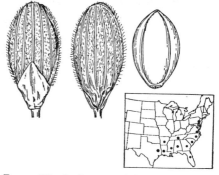

FIGURE 927.—*Panicum angustifolium.* Two views of spikelet, and floret, × 10. (Type.)

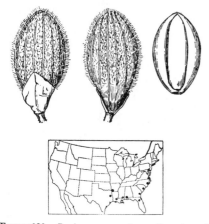

FIGURE 926.—*Panicum consanguineum.* Two views of spikelet, and floret, × 10. (Type.)

17. Panicum consanguíneum Kunth. (Fig. 926.) Vernal culms ascending or spreading, 20 to 50 cm. tall, densely felty-villous below, the nodes bearded; sheaths villous, especially the lower; blades 7 to 11 cm. long, 5 to 8 mm. wide, villous, or nearly glabrous above; panicle 4 to 8 cm. long, the lower branches narrowly ascending; spikelets 2.6 to 2.8 mm. long, obovate, papillose-villous. Autumnal phase spreading or decumbent, the numerous branches somewhat flabellately fascicled, the blades 3 to 4 cm. long, 2 to 3 mm. wide, flat, thin, papery. ♃ — Sandy pine woods, Coastal Plain, Virginia to northern Florida, west to Arkansas and Texas.

18. Panicum angustifólium Ell. 927.) Vernal culms erect or nearly so, 30 to 50 cm. tall, the lowermost internodes gray crisp-villous; lower sheaths

appressed-villous, the upper glabrous; blades stiffly ascending, 8 to 15 cm. long, 4 to 8 mm. wide, long-acuminate; panicle long-exserted, 4 to 10 cm. long, loosely flowered, the branches widely spreading at anthesis, the lower often reflexed; spikelets 2.5 to 2.8 mm. long, elliptic-obovate, papillose-villous. Autumnal phase ascending or somewhat top-heavy-reclining, not spreading or mat-like; blades very numerous, flat, appressed, rather thin and papery. ♃ —Sandy pine woods, Coastal Plain, New Jersey to northern Florida and Texas; Tennessee (Knoxville), Arkansas; Nicaragua.

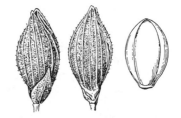

FIGURE 928.—*Panicum pinetorum.* Two views of spikelet, and floret, × 10. (Type.)

19. Panicum pinetórum Swallen. (Fig. 928.) Vernal culms slender, wiry, 55 to 90 cm. tall; sheaths glabrous or the lowermost appressed-pilose; blades 6 to 9 cm. long, 2 to 3 mm. wide, involute in drying, glabrous; panicle 7 to 9 cm. long, narrow, the branches not more than 3 cm. long, ascending; spikelets 2.3 to

2.5 mm. long, (or before maturity to 3 mm. long), commonly somewhat twisted, pointed beyond the fruit, minutely pubescent; fruit 1.6 to 1.7 mm. long. Autumnal phase erect or top-heavy reclining, freely branching, the slender involute blades scarcely reduced; panicles reduced, few-flowered, obscured by the foliage. ⚇ —Known only from open pine woods near Bonita Springs, Lee County, Fla.

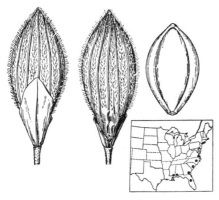

FIGURE 929.—*Panicum fusiforme*. Two views of spikelet, and floret, X 10. (Type.)

20. Panicum fusifórme Hitchc. (Fig. 929.) Vernal culms erect, 30 to 70 cm. tall, the basal and lower sheaths and lower surface of blades softly pubescent; panicle loose, the lower branches spreading or drooping; spikelets 3.3 to 3.5 mm. long, elliptic, acutish or beaked beyond the fruit, long-attenuate at base, papillose-villous. Autumnal phase bushy, the blades soon involute, 3 to 5 cm. long. ⚇ —Sandy pine woods, Virginia to Florida and Mississippi; West Indies; British Honduras.

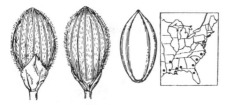

FIGURE 930.—*Panicum arenicoloides*. Two views of spikelet, and floret, X 10. (Type.)

21. Panicum arenicoloídes Ashe. (Fig. 930.) Vernal phase intermediate between that of *P. angustifolium* and *P. aciculare;* culms 30 to 50 cm. tall; lower sheaths and blades softly villous; blades 7 to 12 cm. long, 3 to 4 mm. wide, apex subinvolute; panicle 4 to 6 cm. long, the lower branches ascending; spikelets 2.1 to 2.5 mm. long, obovate, papillose-pilose. Autumnal phase bushy-branching, erect or top-heavy, the blades involute. ⚇ —Sandy pine woods, Coastal Plain, North Carolina to Florida, Arkansas, and Texas; Cuba; Guatemala; northern South America.

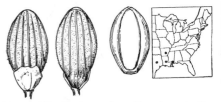

FIGURE 931.—*Panicum ovinum*. Two views of spikelet, and floret, X 10. (Type.)

22. Panicum ovínum Scribn. and Smith. (Fig. 931.) Vernal culms erect or nearly so, not densely tufted, glabrous, 30 to 50 cm. tall; sheaths glabrous or the lowermost appressed-pubescent; blades erect or ascending, 10 to 15 cm. long, 3 to 6 mm. wide, glabrous; panicle 5 to 9 cm. long, the lower branches ascending; spikelets 2.1 to 2.2 mm. long, papillose-pubescent, sometimes minutely so. Autumnal phase erect or nearly so, the blades loosely involute. ⚇ —Dry or moist open ground, Mississippi to Arkansas and eastern Texas; Mexico.

FIGURE 932.—*Panicum neuranthum*. Two views of spikelet, and floret, X 10. (Type.)

23. Panicum neuránthum Griseb. (Fig. 932.) Vernal phase glabrous as a whole; culms 30 to 60 cm. tall; blades erect or ascending, the short basal blades few or wanting; panicle 5 to 9 cm. long, narrow, the flexuous branches narrowly ascending, the branchlets appressed, the short-pediceled spikelets more or less secund along the branches; spikelets 2 mm. long, finely papillose-pubescent. Autumnal culms erect, about as tall as the vernal phase; blades involute. ♃ —Savannas and open ground, southern Florida; Mississippi (Horn Island); Texas; British Honduras; Cuba.

4. Bicknelliána.—In small tufts, erect or ascending; sheaths glabrous; ligules nearly obsolete; panicles few-flowered; spikelets long-pedi-

celed, 7-nerved. Autumnal culms sparingly branching from upper or middle nodes, the blades not much reduced. Intermediate in habit between Depauperata and Dichotoma.

24. Panicum bicknéllii Nash. (Fig. 933.) Vernal phase bluish green; culms 30 to 50 cm. tall; nodes sparsely bearded or glabrous; blades stiffly ascending, 8 to 15 cm. long, 3 to 8 mm. wide, the uppermost usually the longest, narrowed toward the usually ciliate base; panicle 5 to 8 cm. long, the branches ascending; spikelets 2.3 to 2.8 mm. long, sparsely pubescent or rarely glabrous. Autumnal culms erect, forming a loose bushy tuft, the stiffly ascending blades not much reduced, overtopping the narrow few-flowered panicles. ♃ —Dry sterile or rocky woods, Connecticut and Michigan to Georgia and Arkansas.

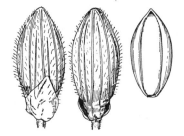

FIGURE 934.—*Panicum calliphyllum*. Two views of spikelet, and floret, × 10. (Type.)

25. Panicum calliphýllum Ashe. (Fig. 934.) Vernal phase yellowish green; culms 35 to 50 cm. tall; nodes sparsely villous; blades ascending, 8 to 12 cm. long, 9 to 12 mm. wide, ciliate at the rounded base; panicle 7 to 9 cm. long, with a few ascending branches; spikelets mostly 3 mm. long, elliptic, sparsely pubescent. Autumnal culms sparingly branching from the middle nodes, the branches about as long as the internodes, erect. ♃ —Woods, rare and local, Ontario, Massachusetts, New York, Ohio, Michigan, and Missouri.

5. Nudicaúlia.—A single rare and local species.

FIGURE 933.—*Panicum bicknellii*. Plant, × 1; two views of spikelet, and floret, × 10. (Porter, Pa.)

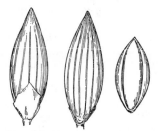

FIGURE 935.—*Panicum nudicaule.* Two views of spikelet, and floret, × 10. (Type.)

26. Panicum nudicaúle Vasey.
(Fig. 935.) Vernal culms erect from
a somewhat spreading base, 40 to 60
cm. tall, glabrous; sheaths glabrous;
blades erect, rather thick, 4 to 10 cm.
long, 5 to 8 mm. wide, the uppermost
reduced, giving the culm a naked appearance; panicle long-exserted, 4 to
7 cm. long, few-flowered, the branches
ascending; spikelets 2.7 to 2.9 mm.
long, narrowly ovate, acuminate, glabrous. Autumnal phase unknown.
2⟁ —Swamps, rare, western Florida,
southern Alabama, and Mississippi.

6. Dichotóma.—Culms few to many
in a tuft, glabrous, or only the
nodes pubescent; sheaths mostly
glabrous or nearly so; ligules
minute; panicles open; spikelets
5- to 7-nerved. Autumnal culms
usually freely branching; leaves
and panicles usually much reduced.

27. Panicum microcárpon Muhl. ex
Ell. (Fig. 936.) Vernal culms tufted,
erect or sometimes geniculate at base,
60 to 100 cm. tall, the nodes densely
bearded with reflexed hairs; sheaths
often mottled with white spots between the nerves; blades spreading,
the upper often reflexed, 10 to 12 cm.
long, 8 to 15 mm. wide, glabrous,
sparsely papillose-ciliate at base; panicle many-flowered, 8 to 12 cm. long;
spikelets 1.6 mm. long, elliptic, glabrous (rarely minutely pubescent).
Autumnal phase much branched from
all the nodes, reclining from the
weight of the dense mass of branches;
blades flat, mostly 2 to 4 cm. long.
2⟁ —Wet woods and swampy places,
Massachusetts to Illinois, south to
northern Florida and eastern Texas.

28. Panicum nítidum Lam. (Fig.
937.) Vernal culms tufted, erect, 30 to
60 cm. tall, the nodes bearded with
reflexed hairs; upper sheaths often
glandular-mottled; blades glabrous, 5
to 10 mm. wide, the upper usually reflexed; panicle ovoid, 5 to 8 cm. long,
many-flowered; spikelets elliptic, 2
mm. long, pubescent. Autumnal

FIGURE 936.—*Panicum microcarpon.* Plant, × 1; two views of spikelet, and floret, × 10. (Maxon and Standley
86, Md.)

phase erect or reclining, the branchlets and foliage forming large clusters from the nodes of the primary culms. ♃ —Low moist or marshy ground, Coastal Plain, New Jersey; Virginia to Florida and Texas; Missouri (Carter County); Bahamas, Cuba.

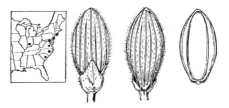

FIGURE 939.—*Panicum mattamuskeetense.* Two views of spikelet, and floret, X 10. (Type coll.)

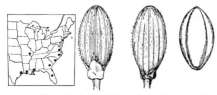

FIGURE 937.—*Pdnicum nitidum.* Two views of spikelet, and floret, X 10. (Type.)

29. Panicum ánnulum Ashe. (Fig. 938.) Vernal phase usually purplish, in small tufts or solitary; culms 35 to 60 cm. tall, the nodes densely bearded; sheaths velvety-pubescent or the upper nearly glabrous; blades densely velvety-pubescent on both surfaces; panicle 6 to 8 cm. long; spikelets 2 mm. long, elliptic, pubescent. Autumnal phase suberect, bearing in late autumn a few short erect branches at the upper nodes. ♃ —Dry woods, Coastal Plain, rare, Massachusetts to Florida and Mississippi; Michigan; Missouri.

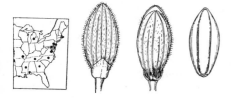

FIGURE 938.—*Panicum annulum.* Two views of spikelet, and floret, X 10. (Type.)

30. Panicum mattamuskeeténse Ashe. (Fig. 939.) Vernal phase olivaceous, usually tinged with purple; culms erect, often 1 m. tall, the nodes bearded or the upper puberulent only; sheaths velvety-pilose or the upper sometimes glabrous; blades horizontally spreading, 8 to 12 cm. long, 8 to 12 mm. wide, velvety-pubescent, or the upper glabrous; panicle 8 to 10 cm. long, many-flow-

ered; spikelets about 2.5 mm. long, elliptic, pubescent. Autumnal phase erect or leaning, branching rather sparingly from the middle nodes, ♃ —Low moist ground, Coastal Plain, New York to South Carolina; Indiana.

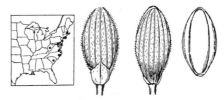

FIGURE 940.—*Panicum clutei.* Two views of spikelet, and floret, X 10. (Type.)

31. Panicum clútei Nash. (Fig. 940.) Similar to *P. mattamuskeetense* but less pubescent, only the lowermost nodes, sheaths, and blades velvety; spikelets 2.2 to 2.3 mm. long. ♃ —Low moist ground and cranberry bogs, Massachusetts to South Carolina; West Virginia. Intergrades with *P. mattamuskeetense.*

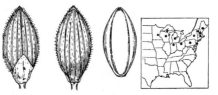

FIGURE 941.—*Panicum boreale.* Two views of spikelet, and floret, X 10. (Type.)

32. Panicum boreále Nash. (Fig. 941.) Vernal culms usually erect, 30 to 50 cm. tall, the nodes mostly glabrous; blades erect or sometimes spreading, 7 to 12 mm. wide, sparsely ciliate at the rounded base; panicle loosely rather few-flowered, 5 to 10

FIGURE 942.—*Panicum dichotomum*. Plant, × ½; two views of spikelet, and floret, × 10. (Bissell 5576, Conn.)

branches erect, the leaves and panicles not greatly reduced. ♃ — Moist open ground or woods, Newfoundland to Minnesota, south to New Jersey and Indiana.

33. Panicum dichótomum L. (Fig. 942.) Vernal phase often purplish; culms slender, erect from a knotted crown, 30 to 50 cm. tall, the lower nodes sometimes with a few spreading hairs; blades spreading, 4 to 8 mm. wide, glabrous; panicle 4 to 9 cm. long, the axis and spreading branches flexuous; spikelets 2 mm. long, elliptic, glabrous (very rarely pubescent); second glume shorter than the fruit at maturity. Autumnal phase much branched at the middle nodes, the lower part usually erect and devoid of blades, giving the plants the appearance of diminutive trees; blades numerous, often involute. ♃ —Dry or sterile woods, New Brunswick to Illinois, south to Florida and eastern Texas.

cm. long; spikelets 2 to 2.2 mm. long, elliptic, pubescent. Autumnal phase erect or leaning, sparingly branching from all the nodes in late summer, the

34. Panicum barbulátum Michx. (Fig. 943.) Vernal phase, resembling that of *P. dichotomum*, the culms 50 to 80 cm. tall, the lower nodes usually bearded; blades slightly wider, panicle slightly larger, spikelets 2 mm. long, glabrous; second glume as long as the fruit at maturity. Autumnal phase diffusely branched, forming very large topheavy reclining bunches, the slender branches recurved, the numerous flat blades horizontally spreading. ♃ —Sterile or rocky woods, Massachusetts to Michigan and Missouri, south to Georgia and eastern Texas. This species seems to intergrade with *P. dichotomum*, but typically the autumnal phases are distinctly different. The vernal culms of *P. barbulatum* are usually more robust and the lower nodes are rather strongly bearded.

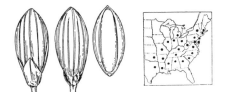

FIGURE 943.—*Panicum barbulatum.* Two views of spikelet, and floret, X 10. (Type.)

35. Panicum yadkinénse Ashe. (Fig. 944.) Vernal phase similar to that of *P. dichotomum*, the culms sometimes 1 m. tall; sheaths bearing pale glandular spots; blades longer and 8 to 11 mm. wide; panicle 10 to 12 cm. long; spikelets 2.3 to 2.5 mm. long, elliptic to subfusiform, pointed a little beyond the fruit, glabrous. Autumnal phase erect or leaning, loosely branching from the middle nodes, the blades not conspicuously reduced. ♃ —Moist woods and thickets, Pennsylvania to Michigan and Illinois, south to Georgia and Texas. Named from Yadkin River, N. C.

36. Panicum roanokénse Ashe. (Fig. 945.) Vernal phase somewhat glaucous olive green; culms erect or ascending, 50 to 100 cm. tall; blades

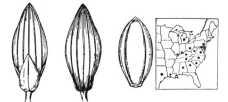

FIGURE 944.—*Panicum yadkinense.* Two views of spikelet, and floret, X 10. (Type coll.)

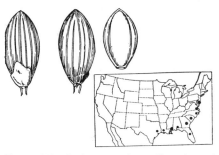

FIGURE 945.—*Panicum roanokense.* Two views of spikelet, and floret, X 10. (Ashe, N. C.)

at first stiffly erect, later somewhat spreading, 3 to 8 mm. wide, glabrous; panicle 4 to 8 cm. long; spikelets 2 mm. long, turgid, elliptic, glabrous, the second glume often purple at base. Autumnal phase erect or decumbent, branching at the middle and upper nodes, the branches numerous but not in tufts, the reduced blades subinvolute. ♃ —Open swampy woods or wet peaty meadows, Coastal Plain, southeastern Delaware to Florida and Texas; Jamaica.

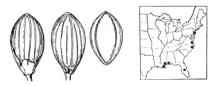

FIGURE 946.—*Panicum caerulescens.* Two views of spikelet, and floret, X 10. (Type.)

37. Panicum caeruléscens Hack. ex Hitchc. (Fig. 946.) Vernal phase similar to that of *P. roanokense;* culms more slender; blades ascending or spreading, commonly purplish beneath; panicle 3 to 7 cm. long; spikelets 1.5 to 1.6 mm. long, obovoid, turgid, glabrous. Autumnal phase erect or leaning, producing short densely fascicled branches at the middle and upper nodes, these tufts scarcely as long as the primary internodes. ♃ —Marshes and swampy woods, Coastal Plain, southern New Jersey to Florida and Louisiana; Cuba.

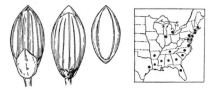

Figure 947.—*Panicum lucidum.* Two views of spikelet, and floret, X 10. (Type.)

38. Panicum lúcidum Ashe. (Fig. 947.)

Vernal phase at first erect and resembling that of *P. dichotomum,* but the weak culms soon decumbent; blades thin, shining, bright green, glabrous, at first erect but soon widely spreading, 4 to 6 mm. wide; panicle resembling that of *P. dichotomum* but fewer-flowered; spikelets 2 to 2.1 mm. long, elliptic, glabrous (rarely pubescent), the tip of the fruit exposed at maturity. Autumnal phase repeatedly branching, forming large clumps or mats of slender weak vinelike culms, the branches elongate and diverging at a wide angle, not fascicled, the blades waxy, flat, spreading. ♃ —Wet woods and sphagnum swamps, Coastal Plain, Massachusetts to Florida, Arkansas, and Texas; Indiana (near Lake Michigan), Michigan (Port Huron). P. LUCIDUM var. OPÁCUM Fernald. Blades not glossy. Virginia.

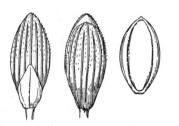

Figure 948.—*Panicum sphagnicola.* Two views of spikelet, and floret, X 10. (Type.)

39. Panicum sphagnícola Nash.

(Fig. 948.) Vernal phase grayish olive green; culms strongly flattened, erect or reclining, 50 to 100 cm. tall; sheaths soon divaricate; blades glabrous, 3 to 7 mm. wide; panicle narrow, 5 to 6 cm. long; spikelets 2.5 mm. long, elliptic, glabrous or minutely pubescent toward the summit. Autumnal phase decumbent or finally prostrate-spreading, divaricately branching from all the nodes, the branches slender, elongate. ♃ — Edges of cypress swamps, in sphagnum bogs, and in similar moist shady places, southern Georgia and Florida.

7. Spréta.—Culms tufted, rather stiff, mostly glabrous or nearly so; ligules densely hairy, 2 to 5 mm. long; blades mostly firm; spikelets 5- to 7-nerved, mostly pubescent. Autumnal culms with rather short-tufted branchlets and greatly reduced leaves and panicles.

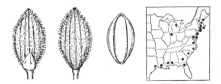

Figure 949.—*Panicum spretum.* Two views of spikelet, and floret, X 10. (Type.)

40. Panicum sprétum Schult. (Fig. 949.)

Vernal culms 30 to 90 cm. tall, erect; sheaths glabrous; ligule 2 to 3 mm. long; blades firm, ascending to reflexed, 4 to 8 mm. wide, sparingly ciliate around the base; panicle 8 to 12 cm. long, the branches ascending or appressed; spikelets about 1.5 mm. long, elliptic, rarely glabrous. Autumnal phase mostly reclining, the early branches elongate, the subsequent branches in short fascicles. ♃ —Wet usually sandy soil, Coastal Plain, Nova Scotia to Florida and Texas; Indiana and Michigan.

41. Panicum lindheímeri Nash.

(Fig. 950.) Vernal culms ascending or spreading, 30 to 100 cm. tall, the lower internodes and sheaths some-

times ascending-pubescent; ligule 4
to 5 mm. long; blades 6 to 8 mm. wide,
glabrous; panicle 4 to 7 cm. long,
about as wide; spikelets 1.4 to 1.6
mm. long, obovate. Autumnal phase
usually stiffly spreading or radiate-
prostrate, with elongate internodes
and tufts of short appressed branches;
blades involute-pointed, often conspicuously ciliate at base. ♃ —
Dry sandy or sterile woods or open
ground, Quebec and Maine to Minnesota, south to northern Florida and
New Mexico; California.

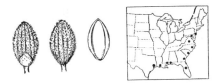

FIGURE 951.—*Panicum leucothrix*. Two views of spikelet, and floret, × 10. (Type.)

middle nodes long branches similar
to primary culms, later producing
more or less fascicled branches. ♃
—Low pinelands, Coastal Plain, New
Jersey to Florida and Texas; Tennessee; West Indies; Colombia.

43. Panicum longiligulátum Nash.
(Fig. 952.) Vernal culms 30 to 70
cm. tall; sheaths glabrous; ligule 2
to 3 mm. long; blades 4 to 8 mm.
wide, glabrous on the upper surface,
puberulent beneath; panicle 3 to 8
cm. long, the slender branches stiffly
ascending; spikelets 1.1 to 1.2 mm.
long. Autumnal culms reclining, the
branches spreading, the branchlets
crowded, the blades subinvolute. ♃
—Low pine barrens and swamps,
Coastal Plain, Pennsylvania (Bucks
County), Delaware to Florida and
Texas; Tennessee; Central America.

FIGURE 950.—*Panicum lindheimeri*. Plant, × 1; two
views of spikelet, and floret, × 10. (Chase 4449,
Miss.)

42. Panicum leucóthrix Nash. (Fig.
951.) Vernal phase light olive green;
culms 25 to 45 cm. tall, erect or ascending, appressed papillose-pilose,
the nodes pubescent; sheaths papillose-pilose; ligule 3 mm. long; blades
3 to 7 mm. wide, glabrous or sparsely
villous on the upper surface, velvety-
puberulent beneath; panicle 3 to 8
cm. long, rather densely flowered;
spikelets 1.2 to 1.3 mm. long, densely
papillose-pubescent. Autumnal culms
at first sending out from lower and

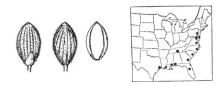

FIGURE 952.—*Panicum longiligulatum*. Two views of
spikelet, and floret, × 10. (Type.)

44. Panicum wrightiánum Scribn.
(Fig. 953.) Vernal culms weak, slender, ascending from a decumbent
base, 15 to 60 cm. tall, minutely
puberulent; sheaths glabrous or puberulent; ligule 2 to 3 mm. long;
blades 2 to 4 cm. long, 3 to 5 mm.
wide, glabrous or puberulent beneath
and minutely pilose above; panicle
3 to 6 cm. long; spikelets 1 mm.
long. Autumnal culms decumbent-
spreading, sending out from lower
and middle nodes numerous ascending branches, becoming bushy-
branched, the flat or subinvolute

blades and secondary panicles not greatly reduced. ♃ —Margins of streams and ponds in sandy or mucky soil, Coastal Plain, Massachusetts to Florida and Mississippi; Cuba and Central America.

FIGURE 953.—*Panicum wrightianum.* Two views of spikelet, and floret, × 10. (Type.)

8. Lanuginósa.—Mostly pubescent throughout; ligules densely hairy, 2 to 5 mm. long; spikelets 5- to 9-nerved, pubescent. Autumnal culms usually freely branching, the leaves and panicles mostly greatly reduced.

45. Panicum meridionále Ashe. (Fig. 954.) Vernal culms 15 to 40 cm. tall, the lower internodes and sheaths pilose, the upper minutely appressed-pubescent; ligule 3 to 4 mm. long; blades 1.5 to 3 cm. long, 2 to 4 mm. wide, long-pilose on the upper surface, the hairs erect; panicle 1.5 to 4 cm. long, the axis appressed-pubescent to glabrous; spikelets 1.3 to 1.4 mm. long. Autumnal culms erect, with fascicled branchlets from all the nodes; leaves and panicles not greatly reduced. ♃ —Sandy or sterile woods and clearings, Nova Scotia to Minnesota, south to Alabama.

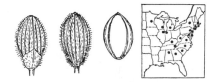

FIGURE 954.—*Panicum meridionale.* Two views of spikelet, and floret, × 10. (Type.)

46. Panicum albemarlénse Ashe. (Fig. 955.) Vernal phase olivaceous, grayish-villous throughout; culms 25 to 45 cm. tall, at first erect, soon geniculate and spreading; blades 3 to 6 mm. wide, the upper surface puberulent as well as long-villous; panicle 3 to 5 cm. long, the axis puberulent; spikelets 1.4 mm. long, pilose. Autumnal culms widely decumbent, spreading or ascending, freely branching at all but the uppermost nodes, the branches narrowly ascending. ♃ —Low sandy woods or open ground, Coastal Plain, Massachusetts to North Carolina; Indiana to Minnesota; West Virginia; Tennessee.

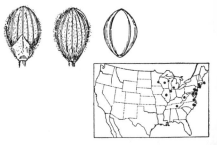

FIGURE 955.—*Panicum albemarlense.* Two views of spikelet, and floret, × 10. (Type.)

47. Panicum implicátum Scribn. (Fig. 956.) Vernal culms slender, 20 to 55 cm. tall, erect or ascending, papillose-pilose with spreading hairs; sheaths papillose-pilose; ligule 4 to

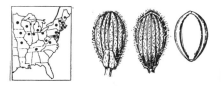

FIGURE 956.—*Panicum implicatum.* Two views of spikelet, and floret, × 10. (Type.)

5 mm. long; blades more or less involute-acuminate, the upper surface pilose with erect hairs 3 to 4 mm. long, appressed-pubescent beneath; panicle 3 to 6 cm. long, the axis long-pilose, the branches flexuous, in typical specimens tangled or implicate; spikelets 1.5 mm. long, papillose-pilose. Autumnal culms erect or spreading, loosely branching from the lower and middle nodes. ♃ —Wet meadows, bogs, and sandy soil, cedar and hemlock swamps,

Newfoundland to Minnesota, south to Delaware, Tennessee, and Missouri.

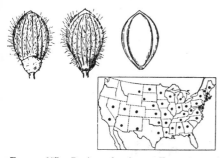

FIGURE 957.—*Panicum huachucae.* Two views of spikelet, and floret, × 10. (Type.)

48. Panicum huachúcae Ashe.

(Fig. 957.) Vernal phase light olivaceous, often purplish, harsh to the touch from copious spreading papillose pubescence; culms usually stiffly upright, 20 to 60 cm. tall, the nodes bearded with spreading hairs; ligule 3 to 4 mm. long; blades firm, stiffly erect or ascending, 4 to 8 cm. long, 6 to 8 mm. wide, the upper surface copiously short-pilose, the lower densely pubescent; panicle 4 to 6 cm. long, the axis and often the branches pilose; spikelets 1.6 to 1.8 mm. long, obovate, papillose-pubescent. Autumnal culms stiffly erect or ascending, the branches fascicled, the crowded blades ascending, 2 to 3 cm. long, much exceeding the panicles. ♃ —Prairies and open ground, Nova Scotia to Montana, south to North Carolina and Texas, westward here and there to southern California. Naturalized in China and Japan.

PANICUM HUACHUCAE var. FASCICULÁTUM (Torr.) Hubb. Vernal culms taller, more slender, less pubescent, the culms 30 to 75 cm. tall; blades thin, lax, spreading, 5 to 10 cm. long, 6 to 12 mm. wide, the upper surface sparsely short-pilose or with copious long hairs toward the base, the lower surface pubescent and with a satiny luster. Autumnal culms more or less decumbent with numerous fascicled branches. ♃ (*P. huachucae* var.

silvicola Hitchc. and Chase.)—Open woods and clearings, Quebec to Minnesota and Nebraska, south to northern Florida and Texas; Arizona (Tucson).

Panicum huachucae, P. huachucae var. *fasciculatum, P. tennesseense,* and *P. pacificum* intergrade more or less. The descriptions apply to the great bulk of specimens, but the distinctions fail to hold for occasional specimens.

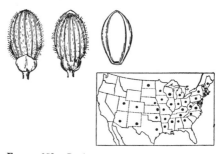

FIGURE 958.—*Panicum tennesseense.* Two views of spikelet, and floret, × 10. (Type.)

49. Panicum tennesseénse Ashe.

(Fig. 958.) Vernal phase bluish green; culms suberect or stiffly spreading, 25 to 60 cm. tall, papillose-pilose or the upper portion glabrous; ligule dense, 4 to 5 mm. long; blades firm, with a thin white cartilaginous margin, 5 to 8 mm. wide, the upper surface glabrous or with a few long hairs toward the base, the lower surface appressed-pubescent or nearly glabrous; panicle 4 to 7 cm. long; spikelets 1.6 to 1.7 mm. long. Autumnal culms widely spreading or decumbent, with numerous fascicled somewhat flabellate branches, often forming prostrate mats; blades usually ciliate at base. ♃ —Open rather moist ground and borders of woods, Quebec to North Dakota, south to Georgia and Texas, and also at a few points west to Utah and Arizona.

50. Panicum lanuginósum Ell.

(Fig. 959.) Vernal phase grayish olive green, velvety-villous throughout; culms usually in large clumps, 40 to 70 cm. tall, lax, spreading, often

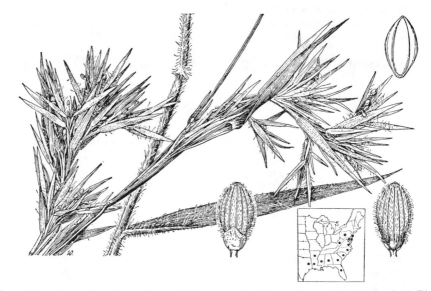

FIGURE 959.—*Panicum lanuginosum.* Plant, ✕ 1; two views of spikelet, and floret, ✕ 10. (Hitchcock, N. C.)

with a glabrous ring below the villous nodes; ligule 3 to 4 mm. long; blades thickish but not stiff, somewhat incurved or spoon-shaped (when fresh), 5 to 10 cm. long, 5 to 10 mm. wide; panicle 6 to 12 cm. long, the axis pubescent; spikelets 1.8 to 1.9 mm. long. Autumnal culms widely spreading or decumbent, freely branching from the middle nodes, the branches repeatedly branching and much exceeding the internodes, the ultimate branchlets forming flabellate fascicles. ♃ —Moist sandy woods, Coastal Plain, New Jersey to Florida, Tennessee, Arkansas, and Texas. The plants have much the habit and pubescence of *P. scoparium,* but much smaller and more slender.

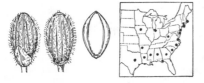

FIGURE 960.—*Panicum auburne.* Two views of spikelet, and floret, ✕ 10. (Type.)

51. Panicum aubúrne Ashe. (Fig. 960.) Vernal phase grayish velvety-villous throughout; culms 20 to 50

cm. tall, geniculate, widely spreading, soon becoming branched and decumbent; ligule 3 to 4 mm. long; blades 3 to 7 cm. long, 3 to 5 mm. wide; panicle 3 to 5 cm. long, the axis velvety; spikelets 1.3 to 1.4 mm. long. Autumnal culms early becoming diffusely branched at all the nodes, prostrate-spreading, forming large mats, the branches curved upward at the ends. ♃ —Sandy pine and oak woods, Coastal Plain, Massachusetts to northern Florida, West Virginia; Arkansas and Texas; Indiana, near Lake Michigan, and Emmet County, Iowa.

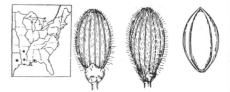

FIGURE 961.—*Panicum thurowii.* Two views of spikelet, and floret, ✕ 10. (Type.)

52. Panicum thurówii Scribn. and Smith. (Fig. 961.) Vernal phase bluish green but drying olive; culms 35 to 70 cm. tall, erect or ascending, villous, the nodes bearded, usually with a glabrous ring below; sheaths sparsely to densely villous; ligule 4 mm. long; blades rather stiff, 6 to 10 mm. wide, the upper surface sparingly pilose toward the base and margins, otherwise glabrous, the lower surface velvety-villous; panicle 7 to 11 cm. long; spikelets 2 mm. long. Autumnal culms erect, bearing at the middle nodes a few appressed fascicles of branches. ♃ —Prairies and dry open woods, Alabama (Mobile) to Texas and Arkansas.

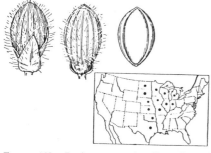

FIGURE 962.—*Panicum praecocius.* Two views of spikelet, and floret, × 10. (Type.)

53. Panicum praecócius Hitchc. and Chase. (Fig. 962.) Vernal culms 15 to 25 cm. tall, at first erect and simple, soon branching and geniculate, becoming 30 to 45 cm. long, papillose-pilose with weak spreading hairs 3 to 4 mm. long; sheaths pilose; ligule 3 to 4 mm. long; blades 5 to 9 cm. long, 4 to 6 mm. wide, long-pilose on both surfaces, the hairs on the upper surface 4 to 5 mm. long, erect; panicle 4 to 6 cm. long, the axis pilose; spikelets 1.8 to 1.9 mm. long, pilose. Autumnal culms in close bunches, 10 to 20 cm. tall, the branches appressed, the scarcely reduced blades erect. ♃ —Dry prairies and clearings, Michigan to North Dakota, south to Arkansas and eastern Texas.

54. Panicum subvillósum Ashe. (Fig. 963.) Vernal culms leafy below, 10 to 45 cm. tall, ascending or spreading, pilose, the nodes short-bearded; sheaths sparsely pilose with ascending hairs; ligule 3 mm. long; blades 4 to 6 cm. long, 4 to 6 mm. wide, both

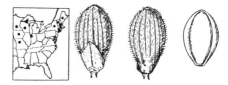

FIGURE 963.—*Panicum subvillosum.* Two views of spikelet, and floret, × 10. (Type.)

surfaces pilose, the hairs on the upper surface 3 to 5 mm. long; panicle long-exserted, 3 to 5 cm. long; spikelets 1.8 to 1.9 mm. long. Autumnal culms widely spreading or prostrate, sparingly branching from the lower nodes, the leaves and panicles not greatly reduced. ♃ —Dry woods and sandy ground, Nova Scotia to Minnesota, south to Connecticut, Indiana, and Missouri.

55. Panicum occidentále Scribn. (Fig. 964.) Vernal culms yellowish green, leafy toward base, 15 to 40 cm. tall, spreading, sparsely pubescent; sheaths sparsely pubescent; ligule 3 to 4 mm. long; blades firm, erect, or ascending, 4 to 8 cm. long, 5 to 7

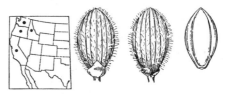

FIGURE 964.—*Panicum occidentale.* Two views of spikelet, and floret, × 10. (Type.)

mm. wide, the upper surface nearly glabrous, the undersurface appressed-pubescent; panicle 4 to 7 cm. long; spikelets 1.8 mm. long. Autumnal culms branching from the lower nodes, forming a spreading tussock 10 to 15 cm. high; leaves and panicles reduced. ♃ —Peat bogs and moist sandy ground, British Columbia and Idaho to southern California.

56. Panicum pacíficum Hitchc. and Chase. (Fig. 965.) Vernal phase light green; culms 25 to 50 cm. tall, ascending or spreading, leafy, pilose, the nodes short-bearded; sheaths pilose; ligule 3 to 4 mm. long; blades erect or ascending, 5 to 10 cm. long, 5 to 8

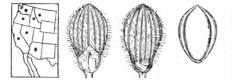

FIGURE 965.—*Panicum pacificum.* Two views of spikelet, and floret, × 10. (Type.)

mm. wide, the upper surface pilose, the lower surface appressed-pubescent; panicle 5 to 10 cm. long; spikelets 1.8 to 2 mm. long. Autumnal culms prostrate spreading, repeatedly branching from the middle and upper nodes. ♃ —Sandy shores and slopes, and moist crevices of rocks, ascending to 1,600 m., British Columbia and Montana to southern California and Arizona.

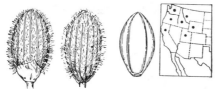

FIGURE 966.—*Panicum thermale.* Two views of spikelet, and floret, × 10. (Type.)

57. Panicum thermále Boland.
(Fig. 966.) Vernal phase grayish green, densely tufted, velvety-villous; culms 10 to 30 cm. tall, ascending or spreading, the nodes with a dense ring of short hairs; ligule 3 mm. long; blades thick, 3 to 8 cm. long, 5 to 12 mm. wide; panicle 3 to 6 cm. long, the axis villous; spikelets 1.9 to 2 mm. long, pilose. Autumnal culms widely spreading, repeatedly branching, the whole forming a dense cushion. ♃ —Wet saline soil in the immediate vicinity of geysers and hot springs, ascending to 2,500 m., Alberta to Washington, south to Wyoming, Utah, and California.

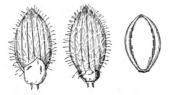

58. Panicum lánguidum Hitchc.
and Chase. (Fig. 967.) Vernal culms 25 to 40 cm. tall, weak, slender, ascending or spreading, pilose; sheaths pilose; ligule 3 mm. long; blades thin, lax, ascending or spreading, 4 to 7 cm. long, 4 to 9 mm. wide, sparsely pilose on the upper surface, minutely appressed-pubescent beneath; panicle 3 to 6 cm. long, the axis and branches sparsely long-pilose; spikelets 2 mm. long, pilose. Autumnal culms decumbent, branching from all the nodes, forming a large loose straggling clump, the ultimate blades and panicles scarcely reduced. ♃ — Dry or sandy open woods, Maine, Massachusetts, Vermont, and eastern New York, apparently rare.

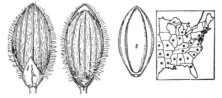

FIGURE 968.—*Panicum villosissimum.* Two views of spikelet, and floret, × 10. (Type.)

59. Panicum villosíssimum Nash.
(Fig. 968.) Vernal phase light olive green; culms 25 to 45 cm. tall, erect or ascending, pilose with spreading hairs 3 mm. long; sheaths pilose; ligule 4 to 5 mm. long; blades rather firm, 6 to 10 cm. long, 5 to 10 mm. wide, pilose on both surfaces; panicle 4 to 8 cm. long, the branches stiffly ascending or spreading; spikelets 2.2 to 2.3 mm. long, pilose. Autumnal culms finally prostrate, the leaves of the fascicled branches appressed, giving the cluster or mat a combed-out appearance. ♃ —Dry sandy or sterile soil, open woods, and hillsides, Massachusetts to Michigan and Kansas, south to Florida and Texas; Guatemala.

60. Panicum bénneri Fernald. (Fig.
969.) Vernal phase light olive green; culms 20 to 35 cm. tall, papillose-

FIGURE 967.—*Panicum languidum.* Two views of spikelet, and floret, × 10. (Type.)

pilose with ascending hairs; nodes inconspicuously bearded; sheaths papillose-pilose; ligule 2 to 3 mm. long; blades 4 to 6 cm. long, 4 to 8 mm. wide, glabrous or with a few long hairs toward the base on the upper surface, very sparsely appressed-pubescent beneath; panicle short-exserted, 3 to 6 cm. long, the axis and flexuous spreading branches pubescent; spikelets 2.2 to 2.5 mm. long, pilose. Autumnal phase unknown, young branches appearing before maturity of primary panicle. ♃ — Only known from an old field along the Delaware River, about 1.5 miles east of Raven Rock, Hunterdon County, N. J. Insufficiently known, may be an exceptional specimen of *P. pseudopubescens.*

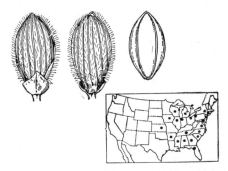

FIGURE 970.—*Panicum pseudopubescens.* Two views of spikelet, and floret, × 10. (Type.)

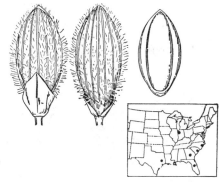

FIGURE 971.—*Panicum ovale.* Two views of spikelet, and floret, × 10. (Type.)

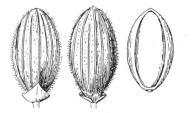

FIGURE 969.—*Panicum benneri.* Two views of spikelet, and floret, × 10. (Type.)

61. Panicum pseudopubéscens Nash. (Fig. 970.) Vernal phase similar to that of *P. villosissimum;* ligule 2 to 3 mm. long; blades with the pubescence on the upper surface short, sparse or wanting down the center, occasionally glabrous; spikelets 2.2 to 2.4 mm. long, pilose. Autumnal culms stiffly spreading, sometimes prostrate, sparingly branching from the middle and lower nodes. ♃ —Sandy open woods, Connecticut to Wisconsin and Iowa south to Florida, Kansas, and Mississippi; Mexico.

62. Panicum ovále Ell. (Fig. 971.) Vernal culms 20 to 50 cm. tall, erect or ascending, rather stout, long-pilose below with ascending or appressed hairs, often nearly glabrous above, the nodes bearded; sheaths ascending-pilose; ligule 2 to 3 mm. long, rather sparse; blades 5 to 10 mm. wide, the

upper surface nearly glabrous except for long hairs near the base and margins, the lower surface appressed-pubescent; panicle 5 to 9 cm. long; spikelets, 2.7 to 2.9 mm. long. Autumnal phase spreading-decumbent, the stiff culms rather loosely branching from the middle and upper nodes. ♃ —Dry sandy woods, Coastal Plain, North Carolina to Florida; Indiana (near Lake Michigan), Illinois (Mason County), and Texas (Waller County).

63. Panicum scoparióides Ashe. (Fig. 972.) Vernal phase light green; culms 30 to 50 cm. tall, erect or ascending, pilose with ascending hairs or nearly glabrous; sheaths pilose to nearly glabrous; ligule 2 to 3 mm. long; blades 6 to 10 mm. wide, sparsely hispid on the upper surface, appressed-pubescent beneath; panicle 4 to 7 cm. long; spikelets 2.2 to 2.3 mm. long, pubescent. Autumnal

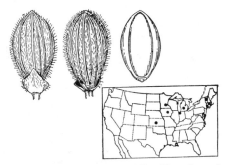

FIGURE 972.—*Panicum scoparioides.* Two views of spikelet, and floret, × 10. (Type.)

culms erect or spreading, sparingly branching from the upper and middle nodes. ♃ —Dry sandy or gravelly soil, Vermont to Delaware; Indiana and Michigan to Minnesota, Iowa, and Kansas.

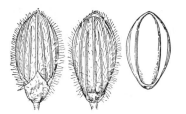

FIGURE 973.—*Panicum shastense.* Two views of spikelet, and floret, × 10. (Type.)

64. Panicum shasténse Scribn. and Merr. (Fig. 973.) Vernal culms 30 to 50 cm. tall, pilose with ascending hairs, the nodes short-bearded; sheaths papillose-pilose, the hairs spreading; ligule sparse, 2 to 3 mm. long; blades 6 to 8 mm. wide, sparsely pilose on the upper surface, pilose beneath; panicle 6 to 8 cm. long; spikelets 2.4 to 2.6 mm. long. Autumnal culms spreading, with geniculate nodes and elongate arched internodes, rather sparingly branched from the middle nodes. ♃ —Moist meadows. Known only from Castle Crag, Shasta County, Calif.

9. Columbiána.—Culms and sheaths appressed-pubescent to crisp-puberulent, the culms stiff; ligules mostly less than 1 mm. long (sometimes to 1.5 mm. in

P. tsugetorum and *P. oricola*); blades firm, thick, stiffly ascending; spikelets 5- to 9-nerved, pubescent, the first glume mostly one-third to half as long as the spikelet. Autumnal culms freely branching, the branches and stiff blades mostly appressed.

65. Panicum málacon Nash. (Fig. 974.) Vernal culms erect to stiffly spreading, purplish olive green; culms and sheaths appressed-pubescent, the culms 30 to 50 cm. tall; blades 3 to 5 mm. wide, sharply acuminate, pu-

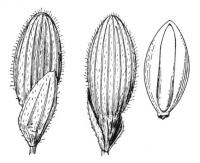

FIGURE 974.—*Panicum malacon.* Two views of spikelet, and floret, × 10. (Type.)

berulent beneath, puberulent to glabrous above; panicle 4 to 7 cm. long, the branches few, stiffly ascending, the pedicels long and stiff; spikelets 3 to 3.2 mm. long, obovate, the first glume distant, about half as long as the spikelet. Autumnal culms subdecumbent-spreading, branching from the lower and middle nodes, the branches appressed. ♃ —Dry pine woods, high pineland, North Carolina (Wilmington); Georgia and Florida; Texas.

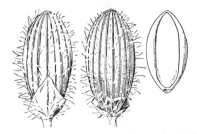

FIGURE 975.—*Panicum deamii.* Two views of spikelet, and floret, × 10. (Type.)

66. Panicum deámii Hitchc. and Chase. (Fig. 975.) Vernal phase yellowish green; culms 25 to 35 cm. tall, erect or ascending, papillose-pilose; sheaths papillose-villous, densely so at base and summit; blades suberect, 8 to 15 cm. long, 4 to 6 mm. wide, sparsely villous on the upper surface, appressed-pilose beneath; panicle rather short-exserted, 6 to 10 cm. long, the branches ascending; spikelets 2.8 to 2.9 mm. long, pilose. Autumnal culms branching from the middle and upper nodes, forming a somewhat bushy summit, the culms sprawling. ♃ —Sand dunes and sandy woods, northern Indiana and Iowa.

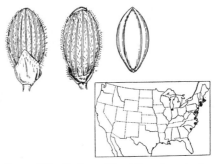

FIGURE 977.—*Panicum addisoni.* Two views of spikelet, and floret, X 10. (Type.)

surface, pubescent or glabrous beneath; panicle 2 to 6 cm. long, more densely flowered than in *P. commonsianum;* spikelets about 2 mm. long. Autumnal culms more or less spreading, rather freely branching from all the nodes, the branches appressed. ♃ —Sand barrens, Coastal Plain, Massachusetts to South Carolina; Indiana. Closely approaching *P. commonsianum* but having smaller spikelets.

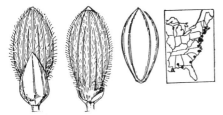

FIGURE 976.—*Panicum commonsianum.* Two views of spikelet, and floret, X 10. (Type.)

67. Panicum commonsiánum Ashe. (Fig. 976.) Vernal phase greenish olive, drying brownish; culms and sheaths appressed-pilose, the culms 20 to 50 cm. tall, ascending or spreading, appressed-pilose; blades 4 to 7 mm. wide, broadest near the rounded base, glabrous or nearly so on the upper surface, strigose or glabrous beneath; panicle 4 to 8 cm. long, the branches stiffly spreading; spikelets 2.2 to 2.4 mm. long. Autumnal culms branching from the middle and upper nodes, finally spreading or prostrate in mats. ♃ —Dunes and sandy woods near the coast, Massachusetts to northern Florida and Alabama.

68. Panicum addisóni Nash. (Fig. 977.) Vernal phase similar to that of *P. commonsianum;* culms usually less than 40 cm. tall, appressed-pilose below, puberulent above; sheaths sparsely ascending-pilose, blades 3 to 6 mm. wide, glabrous on the upper

69. Panicum wilmingtonénse Ashe. (Fig. 978.) Vernal phase bluish green, culms solitary or in small tufts, slender, erect from an ascending base, 20

FIGURE 978.—*Panicum wilmingtonense.* Two views of spikelet, and floret, X 10. (Type.)

to 40 cm. tall, pilose with soft ascending hairs; sheaths pubescent like the culms, densely villous-ciliate at the summit; blades 3 to 7 cm. long, glabrous on the upper surface, softly pubescent or nearly glabrous beneath, strongly ciliate near the base, the thick cartilaginous margin white when dry; panicle 5 to 8 cm. long; spikelets 2 mm. long. Autumnal culms spreading, branching from the middle and upper nodes. ♃ —Sandy woods, North Carolina, South Carolina, and Alabama, rare.

FIGURE 979.—*Panicum tsugetorum.* Two views of spikelet, and floret, × 10. (Type.)

70. Panicum tsugetórum Nash. (Fig. 979.) Vernal phase usually pale bluish green; culms 30 to 50 cm. tall, spreading or ascending, the lower nodes often geniculate, densely appressed-pubescent with short crisp hairs, long hairs more or less intermixed; sheaths pubescent like the culm; ligule 1 to 1.5 mm. long; blades 4 to 7 mm. wide, glabrous or nearly so on the upper surface, appressed-pubescent beneath; panicle 3 to 7 cm. long; spikelets 1.8 to 1.9 mm. long. Autumnal culms decumbent-spreading, branching from the lower and middle nodes. ♃ —Sandy woods, Maine to Wisconsin, south to Georgia and Tennessee.

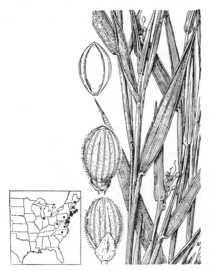

FIGURE 980.—*Panicum columbianum.* Plant, × 1; two views of spikelet, and floret, × 10. (Type.)

71. Panicum columbiánum Scribn. (Fig. 980.) Vernal culms 15 to 50 cm. tall, ascending, densely crisp-puberulent; sheaths less pubescent than the culms; blades 3 to 6 cm. long, 3 to 5 mm. wide, usually glabrous on the upper surface, appressed-puberulent or glabrous beneath; panicle 2 to 4 cm. long; spikelets 1.5 to 1.6 mm. long. Autumnal culms branching from the middle and upper nodes, becoming widely spreading or decumbent at base. ♃ —Sandy woods and open ground, Maine to North Carolina; Indiana and Michigan.

PANICUM COLUMBIANUM var. THÍNIUM Hitchc. and Chase. Vernal culms more slender, usually about 20 cm. tall; blades rarely more than 3 cm. long, sparsely pilose with long hairs on the upper surface; panicle 1.5 to 4 cm. long; spikelets 1.3 to 1.4 mm. long. Autumnal culms with branches crowded and aggregate toward the summit. ♃ —Dry sand, Massachusetts to North Carolina; Tennessee.

FIGURE 981.—*Panicum oricola.* Two views of spikelet, and floret, × 10. (Type.)

72. Panicum orícola Hitchc. and Chase. (Fig. 981.) Vernal phase grayish, often purplish; culms and sheaths appressed-pilose, the culms 10 to 30 cm. tall, spreading; ligule 1 to 1.5 mm. long; blades 2 to 5 cm. long, 2 to 4 mm. wide, the upper surface pilose with hairs 3 to 5 mm. long, the lower surface appressed-pilose; panicle short-exserted, ovoid, 1.8 to 3 cm. long, rather densely flowered; spikelets 1.5 mm. long, broadly obovate, turgid. Autumnal culms prostrate, forming mats, with short fascicled branches at all the nodes. ♃ — Sand barrens along the coast, Massachusetts to Virginia.

10. Sphaerocárpa.—Glabrous as a whole; culms few in a tuft, relatively stout; ligules obsolete or nearly so; blades mostly thick, firm, cartilaginous-margined, cordate and ciliate at base, panicle branches mostly viscid; spikelets obovoid-spherical at maturity, oval when young, 5- to 7-nerved, puberulent. Autumnal culms remaining simple or only sparingly branching, the thick white-margined blades of the winter rosette conspicuous.

73. Panicum sphaerocárpon Ell. (Fig. 982.) Vernal phase light green; culms 20 to 80 cm. tall, radiate-spreading, sometimes nearly erect, the nodes appressed-pubescent; blades 7 to 14 mm. wide; panicle 5 to 10 cm. long, about as wide; spikelets 1.6 to 1.8 mm. long. Autumnal phase prostrate-spreading, sparingly branched late in the season from the lower and middle nodes, the branches short, mostly simple. ♃ —Sandy soil, Vermont to Kansas, south to north-

ern Florida and Texas; Mexico to Venezuela. PANICUM SPHAEROCARPON var. INFLÁTUM (Scribn. and Smith) Hitchc. and Chase. Differing from *P. sphaerocarpon* in having a ligule as much as 1 mm. long, spikelets 1.4 to 1.5 mm. long, and more freely branching autumnal culms; many intergrades occur. ♃ —Moist sandy soil, Coastal Plain, Delaware to Florida and Texas, north to Oklahoma and Missouri.

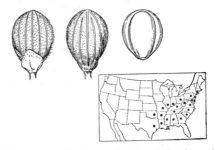

FIGURE 983.—*Panicum polyanthes.* Two views of spikelet, and floret, × 10. (Type.)

74. Panicum polyánthes Schult. (Fig. 983.) Vernal culms erect, 30 to 90 cm. tall, the nodes glabrous or nearly so; blades 12 to 23 cm. long, 15 to 25 mm. wide, the upper scarcely reduced; panicle 8 to 25 cm. long, one-fourth to half as wide, densely flowered; spikelets 1.5 to 1.6 mm. long, minutely puberulent. Autumnal phase remaining erect, producing simple branches from the lower and middle nodes. ♃ —Damp ground, woods, and openings, Connecticut to Oklahoma, south to Georgia and Texas.

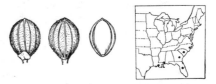

FIGURE 984.—*Panicum erectifolium.* Two views of spikelet, and floret, × 10. (Type.)

75. Panicum erectifólium Nash. (Fig. 984.) Vernal culms 30 to 70 cm. tall, erect or ascending; sheaths usually crowded at base; ligule very

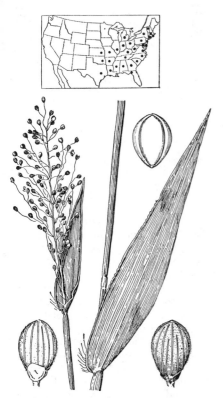

FIGURE 982.—*Panicum sphaerocarpon.* Plant, × 1; two views of spikelet, and floret, × 10. (Deam, Ind.)

short; blades 7 to 13 cm. long, 6 to 12 mm. wide, the crowded lower ones usually much larger than the others; panicle 6 to 12 cm. long, rather narrow, densely flowered, spikelets 1 to 1.2 mm. long, nearly spherical, densely puberulent. Autumnal culms remaining erect, late in the season producing branches from the third or fourth node, the branches nearly as long as the primary culms. ♃ — Moist pine barrens, swamps, and borders of ponds, North Carolina to Florida and Louisiana; Cuba.

11. Ensifólia.—Low and slender, mostly glabrous throughout (except in *P. curtifolium* and *P. tenue*); ligules nearly obsolete; spikelets 5- to 7-nerved. Autumnal culms simple to freely branching.

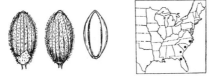

FIGURE 985.—*Panicum tenue.* Two views of spikelet, and floret, × 10. (Type.)

76. Panicum ténue Muhl. (Fig. 985.) Vernal phase olive green; culms 20 to 55 cm. tall, sometimes sparsely appressed-pubescent below; sheaths puberulent between the nerves or sparsely appressed-pilose, or the upper glabrous; blades distant, 2 to 5 cm. long, 3 to 4 mm. wide, rather thick, the margin cartilaginous, puberulent beneath, glabrous on the upper surface; panicle 3 to 5 cm. long; spikelets 1.6 to 1.7 mm. long, puberulent. Autumnal culms erect or leaning, sparingly branching from the middle nodes, the branches in small fascicles. ♃ —Moist sandy woods, eastern North Carolina to northern Florida.

77. Panicum albomarginátum Nash. (Fig. 986.) Vernal culms 15 to 40 cm. tall, ascending or spreading; leaves crowded at the base; blades thick and firm, those of the midculm

4 to 6 cm. long, 4 to 6 mm. wide, with a prominent white cartilaginous margin, the uppermost much reduced; panicle 3 to 6 cm. long; spikelet 1.4 to 1.5 mm. long, puberulent. Autumnal culms spreading, branching at the base, forming bushy tufts. ♃ — Low sandy soil, Coastal Plain, southeastern Virginia to Florida, Tennessee, Arkansas, and Louisiana; Cuba; Guatemala.

78. Panicum trifólium Nash. (Fig. 987.) Vernal phase similar to that of *P. albomarginatum*, the culms more slender, 20 to 50 cm. tall, the blades less crowded at the base, the upper blade not reduced. Autumnal culms erect or leaning, sparingly branching from the middle and upper nodes. ♃— Low, mostly moist, sandy woods, New Jersey to Florida and Texas; Tennessee.

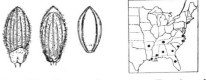

FIGURE 986.—*Panicum albomarginatum.* Two views of spikelet, and floret, × 10. (Type.)

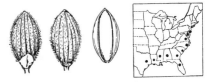

FIGURE 987.—*Panicum trifolium.* Two views of spikelet, and floret, × 10. (Type.)

79. Panicum flavóvirens Nash. (Fig. 988.) Vernal phase bright glossy green; culms very slender, ascending or spreading, 15 to 30 cm. tall; blades 2 to 5 cm. long, 3 to 4 mm. wide, thin; panicle few-flowered; spikelets 1.3 to 1.4 mm. long, pubescent. Autumnal culms spreading, decumbent or prostrate, branching from the lower and middle nodes. ♃ — Moist, shady, or mucky soil, North Carolina to Florida and Mississippi. *Panicum albomarginatum, P. trifolium,* and *P. flavovirens* form a series of closely allied species.

80. Panicum concínnius Hitchc. and Chase. (Fig. 989.) Vernal phase bright green; culms very slender, 12 to 50 cm. tall; blades 5 to 7 cm. long, 5 to 6 mm. wide; panicle 3 to 6 cm. long; spikelets 1.1 mm. long, pubescent. Autumnal culms radiate-spreading, late in the season bearing a few branches, with somewhat reduced blades. ⨞ —Moist sandy ground, northern Georgia, Florida, and northern Alabama, rare.

81. Panicum ensifólium Baldw. ex Ell. (Fig. 990.) Vernal culms 20 to 40 cm. tall, erect or reclining; blades distant, often reflexed, 1 to 3 cm. long, 1.5 to 3 mm. wide, puberulent beneath, at least toward the tip; panicle 1.5 to 4 cm. long; spikelets 1.3 to 1.5 mm. long, glabrous or puberulent. Autumnal culms spreading or reclining, sparingly branching

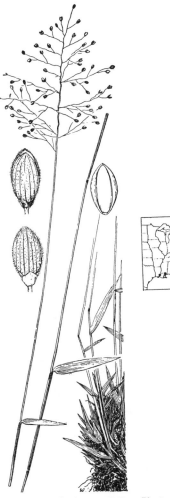

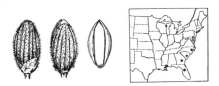

FIGURE 988.—*Panicum flavovirens.* Two views of spikelet, and floret, × 10. (Type.)

FIGURE 989.—*Panicum concinnius.* Two views of spikelet, and floret, × 10. (Type.)

from the middle nodes, the branches mostly simple. ⨞ —Wet places, mostly sphagnum bogs or swamps, Coastal Plain, New Jersey to Florida and Louisiana.

82. Panicum vernále Hitchc. and Chase. (Fig. 991.) Vernal phase light green, soft in texture; culms 15 to 30 cm. tall, very slender, ascending or spreading; leaves clustered at the base; blades thin, 2 to 7 cm. long, 3 to 5 mm. wide, the culm blades smaller; panicle 1.5 to 3 cm. long, few-flowered; spikelets 1.4 to 1.5 mm. long, elliptic, subacute, pubescent.

FIGURE 990.—*Panicum ensifolium.* Plant, × 1; two views of spikelet, and floret, × 10. (Biltmore Herb., N. C.)

FIGURE 991.—*Panicum vernale.* Two views of spikelet, and floret, × 10. (Type.)

Autumnal phase like the vernal in appearance, branching from the base, these culms simple and soon dying to the ground, rarely late in the season producing a few short fascicled branchlets from the nodes, the scarcely reduced flat blades spreading. ♃ —Moist places, especially sphagnum bogs, Florida to Mississippi.

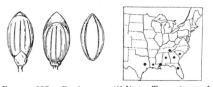

FIGURE 992.—*Panicum curtifolium.* Two views of spikelet, and floret, × 10. (Type.)

83. Panicum curtifólium Nash. (Fig. 992.) Vernal culms 10 to 30 cm. tall, slender, weak, angled, erect or spreading, sheaths striate-angled, sparsely pilose; ligule about 1 mm. long; blades spreading or reflexed, 1.5 to 3 cm. long, 2 to 5 mm. wide, thin, soft, sparsely pilose on both surfaces or nearly glabrous above; panicle 2 to 3 cm. long; spikelets 1.4 mm. long, glabrous or minutely pubescent. Autumnal culms weakly spreading, branching from the middle nodes, the ultimate branches in small fascicles toward the summit of the culm. ♃ —Boggy soil and shady moist places, sometimes forming a rather dense carpet, South Carolina to Tennessee, south to Florida and Texas.

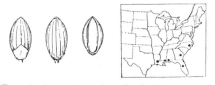

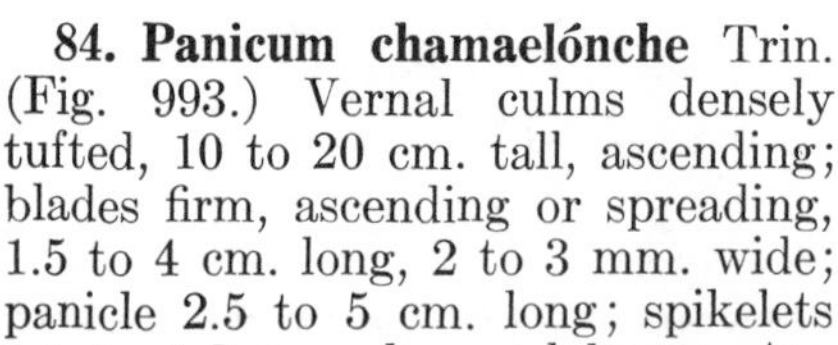

FIGURE 993.—*Panicum chamaelonche.* Two views of spikelet, and floret, × 10. (Type.)

84. Panicum chamaelónche Trin. (Fig. 993.) Vernal culms densely tufted, 10 to 20 cm. tall, ascending; blades firm, ascending or spreading, 1.5 to 4 cm. long, 2 to 3 mm. wide; panicle 2.5 to 5 cm. long; spikelets 1.1 to 1.2 mm. long, glabrous. Autumnal culms freely branching from the base and lower nodes, forming dense cushions as much as 50 cm. across. ♃ —Open sandy soil in low pineland, North Carolina to Florida and Louisiana; Isla de Pinos.

85. Panicum glabrifólium Nash. (Fig. 994.) Vernal phase similar to that of *P. chamaelonche;* culms stouter, 15 to 50 cm. tall, mostly erect; blades erect, 4 to 12 cm. long, 2 to 4 mm. wide, usually involute; panicle 4 to 9 cm. long; spikelets 1.2 to 1.4 mm. long, glabrous. Autumnal culms wiry, elongate, spreading, freely branching from the middle and upper nodes, the blades long and narrow. ♃ —Low sandy woods, peninsular Florida. Closely allied to *P. chamaelonche,* but taller and with different autumnal phase.

FIGURE 994.—*Panicum glabrifolium.* Two views of spikelet, and floret, × 10. (Type.)

FIGURE 995.—*Panicum breve.* Two views of spikelet, and floret, × 10. (Type.)

86. Panicum bréve Hitchc. and Chase. (Fig. 995.) Vernal phase purplish; culms 5 to 15 cm. tall, erect, stiff and wiry; sheaths crowded at the base; blades erect, 3 to 6 cm. long, strongly involute, with a few stiff hairs at the base; panicle 1.5 to 4 cm. long; spikelets 1.3 to 1.4 mm. long, puberulent. Autumnal phase erect, branching from the middle nodes, the fascicled branches strict. ♃ —Low pine woods and hammocks, east coast of southern Florida.

12. Lanceária.—Olive green, often purplish; vernal culms usually wiry; ligules nearly obsolete; blades usually ciliate toward the base; spikelets asymmetrically pyriform, strongly 7- to 9-nerved. Autumnal culms spreading, freely branching.

87. Panicum portoricénse Desv. ex Hamilt. (Fig. 996.) Vernal culms 15 to 30 cm. tall, slender, crisp-puberulent to nearly glabrous; sheaths glabrous or crisp-puberulent; blades firm, 2 to 5 cm. long, 3 to 6 mm. wide, glabrous to puberulent; panicle 2 to

4 cm. long; spikelets 1.5 to 1.6 mm. long, puberulent. Autumnal culms branching from all but the uppermost node, the reduced blades involute-pointed. ♃ (*P. pauciciliatum* Ashe.)—Sandy woods of the Coastal Plain, mostly in moist places, North Carolina to Florida and Texas; Cuba; Puerto Rico.

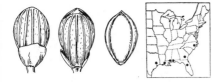

FIGURE 996.—*Panicum portoricense.* Two views of spikelet, and floret, × 10. (Ashe, N. C.)

88. Panicum lanceárium Trin. (Fig. 997.) Vernal culms 20 to 50

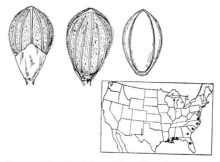

FIGURE 998.—*Panicum patulum.* Two views of spikelet, and floret, × 10. (Type.)

cm. tall, minutely grayish crisp-puberulent; sheaths puberulent; blades firm, 2 to 6 cm. long, 3 to 7 mm. wide, usually glabrous on the upper surface, puberulent or nearly glabrous beneath; panicle 3 to 6 cm. long; spikelets 2 to 2.1 mm. long, glabrous or usually puberulent. Autumnal culms geniculate-spreading, branching from the middle nodes. ♃ —Low sandy woods, Coastal Plain, southeastern Virginia to Florida and Texas; Cuba; Hispaniola; British Honduras.

89. Panicum pátulum (Scribn. and Merr.) Hitchc. (Fig. 998.) Vernal phase grayish olive green; culms geniculate-decumbent, as much as 50 cm. long, internodes and sheaths densely velvety-puberulent; blades rather lax, spreading, 4 to 8 cm. long, 4 to 8 mm. wide, velvety-puberulent beneath, pubescent above, ciliate at least half their length; spikelets as in *P. lancearium* but densely pubescent. Autumnal culms more freely branching than in *P. lancearium*, often forming large mats. ♃ —Low moist woods, Coastal Plain, southeastern Virginia to Florida and Louisiana; British Honduras and Hispaniola.

90. Panicum webberiánum Nash. (Fig. 999.) Vernal phase usually purplish; culms rather stout, erect or ascending, 20 to 50 cm. tall, minutely puberulent to glabrous; leaves somewhat crowded below; sheaths glabrous or nearly so; blades firm, ascending, often incurved or spoon-

FIGURE 997.—*Panicum lancearium.* Plant, × 1; two views of spikelet, and floret, × 10. (Chase 4545, S. C.)

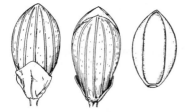

FIGURE 999.—*Panicum webberianum*. Two views of spikelet, and floret, × 10. (Type.)

shaped, 3 to 9 cm. long, 4 to 12 mm. wide, usually ciliate at the subcordate base, glabrous; panicle 4 to 10 cm. long; spikelets 2.3 to 2.5 mm. long, purple-stained at base, glabrous or minutely pubescent. Autumnal culms spreading or decumbent, flabellately branched at the middle and upper nodes. ♃ —Low pineland, North Carolina, Georgia, and Florida.

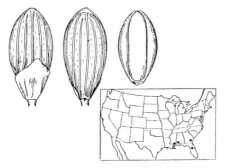

FIGURE 1000.—*Panicum patentifolium*. Two views of spikelet, and floret, × 10. (Type.)

91. Panicum patentifólium Nash. (Fig. 1000.) Vernal culms widely decumbent-ascending, slender, 25 to 55 cm. tall, minutely puberulent to nearly glabrous; blades stiffly spreading, 2.5 to 8 cm. long, 2 to 5 mm. wide, glabrous; panicle 3 to 7 cm. long; spikelets 2.4 to 2.6 mm. long, obovate, turgid, puberulent to nearly glabrous. Autumnal phase, decumbent or spreading, branching from the middle and upper nodes, the branches appressed. ♃ —Dry sand, especially in "scrub," Georgia and Florida to Mississippi.

13. Oligosánthia. — Culms mostly relatively stout, usually erect;

ligules inconspicuous except in *P. ravenelii;* blades firm; spikelets turgid, strongly 7- to 9-nerved. Autumnal culms with branches more or less crowded toward the summit.

92. Panicum wilcoxiánum Vasey. (Fig. 1001.) Vernal culms 10 to 25 cm. tall, copiously papillose-hirsute, as are sheaths and blades; ligule 1 mm. long; blades firm, erect, 5 to 8

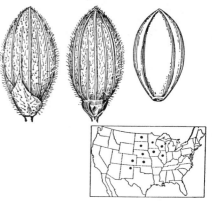

FIGURE 1001.—*Panicum wilcoxianum*. Two views of spikelet, and floret, × 10. (Type.)

cm. long, 3 to 6 mm. wide, usually involute-acuminate; panicle 2 to 5 cm. long; spikelets 2.7 to 3 mm. long, papillose-pubescent. Autumnal culms branching from all the nodes, forming bushy tufts with rigid erect blades. ♃ —Prairies, Alberta and Manitoba; Wisconsin and North Dakota to Illinois; Tennessee; Colorado and New Mexico.

93. Panicum malacophýllum Nash. (Fig. 1002.) Vernal phase velvety or velvety-pilose throughout; culms slender, 25 to 70 cm. tall, ascending or spreading, the nodes retrorsely bearded; ligule 1 to 1.5 mm. long; blades 7 to 10 cm. long, 6 to 12 mm. wide; panicle 3 to 7 cm. long; spikelets 2.9 to 3 mm. long, papillose-pilose. Autumnal phase spreading, forming bushy topheavy clumps with reduced blades. ♃ —Sandy woods, Tennessee to Kansas and Texas.

94. Panicum helléri Nash. (Fig. 1003.) Vernal culms 25 to 60 cm.

FIGURE 1002.—*Panicum malacophyllum.* Two views of spikelet, and floret, × 10. (Type.)

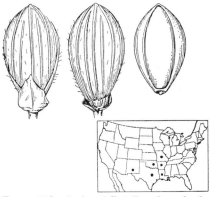

tall, ascending or spreading, appressed-pilose below, often glabrous above; sheaths sparsely papillose-hispid to glabrous; blades rather thin, glabrous on both surfaces or pubescent beneath, ciliate toward the base; panicle 6 to 12 cm. long; spikelets 2.9 to 3 mm. long, glabrous or with a few scattered hairs. Autumnal phase branching at all but the lowest nodes, forming loose sprawling tufts, the blades widely spreading, not much reduced, the long-pediceled spikelets rather conspicuous among the foliage. ♃ —Open woods and prairies, Missouri and Oklahoma to

FIGURE 1003.—*Panicum helleri.* Two views of spikelet, and floret, × 10. (Type.)

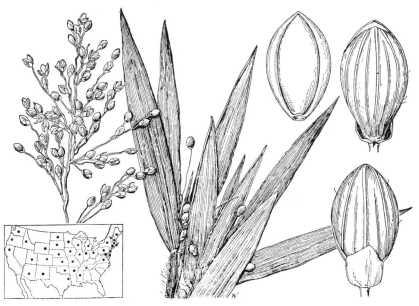

FIGURE 1004.—*Panicum scribnerianum.* Plant, × 1; two views of spikelet, and floret, × 10. (Vernal phase, McDonald 32, Ill.; autumnal phase, Umbach 2365, Ill.)

Louisiana and New Mexico. Closely related to *P. scribnerianum*.

95. Panicum scribneriánum Nash. (Fig. 1004.) Vernal culms 20 to 50 cm. tall, glabrous or harshly puberulent or sometimes ascending-pilose; sheaths striate, papillose-hispid to nearly glabrous; blades ascending or erect, 5 to 10 cm. long, 6 to 12 mm. wide, firm, rounded at the ciliate base, glabrous on the upper surface, appressed-pubescent

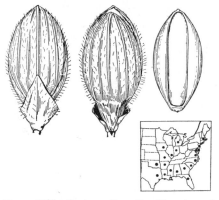

FIGURE 1005.—*Panicum oligosanthes.* Two views of spikelet, and floret, × 10. (Type.)

96. Panicum oligosánthes Schult. (Fig. 1005.) Vernal culms 35 to 80 cm. tall, appressed-pubescent, especially below; sheaths with ascending papillose pubescence; blades stiffly spreading or ascending, 6 to 14 cm. long, 5 to 8 mm. wide, glabrous or nearly so on the upper surface, harshly puberulent beneath; panicle 6 to 12 cm. long; spikelets long-pediceled, 3.5 to 4 mm. long, subacute, sparsely hirsute. Autumnal phase erect to spreading, branching freely from the upper nodes. ♃ — Sandy, usually moist woods, Massachusetts and Michigan to Iowa, south to Florida and Texas.

97. Panicum ravenélii Scribn. and Merr. (Fig. 1006.) Vernal culms 30 to 70 cm. tall, densely papillose-hirsute with ascending hairs, the nodes short-bearded; sheaths hirsute like the culm; ligule 3 to 4 mm. long; blades thick, 8 to 15 cm. long, 1 to 2 cm. wide, glabrous on the upper surface, densely velvety-hirsute beneath; panicle 7 to 12 cm. long; spikelets 4 to 4.3 mm. long, sparsely papillose-pubescent. Autumnal phase

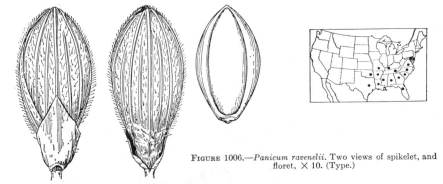

FIGURE 1006.—*Panicum ravenelii.* Two views of spikelet, and floret, × 10. (Type.)

to glabrous beneath; panicle 4 to 8 cm. long; spikelets 3.2 to 3.3 mm. long, obovate, blunt, sparsely pubescent to nearly glabrous. Autumnal phase branching from the middle and upper nodes. ♃ —Sandy soil or dry prairies, Maine to British Columbia and Washington, south to Virginia, Mississippi, Texas, and Arizona; Mexico.

more or less spreading, branching from the middle and upper nodes, the short branches crowded at the summit. ♃ —Sandy or gravelly woods or open ground, Delaware to Missouri, south to Florida and Texas.

98. Panicum leibérgii (Vasey) Scribn. (Fig. 1007.) Vernal culms slender, 25 to 75 cm. tall, erect from a more or less geniculate base, pilose

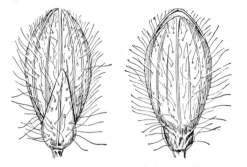

FIGURE 1007.—*Panicum leibergii*. Two views of spikelet, and floret, × 10. (Type.)

or scabrous; sheaths papillose-hispid with spreading hairs; ligule obsolete or nearly so; blades ascending or erect, rather thin, 6 to 15 cm. long, 7 to 15 mm. wide, papillose-hispid on both surfaces, often sparsely so above; panicle 8 to 15 cm. long, less than half as wide; spikelets 3.7 to 4 mm. long, strongly papillose-hispid. Autumnal phase leaning, sparingly branching from the middle and lower nodes. ⚐ —Prairies, New York and Pennsylvania to Manitoba and North Dakota, south to Ohio and Kansas; Texas.

99. Panicum xanthophýsum A. Gray. (Fig. 1008.) Vernal phase yellowish green; culms 20 to 55 cm. tall, more or less scabrous; sheaths sparsely papillose-pilose; blades erect or nearly so, rather thin, prominently nerved, 10 to 15 cm. long, 1 to 2 cm. wide, glabrous except the ciliate base; panicle 5 to 12 cm. long, very narrow, few-flowered, the stiff branches erect or nearly so; spikelets 3.7 to 4 mm. long, blunt, pubescent. Autumnal

phase erect or ascending, branching from the second and third nodes, the branches erect, mostly simple. ⚐ —Sandy or gravelly soil, Quebec to Manitoba, south to Pennsylvania, West Virginia, and Minnesota.

14. Pedicelláta.—Culms slender from a knotted crown; sheaths papillose-hirsute; ligules about 1 mm. long; blades long-ciliate at least toward base; spikelets attenuate at base, 7- to 9-nerved, papillose-pubescent. Autumnal culms freely branching, the branches appearing before the maturity of the primary panicle; no distinct winter rosette formed.

100. Panicum pedicellátum Vasey. (Fig. 1009.) Vernal culms erect or ascending, 20 to 50 cm. tall, usually ascending-hirsute, at least below; blades 5 to 9 cm. long, 3 to 6 mm. wide, glabrous or sometimes minutely hispid; panicle 3 to 6 cm. long; spikelets 3.5 to 3.7 mm. long, elliptic; first glume about half as long as the spikelet, acute, the

FIGURE 1008.—*Panicum xanthophysum*. Two views of spikelet, and floret, × 10. (Type.)

second shorter than the fruit. Autumnal culms erect or leaning, branching from all but the uppermost nodes, the branches spreading. ♃ —Dry woods and prairies, central and southern Texas.

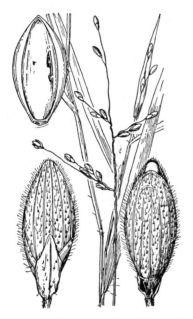

FIGURE 1009.—*Panicum pedicellatum*. Plant, X 1; two views of spikelet, and floret, X 10. (Heller, Tex.)

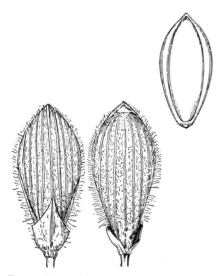

FIGURE 1010.—*Panicum nodatum*. Two views of spikelet, and floret, X 10. (Type.)

101. Panicum nodátum Hitchc. and Chase. (Fig. 1010.) Vernal culms tufted, ascending or spreading, hard and wiry, 25 to 35 cm. tall, finely papillose, crisp-puberulent; blades firm, ascending, 3 to 5 cm. long, 3 to 6 mm. wide, puberulent on both surfaces; panicle 4 to 5 cm. long, few-flowered; spikelets 4 mm. long, pyriform. Autumnal culms widely geniculate-decumbent, branching from all but the uppermost node, the branches somewhat divaricate, the nodes of the main culm swollen. ♃ —Oak woods in sand dunes, southern Texas and northern Mexico.

15. Scopária.—Species of various habit, vernal culms tall; ligules 1 mm. long or less; blades elongate; spikelets abruptly pointed, 7- to 9-nerved; autumnal culms branching from the middle or upper nodes.

102. Panicum scopárium Lam. (Fig. 1011.) Vernal phase grayish olive green, velvety-pubescent throughout except on a viscid ring below the nodes and at the summit of the sheath; culms 80 to 130 cm. tall, stout, erect or ascending, usually geniculate at base; blades rather thick, 12 to 20 cm. long, 10 to 18 mm. wide; panicle 8 to 15 cm. long, the axis and branches with viscid blotches; spikelets 2.4 to 2.6 mm. long, obovate, turgid, papillose-pubescent. Autumnal phase leaning or spreading, freely branching from the middle nodes, forming flabellate fascicles. ♃ —Wet or damp soil, Massachusetts to Florida, west through Kentucky to Missouri, Oklahoma, and Texas; Cuba.

103. Panicum aculeátum Hitchc. and Chase. (Fig. 1012.) Vernal culms in large clumps, slender, 70 to 100 cm. tall, ascending, scabrous, harshly pubescent below; sheaths papillose-hispid with stiff sharp-pointed hairs, a puberulent ring at the summit, the uppermost usually glabrous; blades firm, stiffly ascending or spreading, 12 to 20 cm. long, 9 to 13 mm. wide,

FIGURE 1011.—*Panicum scoparium*. Plant, × 1; two views of spikelet, and floret, × 10. (McGregor 212, S. C.)

scabrous on the upper surface and toward the apex beneath; panicle 8 to 12 cm. long, few-flowered; spikelets 3 mm. long, elliptic, minutely pubescent, pointed beyond the fruit. Autumnal culms branching from the middle nodes, the branches more or less divaricate, the ultimate panicles wholly or partly included in the sheaths. ♃ —Swampy woods, Connecticut to North Carolina, rare.

104. Panicum recógnitum Fernald. (Fig. 1013.) Culms 60 to 150 cm. tall, with elongate internodes, glabrous; sheaths much shorter than the internodes, papillose-pilose to glabrate; ligule minute; blades 6 to 13 cm. long, 8 to 15 mm. wide, lanceolate, acuminate, glabrous, or sometimes pilose on the lower surface, the margins ciliate toward the cordate base, pubescent on the collar; primary panicle 8 to 13 cm. long, the branches broadly ascending, few-flowered; pulvini pubescent; spikelets 2.2 to 2.8 mm. long, elliptic, rather sparsely pubescent; first glume 0.8 to 1 mm. long, ovate, acute, the second glume and sterile lemma subequal, scarcely

FIGURE 1012.—*Panicum aculeatum*. Two views of spikelet, and floret, × 10. (Type.)

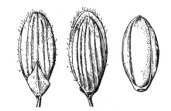

FIGURE 1013.—*Panicum recognitum*. Two views of spikelet, and floret, × 10. (Long 7672, N. J.)

covering the fruit. Autumnal phase sparingly branched, the branches elongate, ascending, the panicles 1.5 to 5 cm. long. ♃ —Open sandy ground, swamps, and moist places, Rhode Island, New Jersey, and eastern Pennsylvania.

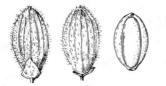

Figure 1014.—*Panicum mundum.* Two views of spikelet, and floret, × 10. (Fernald and Long 6017, Va.)

105. Panicum múndum Fernald.

(Fig. 1014.) Culms 50 to 140 cm. tall, densely tufted, pilose or papillose-pilose with ascending hairs, the nodes retrorsely bearded, with a glabrous glandular ring below; sheaths much shorter than the internodes, viscid-spotted, ascending-pilose or glabrate; ligule about 1 mm. long; blades 6 to 15 cm. long, 8 to 13 mm. wide, lanceolate, subcordate, papillose-ciliate toward the base; primary panicle 7 to 12 cm. long, 5 to 10 cm. wide, the branches ascending; spikelets 1.8 to 2.2 mm. long, subglobose or ellipsoid, densely pubescent, first glume about one-fourth the length of the spikelet, subacute. Autumnal phase sparingly branched, the panicles 1 to 6 cm. long. ♃ — Borders of swamps and sandy, peaty meadows, southeastern Virginia (Princess Anne and Sussex Counties) and Durham County, N. C.

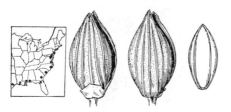

Figure 1015.—*Panicum scabriusculum.* Two views of spikelet, and floret, × 10. (Type.)

106. Panicum scabriúsculum Ell.

(Fig. 1015.) Vernal phase grayish olive green; culms erect, 1 to 1.5 m. tall, scabrous at least below the nodes, sometimes puberulent; sheaths glabrous or more or less hispid at least toward the summit, often mottled or white-spotted, commonly swollen at the base and contracted toward the summit; blades stiffly ascending or spreading, often reflexed, 15 to 25 cm. long, 9 to 12 mm. wide, glabrous or scabrous, often more or less pubescent beneath, tapering to an involute point; panicle 10 to 20 cm. long; spikelets 2.3 to 2.6 mm. long, ovate, glabrous or obscurely puberulent. Autumnal culms erect, branching from the middle and upper nodes, the branches appressed, finally forming dense oblong masses along the upper part of the primary culm, the panicles partly or entirely enclosed in the sheaths. ♃ —Moist ground, especially along ditches, streams, and swamps, Coastal Plain, New Jersey to Florida and Texas.

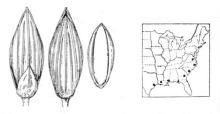

Figure 1016.—*Panicum cryptanthum.* Two views of spikelet, and floret, × 10. (Type.)

107. Panicum cryptánthum Ashe.

(Fig. 1016.) Vernal culms erect, 80 to 100 cm. tall, glabrous except the usually bearded nodes; sheaths glabrous or the lowermost sparsely hirsute, the upper somewhat inflated; blades stiff, glabrous, sparingly ciliate at base, 10 to 15 cm. long, 7 to 9 mm. wide; panicle 6 to 10 cm. long, the axis and ascending branches viscid-spotted; spikelets 2.2 to 2.4 mm. long, lanceolate-elliptic, pointed. Autumnal culms erect, sparingly branching from the middle nodes, the panicles partly hidden in the sheaths. ♃ —Low swampy ground, Virginia to Florida and Texas; infrequent.

16. Commutáta.—Culms relatively stout, glabrous or puberulent; ligules obsolete or nearly so; blades cordate and more or less ciliate at base; spikelets elliptic, not very turgid, 7- to 9-nerved, pubescent. Autumnal culms usually rather sparingly branching.

108. Panicum áshei Pearson. (Fig. 1017.) Vernal phase usually purplish, from a knotted crown; culms 25 to 50 cm. tall, erect, stiff and wiry, densely crisp-puberulent; sheaths less densely puberulent; blades rather thick and firm, 4 to 8 cm. long, 5 to 10 mm. wide, glabrous; panicle 5 to 8 cm. long, loosely flowered; spikelets 2.4 to 2.7 mm. long. Autumnal culms erect or topheavy-reclining, bearing divergent branches from the middle and upper nodes or from the upper nodes only. ♃ —Dry, especially rocky

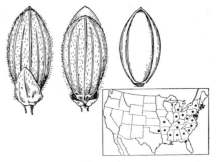

FIGURE 1017.—*Panicum ashei.* Two views of spikelet, and floret, X 10. (Type coll.)

woods, Massachusetts to Michigan and Missouri, south to northern Florida, Mississippi, Arkansas, and Oklahoma.

109. Panicum commutátum Schult. (Fig. 1018.) Vernal culms 40 to 75 cm. tall, erect; sheaths glabrous or nearly so; blades 5 to 12 cm. long, 12 to 25 mm. wide, glabrous on both surfaces or puberulent beneath; pan-

FIGURE 1018.—*Panicum commutatum.* Plant, X 1; two views of spikelet, and floret, X 10. (Bock and Chase 118, Ill.)

icle 6 to 12 cm. long; spikelets 2.6 to 2.8 mm. long. Autumnal culms erect or leaning, branching from the middle nodes, the secondary branches crowded toward the summit.   ♃

culms widely spreading, bearing more or less divaricate branches from all the nodes, the ultimate branches in short dense fascicles.   ♃   —Low or swampy woods, Coastal Plain, south-

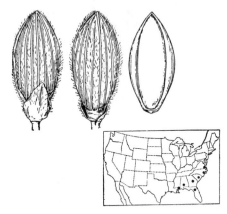

FIGURE 1019.—*Panicum mutabile.* Two views of spikelet, and floret, × 10. (Type.)

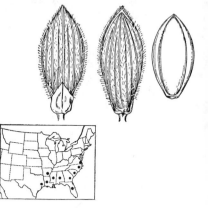

FIGURE 1020.—*Panicum joorii.* Two views of spikelet, and floret, × 10. (Type.)

—Woods and copses, Massachusetts to Michigan and Oklahoma, south to Florida and Texas.

110. Panicum mutábile Scribn. and Smith ex Nash. (Fig. 1019.) Vernal phase blue green, glaucous; culms solitary or few in a tuft, erect, 30 to 70 cm. tall; sheaths glabrous; blades horizontally spreading, 6 to 15 cm. long, 8 to 20 mm. wide, tapering to both ends, glabrous, ciliate toward the cordate base or the lower ciliate nearly to apex; panicle 7 to 15 cm. long; spikelets 2.9 to 3 mm. long. Autumnal culms erect or reclining, sparingly branched from the middle and upper nodes.   ♃   —Sandy pine woods or hammocks, Coastal Plain, southeastern Virginia to Florida and Mississippi.

111. Panicum joórii Vasey. (Fig. 1020.) Vernal culms 20 to 55 cm. tall, slender, spreading or ascending from a decumbent base, at least the lower internodes purplish red; sheaths glabrous; blades 6 to 15 cm. long, 7 to 18 mm. wide, thin, often subfalcate, glabrous on both surfaces; panicle loosely flowered, 5 to 9 cm. long; spikelets 3 to 3.1 mm. long. Autumnal

eastern Virginia to Florida, west to Arkansas and Texas; Mexico.

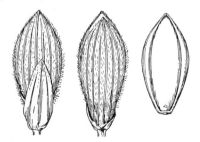

FIGURE 1021.—*Panicum equilaterale.* Two views of spikelet, and floret, × 10. (Type.)

112. Panicum equilaterále Scribn. (Fig. 1021.) Vernal culms 25 to 70 cm. tall, stiff and erect; sheaths glabrous, the upper two often approximate; blades firm, widely spreading, 6 to 17 cm. long, 6 to 14 mm. wide, the margins nearly parallel, glabrous, often ciliate at the rounded or subcordate base; panicle 5 to 10 cm. long; spikelets 3.2 mm. long. Autumnal culms erect or leaning, branching from the upper and middle nodes.   ♃   —Pinelands, hammocks, and sandy woods, Coastal Plain, North

Carolina, South Carolina, and Florida.

17. Latifólia.—Culms rather stout, erect or suberect; ligules not more than 1 mm. long; blades cordate, clasping; spikelets rather turgid, 7- to 9-nerved, pubescent. Autumnal phase usually rather sparingly branching.

113. Panicum clandestínum L. (Fig. 1022.) Vernal culms in large dense clumps, sometimes with strong rhizomes 5 to 10 cm. long, 70 to 150 cm. tall, scabrous to papillose-hispid,

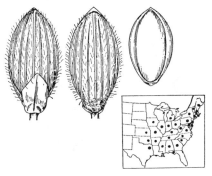

FIGURE 1022.—*Panicum clandestinum.* Two views of spikelet, and floret, × 10. (Torrey, N. Y.)

FIGURE 1023.—*Panicum latifolium.* Plant, × 1; two views of spikelet, and floret, × 10. (Schenck, Ill.)

at least below the nodes; sheaths strongly papillose-hispid to nearly glabrous; blades spreading or finally reflexed, 10 to 20 cm. long, 1.2 to 3 cm. wide, scabrous on both surfaces, at least toward the end, usually ciliate at base; panicle 8 to 15 cm. long; spikelets 2.7 to 3 mm. long. Autumnal culms erect or leaning, the branches leafy, the swollen bristly sheaths overlapping and wholly or partly enclosing the panicles. ♃ —Moist mostly sandy ground, Nova Scotia, Quebec, and Maine to Kansas, south to northern Florida and Texas.

114. Panicum latifólium L. (Fig. 1023.) Vernal culms from a knotted crown; culms 45 to 100 cm. tall, glabrous or the lower part sparsely pu-

bescent; sheaths ciliate; blades 8 to 18 cm. long, 1.5 to 4 cm. wide, glabrous; panicle 7 to 15 cm. long; spikelets 3.4 to 3.7 mm. long. Autumnal culms more or less spreading, branching from the middle nodes, the upper leaves of the branches crowded and spreading, not much reduced. ♃ —Rocky or sandy woods, Maine and Quebec to Minnesota, south to Georgia, Kansas and Arkansas.

115. Panicum bóscii Poir. (Fig. 1024.) Vernal phase resembling that of *P. latifolium;* culms 40 to 70 cm. tall, glabrous or minutely puberulent, the nodes retrorsely bearded; sheaths glabrous or nearly so; blades spreading, 7 to 12 cm. long, 1.5 to 3 cm. wide, sparsely ciliate at base, gla-

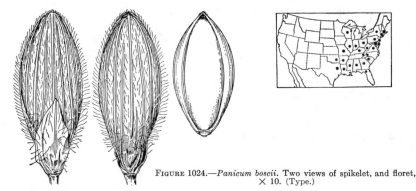

FIGURE 1024.—*Panicum boscii*. Two views of spikelet, and floret, X 10. (Type.)

brous or nearly so; panicle 6 to 12 cm. long; spikelets 4 to 4.5 mm. long, about half as wide, papillose-pubescent. Autumnal phase about as in *P. latifolium*, finally top-heavy-reclining. ♃ —Woods, Massachusetts to Wisconsin and Oklahoma, south to northern Florida and Texas. PANICUM BOSCII var. MÓLLE (Vasey) Hitchc. and Chase. Differing from *P. boscii* in the downy-villous culms and sheaths and the velvety blades. ♃ — About the same range as the species.

SUBGENUS 3. EUPÁNICUM Godr.

Spikelets in open or condensed panicles or in spikelike racemes, the branchlets not produced as bristles (the naked tip forming a short point in Geminata); not presenting vernal and autumnal phases of a distinctive character, with winter rosettes of leaves different from the culm leaves.

1. **Geminàta.**—Subaquatic glabrous perennials; inflorescence of several erect, spikelike racemes distant on an elongate axis; rachis ending in a short naked point; spikelets subsessile, abruptly pointed, glabrous, first glume truncate; fruit transversely rugose.

116. **Panicum geminàtum** Forsk. (Fig. 1025.) Culms tufted, 25 to 80 cm. tall, scarcely succulent, often decumbent at base or with stolons rooting at the nodes; blades 10 to 20 cm. long, 3 to 6 mm. wide, flat, or in-

volute toward the apex; panicle 12 to 30 cm. long, the appressed racemes 12 to 18, the lower 2.5 to 3 cm. long, the upper gradually shorter; spikelets 2.2 to 2.4 mm. long, 5-nerved. ♃ — Moist ground or shallow water, mostly near the coast, southern Florida, Louisiana, Texas, and Oklahoma; warmer regions of both hemispheres.

117. **Panicum paludívagum** Hitchc. and Chase. (Fig. 1026.) Resembling *P. geminatum*, but the culms elongate from a long creeping rooting base, rather succulent, as much as 2 m. long, the lower part submerged, loosely branching; blades 15 to 40 cm. long, scabrous on the upper surface; spikelets 2.8 to 3 mm. long, faintly 3-nerved; fruit obscurely rugose. ♃ —More or less submerged in fresh-water rivers and lakes, Florida, Texas; Mexico, Guatemala.

2. **Purpurascéntia.**—Stoloniferous robust perennial; a single species introduced.

118. **Panicum purpuráscens** Raddi. PARA GRASS. (Fig. 1027.) Culms decumbent and rooting at base, 2 to 5 m. long, the nodes densely villous; sheaths villous or the upper glabrous, densely pubescent on the collar; blades 10 to 30 cm. long, 10 to 15 mm. wide, flat, glabrous; panicle 12 to 20 cm. long, the rather distant subracemose densely flowered branches ascending or spreading; spikelets subsessile, 3 mm. long, elliptic, 5-nerved, glabrous; fruit minutely transversely

FIGURE 1025.—*Panicum geminatum.* Plant, × ½; two views of spikelet, and floret, × 10. (Tracy 9395, Fla.)

rugose. ♃ (*P. barbinode* Trin.)—
Cultivated and waste ground in moist

soil, borders of rivers, marshes, and
swamps, Florida, Alabama (Mobile),

Texas; Oregon (Linnton); throughout tropical America at low altitudes. Commonly cultivated in tropical America as a forage grass, being cut for green feed. It probably was introduced into Brazil at an early date from Africa.

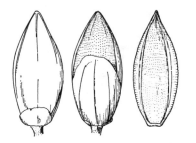

FIGURE 1026.—*Panicum paludivagum.* Two views of spikelet, and floret, × 10. (Type.)

3. **Fasciculáta.**—Branching annuals; blades flat; ligules not more than 1 mm. long; panicles of ascending spikelike racemes along an angled axis; spikelets subsessile, abruptly pointed, strongly 5- to 7-nerved; fruit transversely rugose.

119. Panicum réptans L. (Fig. 1028.) Culms ascending 10 to 30 cm. above the creeping base; blades 1.5 to 6 cm. long, 4 to 12 mm. wide, cordate, usually glabrous, ciliate on the undulate margin at base; panicle 2 to 6 cm. long, the 3 to 12 ascending or spreading racemes 2 to 3 cm. long, aggregate, the rachis usually pilose with long weak hairs; spikelets secund, about 2 mm. long, glabrous, on pubescent or pilose pedicels about 1 mm.; first glume very short, truncate or rounded. ⊙ —Moist open ground, or a weed in cultivated fields, Florida to Texas; tropical regions of both hemispheres.

120. Panicum fasciculátum Swartz. BROWNTOP PANICUM. (Fig. 1029.) Culms erect or spreading from a decumbent base, 30 to 100 cm. tall, sometimes pubescent below the panicle or hispid below the appressed-pubescent nodes, the more robust freely branched from the lower nodes; sheaths glabrous to papillose-hispid; blades 4 to 30 cm. long, 6 to 20 mm. wide, glabrous; panicle 5 to 15 cm. long; the racemes 5 to 10 cm. long; spikelets yellow or bronze brown, 2.1 to 2.5 mm. long, rarely 3 mm., obovate, turgid, glabrous, strongly transversely wrinkled or veined. ⊙ —Moist open ground, often a weed in fields, southern Florida, southern Texas; tropical America, at low altitudes.

PANICUM FASCICULATUM var. RETICULÁTUM (Torr.) Beal. Differing from *P. fasciculatum* in having smaller more compact panicles, narrower pubescent blades, less regular suberect racemes and larger, mostly more yellowish spikelets 2.6 to 3 mm. long. Many intergrades occur. ⊙ (This has been erroneously referred to *P. fasciculatum* var. *chartaginense* (Swartz) Doell.)—Prairies, fields, and waste ground; New Mexico and Arizona; Mexico.

121. Panicum adspérsum Trin. (Fig. 1030.) Culms ascending or spreading from a decumbent base, rooting at the lower nodes, 30 to 100 cm. tall; blades 5 to 15 cm. long, 8 to 20 mm. wide; panicle 6 to 15 cm. long, the racemes 3 to 10 cm. long; spikelets 3.2 to 4 mm. long, fusiform, abruptly acuminate, hispid or hispidulous, sometimes only at the summit, rarely glabrous, obscurely reticulate-veined. ⊙ —Moist open ground, often on coral limestone, Florida; ballast, Philadelphia and Camden; Mobile; West Indies. The Florida specimens, commonly more robust than the typical form from the West Indies, have been described as *P. keyense* Mez.

122. Panicum ramósum L. BROWNTOP MILLET. (Fig. 1031.) Resembling *P. fasciculatum* var. *reticulatum;* pedicels bristly; spikelets glabrous to finely pubescent, about 3 mm. long, tawny or dull brown. ⊙ —Waste ground, North Carolina to Florida, Arkansas, and Louisiana; tropical Asia. Cultivated for bird food.

123. Panicum arizónicum Scribn. and Merr. ARIZONA PANICUM. (Fig.

FIGURE 1027.—*Panicum purpurascens.* Plant, × ½; two views of spikelet, and floret, × 10. (Hitchcock 9693, Jamaica.)

1032.) Culms erect or sometimes decumbent at base, 20 to 60 cm. tall; sheaths glabrous to papillose-hispid; blades 5 to 15 cm. long, 6 to 12 mm. wide, glabrous or papillose-hispid beneath, ciliate near base; panicle 7 to 20 cm. long, the branches rather loosely flowered, finely pubescent

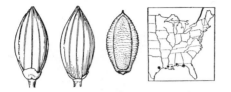

FIGURE 1028.—*Panicum reptans.* Two views of spikelet, and floret, × 10. (Type of *P. prostratum* Lam.)

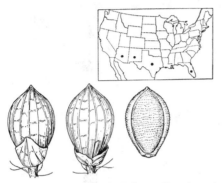

FIGURE 1029.—*Panicum fasciculatum.* Two views of spikelet, and floret, × 10. (Type.)

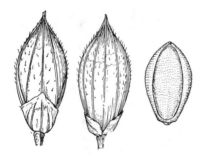

FIGURE 1030.—*Panicum adspersum.* Two views of spikelet, and floret, × 10. (Type.)

and papillose-hirsute; spikelets 3.5 to 3.8 mm. long, obovate-elliptic, densely hirsute to glabrous. ⊙ — Open sandy or stony ground, or in cultivated soil, western Texas to southern California; Mexico.

124. Panicum texánum Buckl. TEXAS MILLET. (Fig. 1033.) Culms erect or ascending, often decumbent and rooting at the lower nodes, 50 to 150 cm. or even to 3 m. long, softly pubescent, at least below the nodes and below the panicles; sheaths softly pubescent, often papillose; blades 8 to 20 cm. long, 7 to 15 mm. wide, softly pubescent; panicle 8 to 20 cm. long, the branches short, appressed, loosely flowered, the axis and rachises pubescent, with long hairs intermixed; spikelets 5 to 6 mm. long, fusiform, pilose, often obscurely reticulate. ⊙ —Prairies and open ground, especially on low land along streams, often a weed in fields, Texas; introduced at several localities, North Carolina to Florida and Oklahoma; Arizona; northern Mexico.

4. Dichotomiflóra.—Somewhat succulent branching annuals (a few species perennial); blades flat, panicles many-flowered, the branchlets short and appressed along the rather stiff main branches; spikelets short-pediceled, 7-nerved, glabrous; first glume short, broad; fruit smooth and shining.

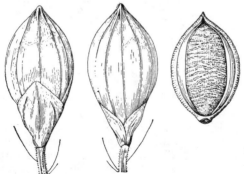

FIGURE 1031.—*Panicum ramosum.* Two views of spikelet, and floret, × 10. (Handley A–75, Fla.)

125. Panicum dichotomiflórum Michx. FALL PANICUM. (Fig. 1034.) Culms ascending or spreading from a geniculate base, 50 to 100 cm. long, or in robust specimens as much as 2 m. long; ligule a dense ring of white hairs 1 to 2 mm. long; blades scaberulous and sometimes sparsely pilose on the upper surface, 10 to 50 cm. long, 3 to 20 mm. wide, the white midrib usually prominent; panicles terminal and axillary, mostly included at base, 10 to 40 cm. long or more, the main branches ascending; spikelets narrowly oblong-ovate, 2 to 3 mm., usually about 2.5 mm., long, acute. ⊙ —Moist ground, along streams, and a weed in waste places and cultivated soil, Nova Scotia and

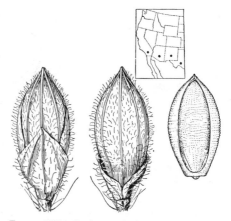

FIGURE 1032.—*Panicum arizonicum.* Two views of spikelet, and floret, × 10. (Palmer 159, Mexico.)

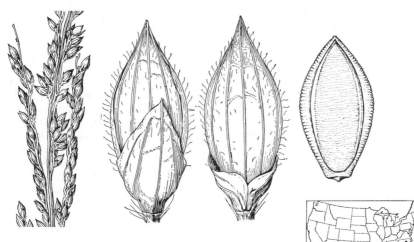

FIGURE 1033.—*Panicum texanum.* Panicle, × 1; two views of spikelet, and floret, × 10. (Hitchcock 3187, Tex.)

Maine to Minnesota, south to Florida and Texas, occasionally introduced farther west; here and there in the West Indies. PANICUM DICHOTOMIFLORUM var. PURITANÓRUM Svenson. Differing in the shorter, more slender culms and looser panicles and in the rather less-pointed spikelets about 2 mm. long. Intergrades with the species. ⊙ —Wet sandy or boggy shores of ponds, Massachusetts, Connecticut, Long Island; Indiana.

126. Panicum bartowénse Scribn. and Merr. (Fig. 1035.) Resembling

P. dichotomiflorum, mostly larger; culms erect, simple or sparingly branched, as much as 2 m. tall and 7 mm. thick; sheaths papillose-hispid; ligule 2 to 3 mm. long. ⊙ —Low ground, often in shallow water, Florida; Bahamas; Cuba, Jamaica.

127. Panicum lacústre Hitchc. and Ekman. (Fig. 1036.) Aquatic or terrestrial perennial; culms nearly simple, those of terrestrial plants erect, about 1 m. tall, with short innovations with pilose sheaths and flat blades, 1 to 10 cm. long, 2 to 4

FIGURE 1034.—*Panicum dichotomiflorum.* Panicle, X 1; two views of spikelet, and floret, X 10. (Deam, Ind.)

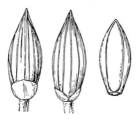

FIGURE 1035.—*Panicum bartowense.* Two views of spikelet, and floret, X 10. (Type.)

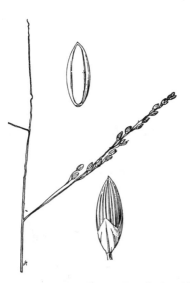

FIGURE 1036.—*Panicum lacustre.* Branch of panicle, X 1; (Brass 15910, Fla.); spikelet and floret, X 10. (Type.)

mm. wide, pilose on the upper surface; culm sheaths mostly longer than the internodes, but narrow, exposing the nodes, glabrous; ligule membranaceous, densely ciliate; blades flat or folded, 15 to 30 cm. long, 2 to 3 mm. wide, sparsely pilose on the upper surface; panicle erect, 10 to 25 cm. long, the rather distant branches ascending, with appressed branchle s except toward the base; spikelets 2 to 2.2 mm. long, subacute, glabrous; first glume one-fourth to one-third as long as the spikelet. ♃ —Edges of cypress ponds, west of Miles City, Collier County, Fla. The type, from western Cuba, is an aquatic plant with a succulent base rooting at the nodes and with loose papery lower sheaths.

5. **Capillária.** — Branching annuals, papillose-hispid, at least on the sheaths; ligules 1 to 3 mm. long; panicles many-flowered, mostly diffuse; spikelets pointed, 7- to 9-nerved, glabrous; first glume large, clasping; fruit smooth and shining, usually olive brown at maturity.

128. **Panicum fléxile** (Gattinger) Scribn. (Fig. 1037.) Culms slender, erect, much-branched from the base, 20 to 70 cm. tall, somewhat hispid below, the nodes pubescent; blades erect but not stiff, glabrous or sparse-

ly hispid, as much as 30 cm. long, 2 to 6 mm. wide; panicles relatively few-flowered, oblong, narrow, 10 to 20 cm. long, about one-third as wide; spikelets 3.1 to 3.5 mm. long. ☉ —Sandy, mostly damp soil, meadows and open woods, eastern Canada and New York to North Dakota, south to Florida and Texas; introduced in Utah.

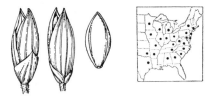

FIGURE 1039.—*Panicum philadelphicum.* Two views of spikelet, and floret, × 10. (Type coll.)

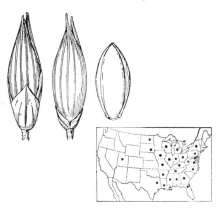

FIGURE 1037.—*Panicum flexile.* Two views of spikelet, and floret, × 10. (Type.)

129. Panicum gattingéri Nash. (Fig. 1038.) Culms at first erect, soon decumbent and rooting at the lower nodes, papillose-hispid, in robust specimens as much as 1 m. long; blades 6 to 10 mm. wide, more or less hispid or nearly glabrous; panicles numerous, terminal and axillary, oval or elliptic in outline, the terminal 10 to 15 cm. long, the lateral smaller; spikelets 2 mm. long. ☉ —Open ground and waste places, often a weed in cultivated soil, New York and Ontario to Minnesota, south to North Carolina, Tennessee, and Arkansas.

130. Panicum philadélphicum Bernh. ex Trin. (Fig. 1039.) Plants light yellowish green; culms slender, usually erect, 15 to 50 cm. tall, papillose-hispid to nearly glabrous, more or less zigzag at base; blades usually erect, 5 to 15 cm. long, 2 to 6 mm. wide, rather sparsely hirsute; panicles 10 to 20 cm. long, few-flowered, the branches solitary, rather stiffly ascending, the axillary pulvini hispid; spikelets 1.7 to 2 mm. long, mostly in twos at the ends of the branchlets. ☉ —Dry open or sandy ground, Connecticut to Minnesota, south to Georgia and Texas.

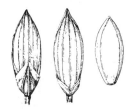

FIGURE 1040.—*Panicum lithophilum.* Two views of spikelet, and floret, × 10. (Type.)

131. Panicum lithóphilum Swallen. (Fig. 1040.) Culms 10 to 30 cm. tall, in small tufts, glabrous or sparsely hispid; sheaths papillose-hispid; blades erect, 6 to 8 cm. long, 2 to 4 mm. wide, conspicuously tinged with purple; panicles 7 to 15 cm. long, the branches stiffly spreading, few-flow-

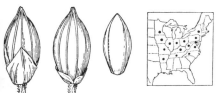

FIGURE 1038.—*Panicum gattingeri.* Two views of spikelet, and floret, × 10. (Type.)

FIGURE 1041.—*Panicum tuckermani.* Two views of spikelet, and floret, × 10. (Type coll.)

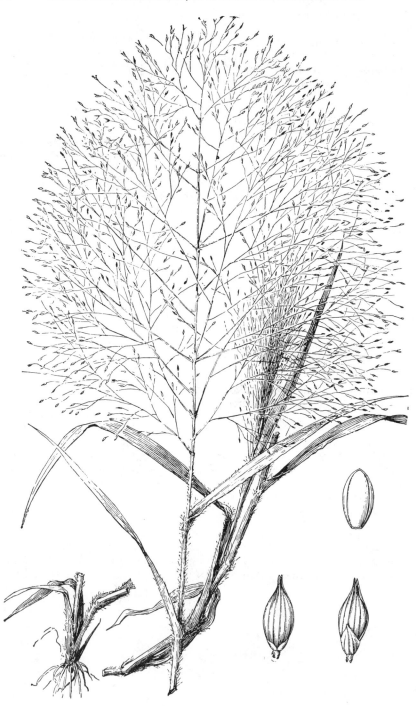

FIGURE 1042.—*Panicum capillare.* Plant, × ½; two views of spikelet, and floret, × 10. (V. H. Chase 774, Ill.)

ered; pulvini glabrous or very sparsely pilose; spikelets 2.1 to 2.2 mm. long, short-pediceled, appressed, in pairs at the ends of the branchlets. ⊙ —Granite outcrops, Georgia.

132. Panicum tuckermáni Fernald. (Fig. 1041.) Resembling *P. philadelphicum* and intergrading with it; often spreading or prostrate and much branched at base; panicles more densely flowered, the branches more spreading, the axillary pulvini glabrous; spikelets somewhat racemosely arranged, rather than in twos at the end. ⊙ —Sandy or gravelly shores and open ground, Maine and eastern Canada to Connecticut and New York; Ohio and Indiana to Minnesota.

133. Panicum capilláre L. WITCH-GRASS. (Fig. 1042.) Culms erect or somewhat spreading at base, 20 to 80 cm. tall, papillose-hispid to nearly glabrous; sheaths hispid; blades 10 to 25 cm. long, 5 to 15 mm. wide, hispid on both surfaces; panicles densely flowered, very diffuse, often half the length of the entire plant, included at the base until maturity, the branches finally divaricately spreading, the whole panicle breaking away and rolling before the wind; spikelets 2 to 2.5 mm. long. ⊙ —Open ground and waste places, a weed in cultivated ground, Maine to Montana, south to Florida and Texas, and occasionally west of this area.

PANICUM CAPILLARE var. OCCIDENTÁLE Rydb. Usually with short flowering branches at the base; blades shorter, less pubescent, crowded toward the base, panicles more exserted and divaricate; spikelets usually about 3 mm. long (2.5 to 3.3 mm.), attenuate at tip; fruit 1.7 to 1.8 mm. long. ⊙ (*P. barbipulvinatum* Nash.)—Open ground and waste places, Prince Edward Island and Quebec to British Columbia, south to New Jersey, Missouri, Texas, and California, more common westward.

134. Panicum hillmáni Chase. (Fig. 1043.) Resembling *P. capillare*, espe-

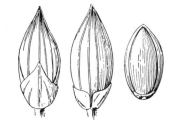

FIGURE 1043.—*Panicum hillmani*. Two views of spikelet, and floret, × 10. (Type.)

cially the var. *occidentale*, differing from this in having no short flowering branches at the base, in the stouter culms, firmer foliage, stiffer panicle branches with the lateral spikelets on shorter more appressed pedicels, in the well-developed sterile palea, and especially in the larger darker fruit (2 mm. long) with a prominent lunate scar at the base. ⊙ —Prairies and plains, Kansas to Texas; California.

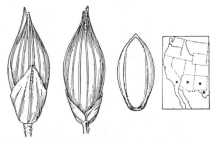

FIGURE 1044.—*Panicum hirticaule*. Two views of spikelet, and floret, × 10. (Type.)

135. Panicum hirticáule Presl. (Fig. 1044.) Culms usually simple or nearly so, 15 to 70 cm. tall, papillose-hispid to nearly glabrous; blades 5 to 15 cm. long, 4 to 13 mm. wide, often cordate at base, sparsely hispid or nearly glabrous, ciliate toward base; panicles 5 to 15 cm. long, scarcely one-third the entire height of the plant; spikelets 2.7 to 3.3 mm. long, lanceolate-fusiform, acuminate, usually reddish brown; first glume half to three-fourths the length of the spikelet; fruit 2 mm. long. ⊙ — Rocky or sandy soil, Arkansas and western Texas to Southern California; Mexico to western South America; Argentina.

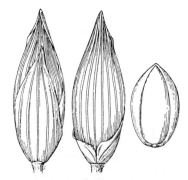 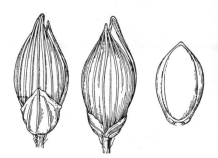

FIGURE 1045.—*Panicum pampinosum*. Two views of spikelet, and floret, × 10. (Type.)

FIGURE 1046.—*Panicum stramineum*. Two views of spikelet, and floret, × 10. (Type.)

136. Panicum pampinósum Hitchc. and Chase. (Fig. 1045.) Resembling *P. hirticaule*, but freely branching and with larger spikelet; spikelets very turgid, about 4 mm. long; first glume more than three-fourths the length of the spikelet; second glume and sterile lemma equal; fruit 2.2 mm. long. ☉ —Mesas, Texas to Arizona; Mexico.

137. Panicum stramíneum Hitchc. and Chase. (Fig. 1046.) Resembling *P. hirticaule*, but freely branching and nearly glabrous throughout; blades longer; spikelets more turgid, less pointed, 3.2 to 3.7 mm. long, the first glume about one-third the length of the spikelet; fruit 2.2 mm. long, with a prominent lunate scar at base. ☉ —Rich bottom lands and damp soil, southern Arizona; northwestern Mexico.

Panicum sonórum Beal. Robust, 60 cm. to 1 m. or more tall; sheaths mostly papillose-hispid; blades elongate, 15 to 30 mm. wide; panicles large, drooping, brownish, densely flowered; spikelets 3 to 3.3 mm. long, lanceolate; first glume half to two-thirds as long as the spikelet; second glume slightly exceeding the sterile lemma. ☉ —Yuma, Ariz., possibly introduced. Northern Mexico. Cultivated by Cocopa Indians, the seed used for food.

138. Panicum miliáceum L. BROOMCORN MILLET. (Fig. 1047.) Culms stout, erect or decumbent at base, 20 to 100 cm. tall; blades more or less pilose on both surfaces or glabrate, as much as 30 cm. long and 2 cm. wide, rounded at base; panicles usually more or less included at base, 10 to 30 cm. long, usually nodding, rather

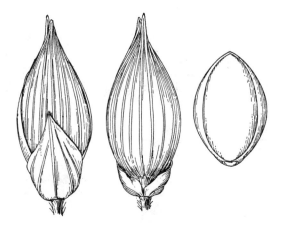 

FIGURE 1047.—*Panicum miliaceum*. Two views of spikelet, and floret, × 10. (Griffith 6490, India.)

compact, the numerous branches ascending, very scabrous, spikelet-bearing toward the ends; spikelets 4.5 to 5 mm. long, ovate, acuminate, strongly many-nerved; fruit 3 mm. long, stramineous to reddish brown. ⊙ —Waste places, introduced or escaped from cultivation, Northeastern States and occasional in other parts of the United States; temperate parts of the Old World. Broomcorn millet is cultivated in the cooler parts of the United States to a limited extent for forage and occasionally the seed is used for feed for hogs, hence it is sometimes known as hog millet. Also called proso. Commonly cultivated in Europe and western Asia.

6. **Diffúsa.**—Perennials; culms stiff, mostly tufted; sheaths mostly hirsute; ligules membranaceous, ciliate; spikelets pointed, 7- to 9-nerved, glabrous; fruit smooth and shining.

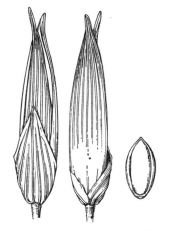

FIGURE 1048.—*Panicum capillarioides.* Two views of spikelet, and floret, × 10. (Type.)

139. Panicum capillarioídes Vasey. (Fig. 1048.) Culms erect or ascending from a knotted crown, 30 to 55 cm. tall, appressed-pubescent or glabrate, the nodes densely ascending-pubescent; blades rather stiff, 10 to 30 cm. long, 2 to 10 mm. wide, flat, harshly papillose-pubescent; panicle diffuse, few-flowered, 10 to 20 cm. long, the capillary branches stiffly spreading at maturity; spikelets 5 to 6 cm. long, lanceolate, long-acuminate, fruit 1.6 to 1.8 mm. long. ♃ —Prairies and plains, southern Texas and northern Mexico. This species is readily distinguished from all others by the peculiar elongated second glume and sterile lemma.

Panicum bérgi Arech. Tufted, with numerous leaves clustered at base; sheaths hispid; blades involute; panicle very diffuse, a third or more the entire height of the plant, the lower branches verticillate, conspicuously pilose in the axils; spikelets short-pointed, 2.2 to 2.6 mm. long. ♃ —Weed in grass plots, Experiment Station, Tifton, Ga. Adventive from South America.

PANICUM PILCOMÁYENSE Hack. Culms robust, few together, 70 to 100 cm. tall, at least the lower nodes with a ring of ere t hairs; blades flat, elongate, 4 to 8 mm. wide; panicle very diffuse, nearly half the height of the plant, the branches to 30 cm. long, in fascicles of 2 to 4 or solitary, scabrous, naked below, loosely branched toward the ends, at least the lower axils pilose; spikelets about 3 mm. long, on appressed pedicels. ♃ —Collegeport, Matagorda County, Tex. Probably introduced from Paraguay.

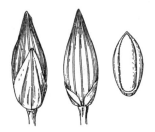

FIGURE 1049.—*Panicum filipes.* Two views of spikelet, and floret, × 10. (Type.)

140. Panicum fílipes Scribn. (Fig. 1049.) Culms 30 to 80 cm. tall, erect or ascending; blades laxly ascending or spreading, 10 to 25 cm. long, 3 to 8 mm. wide, flat, glaucous, glabrous or sometimes sparseıy hirsute beneath; panicles 7 to 25 cm. long, usu-

FIGURE 1050.—*Panicum hallii.* Plant, × 1; two views of spikelet, and floret, × 10. (Type.)

pilose to scabrous; blades suberect, 7 to 30 cm. long, 5 to 10 mm. wide, sparsely papillose-pilose to nearly glabrous; panicle 7 to 20 cm. long, usually scarcely half as wide, branches ascending with short spreading branchlets with 1 to 3 spikelets; spikelets 4 to 4.2 mm. long, turgid. ♀ —Moist places mostly in the uplands, Utah, New Mexico, Arizona, and Mexico.

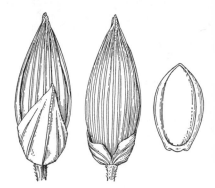

FIGURE 1051.—*Panicum lepidulum.* Two views of spikelet, and floret, × 10. (Type.)

ally equaled or exceeded by the upper blades, the distant branches spreading; spikelets 2 to 2.6 mm. long. ♀ —Low open ground or among chaparral, Louisiana (Shreveport) and Texas; northeastern Mexico. Distinguished from *P. hallii* by the longer blades, looser panicle, and smaller spikelets.

141. Panicum hállii Vasey. HALL'S PANICUM. (Fig. 1050.) Somewhat glaucous green, leaves usually crowded toward the base, the blades curling like shavings with age; culms erect, 15 to 60 cm. tall; sheaths sparsely papillose-hispid to glabrous; blades erect or nearly so, flat, 4 to 15 cm. long, 2 to 6 mm. wide, sparsely ciliate toward base, otherwise glabrous or nearly so; panicle 6 to 20 cm. long, the few branches stiffly ascending; spikelets 3 to 3.7 mm. long. ♀ —Dry prairie, rocky and gravelly hills and canyons, and in bottom lands and irrigated fields, Oklahoma and Colorado to Texas and Arizona; Mexico.

142. Panicum lepídulum Hitchc. and Chase. (Fig. 1051.) Culms 25 to 70 cm. tall, erect, usually sparingly branching from lower nodes, sparsely

143. Panicum ghiesbréghtii Fourn. (Fig. 1052.) Culms erect, rather robust, ascending-hirsute, 60 to 80 cm. tall, the nodes densely hirsute; blades as much as 60 cm. long and 12 mm. wide, flat, papillose-hirsute to glabrescent; panicles 20 to 30 cm. long, usually less than half as wide, the branches ascending, naked at base, the branchlets more or less appressed; spikelets 3 mm. long, 1 mm. wide. ♀ —Low moist ground, southern Texas; tropical America.

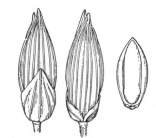

FIGURE 1052.—*Panicum ghiesbreghtii.* Two views of spikelet, and floret, × 10. (Type.)

FIGURE 1054.—*Panicum trichoides*. Panicle, × 1. (Runyon 1873, Tex.); two views of spikelet, and floret, × 10 (Type.)

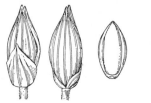

FIGURE 1053.—*Panicum hirsutum*. Two views of spikelet, and floret, × 10. (Type.)

144. Panicum hirsútum Swartz. (Fig. 1053.) Culms robust, erect, as much as 1.5 m. tall and 1 cm. thick, simple or branched at base only; nodes appressed-pubescent; sheaths papillose-hirsute, the hairs stiff, spreading, fragile, causing mechanical irritation to the skin when handled; blades flat, as much as 60 cm. long and 3.5 cm. wide, glabrous; panicle 20 to 35 cm. long, at first condensed, finally open, the branches ascending; spikelets 2 to 2.2 mm. long. ♃ — Open moist ground, southern Texas; tropical America at low altitudes.

7. Trichoídea.—Decumbent, spreading, freely branching annual; blades ovate to ovate-lanceolate; panicles diffuse, with capillary branches; spikelets minute.

145. Panicum trichoídes Swartz. (Fig. 1054.) Culms slender, widely creeping; freely branching; sheaths mostly longer than the internodes,

FIGURE 1055.—*Panicum maximum.* Plant, × ½; two views of spikelet, and floret, × 10. (Combs and Baker 1170, Fla.)

pilose; blades 4 to 7 cm. long, 8 to 15 mm. wide, thin, ovate-lanceolate, asymmetrical, ciliate at the base; panicles 8 to 15 cm. long, the slender ascending to spreading branches with capillary, spreading, few-flowered branchlets; spikelets about 1.3 mm. long, acute, sparsely pubescent. ⊙ —Waste places, woods and open ground, Texas (Brownsville); Mexico and the West Indies to Peru and Brazil; southeastern Asia and the Philippines.

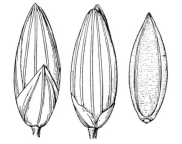

FIGURE 1056.—*Panicum plenum*. Two views of spikelet, and floret, X 10. (Type.)

8. Máxima.—Tall robust perennials; ligules membranaceous, ciliate; blades linear, flat; panicles large; many-flowered; spikelets ellipsoid, faintly nerved, glabrous; fruit transversely rugose.

146. Panicum máximum Jacq. GUINEA GRASS. (Fig. 1055.) Plants light green, in large bunches from short stout rhizomes; culms mostly erect, the nodes usually densely hirsute; sheaths papillose-hirsute to glabrous, usually densely pubescent on the collar; ligule 4 to 6 mm. long; blades 30 to 75 cm. long, as much as 3.5 cm. wide, glabrous, very scabrous on the margins, sometimes hirsute on the upper surface near the base; panicles 20 to 50 cm. long, about one-third as wide, the long rather stiff branches ascending, naked at base, the lower in whorls, the axils pilose, the branchlets short, appressed, bearing more or less clustered short-pediceled spikelets; spikelets 3 to 3.3 mm. long; first glume about one-third the length of the spikelet. ♃ —Fields and waste places, southern Florida, and southern Texas, introduced from Africa; tropical regions of both hemispheres at low altitudes. Guinea grass is the most important cultivated forage grass of tropical America. It grows in moderately dry ground and can be used for pasture or for soiling. Much of the green feed cut for forage is this species.

147. Panicum plénum Hitchc. and Chase. (Fig. 1056.) Plants mostly in large clumps, mostly glaucous from a stout rhizome; culms 1 to 2 m. tall, erect from a usually decumbent base, compressed; sheaths glabrous, somewhat keeled; blades 20 to 35 cm. long, 7 to 17 mm. wide, glabrous or nearly so; panicle 20 to 50 cm. long, open; spikelets 3 to 3.4 mm. long. ♃ —Moist places in rocky hills and canyons, Texas to Arizona; Mexico. Differs from *P. bulbosum* in the absence of the basal corm.

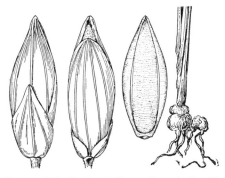

FIGURE 1057.—*Panicum bulbosum*. Base of culm, X ½; two views of spikelet, and floret, X 10. (Lemmon 2914, Ariz.)

148. Panicum bulbósum H. B. K. BULB. PANICUM. (Fig. 1057.) Culms in tufts, 1 to 2 m. tall, erect, the lowest internode thickened to a hard cormlike base 1 to 2 cm. thick, budding at base, sometimes with one or more corms of previous years attached; sheaths glabrous or pilose toward the summit; blades 25 to 60 cm. long, 3 to 12 mm. wide, scabrous above, glabrous beneath; panicle 20 to 50 cm. long, open; spikelets

FIGURE 1058.—*Panicum repens*. Plant, × ½; two views of spikelet, and floret, × 10. (Hitchcock 14145, Hawaii.)

3.5 to 4.2 mm. long. ♃ —Moist places in canyons and valleys, western Texas to Arizona; Mexico. PANICUM BULBOSUM var. MÍNUS Vasey. Culms slender, mostly less than 1 m. tall, the corms smaller than in the species; blades mostly 2 to 4 mm. wide; spikelets 2.8 to 3.2 mm. long. ♃ (*P. bulbosum* var. *sciaphilum* Hitchc. and Chase.)— Same range as the species and more common in the United States.

9. Virgáta.—Perennials from stout rhizomes; culms mostly stout; ligules membranaceous, ciliate; blades linear, mostly firm; spikelets turgid, usually gaping, strongly 5- to 9-nerved, glabrous, pointed; lower floret usually staminate; fruit smooth and shining.

149. Panicum répens L. (Fig. 1058.) Culms rigid, 30 to 80 cm. tall, erect from the nodes of strong horizontal often extensively creeping rhizomes, clothed at base with bladeless sheaths; sheaths more or less pilose; blades flat or folded, 2 to 5 mm. wide, sparsely pilose to glabrous; panicle open, 7 to 12 cm. long, the somewhat distant branches stiffly ascending; spikelets 2.2 to 2.5 mm. long, ovate; first glume about one-fifth as long as the spikelet, loose, truncate. ♃ —Sea beaches along the Gulf coast, Florida to Texas. Tropical and subtropical coasts of both hemispheres, possibly introduced in America.

150. Panicum gouíni Fourn. (Fig. 1059.) Resembling *P. repens*, but the culms usually less than 30 cm. tall; sheaths and blades usually glabrous; panicle smaller, more densely flowered; first glume longer. ♃ —Sea beaches, Alabama to Louisiana; Gulf coast of Mexico.

151. Panicum virgátum L. SWITCHGRASS. (Fig. 1060.) Plants usually in large bunches, green or glaucous, with numerous scaly creeping rhizomes; culms erect, tough and hard, 1 to 2 m., rarely to 3 m., tall; sheaths gla-

FIGURE 1059.—*Panicum gouini*. Two views of spikelet, and floret, × 10. (Type.)

brous; blades 10 to 60 cm. long, 3 to 15 mm. wide, flat, glabrous, or sometimes pilose above near the base, rarely pilose all over; panicle 15 to 50 cm. long, open, sometimes diffuse; spikelets 3.5 to 5 mm. long, acuminate; first glume clasping, two-thirds to three-fourths as long as the spikelet, acuminate or cuspidate; fruit narrowly ovate, the margins of the lemma inrolled only at base. ♃ —Prairies and open ground, open woods, and brackish marshes, Nova Scotia and Ontario, Maine to North Dakota and Wyoming, south to Florida, Nevada, and Arizona; Mexico and Central America.

PANICUM VIRGATUM var. CUBÉNSE Griseb. Culms more slender, usually solitary or few in a tuft; panicle narrower, with ascending branches; spikelets 2.8 to 3.2 mm. long, the second glume and sterile lemma not extending much beyond the fruit. ♃ —Pine woods, Coastal Plain, Massachusetts to Florida, Michigan, Wisconsin, Tennessee (Coffee County), and Mississippi; Michigan; Cuba.

PANICUM VIRGATUM var. SPÍSSUM Linder. Culms from short stout knotty rhizomes. ♃ —Nova Scotia to Pennsylvania.

152. Panicum havárdii Vasey. (Fig. 1061.) Pale green, glaucous, glabrous throughout; culms robust, solitary, 1 m. tall or more, erect from creeping rhizomes; blades 5 to 10 mm. wide, tapering into long involute-setaceous tips; panicle as much as 40 cm. long; spikelets 6 to 8 mm. long. ♃ — Arroyos and sand hills, western Texas and southern New Mexico; northern Mexico.

FIGURE 1060.—*Panicum virgatum.* Plant, × ½; two views of spikelet, and floret, × 10. (V. H. Chase, Ill.)

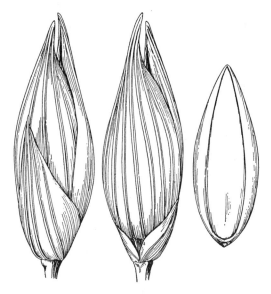

FIGURE 1061.—*Panicum havardii.* Two views of spikelet, and floret, × 10. (Type.)

153. Panicum amárum Ell. (Fig. 1062.) Glaucous and glabrous throughout; culms solitary from extensively creeping rhizomes, 30 to 100 cm. tall; blades thick, 10 to 30 cm. long, 5 to 12 mm. wide, flat, involute toward the tip, the margins smooth; panicle one-fourth to one-third the height of the plant, not more than 3 cm. wide, the branches appressed; spikelets 5 to 6.5 mm. long, acuminate. ♃ — Sandy seashores and coast dunes, Connecticut to Georgia; southern Mississippi; Texas.

154. Panicum amárulum Hitchc. and Chase. (Fig. 1063.) Culms as much as 1 cm. thick, in large bunches as much as 1 m. across, 1 to 2 m.

FIGURE 1062.—*Panicum amarum.* Two views of spikelet, and floret, × 10. (Vasey, Va.)

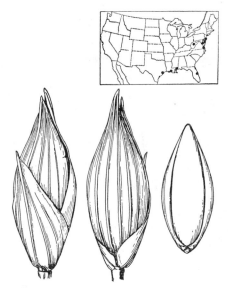

FIGURE 1063.—*Panicum amarulum*. Two views of spikelet, and floret, × 10. (Type.)

tall, glaucous; rhizomes vertical or ascending; blades 20 to 50 cm. long, 5 to 12 mm. wide, more or less involute, pilose on the upper surface near the base; panicle large, rather compact, 5 to 10 cm. wide, slightly nodding, densely flowered; spikelets 4.3 to 5.5 mm. long, acuminate. ♃ —Sandy shores and coast dunes, New Jersey to Virginia; Florida; Louisiana and Texas; introduced in West Virginia; Yucatan; Bahamas; Cuba.

10. Ténera.—Perennials; culms subcompressed, wiry; ligules minute; spikelets short-pediceled; fruit smooth and shining.

155. Panicum ténerum Beyr. (Fig. 1064.) Culms in small tufts from a

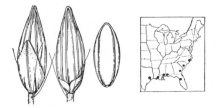

FIGURE 1064.—*Panicum tenerum*. Two views of spikelet, and floret, × 10. (Type.)

knotted crown, erect, 40 to 90 cm. tall; lower sheaths pubescent toward the summit, with spreading hairs; blades 4 to 15 cm. long, 2 to 4 mm. wide, erect, firm, subinvolute, pilose on upper surface toward base; panicles 3 to 8 cm. long, very slender, terminal and axillary; spikelets 2.2 to 2.8 mm. long, pointed, glabrous, the pedicel usually with a few long hairs. ♃ —Margins of swamps and wet places in pine barrens near the coast, North Carolina to Florida and Texas; West Indies.

11. Agrostoídea.—Tufted perennials; culms erect, compressed; sheaths keeled; ligules membranaceous, mostly about 1 mm. long; spikelets short-pediceled, lanceolate, pointed, 5- to 7-nerved, glabrous; glumes and sterile lemma mostly keeled; fruit smooth and shining, with a minute tuft of thickish hairs at apex.

156. Panicum abscíssum Swallen. (Fig. 1065.) Culms 50 to 70 cm. tall, densely tufted, compressed; lower

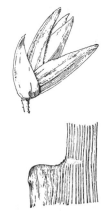

FIGURE 1065.—*Panicum abscissum*. Spikelet and summit of sheath, × 10. (Type.)

sheaths broad, strongly keeled, crowded, 3 to 4 mm. wide from keel to margin, truncate or extended at the summit into short, broad, obtuse auricles; blades 15 to 25 cm. long, 1 to 2 mm. wide, folded, glabrous or scabrous; panicles terminal and axillary, 7 to 15 cm. long, the branches ascending or appressed; spikelets 2.8 to 3 mm. long, obliquely set on the pedicels. ♃ —Sandy or swampy woods, central Florida.

157. Panicum agrostoídes Spreng. (Fig. 1066.) In dense clumps from a short crown, with numerous short-leaved innovations at base; culms 50 to 100 cm. tall; blades erect, folded at base, flat above, 20 to 50 cm. long, 5 to 12 mm. wide; panicles terminal and axillary, 10 to 30 cm. long, half to two-thirds as wide, sometimes more diffuse, the densely flowered branchlets mostly on the under side of the branches, the pedicels usually bearing at the summit one to several delicate hairs; spikelets about 2 mm. long. ♃ —Wet meadows and

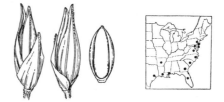

FIGURE 1067.—*Panicum condensum.* Two views of spikelet, and floret, × 10. (Type.)

more pointed, resembling *P. stipitatum* Nash. ♃ —Virginia to Florida and Texas.

158. Panicum condénsum Nash. (Fig. 1067.) Resembling *P. agrostoides;* culms on the average taller; blades often sparsely pilose on the upper side at the folded base; panicles 10 to 25 cm. long, rarely more than 5 cm. wide, the long branches erect, naked at base, with appressed branchlets bearing crowded spikelets, the pedicels not pilose; spikelets 2.2 to 2.5 mm. long. ♃ —Borders of streams and ponds and in wet places, Coastal Plain, Pennsylvania to Florida, Arkansas, and Texas; West Indies.

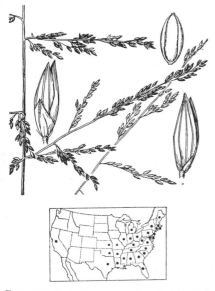

FIGURE 1066.—*Panicum agrostoides.* Panicle, × 1; two views of spikelet, and floret, × 10. (Fisher 30, N. J.)

shores, Maine to Kansas, south to Florida and Texas; Vancouver Island; California; British Honduras.

PANICUM AGROSTOIDES var. RA-MÓSIUS (Mohr) Fernald. Panicles more open and loosely flowered than in the species; spikelets more or less secund on the branchlets, slender and

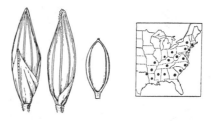

FIGURE 1068.—*Panicum stipitatum.* Two views of spikelet, and floret, × 10. (Commons 305, Del.)

159. Panicum stipitátum Nash. (Fig. 1068.) Resembling *P. agrostoides;* often purple-tinged throughout, especially the panicles; sheaths much overlapping, the blades usually equaling or exceeding the terminal panicle; panicles usually several to a culm, 10 to 20 cm. long, narrow, densely flowered, the numerous stiff branches ascending, with numerous divaricate branchlets, mostly on the lower side; spikelets 2.5 to 2.8 mm. long, often curved at the tip. ♃ —

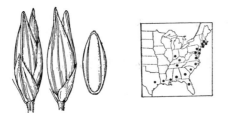

FIGURE 1069.—*Panicum longifolium.* Two views of spikelet, and floret, X 10. (Type.)

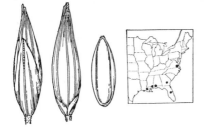

FIGURE 1070.—*Panicum combsii.* Two views of spikelet, and floret, X 10. (Type.)

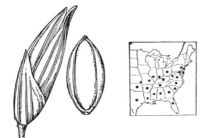

FIGURE 1071.—*Panicum anceps.* Spikelet and floret, X 10. (Type.)

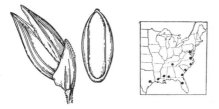

FIGURE 1072.—*Panicum rhizomatum.* Spikelet and floret, X 10. (Type.)

Moist soil, Connecticut to Missouri, south to Georgia and Texas.

160. Panicum longifólium Torr. (Fig. 1069.) Culms rather slender, 35 to 80 cm. tall, in dense tufts, usually surrounded by basal leaves nearly half as long; sheaths usually villous near the summit; ligule fimbriate-ciliate, 2 to 3 mm. long; blades elongate, 2 to 5 mm. wide, pilose on the upper surface near the base; lateral panicles few or none, the terminal 10 to 25 cm. long, the branches slender, ascending; spikelets 2.4 to 2.7 mm. long. ♃ —Moist sandy ground, Massachusetts to Florida and Texas to Indiana and Tennessee.

161. Panicum cómbsii Scribn. and Ball. (Fig. 1070.) Resembling *P. longifolium;* sheaths glabrous or nearly so; ligule less than 1 mm. long; blades on the average shorter; spikelets 3 to 3.5 mm. long, acuminate. ♃ —Margins of ponds and wet woods, southeastern Virginia; Georgia and Florida to Louisiana.

162. Panicum ánceps Michx. (Fig. 1071.) Culms 50 to 100 cm. tall, with numerous scaly rhizomes; sheaths glabrous or pilose; blades elongate, 4 to 12 mm. wide, pilose above near the base; panicles 15 to 40 cm. long, the slender, remote branches somewhat spreading, bearing short mostly appressed branchlets with rather crowded somewhat curved subsecund spikelets, set obliquely on their pedicels; spikelets 3.4 to 3.8 mm. long. ♃ —Moist sandy soil, New Jersey to Kansas, south to Florida and Texas.

163. Panicum rhizomátum Hitchc. and Chase. (Fig. 1072.) Resembling *P. anceps;* culms less robust, the rhizomes more slender and numerous; sheaths densely to sparsely villous, especially at the summit; blades usually pubescent on both surfaces; panicles more or less contracted; spikelets 2.4 to 2.8 mm. long. ♃ — Moist sandy woods and savannas, Coastal Plain, Maryland to Florida and Texas; Tennessee.

12. Láxa.—Slender perennials; culms compressed; ligules minute; spikelets short-pediceled, 5-nerved, glabrous, the palea of the sterile floret becoming enlarged and indurate, expanding the spikelet at maturity; fruit minutely papillose-roughened, relatively thin in texture.

164. Panicum híans Ell. (Fig. 1073.) Culms 20 to 60 cm. tall, mostly erect, sometimes more or less decumbent or prostrate with erect branches; blades 5 to 15 cm. long, 1 to 5 mm. wide, flat or folded, pilose on the upper surface near base; panicles 5 to 20 cm. long, usually loose and open, the primary branches few, slender, distant, spreading or drooping, the branchlets borne on the upper half or towards the ends only; spikelets in more or less secund clusters, 2.2 to 2.4 mm. long, at maturity about twice as thick as wide. ♃ —Damp soil along ponds and streams, Virginia to Florida and New Mexico; Tennessee; Oklahoma and southern Missouri; Mexico.

13. Verrucósa.—Glabrous branching annuals; culms slender, weak, decumbent at base, usually with stilt-roots; ligules minute; panicles with divaricate capillary branches, spikelet-bearing toward the ends, the spikelets mostly in twos; spikelets tuberculate, nerves obscure or obsolete; first glume minute; fruit minutely papillose, margin of the lemma inrolled only at base.

165. Panicum verrucósum Muhl. (Fig. 1074.) Bright green, at first erect, later widely spreading; culms 20 to 150 cm. long; blades thin, flat, lax, 5 to 20 cm. long, 4 to 10 mm. wide; panicles 5 to 30 cm. long, about as wide, diffuse, small panicles often produced at the lower nodes; spikelets 1.8 to 2.1 mm. long, ellipticobovate, subacute, roughened with small warts. ☉ —Wet, mostly shady soil, Massachusetts to Florida, west to Michigan, Kentucky, Arkansas and Texas.

166. Panicum brachyánthum Steud. (Fig. 1075.) Culms 30 to 100 cm. tall; blades 5 to 15 cm. long, 2 to 3 mm. wide; panicles 5 to 15 cm. long, the branches few; spikelets 3.2 to 3.6 mm. long, fusiform, acute, tu-

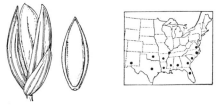

FIGURE 1073.—*Panicum hians.* Spikelet and floret, × 10. (Type.)

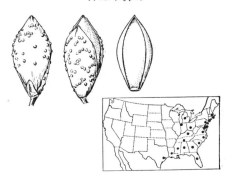

FIGURE 1074.—*Panicum verrucosum.* Two views of spikelet, and floret, × 10. (Type.)

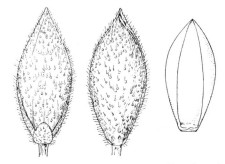

FIGURE 1075.—*Panicum brachyanthum.* Two views of spikelet, and floret, × 10. (Type.)

berculate-hispid. ☉ —Sandy soil, Arkansas, Louisiana, Texas, and Oklahoma.

14. Urvilleána.—Robust perennials; spikelets large, densely villous; fertile lemma long-villous on the margin.

167. Panicum urvilleánum Kunth. (Fig. 1076.) Culms solitary or few in a tuft, 50 to 100 cm. tall, erect from a creeping rhizome; nodes densely bearded; sheaths overlapping, densely retrorse-villous; blades elongate, 4 to 7 mm. wide, tapering from a flat base

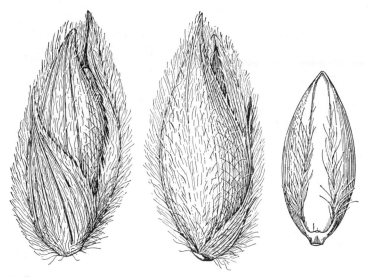

FIGURE 1076.—*Panicum urvilleanum.* Two views of spikelet, and floret, × 10. (Type.)

to a long involute setaceous point, strigose or glabrous; panicle 25 to 30 cm. long, the slender branches ascending; spikelets 6 to 7 mm. long, densely silvery- or tawny-villous; first glume clasping, from two-thirds to nearly as long as the spikelet. ♃ —Sandy deserts, Arizona and southern California; Argentina, Chile.

15. **Obtúsa.**—Stoloniferous wiry perennial; ligules about 1 mm. long; panicles narrow, the few appressed branches densely flowered; spikelets short-pediceled, secund, glabrous; fruit smooth and shining.

168. Panicum obtúsum H. B. K. VINE-MESQUITE. (Fig. 1077.) Tufted from a knotted crown, the stolons sometimes 2 m. long or more, with long internodes and geniculate, swollen, conspicuously villous nodes; culms compressed, 20 to 80 cm. tall; blades mostly elongate, 2 to 7 mm. wide, glabrous or nearly so; panicles 3 to 12 cm. long, about 1 cm. wide; spikelets 3 to 3.8 mm. long, obovoid, brownish, obtuse; first glume nearly as long as the spikelet. ♃ —Sandy or gravelly soil, mostly along banks

of rivers, arroyos, and irrigation ditches, western Missouri to Colorado, south to Arkansas, Texas, Utah, and Arizona; Mexico.

16. **Hemítoma.**—Aquatic or subaquatic perennial; panicles elongate, very narrow; spikelets subsessile, 3- to 5-nerved, glabrous.

169. Panicum hemítomon Schult. MAIDENCANE. (Fig. 1078.) With extensively creeping rhizomes, often producing numerous sterile shoots with overlapping, sometimes densely hirsute, sheaths; culms 50 to 150 cm. tall, usually hard; sheaths of fertile culms usually glabrous; blades 10 to 25 cm. long, 7 to 15 mm. wide, usually scabrous on the upper surface and smooth beneath; panicle 15 to 30 cm. long, the branches erect, the lower distant, the upper approximate, 2 to 10 cm. long; spikelets 2.4 to 2.7 mm. long, lanceolate, acute; first glume about half the length of the spikelet; fruit less rigid than usual in the genus, the apex of the palea scarcely enclosed. ♃ —Moist soil along river banks and ditches, borders of lakes and ponds, often in the water, sometimes a weed in moist cultivated

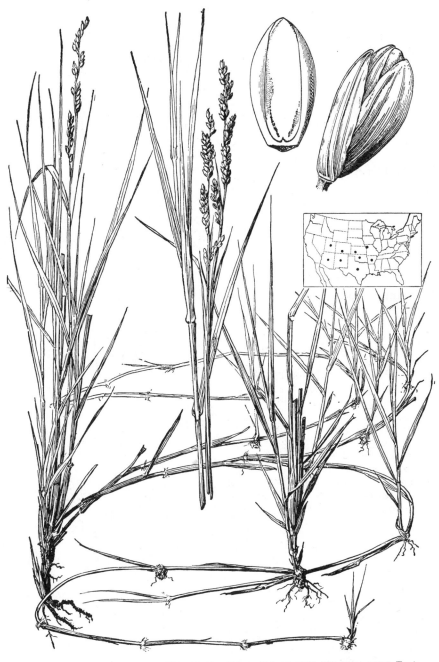

FIGURE 1077.—*Panicum obtusum.* Plant, × ½; spikelet and floret, × 10. (Hitchcock 13412, **Tex.**)

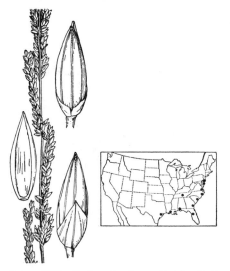

FIGURE 1078.—*Panicum hemitomon*. Panicle, × 1; spikelet and floret, × 10. (Tracy 6731, Fla.)

fields, Coastal Plain, New Jersey to Florida and Texas; Tennessee; Brazil.

17. Gymnocárpa.—Succulent glabrous perennial; panicles of several to many long stiffly ascending racemes along a main axis; spikelets strongly 3- to 5-nerved, glabrous.

170. Panicum gymnocárpon Ell. (Fig. 1079.) Creeping, the base as much as 2 m. long, rooting at the nodes; culms 60 to 100 cm. tall; blades elongate, 15 to 25 mm. wide, flat, scarcely narrowed at the cordate, sparingly ciliate base, the margin very scabrous; panicle 20 to 40 cm. long; spikelets 6 to 7 mm. long; first glume nearly as long as the sterile lemma, the second glume exceeding the sterile lemma, all acuminate-pointed, much exceeding the obovate, stipitate fruit, this 2 mm. long, smooth and shining. ♃ —Ditches and muddy banks of streams and lakes, South Carolina to Florida, Arkansas, and Texas.

PANICUM ANTIDOTÁLE Retz. Robust glabrous, branching, leafy perennial, to 3 m. tall, with strong rhizomes; blades elongate, flat, 5 to 12 mm. wide; panicle 20 to 30 cm. long, the many-flowered branches ascending; spikelets 2.5 to 3 mm. long, strongly nerved, pointed, the first glume one-third to scarcely half as long as the spikelet. ♃ — Cultivated in experiment stations in Missouri, Texas, Oklahoma, Arizona (spreading in Cochise County), and California. India

138. LASÍACIS (Griseb.) Hitch

Spikelets subglobose, placed obliquely on their pedicels; first glume broad, somewhat inflated-ventricose, usually not more than one-third the length of the spikelet, several-nerved; second glume and sterile lemma about equal, broad, abruptly apiculate, papery-chartaceous, shining, many-nerved, glabrous, or lanose at the apex only, the lemma enclosing a membranaceous palea and sometimes a staminate flower; fertile lemma white, bony-indurate, obovoid, obtuse, this and the palea of the same texture, bearing at the apex in a slight depression a tuft of woolly hairs, the palea concave below, gibbous above, the apex often free at maturity. Large branching peren-

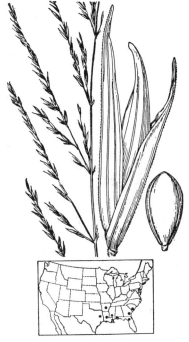

FIGURE 1079.—*Panicum gymnocarpon*. Panicle, × 1; spikelet and floret, × 10. (Type.)

FIGURE 1080.—*Lasiacis divaricata*. Plant, × ½; spikelet and floret, × 10. (Curtiss 5530, Fla.)

nials, with woody culms often clambering several meters high into shrubs or trees, the blades firm, flat, usually lanceolate and narrowed into a petiole, the spikelets in an open panicle. Type species, *Lasiacis divaricata*. Name from Greek *lasios*, woolly, and *akis*, point, alluding to the tuft of wool at the tip of the fruit.

1. Lasiacis divaricáta (L.) Hitchc. TIBISEE. (Fig. 1080.) Glabrous throughout except the margins of the sheaths; culms much-branched, clambering over shrubs to the height of 3 or 4 m., the main culm (cane) strong, as much as 6 mm. in diameter, the main branches often fascicled, the vigorous secondary sterile shoots usually strongly divaricate or zigzag; blades narrowly lanceolate, 5 to 20 cm. long, 5 to 15 mm. wide, or larger on vigorous sterile shoots; panicles terminating the main culm and branches, 5 to 20 cm. long, loosely few-flowered, the branches distant, spreading or reflexed; spikelets ovoid, about 4 mm. long, black at maturity. ♃ —Copses and edges of woods,

southern Florida; tropical America, at low altitudes, especially near the seacoast.

139. SACCIÓLEPIS Nash

Spikelets oblong-conic; first glume much shorter than the spikelet; second glume broad, inflated-saccate, strongly many-nerved; sterile lemma narrower, flat, fewer nerved, its palea nearly as long, often subtending a staminate flower; fertile lemma stipitate, elliptic, chartaceous-indurate, the margins inrolled, the palea not enclosed at the summit. Annuals or perennials, of wet soil, usually branching, the inflorescence a dense, usually elongate, spikelike panicle. Type species, *Panicum gibbum* Ell. (*Sacciolepis striata.*) Name from Greek *sakkion*, a small bag, and *lepis*, scale, alluding to the saccate second glume.

1. Sacciolepis striáta (L.) Nash. (Fig. 1081.) Perennial, glabrous, often decumbent and rooting at base; culms as much as 1 to 2 m. tall; sheaths glabrous to more or less

FIGURE 1081.—*Sacciolepis striata.* Plant, × ½; two views of spikelet, and floret, × 10. (Chase 4240, Fla.)

papillose-hirsute; blades lanceolate, 4 to 20 cm. long; panicles 6 to 30 cm. long; spikelets about 4 mm. long. ♃ (*Sacciolepis gibba* Nash.)—Marshes, ditches, and wet places, Coastal Plain, New Jersey (Cape May) to Florida; Tennessee, Texas, and Oklahoma; West Indies.

FIGURE 1082.—*Oplismenus setarius.* Plant, × ½; two views of spikelet, and floret, × 10. (Curtiss 5553, Fla.)

Sacciolepis índica (L.) Chase. Annual; culms slender, spreading, 20 to 60 cm. tall; blades 2 to 4 mm. wide; panicle spikelike, 1 to 4 cm. long; spikelets about 2.5 mm. long, glabrous or pilose near the summit. ♃ —Introduced in a Government pecan orchard, Thomasville, Ga.; India.

140. OPLÍSMENUS Beauv.

Spikelets terete or somewhat laterally compressed, subsessile, solitary or in pairs, in 2 rows crowded or approximate on one side of a narrow scabrous or hairy rachis; glumes about equal, entire, or emarginate, awned from the apex or from between the lobes; sterile lemma exceeding the glumes and fruit, notched or entire, mucronate or short-awned, enclosing a hyaline palea; fertile lemma elliptic, acute, convex or boat-shaped, the firm margins clasping the palea, not inrolled. Freely branching, creeping, shade-loving annuals or perennials, with erect flowering shoots, flat, thin, lanceolate or ovate blades, and several one-sided, thickish, short racemes rather distant on a slender axis. Type species, *Oplismenus africanus* Beauv. Name from Greek *hoplismenos*, armed, alluding to the awned spikelets.

Rachis of racemes mostly 2 to 3 mm. long, bearing usually not more than 5 spikelets; blades 1 to 3 cm. long.   1. O. SETARIUS.
Rachis of lower racemes 10 to 30 mm. long, bearing more than 8 spikelets; blades mostly 5 cm. or more long.   2. O. HIRTELLUS.

1. **Oplismenus setárius** (Lam.) Roem. and Schult. (Fig. 1082.) Perennial; culms slender, lax, ascending or prostrate, 10 to 20 cm. long, sometimes as much as 30 cm.; blades ovate to ovate-lanceolate, thin, 1 to 3 cm. long, 4 to 10 mm. wide; panicle long-exserted, usually not more than 5 cm. long; racemes usually 3 to 5, subglobose, distant or the upper approximate, the lower internodes sometimes as much as 2 cm. long, the rachis 2 to 3 mm. long, sometimes to 6 mm.; spikelets about 5 (4 to 8) on each rachis; awn of first glume 4 to 8 mm. long. ♃ —Shaded places along the coast, North Carolina to Florida, Arkansas, and Texas; tropical America at low altitudes.

2. **Oplismenus hirtéllus** (L.)Beauv. (Fig. 1083.) Perennial; culms widely creeping and branching, the fertile culms erect from an ascending base, commonly 20 to 30 cm. tall; sheaths glabrous to papillose-hispid; blades 5 to 10 cm. long, 1 to 2 cm. wide; panicle 5 to 10 cm. long; racemes 3 to 7, rather distant, the rachis 1 to 3 cm. long, the spikelets green with erect purple awns, the awn of the

FIGURE 1083.—*Oplismenus hirtellus*, × ½. (Amer. Gr. Natl. Herb. 602, Trinidad.)

first glume 5 to 10 mm. long. ♃ —Shady places, Texas (Cameron County); Mexico, and the West Indies to Argentina. Sometimes cultivated by florists as a basket plant and for edging, under the name *Panicum variegatum*. It has been incorrectly referred to *Oplismenus burmanni* (Retz.) Beauv. The common form in cultivation is variegated, the blades striped with white.

141. ECHINÓCHLOA Beauv.

Spikelets planoconvex, often stiffly hispid, subsessile, solitary or in irregular clusters on one side of the panicle branches; first glume about half the length of the spikelet, pointed; second glume and sterile lemma equal, pointed, mucronate, or the glume short-awned and the lemma long-awned, sometimes conspicuously so, enclosing a membranaceous palea and sometimes a staminate flower; fertile lemma planoconvex, smooth and shining, acuminate-pointed, the margins inrolled below, flat above, the apex of the palea not enclosed. Coarse, often succulent, annuals or perennials, with compressed sheaths, linear flat blades, and rather compact panicles composed of short, densely flowered racemes along a main axis. Type species, *Echinochloa crusgalli*. Name from Greek *echinos*, hedgehog, and *chloa*, grass, alluding to the echinate spikelets.

All the species are grazed by stock but usually grow in sparse stands or in situations where they cannot well be utilized. *E. crusgalli* is occasionally cut for hay. *Echinochloa crusgalli* var. *frumentacea*, Japanese millet, has been advertised by seedsmen in this country as billion-dollar grass and recommended for forage. It has some forage value, but requires considerable moisture to produce abundantly, and is rather too succulent for hay. This and forms of *E. colonum* are cultivated in tropical Asia and tropical Africa for the seeds which are used for food.

Ligule a dense line of stiff yellowish hairs; plants perennial................ 1. E. POLYSTACHYA.
Ligule wanting; plants annual.
Racemes simple, rather distant, 1 to 2 cm. long; spikelets crowded in about 4 rows, the
    awn of the sterile lemma reduced to a short point; blades 3 to 6 mm. wide.
                                                                            2. E. COLONUM.
Racemes more or less branched, usually more than 2 cm. long; spikelets irregularly
    crowded and fascicled, usually not arranged in rows, the awn of the sterile lemma
    variable; blades usually more than 5 mm. wide.
    Sterile floret staminate.. 5. F. PALUDIGENA.
    Sterile floret neuter.
        Sheaths smooth; awns variable, but the panicle not a dense mass of long-awned
            spikelets.
            Panicles erect and rather stiff (heavy panicles somewhat nodding); spikelets con-
                spicuously hispid.. 3. E. CRUSGALLI.
            Panicles soft and nodding; spikelets inconspicuously hispid.
                                                                            4. E. CRUS-PAVONIS.
        Sheaths, at least the lower, hispid or scabrous (glabrous in forma *laevigata*); panicle
            dense, the spikelets long-awned.................................... 6. E. WALTERI.

1. Echinochloa polystáchya (H. B. K.) Hitchc. (Fig. 1084.) Aquatic or subaquatic; culms coarse, 1 to 2 m. tall, from a long creeping base, glabrous; nodes glabrous or obscurely pubescent; sheaths glabrous or very sparsely papillose; ligule a dense line of stiff yellow hairs as much as 4 mm. long; blades 30 to 40 mm. long, 1.5 to 2.5 cm. wide, scabrous on the margin; panicles mostly 15 to 25 cm. long, dense, the short thick branches ascending; pedicels with stiff hairs 3 to 5 mm. long; spikelets about 5 mm. long, the nerves papillose-hispid, the sterile floret staminate; awns 2 to

15 mm. long. ♃ —Swamps and ditches near the coast, Louisiana and Brownsville, Tex.; West Indies to Argentina.

2. Echinochloa colónum (L.) Link. JUNGLE-RICE. (Fig. 1085.) Culms prostrate to erect, 20 to 40 cm. long; blades rather lax, 3 to 6 mm. wide, occasionally transversely zoned with purple; panicle 5 to 15 cm. long; racemes several, 1 to 2 cm. long, appressed or ascending, single or occasionally two approximate, the lower usually distant as much as 1 cm.; spikelets about 3 mm. long,

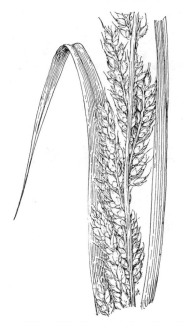

FIGURE 1084.—*Echinochloa polystachya*, × 1. (Chase 6319, P. R.)

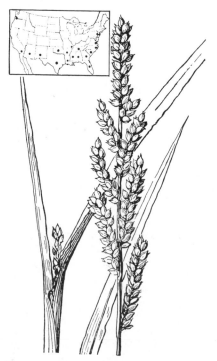

FIGURE 1085.—*Echinochloa colonum*, × 1. (Bentley, Tex.)

crowded, nearly sessile; second glume and sterile lemma short-pointed, rather soft, faintly nerved, the nerves weakly hispid-scabrous. ⊙ — Ditches and moist places, Virginia to Missouri, south to Florida, Texas, and southeastern California; ballast, Camden, N. J., Philadelphia, Pa., and Portland, Oreg.; tropical regions of both hemispheres; introduced in America.

3. **Echinochloa crusgálli** (L.) Beauv. BARNYARD GRASS. (Fig. 1086.) Culms erect to decumbent, stout, as much as 1 m. or even 1.5 m. tall, often branching at base; sheaths glabrous; blades elongate, 5 to 15 mm. wide; panicle erect or nodding, purple-tinged, 10 to 20 cm. long; racemes spreading, ascending or appressed, the lower somewhat distant, as much as 10 cm. long, sometimes branched, the upper approximate; spikelets crowded, about 3 mm. long, excluding the awns; internerves hispidulous; nerves strongly tuberculate-hispid; awn variable, mostly 5 to 10 mm. long on at least some of the spikelets, sometimes as much as 3 cm. ⊙ — Moist open places, ditches, cultivated fields, and waste ground, New Brunswick to Washington, south to Florida and California, mostly at low and medium altitudes; Eastern Hemisphere. *Echinochloa pungens* (Poir.) Rydb. (*E. muricata* (Michx.) Fernald) has been differentiated from *E. crusgalli* by the papillae at the base of the stiff hairs on the spikelets; true *E. crusgalli*, as understood by Fernald and by Rydberg, having hairs that lack the papillose base. But the European specimens have on the average about as strongly tuberculate spikelets as the American. The three following varieties intergrade and can sometimes be only arbitrarily distinguished.[14]

[14] For various treatments of the *Echinochloa crusgalli* complex, and for names here cited in Synonymy see FERNALD, F. L., Rhodora 17: 105–107. 1915; FERNALD, F. L. and GRISCOM, L., Rhodora 37: 136–137. 1935. WIEGAND, K. M., 23: 49–65. 1921. For FARWELL, FASSETT, GLEASON, RYDBERG, and others, see references in Synonymy.

FIGURE 1086.—*Echinochloa crusgalli.* Plant, × ½; two views of spikelet, and floret, × 10. (Somes 3725, Iowa.)

FIGURE 1087.—*Echinochloa crusgalli* var. *mitis,* × 1.
(Pammel and Cratty 791, Iowa.)

ECHINOCHLOA CRUSGALLI var. MÍTIS (Pursh) Peterm. (Fig. 1087.) Racemes dense, mostly somewhat spreading-flexuous; spikelets awnless or nearly so, the awns less than 3 mm. long; basal sheaths occasionally hirsute. ⊙ —Moist places over about the same area as the species and nearly as common.

ECHINOCHLOA CRUSGALLI var. ZELAYÉNSIS (H. B. K.) Hitchc. (Fig. 1088.) Differs from *E. crusgalli* var. *mitis* in having less succulent culms, mostly simple, more or less appressed racemes, the spikelets less strongly hispid but papillose, usually green. Small plants resemble *E. colonum,* but differ in the more distinctly pointed spikelets, more spreading racemes, and erect more robust culms. ⊙ —Moist, often alkaline places, Oklahoma to Oregon, south to Texas and California; Mexico to Argentina, in the tablelands. (Type from Zelaya, Mexico.)

ECHINOCHLOA CRUSGALLI var. FRUMENTÁCEA (Roxb.) W. F. Wight. JAPANESE MILLET. (Fig. 1089.) Ra-

FIGURE 1088.—*Echinochloa crusgalli* var. *zelayensis,* × 1. (Mearns 744, Mex.)

FIGURE 1089.—*Echinochloa crusgalli* var. *frumentacea,* × 1. (Piper, Tex.)

cemes thick, appressed, incurved; spikelets more turgid, awnless, mostly purple, the nerves hispid, but not, or only slightly, tuberculate. ⊙ (Var. *edulis* Hitchc.)—Occasionally cultivated as a forage grass and escaped here and there. Exploited at one time under the name, "billion-dollar grass."

4. Echinochloa crus-pavónis (H. B. K.) Schult. (Fig. 1090.) Culms erect or sometimes decumbent at base, as much as 1 m. tall; blades 5 to 15 mm. wide; panicle 10 to 20 cm. long, nodding, rather soft, pinkish or pale purple; racemes mostly ascending or appressed, the lower somewhat distant; spikelets about 3 mm. long, hispid on the nerves, hispidulous on the internerves, the awn usually about 1 cm. long. ⊙ (*E. crusgalli crus-pavonis* Hitchc.)—Marshes and wet places, often in the water, Virginia, Alabama, Louisiana, southern Texas, and through tropical America at low altitudes.

5. Echinochloa paludígena Wiegand. (Fig. 1091.) Culms mostly soli-

FIGURE 1091.—*Echinochloa paludigena,* ✕ 1. (Fredholm 6390, Fla.)

tary, erect, rather stout, usually 1 to 1.5 m. tall; blades elongate, 8 to 20 mm. wide; panicle narrow, usually 20 to 30 cm. long; racemes ascending, usually simple, rather evenly distributed on the axis, not closely crowded, sometimes remote; spikelets about as in *E. crusgalli*, but on the average less strongly tuberculate; sterile floret staminate. ⊙ — Ditches, marshes, and wet places, often in shallow water, south and central Florida.

6. Echinochloa waltéri (Pursh) Heller. (Fig. 1092.) Culms usually stout, erect, 1 to 2 m. tall; sheaths papillose-hispid or papillose only, sometimes only the lower sheaths hispid or the hairs on the margins only; blades elongate; panicle dense, nodding, mostly 20 to 30 cm. long, purplish; spikelets about 4 mm. long, less turgid than in *E. crusgalli;* the

FIGURE 1090.—*Echinochloa crus-pavonis,* ✕ 1. (Sintenis 1889, P. R.)

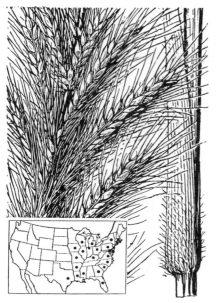

FIGURE 1092.—*Echinochloa walteri*, × 1. (Chase 1426, Ill.)

stiff hairs on the nerves not tuberculate; awns mostly 1 to 2.5 cm. long. ⊙ —Wet places, often in shallow water or brackish marshes, Coastal Plain, Massachusetts to Florida and Texas; Wisconsin, Iowa, and Arkansas. Short-awned specimens have been segregated as forma *breviseta* Fern. and Grisc. ECHINOCHLOA WALTERI forma LAÉVIGATA Wiegand. Sheaths glabrous. (*E. longearistata* Nash.) Wisconsin, Virginia, South Carolina, Arkansas, Louisiana, and Texas.

142. RHYNCHELÝTRUM Nees

(Included in *Tricholaena* Schrad. in Manual, ed. 1)

Spikelets on short capillary pedicels; first glume minute, villous; second glume and sterile lemma equal, gibbous below, raised on a stipe above the first glume, emarginate or slightly lobed, short-awned, covered, except toward the slightly spreading apex, with long silky hairs, the palea of the sterile lemma well developed; fertile lemma shorter than the spikelet, cartilaginous, smooth, boat-shaped, ob-

tuse, the margin thin, not inrolled, enclosing the margins of the palea. Perennials or annuals, with rather open panicles of silky spikelets, the fruit not falling from the spikelet at maturity. Type species *Rhynchelytrum dregeanum* Nees. Name from Greek, *rhychos*, beak, and *elytron*, scale, alluding to the beaked second glume and sterile lemma. This genus has, until recently, generally been included in *Tricholaena* Schrad. The type species of the two are sufficiently different to recognize this as generically distinct.

1. Rhynchelytrum róseum (Nees) Stapf and Hubb. NATAL GRASS. (Fig. 1093.) Short-lived perennial, sometimes apparently annual; culms slender, about 1 m. tall; blades flat, 2 to 5 mm. wide; panicle rosy purple, fading to pink, silvery in age, 10 to 15 cm. long, the branches slender, ascending; spikelets about 5 mm. long, the capillary pedicels flexuous or recurved. ♃ (*Tricholaena rosea* Nees.)—Sandy prairies, open woods, fields, and waste places, Florida, Texas, and Arizona, naturalized from South Africa; drier parts of tropical America at low altitudes. Cultivated as a meadow grass in sandy soil in Florida and more rarely along the Gulf coast.

CORIDÓCHLOA Nees

Spikelets flattened, ovate, in 2's or 3's, subsessile along a slender rachis; glumes and sterile lemma papery, the second glume stiffly ciliate; fruit stipitate, concavo-convex, awned. Annual, with several digitate racemes naked at base.

Coridochloa cimicína (L.) Nees ex Jacks. Culms 20 to 60 cm. tall; sheaths hispid; blades 3 to 8 cm. long, 1.5 to 2.5 cm. wide, subcordate; racemes mostly 4 to 8, digitate, sometimes a second whorl below; spikelets about 3 mm. long, the awn of the fruit curved, about 1 mm. long. ⊙ —Sparingly introduced in Florida. Southern Asia.

FIGURE 1093.—*Rhynchelythrum roseum.* Plant, × ½; spikelet and floret, × 10. (Tracy 9365, Fla.)

143. SETÁRIA Beauv.

(*Chaetochloa* Scribn.)

Spikelets subtended by one to several bristles (sterile branchlets), falling free from the bristles, awnless; first glume broad, usually less than half the length of the spikelet, 3- to 5-nerved; second glume and sterile lemma equal, or the glume shorter, several-nerved; fertile lemma coriaceous-indurate, transversely rugose or smooth. Annual or perennial grasses, with narrow terminal panicles, these dense and spikelike or somewhat loose and open. Type species, *Setaria viridis*. Name from Latin *seta*, a bristle, alluding to the numerous bristles of the inflorescence. The species are, in general, palatable and nutritious. A few species, especially *S. macrostachya*, form an appreciable part of the forage on southwestern ranges. Primitive peoples have cultivated *S. italica*, Italian or foxtail millet, since prehistoric times. The seed has been found in early remains such as those of the Swiss lake dwellings of the stone age. In America this species is used for hay. Another species, *S. palmifolia*, cultivated for ornament in greenhouses.

Bristles below each spikelet numerous, at least more than 5. Panicle dense, cylindric, spikelike.
    Plants annual; spikelets 3 mm. long; lower floret staminate, the palea well developed.
                                                 1. S. LUTESCENS.
    Plants perennial; spikelets 2 to 2.5 mm. long; lower floret neuter, the palea reduced.
                                                2. S. GENICULATA.
Bristles below each spikelet 1 to 3, or, by the abortion of the spikelets, 4 or 6. (See also *S. faberii*.)
    Bristles more or less retrorsely scabrous (antrorsely in var. *ambigua*).
                                          3. S. VERTICILLATA.
    Bristles antrorsely scabrous only.
      Plants perennial.
        Spikelets 3 mm. long.
          Blades scabrous_____ 4. S. MACROSPERMA.
          Blades villous_____ 5. S. VILLOSISSIMA.
        Spikelets 2 to 2.5 mm. long.
          Blades mostly less than 1 cm. wide, often folded; panicles usually loosely or interruptedly spikelike, the branches usually not more than 1 cm. long.
                                      6. S. MACROSTACHYA.
          Blades flat, as much as 1.5 cm. wide; panicles tapering from near the base, the lower branches as much as 3 cm. long_____ 7. S. SCHEELEI.
      Plants annual.
        Fertile lemma coarsely transversely rugose.
          Panicle densely cylindric_____ 8. S. CORRUGATA.
          Panicle loosely flowered_____ 9. S. LIEBMANNI.
        Fertile lemma finely cross-lined or nearly smooth.
          Panicle loosely flowered, tapering above_____ 10. S. GRISEBACHII.
          Panicle compactly flowered, sometimes interrupted at base.
            Culms as much as 3 m. tall; bristles 1 to 2 cm. long; fertile lemma smooth or nearly so_____ 11. S. MAGNA.
           Culms mostly less than 1 m. tall.
              Panicle cylindric, tapering above, green; spikelets falling entire.
               Spikelets 2 to 2.5 mm. long; bristles 1 to 3 below each spikelet; panicle erect or somewhat nodding_____ 12. S. VIRIDIS.
                Spikelets 2.8 to 3 mm. long; bristles 3 to 6 below each spikelet; panicle conspicuously nodding_____ 13. S. FABERII.
              Panicle lobed or interrupted, often large and heavy, purple or yellow; fruit deciduous from glumes and sterile lemma_____ 14. S. ITALICA.

1. Setaria lutéscens (Weigel) Hubb. YELLOW BRISTLEGRASS. (Fig. 1094.) Annual, branching at base; culms erect to prostrate, mostly 50 to 100 cm. tall, compressed; sheaths keeled; blades as much as 25 cm. long and 1 cm. wide, flat, twisted in a loose spiral, villous toward the base above; panicle dense, evenly cylindric, spikelike, yellow at maturity, mostly 5 to 10 cm. long, about 1 cm. thick, the

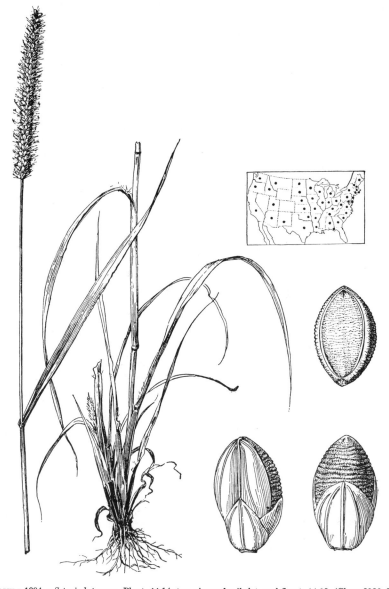

FIGURE 1094.—*Setaria lutescens*. Plant, × ½; two views of spikelet, and floret, × 10. (Chase 2986, D. C.)

axis densely pubescent; bristles 5 to 20 in a cluster, the longer 2 to 3 times as long as the spikelet; spikelets 3 mm. long; fruit strongly rugose. ⊙ —Cultivated soil and waste places, New Brunswick to North Dakota, south to northern Florida and Texas, occasional from British Columbia to California, New Mexico, and Arizona; Jamaica, at high altitudes; introduced from Europe; widely distributed in temperate regions. This species has been erroneously referred to *S. glauca* (L.) Beauv.

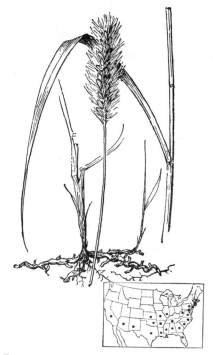

FIGURE 1095.—*Setaria geniculata*, × 1. (Chase 2981, Md.)

2. Setaria geniculáta (Lam.) Beauv. KNOTROOT BRISTLEGRASS. (Fig. 1095.) Resembling *S. lutescens* but perennial, producing short knotty branching rhizomes as much as 4 cm. long; base of plant slender, wiry; blades mainly straight (not twisted as in *S. lutescens*); bristles yellow or purple, 1 to 3 times or even 6 times as long as the spikelet; spikelets 2 to 2.5 or even 3 mm. long. ♃ —Open ground, pastures, cultivated soil, salt marshes, and moist ground along the coast, Massachusetts to Florida and Texas, in the interior north to West Virginia, Illinois, and Kansas, west to California; tropical America to Argentina and Chile.

Setaria nigriróstris (Nees) Dur. and Schinz. Perennial; resembling *S. lutescens*, but the dense spikelike racemes purple or dark brown. ♃ — Ballast, near Portland, Oreg.; South Africa.

SETARIA SPHACELÁTA (Schum.) Stapf and C. E. Hubb. Tufted perennial, glabrous or nearly so, often with stout rhizomes; culms 0.5 to 1.5 m. tall, flattened; blades flat, rather lax, 4 to 10 mm. wide; panicle dense, cylindric, 8 to 15 cm. long, usually orange to purple, bristles mostly 5 or more, 3 to 6 mm. long; spikelets 2.5 to 3 mm. long; fruit finely rugose. ♃ —Cultivated in experiment stations and escaped along irrigation ditches, Stanislaus and Kern Counties, Calif. Introduced from Africa.

3. Setaria verticilláta (L.) Beauv. BUR BRISTLEGRASS. (Fig. 1096.) Annual, often much branched at base and geniculate-spreading, as much as 1 m. long; blades flat, rather thin, scabrous and often more or less pilose, 10 to 20 cm. long, 5 to 10 mm. wide; panicle erect but not stiff, cylindric or somewhat tapering upward, more or less lobed or interrupted, especially toward base, 5 to 15 cm. long, 7 to 15 mm. wide; bristles single below each spikelet, 1 to 3 times as long as the spikelet, retrorsely scabrous; spikelets 2 mm. long; fruit finely rugose. ⊙ —Cultivated soil and waste places, Massachusetts to North Dakota, south to Alabama, Louisiana, and Missouri, occasional west to

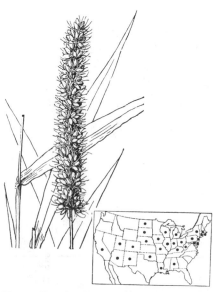

FIGURE 1096.—*Setaria verticillata*, × 1. (Steele, D. C.)

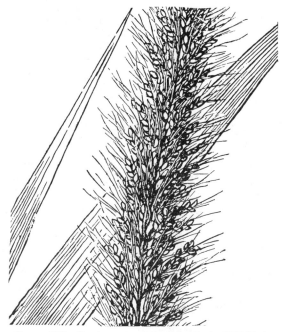

FIGURE 1097.—*Setaria macrosperma*, × 1. (Curtiss 3617, Fla.)

California; introduced from Europe; tropical America at medium altitudes. SETARIA VERTICILLATA var. AMBÍGUA (Guss.) Parl. Differing from *S. verticillata* in the scabrous but not pilose axis of the panicle and the antrorsely scabrous bristles, mostly 2 to 3 times as long as the spikelets, at maturity spreading and more or less implicate. ⊙ —Sparingly introduced in the United States, ballast, and waste places, Albany, N. Y., Philadelphia, District of Columbia, and Mobile, Ala.; Europe.

Setaria cárnei Hitchc. Resembling *S. verticillata* (L.) Beauv., but having looser panicles and larger spikelets, brown at maturity. ⊙ —A rapidly spreading weed in vineyards, Fresno County, Calif.; introduced from Western Australia.

4. Setaria macrospérma (Scribn. and Merr.) Schum. (Fig. 1097.) Perennial, often in large tufts, 1 to 1.5 m. tall; sheaths keeled; blades elongate, 1 to 2 cm. wide, scabrous on upper surface; panicle 15 to 30 cm. long, 2 to 4 cm. wide, tapering to both ends, rather loose, the secondary panicles smaller, compact, the branches of the terminal panicle as much as 2 cm. long, about equally distributed; bristles single below each spikelet, 1.5 to 3 cm. long; spikelets 3 mm. long. ♃ —Open ground, mostly on coral rock or coral sand, Florida; Bahamas.

5. Setaria villosíssima (Scribn. and Merr.) Schum. (Fig. 1098.) Perennial, as much as 1 m. tall; blades flat, villous or scabrous only, 15 to 30 cm. long, 5 to 10 mm. wide; panicle rather loose, more or less interrupted, tapering above, as much as 27 cm. long, the branches ascending, the axis villous; bristles 1.5 to 2.5 cm. long; spikelets about 3 mm. long, the second glume slightly shorter; fruit finely rugose. ♃ —Open or wooded rocky places, Texas and Arizona (locality unknown). Differing from *S. macrosperma* in the villous blades and looser panicles.

6. Setaria macrostáchya H. B. K. PLAINS BRISTLEGRASS. (Fig. 1099.) Perennial, densely tufted, usually pale or glaucous, 40 to 120 cm. tall;

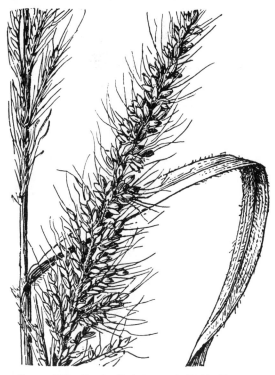

FIGURE 1098.—*Setaria villosissima*, × 1. (Smith, Tex.)

blades flat or folded, scabrous on the upper surface, rarely pubescent on both surfaces, 15 to 40 cm. long, 3 to 10 mm. wide; panicle spikelike, 10 to 25 cm. long, mostly 5 to 10 mm. thick, somewhat tapering but not attenuate, more or less interrupted or lobed; bristles 10 to 15 mm. long; spikelets 2 to 2.5 mm. long, very turgid; fruit rugose. ♃ —Open dry ground and dry woods, Texas to Colorado and Arizona; Mexico. Variable, especially in the thickness of the panicle, sometimes very slender, occasionally to 15 mm. thick. The type, from Mexico, is the robust form with thick panicles.

Setaria setósa (Swartz) Beauv. Panicle interrupted, attenuate at apex. ♃ —Ballast, Camden, N. J., and Key West, Fla.; adventive from the West Indies.

Setaria rariflóra Mikan ex Trin. Similar to *S. setosa*, the panicle and blades more slender. ♃ —Mobile, Ala.; adventive from South America.

7. Setaria scheélei (Steud.) Hitchc. (Fig. 1100.) Perennial, 60 to 120 cm. tall; sheaths compressed-keeled, glabrous or more or less hispid, the collar hispid; blades flat, elongate, as much as 1.5 cm. wide, scabrous or more or less pubescent; panicle rather loose, mostly 15 to 20 cm. long, tapering from near the base, the lower branches as much as 3 cm. long, ascending, the axis scabrous-pubescent and rather sparsely villous; bristles 1 to 1.5 cm. long, rather numerous, flexuous; spikelets about 2 mm. long; fruit rugose. ♃ —Open or rocky woods, Texas and Arizona. Differing from *S. macrostachya* in the looser panicle and the longer lower branches.

8. Setaria corrugáta (Ell.) Schult. (Fig. 1101.) Annual, erect or geniculate-spreading; culms freely branching, as much as 1 m. tall; sheaths scabrous to appressed-hirsute; blades flat, scabrous, as much as 30 cm.

long and 1 cm. wide (commonly less than 5 mm.); panicle dense, cylindric, usually 5 to 10 cm. long, the axis densely hispid-scabrous and also villous; bristles much exceeding the spikelets, sometimes as much as 2 cm. long, green or purple; spikelets 2 mm. long; fruit coarsely rugose. ⊙ —Sandy woods, cultivated fields, and waste places, along the coast, North Carolina to Florida and Louisiana; Cuba.

9. **Setaria liebmánni** Fourn. (Fig. 1102.) Annual, branching below, 30 to 100 cm. tall; blades flat, rather thin, 10 to 20 cm. long, 1 to 2 cm. wide, scabrous; panicle loosely flowered, tapering at each end, often nodding, usually 10 to 25 cm. long,

FIGURE 1100.—*Setaria scheelei*, × 1. (Bush 1244, Tex.)

FIGURE 1099.—*Setaria macrostachya*, × 1. (Hitchcock 13605, Tex.)

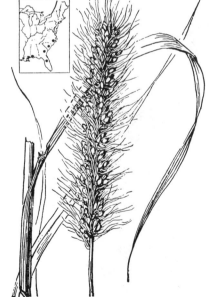

FIGURE 1101.—*Setaria corrugata*, × 1. (Pollard and Collins 253, Fla.)

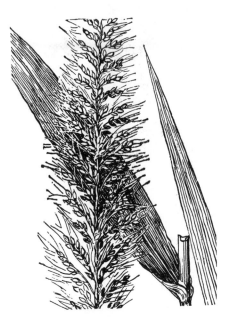

FIGURE 1102.—*Setaria liebmanni*, × 1. (Palmer 52, Mex.)

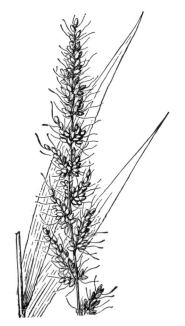

FIGURE 1103.—*Setaria grisebachii*, × 1. (Metcalf 1262, N. Mex.)

from slender to 25 mm. wide; bristles 7 to 15 mm. long; spikelets about 2 mm. long; fruit coarsely and strongly rugose. ☉ —Open sandy or rocky soil, Arizona (Tucson); Mexico to Nicaragua.

10. Setaria grisebáchii Fourn. GRISEBACH BRISTLEGRASS. (Fig. 1103.) Resembling *S. liebmanni;* blades smaller, panicle branches densely flowered; fruit finely rugose. ☉ — Open ground, often a weed in fields, Texas to Arizona; Mexico.

11. Setaria mágna Griseb. GIANT BRISTLEGRASS. (Fig. 1104.) Annual, robust, erect; culms sparingly branching, as much as 4 m. tall and 2 cm.

FIGURE 1104.—*Setaria magna*, × ⅛. (Nash 1279, Fla.)

thick at base; blades flat, scabrous, as much as 50 cm. long and 3.5 cm. wide; panicles densely flowered, nodding, often interrupted at base, tapering at each end, as much as 50 cm. long and 3 cm. thick, those of the branches much smaller; bristles 1 to 2 cm. long; spikelets about 2 mm. long; fruit smooth or nearly so, brown and shining at maturity. ☉ —Marshes and wet places along the coast, New Jersey to Florida; Arkansas and Texas; West Indies.

12. Setaria víridis (L.) Beauv. GREEN BRISTLEGRASS. (Fig. 1105.)

Annual, branching at base, sometimes geniculate-spreading, 20 to 40 cm. tall (or even 1 m.); blades flat, usually less than 15 cm. long and 1 cm. wide; panicle erect or somewhat nodding, densely flowered, green or purple, cylindric but tapering a little at the summit, usually less than 10 cm. long; bristles 1 to 3 below each spikelet, mostly 3 to 4 times their length, spikelets 2 to 2.5 mm. long; fruit very finely rugose. ⊙ —A weed in cultivated soil and waste places, common throughout the cooler parts of the United States, Newfoundland to British Columbia, south to Florida and California, infrequent in the Southern States and in the mountains; Mexico; introduced from Europe.

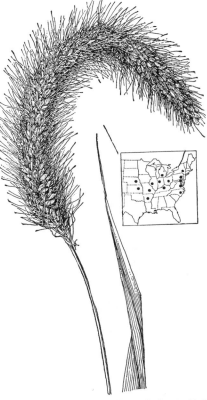

FIGURE 1106.—*Setaria faberii.* × 1; floret, × 5.
(V. H. Chase 8395, Va.)

FIGURE 1105.—*Setaria viridis,* × 1. (Thompson 129, Kans.)

13. Setaria fabérii Herrm. (Fig. 1106.) Similar to *S. viridis,* usually taller; blades softly pubescent to glabrescent; panicle conspicuously nodding; spikelets about 3 mm. long, the second glume shorter than the more rugose fruit. ⊙ —Becoming a weed in waste and cultivated ground, apparently spreading rapidly, New York to Nebraska and Arkansas, North Carolina, Kentucky, and Ten-

nessee. Introduced from China, probably in seed of Chinese millet.

14. Setaria itálica (L.) Beauv. FOX-TAIL MILLET. (Fig. 1107.) Cultivated form of *S. viridis,* more robust, with broader blades and larger lobed panicles, the fruit smooth or nearly so, shining at maturity, falling away from the remainder of the spikelet. In the larger forms the culms may be as much as 1 cm. thick and the panicles as much as 30 cm. long and 3 cm. thick, yellow or purple; bristles from scarcely longer than the spikelets to 3 to 4 times as long; fruit tawny to red, brown, or black. The smaller forms are known as Hungarian grass. ⊙ —Cultivated in the warmer parts of the United States, especially from Nebraska to Texas; escaped from cultivation in waste plicate, 1 to 2.5 cm. wide; panicles narrow, loose; bristles 5 to 10 mm. long. ⊙ —Ballast, Apalachicola and Miami, Fla.; adventive from East Indies. Many cultivated varieties and forms of this species have been described. For a study of these variants and a key to them see Hubbard, F. T., Rhodora 2: 187–196. 1915.

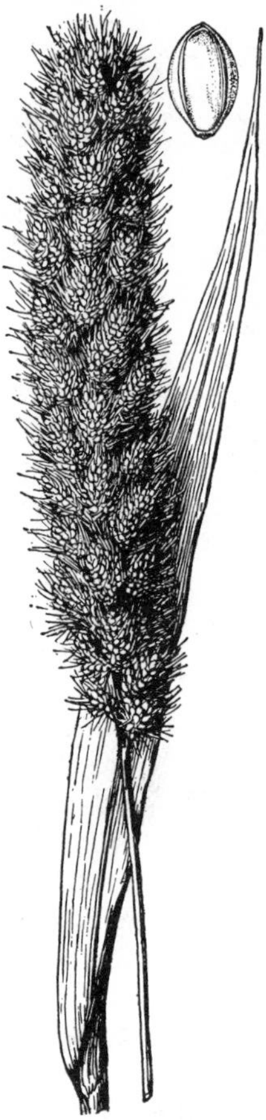

SETARIA PALMIFÓLIA (Koen.) Stapf. PALM-GRASS. (Fig. 1108.) Tall perennial; blades plicate, as much as 50 cm. long and 6 cm. wide; panicle loose, 20 to 40 cm. long; bristles inconspicuous. ♃ —Cultivated in the South and in greenhouses for ornament. (Sometimes called *Panicum plicatum*.) Native of India.

SETARIA POIRETIÁNA (Schult.) Kunth. Differing from *S. palmifolia* in having a narrow panicle about 30 cm. long with numerous ascending branches. ♃ —Occasionally cultivated for ornament. (Sometimes called *Panicum sulcatum*.) Tropical America. The last three species belong to the section *Ptychophyllum*.

FIGURE 1107.—*Setaria italica,* × 1; floret, × 5. (Williams 82, D. C.)

places throughout the United States; Eurasia.

Setaria barbáta (Lam.) Kunth. Decumbent annual; blades thin, lightly

FIGURE 1108.—*Setaria palmifolia,* × 1. (Hitchcock 9727, Jamaica.)

144. PENNISÉTUM L. Rich.

Spikelets solitary or in groups of 2 or 3, surrounded by an involucre of bristles (sterile branchlets), these not united except at the very base, often plumose, falling attached to the spikelets; first glume shorter than the spikelet, sometimes minute or wanting; second glume shorter than or equaling the sterile lemma; fertile lemma chartaceous, smooth, the margin thin, enclosing the palea. Annuals or perennials, often branched, with usually flat blades and dense spikelike panicles. Type species, *Pennisetum typhoideum* L. Rich. (*P. glaucum*). Name from Latin *penna*, feather, and *seta*, bristle, alluding to the plumose bristles of some species.

The most important species is *P. glaucum*, pearl millet, which is widely cultivated in tropical Africa and Asia, the seed being used for human food. It has been cultivated since prehistoric times, its wild prototype being unknown. In the United States pearl millet is used to a limited extent in the Southern States for forage, especially for soiling. Two species, *P. villosum* and *P. setaceum*, are cultivated for ornament. An African species, *P. purpureum*, elephant or Napier grass, is used in Florida as a forage plant.

Plants annual; bristles of involucre about as long as the spikelets. Cultivated.
                                          1. P. GLAUCUM.
Plants perennial; bristles much longer than the spikelets.
   Culms extensively creeping; spikelets few, hidden in the upper sheath.
                                         6. P. CLANDESTINUM.
   Culms not creeping; panicle exserted.
     Longer bristles 1 cm. long.
       Bristles unlike, the inner silky, plumose............................ 2. P. SETOSUM.
       Bristles all scabrous............ 3. P. NERVOSUM.
     Longer bristles 3 to 4 cm. long, the panicles feathery.
       Panicle oval, tawny............ 4. P. VILLOSUM.
       Panicle elongate, purple or rosy............ 5. P. SETACEUM.

1. Pennisetum glaúcum (L.) R. Br. PEARL MILLET. (Fig. 1109.) Annual; culms robust, as much as 2 m. tall, densely villous below the panicle; blades flat, cordate, sometimes as much as 1 m. long and 5 cm. wide; panicle cylindric, stiff, very dense, as much as 40 to 50 cm. long and 2 to 2.5 cm. thick, pale, bluish-tinged, or sometimes tawny, the stout axis densely villous; fascicles peduncled, spikelets short-pediceled, 2 in a fascicle, 3.5 to 4.5 mm. long, obovate, turgid, the grain at maturity protruding from the hairy-margined lemma and palea. (*P. typhoideum* L. Rich.; *Penicillaria spicata* Willd.)— Cultivated to a limited extent in the Southern States for forage; Eastern Hemisphere.

Pennisetum purpúreum Schumach. NAPIER GRASS. Robust leafy perennial, 2 to 4 m. tall; blades elongate, 2 to 3 cm. wide; panicle dense, elongate, stiff, tawny or purplish, with sparsely plumose bristles about 1 cm. long. ♃ —Introduced from Africa; used as a forage plant from central to southern Florida; grown in the West Indies and South America. Also called elephant grass.

2. Pennisetum setósum (Swartz) L. Rich. (Fig. 1110.) Perennial; culms sometimes 30 or more in loose clumps, 1 to 2 m. tall, geniculate, sometimes rooting at the lower nodes, bearing 1 to several flowering branches from the lower and middle nodes, scabrous below the panicle; blades elongate, 4 to 18 mm. wide; panicle 10 to 25 cm. long, 8 to 10 mm. thick, excluding the bristles, rather dense, yellow to purple; fascicles reflexed at maturity; bristles unequal, the outer delicate, mostly shorter than the spikelet, the inner densely silky-plumose below, as much as 1 cm. long, the hairs beautifully crimped; spikelet solitary, 3.2 to 4 mm. long; fruit subindurate, smooth and shining. ♃ —Open slopes and savannas, southern Florida; tropical America.

3. Pennisetum nervósum (Nees) Trin. (Fig. 1111.) Perennial; culms robust, branching, as much as 3 m.

tall; blades elongate, 5 to 10 mm. wide, scabrous; panicle dense, somewhat flexuous, 10 to 20 cm. long; fascicles spreading to reflexed; bristles scabrous, the outer about as long as the spikelet, the inner about 10 mm. long; spikelet solitary, 5 to 6 mm. long. ♃ —Moist open or brushy places, Brownsville, Tex., along the Rio Grande; apparently introduced; Ecuador to Brazil and Argentina.

4. Pennisetum villósum R. Br. Feathertop. (Fig. 1112.) Perennial; culms tufted, 30 to 60 cm. tall, pubescent below the panicle; blades 3 to 5 mm. wide; panicle tawny, ovoid or oblong, 3 to 10 cm. long, 1 to 5 cm. wide including bristles, dense, feathery; spikelets 1 to 4 in a fascicle; fascicles short-peduncled, a tuft of white hairs at base of peduncle; bristles numerous, spreading, the inner very plumose, the longer 4 to 5 cm. long. ♃ (*P. longistylum* of

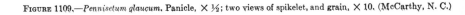

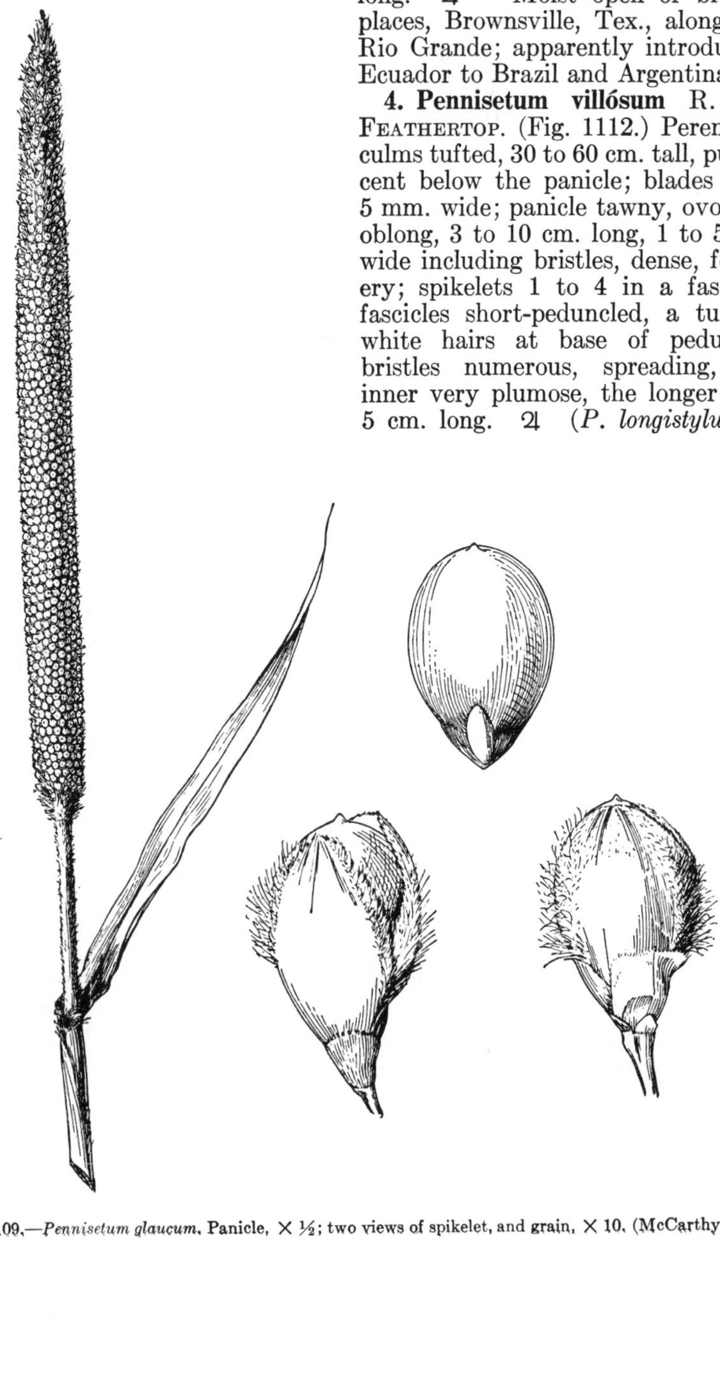

Figure 1109.—*Pennisetum glaucum*. Panicle, × ½; two views of spikelet, and grain, × 10. (McCarthy, N. C.)

florists, not Hochst.)—Cultivated for ornament, sparingly escaped in dry ground, Michigan, Texas, and California; introduced from Africa.

5. Pennisetum setáceum (Forsk.) Chiov. FOUNTAIN GRASS. (Fig. 1113.) Perennial, culms tufted, simple, about 1 m. tall; blades narrow, elongate, scabrous; panicle 15 to 35 cm. long, nodding, pink or purple; fascicles peduncled, rather loosely arranged, containing 1 to 3 spikelets; bristles plumose toward base, unequal, the longer 3 to 4 cm. long. ♃ (*P. ruppelii* Steud.)—Cultivated for ornament, especially as a border plant or around fountains; introduced from Africa.

6. Pennisetum clandestínum Hochst. ex Chiov. KIKUYU GRASS. (Fig. 1114.) Low-growing, rhizomatous, stoloniferous perennial, the stolons with short internodes; inflorescence consisting of 2 to 4 spikelets almost entirely enclosed in the upper sheath of the short culms. ♃ — A troublesome weed in orchards and

FIGURE 1111.—*Pennisetum nervosum,* × ½. (Ferris and Duncan 3198, Tex.)

gardens in southern California. Introduced from Africa. A good forage grass in tropics and subtropics.

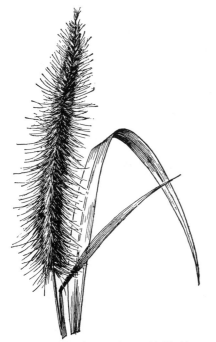

FIGURE 1110.—*Pennisetum setosum,* × ½. (Amer. Gr. Natl. Herb. 611, Trinidad.

FIGURE 1112.—*Pennisetum villosum,* × ½. (Eastwood 172, Calif.)

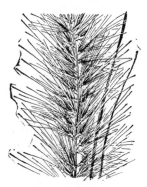

FIGURE 1113.—*Pennisetum setaceum,* × ½. (Hitchcock, D. C.)

PENNISETUM CILIÁRE (L.) Link. Culms geniculate, from a knotted crown, 10 to 50 cm. tall; panicle 2 to 10 cm. long; bristles united at the very base, flexuous, purple, 5 to 10 mm. long, the inner plumose. ♃ —Occasionally cultivated in the Southern States; adventive in wool waste, Yonkers, N. Y. In the West Indies said to be good forage. India.

PENNISETUM ALOPECUROÍDES (L.) Spreng. Perennial; culms compressed, to 1 m. tall, with elongate scabrous blades and softly bristly panicles 8 to 15 cm. long; bristles of the fascicles to 2 cm. long. ♃ —Sparingly cultivated; escaped in Berks County, Pa.; Asia.

PENNISETUM MACROSTÁCHYUM (Brongn.) Trin. Resembling *P. setaceum,* blades as much as 2.5 cm. wide; panicle denser, brownish purple, fascicles smaller; bristles not plumose. ♃ —Sparingly cultivated for ornament. East Indies.

PENNISETUM LATIFÓLIUM Spreng. Perennial; culms 100 to 150 cm. tall, the nodes appressed-pubescent; blades 2 to 3 cm. wide, tapering to a long point; panicles terminal and axillary, nodding, 5 to 8 cm. long, the bristles prominent. ♃ —Occasionally cultivated for ornament. South America.

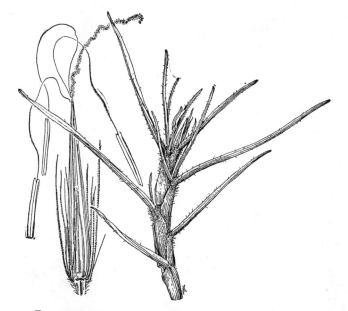

FIGURE 1114.—*Pennisetum clandestinum,* × 1. (Chase 10181, Brazil.)

145. CÉNCHRUS L. SANDBUR

Spikelets solitary or few together, surrounded and enclosed by a spiny bur composed of numerous coalescing bristles (sterile branchlets), the bur subglobular, the peduncle short and thick, articulate at base, falling with the spikelets and permanently enclosing them, the seed germinating within the old involucre, the spines usually retrorsely barbed. Annuals or sometimes perennials, commonly low and branching, with flat blades and racemes of burs, the burs readily deciduous. Type species, *Cenchrus echinatus*. Name from Greek *kegchros*, a kind of millet.

The species are excellent forage grasses before the burs are formed. Several species are weeds and become especially troublesome after the maturity of the burs.

Involucral lobes united at the base only; racemes dense; plants perennial.
1. C. MYOSUROIDES.
Involucral lobes united above the base.
 Involucre with a ring of slender bristles at base; plants annual.
  Burs, excluding the bristles, not more than 4 mm. wide, numerous, crowded in a long raceme; lobes of the involucre interlocking, not spinelike............ 2. C. BROWNII.
  Burs, excluding the bristles, 5 to 7 mm. wide, not densely crowded; lobes of the involucre erect or nearly so or rarely one or two lobes loosely interlocking, the tips spinelike.
3. C. ECHINATUS.
 Involucre with flattened spreading spines, no ring of slender bristles at base.
  Body of bur ovate, usually not more than 3.5 mm. wide, tapering at base; plants perennial.
   Burs glabrous; spines 4 to 6 mm. long............................... 4. C. GRACILLIMUS.
   Burs pubescent; spines rarely more than 4 mm. long, usually shorter.
5. C. INCERTUS.
  Body of bur globose, 5 mm. wide or more, not tapering at base; plants annual.
   Burs, including spines, 7 to 8 mm. wide, finely pubescent........ 6. C. PAUCIFLORUS.
   Burs, including spines, 10 to 15 mm. wide, densely woolly........ 7. C. TRIBULOIDES.

1. Cenchrus myosuroídes H. B. K.

(Fig. 1115.) Stout glaucous woody perennial; culms erect from an often decumbent base, 1 to 1.5 m. tall, branching below; blades 5 to 12 mm. wide; raceme 10 to 25 cm. long, strict, erect, dense; burs 1-flowered, about 5 mm. wide, the bristles united at the base only, the outer shorter, the inner about as long as the spikelet; spikelet 4.5 to 5.5 mm. long. ♃ —

FIGURE 1115.—*Cenchrus myosuroides.* Bur, two views of spikelet, and floret, × 5. (Léon 835, Cuba.)

Moist sandy open ground or scrubland near the coast, Georgia and Florida, southern Louisiana and southern Texas; tropical America.

Cenchrus biflórus Roxb. Annual; culms 30 to 100 cm. tall; raceme 8 to 10 cm. long, the burs usually 2-flowered, 4 to 6 mm. long, the outer row of bristles short, spreading, the inner flattened, rigid, erect. ☉ (*C. barbatus* Schum., *C. catharticus* Del.) —Ballast, Mobile, Ala.; wool waste, Yonkers, N. Y. Native of India and north Africa.

2. Cenchrus brównii Roem. and

Schult. (Fig. 1116.) Annual, mostly erect, 30 to 100 cm. tall; blades thin, flat, lax, 6 to 12 mm. wide; raceme 4 to 10 cm. long, dense; burs depressed globose, about 4 mm. high, the outer bristles numerous, very slender, the inner somewhat exceed-

FIGURE 1116.—*Cenchrus brownii.* Bur, two views of spikelet, and floret, × 5. (Type.)

ing the body, the lobes interlocking at maturity; spikelets usually 3. ☉ (*C. viridis* Spreng.)—Open ground, often a weed in waste places, Florida Keys; adventive in North Carolina; tropical America at low altitudes; introduced in Malaysia.

3. Cenchrus echinâtus L. (Fig.

1117.) Annual; culms compressed,

Figure 1117.—*Cenchrus echinatus.* Bur, two views of spikelet, and floret, × 5. (Hitchcock 9397, Jamaica.)

usually geniculate, branching at base, 25 to 60 cm. long; blades 3 to 8 mm. wide, pilose on the upper surface near the base; raceme 3 to 10 cm. long, the burs larger, fewer, and less crowded than in *C. brownii;* bur 4 to 7 mm. high, as broad or broader, pubescent, the lobes of the involucre erect or

forming dense clumps, glabrous as a whole; culms slender, wiry, erect or ascending, 20 to 80 cm. tall; blades usually folded, 2 to 3 mm. wide; raceme 2 to 6 cm. long, the burs relatively distant, about 3.5, rarely as much as 5 mm., wide (excluding spines), tapering at base, glabrous;

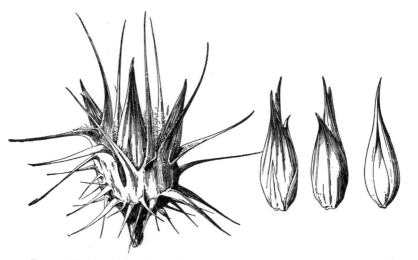

Figure 1118.—*Cenchrus gracillimus.* Bur, two views of spikelet, and floret, × 5. (Type coll.)

bent inward but not interlocking; spikelets usually 4 in each bur. ☉ —Open ground and waste places, South Carolina to southern California; a common weed in tropical America; sparingly introduced in Hawaii and Malaysia.

4. Cenchrus gracíllimus Nash. (Fig. 1118.) Perennial, at length

spines spreading or reflexed, flat, 4 to 6 mm. long, the lobes about 8; spikelets 2 or 3 in each bur. ♃ —Sandy open ground and high pineland, Florida, southern Alabama and Mississippi; Cuba, Jamaica.

5. Cenchrus incértus M. A. Curtis. COAST SANDBUR. (Fig. 1119.) Perennial, glabrous as a whole; culms 25 to

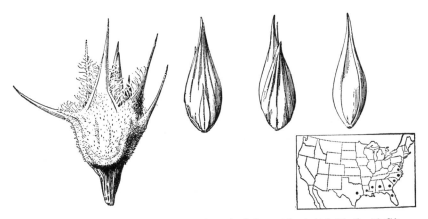

FIGURE 1119.—*Cenchrus incertus*. Bur, two views of spikelet, and floret, × 5. (Curtiss, N. C.)

100 cm. tall; blades commonly folded but sometimes flat, 2 to 5 mm. wide; raceme 4 to 10 cm. long, the burs not crowded; burs about 3.5 (3 to 5) mm. wide, the body finely and densely pubescent, the base glabrous; spines few, mostly less than 5 mm. long, the lower often reduced or obsolete; spike-

spreading, 20 to 90 cm. long, rather stout; blades usually flat, 2 to 7 mm. wide; raceme usually 3 to 8 cm. long, the burs somewhat crowded; burs (excluding spines) mostly 4 to 6 mm. wide, pubescent, often densely so; spines numerous, spreading or reflexed, flat, broadened at base, the

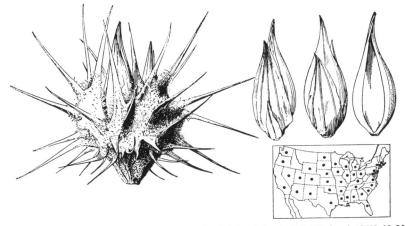

FIGURE 1120.—*Cenchrus pauciflorus*. Bur, two views of spikelet, and floret, × 5. (Hitchcock 13582, N. Mex.)

lets 1 to 3 in each bur. ♃ —Open sandy soil, Coastal Plain, Virginia to Florida and Texas.

6. Cenchrus pauciflórus Benth. FIELD SANDBUR. (Fig. 1120.) Annual, at times a short-lived perennial, sometimes forming large mats; culms

lowermost shorter and relatively slender, some of the upper ones commonly 4 to 5 mm. long, usually villous at the base; spikelets usually 2 in each bur. ☉ (Confused with *C. tribuloides* in early manuals; *C. carolinianus* of recent manuals, not of Walt.)

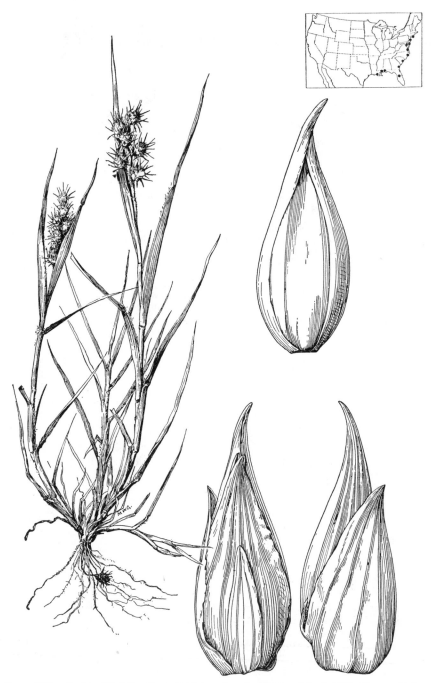

FIGURE 1121.—*Cenchrus tribuloides*. Plant, × ½; two views of spikelet, and floret, × 10. (Kearney, Va.)

—Sandy open ground, often a weed in sandy fields, Ontario to Oregon, south to Florida, Texas, and California; Mexican plateau; coastal region of tropical America; southern South America. The type, from Baja California, is a small arid-ground specimen, the burs smaller than those of plants of more favorable situations. Specimens with long spines have been differentiated as *C. albertsonii* Runyon and *C. longispinus* (Hack.) Fernald. The spikelets are identical except in size.

7. Cenchrus tribuloídes L. DUNE SANDBUR. (Fig. 1121.) Stouter than *C. pauciflorus;* soon branching and radiate-decumbent, rooting at the nodes; sheaths usually much overlapping; burs (excluding spines) 5 to 6 mm. wide and 8 to 9 mm. high, usually conspicuously villous. ⊙ —In loose sands of the coast, Staten Island, N. Y., to Florida and Louisiana; West Indies.

146. AMPHICÁRPUM Kunth

(Amphicarpon Raf.)

Spikelets of 2 kinds on the same plant, one in a terminal panicle, perfect but not fruitful, the other cleistogamous on slender leafless subterranean branches from the base of the culm or sometimes also from the lower nodes; first glume of the aerial spikelets variable in size, sometimes obsolete; second glume and sterile lemma about equal; lemma and palea indurate, the margins of the lemma thin and flat; fruiting spikelets much larger, the first glume wanting; second glume and sterile lemma strongly nerved, subrigid, exceeded at maturity by the turgid, elliptic, acuminate fruit with strongly indurate lemma and palea, the margins of the lemma thin and flat; stamens with small anthers on short filaments. Annual or perennial erect grasses, with flat blades and narrow terminal panicles. Type species, *Milium amphicarpon* Pursh (*Amphicarpum purshii*). Name from Greek *amphikarpos*, doubly fruit-bearing, alluding to the two kinds of spikelets.

Blades conspicuously hirsute.. 1. A. PURSHII.
Blades glabrous or nearly so.. 2. A. MUHLENBERGIANUM.

1. Amphicarpum púrshii Kunth. (Fig. 1122.) Annual; culms erect, 30 to 80 cm. tall, the leaves crowded toward the base, hirsute; blades erect, 10 to 15 cm. long, 5 to 15 mm. wide, sharp-pointed; panicle 3 to 20 cm. long; spikelets elliptic, 4 to 5 mm. long; subterranean spikelets 7 to 8 mm. long, plump, acuminate. ⊙ (*Amphicarpon amphicarpon* Nash.)— Sandy pinelands, New Jersey to Georgia.

2. Amphicarpum muhlenbergiánum (Schult.) Hitchc. (Fig. 1123.) Perennial; culms usually decumbent at base, 30 to 100 cm. tall; leaves evenly distributed; blades firm, white-margined when dry, mostly less than 10 cm. long, 5 to 10 mm. wide; panicle long-exserted, few-flowered; spikelets narrowly lanceolate, 6 to 7 mm. long; subterranean spikelets 6 to 9 mm. long. ♃ (*A. floridanum* Chapm.)—Low pinelands, South Carolina and Florida.

147. OLÝRA L.

Plants monoecious; inflorescence paniculate; pistillate spikelets borne on the ends of the branches of loose panicles, the smaller staminate spikelets pedicellate below the pistillate ones, sometimes the upper branches all pistillate and the lower ones all staminate; pistillate spikelets rather large; first glume wanting; second glume and sterile lemma herbaceous, often caudate-acuminate; fruit bony-indurate; staminate spikelets readily deciduous; glumes and sterile lemma wanting, the lemma and palea mem-

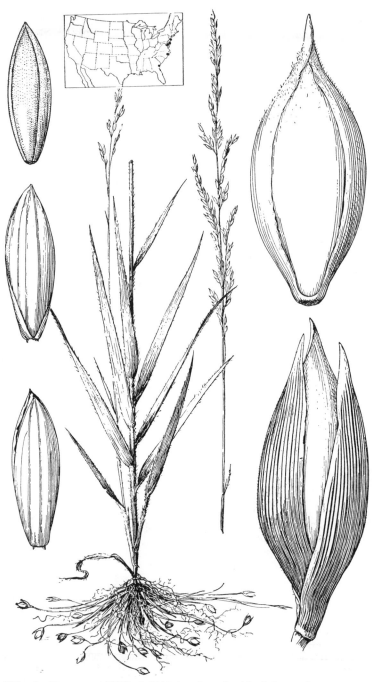

FIGURE 1122.—*Amphicarpum purshii*. Plant, × ½; two views of aerial spikelet and floret, and subterranean spikelet and floret, × 10. (Brinton, N. J.)

branaceous. Mostly tall perennials with broad flat blades, contracted into a petiole, and open or contracted panicles of glabrous spikelets. Type species, *Olyra latifolia*. Name from *olura*, an old Greek name for a kind of grain.

1. Olyra latifólia L. (Fig. 1124.) Glabrous perennial, bamboolike in aspect, commonly 3 m. tall, with flat, firm, asymmetrically lanceolate-oblong, abruptly acuminate blades commonly 20 cm. long and 5 cm. wide, and ovoid panicles 10 to 15 cm. long, the branches stiffly ascending or spreading, each bearing a single large long-acuminate pistillate spikelet at the thickened summit and several small slender-pediceled staminate spikelets along the branches. ♃ — Said to occur in the region of Tampa Bay, Fla., but the record is doubtful; tropical America; Africa.

FIGURE 1123.—*Amphicarpum muhlenbergianum*, × 1. (Chapman, Fla.)

TRIBE 13. ANDROPOGONEAE

148. IMPERÁTA Cyrillo

Spikelets all alike, awnless, in pairs, unequally pedicellate on a slender continuous rachis, surrounded by long silky hairs; glumes about equal, membranaceous; sterile lemma, fertile lemma, and palea thin and hyaline. Perennial, slender, erect grasses, from hard scaly rhizomes, with terminal narrow silky panicles. Type species, *Imperata cylindrica*. Named for Ferrante Imperato.

Spikelets 4 mm. long, the hairs at base twice as long; panicle oblong, rather lax.
                                                     1. I. BRASILIENSIS.
Spikelets 3 mm. long, the hairs three times as long; panicle elongate.. 2. I. BREVIFOLIA.

1. Imperata brasiliénsis Trin. (Fig. 1125.) Culms 50 to 100 cm. tall, from scaly rhizomes; leaves crowded below, 3 to 8 mm. wide, the lower blades elongate, those of the culm short, the uppermost much reduced; panicle dense, pale or silvery, mostly 10 to 12 cm. long; spikelets 4 mm. long. ♃ —Pinelands, prairies, and Everglades, southern Florida and Alabama; tropical America at low altitudes.

2. Imperata brevifólia Vasey. SATINTAIL. (Fig. 1126.) Resembling *I. brasiliensis;* culms 1 to 1.5 m. tall; leaves less crowded at base, all but the uppermost elongate; panicle 15 to 30 cm. long; spikelets 3 mm. long, the hairs three times as long. ♃ (*I. hookeri* Rupr. ex Hack.)—Desert regions, western Texas to southern California, Utah, and Nevada; Mexico.

Imperata cylíndrica (L.) Beauv. COGON GRASS. Spikelets 4 to 5 mm. long, the hairs as long as in *I. brevifolia*. ♃ —Ballast, Portland, Oreg.; recently introduced in Florida and spreading in the west central part of the State. It is fairly good forage, but because of the strong creeping rhizomes it spreads into cultivated ground and is difficult to eradicate.

FIGURE 1124.—*Olyra latifolia*. Plant, × ½; pistillate and staminate spikelets, and fertile floret, × 5. (Chase 6416, P. R.)

FIGURE 1125.—*Imperata brasiliensis*. Plant, ✕ ½; spikelet, ✕ 5. (Chapman, Fla.)

149. MISCÁNTHUS Anderss.

Spikelets all alike, in pairs, unequally pedicellate along a slender continuous rachis; glumes equal, membranaceous or somewhat coriaceous; sterile lemma a little shorter than the glumes, hyaline; fertile lemma hyaline, smaller than the sterile lemma, extending into a delicate

FIGURE 1126.—*Imperata brevifolia*. Plant, × ½.
(Toumey 782, Ariz.)

bent and flexuous awn; palea small and hyaline. Robust perennials, with long flat blades and terminal panicles of aggregate spreading slender racemes. Type species, *Miscanthus japonicus* Anderss. (*M. floridulus* (Labill.) Warb.) Name from Greek *mischos*, pedicel, and *anthos*, flower, both spikelets of the pair being pedicellate.

1. Miscanthus sinénsis Anderss. EULALIA. (Fig. 1127.) Culms robust in large bunches, erect, 2 to 3 m. tall; leaves numerous, mostly basal, the blades flat, as much as 1 m. long,

about 1 cm. wide, tapering to a slender point, the margin sharply serrate; panicle somewhat fan-shaped, consisting of numerous silky aggregate racemes, 10 to 20 cm. long; spikelets with a tuft of silky hairs at base surrounding them and about as long as the glumes. ♃ —Cultivated for ornament and now growing wild in some localities in the Eastern States; native of eastern Asia. There are three varieties in cultivation besides the usual form described above: M. SINENSIS var. VARIEGÁTUS Beal, with blades striped with white, M. SINENSIS var. ZEBRÍNUS Beal, with blades banded or zoned with white, and M. SINENSIS var. GRACÍLLIMUS Hitchc., with very narrow blades.

Miscanthus nepalénsis (Trin.) Hack. Panicles yellowish brown; spikelets about one-fourth as long as the hairs at their base. ♃ —Occasionally cultivated under the name of Himalaya fairy grass. Nepal, India.

MISCANTHUS SACCHARIFLÓRUS (Maxim.) Hack. Perennial with thick horizontal rhizomes; culms 1.5 to 2 m. tall; blades 1 to 1.8 cm. wide; panicle more silky than in *M. sinensis*, the spikelets awnless. ♃ —Sparingly cultivated for ornament; escaped in Clinton County, Iowa; Asia.

150. SÁCCHARUM L.

Spikelets in pairs, one sessile, the other pedicellate, both perfect, awnless, arranged in panicled racemes, the axis disarticulating below the spikelets; glumes somewhat indurate, sterile lemma similar but hyaline; fertile lemma hyaline, sometimes wanting. Robust perennials of tropical regions. Type species, *Saccharum officinarum*. Name from Latin *saccharum* (*saccharon*), sugar, because of the sweet juice.

1. Saccharum officinárum L. SUGARCANE. (Fig. 1128.) Culms 3 to 5 m. tall, 2 to 3 cm. thick, solid, juicy, the lower internodes short, swollen; sheaths greatly overlapping, the lower usually falling from the culms; blades elongate, mostly 4 to 6 cm. wide, with a very thick midrib; panicle plume-

FIGURE 1127.—*Miscanthus sinensis*. Plant, much reduced; raceme, × ½; spikelet, × 5. (Cult.)

Figure 1128.—*Saccharum officinarum.* Plant, much reduced; racemes, × ½; spikelet with pedicel and rachis joint, × 5. (Pringle, Cuba.)

like, 20 to 60 cm. long, the slender racemes drooping; spikelets about 3 mm. long, obscured in a basal tuft of silky hairs 2 to 3 times as long as the spikelet. ♃ —Cultivated in the Southern States, especially Louisiana, for sugar and byproducts, and for sirup, and also used for forage; commonly cultivated in tropical regions.

The sugarcanes cultivated in the United States are derived chiefly from four species and their hybrids. In the Noble canes (*S. officinarum*, chromosomes 40), described above, the axis of inflorescence is without long hairs. Chinese canes (*S. sinensis* Roxb., chromosomes about 58 to 60), with long hairs on the axis of inflor-escence, are cultivated chiefly for sirup. *Saccharum barberi* Jeswiet (chromosomes about 45 or 46) from northern India, differs from the last in having narrower blades and more slender canes. Varieties of this species do not form an entirely homogeneous group and may later be separated into two or more species. The wild cane of Asia (*S. spontaneum* L., chromosomes 56), is used as a basis for hybrids with other species. There are numerous hybrids and varieties of the species mentioned.

SACCHARUM BENGALÉNSE Retz. MUNJ. Tall cane; blades very scabrous; panicle 70 to 80 cm. long, narrow, dense, silvery. ♃ —Sometimes cultivated for ornament. India.

151. ERIÁNTHUS Michx. PLUMEGRASS

Spikelets all alike, in pairs along a slender axis, one sessile, the other pedicellate, the rachis disarticulating below the spikelets, the rachis joint and pedicel falling attached to the sessile spikelet; glumes coriaceous, equal, usually copiously clothed, at least at the base, with long silky spreading hairs; sterile lemma hyaline; fertile lemma hyaline, the midnerve extending into a slender awn; palea small, hyaline. Perennial reedlike grasses, with elongate flat blades and terminal oblong, usually dense silky panicles. Type species, *Erianthus saccharoides* (*E. giganteus*). Name from Greek *erion*, wool, and *anthos*, flower, alluding to the woolly glumes.

Spikelets naked, or nearly so, at base... 1. E. STRICTUS.
Spikelets with a conspicuous tuft of hairs at base.
  Awn flat, spirally coiled at base, the upper portion more or less bent and flexuous or
    loosely spiral.
    Basal hairs nearly as long as the brownish spikelets; panicle not conspicuously hairy,
      the main axis and branches visible; culms usually glabrous below panicle.
                                            2. E. CONTORTUS.
    Basal hairs copious, about twice as long as the yellowish spikelets; panicle conspicu-
      ously woolly, the hairs hiding the main axis and branches; culms villous below
      panicle... 3. E. ALOPECUROIDES.
  Awn terete, or flattened at base, not coiled, the upper portion straight or slightly flexuous.
    Basal hairs copious, much longer then the spikelet; panicle conspicuously woolly.
                                            6. E. GIGANTEUS.
    Basal hairs rather sparse, shorter than the spikelet; panicle not woolly.
      Uppermost blade not reduced, reaching the summit of the panicle; rachis joint and
        pedicel terete, sparsely long-pilose............................ 4. E. BREVIBARBIS.
      Uppermost blade usually much reduced; rachis joint and pedicel somewhat angled,
        sparsely short-pilose... 5. E. COARCTATUS.

1. Erianthus stríctus Baldw. NARROW PLUMEGRASS. (Fig. 1129.) Culms 1 to 2 m. tall, relatively slender, glabrous; nodes hirsute with stiff erect deciduous hairs; foliage glabrous, the lower sheaths narrow, crowded, the blades mostly 8 to 12 mm. wide; panicle 20 to 40 cm. long, strict, the branches closely appressed; spikelets brown, about 8 mm. long, scabrous, nearly naked to sparsely short-hairy at base; awn straight, about 15 mm. long; rachis joint and pedicel scabrous. ♃ —Marshes and wet places, Coastal Plain, Vir-

and Texas, north to Tennessee and Oklahoma.

3. Erianthus alopecuroídes (L.) Ell. SILVER PLUMEGRASS. (Fig. 1131.) Culms robust, 1.5 to 3 m. tall, appressed-villous below the panicle, and usually on the nodes; sheaths pilose at the summit; blades 1.2 to 2 cm. wide, scabrous, pilose on upper surface toward the base; panicle 20 to 30 cm. long, silvery to tawny or purplish; spikelets 5 to 6 mm. long, pale, sparsely villous, shorter than the copious basal hairs; awn 1 to 1.5 cm. long, flat, loosely twisted; rachis joint and pedicel long-villous, ♃ (*E. divaricatus* Hitchc.)—Damp woods, open ground, and borders of fields,

FIGURE 1129.—*Erianthus strictus*, × ½. (Curtiss 6936, Fla.)

FIGURE 1130.—*Erianthus contortus*, × ½. (Amer. Gr. Natl. Herb. 234, S. C.)

ginia to Florida and Texas, north to Tennessee and southern Missouri.

2. Erianthus contórtus Baldw. ex Ell. BENT-AWN PLUMEGRASS. (Fig. 1130.) Culms 1 to 2 m. tall, glabrous or sometimes sparsely appressed-pilose below the panicle; nodes glabrous or pubescent with erect deciduous hairs; sheaths sparsely pilose at summit or glabrous; blades 1 to 1.5 cm. wide, scabrous; panicle 15 to 30 cm. long, narrow, the branches ascending but not closely appressed; spikelets 6 to 8 mm. long, brownish, basal hairs nearly or about as long as the spikelet, awn about 2 cm. long, spirally coiled at base; rachis joints and pedicels villous. ♃ —Moist sandy pinelands or open ground, Coastal Plain, Maryland to Florida

FIGURE 1131.—*Erianthus alopecuroides*, × ½. (Chase 4213, Fla.)

southern New Jersey to southern
Illinois, southern Missouri, and Oklahoma, south to Florida and Texas.
ERIANTHUS ALOPECUROIDES var. HIR-
SÚTUS Nash. Sheaths and lower surface of the blades appressed-hirsute.
♃ —North Carolina and Florida.
4. Erianthus brevibárbis Michx.
(Fig. 1132.) Culms stout, nearly 2
m. tall, with 9 or 10 nodes; glabrous;
sheaths glabrous or sparingly pubescent at the summit; blades scabrous
on the upper surface, pilose at the
base, 1 to 1.5 cm. wide, the upper
not reduced; panicle 35 cm. long,
tawny brown, not conspicuously woolly; spikelets 6 to 7 mm. long; glumes

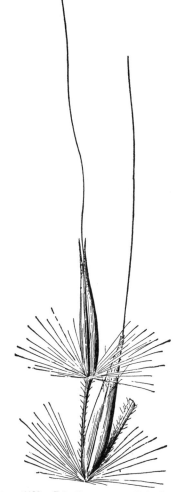

FIGURE 1133.—*Erianthus coarctatus.* Pair of spikelets
with pedicel and rachis joint, × 5. (Type collection.)

acuminate, glabrous or with a few
long hairs on the inflexed margins,
the spreading basal hairs about two-
thirds as long as the spikelet; awn
terete, straight or subflexuous, 1.5
to 1.6 cm. long; rachis joint and
pedicel sparsely long-pilose. ♃ —
Dry hills, southern Illinois (type)
and Arkansas (Pulaski County); rare.
5. Erianthus coarctátus Fernald.
(Fig. 1133.) Culms relatively slender,
75 to 150 cm. tall, subcompressed,
the nodes bearded, appressed-pubes-

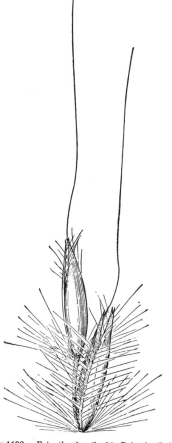

FIGURE 1132.—*Erianthus brevibarbis.* Pair of spikelets
with pedicel and rachis joint, × 5. (Demaree 8228,
Ark.)

cent, or glabrescent; sheaths glabrous, the lower narrow, somewhat keeled; blades 3 to 10 mm. wide, scaberulous, the upper reduced; panicle 10 to 27 cm. long, 2.5 to 4 cm. wide, purplish brown, not conspicuously woolly; spikelets 7 to 8 mm. long; glumes acuminate, scaberulous, the first sometimes with a few long hairs on the back, the second without hairs on the inflexed margins, the basal hairs about half as long as the spikelet; awn terete, straight, 1.5 to 2.3 cm. long, straight; rachis joint and pedicel somewhat angled, very sparsely short-pilose. ♃ —Peaty, sandy, moist meadows and swales and margin of swamps, Delaware, Maryland, Virginia, Georgia, and Florida (near Gainesville); insufficiently known, apparently rare.

FIGURE 1134.—*Erianthus coarctatus* var. *elliottianus.* Racemes, × ½. (Hitchcock, N. C.)

ERIANTHUS COARCTATUS var. ELLIOTTIÁNUS Fernald. (Fig. 1134.) Taller and more robust, resembling *E. brevibarbis,* but nodes appressed-pubescent, upper blades mostly reduced, the brownish panicle mostly smaller; spikelets 7 mm. long, more slender, as in *E. coarctatus,* the first glume usually with a few long hairs on the back, occasionally the second glume likewise, the margins without long hairs; awns, rachis joints, and pedicels as in *E. coarctatus.* ♃ — Wet ground, swales, and pond borders, North Carolina to Florida; Louisiana.

This group is insufficiently known;

the size of upper blade and pubescence on spikelets is not constant.

6. Erianthus gigantéus (Walt.) Muhl. SUGARCANE PLUMEGRASS. (Fig. 1135.) Culms 1 to 3 m. tall, appressed-villous below the panicle, the nodes appressed-hispid, the hairs deciduous; sheaths and blades from nearly glabrous to shaggy appressed-villous, the blades 8 to 15 mm. wide; panicle 10 to 40 cm. long, oblong or ovoid, tawny to purplish; spikelets 5 to 7 mm. long, sparsely long-villous on the upper part, shorter than the copious basal hairs; awn 2 to 2.5 cm. long, terete, straight or slightly flexuous; rachis joint and pedicel long-pilose. ♃ (*E. saccharoides* Michx.)—Moist soil, Coastal Plain, New York to Florida and Texas, north to Kentucky; Cuba. A common form with relatively small compact panicles has been segregated as *E. compactus* Nash; a robust form with long, copiously silky, tawny panicle, as *E. tracyi* Nash; and a form with rather looser panicle, the lower rachis joints longer than the spikelets, and pubescent foliage was described from Florida as *E. laxus* Nash.

Erianthus ravénnae (L.) Beauv. RAVENNA GRASS. (Fig. 1136.) Culms stout, as much as 4 m. tall; panicle as much as 60 cm. long, silvery (purplish in var. *purpuráscens* (Anderss.) Hack.); spikelets awnless or nearly so. ♃ —Cultivated for ornament; hardy as far north as New York City; native of Europe. Established along irrigation ditches near Phoenix, Ariz.

152. MICROSTÉGIUM Nees
(Included in *Eulalia* Kunth in Manual, ed. 1)

Spikelets in pairs, alike, perfect, on an articulate rachis, 1 sessile, 1 pedicellate; racemes 1 to several, digitate or approximate; first glume sulcate. Straggling annuals with flat lanceolate blades. Type species, *M. willdenovianum* Nees (*M. vimineum* (Trin.) A. Camus). Name from Greek *micros*, small, and *stege*, cover, probably alluding to the minute lemma.

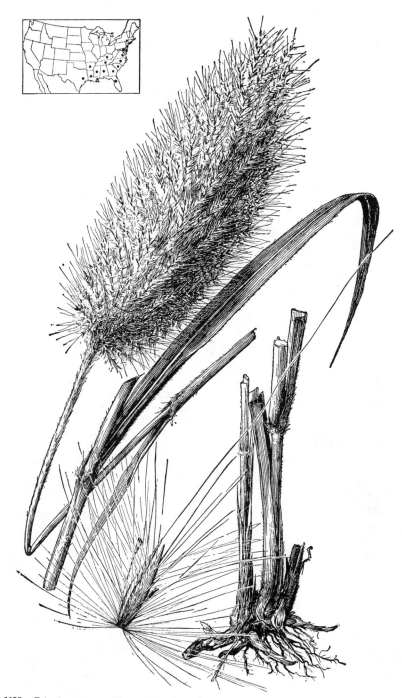

FIGURE 1135.—*Erianthus giganteus.* Plant, × ½; spikelet with pedicel and rachis joint, × 5. (Langlois 96, La.)

FIGURE 1136.—*Erianthus ravennae.* Racemes, × 1. (Cult.)

FIGURE 1137.—*Microstegium vimineum,* × 1. (Wilkins 3716, Tenn.)

FIGURE 1138.—*Arthraxon hispidus* var. *cryptatherus,* × 1. (Cult.)

1. Microstegium vimíneum (Trin.) A. Camus. (Fig. 1137.) Annual; culms slender, straggling, rooting at the nodes, 50 to 100 cm. long, freely branching; blades lanceolate, 3 to 8 cm. long, 5 to 10 mm. wide; racemes 2 to 6, sometimes only 1, approximate; spikelets about 5 mm. long. ☉ (*Eulalia viminea* (Trin.) Kuntze.) —Shaded banks and roadsides, Ohio, Virginia, North Carolina, Kentucky, Tennessee, and Alabama. Introduced from Asia.

MICROSTEGIUM VIMINEUM var. IMBÉRBE (Nees) Honda, an awned form, found in Berks County, Pa., and Greenville, Va.

153. ARTHRÁXON Beauv.

Perfect spikelets usually awned, sessile, the secondary spikelet and its pedicel wanting or the pedicel (rarely a spikelet) developed only at the lower joints of the filiform articulate rachis; racemes terminating the branches of a dichotomously forking panicle, in appearance subdigitate or fascicled. Usually low creeping grasses with broad cordate-clasping blades and subflabellate panicles. Type species, *Arthraxon ciliaris* Beauv. Name from Greek *arthron*, joint, and *axon*, axis, alluding to the jointed rachis.

1. Arthraxon híspidus (Thunb.) Makino. Annual; culms slender, branching, decumbent or creeping, 20 to 100 cm. long; sheaths hispid; blades ovate to ovate-lanceolate, 2 to 5 cm. long, 5 to 15 mm. wide, ciliate toward base; panicles of few to several racemes, flabellate, contracting toward maturity, on filiform peduncles; rachis joints glabrous; spikelets 4 to 5 mm. long, the strong nerves aculeate-scabrous; sterile lemma with a slender geniculate awn. ☉ —Waste ground, rare, Maryland (near Washington, D. C.), Missouri (St. Louis), and Louisiana (Richland County). A. HISPIDUS var. CRYPTÁTHERUS (Hack.) Honda. (Fig. 1138.) Spikelets slightly smaller; awn wanting or included in the glumes. ☉ —Pastures, lawns, and open ground in a few localities, Pennsylvania to Florida and Tennessee; Arkansas and Washington; introduced from the Orient.

154. ANDROPÓGON L. Beardgrass

Spikelets in pairs at each node of an articulate rachis, one sessile and perfect, the other pedicellate and either staminate, neuter, or reduced to the pedicel, the rachis and pedicels of the sterile spikelets often villous, sometimes conspicuously so; glumes of fertile spikelet coriaceous, narrow, the first rounded, flat, or concave on the back, the median nerve weak or wanting, the second laterally compressed; sterile lemma shorter than the glumes, empty, hyaline; fertile lemma hyaline, narrow, entire or bifid, usually bearing a bent and twisted awn from the apex or from between the lobes; palea hyaline, small or wanting; pedicellate spikelet awnless, sometimes staminate and about as large as the sessile spikelet, sometimes consisting of 1 or 2 reduced glumes, sometimes wanting, only the pedicel present. Rather coarse grasses (perennial in the United States), with solid culms, the spikelets arranged in racemes, these numerous, aggregate on an exserted peduncle, or single, in pairs, or sometimes in threes or fours, the common peduncle usually enclosed by a spathelike sheath, these sheaths often numerous, the whole forming a compound inflorescence, usually narrow, but sometimes in dense subcorymbose masses. Standard species, *Andropogon distachyus* L. Name from Greek *aner* (*andr-*), man, and *pogon*, beard, alluding to the villous pedicels of the staminate or sterile spikelets.

Several of the species, especially in the Southwest, are regarded as good forage grasses but may soon become woody toward maturity and thus decrease in value. *Andropogon gerardi*, big bluestem, is the most important constituent of the wild hay of the prairie States. The amount is decreasing rapidly because the rich land upon which it grows is being converted into cultivated fields. Little bluestem (*A. scoparius*) is also a common constituent of wild hay.

Racemes solitary on each peduncle; rachis joints oblique and hollow at the summit.
            SECTION 1. SCHIZACHYRIUM.
Racemes 2 to numerous on each peduncle.
Racemes 2 to several on each peduncle, digitate; joints of rachis slender, sometimes with
 a shallow groove on one side.................................... SECTION 2. ARTHROLOPHIS.
Racemes several to numerous (rarely few) in a leafless panicle usually on a relatively long
 axis, the joints of the rachis flat, the margins thick and ciliate, the center very thin.
            SECTION 3. AMPHILOPHIS.

Section 1. Schizachyrium

Blades slender, terete, the upper surface a mere groove................ 1. A. GRACILIS.
Blades flat or folded, not terete.
 First glume of sessile spikelet pubescent........................... 3. A. HIRTIFLORUS.
 First glume of sessile spikelet glabrous.
  Internodes of rachis relatively thick, glabrous or ciliate at base and near apex only;
   racemes straight.
   Sessile spikelet 4 mm. long; blades about 1 mm. wide.......... 2. A. TENER.
   Sessile spikelet 6 to 9 mm. long; blades mostly 2 to 3 mm. wide.
    Sterile pedicel ciliate from below the middle to the apex; sterile spikelet about 3
     mm. long, the awn somewhat exserted.................... 4. A. SEMIBERBIS.
    Sterile pedicel ciliate only at the apex; sterile spikelet about 5 mm. long, the awn
     wanting or included.................................. 5. A. CIRRATUS.
  Internodes of rachis and sterile pedicels slender, villous throughout or nearly so;
   racemes flexuous.
   Culms tufted; rhizomes wanting (base sometimes slightly rhizomatous in *A. littoralis*).
    Sheaths and blades glabrous or nearly so (occasionally sparsely to conspicuously
     pilose in *A. scoparius*); pedicellate spikelet usually much reduced.
     Racemes nearly straight, densely villous, the hairs obscuring the rachis and
      spikelets; blades 5 to 9 cm. long, spreading................ 6. A. NIVEUS,

Racemes flexuous, the hairs not obscuring the rachis and spikelets; blades more than 10 cm. long, usually elongate.

Racemes numerous in a dense flabellate but delicate inflorescence; sessile spikelet 5 mm. long_____ 7. A. SERICATUS.

Racemes relatively few in a narrow elongate inflorescence; sessile spikelet 6 to 10 mm. long.

Culms strictly erect; sessile spikelet 6 to 8 mm. long; hairs on the rachis and sterile pedicel inconspicuous_____ 8. A. SCOPARIUS.

Culms decumbent at the base, usually very glaucous; sessile spikelet about 1 cm. long; hairs on the rachis and sterile pedicel rather prominent.
9. A. LITTORALIS.

Sheaths and blades villous; pedicellate spikelet prominent____ 10. A. DIVERGENS.

Culms solitary or few together; creeping rhizomes developed.

Sessile spikelet 8 to 10 mm. long; sterile spikelet mostly not much reduced.
11. A. MARITIMUS.

Sessile spikelet 5 to 7 mm. long; sterile spikelet much reduced.

Rachis tortuous, the joints as long as the sessile spikelets; blades 1 to 3 mm. wide, at least some of them involute_____ 12. A. RHIZOMATUS.

Rachis somewhat flexuous, but not conspicuously tortuous; blades mostly 3 to 5 mm. wide, flat_____ 13. A. STOLONIFER.

Section 2. Arthrolophis

1a. Pedicellate spikelet staminate, similar to the sessile spikelet, but awnless.

Rhizomes short or wanting; rachis joint and sterile pedicel ciliate, the joints short-hispid at base; awn of sessile spikelet 1 to 2 cm. long_____ 14. A. GERARDI.

Rhizomes well developed; rachis joint and sterile pedicel densely long-villous; awn of sessile spikelet rarely more than 5 mm. long, often obsolete_____ 15. A. HALLII.

1b. Pedicellate spikelet reduced to 1 or 2 glumes, or obsolete, the pedicel only developed; racemes silky-villous.

2a. Inflorescence very decompound, the profuse pairs of racemes aggregate in an elongate or corymbose mass; spathes rarely more than 2 mm. wide; pedicellate spikelet obsolete (see also A. virginicus var. hirsutior)_____ 27. A. GLOMERATUS.

2b. Inflorescence not conspicuously decompound nor dense (rather dense in A. virginicus var. hirsutior).

3a. Peduncle not more than 1 cm. long, the dilated spathes exceeding the 2 (occasionally 3 or 4) racemes.

Upper sheaths inflated spathelike, aggregate, the late inflorescence a flabellate tuft.
28. A. ELLIOTTII.

Upper sheaths not inflated and aggregate.

Blades of the innovations subfiliform; ligule acute, protruding from the folded blade; foliage usually glabrous_____ 23. A. PERANGUSTATUS.

Blades 2 to 5 mm. wide; ligule minute, concealed within the folded blade; foliage from obscurely to conspicuously pubescent.

Hairs of the racemes copious_____ 22. A. LONGIBERBIS.

Hairs of the racemes comparatively sparse.

Rachis joints shorter than the spikelets; branches glabrous below the spathes.
25. A. CAPILLIPES.

Rachis joints usually as long as the spikelets; branches, at least some of them, bearded below the spathes_____ 26. A. VIRGINICUS.

3b. Peduncles 2 cm. long or more.

4a. Peduncles not more than 5 cm. long, enclosed in the spathe or only slightly exserted (see also A. perangustatus).

Racemes usually not more than 15 mm. long; ultimate branchlets capillary, spreading or recurved, long-villous at summit_____ 24. A. BRACHYSTACHYS.

Racemes 2 to 5 cm. long.

Racemes 4 to 6 to a peduncle, tawny; sheaths villous_____ 16. A. MOHRII.

Racemes 2 to a peduncle, silvery or creamy white; sheaths glabrous or nearly so.

Pairs of racemes numerous; spathes inconspicuous, at least some of the peduncles as much as 5 cm. long_____ 20. A. FLORIDANUS.

Pairs of racemes not more than 10 to a culm; spathes dilated; peduncles 1 to 3 cm. long_____ 21. A. TRACYI.

4b. Peduncles or most of them 5 to 15 cm. long, long-exserted (short-exserted peduncles intermixed with long in A. elliottii and A. subtenuis).

Rachis joints longer than the spikelets; racemes 5 to 10 cm. long, conspicuously slender and flexuous_____ 30. A. CAMPYLORACHEUS.

Rachis joints not longer than the spikelets; racemes not more than 7 cm. long, usually not more than 5 cm.

Upper sheaths inflated, overlapping, conspicuous................... 29. A. ELLIOTTII.
Upper sheaths not inflated, overlapping, nor conspicuous.
Spikelets 4 mm. long; racemes very flexuous, the rachis joints nearly as long
    as the spikelets... 29. A. SUBTENUIS.
Spikelets 5 to 7 mm. long; racemes slightly or not at all flexuous, the rachis
    joints distinctly shorter than the spikelets.
Sessile spikelets about 5 mm. long, about 0.5 mm. wide, the glume deeply
    grooved; hairs of racemes not obscuring the spikelets.
                                                                    19. A. ARCTATUS.
Sessile spikelets somewhat more than 5 mm. long, 1 to 1.5 mm. wide, the
    glume concave but not grooved; hairs of racemes conspicuous to
    copious.
Racemes copiously long-villous, the hairs about twice as long as the
    spikelet and obscuring it; first glume of sessile spikelet nerveless and
    glabrous between the keels.................................. 18. A. TERNARIUS.
Racemes not copiously villous, the hairs about as long as the spikelet, not
    obscuring it; first glume of sessile spikelet scabrous and often 2-
    nerved between the keels.................................... 17. A. CABANISII.

Section 3. Amphilophis

Racemes 3 to 7, not conspicuously woolly; pedicellate spikelet about as large as the sessile
    one. Sessile spikelet often pitted.. 31. A. WRIGHTII.
Racemes few to many, conspicuously woolly; pedicellate spikelet reduced.
Panicle subflabellate, often short-exserted or included at base in a dilated sheath; racemes
    few to many on a relatively short axis; spikelets 5 to 6 mm. long.
                                                                    33. A. BARBINODIS.
Panicle oblong, usually long-exserted; racemes numerous on a long axis; spikelets 3.5 to
    6 mm. long.
First glume of sessile spikelet pitted.. 32. A. PERFORATUS.
First glume of sessile spikelet not pitted.
Spikelets awned.. 34. A. SACCHAROIDES.
Spikelets awnless.. 35. A. EXARISTATUS.

SECTION 1. SCHIZACHÝRIUM (Nees) Trin.

Branching perennials; racemes solitary on each peduncle; rachis
joints tapering to base, the apex
oblique and hollow; sessile spikelets awned, the awns twisted,
geniculate.

1. Andropogon grácilis Spreng.
(Fig. 1139.) Culms slender, wiry,
densely tufted, erect, glabrous, 20 to
0 cm. tall; blades terete, filiform;
peduncles few to several, filiform,
long-exserted, with a tuft of long
white hairs at summit; raceme 2 to
cm. long, silvery white; rachis
slender, flexuous, copiously long-
villous; sessile spikelet about 5 mm.
long, the awn 1 to 2 cm. long; pedicellate spikelet reduced to an awned
or awnless glume, the pedicel very
villous. 2⟂ —Rocky pine woods
southern Florida; West Indies.
2. Andropogon téner (Nees) Kunth.
(Fig. 1140.) Culms slender, tufted,
sometimes reclining or decumbent,
0 to 100 cm. long, the upper half

FIGURE 1139.—*Andropogon gracilis*, × 1. (Hitchcock
682, Fla.)

rather sparingly branching; blades
scarcely 1 mm. wide, flat or loosely
involute, often sparingly long-pilose
on upper surface near base; raceme
finally long-exserted, slender, sub-
terete, glabrous, 2 to 6 cm. long;

FIGURE 1140.—*Andropogon tener*, X 1. (Rolfs 986, Fla.)

base often included in the somewha
dilated sheath, the rachis joints
pedicels, and first glume of sessil
spikelet pubescent, the rachi
straight; sessile spikelet about 6 mn
long, the awn 10 to 15 mm. long
pedicellate spikelets much reducec
short-awned. ♃ (*A. oligostachyu*
Chapm.)—Pine woods, souther
Georgia and Florida; tropical Amer
ica. ANDROPOGON HIRTIFLORUS va
FꝛꝛÉNSIS (Fourn.) Hack. Blades sca
brous; sessile spikelet as much as
mm. long, the first glume minutel
papillose, the pubescence less copiou:
♃ —Canyons and rocky slope:
western Texas to Arizona; Mexicc

sessile spikelet about 4 mm. long,
the awn 7 to 10 mm. long; ♃ —
Dry pine woods and prairies, Coastal
Plain, Georgia to Florida, Texas,
and Oklahoma; tropical America.

3. **Andropogon hirtiflórus** (Nees)
Kunth. (Fig. 1141.) Culms tufted,
60 to 120 cm. tall, erect, reddish, the
upper half sparingly branching; foliage
often glaucous, the blades 2 to 4 mm.
wide; raceme 6 to 10 cm. long, the

FIGURE 1141.—*Andropogon hirtiflorus*, X 1. (Chase 4193, Fla.)

FIGURE 1142.— *Andro-*
pogon semiberbis, X 1.
(C. H. Baker 327,
Fla.)

FIGURE 1143.—*Andropog*
cirratus, X 1. (Greene 40
N. Mex.)

4. **Andropogon semibérbis** (Nee
Kunth. (Fig. 1142.) Culms usuall
in rather small tufts, 60 to 120 cn
tall, erect, pinkish, compressed, th
upper third to half freely branchinₑ
blades 2 to 4 mm. wide, glabrou:
raceme 5 to 8 cm. long, the ba:
often included in the sheath, th
rachis straight, the joints shor

hispid at base with erect hairs; sessile spikelet about 6 mm. long, the awn 10 to 15 mm. long; pedicellate spikelet much reduced, short-awned, the pedicel more or less ciliate on one margin. ⚥ —Pine woods, Florida; tropical America.

5. Andropogon cirrátus Hack. TEXAS BEARDGRASS. (Fig. 1143.) Plants pale, glaucous to purplish; culms slender, tufted, 30 to 70 cm. tall, erect, the upper half sparingly branching; blades flat, 1 to 4 mm. wide, usually scabrous; raceme exserted, 3 to 6 cm. long, the rachis straight; sessile spikelet 8 to 9 mm. long, the awn 5 to 10 mm. long; pedicellate spikelet scarcely reduced, awnless, the pedicel stiffly ciliate on one side near the summit. ⚥ — Canyons and rocky slopes, western Texas to Arizona and southern California (Jamacha); northern Mexico.

6. Andropogon níveus Swallen. (Fig. 1144.) Culms 50 to 65 cm. tall, slender, erect in small tufts; sheaths narrow, keeled, glabrous; blades 5 to 9 cm. long, 1 to 2 mm. wide, flat, spreading or reflexed; raceme 3 to 4 cm. long, the rachis nearly straight or somewhat flexuous, the joints very densely villous; sessile spikelet 5 to 6 mm. long, the first glume glabrous, obscurely bifid at the summit, 2-nerved between the keels; awn about 1 cm. long, tightly twisted below the bend; pedicellate spikelet 3 mm. long, the pedicel densely villous. ⚥ —Open sandy woods, central Florida.

7. Andropogon sericátus Swallen. (Fig. 1145.) Culms 50 to 80 cm. tall, slender, tufted, erect, profusely branching in the upper half; sheaths keeled, glabrous, mostly shorter than the internodes; blades of the innovations subfiliform, 10 to 20 cm. long, the culm blades broader, 2 to 3 mm. wide, folded; spathes very inconspicuous; peduncles filiform, 4 to 6 cm. long; raceme 3 cm. long, scarcely exserted, the rachis flexuous, conspicuously hairy; sessile spikelet 5 mm. long, the first glume prominently 2-keeled, sulcate; awn 15 to

FIGURE 1144.—*Andropogon niveus,* × 1. (Type.)

FIGURE 1145.—*Andropogon sericatus,* × 1. (Type.)

20 mm. long, geniculate, tightly twisted below the bend; pedicellate spikelet 3 to 4 mm. long, including the short awn. ⚥ —Ramrod Key, Fla.

8. Andropogon scopárius Michx. LITTLE BLUESTEM. (Fig. 1146.) Plants green or glaucous, often purplish, culms tufted, from slender to robust, compressed, 50 to 150 cm. tall, erect, the upper half freely branching; sheaths and blades commonly glabrous or nearly so, frequently sparsely pilose at their junction, rarely pubescent to villous throughout, the

FIGURE 1146.—*Andropogon scoparius*. Plant, × ½; pair of spikelets, × 5. (Amer. Gr. Natl. Herb. 268, D. C.)

blades 3 to 6 mm. wide, flat; raceme 3 to 6 cm. long, mostly curved, the filiform peduncles mostly wholly or partly included in the sheaths, commonly spreading, the rachis slender, flexuous, pilose, sometimes copiously so; sessile spikelet mostly 6 to 8 mm. long, scabrous, the awn 8 to 15 mm. long; pedicellate spikelet usually reduced, short-awned, spreading, the pedicel pilose. ♃ —Prairies, open woods, dry hills, and fields, Quebec and Maine to Alberta and Idaho, south to Florida and Arizona. A form with villous foliage has been segregated as *A. scoparius* var. *villosissimus* Kearney (*Schizachyrium villosissimum* Nash). *Schizachyrium acuminatum* Nash was described from a specimen, otherwise typical, having spikelets 10 mm. long. Specimens with spikelets 4.5 to 6 mm. long and reduced sterile spikelets have been differentiated as var. *frequens* Hubb., and northern specimens with few racemes, relatively distant spikelets 7 to 8 mm. long, and sterile spikelets, including awn, 6.5 to 10 mm. long, as var. *septentrionalis* Fern. and Grisc. Specimens from Virginia to South Carolina, collected from June 8 to September 13, have been segregated as *Andropogon praematurus* Fernald.[15]

ANDROPOGON SCOPARIUS var. NEO-MEXICÁNUS (Nash) Hitchc. (Fig. 1147.) Rachis and pedicels copiously villous, the rachis mostly nearly straight. In the Southwest the species verges into this variety. ♃ (*Schizachyrium neomexicanum* Nash.)—Sandy soil and rocky hills, Texas to Arizona.

9. Andropogon littorá is Nash. (Fig. 1148.) Resembling *A. scoparius*, but culms more compressed, with broad, keeled, overlapping lower sheaths, often bluish-glaucous, the flat tufts crowded on a slender rhizome, decumbent or bent at base; blades 4

FIGURE 1147.—*Andropogon scoparius* var. *neomexicanus*, × 1 (Wooton, N. Mex.)

to 6 mm. wide; rachis joints and pedicels copiously long-villous. ♃ —Sandy shores, Ontario; Massachusetts and Staten Island, N. Y., to North Carolina; Ohio (Sandusky); Indiana (sand dunes of Lake Michigan); southeastern Texas. A short specimen without rhizomes and with a rather crowded inflorescence, from Elizabeth Islands, Mass., has been described as *A. scoparius* var. *ducis* Fern. and Grisc.

10. Andropogon divérgens (Hack.) Anderss. ex Hitchc. (Fig. 1149.) Culms rather robust, 80 to 120 cm. tall, sparingly branching toward the summit; sheaths grayish villous, the lower crowded, compressed-keeled; blades rather firm, 3 to 6 mm. wide, villous, elongate, flat or folded; raceme mostly 3 to 4 cm. long, mostly 6- to 8-jointed, rather stout, usually partly included, the rachis slightly to strongly flexuous, rather stout, the joints long-ciliate on the upper half, rarely throughout, and with a short tuft of hairs at the summit, the pedicel long-ciliate on the upper half; sessile spikelet 6 to 8 mm. long, minutely roughened, the awn 5 to 10 mm. long; pedicellate spikelet about as long as the sessile one, the first

---

[15] There are numerous collections of *A. scoparius* throughout its range made in June, July, and August. *A. praematurus*, with a single pedicellate spikelet at each joint of the rachis, was differentiated from *A. scoparius*, which is said to have 2 pedicels. In *A. scoparius*, as in all species of *Andropogon*, each rachis joint bears 1 sessile and 1 pedicellate spikelet. The second "truncate" pedicel described was undoubtedly a rachis joint from which a sessile spikelet had fallen.

8 to 10 mm. long, the awn 8 to 12 mm. long; pedicellate spikelet scarcely reduced, short-awned. ♃ —

FIGURE 1148.—*Andropogon littoralis,* × 1. (Burk, N. J.)

FIGURE 1149.—*Andropogon divergens,* × 1. (Tharp 3094, Tex.)

glume awn-tipped. ♃ —Pinelands, Mississippi to Arkansas and Texas.

11. Andropogon marítimus Chapm. (Fig. 1150.) Culms solitary, compressed, ascending from a decumbent, short-noded base, 50 to 60 cm. long, branching toward the ends, and with long creeping rhizomes; sheaths overlapping on the short internodes, strongly keeled, commonly reddish; blades 3 to 5 mm. wide, often folded and reflexed, the midnerve deeply impressed; raceme 4 to 6 cm. long, the base included in the dilated sheath, the rachis very flexuous, the joints and pedicels copiously long-ciliate except at base; sessile spikelet

FIGURE 1150.—*Andropogon marítimus,* × 1. (Chapman, Fla.)

Sandy ground along the Gulf coast, western Florida, Mississippi (Horn Island), and Louisiana (Last Island).

12. Andropogon rhizomátus Swallen. (Fig. 1151.) Culms 50 to 70 cm. tall, scattered or in small dense tufts, erect from short rhizomes, sparingly branching above the middle; sheaths rounded or obscurely keeled, much longer than the internodes; blades 10 to 25 cm. long, 1 to 3 mm. wide, flat or loosely involute, glabrous; raceme 2 to 3 cm. long, strongly flexuous, partly enclosed or exserted from the very inconspicuous spathe; peduncles 3 to 7 cm. long; sessile spikelet 5 to 6 mm. long, the first glume rounded on the back, obscurely keeled near the summit; awn 8 to 10 mm. long, geniculate, twisted below the bend; pedicellate spikelet 2 to 3 mm. long. ♃ — Rocky ground, southern Florida.

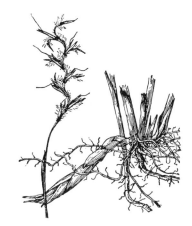

FIGURE 1151.—*Andropogon rhizomatus,* × 1. (Type.)

13. Andropogon stolónifer (Nash) Hitchc. (Fig. 1152.) Resembling *A. scoparius;* culms as much as 1.5 m. tall, solitary or few in a tuft, with slender, creeping scaly rhizomes; foliage glabrous to villous, the blades flat, as much as 5 mm. wide; racemes 3 to 4 cm. long, the slender rachis joints and pedicels silky villous; first glume of both sessile and pedicellate spikelets sometimes bifid at apex; sessile spikelet 5 to 7 mm. long, scabrous, especially toward the summit and on the margins. ♃ (*Schizachyrium triaristatum* Nash.)—Sandy woods, southern Georgia, Florida, and Alabama.

SECTION 2. ARTHRÓLOPHIS Trin.

Branching perennials; racemes 2 to few on each peduncle; rachis joints slender, mostly pubescent; sessile spikelet awned.

14. Andropogon gerárdi Vitman. BIG BLUESTEM. (Fig. 1153.) Plants often glaucous; culms robust, often in large tufts, sometimes with short rhizomes, 1 to 2 m. tall, usually sparingly branching toward the summit; lower sheaths and blades sometimes villous, occasionally densely so, the blades flat, elongate, mostly 5 to

FIGURE 1152.—*Andropogon stolonifer,* × 1. (Fredholm, 6122 Fla.)

10 mm. wide, the margins very scabrous; racemes on the long-exserted terminal peduncle mostly 3 to 6, fewer on the branches, 5 to 10 cm. long, usually purplish, sometimes yellowish; rachis straight, the joints and pedicels stiffly ciliate on one or both margins, the joints hispid at base; sessile spikelet 7 to 10 mm. long, the first glume slightly sulcate, usually scabrous, the awn geniculate and tightly twisted below, 1 to 2 cm. long; pedicellate spikelet not

FIGURE 1153.—*Andropogon gerardi*. Plant, × ½; pair of spikelets, × 5. (Amer. Gr. Natl. Herb. 255, D. C.)

reduced, or but slightly so, awnless, staminate. ♃ —(*A. provincialis* Lam. not Retz., *A. furcatus Muhl.*)— Dry soil, prairies and open woods, Quebec and Maine to Saskatchewan and Montana, south to Florida, Wyoming, Utah, and Arizona; Mexico. An important forage grass in the prairie States of the Mississippi Valley, and a constituent of prairie hay.

15. Andropogon hállii Hack. SAND BLUESTEM. (Fig. 1154.) Resembling *A. gerardi*, but with creeping rhizomes; racemes conspicuously villous, the hairs grayish to pale golden; awn of sessile spikelet rarely more than 5 mm. long, often obsolete. ♃ —Sand hills and sandy soil, North Dakota and eastern Montana to Texas, Wyoming, Utah, and Arizona; Iowa. Intergrades with *A. gerardi*. A form with yellow-villous racemes and awns 5 to 10 mm. long has been segregated as *A. chrysocomus* Nash.

16. Andropogon móhrii (Hack.) Hack. ex Vasey. (Fig. 1155.) Culms stout, compressed, tufted, erect, 80 to 130 cm. tall, the upper half sparingly to rather freely branching; leaves villous, the lower sheaths

FIGURE 1155.—*Andropogon mohrii*, × 1. (Mohr, Ala.)

strongly keeled and glabrous at base, the blades elongate, 3 to 5 mm. wide; inflorescence narrow, the branches approximate, the ultimate branchlets short, densely bearded at summit, the purplish spathes 4 to 6 cm. long;

FIGURE 1154.—*Andropogon hallii*, × 1. (Hitchcock 584, Kans.)

FIGURE 1156.—*Andropogon cabanisii*, × 1. (Fredholm 6416, Fla.)

racemes mostly 4, tawny, 2 to 4 cm. long, on peduncles mostly about 2 cm. long, or the terminal ones sometimes long-exserted; rachis scarcely flexuous, the joints shorter than the spikelets, copiously long-villous; sessile spikelet 4 to 5 mm. long, the awn loosely twisted below, 1.5 to 2 cm. long; pedicel long-villous, the spikelet reduced to a minute glume. ♃ —Wet pine woods and sandy seacoast, Virginia to Georgia and Louisiana.

FIGURE 1157.—*Andropogon ternarius,* ✕ 1. (Chase 4557, N. C.)

17. Andropogon cabanísii Hack.

(Fig. 1156.) Culms in small tufts, erect, 80 to 150 cm. tall, the upper half bearing long slender branches; sheaths villous to nearly glabrous; blades 2 to 3 mm. wide; inflorescence loose; racemes 2, pale grayish tawny, with about 15 joints, 4 to 7 cm. long, on slender long-exserted peduncles, the spathes narrow, inconspicuous, or a few occasionally dilated; rachis not flexuous or but slightly so, the joints

shorter than the spikelets, long-villous; sessile spikelets 6 to 7 mm. long, the first glume firm, scabrous and often 2-nerved between the keels, the awn twisted below, about 1.5 cm. long; pedicel long-villous, the spikelet reduced to a slender glume or obsolete. ♃ —Dry pine woods, peninsular Florida.

18. Andropogon ternárius Michx.

(Fig. 1157.) Culms tufted, erect, 80 to 120 cm. tall, the upper half to two-thirds branching, the branches usually long, slender and erect; leaves often purplish-glaucous, glabrous, or the lower loosely villous, the blades 2 to 4 mm. wide; inflorescence elongate, loose, of few to many pairs of silvery to creamy or grayish feathery racemes, usually on long-exserted peduncles from slender inconspicuous spathes, some of the lateral peduncles often short, from dilated spathes, rarely most of them so; racemes 3 to 6 cm. long, with mostly less than 12 joints, the rachis not flexuous, the joints shorter than the spikelets, copiously long-villous; sessile spikelets 5 to 7 mm. long, glabrous and nerveless between the keels, the awn twisted below, 1.5 to 2 cm. long;

FIGURE 1158.—*Andropogon arctatus,* ✕ 1. (Chapman, Fla.)

stamens 3; pedicel long-villous, the spikelet obsolete or nearly so. ♃ —Dry sandy soil, open woods, mostly Coastal Plain, Delaware to Kentucky and Kansas, south to Florida and Texas. Variable in the density and length of pubescence on the rachis and pedicels, the less hairy specimens verging toward *A. arctatus.*

19. Andropogon arctátus Chapm. (Fig. 1158.) Resembling *A. ternarius;* culms 1 to 1.5 m. tall; the blades often wider and firmer; branches of the inflorescence rather more slender; racemes 3 to 5 cm. long, tawny; sessile spikelets 4 to 5 mm. long, brown, the awn 1 to 5 cm. long; first glume concave, the pale or tawny hairs of rachis and pedicels shorter and less copious than in *A. ternarius;* sessile spikelet 5 mm. long, 0.5 mm. wide, the glume grooved; stamen 1. ♃ —Low pine woods, Florida.

20. Andropogon floridánus Scribn. (Fig. 1159.) Culms often stout, 1 to 1.8 m. tall; the upper one-third to half bearing long slender branches; blades elongate, 2 to 6 mm. wide; inflorescence loosely subcorymbose of usually numerous pairs of silvery-white to creamy racemes on sub-capillary peduncles, mostly 2 to 8 cm. long, included in very slender spathes or exserted, the ultimate branchlets filiform, often long-ciliate toward the summit; racemes 3 to 4 cm. long, the slender rachis not flexuous, the joints a little shorter than the spikelets, rather copiously long-villous; sessile spikelets 4 to 4.5 mm. long, the delicate awn straight, 6 to 10 mm. long; pedicel long-villous, the spikelet obsolete. ♃ —

FIGURE 1159.—*Andropogon floridanus,* X 1. (Type coll.)

FIGURE 1160.—*Andropogon tracyi,* X 1. (Type.)

Low pine woods, Florida. An occasional peduncle bears 3 racemes.

21. Andropogon trácyi Nash. (Fig. 1160.) Culms in small tufts, slender, erect, the upper third sparingly branching; sheaths keeled, narrow, glabrous or nearly so; blades 2 to 3 mm. wide, sometimes ciliate toward base; inflorescence of 8 to 10 relatively distant racemes, the slender ultimate branches often recurved, cence on the average less compound, the racemes mostly 3, more copiously long-villous, the spikelets 4 to 4.5 mm. long. ♃ —Pine woods, Georgia and Florida. Intergrades with *A. virginicus.*

23. Andropogon perangustátus Nash. (Fig. 1162.) Culms in small tufts, slender, wiry, erect, the upper third to half sparingly branching; lower sheaths keeled, very narrow,

FIGURE 1161.—*Andropogon longiberbis,* × 1. (Garber, Fla.)

FIGURE 1162.—*Andropogon perangustatus,* × 1. (Fredholm 6072, Fla.)

the dilated spathes 4 to 6 cm. long, attenuate below, the enclosed peduncle 1 to 3 cm. long; ultimate branchlets long-bearded toward the summit; racemes 2 or 3, feathery, 2 to 4 cm. long, the very slender flexuous rachis and the pedicel copiously long-villous; sessile spikelet about 4 mm. long, the awn loosely twisted below, 1 to 2 cm. long; pedicellate spikelet obsolete. ♃ — Pine woods, Georgia and Florida to Louisiana. Resembling *A. longiberbis,* mostly more slender and with nearly glabrous foliage.

22. Andropogon longibérbis Hack. (Fig. 1161.) Resembling *A. virginicus;* sheaths, especially of the innovations, appressed grayish-villous; inflores-

occasionally sparsely villous; ligule about 1.5 mm. long, firm; blades mostly folded, subfiliform, flexuous, glabrous or rarely pilose; inflorescence slender, of few to several racemes, resembling that of slender specimens of *A. virginicus,* the peduncles usually short but the spathes sometimes attenuate to base, the peduncle 1 to 2 cm. long; racemes as in *A. virginicus.* ♃ —Bogs and moist pine woods, Florida and Mississippi.

24. Andropogon brachystachys Chapm. (Fig. 1163.) Culms tufted, erect, 1 to 1.5 m. tall, the upper half loosely branching; sheaths crowded at base, broad, strongly keeled; blades mostly folded, 4 to 6 mm.

wide; inflorescence decompound, loose, the ultimate capillary branchlets commonly recurved, long-villous toward the summit; spathes slender, the long peduncles often exserted from the summit; racemes 2, flexuous, mostly 1 to 1.5 cm. long, the rachis joint and pedicel long-villous; sessile spikelet about 4 mm. long, the awn scarcely 1 cm. long. ♃ —Moist pine woods, southern Georgia and Florida. The racemes are frequently affected by a smut, making them shorter and denser, reducing the size of the spikelet and the awn. The inflorescence resembles that of *A. capillipes*, but the racemes mostly more numerous; the ultimate branchlets are long-villous toward the summit and the spikelets larger.

25. Andropogon capíllipes Nash. (Fig. 1164.) Plants conspicuously glaucous; culms tufted, slender, erect, 60 to 100 cm. tall, the upper third to half with few to several slender branches; sheaths crowded at base,

FIGURE 1164.—*Andropogon capillipes*, × 1. (Curtiss 3638b, Fla.)

keeled, chalky-glaucous; blades mostly folded, 2 to 4 mm. wide; inflorescence narrow but loose, the branches often flexuous to zigzag, the ultimate capillary branchlets finally spreading or recurved, glabrous, the dilated purplish-brown spathes 2 to 3.5 cm. long, glabrous; racemes 2, less flexuous than in *A. virginicus*, 1 to 2.5 cm. long; rachis joint about half as long as the sessile spikelet, the pedicel about equaling the spikelet, both copiously long-villous; sessile spikelet 3 mm. long, the delicate straight awn about 1 cm. long. ♃ — Sandy pine and oak woods, southern North Carolina, South Carolina, and Florida.

26. Andropogon virgínicus L. BROOMSEDGE. (Fig. 1165.) Culms erect, 50 to 100 cm. tall, usually in rather small tufts, the upper two-thirds mostly freely branching; lower sheaths compressed, keeled, equitant; sheaths glabrous or more or less pilose along the margins, occasionally conspicuously so; ligule strongly ciliate; blades flat or folded, 2 to 5 mm. wide, pilose on the upper surface toward base; inflorescence elongate, narrow, the 2 to 4 racemes 2 to 3 cm. long, partly included and shorter than the inflated tawny to bronze

FIGURE 1163.—*Andropogon brachystachys*, × 1. (Curtiss 3632, Fla.)

FIGURE 1165.—*Andropogon virginicus.* Plant, × ½; spikelet with rachis joint and pedicel, × 5. (Earle 4, Ala.)

spathes; rachis very slender, flexuous, long-villous; sessile spikelet about 3 mm. long, the delicate straight awn 1 to 2 cm. long; pedicel long-villous, its spikelet obsolete or nearly so. ♃ —Open ground, old fields, open woods, sterile hills, and sandy soil, Massachusetts, New York, Michigan, and Kansas, south to Florida and Texas; California; Mexico, Central America, West Indies. ANDROPOGON VIRGINICUS var. HIRSÚTIOR (Hack.) Hitchc. Flowering branches more numerous than in the species, the inflorescence often rather dense, resembling that of *A. glomeratus*, but the spathes mostly larger and the peduncles usually shorter. ♃ — Moist meadows and old fields, Florida to Texas; Tennessee; Oklahoma; Mexico. Intergrades with *A. virginicus* and appears to be intermediate between that and *A. glomeratus*. ANDROPOGON VIRGINICUS var. GLAU-CÓPSIS (Ell.) Hitchc. Resembling the species, but foliage, especially the lower sheaths, very glaucous; inflorescence sometimes as dense as in var. *hirsutior*, the spathes dull purple. ♃ (*A. glaucopsis* Nash.)—Moist sandy soil and low pine barrens, Virginia to Florida and Mississippi.

27. Andropogon glomerátus (Walt.) B. S. P. BUSHY BEARDGRASS. (Fig. 1166.) Culms erect, 50 to 150 cm. tall, compressed, with broad keeled overlapping lower sheaths, the flat tufts often forming dense, usually glaucous clumps, the culms from freely to bushy-branching toward the summit; sheaths occasionally villous; blades elongate, 3 to 8 mm. wide; inflorescence dense, feathery, from flabellate to oblong, the paired racemes 1 to 3 cm. long, about equaling the slightly dilated spathes, the enclosed peduncle and ultimate branchlets long-villous, the peduncle at least 5 mm. long, often longer; rachis very slender, flexuous, long-villous; sessile spikelets 3 to 4 mm. long, the awn straight, 1 to 1.5 cm. long; sterile spikelet reduced to a subulate

FIGURE 1166.—*Andropogon glomeratus*, branchlet of inflorescence, × 1. (Hitchcock 437, Fla.)

glume or wanting, the pedicel slender, long-villous. ♃ —Low moist ground, marshes, and swamps, Massachusetts to Florida, west to Ken-

FIGURE 1167.—*Andropogon elliottii*, × 1. (Commons 115, Del.)

tucky, southern California, and Nevada; West Indies, Yucatan, Central America.

28. Andropogon ellióttii Chapm. ELLIOTT BEARDGRASS. (Fig. 1167.) Culms tufted, erect, 30 to 80 cm. tall, at first nearly simple, later branching toward the summit; lower sheaths keeled, rather narrow, commonly loosely pilose, those near the summit inflated and spathelike, crowded, the very short internodes densely bearded; blades flat, 3 to 4 mm. wide; primary inflorescence of few to several racemes, mostly in pairs, rarely threes or fours, on filiform, often strongly flexuous peduncles, long-exserted from inconspicuous spathes, these on slender branchlets borne in the axils of the broad spathelike sheaths of the main culm; secondary inflorescence of numerous pairs of racemes on short peduncles subtended by broad spathes, these on short, bearded, often fascicled, branchlets borne in the axils of the spathelike sheaths of the main culm and short primary branches, the whole forming a series of flabellate tufts with conspicuous purplish to copper-brown spathes, 5 to 10 mm. wide, much exceeding the feathery racemes; racemes flexuous, 3 to 4 rarely to 5 cm. long, the slender rachis joints and pedicels long-villous; sessile spikelets 4 to 5 mm. long, those of the late enclosed racemes cleistogamous, the awn loosely twisted, 10 to 15 mm. long; pedicellate spikelets obsolete or nearly so. ♃ —Open ground, old fields, and open woods, mostly on the Coastal Plain, New Jersey to Florida and Texas, north to southern Missouri, Illinois, Indiana, and Ohio; British Honduras. The flattened ferruginous upper sheaths are conspicuous in winter. The characteristic plant is very striking, but occasional individuals occur with less aggregate upper sheaths, and others with scarcely dilated sheaths, aggregate or scarcely aggregate. This form, which has been distinguished as *A. elliottii* var. *graci-*

lior Hack., appears to merge into *A. subtenuis* Nash.

29. Andropogon subténuis Nash. (Fig. 1168.) Culms in small tufts, slender, erect, 40 to 70 cm. tall, the upper third sparingly branching; foliage glabrous or nearly so, the blades 1.5 to 2 mm. wide; inflorescence narrow, of few to several pairs of racemes on elongate filiform peduncles short-exserted from near the

FIGURE 1168.—*Andropogon subtenuis*, × 1. (Tracy 4701, Miss.)

summit of the elongate slender spathe, the ultimate branches sometimes long-villous toward the summit; racemes 2, flexuous, 2 to 3 cm. long, very like the primary racemes of *A. elliottii;* spikelets 4 mm. long. ⚇ —Dry sandy soil, northern Florida to Louisiana. Possibly a form of *A. elliottii* in which the enlarged sheaths and cleistogamous inflorescence are not developed.

30. Andropogon campylorácheus Nash. (Fig. 1169.) Culms tufted, erect, 40 to 80 cm. tall, simple or with a few branches about the middle;

FIGURE 1169.—*Andropogon campyloracheus,* × 1. (Combs 677, Fla.)

sheaths and lower part of the blades appressed-villous, the blades about 2 mm. wide; racemes 2 to 4, mostly 2, on long flexuous peduncles exserted from long narrow spathes, the slender rachis very flexuous, the joints and pedicels much longer than the sessile spikelet, long-villous, the lowermost rachis joint often elongate; sessile spikelet 5 to 6 mm. long, slender, the awn loosely twisted, mostly about 2 cm. long; pedicellate spikelet reduced to a slender glume or obsolete. ⚇ —Dry sandy pine woods, Florida, Mississippi, and Louisiana.

SECTION 3. AMPHÍLOPHIS Trin.

Perennials, simple or sparingly branching; racemes several to numerous in a leafless panicle, at least the lower racemes short-peduncled, mostly on a relatively long axis, rachis straight, the joints and pedicels flat, with thick bearded margins, the center subhyaline.

31. Andropogon wríghtii Hack. (Fig. 1170.) Plants somewhat glaucous; culms tufted, 50 to 100 cm. tall, simple, the nodes usually hispid; blades flat, 3 to 5 mm. wide, tapering to a fine point; racemes 3 to 7, suberect, mostly 3 to 6 cm. long, green or tawny, not conspicuously woolly, the hairs of rachis joints and pedicels much shorter than the spikelets; peduncle usually long-exserted; sessile spikelet about 6 mm. long, short-pilose at base, the first glume several-nerved toward the summit, stiffly short-ciliate on the keels above; awn twisted below, geniculate, 10 to 15 mm. long; pedicellate spikelet about as large as the sessile one, awnless. ⚇ —Rocky hills and mesas, southern New Mexico, and northern Mexico. An occasional spikelet is found with a pitted first glume. In Mexican specimens the glumes are commonly pitted.

32. Andropogon perforátus Trin. ex Fourn. (Fig. 1171.) Culms densely tufted, geniculate at base, 50 to 100 cm. tall, simple or with a few leafy shoots at base; nodes from obscurely appressed-pubescent to densely short-bearded; blades 2 to 4 mm. wide, the apex attenuate; racemes few to several, mostly 5 to 7 cm. long, one or more of them on slender individual peduncles aggregate on a short axis, the common peduncle usually long-exserted; margins of rachis joints and pedicels densely long-villous; sessile spikelet 4 to 6 mm. long, short-pilose at base, the first glume sparsely hairy and with a small pit like a pinhole; awn twisted below, geniculate, 2 to 2.5 cm. long; pedicellate spikelet reduced. ⚇ —Mesas, rocky

FIGURE 1170.—*Andropogon wrightii,* × 1. (Metcalfe 1371, N. Mex.)

hills, and dry woods, southern Texas; Mexico.

33. Andropogon barbinódis Lag. (Fig. 1172.) Culms tufted, 40 to 120 cm. tall, spreading to ascending, often branching below, the nodes bearded with short spreading hairs; sheaths sparsely hairy in the throat, foliage

FIGURE 1171.—*Andropogon perforatus,* × 1. (Hitchcock 5218, Tex.)

otherwise glabrous or nearly so, the blades 2 to 7 mm. wide, scabrous; panicles from rather long-exserted to included at base, those of the branches often partly included in dilated sheaths, silvery to creamy white, silky, subflabellate, mostly 7 to 10 cm. long; racemes several to many, or sometimes few on the branches, 2 to 6 cm. long, the common axis usually shorter than the racemes, rarely longer; rachis joints and pedicels copiously long-villous, the hairs on the average longer than in *A. saccharoides;* spikelets 5 to 6 mm. long, the awn twisted below, geniculate, 20 to 25 mm. long; pedicellate spikelet reduced. ♃ —Mesas, rocky slopes, and open ground, Oklahoma and Texas to California and Arizona, south through Mexico. Has been confused with *A. saccharoides,* differing chiefly in the subflabellate panicle and larger spikelets.

34. Andropogon saccharoídes Swartz. SILVER BEARDGRASS. (Fig. 1173.) Culms tufted, 60 to 130 cm. tall, erect or ascending, often branching below, the nodes from appressed hispid to glabrous; foliage commonly glaucous, glabrous or nearly so, the blades 3 to 6 mm. wide; panicle long-exserted or those of the branches short-exserted, silvery white, silky, dense, oblong, mostly 7 to 15 cm. long; racemes 2 to 4 cm. long, the common axis mostly at least twice as long, but readily breaking; rachis joints and pedicels long-villous; spikelets about 4 mm. long, the delicate awn twisted below, geniculate, 10 to 15 mm. long; pedicellate spikelet reduced. ♃ —Prairies and rocky slopes, especially in limestone areas, Missouri to Colorado, and Alabama to Arizona; Mexico and West Indies to Brazil. Our plants, which have been differentiated as *A. torreyanus* Steud., are more freely branching than the typical form of the West Indies.

35. Andropogon exaristátus (Nash) Hitchc. (Fig. 1174.) Resembling *A. saccharoides;* panicle slender, spike-

FIGURE 1172.—*Andropogon barbinodis.* Plant, X ½; pair of spikelets, X 5. (Amer. Gr. Natl. Herb. 549, Ariz.)

FIGURE 1173.—*Andropogon saccharoides*, × 1. (Hitchcock 5370, Tex.)

lets slightly smaller, awnless or nearly so; rare. ♃ —Low open ground, southern Louisiana and eastern Texas.

Andropogon pertúsus (L.) Willd. Culms ascending, branching; racemes few to several, aggregate on a short axis, sparsely villous; first glume pitted; awn of sessile spikelet geniculate, 10 to 15 mm. long. ♃ —A troublesome weed in lawns and pastures, State College, Miss.; West Indies; introduced from the Old World.

Andropogon seríceus R. Br. Culms slender, leafy, 50 to 80 cm. tall, branching; nodes bearded; racemes 2 to 7, aggregate, 3 to 5 cm. long, nodding from a very slender peduncle,

conspicuously silky; sessile and pedicellate spikelets about equal, the first glumes strongly several-nerved; awn of sessile spikelet twisted, geniculate, 20 to 30 mm. long. ♃ — Spontaneous on roadside banks, Cameron County, Tex. Introduced from Australia.

Andropogon nodósus (Willem.) Nash. Culms ascending from a decumbent base, leafy, branching; nodes bearded; peduncle villous below the inflorescence; racemes 1 to 4, approximate, the sterile spikelets as conspicuous as the fertile, giving the appearance of a flat 2-ranked scaly spike, the first glume broad, obtuse, many-nerved; awns slender, twisted and bent, 15 to 25 mm. long. ♃ — Cultivated at experiment stations, spontaneous along ditches in southeast Texas, and formerly near Miami, Fla. Has been confused with *A. annulatus* Forsk., cultivated under that name, and called "Angleton grass." Established in a few of the West Indian islands. Introduced from Old World tropics.

Andropogon ischaémum L. Culms ascending, 70 to 100 cm. tall; nodes glabrous; racemes nodding, few to several, on slender peduncles aggregate or somewhat distant on a slender axis 3 to 5 cm. long, the sterile spikelets as conspicuous as the fertile, the rachis and pedicels silky-ciliate; awns slender, twisted and bent, about 15 mm. long. ♃ —Cultivated at experiment stations, reported to be a promising pasture grass in southern

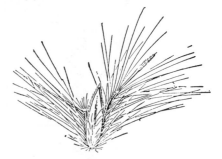

FIGURE 1174.—*Andropogon exaristatus*, × 5. (Type.)

Texas; adventive in Kansas, Knoxville, Tenn., and in wool waste, Yonkers, N. Y.

CYMBOPÓGON Spreng. OILGRASS

Closely allied to *Andropogon;* the pairs of racemes included in an inflated spathe, the spathes in a large compound inflorescence; sessile and pedicellate spikelets of lower pair alike, well developed, but staminate or neuter. Robust mostly aromatic perennials, including the oilgrasses of commerce. The most important are

FIGURE 1175.—*Vetiveria zizanioides,* × ½. (Hitchcock 9435, Jamaica.)

CYMBOPOGON NÁRDUS (L.) Rendle, citronella grass, nard grass, in which the first glume of the sessile spikelet is flat on the back and, C. CITRÁTUS (DC.) Stapf, lemon grass, in which the first glume is concave on the back. These species are sometimes cultivated in gardens in southern Florida and southern California but do not flower there. Name from Greek *kumbe*, boat, and *pogon*, beard, alluding to the boat-shaped spathes.

VETIVÉRIA Bory

Vetiveria zizanioídes (L.) Nash. VETIVER. (Fig. 1175.) Robust densely tufted perennial with simple culms and large erect panicles, the slender whorled branches ascending, naked at base, the awnless spikelets muricate. Also called khus-khus and khas-khas. ♃ —Native of the Old World, frequently cultivated in tropical America for hedges and for the aromatic roots, these being used for making screens and mats which are fragrant when wet. Vetiver oil is much used in perfumery. Escaped

from cultivation in Louisiana. Name from *vettiver*, the native Tamil name.

155. HYPARRHÉNIA Anderss. ex Stapf

Spikelets in pairs as in *Andropogon*, but spikelets of the lower pairs alike, sterile, and awnless; fertile spikelets 1 to few in each raceme, terete or flattened on the back (keeled toward the summit in *Hyparrhenia rufa*), the base usually elongate into a sharp callus, the fertile lemma with a strong geniculate awn; sterile spikelets awnless; racemes in pairs, on slender peduncles, and subtended by a spathe. Tall perennials, the pairs of racemes and their spathes more or less crowded, forming a rather large elongate inflorescence. Type species, *Hyparrhenia pseudocymbaria* (Steud.) Stapf. Name from Greek *hypo*, under, and *arren*, masculine, alluding to the pair of staminate spikelets at the base of the raceme.

1. Hyparrhenia rúfa (Nees) Stapf. (Fig. 1176.) Culms erect, rather stout, 1 to 2.5 m. tall; blades flat,

FIGURE 1176.—*Hyparrhenia rufa*, × 1. (Moldenke 243, Fla.)

elongate, 2 to 8 mm. wide, sometimes wider, very scabrous on the margins; inflorescence 20 to 40 cm. long, the pairs of racemes on long slender flexuous peduncles; racemes about 2 cm. long, reddish brown; fertile spikelets mostly 5 to 7 in each raceme, 3 to 4 mm. long, flattened from the back, pubescent with dark-red hairs, the pedicels and rachis joint ciliate with red hairs; awn 15 to 20 mm. long, twice geniculate, twisted, red brown, hispidulous. ♃ — Tropics of the Old World; introduced in tropical America; sparingly culti-

vated in Florida (where it has escaped) and along the Gulf coast. Adapted to conditions in the regions mentioned, but only moderately valuable as a forage grass. The native name in Brazil is jaraguá.

Hyparrhenia hírta (L.) Stapf. Usually not more than 1 m. tall; blades usually less than 3 mm. wide, more or less involute, flexuous; racemes whitish or grayish silky-villous. ♃ —Warmer parts of the Old World; cultivated at the Florida State Experiment Station and probably elsewhere. Appears to have little forage value.

156. SÓRGHUM Moench

Spikelets in pairs, one sessile and fertile, the other pedicellate, sterile but well developed, usually staminate, the terminal sessile spikelet with two pedicellate spikelets. Tall or moderately tall annuals or perennials, with flat blades and terminal panicles of 1- to 5-jointed tardily disarticulating racemes. Type species, *Sorghum saccharatum* (L.) Moench. Name from *Sorgho*, the Italian name of the plant.

The sorghums and Johnson grass sometimes produce cyanogenetic compounds in sufficient abundance, especially in second growth, to cause prussicacid poisoning in grazing animals. The leaves are often splotched with purple, due to a bacterial disease.

Plants perennial.. 1. S. HALEPENSE.
Plants annual.. 2. S. VULGARE.

1. Sorghum halepénse (L.) Pers. JOHNSON GRASS. (Fig. 1177.) Culms 50 to 150 cm. tall, from extensively creeping scaly rhizomes; blades mostly less than 2 cm. wide; panicle open, 15 to 50 cm. long; sessile spikelet 4.5 to 5.5 mm. long, ovate, appressed-silky, the readily deciduous awn 1 to 1.5 cm. long, geniculate, twisted below; pedicellate spikelet 5 to 7 mm. long, lanceolate. ♃ (*Holcus halepensis* L.)—Open ground, fields, and waste places, Massachusetts to Iowa and Kansas, south to Florida and Texas, west to southern California; native of the Mediterranean region found in the tropical and warmer regions of both hemispheres. Cultivated for forage, but on account of the difficulty of eradication it becomes a troublesome weed.

2. Sorghum vulgáre Pers. SORGHUM.[16] Differing from *S. halepense*

in being annual and more robust. ⊙ (*Holcus sorghum* L.)—This species has been cultivated in warmer regions since prehistoric times for the seed, which has been used for food, for the sweet juice, and for forage. In the United States it is cultivated under the general name of sorghum. There are many varieties or races of cultivated sorghums, all of which have the same chromosome number (10) and which fall naturally into distinct groups, the chief of which (in the United States) are sorgo, kafir, durra, milo, feterita, shallu, kaoliang, and broomcorn. Sorgo includes the varieties known collectively as sweet or saccharine sorghums, in which the juice in the stems is

[16] For elaboration of cultivated sorghums see SNOW-DEN, J. D., THE CULTIVATED RACES OF SORGHUM. vii + 272 pp. 1936. London.

FIGURE 1177.—*Sorghum halepense.* Plant, × ½; two views of terminal raceme, × 5. (Small, Ga.)

abundant and very sweet. In this country sorgo is cultivated chiefly in the region from Kansas and Texas to North Carolina for forage and for the juice which is made into sirup. The large panicles of broomcorn, grown especially in Oklahoma and Illinois, furnish the material for brooms. The other forms are grown for forage or for the seed which is used for feed. Chicken corn (S. VULGARE var. DRUMMÓNDII (Nees) Hack. ex Chiov.), described from New Orleans, La., was early introduced from Africa and became naturalized in Mississippi and Louisiana, but is apparently dying out. Culms up to 2 m. tall; blades to 5 cm. wide; panicle elongate, narrow but loose. Near railway, Illinois; weed in cotton field, Alabama; Mississippi; California; rare.

The differences between most of the varieties are so indistinct and so unstable because of intercrossing as to make it very difficult to assign descriptive limits. The application of botanical names is uncertain, and it seems best, therefore, not to assign to them definite varietal or specific Latin names.

The following names have been applied in American literature to some of the more important varieties.

Kafir. *S. vulgare* var. *caffrorum* (Retz.) Hubb. and Rehder.

Shallu. *S. vulgare* var. *roxburghii* (Stapf) Haines.

Durra. *S. vulgare* var. *durra* (Forsk.) Hubb. and Rehder.

Broomcorn. *S. vulgare* var. *technicum* (Koern.) Jav.

Sorgo. *S. vulgare* var. *saccharatum* (L.) Boerl.

Tunis grass (*S. virgatum* (Hack.) Stapf) is a tall annual with a narrow slender open panicle and narrowly-lanceolate green finely awned spikelets. Africa. Has been tried at experiment stations, but has not been brought into commercial cultivation, being inferior to Sudan grass.

Sorghum lanceolátum Stapf. Robust annual to 1.5 m. tall; blades 30 to 60 cm. long, 2 to 3.5 cm. wide; panicle 25 to 40 cm. long with ascending branches; rachis joints and pedicels ciliate; spikelets about 6 mm. long, silky-pubescent, becoming glabrous and shining on the lower half; awn about 1 cm. long. ☉ — Becoming a weed at Yuma, Calif. Introduced from tropical Africa.

Sorghum sudanénse (Piper) Stapf. SUDAN GRASS. Annual, branching from the base, 2 to 3 m. tall; blades 15 to 30 cm. long, 8 to 12 mm. wide; panicle erect, loose, 15 to 30 cm. long, about half as wide, the branches subverticillate, the lower half or third naked; sessile spikelet 6 to 7 mm. long, lanceolate-ovate, a ring of hairs at base, sparsely appressed-silky toward the apex; awn persistent, 10 to 15 mm. long, geniculate, twisted below; pedicellate spikelet narrow, about as long as the sessile spikelet, strongly nerved. ☉ (*Sorghum vulgare* var. *sudanense* Hitchc.)—Extensively cultivated for pasture and hay and escaped in the Southern and Midwestern States and in Arizona and California. Originally from Anglo-Egyptian Sudan.

157. SORGHÁSTRUM Nash

Spikelets in pairs, one nearly terete, sessile, and perfect, the other wanting, only the hairy pedicel being present; glumes coriaceous, brown or yellowish, the first hirsute, the edges inflexed over the second; sterile and fertile lemmas thin and hyaline, the latter extending into a usually well-developed bent and twisted awn. Perennial, erect, rather tall grasses, with auricled sheaths, narrow flat blades, and narrow terminal panicles of 1- to few-jointed racemes. Type species, *Sorghastrum avenaceum* (Michx.) Nash (*S. nutans*). Name from *Sorghum* and the Latin suffix *astrum*, a poor imitation of, alluding to the resemblance to *Sorghum*.

The most important species, *S. nutans*, is a common constituent of wild or prairie hay in the eastern part of the Great Plains region.

Awn usually 15 mm. long or less, once geniculate. Panicle rather dense, yellowish.

1. S. NUTANS.

FIGURE 1178.—*Sorghastrum nutans.* Plant, × ½; spikelet with pedicel and rachis joint, × 5. (Deam, Ind.)

Awn 20 to 35 mm. long, twice-geniculate, twisted below the second bend.
 Spikelets chestnut-brown, the ultimate branchlets with a few long hairs at the tip only;
  panicle loose, not unilateral... 2. S. ELLIOTTII.
 Spikelets yellowish brown, the upper portion of the ultimate branchlets conspicuously
  long-hairy toward the tip; panicle distinctly unilateral................... 3. S. SECUNDUM.

1. Sorghastrum nútans (L.) Nash.
INDIAN GRASS. (Fig. 1178.) Culms 1
to 2.5 m. tall from short scaly rhizomes; blades elongate, flat, mostly
5 to 10 mm. wide, tapering to a
narrow base, scabrous; panicle narrow, yellowish, rather dense, 15 to
30 cm. long, contracted and darker
at maturity; summit of branchlets,
rachis joints, and pedicels grayish-
hirsute; spikelets 6 to 8 mm. long,
lanceolate, hirsute, the awn 1 to 1.5
cm. long, once-geniculate. 4 —
Prairies, open woods, and dry slopes,
Quebec and Maine to Manitoba and
North Dakota, south to Florida and
Arizona; Mexico.

2. Sorghastrum ellióttii (Mohr)
Nash. (Fig. 1179.) Culms 1 to 1.5
m. tall, more slender than in *S.
nutans*, without rhizomes; the base
comparatively delicate, smooth or
nearly so; blades on the average
narrower; panicle loose, 15 to 30
cm. long, nodding at apex, the filiform branchlets and pedicels flexuous
but not recurved, with a few long
hairs at the tip; spikelets 6 to 7 mm.
long, chestnut brown at maturity,
with a short blunt bearded callus,
the first glume hirsute or glabrescent
on the back; awn 2.5 to 3.5 cm. long,
twice-geniculate. 4 —Open woods
dry hills, and sandy fields, eastern
Maryland to Tennessee, south to
Florida and Texas.

3. Sorghastrum secúndum (Ell.)
Nash. (Fig. 1180.) Culms 1 to 2 m.
tall, without rhizomes, the base
robust and felty-pubescent; blades
mostly less than 5 mm. wide, flat
or subinvolute; panicle narrow, 20
to 40 cm. long, 1-sided, the branches
mostly in separated fascicles, the
capillary branchlets and pedicels
strongly curved or circinately recurved, stiffly long-pilose below the
tip; spikelets about 7 mm. long,
brownish, pilose, with an acute dense-

FIGURE 1179.—*Sorghastrum elliottii,* × 1. (Harper 1718, Ga.)

ly bearded callus 1 to 1.5 mm. long.
4 —Pine barrens, South Carolina
to Florida and Texas.

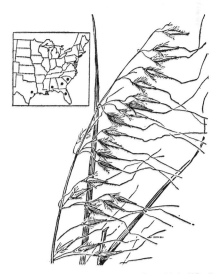

FIGURE 1180.—*Sorghastrum secundum,* × 1. (Hood, Fla.)

FIGURE 1181.—*Chrysopogon pauciflorus*. Plant, × ½; fruiting spikelet, × 5. (Combs 1359, Fla.)

158. CHRYSOPÓGON Trin.

(Rhaphis Lour.)

Spikelets in threes, one sessile and perfect, the other two pedicellate and sterile, or sometimes a pair below, one fertile and one sterile; fertile spikelet terete, the glumes coriaceous; sterile and fertile lemmas thin and hyaline, the latter awned. Perennial grasses, or, our species, annual, with open panicles, the three spikelets (reduced raceme) borne at the ends of long, slender, naked branches. Type species, *Andropogon gryllus* L. Name from Greek *chrysos*, golden, *pogon*, beard.

1. Chrysopogon pauciflórus (Chapm.) Benth. ex Vasey. (Fig.

1181.) Annual; culms 60 to 120 cm. tall, erect or somewhat decumbent at base; blades flat, mostly 4 to 8 mm. wide; panicle loose, the axis 5 to 10 cm. long, the branches few, very slender, 5 to 8 cm. long; sessile spikelet about 1.5 cm. long, including the slender villous callus about 7 mm. long, this disarticulating by a long-oblique line, the tip of the pedicel thus villous on one side; awn stout, brown, geniculate, twisted below, about 15 cm. long. ⊙ —Sandy pine woods, open ground, and fields, Florida; Cuba. The fertile spikelets resemble the fruits of certain species of *Stipa*, such as *S. spartea* L.

159. HETEROPÓGON Pers.

Spikelets in pairs, one sessile, the other pedicellate, both of the lower few to several pairs staminate or neuter, the remainder of the sessile spikelets perfect, terete, long-awned, the pedicellate spikelets, like the lower, staminate, flat, conspicuous, awnless; glumes of the fertile spikelet equal, coriaceous, the first brown-hirsute, infolding the second; lemmas thin and hyaline, the fertile one narrow, extending into a strong bent and twisted brown awn; palea wanting; glumes of the staminate spikelet membranaceous, the first green, faintly many-nerved, asymmetric, one submarginal keel rather broadly winged, the other wingless, the margins inflexed, the second glume narrower, symmetric; lemmas hyaline; palea wanting. Annual or perennial, often robust grasses, with flat blades and usually solitary terminal racemes; rachis slender, the lower part, bearing the pairs of staminate spikelets, continuous, the remainder disarticulating obliquely at the base of each joint, the joint forming a sharp-barbed callus below the fertile spikelet, the pedicellate spikelet readily falling, its pedicel remaining obscured in the hairs of the callus. Type species, *Heteropogon glaber* Pers. (*H. contortus*). Name from Greek *heteros*, different, and *pogon*, beard, alluding to the difference between the awnless-staminate and awned-pistillate spikelets.

One species, *H. contortus*, has a world-wide distribution. It is a good forage grass in the Southwest; if grazed constantly the troublesome awns do not develop. In the Hawaiian Islands, where it is called pili, it is an important range grass on the drier areas; also used there by the natives to thatch their grass huts. The mature fruits are injurious to sheep.

Plants perennial, less than 1 m. tall; first glume of staminate spikelet usually papillose-hispid.. 1. H. CONTORTUS.
Plants annual, usually more than 1 m. tall; first glume of staminate spikelet with a row of glands along the back, glabrous..................................... 2. H. MELANOCARPUS.

1. Heteropogon contórtus (L.) Beauv. ex Roem. and Schult. TANGLEHEAD. (Fig. 1182.) Plants perennial, tufted; culms 20 to 80 cm. tall, branched above, the branches erect; sheaths smooth, compressed-keeled; blades flat or folded, 3 to 7 mm. wide; raceme 4 to 7 cm. long, 1-sided; sessile spikelets about 7 mm. long,

FIGURE 1182.—*Heteropogon contortus.* Plant, × ½; fruiting spikelet, × 5. (Griffiths 1844, Ariz.)

slender, nearly hidden by the im-   5 to 12 cm. long, bent and flexuous,
bricate pedicellate spikelets, the awns   commonly tangled; pedicellate spike-

let about 1 cm. long, the first glume papillose-hispid toward the tip and margins, sometimes nearly glabrous. ♃ —Rocky hills and canyons, Texas to Arizona; tropical and warmer regions of both hemispheres.

2. Heteropogon melanocárpus (Ell.) Benth. SWEET TANGLEHEAD. (Fig. 1183.) Plants annual, 1 to 2 m. tall, freely branching; sheaths smooth, the upper part of the keel, especially of the upper sheaths, with a row of concave glands; blades 5 to 10 mm. wide; raceme 3 to 6 cm. long; looser than in *H. contortus;* sessile spikelets 9 to 10 mm. long, relatively thick, the awns 10 to 15 cm. long; pedicellate spikelet 1.5 to 2.5 cm. long, the first glume with a line of punctate glands along the middle. ⊙ —Pine woods, fields, and waste places, Georgia, Florida, and Alabama; Arizona; tropical regions of both hemispheres. The plant when fresh emits an odor like that of citronella oil.

FIGURE 1183.—*Heteropogon melanocarpus,* × 1 (Fredholm 6405, Fla.)

160. TRACHYPÓGON Nees

Spikelets in pairs, along a slender continuous rachis, one nearly sessile, staminate, awnless, the other pedicellate, perfect, long-awned; the pedicel of the perfect spikelet obliquely disarticulating near the base, forming a sharp-barbed callus below the spikelet; first glume firm-membranaceous, rounded on the back, several-nerved, obtuse; second glume firm, obscurely nerved; fertile lemma narrow, extending into a stout twisted and bent or flexuous awn; palea obsolete; sessile spikelet persistent, as large as the fertile spikelet and similar but awnless. Perennial, moderately tall grasses, with terminal spikelike solitary or fascicled racemes. Type species, *Trachypogon montufari.* Name from Greek *trachus,* rough, and *pogon,* beard, alluding to the plumose awn of the fertile spikelet.

1. Trachypogon secúndus (Presl) Scribn. CRINKLE-AWN. (Fig. 1184.) Culms tufted erect, slender, 60 to 120 cm. tall, the nodes appressed hirsute; sheaths with erect auricles 2 to 5 mm. long; blades flat to subinvolute, 3 to 8 mm. wide; raceme solitary, 10 to 18 cm. long, the rachis glabrous; spikelets 6 to 8 mm. long, pubescent, the awns of perfect spikelets 4 to 6 cm. long, short-plumose below, nearly glabrous toward the tip. ♃ (Included in *T. montufari* (H. B. K.) Nees in the Manual, ed. 1.)—Rocky hills and canyons, southern Texas, southwestern New Mexico, and southern Arizona; Mexico to Argentina.

161. ELYONÚRUS Humb. and Bonpl. ex Willd.

Spikelets in pairs along a somewhat tardily disarticulating rachis, the joints and pedicels short, thickened, and parallel, the sessile spikelets perfect, appressed to the concave side, the pedicellate spikelet staminate, similar to the sessile one, both awnless, the pair falling with a joint of the rachis; first glume

Figure 1184.—*Trachypogon secundus.* Plant, × ½; fertile spikelet, × 5. (Griffiths and Thornber 300, Ariz.)

firm, somewhat coriaceous, dorsally flattened, the margins inflexed around the second glume, a line of balsam glands on the marginal nerves, the apex entire and acute or acuminate, or bifid with aristate teeth; second glume similar to the first; sterile and fertile lemmas thin and hyaline; palea obsolete. Erect, moderately tall perennials, with solitary spikelike, often woolly racemes. Type species, *Elyonurus tripsacoides*. Name from Greek *eluein*, to roll, and *oura*, tail, alluding to the cylindric inflorescence.

The species are important grazing grasses in the savannas and plains of tropical America, but they extend only a short distance into the United States.

Rhizomes wanting; culms hirsute below the nodes; racemes conspicuously woolly.
                                                                    1.   E. BARBICULMIS.
Rhizomes present; culms glabrous; racemes slightly pubescent, the first glume glabrous or
  nearly so on the back.. 2.   E. TRIPSACOIDES.

1. Elyonurus barbicúlmis Hack.

(Fig. 1185.) Culms tufted, erect, simple or sparingly branching, 40 to 60 cm. tall, pubescent below the nodes; blades involute, striate, about 1 mm. thick, the upper surface usually long-pilose; raceme mostly 5 to 10 cm. long, pale; rachis joints, pedicels, and spikelets densely woolly, the spikelets 6 to 8 mm. long; first glume acuminate. ♃ —Mesas, rocky hills, and canyons, western Texas to southern Arizona; northern Mexico.

2. Elyonurus tripsacoídes Humb.

and Bonpl. ex Willd. (Fig. 1186.) Culms 60 to 120 cm. tall, glabrous, rather freely branching and with short rhizomes; blades flat or involute, 2 to 4 mm. wide, slightly pilose on the upper surface near the base; raceme 7 to 15 cm. long; rachis joints ciliate, the pedicels pilose; spikelets 6 to 8 mm. long, the first glume ciliate toward the acuminate 2-toothed apex, usually glabrous on the back. ♃ —Moist pine woods and low prairies, Georgia, Florida, southern Mississippi, and southern Texas; Mexico to Argentina.

162. ROTTBOÉLLIA L. f.

Spikelets awnless, in pairs at the nodes of a thickened articulate rachis, one sessile and perfect, the other pedicellate, sterile; rachis joints hollow above, the thickened pedicel adnate to it, the pedicellate spikelet appearing to be sessile; sessile spike-

FIGURE 1185.—*Elyonurus barbiculmis*, × 1. (Type coll.)

let fitting closely against the concave side of the rachis joint, the first glume coriaceous, the second less coriaceous; sterile and fertile lemmas and palea hyaline. Coarse branching annual, with broad flat blades and subcylindric racemes, dwindling toward the summit and bearing abortive spikelets only. Type species, *Rottboellia exaltata*. Named for C. F. Rottboell.

1. Rottboellia exaltáta L. f. (Fig.

1187.) Culms robust, 1 to 3 m. tall, branching; sheaths papillose-hispid, especially toward the summit; blades flat, in robust specimens as much as 3 cm. wide; racemes mostly 8 to 12 cm. long, 3 to 4 mm. thick, dwindling at the summit; sessile spikelet 5 to

FIGURE 1186.—*Elyonurus tripsacoides*. Plant, ✕ ½; two views of pair of spikelets with rachis joint, ✕ 5. (Chase 4144, Fla.)

7 mm. long; first glume finely papillose; pedicellate spikelet scarcely as long as the sessile one. ⊙ (*Manisuris exaltata* Kuntze.)—Introduced at Miami, Fla.; West Indies; native of tropical Asia. The fragile hairs of the sheaths are irritating to the skin of persons handling the plant.

163. MANISÚRIS L.

Spikelets awnless, in pairs at the nodes of a thickened articulate rachis, one sessile and perfect, the other pedicellate, rudimentary (developed but sterile in *M. altissima*), the pedicel thickened and appressed to the rachis, the sessile spikelet fitting closely against the rachis (racemes partly adnate in *M. altissima*), forming a cylindric or flattened raceme; glumes mostly obtuse, the first coriaceous, fitting over the hollow containing the spikelet, the keels winged at the summit, the second less coriaceous than the first; sterile lemma, fertile lemma, and palea thin and hyaline. Perennial, slender, moderately tall, or tall grasses, with usually numerous glabrous cylindric or flattened solitary racemes. Type species, *Manisuris myuros* L. Name from Greek *manos*, necklace, and *oura*, tail, presumably alluding to the jointed racemes. The species probably have some forage value but they are nowhere abundant.

Racemes flattened, tardily disarticulating; first glume of sessile spikelet smooth.
                        1. M. ALTISSIMA.
Racemes cylindric, readily disarticulating at maturity; first glume of sessile spikelet marked with pits or wrinkles (sometimes smooth in *M. tuberculosa*).
    Sheaths not compressed-keeled; first glume more or less pitted........ 2. M. CYLINDRICA.
    Sheaths compressed-keeled; first glume tessellated, wrinkled, tubercled, or smooth.
        First glume tessellated, the depressions rectangular...................... 3. M. TESSELLATA.
        First glume with prominent transverse wrinkles.... .... 4. M. RUGOSA.
        First glume with a few low tubercles or smooth.. .... 5. M. TUBERCULOSA.

1. Manisuris altíssima (Poir.) Hitchc. (Fig. 1188.) Perennial; culms ascending from a long creeping base, compressed and 2-edged, 40 to 80 cm. long, freely branching toward the ends; blades flat, 3 to 8 mm. wide; flowering branches often short and fascicled, the racemes 3 to 5 cm., sometimes 10 cm. long, compressed; pedicel free or partly adnate to the rachis joint; sessile spikelet 5 to 7 mm. long, the keels of the first glume very narrowly winged toward the apex; pedicellate spikelet 5 to 6 mm. long, acute. ♃ (*M. fasciculata* Hitchc.)—Ponds and ditches, southern Texas; warm-temperate and tropical regions of both hemispheres; introduced in America.

2. Manisuris cylíndrica (Michx.) Kuntze. (Fig. 1189.) Culms tufted, with short rhizomes, erect, rather slender, 30 to 100 cm. tall, simple or

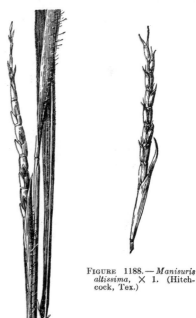

FIGURE 1188.— *Manisuris altissima*, × 1. (Hitchcock, Tex.)

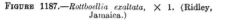

FIGURE 1187.—*Rottboellia exaltata*, × 1. (Ridley, Jamaica.)

FIGURE 1189.—*Manisuris cylindrica*. Plant, × ½; two views of rachis joint with fertile and sterile spikelets attached, × 5. (Harvey, Ark.)

with a few branches; sheaths not compressed-keeled; blades flat or folded, 2 to 3 mm. wide; raceme cylindric, 5 to 15 cm. long, slightly curved; sessile spikelet 4 to 5 mm. long, the first glume pitted along the nerves. ♃ —Pine woods and prairies, Coastal Plain, North Carolina to Florida and Texas, north to Missouri and Oklahoma.

3. Manisuris tesselláta (Steud.) Scribn. (Fig. 1190.) Culms 80 to 120 cm. tall, rather stout, branching; sheaths, especially the basal ones, compressed-keeled; blades elongate, flat, mostly 5 to 8 mm. wide; raceme 5 to 12 cm. long; sessile spikelets 4 to 5 mm. long; first glume tessellated with rectangular depressions, the keels narrowly winged at the apex. ♃ —Moist pine woods, Coastal Plain, Florida to Louisiana.

4. Manisuris rugósa (Nutt.) Kuntze. (Fig. 1191.) Culms mostly rather stout, 70 to 120 cm. tall, freely branching; sheaths compressed-keeled; blades commonly folded, 3 to 8 mm. wide; flowering branches often numerous, the racemes 4 to 8 cm. long, partly included in brownish sheaths; rachis joint and pedicel contracted in the middle; sessile spikelet 3.5 to 5 mm. long, the first glume strongly and irregularly transversely ridged, the keels narrowly winged toward the summit. ♃ —Wet pine woods, Coastal Plain, southern New Jersey to Florida, Arkansas, and Texas.

5. Manisuris tuberculósa Nash. (Fig. 1192.) Differing from *M. rugosa* chiefly in the straight rachis joints, not contracted in the middle, and in the smooth to obscurely ridged or tuberculate first glume of the sessile spikelet, varying in a single raceme. ♃ —Moist ground along lakes, central peninsular Florida. Apparently rare.

Eremóchloa ophiuroídes (Munro) Hack. CENTIPEDE GRASS. Low perennial, creeping by thick short-noded leafy stolons; racemes spikelike, smooth, subcylindric, terminal and axillary on slender peduncles, 2 to 6 cm. long; rachis flat, not thickened as in *Manisuris*, the first glume of

FIGURE 1190.—*Manisuris tessellata*, × 1. (Tracy and Ball 1, Miss.)

sessile spikelet winged at summit. ♃ —Southeastern Asia; valuable as a lawn grass from South Carolina to Florida, and the Gulf States. It is commonly used in northern Florida, replacing to a large extent carpet grass and St. Augustine grass. It is easily established and quickly forms a dense turf.

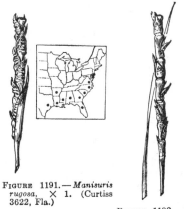

FIGURE 1191.—*Manisuris rugosa*, × 1. (Curtiss 3622, Fla.)

FIGURE 1192.—*Manisuris tuberculosa*, × 1. (Nash 1074, Fla.)

EREMOCHLOA CILIÁRIS (L.) Merr. Found near a Chinese warehouse in San Francisco. Southeastern Asia. Mentioned in the Botany of California (2:262. 1880) under *Ischaemum leersioides* Munro. Not since collected in the United States.

nate. Freely branching annual with flat blades, the numerous racemes solitary and more or less enclosed in the spathes, these usually fascicled in the axils of the leaves. Type species, *Hackelochloa granularis*. Named

FIGURE 1193.—*Hackelochloa granularis*. Plant, × ½; single raceme, × 2; two views of spikelets with rachis joint, × 5. (Pringle, Ariz.)

164. HACKELÓCHLOA Kuntze

(*Rytilix* Raf.)

Spikelets awnless, in pairs, the rachis joint and pedicel grown together, the two clasped between the edges of the globose alveolate first glume of the sessile spikelet; pedicellate spikelet conspicuous, stami-

for Eduard Hackel and Greek *chloa*, grass.

1. Hackelochloa granuláris (L.) Kuntze. (Fig. 1193.) Culms 30 to 100 cm. tall; sheaths papillose-hispid; blades flat, 5 to 15 cm. long, 3 to 15 mm. wide, papillose-hirsute, ciliate; racemes 1 to 2 cm. long; sessile spikelet about 1 mm. thick; pedicellate spikelet about 2 mm. long. ☉ — Open ground, fields, and waste places, Georgia and Florida to Louisiana; New Mexico to Arizona; tropics of both hemispheres, introduced in America. Furnishes some forage in the Southwest.

THEMÉDA Forsk.

Inflorescence a flabellate cluster of several short racemes, each subtended by a spathe, the entire cluster subtended by a larger spathe; racemes consisting of 2 approximate pairs of sessile awnless staminate or neuter spikelets and a single fertile awned spikelet with a pair of sterile pedicellate ones, the rachis disjointing above the pairs of sessile staminate spikelets and forming a pointed callus below the fertile one. Annuals or perennials. Name from the Arabian, *Thaemed*.

Themeda quadriválvis (L.) Kuntze. KANGAROO GRASS. Robust annual, 1 to 1.5 m. tall; blades flat, elongate, 5 to 10 cm. wide; inflorescence often elongate, narrow, loose to dense, with conspicuous bent brown awns 4 to 5 cm. long. ☉ —Established on bottom land, near Opelousas, St. Landry Parish, La. Introduced in the West Indies. East Indies.

TRIBE 14. TRIPSACEAE

165. CÓIX L. JOBS-TEARS

Spikelets unisexual; staminate spikelets 2-flowered, in twos or threes on the continuous rachis, the normal group consisting of a pair of sessile spikelets with a single pedicellate spikelet between, the latter sometimes reduced to a pedicel or wanting; glumes membranaceous, obscurely nerved; lemma and palea hyaline; stamens 3; pistillate spikelets 3 together, 1 fertile and 2 sterile at the base of the inflorescence; glumes of fertile spikelet several-nerved, hyaline below, chartaceous in the upper narrow pointed part, the first very broad, infolding the spikelet, the margins infolded beyond the 2 lateral stronger

pair of nerves; second glume narrower than the first, keeled, sterile lemma similar but a little narrower; fertile lemma and palea hyaline; sterile spikelets consisting of a single narrow tubular glume as long as the fertile spikelet, somewhat chartaceous. Tall branched grasses with broad flat blades, the monoecious inflorescences numerous on long, stout peduncles, these clustered in the axils of the leaves, each inflorescence consisting of an ovate or oval pearly-white or drab, beadlike, very hard, tardily deciduous involucre (much modified sheathing bract) containing the pistillate lower portion of the inflorescence, the points of the pistillate spikelets and the slender axis of the staminate portion of the inflorescence protruding through the orifice at the apex, the staminate upper portion of the inflorescence 2 to 4 cm. long, soon deciduous, consisting of several clusters of staminate spikelets. Type

FIGURE 1194.—*Coix lacryma-jobi*, × 1. (Cult.)

species, *Coix lacryma-jobi*. Name from Greek *koix*, a kind of palm, applied by Linnaeus to this grass.

1. Coix lácryma-jóbi L. JOBS-TEARS. (Fig. 1194.) Annual; culms usually about 1 m. tall; blades as much as 4 cm. wide; beads white to bluish gray, globular or ovoid, 6 to 12 mm. long. ⊙ —Occasionally

cultivated for ornament, escaped into waste places in the Southern States; all tropical countries; introduced in America. The beadlike fruits are used as beads and for rosaries. A garden form (called by gardeners var. *aurea zebrina*) has yellow-striped blades.

166. TRÍPSACUM L. GAMAGRASS

Spikelets unisexual; staminate spikelets 2-flowered, in pairs on one side of a continuous rachis, one sessile, the other sessile or pedicellate, similar to those of *Zea*, the glumes firmer; pistillate spikelets solitary (a minute rudiment of a sterile spikelet sometimes found), on opposite sides at each joint of the thick, hard articulate lower part of the same rachis, sunken in hollows in the joints, consisting of one perfect floret and a sterile lemma; first glume coriaceous, nearly infolding the spikelet, fitting into and closing the hollow of the rachis; second glume similar to the first but smaller, infolding the remainder of the spikelet; sterile lemma, fertile lemma, and palea very thin and hyaline, these progressively smaller. Robust perennials, with usually broad flat blades and monoecious terminal and axillary inflorescences of 1 to 3 racemes, the pistillate part below, breaking up into bony, seedlike joints, the staminate above on the same rachis, deciduous as a whole. Type species, *Tripsacum dactyloides*. Name of unknown origin, said by some to come from Greek *tribein*, to rub, alluding to the smooth joints.

The species are good forage grasses, but even the more widely spread *T. dactyloides* is not common enough to be of importance. Two large species not found in the United States, *T. laxum* Nash and *T. latifolium* Hitchc., of Central America, are occasionally cultivated for forage in that region. The genus is of interest because it is related to maize. Hybrids between *T. dactyloides* and maize have been made.[17]

Staminate spikelets membranaceous, the members of the pair unequally pedicellate, one
   nearly sessile, the other with a distinct pedicel.............................. 3. T. LANCEOLATUM.
Staminate spikelets rather chartaceous, both members of the pair nearly sessile.
   Blades 1 to 2 cm. wide, flat; plants 1 to 2 m. tall; terminal racemes usually more than one.
                                                         1. T. DACTYLOIDES.
   Blades 1 to 4 mm. wide, subinvolute; plants less than 1 m. tall; all racemes usually solitary.
                                                         2. T. FLORIDANUM.

1. Tripsacum dactyloídes (L.) L. EASTERN GAMAGRASS. (Fig. 1195.) Plants in large clumps, with thick knotty rhizomes, 2 to 3 m. tall or sometimes taller, glabrous throughout; blades usually 1 to 2 cm. wide, flat, scabrous on the margin; inflorescence 15 to 25 cm. long, the pistillate part one-fourth the entire length or less, the terminal racemes usually 2 or 3, sometimes only 1, those of the branches usually solitary; pistillate

spikelets 7 to 10 mm. long, the joints rhombic; staminate spikelets 7 to 11 mm. long, both of a pair nearly sessile, the glumes rather chartaceous. ♃ —Swales, banks of streams, and moist places, Massachusetts to Michigan, Iowa, and Nebraska, south to Florida, Oklahoma, and Texas; West Indies. TRIPSACUM DACTYLOIDES var. OCCIDENTÁLE Cutler and Anders. Differentiated on softer staminate glumes more than 9 mm. long, tapering to an acute tip. ♃ —Texas. Examination of a large number of specimens shows the length and texture of

---

[17] MANGELSDORF, P. C., and REEVES, R. G. Jour. Hered. 22: 329–343. 1931. Ibid., Texas Agr. Expt. Sta. Bul. 574 (monogr.): 1–315. 1939.

FIGURE 1195.—*Tripsacum dactyloides.* Plant, × ½; pistillate spikelets with rachis joint and pair of staminate spikelets with rachis joint, × 5. (Amer. Gr. Natl. Herb. 229, Va.)

FIGURE 1196.—*Tripsacum floridanum*, ✕ 1. (Hitch-cock 686, Fla.)

staminate glumes to vary greatly, often in a single raceme. Occasional specimens with glumes 10 to 11 mm. long, soft or firm, are found also in Iowa, Missouri, Kansas, Virginia, Tennessee, and Oklahoma, the plants not differing otherwise from the species.

2. Tripsacum floridánum Porter ex Vasey. FLORIDA GAMAGRASS. (Fig. 1196.) Smaller than *T. dactyloides* in all ways, commonly less than 1 m. tall; blades mostly 1 to 4 mm. wide; terminal and axillary racemes usually solitary (rarely 2 or more). ♃ — Low rocky pinelands, southern Florida.

3. Tripsacum lanceolátum Rupr. MEXICAN GAMAGRASS. (Fig. 1197.) Resembling *T. dactyloides;* sheaths, especially the lower, sometimes hispid; blades often hispidulous on the upper surface; racemes more slender with smaller spikelets than in *T.*

FIGURE 1197.—*Tripsacum lanceolatum*, ✕ 1. (Lemmon, Ariz.)

dactyloides, the terminal racemes usually 3 to 5; staminate spikelets membranaceous, one of the pair distinctly pediceled. ♃ (*T. lemmoni* Vasey.) —Rocky hills, Huachuca and Mule Mountains, Ariz.; Mexico to Guatemala.

167. EUCHLAÉNA Schrad.

TEOSINTE

Staminate spikelets as in *Zea;* pistillate spikelets solitary on opposite sides, sunken in cavities in the hardened joints of an obliquely articulate rachis, the indurate first glume covering the cavity; second glume membranaceous, the lemma hyaline. Spikes infolded in foliaceous spathes or husks, 2 to several of these together enclosed in the leaf sheaths. Robust annuals and perennials with

FIGURE 1198.—*Euchlaena mexicana.* Plant, much reduced; pistillate inflorescence enclosed in bract (*a*) and with portion of bract removed (*b*), × 1; lateral view of rachis joint and fertile spikelet (*c*), and dorsal view of same, showing first glume (*d*), × 2. (Cult.)

broad flat blades, terminal panicles of staminate spikelets, and axillary spikes of pistillate spikelets. Type species, *Euchlaena mexicana.* Name from Greek *eu*, well, and *chlaina*, cloak, alluding to the husks hiding the pistillate inflorescence.

1. Euchlaena mexicána Schrad. Teosinte. (Fig. 1198.) Tall annual, resembling maize, the culms branching at base, 2 to 3 or even 5 m. tall; blades as much as 8 cm. wide. ⊙ —Occasionally cultivated in the Southern States for green forage; Mexico. Closely related to maize and readily hybridizing with it.

2. Euchlaena perénnis Hitchc., Mexican teosinte, a perennial species from Mexico, is cultivated at the substation of the agricultural college, Angleton, Tex., Sacaton, Ariz., and probably at other points. Established on James Island, S. C. It propagates by creeping rhizomes.

168. ZÉA L.

Spikelets unisexual; staminate spikelets 2-flowered, in pairs, on one side of a continuous rachis, one nearly sessile, the other pedicellate; glumes membranaceous, acute; lemma and palea hyaline; pistillate spikelets sessile, in pairs, consisting of 1 fertile floret and 1 sterile floret, the latter sometimes developed as a second fertile floret; glumes broad, rounded or emarginate at apex; sterile and fertile lemmas hyaline, the palea developed; style very long and slender, stigmatic along both sides well toward the base. Robust annual, with terminal panicles (tassels) of staminate racemes, and short-peduncled, pistillate, 8- to many-rowed spikes (ears) enclosed in numerous spathes (husks). Type species, *Zea mays.* Name Greek *zea*, or *zeia*, a kind of grain.

1. Zea máys L. Maize, Indian corn. (Fig. 1199.) Tall robust monoe-

cious annual, with overlapping sheaths and broad, conspicuously distichous blades; staminate spikelets in long spikelike racemes, these numerous, forming large spreading terminal panicles; pistillate inflorescence in the axils of the leaves, the spikelets in 8 to 16 or even as many as 30 rows on a thickened, almost woody axis (cob), the whole enclosed in numerous large foliaceous bracts or spathes, the long styles (silk) protruding from the summit as a mass of silky threads; grains at maturity greatly exceeding the glumes. ⊙ —Maize or Indian corn is one of the important economic plants of the world, being cultivated for food for man and domestic animals and for forage. It originated[18] in America, probably on the Mexican Plateau, and was cultivated from prehistoric times by the early races of American aborigines, from Peru to middle North America. Several races of maize are grown in the United States,[19] the most important being dent, the common commercial field sort, flint, sweet, and pop. Pod corn (*Z. mays* var. *tunicata* Larr. ex St. Hil.), occasionally cultivated as a curiosity, is a variety in which each kernel is enveloped in the elongate glumes. A variety with variegated leaves (*Z. mays* var. *japónica* (Van Houtte) Wood) is cultivated for ornament.

[18] For a note on the origin of maize, see Collins, G. N. The origin of maize. Jour. Wash. Acad. Sci. 2: 520–530. 1912.
[19] See the following publications:
Mangelsdorf, P. C. and Reeves, R. G. The origin of Indian corn and its relatives. Tex. Agr. Expt. Sta. Bul. 574 (monogr.): 1–315. 1939. Reeves, R. G. and Mangelsdorf, P. C. Amer. Jour. Bot. 29: 815–817. 1942. Sturtevant, E. L. Varieties of corn. U. S. Dept. Agr., Off. Expt. Sta. Bul. 57, 108 pp. 1899. Tapley, W. T., Enzie, W. D., and Van Eseltine, G. P., N. Y. State Agr. Expt. Sta. Rpt., 1934. 1934. Weatherwax, Paul, Morphology of the flowers of zea mays. Torrey Bot. Club Bul. 43: 127–144. 1916. Development of spikelets of zea mays. Torrey Bot. Club Bul. 43: 483–496. 1917; the evolution of maize. Torrey Bot. Club Bul. 45: 309–342. 1918; the story of the maize plant. 247 pp. illus. Chicago, Ill., 1923; the phylogeny of zea mays. Amer. Midl. Nat. 16: 1–71. 1935.

FIGURE 1199.—*Zea mays.* Pistillate inflorescence (ear) and 2 branches of staminate inflorescence (tassel), × ½; pair of pistillate spikelets attached to rachis (cob) with mature grains, the second glume showing, × 2; single pistillate spikelet soon after flowering, × 4; staminate spikelet, × 2. (Cult.)

SYNONYMY

The following names have appeared in botanical literature as applied to grasses growing in the United States. For grasses introduced into the United States from other countries there are here given only the names appearing in American works. No attempt has been made to present the complex synonymy for these introduced grasses given in foreign works. The synonymy for the generic names will be found in The Genera of Grasses of the United States. Genera not included in this work nor in Manual, ed. 1, and a few changes in generic names will be found in the Appendix, page 1001.

For quick reference the names of genera and valid species are arranged in alphabetic order, the names in blackface type. The synonyms, in italics, are arranged chronologically under the names to which they are referred. The numbers in parentheses are the numbers these genera and species bear in the body of this work.

(44) AEGILOPS L.

(1) **Aegilops cylindrica** Host, Icon. Gram. Austr. 2: 6. pl. 7. 1802. Southern Europe.
Triticum cylindricum Ces., Pass. and Gib., Comp. Fl. Ital. 86. 1867. Presumably based on *Aegilops cylindrica* Host.
(3) **Aegilops ovata** L., Sp. Pl. 1050. 1753. Southern Europe.
Triticum ovatum Raspail, Ann. Sci. Nat., Bot. 5: 435. 1825. Based on *A. ovata* L.
(2) **Aegilops triuncialis** L., Sp. Pl. 1051. 1753. Mediterranean region.
Triticum triunciale Raspail, Ann. Sci. Nat. Bot. 5: 435. 1825. Based on *A. triuncialis* L.

(96) AEGOPOGON Humb. and Bonpl. ex Willd.

(1) **Aegopogon tenellus** (DC.) Trin., Gram. Unifl. 164. 1824. Based on *Lamarckia tenella* DC., though Trinius cites not that but *A. pusillus* Beauv., in Roem. and Schult., Syst. Veg. 2: 805. 1817. Roemer and Schultes cite *L. tenella* DC., obviously the basis of Trinius' name, as synonym of *A. pusillus* Beauv., which, however, is the same as *A. cenchroides* Humb. and Bonpl. (not known from the U. S.).

Lamarckia tenella DC., Cat. Hort. Monsp. 120. 1813. Grown in Montpellier, origin unknown, probably Mexico.
Cynosurus tenellus Cav. ex DC., Cat. Hort. Monsp. 120. 1813, as synonym of *Lamarckia tenella* DC.
Hymenothecium unisetum Lag., Gen. and Sp. Nov. 4. 1816. Grown from Mexican seed sent by Sessé.
Hymenothecium tenellum Lag., Gen. and Sp. Nov. 4. 1816. Based on *Cynosurus tenellus* Cav.
Aegopogon unisetus Roem. and Schult., Syst. Veg. 2: 805. 1817. Based on *Hymenothecium unisetum* Lag.
Schellingia tenera Steud., Flora 33: 232. 1850. Mexico, *Galeotti 5750.*
Aegopogon geminiflorus var. *unisetus* Fourn., Mex. Pl. 2: 71. 1886. Based on *A. unisetus* Roem. and Schult.
Chloris pedicellata Steud. ex Fourn., Mex. Pl. 2: 71. 1886, as synonym of *A. geminiflorus* H. B. K. Misapplied by Fournier.
Aegopogon geminiflorus var. *abortivus* Fourn. Mex. Pl. 2: 71. 1886. *Bourgeau 750 bis* and *Schaffner 7*, both in the Paris Herbarium and cited by Fournier have been examined. Both are short-awned specimens of *A. tenellus.*
Aegopogon tenellus var. *abortivus* Beetle, Wyo. Univ. Pub. 13²: 18. 1948. Based on *A. geminiflorus* var. *abortivus* Fourn.

(42) AGROPYRON Gaertn.

(9) **Agropyron albicans** Scribn. and Smith, U. S. Dept. Agr., Div. Agrost. Bul. 4: 32. 1897. Yogo Gulch, Mont., *Rydberg 3405.*
(20) **Agropyron arizonicum** Scribn. and Smith, U. S. Dept. Agr., Div. Agrost. Bul. 4: 27. 1897. New Mexico, Arizona [type, Rincon Mountains, *Nealley 67*], and Chihuahua, Mexico.
Agropyron caninum var. *majus* Scribn., Torrey Bot. Club Bul. 10: 32. 1883. Santa Rita Mountains, Ariz., *Pringle.*
Agropyron spicatum var. *arizonicum* Jones, West. Bot. Contrib. 14: 19. 1912. Based on *A. arizonicum* Scribn. and Smith.
Elymus arizonicus Gould, Madroño 9: 125. 1947. Based on *Agropyron arizonicum* Scribn. and Smith.
(15) **Agropyron bakeri** E. Nels., Bot. Gaz. 38: 378. 1904. Pagosa Peak, Colo., *Baker 139.*
Agropyron caninum (L.) Beauv., Ess. Agrost. 102, 146. 1812. Based on *Triticum caninum* L.
Triticum caninum L., Sp. Pl. 86. 1753. Europe.
Elymus caninus L., Fl. Suec. ed. 2. 39. 1755. Based on *Triticum caninum* L.
Zeia canina Lunell, Amer. Midl. Nat. 4: 226. 1915. Based on *Triticum caninum* L.

Roegneria canina Nevski in Komarov, Fl. U. R. S. S. 2: 617. 1934. Based on *Triticum caninum* L.

Agropyron cristatum (L.) Gaertn., Nov. Comm. Petrop. 14: 540. 1770. Based on *Bromus cristatus* L.

Bromus cristatus L., Sp. Pl. 78. 1753. Northern Asia.

Triticum cristatum Schreb., Beschr. Gras. 2: 12. pl. 23. f. 2. 1769. Based on *Bromus cristatus* L.

Avena cristata Roem. and Schult., Syst. Veg. 2: 758. 1817, as synonym of *Agropyron cristatum* Gaertn.

Costia cristata Willk. Bot. Ztg. 16: 377. 1858. Based on *Bromus cristatus* L.

Eremopyrum cristatum Willk. and Lange, Prodr. Fl. Hisp. 1: 108. 1870. Presumably based on *Bromus cristatus* L.

Zeia cristata Lunell, Amer. Midl. Nat. 4: 226. 1915. Based on *Agropyron cristatum* Gaertn.

(6) **Agropyron dasystachyum** (Hook.) Scribn. Torrey Bot. Club Bul. 10: 78. 1883. Based on *Triticum repens* var. *dasystachyum* Hook.

Triticum repens var. *dasystachyum* Hook., Fl. Bor. Amer. 2: 254. 1840. Saskatchewan, *Richardson*. The type has villous lemmas.

Triticum repens var. *subvillosum* Hook., Fl. Bor. Amer. 2: 254. 1840. Mackenzie River, Canada, *Richardson*. The type has scabrous-pubescent lemmas.

Triticum dasystachyum A. Gray, Man. 602. 1848. Based on *T. repens* var. *dasystachyum* Hook.

Agropyron dasystachyum var. *subvillosum* Scribn. and Smith, U. S. Dept. Agr., Div. Agrost. Bul. 4: 33. 1897. Based on *Triticum repens* var. *subvillosum* Hook.

Agropyron lanceolatum Scribn. and Smith, U. S. Dept. Agr., Div. Agrost. Bul. 4: 34. 1897. Idaho [type, Blackfoot, *Palmer* 266], Washington and Oregon.

Triticum repens acutum Vasey ex Scribn. and Smith, U. S. Dept. Agr., Div. Agrost. Bul. 4: 34. 1897, as synonym of *A. lanceolatum* Scribn. and Smith.

Agropyron subvillosum E. Nels., Bot. Gaz. 38: 378. 1904. Based on *Triticum repens* var. *subvillosum* Hook.

Zeia dasystachyum Lunell, Amer. Midl. Nat. 4: 226. 1915. Based on *Triticum repens* var. *dasystachyum* Hook.

Elymus subvillosus Gould, Madroño 9: 127. 1947. Based on *Triticum repens* var. *subvillosum* Hook.

Elymus lanceolatus Gould, Madroño 10: 94. 1949. Presumably based on *Agropyron lanceolatum* Scribn. and Smith. Not *Elymus dasystachys* Trin., 1829.

(1) **Agropyron desertorum** (Fisch.) Schult., Mantissa 2: 412. 1824. Based on *Triticum desertorum* Fisch.

Triticum desertorum Fisch. ex Link,

Enum. Pl. 1: 97. 1821. Desert Cumano [along River Kuma, southeastern European Russia].

(7) **Agropyron elmeri** Scribn., U. S. Dept. Agr., Div. Agrost. Bul. 11: 54. pl. 12. 1898. Snake River, Wash., *Elmer* 759.

(10) **Agropyron griffithsi** Scribn. and Smith ex Piper, Biol. Soc. Wash. Proc. 18: 148. 1905. North Fork Clear River, Wyo., *Williams* and *Griffiths* 140.

(19) **Agropyron inerme** (Scribn. and Smith) Rydb., Torrey Bot. Club Bul. 36: 539. 1909. Based on *A. divergens* var. *inerme* Scribn. and Smith.

Agropyron divergens var. *inerme* Scribn. and Smith, U. S. Dept. Agr., Div. Agrost. Bul. 4: 27. 1897. British Columbia to Utah and Idaho [type *Henderson* 3058].

Agropyron spicatum inerme Heller, N. Amer. Pl. Cat. ed. 2. 3. 1900. Based on *Agropyron divergens* var. *inerme* Scribn. and Smith.

Agropyron intermedium (Host) Beauv., Ess. Agrost. 102, 146. 1812. Based on *Triticum intermedium* Host.

Triticum intermedium Host, Gram. Austr. 3: 23. 1805. Austria.

Triticum glaucum Desf. ex DC., Fl. Franc. 5: 281. 1815. Not *T. glaucum* Moench, 1794. France.

Agropyron glaucum Roem. and Schult., Syst. Veg. 2: 752. 1817. Based on *Triticum glaucum* Desf.

Braconotia glauca Godr., Fl. Lorr. 3: 192. 1844. Based on *Triticum glaucum* Desf.

Agropyron repens glaucum Scribn., Torrey Bot. Club Mem. 5: 57. 1894. Based on *Triticum glaucum* Desf., but misapplied to *A. smithii* Rydb.

Zeia glauca Lunell, Amer. Midl. Nat. 4: 226. 1915. Based on *Triticum glaucum* Desf., but misapplied to *A. smithii* Rydb.

Elytrigia intermedia Nevski, Akad. Nauk S. S. S. R. Bot Inst. Trudy I. (Acad. Sci. U. R. S. S. Inst. Bot. Acta I. Flora et Syst. Plant. Vasc.) 1: 14. 1933. Based on *Triticum intermedium* Host.

Agropyron junceum (L.) Beauv., Ess. Agrost. 102, 146, 180. 1812. Based on *Triticum junceum* L.

Triticum junceum L., Mant. Pl. 2: 327. 1771. Europe.

Festuca juncea Moench, Meth. Pl. 190. 1794. Based on *Triticum junceum* L.

Braconotia juncea Godr., Fl. Lorr. 3: 192. 1844. Based on *Triticum junceum* L.

Elytrigia juncea Nevski, Akad. Nauk S. S. S. R. Bot. Inst. Trudy I. (Acad. Sci. U. R. S. S. Inst. Bot. Acta I. Flora et Syst. Plant. Vasc.) 1: 17. 1933; 2: 83. 1936. Based on *Triticum junceum* L.

Elymus multinodus Gould, Madroño 9: 126. 1947. Based on *Triticum junceum* L., not *Elymus junceus* Fisch.

The names *Agropyron junceum* and *A.*

intermedium are here applied in accord with Ascherson and Graebner (Syn. Mitteleur. Fl. 2: 654, 662. 1901) under *Triticum*. *Triticum junceum* L. (Cent. Pl. 1: 6. 1755; Amoen. Acad. 4: 266. 1759), which seems to have been generally ignored, appears to be the same as *T. intermedium* Host. Linnaeus later (Mant. Pl. 2: 327. 1771) published a different species under the same name. This second name is the one used by Ascherson and Graebner and other European botanists. The problem involves study of European types not here available.

(14) **Agropyron latiglume** (Scribn. and Smith) Rydb., Torrey Bot. Club Bul. 36: 539. 1909. Based on *A. violaceum* var. *latiglume* Scribn. and Smith.

Agropyron violaceum var. *latiglume* Scribn. and Smith, U. S. Dept. Agr., Div. Agrost. Bul. 4: 30. 1897. Montana [type, Lone Mountain, Gallatin County, *Tweedy* 1011] to Alaska.

Agropyron biflorum latiglume Piper, Torrey Bot. Club Bul. 32: 547. 1905. Based on *A. violaceum* var. *latiglume* Scribn. and Smith.

Agropyron caninum var. *latiglume* Pease and Moore, Rhodora 12: 73. 1910. Based on *A. violaceum* var. *latiglume* Scribn. and Smith.

Roegneria latiglumis Nevski, Akad. Nauk S. S. S. R. Bot Inst. Trudy I. (Acad. Sci. U. R. S. S. Inst. Bot. Acta I. Flora et Syst. Plant. Vasc.) 2: 55. 1936. Based on *Agropyron latiglume* Rydb.

(21) **Agropyron parishii** Scribn. and Smith, U. S. Dept. Agr., Div. Agrost. Bul. 4: 28. 1897. San Bernardino Mountains, Calif., *Parish* 2054.

Elymus stebbinsii Gould, Madroño 9: 126. 1947. Based on *Agropyron parishii* Scribn. and Smith, not *Elymus parishii* Davy and Merr.

AGROPYRON PARISHII var. LAEVE Scribn. and Smith, U. S. Dept. Agr., Div. Agrost. Bul. 4: 28. 1897. Cuyamaca Mountains, Calif., *Palmer* 414.

Agropyron laeve Hitchc. in Jepson, Fl. Calif. 1: 181. 1912. Based on *A. parishii* var. *laeve* Scribn. and Smith.

Elymus pauciflorus subsp. *laeve* Gould, Madroño 9: 126. 1947. Based on *Agropyron parishii* var. *laeve* Scribn. and Smith.

(16) **Agropyron pringlei** (Scribn. and Smith) Hitchc. in Jepson, Fl. Calif. 1: 183. 1912. Based on *A. gmelini* var. *pringlei* Scribn. and Smith.

Agropyron gmelini[20] var *pringlei* Scribn. and Smith, U. S. Dept. Agr., Div. Agrost. Bul. 4: 31. 1897. Wyoming

and California [type, Summit Valley, *Pringle* in 1882].

Agropyron caninum var. *gmelini* forma *pringlei* Pease and Moore, Rhodora 12: 76. 1910. Based on *A. gmelini* var. *pringlei* Scribn. and Smith.

Agropyron spicatum var. *pringlei* Jones, West. Bot. Contrib. 14: 19. 1912. Based on *A. gmelini* var. *pringlei* Scribn. and Smith.

Elymus sierrus Gould, Madroño 9: 125. 1947. Based on *Agropyron gmelini* var. *pringlei* Scribn. and Smith, not *Elymus pringlei* Scribn. and Merr., 1901.

(4) **Agropyron pseudorepens** Scribn. and Smith, U. S. Dept. Agr., Div. Agrost. Bul. 4: 34. 1897. Texas and Arizona to Nebraska [type, Kearney, *Rydberg* 2018], Montana and British Columbia.

Agropyron pseudorepens var. *magnum* Scribn. and Smith, U. S. Dept. Agr., Div. Agrost. Bul. 4: 35. 1897. Enterprise, Colo., *Rydberg* 2401.

Agropyron tenerum magnum Piper, Torrey Bot. Club Bul. 32: 546. 1905. Based on *A. pseudorepens* var. *magnum* Scribn. and Smith.

Agropyron tenerum var. *pseudorepens* Jones, West. Bot. Contrib. 14: 19. 1912. Based on *A. pseudorepens* Scribn. and Smith.

Zeia pseudorepens Lunell, Amer. Midl. Nat. 4: 226. 1915. Based on *Agropyron pseudorepens* Scribn. and Smith.

Elymus pauciflorus subsp. *pseudorepens* Gould, Madroño 10: 94. 1949. Based on *Agropyron pseudorepens* Scribn. and Smith.

(3) **Agropyron pungens** (Pers.) Roem. and Schult., Syst. Veg. 2: 753. 1817. Based on *Triticum pungens* Pers.

Triticum pungens Pers., Syn. Pl. 1: 109. 1805. England.

Triticum repens var. *pungens* Duby in DC., Bot. Gall. 1: 529. 1828. Based on *T. pungens* Pers.

Braconotia pungens Godr., Fl. Lorr. 3: 192. 1844. Based on *Triticum pungens* Pers.

Agropyron repens subsp. *pungens* Hook. f., Stud. Fl. ed. 3: 504. 1884. Based on *A. pungens* Roem. and Schult.

Agropyron tetrastachys Scribn. and Smith, U. S. Dept. Agr., Div. Agrost. Bul. 4: 32. 1897. Cape Elizabeth, Maine, *Scribner* in 1895.

Elymus pauciflorus subsp. *pseudorepens* Gould, Madroño 10: 94. 1949. Based on *Agropyron pseudorepens* Scribn. and Smith.

(2) **Agropyron repens** (L.) Beauv., Ess. Agrost. 102, 146, 180. pl. 20, f. 2. 1812. Based on *Triticum repens* L.

Triticum repens L., Sp. Pl. 86. 1753. Europe.

Triticum infestum Salisb., Prodr. Stirp. 27. 1796. Based on *T. repens* L.

---

[20] *Triticum caninum* var. *gmelini* Griseb. in Ledeb., Icon. Pl. Ross. 3: 16. pl. 248. 1831, the basis of *Agropyron gmelini* Scribn. and Smith, U. S. Dept. Agr., Div. Agrost. Bul. 4: 30. 1897, and of *A. caninum* var. *gmelini* Pease and Moore, Rhodora 12: 75. 1910, is a Siberian species not known from North America. See note under *A. subsecundum.*

?*Triticum vaillantianum* Wulf. and Schreb. in Schweig. and Körte, Spec. Fl. Erlang. 1: 143. 1804. Germany. [This work not in Washington. From the description in ed. 2. 1: 143. 1811, this appears to be an awned form of *A. repens*.] *Braconotia officinarum* Godr., Fl. Lorr. 3: 192. 1844. Based on *Triticum repens* L. *Elytrigia repens* Desv. ex Jacks., Ind. Kew. 1: 836. 1893. Based on *Triticum repens* L.

Agropyron repens var. *pilosum* Scribn. in Rand and Redfield, Fl. Mt. Desert 183. 1894. Mount Desert, Maine, *Rand*.

Agropyron repens forma *geniculatum* Farwell, Detroit Commr. Parks and Boul. Ann. Rpt. 11: 48. 1900. Detroit, Mich., *Farwell* 1635.

Agropyron repens forma *stoloniferum* Farwell, Detroit Commr. Parks and Boul. Ann. Rpt. 11: 48. 1900. Detroit, *Farwell* 1634.

Zeia repens Lunell, Amer. Midl. Nat. 4: 227. 1915. Based on *Triticum repens* L.

Agropyron repens forma *pilosum* Fernald, Rhodora 35: 184. 1933. Based on *A. repens* var. *pilosum* Scribn.

Agropyron repens var. *subulatum* forma *heberhachis* Fernald, Rhodora 35: 184. 1933. Yarmouth, Nova Scotia, *Long* and *Linder* 20,091.

Agropyron repens var. *subulatum* forma *setiferum* Fernald, Rhodora 35: 184. 1933. Chelsea Beach, Mass., *Boott* in 1868.

?*Agropyron repens* var. *subulatum* forma *vaillantianum* Fernald, Rhodora 35: 184. 1933. Based on *Triticum vaillantianum* Wulf. and Schreb.

Elymus repens Gould, Madroño 9: 127. 1947. Based on *Triticum repens* L.

Agropyron leersianum (Wulf.) Rydb. (Brittonia 1: 85. 1931), based on "*Triticum repens leersianum* Wulfen" (apparently error for *T. leersianum* Wulf.) is applied to awned specimens of *A. repens*. The name, ultimately based on a description and figure named "*Elymus caninus* L." by Leers (Fl. Herborn. 46. pl. 12. f. 4. 1775), is uncertain. The figure, showing paired spikelets, appears to represent a species of *Elymus*.

(8) **Agropyron riparium** Scribn. and Smith, U. S. Dept. Agr., Div. Agrost. Bul. 4: 35. 1897. Montana [type, Garrison, *Rydberg* 2127].

Agropyron smithii var. *riparium* Jones, West. Bot. Contrib. 14: 19. 1912. Based on *A. riparium* Scribn. and Smith.

Zeia riparia Lunell, Amer. Midl. Nat. 4: 227. 1915. Based on *Agropyron riparium* Scribn. and Smith.

Elymus riparius Gould, Madroño 9: 127. 1947. Not *E. riparius* Wiegand, 1918. Based on *Agropyron riparium* Scribn. and Smith.

Elymus rydbergii Gould, Madroño 10: 94. 1949. Based on *Agropyron riparium* Scribn. and Smith. Not *Elymus riparius* Wiegand.

(23) **Agropyron saundersii** (Vasey) Hitchc., Wash. Biol. Soc. Proc. 41: 159. 1928. Based on *Elymus saundersii* Vasey.

Elymus saundersii Vasey, Torrey Bot. Club Bul. 11: 126. 1884. Veta Pass, Colo. [Vasey].

(22) **Agropyron saxicola** (Scribn. and Smith) Piper, U. S. Natl. Herb. Contrib. 11: 148. 1906. Based on *Elymus saxicola* Scribn. and Smith.

Elymus saxicola Scribn. and Smith, U. S. Dept. Agr., Div. Agrost. Bul. 11: 56. pl. 15. 1898. Mt. Chapaca, Wash., *Elmer* 554.

Sitanion flexuosum Piper, Erythea 7: 99. 1899. Wawawai, Wash., *Piper* 3004.

Sitanion lanceolatum J. G. Smith, U. S. Dept. Agr., Div. Agrost. Bul. 18: 20. 1899. Barker, Mont., *Rydberg* 3381.

Agropyron flexuosum Piper, Wash. Biol. Soc. Proc. 18: 149. 1905. Based on *Sitanion flexuosum* Piper.

Agropyron sitanioides J. G. Smith in Piper, Wash. Biol. Soc. Proc. 18: 149. 1905. Rapid City, S. Dak., *Griffiths* 735.

(17) **Agropyron scribneri** Vasey, Torrey Bot. Club Bul. 10: 128. 1883. Montana, *Scribner* in 1883.

Elymus scribneri Jones, West. Bot. Contrib. 14: 20. 1912. Based on *Agropyron scribneri* Vasey.

Agropyron semicostatum (Steud.) Nees ex Boiss., Fl. Orient. 5: 662. 1884. Presumably based on *Triticum semicostatum* Steud.

Triticum semicostatum Steud., Syn. Pl. Glum. 1: 346. 1854. Nepal.

Agropyron japonicum Tracy, U. S. Dept. Agr., Div. Agrost. Ann. Rpt. 1891: 6. 1892. Name only; Vasey ex Wickson, Calif.Agr. Expt. Sta. Rpt. 1895–1897: 275. pl. 14. f. 1. 1898. Erroneously listed as *Brachypodium japonicum* Miq. by Scribner, U. S. Dept. Agr., Div. Agrost. Bul. 14 (rev.): 22. 1900.

Roegneria semicostata Kitag., Manchukuo Inst. Sci. Res. Rpt. 3 (App. I): 91. 1939. Based on *Triticum semicostatum* Steud.

Agropyron sibiricum (Willd.) Beauv., Ess. Agrost. 102, 146, 181. 1812. Based on *Triticum sibiricum* Willd.

Triticum sibiricum Willd., Enum. Pl. 135. 1809. Siberia.

(5) **Agropyron smithii** Rydb., N. Y. Bot. Gard. Mem. 1: 64. 1900. (Feb.) Based on *A. spicatum* as described by Scribner and Smith (U. S. Dept. Agr., Div. Agrost. Bul. 4: 33. 1897), ["type * * * Geyer, upper Missouri"], not *Festuca spicata* Pursh, upon which they based the name.

Agropyron glaucum var. *occidentale* Scribn.,

Kans. Acad. Trans. 9: 110. 1885. Kansas. Scribner later (Torrey Bot. Club Mem. 5: 57. 1894) called this *A. repens glaucum*, but he based that name on *Triticum glaucum* Desf.

Agropyron occidentale Scribn., U. S. Dept. Agr., Div. Agrost. Cir. 27: 9. 1900. (Dec.) Based on *A. glaucum* var. *occidentale* Scribn.

Zeia occidentalis Lunell, Amer. Midl. Nat. 4: 226. 1915. Based on *Agropyron occidentale* Scribn.

Zeia smithii Lunell, Amer. Midl. Nat. 4: 227. 1915. Based on *Agropyron smithii* Rydb.

Agropyron spicatum var. *viride* Farwell, Mich. Acad. Sci. Rpt. 21: 356. 1920. Detroit, Mich., *Farwell* 851e.

Elymus smithii Gould, Madroño 9: 127. 1947. Based on *Agropyron smithii* Rydb.

Agropyron smithii var. *typica* Waterf., Rhodora 51: 21. 1949. Based on *A. smithii* Rydb.

AGROPYRON SMITHII var. MOLLE (Scribn. and Smith) Jones, West. Bot. Contrib. 14: 18. 1912. Based on *A. spicatum* var. *molle* Scribn. and Smith.

Agropyron spicatum var. *molle* Scribn. and Smith, U. S. Dept. Agr., Div. Agrost. Bul. 4: 33. 1897. Saskatchewan to Colorado, New Mexico, Idaho, and Washington. [Type, Montana, *Rydberg* 3193.]

Agropyron molle Rydb., N. Y. Bot. Gard. Mem. 1: 65. 1900. Based on *A. spicatum* var. *molle* Scribn. and Smith.

Agropyron occidentale var. *molle* Scribn., U. S. Dept. Agr., Div. Agrost. Cir. 27: 9. 1900. Based on *A. spicatum* var. *molle* Scribn. and Smith.

Zeia mollis Lunell, Amer. Midl. Nat. 4: 226. 1915. Based on *Agropyron spicatum* var. *molle* Scribn. and Smith.

AGROPYRON SMITHII var. PALMERI Heller, N. Amer. Pl. Cat. ed. 2: 3. 1900. Based on *A. spicatum* var. *palmeri* Scribn. and Smith. (Published as *A. smithii palmeri*.)

Agropyron spicatum var. *palmeri* Scribn. and Smith, U. S. Dept. Agr., Div. Agrost. Bul. 4: 33. 1897. Arizona [type *Palmer* in 1869] and New Mexico.

Agropyron occidentale var. *palmeri* Scribn., U. S. Dept. Agr., Div. Agrost. Cir. 27: 9. 1900. Based on *A. spicatum* var. *palmeri* Scribn. and Smith.

Agropyron palmeri Rydb., Colo. Agr. Expt. Sta. Bul. 100: 55. 1906. Based on *A. spicatum* var. *palmeri* Scribn. and Smith.

(18) **Agropyron spicatum** (Pursh) Scribn. and Smith, U. S. Dept. Agr., Div. Agrost. Bul. 4: 33. 1897. Based on *Festuca spicata* Pursh, but due to misidentification of Pursh's species, misapplied to *Agropyron smithii* Rydb.

Festuca spicata Pursh, Fl. Amer. Sept. 1: 83. 1814. Missouri and Columbia Rivers [type from Columbia River, *Lewis* and *Clark* in 1806].

Schedonorus spicatus Roem. and Schult., Syst. Veg. 2: 707. 1817. Based on *Festuca spicata* Pursh.

Triticum divergens Nees ex Steud., Syn. Pl. Glum. 1: 347. 1854. North America, *Douglas*.

Agropyron divergens Vasey, Descr. Cat. Grasses U. S. 96. 1885. Presumably based on *Triticum divergens* Nees.

Agropyron divergens var. *tenue* Vasey, Descr. Cat. Grasses U. S. 96. 1885. Name only; in Macoun, Can. Pl. Cat. 24: 242. 1888. Name only.

Agropyron divergens var. *tenuispicum* Scribn. and Smith, U. S. Dept. Agr., Div. Agrost. Bul. 4: 27. 1897. Washington and Oregon [type, *Howell* 181] to Wyoming and Montana.

Agropyron vaseyi Scribn. and Smith, U. S. Dept. Agr., Div. Agrost. Bul. 4: 27. 1897. Oregon and Washington to Wyoming and Colorado. [Type, Montana, *Rydberg* 2299.]

Agropyron spicatum tenuispicum Rydb., N. Y. Bot. Gard. Mem. 1: 61. 1900. Based on *A. divergens* var. *tenuispicum* Scribn. and Smith.

Agropyron spicatum var. *vaseyi* E. Nels., Bot. Gaz. 38: 378. 1904. Based on *A. vaseyi* Scribn. and Smith.

Zeia spicata Lunell, Amer. Midl. Nat. 4: 227. 1915. Based on *Festuca spicata* Pursh.

Elymus spicatus Gould, Madroño 9: 125. 1947. Based on *Festuca spicata* Pursh. This is the species called *Triticum strigosum* Less., by Thurber (S. Wats., Bot. Calif. 2: 324. 1880), and *Agropyron strigosum* by Coulter (Rocky Mount. Man. 426. 1885, the name erroneously ascribed to Beauv.). Not *T. strigosum* Less., of the Caspian region, nor *A. strigosum* (Bieb.) Boiss. (1884) of Asia Minor.

AGROPYRON SPICATUM var. PUBESCENS Elmer, Bot. Gaz. 36: 52. 1903. Mount Stuart, Wash., *Elmer* 1158.

Agropyron spicatum puberulentum Piper, U. S. Natl. Herb. Contrib. 11: 147. 1906. Based on *Agropyron spicatum* var. *pubescens* Elmer.

(12) **Agropyron subsecundum** (Link) Hitchc., Amer. Jour. Bot. 21: 131. 1934. Based on *Triticum subsecundum* Link.

Triticum subsecundum Link, Hort. Berol. 2: 190. 1833. Garden plant, seed collected by Richardson in western North America.

Triticum richardsoni Schrad., Linnaea 12: 467. 1838. North America.

Agropyron richardsoni Schrad., Linnaea 12: 467. 1838, as synonym of *Triticum richardsoni* Schrad.

Cryptopyrum richardsoni Heynh., Nom.
2: 174. 1846, as synonym of *Triticum richardsoni* Schrad.
Agropyron unilaterale Cassidy, Colo. Agr.
Expt. Sta. Bul. 12: 63. 1890. Not *A. unilaterale* Beauv., 1812. Colorado.
Agropyron caninum var. *unilaterale* Vasey,
U. S. Natl. Herb. Contrib. 1: 279.
1893. Based on *A. unilaterale* Cassidy,
though Vasey adds: "Type specimen
collected by F. Lamson-Scribner in
Montana in 1883 (no. 422)."
Agropyron violaceum forma *caninoides*
Ramaley, Minn. Bot. Studies 1: 108.
1894. Minnesota, *Macmillan* and *Sheldon* 84.
Agropyron caninum forma *violacescens*
Ramaley, Minn. Bot. Studies 1: 107.
1894. Based on *A. caninum* var. *unilaterale* Vasey.
Agropyron violacescens Beal, Grasses N.
Amer. 2: 635. 1896. Based on *A. caninum* forma *violacescens* Pound
(error for Ramaley).
Agropyron caninoides Beal, Grasses N.
Amer. 2: 640. 1896. Based on *A. violaceum* forma *caninoides* Ramaley.
Agropyron caninum var *pubescens* Scribn.
and Smith, U. S. Dept. Agr., Div.
Agrost. Bul. 4: 29. 1897. British Columbia, *Macoun* 99.
Agropyron richardsoni var. *ciliatum*
Scribn. and Smith, U. S. Dept. Agr.,
Div. Agrost. Bul. 4: 29. 1897. Montana, Belt Mountains, *Scribner* in 1883.
Agropyron caninum forma *glaucum* Pease
and Moore, Rhodora 12: 71. 1910.
Maine, *Fernald* 1367.
Agropyron caninum var. *unilaterale* forma
ciliatum Pease and Moore, Rhodora 12:
76. 1910. Based on *A. richardsoni* var.
ciliatum Scribn. and Smith.
Agropyron caninum var. *richardsoni* Jones,
West. Bot. Contrib. 14: 18. 1912.
Based on *Triticum richardsoni* "Trin."
(error for Schrad.).
Zeia richardsoni Lunell, Amer. Midl. Nat.
4: 227. 1915. Based on *Agropyron richardsoni* Schrad.
Agropyron trachycaulum var. *unilaterale*
Malte, Canada Natl. Mus. Ann. Rpt.
1930 (Bul. 68): 46. 1932. Based on *A. unilaterale* Cassidy.
Agropyron trachycaulum var. *ciliatum*
Malte, Canada Natl. Mus. Ann. Rpt.
1930 (Bul. 68): 47. 1932. Based on *A. richardsoni* var. *ciliatum* Scribn. and
Smith.
Agropyron trachycaulum var. *caerulescens*
Malte, Canada Natl. Mus. Ann. Rpt.
1930 (Bul. 68): 47. 1932. Vancouver
Island, *Malte*.
Agropyron trachycaulum var. *glaucum*
Malte, Canada Natl. Mus. Ann. Rpt.
1930 (Bul. 68): 47. 1932. Based on *A. caninum* forma *glaucum* Pease and
Moore.

Agropyron trachycaulum var. *pilosiglume*
Malte, Canada Natl. Mus. Ann. Rpt.
1930 (Bul. 68): 48. 1932. Victoria,
Vancouver Island, *Macoun*.
Agropyron trachycaulum var. *hirsutum*
Malte, Canada Natl. Mus. Ann. Rpt.
1930 (Bul. 68): 48. 1932. Victoria,
Vancouver Island, *Macoun*.
Elymus pauciflorus subsp. *subsecundus*
Gould, Madroño 9: 126. 1947. Based
on *Triticum subsecundum* Link.
This is the species which has been generally called *Agropyron caninum* (L.) Beauv.
by American authors. Most of the specimens
cited under *A. gmelini* Scribn. and Smith
(U. S. Dept. Agr., Div. Agrost. Bul. 4: 30.
1897) belong to *A. subsecundum*, but the
name was based in *Triticum caninum* var.
gmelini Griseb., a Siberian species.
AGROPYRON SUBSECUNDUM var. ANDINUM
(Scribn. and Smith) Hitchc., Amer.
Jour. Bot. 21: 132. 1934. Based on
A. violaceum andinum Scribn. and
Smith.
Agropyron violaceum var. *andinum* Scribn.
and Smith, U. S. Dept. Agr., Div.
Agrost. Bul. 4: 30. 1897. Colorado.
[Type, Grays Peak, *Jones* 720.]
Agropyron brevifolium Scribn., U. S. Dept.
Agr., Div. Agrost. Bul. 11: 55. pl. 13.
1898. Washington, *Elmer* 676.
Agropyron biflorum andinum Piper, Torrey Bot. Club Bul. 32: 547. 1905.
Based on *A. violaceum* var. *andinum*
Scribn. and Smith.
Agropyron andinum Rydb., Colo. Agr.
Expt. Sta. Bul. 100: 54. 1906. Based
on *A. violaceum* var. *andinum* Scribn.
and Smith.
Agropyron caninum var. *andinum* Pease
and Moore, Rhodora 12: 75. 1910.
Based on *A. violaceum* var. *andinum*
Scribn. and Smith.
(13) **Agropyron trachycaulum** (Link) Malte,
Canada Natl. Mus. Ann. Rpt. 1930.
(Bul. 68): 42. 1932. Based on *Triticum trachycaulum* Link.
Triticum pauciflorum Schwein., in Keat.,
Narr. Exped. St. Peter's River 2: 383.
1824. Prairies of the St. Peter [Minn.],
Say in 1823. Not *A. pauciflorum* Schur,
1859.
?*Triticum missuricum* Spreng., Syst. Veg.
1: 325. 1825. Missouri River. *Festuca
spicata* Pursh erroneously cited as synonym. The type has not been found.
A specimen of *Agropyron trachycaulum*
in the Vienna Herbarium, collected by
Geyer, "Missouri" in 1839 (long after
the name was published), is labeled
T. missuricum Spreng. There are no
rhizomes. Sprengel's description is inadequate, but applies to *A. trachycaulum*. *Triticum repens* and other
species having rhizomes are described
as having "radice repente," while *T.
missuricum* is not so described.

Triticum trachycaulum Link, Hort. Berol.
2: 189. 1833. Grown from seed collected by *Richardson* in North America.
Agropyron trachycaulon Steud., Syn. Pl.
Glum. 1: 344. 1854. Garden name as synonym of *Triticum trachycaulum* Link.
Crithopyrum trachycaulon Steud., Syn. Pl.
Glum. 1: 344. 1854. Garden name, as synonym of *Triticum trachycaulum* Link.
Agropyron tenerum Vasey, Bot. Gaz. 10: 258. 1885. Rocky Mountains. [Type, Fort Garland, Colo., *Vasey* in 1884.]
Agropyron violaceum var. *majus* Vasey, U. S. Natl. Herb. Contrib. 1: 280. 1893. Oregon, *Cusick* 1134.
Agropyron repens var. *tenerum* Beal, Grasses N. Amer. 2: 637. 1896. Based on *A. tenerum* Vasey.
Agropyron tenerum var. *longifolium* Scribn. and Smith, U. S. Dept. Agr., Div. Agrost. Bul. 4: 30. 1897. Oregon, Giant's [error for Grant's] Pass, *Howell* 256.
Agropyron tenerum var. *ciliatum* Scribn. and Smith, U. S. Dept. Agr., Div. Agrost. Bul. 4: 30. 1897. Minnesota [type, Duluth, *Vasey* in 1881] to Nebraska and Utah.
Agropyron novae-angliae Scribn. in Brain., Jones, and Eggl., Fl. Vt. 103. 1900. Westmore, Vt., *Grout* and *Eggleston* in 1894.
Agropyron tenerum majus Piper, Torrey Bot. Club Bul. 32: 543. 1905. Based on *A. violaceum* var. *major* Vasey.
Agropyron tenerum trichocoleum Piper, Torrey Bot. Club Bul. 32: 546. 1905. Based on *A. tenerum* var. *ciliatum* Scribn. and Smith.
Agropyron caninum var. *tenerum* Pease and Moore, Rhodora 12: 71. 1910. Based on *A. tenerum* Vasey.
Agropyron caninum var. *tenerum* forma *ciliatum* Pease and Moore, Rhodora 12: 72. 1910. Based on *A tenerum* var. *ciliatum* Scribn. and Smith.
Agropyron caninum var. *tenerum* forma *fernaldii* Pease and Moore, Rhodora 12: 73. 1910. Quebec, *Macoun* Herb. Geol. Survey Canada 68978.
Agropyron caninum var. *hornemanni* forma *pilosifolium* Pease and Moore, Rhodora 12: 75. 1910. Dead River, Maine, *Fernald* 576.
Zeia tenera Lunell, Amer. Midl. Nat. 4: 227. 1915. Based on *Agropyron tenerum* Vasey.
Agropyron tenerum var. *novae-angliae* Farwell, Mich. Acad. Sci. Rpt. 21: 355. 1920. Based on *A. novae-angliae* Scribn.
Agropyron missuricum Farwell, Amer Midl. Natl. 12: 48. 1930. Based on *Triticum missuricum* Spreng.
Agropyron trachycaulum var. *tenerum*

Malte, Canada Natl. Mus. Ann. Rpt. 1930 (Bul. 68): 44. 1932. Based on *A. tenerum* Vasey.
Agropyron trachycaulum var. *glaucescens* Malte, Canada Natl. Mus. Ann. Rpt. 1930 (Bul. 68): 45. 1932. Saskatchewan, *Malte.*
Agropyron trachycaulum var. *trichocoleum* Malte, Canada Natl. Mus. Ann. Rpt. 1930 (Bul. 68): 45. 1932. Based on *A. tenerum trichocoleum* Piper.
Agropyron trachycaulum var. *fernaldii* Malte, Canada Natl. Mus. Ann. Rpt. 1930 (Bul. 68): 46. 1932. Based on *A. caninum* var. *tenerum* forma *fernaldii* Pease and Moore.
Agropyron trachycaulum var. *majus* Fernald, Rhodora 35: 171. 1933. Based on *A. violaceum* var. *major* Vasey.
Agropyron trachycaulum var. *novae-angliae* Fernald, Rhodora 35: 174. 1933. Based on *A. novae-angliae* Scribn.
Agropyron pauciflorum Hitchc., Amer. Jour. Bot. 21: 132. 1934. Based on *Triticum pauciflorum* Schwein. Not *A. pauciflorum* Schur., 1859.
Roegneria trachycaulon Nevski in Komarov, Fl. U. R. S. S. 2: 599. 1934. Based on *Triticum trachycaulum* Link.
Roegneria pauciflora Hylander, Uppsala Univ. Årsk. 7: 36, 89. 1945. Based on *Triticum pauciflorum* Schwein.
Elymus pauciflorus Gould, Madroño 9: 126. 1947. Based on *Triticum pauciflorum* Schwein. Not *Elymus pauciflorus* Lam., 1791.
Alpine forms of this species have been referred to *Agropyron violaceum* (Hornem.) Lange and to *A. biflorum* (Brign.) Roem. and Schult.
Agropyron trichophorum (Link) Richt., Pl. Eur. 1: 124. 1890. Based on *Triticum trichophorum* Link.
Triticum trichophorum Link, Linnaea 17: 395. 1843. Europe.
Elytrigia trichophora Nevski, Acta Univ. Asiae Med. VIII b, Bot. 17: 57. 1934. Based on *Triticum trichophorum* Link.
Agropyron triticeum Gaertn., Nov. Comm. Petrop. 14¹: 540. 1770. Russia.
Secale prostratum Pall., Reise Prov. Russ. Reich. Anhang 1: 485. 1771. Russia.
Triticum prostratum L. f., Sup. Pl. 114. 1781. Based on *Secale prostratum* Pall.
Agropyron prostratum Beauv., Ess. Agrost. 102, 146. 1812. Based on *Triticum prostratum* L. f.
(11) **Agropyron vulpinum** (Rydb.) Hitchc., Amer. Jour. Bot. 21: 132. 1934. Based on *Elymus vulpinus* Rydb.
Elymus vulpinus Rydb., Torrey Bot. Club Bul. 36: 540. 1909. Grant County, Nebr., *Rydberg* 1617.
Agropyron richardsoni vulpinum Hitchc., Wash. Biol. Soc. Proc. 41: 159. 1928. Based on *Elymus vulpinus* Rydb.

(71) AGROSTIS L.

(3) **Agrostis aequivalis** (Trin.) Trin., Mém. Acad. St. Pétersb. VI. Sci. Nat. 4¹: 362. 1841. Based on *A. canina* var. *aequivalvis* Trin.

Agrostis canina var. *aequivalvis* Trin. in Bong., Acad. St. Pétersb. Mém. VI. Math. Phys. Nat. 2: 171. 1832. Sitka, Alaska.

Deyeuxia aequivalvis Benth. ex Vasey, U. S. Natl. Herb. Contrib. 3: 77. 1892, as synonym of *Agrostis aequivalvis* Trin. ex Jacks., Ind. Kew. 1: 740. 1893. Based on *A. aequivalvis* Trin. (as indicated by the reference to Benth., Linn. Soc. Bot. Jour. 19: 91. 1881, the combination not there made).

Podagrostis aequivalvis Scribn. and Merr., U. S. Natl. Herb. Contrib. 13: 58. 1910. Based on *Agrostis canina* var. *aequivalvis* Trin.

(8) **Agrostis alba** L., Sp. Pl. 63. 1753; ed. 2. 1: 93. 1762. Europe. Linnaeus' diagnosis is inadequate and his original application of the name is uncertain, but the specimen in his herbarium bearing the name in his own script belongs to the species for which the name has been generally used by European and American authors ever since. In recent American works this species has been called *A. palustris* Huds. But this name proves to belong to the creeping species with contracted panicle, the same as *A. maritima* Lam. See U. S. Dept. Agr., Bur. Plant Indus. Bul. 25. 1905, and U. S. Dept. Agr. Bul. 772: 128. 1920, for discussion of *A. alba* L. In the second edition of the Species Plantarum an undoubted reference to this species is added to the original uncertain one.

Agrostis gigantea Roth, Tent. Germ. 1: 31. 1788, described from Germany, is this species according to W. R. Philipson, who examined the type specimen borrowed from Berlin. (See Philipson, Revision of British species of Agrostis L., Linn. Soc. Jour. Bot. 51:90. 1937.)

Agrostis dispar Michx., Fl. Bor. Amer. 1: 52. 1803. South Carolina.

Decandolia alba Bast., Fl. Maine-et-Loire 29. 1809. Based on *Agrostis alba* L.

Vilfa alba Beauv., Ess. Agrost. 16, 146, 181. 1812. Based on *Agrostis alba* L.

Vilfa dispar Beauv., Ess. Agrost. 16, 147, 181. 1812. Based on *Agrostis dispar* Michx.

Agrostis alba var. *major* Gaudin, Fl. Helv. 1: 189. 1828. Switzerland.

Agrostis alba var. *dispar* Wood, Class-book ed. 1861. 774. 1861. Based on *A. dispar* Michx.

Agrestis alba Lunell, Amer. Midl. Nat. 4: 216. 1915. Based on *Agrostis alba* L.

Agrostis stolonifera var. *major* Farwell, Mich. Acad. Sci. Rpt. 21: 351. 1920. Based on *A. alba* var. *major* Gaudin.

Agrostis stolonifera forma *aristigera* Fernald, Rhodora 35· 317. 1933. Granville, Mass., *Seymour*.

Agrostis gigantea var. *dispar* Philipson, Linn. Soc. Jour. Bot. 51: 93. pl. 10. 1937. Based on *A. dispar* Michx.

Agrostis alba L. forma *aristata* Fernald, Rhodora 49: 112. 1947. Based on "*A. stolonifera* forma *aristata* Fernald."

Agrostis alba forma *aristigera* Fernald, Rhodora 51: 192. 1949. Based on *A. stolonifera* forma *aristigera* Fernald.

(33) **Agrostis altissima** (Walt.) Tuckerm., Amer. Jour. Sci. 45: 44. 1843. Based on *Cornucopiae altissima* Walt.

Cornucopiae altissima Walt., Fl. Carol. 74. 1788. South Carolina.

Trichodium elatum Pursh, Fl. Amer. Sept. 1: 61. 1814. New Jersey, Carolina.

Agrostis elata Trin., Acad. St. Pétersb. Mém. VI. Sci. Nat. 4¹: 317. 1841. Based on *Trichodium elatum* Pursh.

Trichodium altissimum Michx. ex Wood, Class-book ed. 2. 599. 1847. Based on *Cornucopiae altissima* Walt.

Agrostis perennans var. *elata* Hitchc., U. S. Dept. Agr., Bur. Plant Indus. Bul. 68: 50. 1905. Based on *Trichodium elatum* Pursh. (Published as *A. perennans elata*.)

Agrostis hyemalis var. *elata* Fernald, Rhodora 23: 229. 1921. Based on *Trichodium elatum* Pursh.

(25) **Agrostis ampla** Hitchc., U. S. Dept. Agr., Bur. Plant Indus. Bul. 68: 38. pl. 20. 1905. Rooster Rock, Oreg., *Suksdorf 135*.

Agrostis exarata var. *ampla* Hitchc., Amer. Jour. Bot. 2: 303. 1915. Based on *A. ampla* Hitchc.

(17) **Agrostis aristiglumis** Swallen, West. Bot. Leaflets 5: 56. 1947. Point Reyes Peninsula, Marin County, Calif., *J. T. Howell 23149*.

(1) **Agrostis avenacea** Gmel., Syst. Nat. 2: 171. 1791. Based on *Avena filiformis* G. Forst.

Avena filiformis G. Forst., Fl. Ins. Austr. Prodr. 9. 1786. New Zealand and Easter Island. Not *Agrostis filiformis* Vill., 1787, nor Willd., 1809.

Agrostis retrofracta Willd., Enum. Pl. 1: 94. 1809. Australia.

Vilfa retrofracta Beauv., Ess. Agrost. 16, 148, 182. 1812. Based on *Agrostis retrofracta* Willd.

Lachnagrostis retrofracta Trin., Fund. Agrost. 128. 1820. Based on *Agrostis retrofracta* Willd.

Lachnagrostis willdenovii Trin., Gram. Unifl. 217. 1824. Based on *Agrostis retrofracta* Willd.

Deyeuxia retrofracta Kunth, Rév. Gram. 1: 77. 1829. Based on *Agrostis retrofracta* Willd.

Calamagrostis retrofracta Link, Hort. Berol. 2: 247. 1833. Based on *Agrostis*

retrofracta Willd.
Calamagrostis willdenovii Steud., Syn. Pl.
Glum. 1: 192. 1854. Based on *Lachnagrostis willdenovii* Trin.
(22) **Agrostis blasdalei** Hitchc., Wash. Biol.
Soc. Proc. 41: 160. 1928. Fort Bragg,
Calif., *Davy* and *Blasdale* 6159.
(36) **Agrostis borealis** Hartm., Handb.
Skand. Fl. ed. 3. 17. 1838. Lapland.
?*Agrostis rubra* L., Sp. Pl. 62. 1753.
Sweden. Identity uncertain.
Agrostis canina var. *alpina* Oakes, Cat.
Vt. Pl. 32. 1842. Name only. Camels
Hump Mountain, Vt., *Robbins, Tuckerman*, and *Macrae.*
Agrostis canina var. *tenella* Torr., Fl.
N. Y. 2: 443. 1843. Northern New
York.
Agrostis pickeringii Tuckerm., Mag. Hort.
Hovey 9: 143. 1843. White Mountains, N. H.
Agrostis concinna Tuckerm., Mag. Hort.
Hovey 9: 143. 1843. Mount Monroe,
White Mountains, N. H.
Agrostis pickeringii var. *rupicola* Tuckerm., Amer. Jour. Sci. 45: 42. 1843.
White Mountains, N. H., *Pickering* and
Oakes; Vermont, Camels Hump.
Trichodium concinnum Wood, Class-book
ed. 2. 600. 1847. Based on *Agrostis
concinna* Tuckerm.
Agrostis rubra var. *americana* Scribn. in
Macoun, Can. Pl. Cat. 2⁵: 391. 1890.
Based on "*A. rupestris* Chapm. (non
All.), found on Roan Mountain, North
Carolina"; Tenn. Agr. Expt. Sta. Bul.
7: 77. f. 100. 1894. (See below.)
Agrostis novae-angliae Vasey, U. S. Natl.
Herb. Contrib. 3: 76. 1892. Not *A.
novae-angliae* Tuckerm. [Mount Washington, N. H., *Pringle.*]
Agrostis rubra var. *alpina* MacM., Met.
Minn. Vall. 65. 1892. Based on *A.
canina* var. *alpina* Oakes.
Agrostis borealis var. *macrantha* Eames,
Rhodora 11: 88. 1909. Blow-me-down
Mountains, Nova Scotia, *Eames* and
Godfrey in 1908 [No. 5833, the spikelets
abnormal].
Agrostis bakeri Rydb., Torrey Bot. Club
Bul. 36: 532. 1909. Pagosa Peak,
Colo., *Baker* 150.
Agrostis borealis var. *typica* Fernald,
Rhodora 35: 204. 1933. Based on *A.
borealis* Hartm.
Agrostis borealis var. *americana* Fernald,
Rhodora 35: 205. 1933. Based on *A.
rubra* var. *americana* Scribn.
Agrostis borealis forma *macrantha* Fernald,
Rhodora 35: 205. 1933. Based on *A.
borealis* var. *macrantha* Eames.
This species was erroneously referred to
Agrostis rupestris All. by A. Gray in a list of
plants from Roan Mountain, N. C., and by
Chapman (Fl. South. U. S. 551. 1860).
(26) **Agrostis californica** Trin., Acad. St.
Pétersb. Mém. VI. Sci. Nat. 4¹: 359.

1841. California. (*Vilfa glomerata* Presl
erroneously cited as synonym.)
Agrostis densiflora Vasey, U. S. Natl.
Herb. Contrib. 3: 72. 1892. Santa
Cruz, Calif., *Anderson.*
Agrostis densiflora var. *arenaria* Vasey,
U. S. Natl. Herb. Contrib. 3: 72. 1892.
Mendocino County, Calif., *Pringle.*
Agrostis arenaria Scribn., U. S. Natl.
Herb. Contrib. 3: 72. 1892. Not *A.
arenaria* Gouan, 1773. As synonym of
A. densiflora var. *arenaria* Vasey.
(35) **Agrostis canina** L., Sp. Pl. 62. 1753.
Europe.
Trichodium caninum Schrad., Fl. Germ.
1: 198. 1806. Based on *Agrostis canina*
L.
Agraulus caninus Beauv., Ess. Agrost. 5,
146, 147. 1812. Based on *Agrostis
canina* L.
Agrostis canina var. *alpina* Wood, Amer.
Bot. and Flor. pt. 2: 384. 1871. Not *A.
canina* var. *alpina* Ducomm., 1869.
Mountains of the Eastern States.
Agrostis alba var. *vulgaris* forma *aristata*
Millsp., Fl. W. Va. 469. 1892. Monangalia, W. Va.
Agrestis canina Bubani, Fl. Pyr. 4: 286.
1901. Based on *Agrostis canina* L.
(21) **Agrostis diegoensis** Vasey, Torrey Bot.
Club Bul. 13: 55. 1886. San Diego,
Calif., *Orcutt.*
Agrostis foliosa Vasey, Torrey Bot. Club
Bul. 13: 55. 1886. Not *A. foliosa*
Roem. and Schult., 1817. Oregon,
Howell [type] and *Bolander.*
Agrostis diegoensis var. *foliosa* Vasey,
U. S. Natl. Herb. Contrib. 3: 74. 1892.
Based on *A. foliosa* Vasey.
Agrostis canina var. *stolonifera* Vasey,
U. S. Natl. Herb. Contrib. 3: 75. 1892.
Not *A. canina* var. *stolonifera* Blytt,
1847. Oregon, *Henderson* [type] and
Howell.
Agrostis multiculmis Vasey ex Beal,
Grasses N. Amer. 2: 328. 1896, as
synonym of *A. diegoensis* Vasey.
Agrostis pallens foliosa Hitchc., U. S.
Dept. Agr., Bur. Plant Indus. Bul. 68:
34, pl. 14, f. 1. 1905. Based on *A.
foliosa* Vasey.
Agrostis pallens var. *vaseyi* St. John, Fl.
Southeast. Wash. and Adj. Idaho 30.
1937. Based on *A. foliosa* Vasey, not *A.
foliosa* Roem. and Schult.
(12) **Agrostis elliottiana** Schult., Mantissa
2: 202. 1824. Based on *A. arachnoides*
Ell.
Agrostis arachnoides Ell., Bot. S. C. and
Ga. 1: 134. 1816. Not *A. arachnoides*
Poir., 1810. Orangeburg, S. C., *Bennett.*
Notonema arachnoides Raf. ex Jacks., Ind.
Kew. 2: 319. 1894, as synonym of
Agrostis arachnoides Ell.
Notonema agrostoides Raf. ex Merrill,
Ind. Rafin. 76. 1949. Error for *N.
arachnoides* Raf.

(24) **Agrostis exarata** Trin., Gram. Unifl. 207. 1824. Unalaska, *Eschscholtz.*
Agrostis exarata var. *minor* Hook., Fl. Bor. Amer. 2: 239. 1839. Rocky Mountains, *Drummond, Douglas.*
Agrostis grandis Trin., Acad. St. Pétersb. Mém. VI. Sci. Nat. 4¹: 316. 1841. "Columbia (*Hooker*)."
Agrostis asperifolia Trin., Acad. St. Pétersb. Mém. VI. Sci. Nat. 4¹: 317. 1841. "Amer. bor.? Chile? (*Hooker*)." Probably collected in the Rocky Mountains and received from Hooker.
Agrostis scouleri Trin., Acad. St. Pétersb. Mém. VI. Sci. Nat. 4¹: 329. 1841. Nootka Sound, Vancouver Island, [received from] *Hooker.*
Agrostis albicans Buckl., Acad. Nat. Sci. Phila. Proc. 1862: 91. 1862. Columbia woods, Oreg , *Nuttall.*
Agrostis oregonensis Nutt. ex A. Gray, Acad. Nat. Sci. Phila. Proc. 1862: 334. 1862, as synonym of *A. albicans* Buckl.
Agrostis exarata forma *asperifolia* Vasey, U. S. Dept. Agr., Div. Bot. Bul. 13¹: pl. 31. 1892. No reference to Trinius, but the original of plate 31 is labeled *A. asperifolia* Trin. in Vasey's script.
Agrostis exarata var. *purpurascens* Hultén, Fl. Aleut. Isl. 73. 1937. Aleutian Islands, Unalaska, *Eyerdam* 2285.
AGROSTIS EXARATA var. PACIFICA Vasey, U. S. Dept. Agr., Div. Bot. Spec. Bul. (new ed.) 1889: 107. pl. 106. 1889. Pacific Coast.
AGROSTIS EXARATA var. MONOLEPIS (Torr.) Hitchc., Amer. Jour. Bot. 21: 136. 1934. Based on *Polypogon monspeliensis* var. *monolepis* Torr.
Polypogon monspeliensis var. *monolepis* Torr., U. S. Expl. Miss. Pacif. Rpt. 5: 366. 1858. Posé Creek, Walkers Pass, Calif. [Blake].
Agrostis ampla forma *monolepis* Beetle, Torrey Bot. Club Bul. 72: 544. 1945. Based on *Polypogon monspeliensis* var. *monolepis* Torr.
(13) **Agrostis exigua** Thurb. in S. Wats., Bot. Calif. 2: 275. 1880. Foothills of Sierras, Calif., *Bolander.*
(18) **Agrostis hallii** Vasey, U. S. Natl. Herb. Contrib. 3: 74. 1892. Oregon [type, *Hall* in 1872], Washington, and California.
Agrostis davyi Scribn., U. S. Dept. Agr., Div. Agrost. Cir. 30: 3. 1901. Point Arena, Calif., *Davy* and *Blasdale* 6062.
Agrostis occidentalis Scribn. and Merr., Torrey Bot. Club Bul. 29: 466. 1902. McMinnville, Oreg., *Shear* 1644.
AGROSTIS HALLII var. PRINGLEI (Scribn.) Hitchc., U. S. Dept. Agr., Bur. Plant Indus. Bul. 68: 33. pl. 12. 1905. Based on *A. pringlei* Scribn. (Published as *A. hallii pringlei*.)
Agrostis pringlei Scribn., U. S. Dept. Agr., Div. Agrost. Bul. 7: 156. f. 138. 1897.

Mendocino County, Calif., *Pringle.*
(14) **Agrostis hendersonii** Hitchc., Wash. Acad. Sci. Jour. 20: 381. 1930. Sams Valley, near Gold Hill, Jackson County, Oreg., *Henderson* 12387.
Agrostis microphylla var. *hendersonii* Beetle, Torrey Bot. Club Bul. 72: 547. f. 8. 1945. Based on *A. hendersonii* Hitchc.
(29) **Agrostis hiemalis** (Walt.) B. S. P., Prel. Cat. N. Y. 68. 1888. Based on *Cornucopiae hyemalis* Walt.
Cornucopiae hyemalis Walt., Fl. Carol. 73. 1788. South Carolina.
Trichodium laxiflorum Muhl., Descr. Gram. 60. 1817. Not *T. laxiflorum* Michx., 1803. Pennsylvania.
Agrostis laxiflora Poir. in Lam., Encycl. Sup. 1: 255. 1810. Carolina, Bosc.
Trichodium laxum Schult., Mantissa 2: 157. 1824. Based on *T. laxiflorum* Muhl.
Agrostis leptos Steud., Syn. Pl. Glum. 1: 169. 1854. Louisiana.
Agrostis canina var. *hyemalis* Kuntze, Rev. Gen. Pl. 3²: 338. 1898. Based on *Cornucopiae hyemalis* Walt.
Agrostis antecedens Bicknell, Torrey Bot. Club Bul. 35: 473. 1908. Nantucket, Bicknell in 1908.
Agrostis hyemalis Lunell, Amer. Midl. Nat. 4: 216. 1915. Based on *Cornucopiae hyemalis* Walt.
(27) **Agrostis hooveri** Swallen, West. Bot. Leaflets 5: 198. 1949. Type collected in sandy soil in open oak woodland, at summit on road between Arroyo Grande and Huasna district, San Luis Obispo County, Calif., June 29, 1948, by *Robert F. Hoover* 7549.
(28) **Agrostis howellii** Scribn., U. S. Natl. Herb. Contrib. 3: 76. 1892. Hood River, Oreg., *Howell* 198.
(10) **Agrostis humilis** Vasey, Torrey Bot. Club Bul. 10: 21. 1883. Mount Paddo [Adams], Wash., *Howell* [85].
(31) **Agrostis idahoensis** Nash, Torrey Bot. Club Bul. 24: 42. 1897. Forest, Idaho, *Heller* 3431.
Agrostis tenuis Vasey, Torrey Bot. Club Bul. 10: 21. 1883. Not *A. tenuis* Sibth. 1794. San Bernardino Mountains, Calif., *Parish Bros.* [1085].
Agrostis tenuiculmis Nash in Rydb., N. Y. Bot. Gard. Mem. 1: 32. 1900. Based on *A. tenuis* Vasey.
Agrostis tenuiculmis recta Nash in Rydb., N. Y. Bot. Gard. Mem. 1: 32. 1900. [Belt Pass, Mont., *Rydberg* 3327½.]
Agrostis tenuis erecta Vasey ex Nash in Rydb., N. Y. Bot. Gard. Mem. 1: 32. 1900, as synonym of *A. tenuiculmis recta* Nash.
Agrostis filiculmis Jones, West. Bot. Contrib. 14: 13. 1912. Little De Motte Park on the Kaibab, northern Arizona, [*Jones* 6056 bb.]

(15) **Agrostis kennedyana** Beetle, Torrey Bot. Club Bul. 72: 547. 1945. California, San Diego, *Grant* 896.

(19) **Agrostis lepida** Hitchc. in Jepson, Fl. Calif. 1: 121. 1912. Siberian Pass, Sequoia National Park, Calif., *Hitchcock* 3455.

(37) **Agrostis longiligula** Hitchc., U.S. Dept. Agr., Bur. Plant Indus. Bul. 68: 54. 1905. Fort Bragg, Calif., *Davy* and *Blasdale* 6110.

AGROSTIS LONGILIGULA var. AUSTRALIS J. T. Howell, West. Bot. Leaflets 4: 246. 1946. Marin County, Calif., *J. T. Howell* 18250.

(16) **Agrostis microphylla** Steud., Syn. Pl. Glum. 1: 164. 1854. North America, *Douglas.*

Agraulus brevifolius Nees ex Torr., U. S. Expl. Miss. Pacif. Rpt. 4: 154. 1857, as synonym of *Agrostis microphylla* Steud.

Polypogon alopecuroides Buckl., Acad. Nat. Sci. Phila. Proc. 1862: 88. 1862. Columbia Plains, Oreg., *Nuttall.*

Agrostis alopecuroides A. Gray, Acad. Nat. Sci. Phila. Proc. 1862: 333. 1862. Not *A. alopecuroides* Lam., 1791. Based on *Polypogon alopecuroides* Buckl.

Deyeuxia alopecuroides Nutt. ex A. Gray, Acad. Nat. Sci. Phila. Proc. 1862: 333. 1862, as synonym of *Polypogon alopecuroides* Buckl.

Agrostis exarata var. *microphylla* S. Wats. ex Vasey, U.S. Natl. Herb. Contrib. 3: 72. 1892, as synonym of *A. microphylla* var. *major* Vasey.

Agrostis virescens microphylla Scribn., U. S. Dept. Agr., Div. Agrost. Cir. 30: 2. 1901. Based on *A. microphylla* Steud.

Agrostis microphylla var. *intermedia* Beetle, Torrey Bot. Club Bul. 72: 547. f. 7. 1945. Lake County, Calif., *J. T. Howell* 18063.

AGROSTIS MICROPHYLLA var. MAJOR Vasey, U.S. Natl. Herb. Contrib. 3: 58, 72. 1892. [Truckee Valley, Nev., *Watson* 1284.]

Agrostis exarata microphylla Hitchc., Amer. Jour. Bot. 2: 303. 1915. Based on *A. microphylla* Steud.

Agrostis nebulosa Boiss. and Reut., Bibl. Univ. Genève (n.s.) 38: 218. 1842. Spain.

(7) **Agrostis nigra** With., Bot. Arr. Veg. Brit. ed. 3. 2: 131. 1796. Europe.

(34) **Agrostis oregonensis** Vasey, Torrey Bot. Club Bul. 13: 55. 1886. Oregon, *Howell* [49].

Agrostis attenuata Vasey, Bot. Gaz. 11: 337. 1886. Mount Hood, Oreg., *Howell* [210].

Agrostis hallii var. *californica* Vasey, U. S. Natl. Herb. Contrib. 3: 74. 1892. California [*Bolander* 6103].

Agrostis schiedeana var. *armata* Suksdorf, Werdenda 1[2]: 1. 1923. Klickitat County, Wash., *Suksdorf* 6310.

(20) **Agrostis pallens** Trin., Acad. St. Pétersb. Mem. VI. Sci. Nat. 4[1]: 328. 1841. "Amer.-borealis? (*Hooker*)."

Agrostis exarata var. *littoralis* Vasey, Torrey Bot. Club Bul. 13: 54. 1886. Oregon, *Howell* [64].

Agrostis densiflora var. *littoralis* Vasey, U. S. Natl. Herb. Contrib. 3: 72. 1892. Based on *A. exarata* var. *littoralis* Vasey.

(6) **Agrostis palustris** Huds., Fl. Angl. 27. 1762. England.

Agrostis polymorpha var. *palustris* Huds., Fl. Angl. 32. 1778. Based on *A. palustris* Huds.

Agrostis maritima Lam., Encycl. 1: 61. 1783. France.

Agrostis alba var. *palustris* Pers., Syn. Pl. 1: 76. 1805. Based on *A. palustris* Huds.

Milium maritimum Clem. y Rubio, Ensay. Vid. Andaluc. 285. 1807. Based on *Agrostis maritima* Lam.

Agrostis decumbens Gaud. ex Muhl., Descr. Gram. 68. 1817. Not *A. decumbens* Host, 1809. Pennsylvania, New Jersey.

Vilfa stolonifera var. *maritima* S. F. Gray, Nat. Arr. Brit. Pl. 2: 146. 1821. Based on *Agrostis maritima* With. (error for Lam.).

Apera palustris S. F. Gray, Nat. Arr. Brit. Pl. 2: 148. 1821. Based on *Agrostis palustris* With. (error for Huds.).

Agrostis alba var. *maritima* G. Meyer, Hannov. Mag. 1823: 138. 1824. Based on *A. maritima* Lam.

Agrostis stolonifera var. *maritima* Koch, Syn. Fl. Germ. Helv. 781. 1837. Based on *A. maritima* Lam.

?*Agrostis alba* var. *decumbens* Eaton and Wright, N. Amer. Bot. ed. 8. 117. 1840. Not *A. alba* var. *decumbens* Gaudin, 1828. Eastern United States.

Agrostis stolonifera var. *compacta* Hartm., Skand. Flora Handb. ed. 4. 24. 1843. Scandinavia.

Agrostis alba forma *maritima* Parl., Fl. Ital. 1: 181. 1848. Based on *A. maritima* Lam.

Agrostis depressa Vasey, Torrey Bot. Club Bul. 13: 54. 1886. Clear Creek Canyon, Colo., *Patterson* in 1885.

Agrostis exarata var. *stolonifera* Vasey, Torrey Bot. Club Bul. 13: 54. 1886. Columbia River, *Suksdorf.*

Agrostis reptans Rydb., Fl. Rocky Mount. 54. 1917. Based on *A. exarata* var. *stolonifera* Vasey.

Agrostis stolonifera var. *palustris* Farwell, Mich. Acad. Sci. Rpt. 21: 351. 1920. Based on *A. polymorpha* var. *palustris* Huds.

New England specimens of this species have been referred to *A. alba* var. *coarctata* Scribn., based on *A. coarctata* Ehrh., of Germany, which appears to be a narrow-panicled form of *A. stolonifera* L.

(32) **Agrostis perennans** (Walt.) Tuckerm., Amer. Jour. Sci. 45: 44. 1843. Based on *Cornucopiae perennans* Walt.

Cornucopiae perennans Walt., Fl. Carol. 74. 1788. South Carolina.

Agrostis cornucopiae Smith, Gentleman's Mag. 59: 873. 1789. Based on *Cornucopiae perennans* Walt.

Agrostis elegans Salisb., Prodr. Stirp. 25. 1796. Based on *Cornucopiae perennans* Walt.

Agrostis anomala Willd., Sp. Pl. 1: 370. 1797. Based on *Cornucopiae perennans* Walt.

Alopecurus carolinianus Spreng., Nachtr. Bot. Gart. Halle 10. 1801. Not *A. carolinianus* Walt., 1788. [Kentucky, *Peter.*]

Trichodium decumbens Michx., Fl. Bor. Amer. 1: 42. 1803. Virginia to Florida, *Michaux.*

Trichodium perennans Ell., Bot. S. C. and Ga. 1: 99. 1816. Based on *Cornucopiae perennans* Walt.

Trichodium muhlenbergianum Schult., Mantissa 2: 159. 1824. Pennsylvania, *Muhlenberg.* Based on Muhlenberg's *Trichodium* No. 4.

Agrostis michauxii Trin., Gram. Unifl. 206. 1824. Not *A. michauxii* Zucc., 1809. Based on *Trichodium decumbens* Michx.

Agrostis noveboracensis Spreng., Syst. Veg. 1: 260. 1825. New York, *Torrey.*

Agrostis decumbens Link, Hort. Berol. 1: 80. 1827. Not *A. decumbens* Host, 1809. Based on *Trichodium decumbens* Michx.

Trichodium noveboracense Schult., Mantissa 3 (Add. 1): 555. 1827. Based on *Agrostis noveboracensis* Spreng.

Trichodium scabrum [Muhl., misapplied by] Darl., Fl. Cestr. 1: 54. 1837. Pennsylvania.

Agrostis schweinitzii Trin., Acad. St. Pétersb. Mém. VI. Sci. Nat. 4¹: 311. 1841. Pennsylvania, *Schweinitz.*

Agrostis oreophila Trin., Acad. St. Pétersb. Mém. VI. Sci. Nat. 4¹: 323. 1841. Bethlehem, Pa., *Moser.* (*Trichodium montanum* Torr. is erroneously cited as synonym.)

Agrostis abakanensis Less. ex Trin., Acad. St. Pétersb. Mém. VI. Sci. Nat. 4¹: 325. 1841, as synonym of *A. michauxii* Trin.

Agrostis schiedeana Trin., Acad. St. Pétersb. Mém. VI. Sci. Nat. 4¹: 327. 1841. Mexico, type received from Schrader.

Agrostis novae-angliae Tuckerm., Mag. Hort. Hovey 9: 143. 1843. White Mountains, N. H.

Agrostis campyla Tuckerm., Amer. Jour. Sci. II. 6: 231. 1848. Based on "*A. scabra*" as described by Tuckerman.

Agrostis scabra var. *perennans* Wood, Class-book ed. 1861. 774. 1861. Presumably based on *A. perennans* Tuckerm.

Agrostis perennans var. *aestivalis* Vasey, U.S. Natl. Herb. Contrib. 3: 76. 1892. Athens, Ill. [*Hall*]. The slender lax form.

Agrostis intermedia Scribn., Torrey Bot. Club Bul. 20: 476. 1893. Not *A. intermedia* Balb., 1801. Pine Mountain, Harlan County, Tenn. *Kearney* 39.

Agrostis pseudointermedia Farwell, Detroit Commr. Parks and Boul. Ann. Rpt. 11: 46. 1900. Based on *A. intermedia* Scribn.

Agrostis scribneriana Nash in Small, Fl. Southeast. U. S. 126. 1903. Based on *A. intermedia* Scribn.

Agrostis hyemalis var. *oreophila* Farwell, Mich. Acad. Sci. Rpt. 6: 202. 1904. Based on *A. oreophila* Trin.

Agrostis perennans var. *humilis* Farwell, Mich. Acad. Sci. Papers 1: 87. 1921. Detroit, *Farwell 5672½.*

Agrostis perennans forma *chaetophora* Fernald, Rhodora 35: 317. 1933. Huntingdon County, Pa., *Lowrie.*

Agrostis perennans var. *aestivalis* forma *atherophora* Fernald, Rhodora 35: 317. 1933. Terrebonne, Quebec, *Churchill.*

(11) **Agrostis rossae** Vasey, U. S. Natl. Herb. Contrib. 3: 76. 1892. Yellowstone Park, Wyo., *Edith Ross* in 1890.

Agrostis exarata var. *rossae* G. N. Jones, Wash. Univ. Pubs. Biol. 5: 113. 1936. Based on *A. rossae* Vasey.

(30) **Agrostis scabra** Willd., Sp. Pl. 1: 370. 1797. North America.

Agrostis laxa Muhl., Amer. Phil. Soc. Trans. 4: 236. 1799. Name only.

Trichodium laxiflorum Michx., Fl. Bor. Amer. 1: 42. 1803. Hudson Bay to Florida, *Michaux.*

Vilfa scabra Beauv., Ess. Agrost. 16, 182. 1812. Based on *Agrostis scabra* Willd.

Trichodium scabrum Muhl., Cat. Pl. 10. 1813. Based on *Agrostis scabra* Willd.

Agrostis laxa Schreb. ex Pursh, Fl. Amer. Sept. 1: 61. 1814, as synonym of *Trichodium laxiflorum* Michx.

Agrostis laxiflora Richards., Bot. App. Franklin Jour. 731. 1823. Based on *Trichodium laxiflorum* Michx.

Trichodium montanum Torr., Fl. North. and Mid. U. S. 84. 1823. Fishkill Mountains, N. Y.

Trichodium album Presl, Rel. Haenk. 1: 244. 1830. Nootka Sound, Vancouver Island, *Haenke.*

Agrostis nutkaensis Kunth, Rév. Gram. 1: Sup. 17. 1830. Based on *Trichodium album* Presl.

Agrostis michauxii var. *laxiflora* A. Gray,

N. Amer. Gram. and Cyp. **1: 17.** 1834. Based on *Trichodium laxiflorum* Michx.
Agrostis nootkaensis Trin., Acad. St. Pétersb. Mém. VI. Sci. Nat. 4¹: 326. 1841. Based on *Trichodium album* Presl.
Agrostis laxiflora var. *montana* Tuckerm., Amer. Jour. Sci. 45: 43. 1843. Based on *Trichodium montanum* Torr.
Agrostis scabra var. *tenuis* Tuckerm., Amer. Jour. Sci. 45: 45. 1843. Lincoln, N. H.
Agrostis laxiflora var. *caespitosa* Torr., Fl. N. Y. 2: 442. 1843. Based on *Trichodium montanum* Torr.
Agrostis laxiflora var. *scabra* Torr., Fl. N. Y. 2: 442. 1843. Based on *A. scabra* Willd.
Agrostis laxiflora var. *tenuis* Torr., Fl. N. Y. 2: 442. 1843. Based on *A. scabra* var. *tenuis* Tuckerm.
Agrostis torreyi Tuckerm., Mag. Hort. Hovey 9: 143. 1843. Not *A. torreyi* Kunth, 1830. Based on *Trichodium montanum* Torr.
Agrostis scabra var. *oreophila* Wood, Classbook ed. 1861. **774.** 1861. Based on *A. [laxiflora* var.] *montana* Tuckerm. (There is no reference to *A. oreophila* Trin.)
Agrostis scabriuscula Buckl., Acad. Nat. Sci. Phila. Proc. 1862: 90. 1862. Columbia Plains, Oreg., *Nuttall.*
Agrostis scabrata Nutt. ex A. Gray, Acad. Nat. Sci. Phila. Proc. 1862: 334. 1862, as synonym of *A. scabriuscula* Buckl.
Agrostis scabra var. *montana* Fernald, Portland Soc. Nat. Hist. Proc. 2: 91. 1895. Based on *Trichodium montanum* Torr. This combination was made by Paine (giving Tuckerm. as author), State Cabinet Nat. Hist., N. Y. Ann. Rpt. 18: 166. 1865, and by Vasey (also giving Tuckerm. as author), U. S. Natl. Herb. Contrib. 3: 76. 1892, erroneously cited as synonym of *A. novae-angliae* Vasey. The basis is not given in either publication.
Agrostis hyemalis var. *keweenawensis* Farwell, Mich. Acad. Sci. Rpt. 6: 203. 1904. Keweenaw County, Mich.
Agrostis hiemalis nutkaensis Scribn. and Merr., U. S. Natl. Herb. Contrib. 13: 56. 1910. Based on *A. nutkaensis* Kunth.
Agrostis scabra forma *tuckermani* Fernald, Rhodora 35: 207. 1933. Braintree, Mass., *Churchill* in 1911.
Agrostis peckii House, Amer. Midl. Nat. 7: 126. 1921. Based on *A. laxiflora* var. *caespitosa* Torr. "*A. caespitosa* Torr. . . . Not Salisb." is erroneously cited. The statement that "Torrey's type was collected on Mt. Beacon, near Fishkill," indicates that *A. peckii* is based on *A. laxiflora* var. *caespitosa* Torr.

Agrostis scabra var. *keweenawensis* Farwell, Mich. Acad. Sci. Papers 23: 125. 1938. Based on *A. hyemalis* var. *keweenawensis* Farwell.
AGROSTIS SCABRA var. GEMINATA (Trin.) Swallen, Wash. Biol. Soc. Proc. 54: 1941. Based on *Agrostis geminata* Trin.
Agrostis geminata Trin., Gram. Unifl. 207. 1824. Unalaska, *Eschscholtz.*
Agrostis hyemalis var. *geminata* Hitchc., U. S. Dept. Agr., Bur. Plant Indus. Bul. 68: 44. 1905. Based on *A. geminata* Trin. (Published as *A. hiemalis geminata.*)
Agrostis geminata forma *exaristata* Fernald, Rhodora 35: 211. 1933. Gaspé County, Quebec, *Fernald, Dodge,* and *Smith* 25, 485.
(4) **Agrostis semiverticillata** (Forsk.) C. Christ., Dansk Bot. Arkiv 4³: 12. 1922. Based on *Phalaris semiverticillata* Forsk.
Phalaris semiverticillata Forsk., Fl. Aegypt. Arab. 17. 1775. Egypt.
Agrostis verticillata Vill., Prosp. Pl. Dauph. 16. 1779. France.
Agrostis alba var. *verticillata* Pers., Syn. Pl. 1: 76. 1805. Based on *A. verticillata* Vill.
Agrostis villarsii Poir. in Lam., Encycl. Sup. 1: 251. 1810. Based on *A. verticillata* Vill.
Vilfa verticillata Beauv., Ess. Agrost. 16, 148, 182. 1812. Based on *Agrostis verticillata* Vill.
Agrostis decumbens Muhl. ex Ell. Bot. S. C. and Ga. 1: 136. 1816. Not *A. decumbens* Host, 1809. Charleston, S. C.
Agrostis stolonifera var. *verticillata* St. Amans, Fl. Agen. 28. 1821. Based on *A. verticillata* Vill.
Agrostis condensata Willd. ex Steud., Nom. Bot. ed. 2. 1: 40. 1840, as synonym of *A. verticillata* Vill.
Agrostis aquatica Buckl., Acad. Nat. Sci. Phila. Proc. 1862: 90. 1862. Not *A. aquatica* Pourr., 1783. San Saba County, Tex.
Agrostis verticillata Bubani, Fl. Pyr. 4: 282. 1901. Based on *Agrostis verticillata* Vill.
Nowodworskya verticillata Nevski, Akad. Nauk S. S. S. R. Bot. Inst. Trudy I. (Acad. Sci. U. R. S. S. Inst. Bot. Acta I. Flora et Syst. Plant. Vasc.) 3: 143. 1936. Based on *Agrostis verticillata* Vill.
Nowodworskya semiverticillata Nevski, Akad. Nauk S. S. S. R. Bot. Inst. Trudy I. (Acad. Sci. U. R. S. S. Bot. Acta I. Flora et Syst. Plant. Vasc.) 4: 339. 1937. Based on *Phalaris semiverticillata* Forsk.
Polypogon semiverticillatus Hylander, Uppsala Univ. Årsk. 7: 74. 1945. Based on *Phalaris semiverticillata* Forsk. The same combination made by Hoover, West. Bot. Leaflets 5: 138. 1948.
(5) **Agrostis stolonifera** L., Sp. Pl. 62. 1753. Europe.

Decandolia stolonifera Bast., Fl. Maine-et-Loire 29. 1809. Based on *Agrostis stolonifera* L.

Vilfa stolonifera Beauv., Ess. Agrost. 16, 148, 182. 1812. Based on *Agrostis stolonifera* L.

Agrostis alba var. *stolonifera* Smith, English Fl. 1: 93. 1824. Based on *A. stolonifera* L.

Agrostis vulgaris var. *stolonifera* Koch, Syn. Fl. Germ. Helv. 782. 1837. Based on *A. stolonifera* L.

(9) **Agrostis tenuis** Sibth., Fl. Oxon. 36. 1794. Based on *A. capillaris* Huds.

Agrostis capillaris Huds., Fl. Angl. ed. 2. 27. 1762. Not *A. capillaris* L., 1753. England.

Agrostis sylvatica Huds., Fl. Angl. ed. 2. 28. 1762. England. A teratological form, the florets abnormally elongated. Name rejected, being based on a monstrosity.

Agrostis vulgaris With., Bot. Arr. Veg. Brit. ed. 3. 2: 132. 1796. Europe.

Vilfa vulgaris Beauv., Ess. Agrost. 16, pl. 5. f. 8. 1812. Based on *Agrostis vulgaris* With.

Agrostis alba var. *sylvatica* Smith, English Fl. 1: 93. 1824. Based on *A. sylvatica* Huds. Published as new by Scribner, Torrey Bot. Club Mem. 5: 40. 1894, the basis given as "*A. sylvatica* L." error for Huds.

Agrostis alba var. *vulgaris* Coss. and Dur., Expl. Sci. Alger. 2: 63. 1854–1855. Based on *A. vulgaris* With.

Agrostis stolonifera var. *vulgaris* Celak., Prodr. Fl. Bohm. 710. 1881. Not *A. stolonifera* var. *vulgaris* Heuff., 1858. Based on *A. vulgaris* With.

Agrostis alba var. *minor* Vasey, U. S. Natl. Herb. Contrib. 3: 78. 1892. [Washington, D.C.]

Agrostis stolonifera var. *minor* Farwell, Mich. Acad. Sci. Rpt. 6: 202. 1904. Based on *A. alba* var. *minor* Vasey.

This species has been referred to *Agrostis capillaris* L., a European species not known from America.

AGROSTIS TENUIS var. ARISTATA (Parnell) Druce, List Brit. Pl. 79. 1908. Presumably based on *A. vulgaris* var. *aristata* Parnell.

Agrostis stricta Willd., Sp. Pl. 1: 366. 1797. Not *A. stricta* Gmel., 1791. North America.

Agrostis stricta Muhl., Descr. Gram. 65. 1817. Not *A. stricta* Gmel., 1791. New England and Carolina.

Trichodium strictum Roem. and Schult., Syst. Veg. 2: 281. 1817. Based on *Agrostis stricta* Willd.

Agrostis diffusa Muhl. ex Spreng., Syst. Veg. 1: 260. 1825. Not *A. diffusa* Host, 1809, nor Muhl., 1817. As synonym of *A. stricta* Muhl.

Agrostis vulgaris var. *aristata* Parnell,

Grasses Scotl. 1¹: 34. pl. 13. 1842. Scotland.

Agrostis alba var. *aristata* A. Gray, Man. 578. 1848. Not *A. alba* var. *aristata* Spenner, 1825. Based on *A. stricta* Willd.

Agrostis stricta Buse, in Miquel, Pl. Jungh. 341. 1854. Not *A. stricta* Gmel., 1791. Based on *Trichodium strictum* Roem. and Schult.

Agrostis alba var. *stricta* Wood, Classbook ed. 1861. 774. 1861. Based on *A. stricta* Willd.

Agrostis tenuis forma *aristata* Wiegand, Rhodora 26: 2. 1924. Based on *A. vulgaris* var. *aristata* Parnell.

Agrostis palustris var. *stricta* House, N. Y. State Mus. Bul. 254: 98. 1924. Based on *Agrostis stricta* Willd.

Agrostis capillaris var. *aristata* Druce, Fl. Oxfordsh. ed. 2. 474. 1927. Presumably based on *A. vulgaris* var. *aristata* Parnell.

Agrostis capillaris aristulata Hitchc. Wash. Biol. Soc. Proc. 41: 160. 1928. Alexandria, Va. *Amer. Gr. Natl. Herb.* 344.

(2) **Agrostis thurberiana** Hitchc., U. S. Dept. Agr., Bur. Plant Indus. Bul. 68: 23. pl. 1. f. 1. 1905. Skamania County, Wash., *Suksdorf* 1021.

Agrostis hillebrandii Thurb. ex Boland. Agr. Soc. Calif. Trans. 1864–1865: 136. 1866. Name only. Sierra Nevada, Calif., *Hillebrand*.

Agrostis atrata Rydb., Torrey Bot. Club Bul. 36: 531. 1909. Yoho Valley, British Columbia, *Macoun* 64787.

(23) **Agrostis variabilis** Rydb., N. Y. Bot. Gard. Mem. 1: 32. 1900. Based on *A. varians* Trin.

Agrostis varians Trin., Acad. St. Pétersb. Mém. VI. Sci. Nat. 4¹: 314. 1841. Not *A. varians* Thuill., 1790. "America boreal? (Hoocker 217)." A duplicate type in the Torrey Herbarium (N. Y. Bot. Gard.) is labeled "Rocky Mountains, Hooker 217."

(59) **AIRA L.**

(2) **Aira caryophyllea** L., Sp. Pl. 66. 1753. Europe.

Avena caryophyllea Wigg., Prim. Fl. Hols. 10. 1780. Based on *Aira caryophyllea* L.

Agrostis caryophyllea Salisb., Prodr. Stirp. 25. 1796. Based on *Aira caryophyllea* L.

Airopsis caryophyllea Fries, Nov. Fl. Suec. ed. 2. Cont. 3: 180. 1842. Based on *Aira caryophyllea* L.

Caryophyllea airoides Opiz, Sezn. Rostl. Ceské 27. 1852. Based on *Aira caryophyllea* L.

Fussia caryophyllea Schur, Enum. Pl.

Transsilv. 754. 1866. Based on *Aira caryophyllea* L.
Airella caryophyllea Dum., Soc. Bot. Belg. Bul. 7¹: 68. 1868. Based on *Aira caryophyllea* L.
Salmasia vulgaris Bubani, Fl. Pyr. 4: 316. 1901. Based on *Aira caryophyllea* L.
Aspris caryophyllea Nash in Britt. and Brown, Illustr. Fl. ed. 2. 1: 214. 1913. Based on *Aira caryophyllea* L.
(3) **Aira elegans** Willd. ex Gaudin, Agrost. Helv. 1: 130, 355. 1811. Pavia, Italy.
Aira capillaris Host, Icon. Gram. Austr. 4: 20. pl. 35. 1809. Not *A. capillaris* Savi, 1798, nor *A. capillaris* Lag., 1805. Europe.
Avena capillaris Mert. and Koch in Roehl., Deut. Fl. ed. 3. 1²: 573. 1823. Based on *Aira capillaris* Host.
Airopsis capillaris Schur, Oesterr. Bot. Ztschr. 9: 328. 1859. Based on *Aira capillaris* Host.
Fussia capillaris Schur, Enum. Pl. Transsilv. 754. 1866. Based on *Aira capillaris* Host.
Airella capillaris Dum., Soc. Bot. Belg. Bul. 7¹: 68. 1868. Based on *Aira capillaris* Host.
Aspris capillaris Hitchc., U. S. Dept. Agr. Bul. 772: 116. 1920. Based on *Aira capillaris* Host.
(1) **Aira praecox** L., Sp. Pl. 65. 1753. Europe.
Agrostis praecox Salisb., Prodr. Stirp. 24. 1796. Based on *Aira praecox* L.
Avena praecox Beauv., Ess. Agrost. 89, 154. 1812. Based on *Aira praecox* L.
Trisetum praecox Dum., Obs. Gram. Belg. 122. pl. 8. f. 30. 1823. Based on *Aira praecox* L.
Airopsis praecox Fries, Nov. Fl. Suec. ed. 2. Cont. 3: 180. 1842. Based on *Aira praecox* L.
Caryophyllea praecox Opiz, Sezn. Rostl. Ceské 27. 1852. Based on *Aira praecox* L.
Fussia praecox Schur, Enum. Pl. Transsilv. 754. 1866. Based on *Aira praecox* L.
Airella praecox Dum., Soc. Bot. Belg. Bul. 7¹: 68. 1868. Based on *Aira praecox* L.
Salmasia praecox Bubani, Fl. Pyr. 4: 316. 1901. Based on *Aira praecox* L.
Aspris praecox Nash, in Britt. and Brown, Illus. Fl. ed. 2. 1: 215. 1913. Based on *Aira praecox* L.

(76) **ALOPECURUS L.**

(5) **Alopecurus aequalis** Sobol., Fl. Petrop. 16. 1799. Greece.
Alopecurus aristulatus Michx., Fl. Bor. Amer. 1: 43. 1803. Canada, *Michaux.*
Alopecurus fulvus J. E. Smith in Sowerby, English Bot. 21: pl. 1467. 1805. England.
Alopecurus subaristatus Pers., Syn. Pl. 1: 80. 1805. Canada.

Alopecurus geniculatus var. *natans* Wahl., Fl. Lapp. 22. 1812. Lapland.
Alopecurus geniculatus var. *aristulatus* Torr., Fl. North. and Mid. U. S. 1: 97. 1823. Based on *A. aristulatus* Michx.
Alopecurus caespitosus Trin., Gram. Icon. 3: pl. 241. 1836. North America, [type, Northwest America, *Douglas*].
Alopecurus geniculatus var. *fulvus* Schrad., Linnaea 12: 424. 1838. Based on *A. fulvus* J. E. Smith.
Alopecurus geniculatus var. *robustus* Vasey, Torrey Bot. Club Bul. 15: 13. 1888. Vancouver Island, *Macoun.*
Alopecurus howellii var. *merrimani* Beal, Grasses N. Amer. ¹2: 278. 1896. Pribilof Islands, Alaska, "C. H. Merriman" [error for *Merriam*].
Alopecurus howellii var. *merriami* Beal, ex Macoun, in Jordan, Fur Seals North Pacif. 3: 573. 1899. (Correction of var. *merrimani* Beal.)
Alopecurus aristulatus var. *natans* Simmons, Arkiv Bot. 6¹⁷: 4. 1907. Based on *A. geniculatus* var. *natans* Wahl.
Tozzettia fulva Lunell, Amer. Midl. Nat. 4: 216. 1915. Based on *Alopecurus fulvus* J. E. Smith.
Alopecurus artistulatus var. *merriami* St. John, Canada Dept. Mines Mem. 126: 42. 1922. Based on *A. howellii* var. *merriami* Beal.
Alopecurus aequalis var. *natans* Fernald, Rhodora 27: 198. 1925. Based on *Alopecurus geniculatus* var. *natans* Wahl.
(3) **Alopecurus alpinus** J. E. Smith in Sowerby, English Bot. pl. 1126. 1803. Scotland.
?Alopecurus borealis Trin., Fund. Agrost. 58. 1820. Asia and North America.
Alopecurus occidentalis Scribn. and Tweedy, Bot. Gaz. 11: 170. 1886. Yellowstone National Park, *Tweedy.*
Alopecurus behringianus Gandog., Soc. Bot. France Bul. 66⁷: 298. 1920. St. Paul Island, Alaska, *Macoun.*
Vasey misapplied the name *Alopecurus pratensis* var. *alpestris* Wahl. to this species in U. S. Natl. Herb. Contrib. 3: 86. 1892.
Alopecurus arundinaceus Poir. in Lam. Encycl. 8: 766. 1808. Cultivated in Botanical Garden, Paris.
Alopecurus ventricosus Pers., Syn. Pl. 1: 80. 1805. Not *A. ventricosus* (Gouan) Huds., 1778. France.
Alopecurus pratensis var. *ventricosus* Coss. and Dur. Expl. Sci. Alger. 2: 56. 1854–55. Based on *Alopecurus ventricosus* Pers.
(7) **Alopecurus carolinianus** Walt., Fl. Carol. 74. 1788. South Carolina.
Alopecurus ramosus Poir. in Lam., Encycl. 8: 776. 1808. Carolina, *Bosc.*
Alopecurus pedalis Bosc. ex Beauv., Ess. Agrost. 4. 1812. Name only. [Carolina, *Bosc.*]
Alopecurus gracilis Willd. ex Trin., Acad.

St. Pétersb. Mém. VI. Sci. Nat. 4¹: 38. 1840. Carolina [Bosc].
Alopecurus macounii Vasey, Torrey Bot. Club Bul. 15: 12. 1888. Oak Bay, Vancouver Island, Macoun.
Alopecurus geniculatus var. caespitosus Scribn., in Macoun, Can. Pl. Cat. 2⁵: 389. 1890. Yale, British Columbia, Macoun.
Alopecurus geniculatus var. ramosus St. John, Rhodora 19: 167. 1917. Based on A. ramosus Poir.
Alopecurus creticus Trin. in Spreng., Neu. Entd. 2: 45. 1821. Crete.
(6) **Alopecurus geniculatus** L., Sp. Pl. 60. 1753. Europe.
Tozzettia geniculata Bubani, Fl. Pyr. 4: 275. 1901. Based on Alopecurus geniculatus L.
(8) **Alopecurus howellii** Vasey, Torrey Bot. Club Bul. 15: 12. 1888. [Medford], Oreg., Howell [215].
Alopecurus californicus Vasey, Torrey Bot. Club Bul. 15: 13. 1888. California [type, Santa Cruz, Anderson] and Oregon.
(1) **Alopecurus myosuroides** Huds., Fl. Angl. 23. 1762. England.
Alopecurus agrestis L., Sp. Pl. ed. 2. 1: 89. 1762. Europe.
Tozzettia agrestis Bubani, Fl. Pyr. 4: 274. 1901. Based on Alopecurus agrestis L.
(4) **Alopecurus pallescens** Piper, Fl. Palouse 18. 1901. Pullman, Wash., Piper 1743.
(2) **Alopecurus pratensis** L., Sp. Pl. 60. 1753. Europe.
Alopecurus rendlei Eig, Brit. and For. Jour. Bot. 75: 187. 1937. Based on Phalaris utriculata L., not Alopecurus utriculatus Banks and Solander.
Phalaris utriculata L. Syst. Nat. ed. 10. 869. 1759.
Alopecurus utriculatus Pers., Syn. Pl. 1: 80. 1805. Not Alopecurus utriculatus Banks and Solander, 1794. Based on Phalaris utriculata L.
(9) **Alopecurus saccatus** Vasey, Bot. Gaz. 6: 290. 1881. Eastern Oregon, Howell.

(68) AMMOPHILA Host

(2) **Ammophila arenaria** (L.) Link, Hort. Berol. 1: 105. 1827. Based on Arundo arenaria L.
Arundo arenaria L., Sp. Pl. 82. 1753. Europe.
Calamagrostis arenaria Roth, Tent. Fl. Germ. 1: 34. 1788. Based on Arundo arenaria L.
Ammophila arundinacea Host, Icon. Gram. Austr. 4: 24. pl. 41. 1809. Based on Arundo arenaria L.
Psamma littoralis Beauv., Ess. Agrost. 144. pl. 6. f. 1. 176. 1812. Europe.
Psamma arenaria Roem. and Schult., Syst. Veg. 2: 845. 1817. Based on Calamagrostis arenaria Roth.
Phalaris maritima Nutt., Gen. Pl. 1: 48.

1818. Based on Arundo arenaria L., but misapplied to Ammophila breviligulata.
Phalaris ammophila Link, Enum. Hort. Berol. 1: 66. 1821. Based on Ammophila arundinacea Host.
Arundo littoralis Beauv. ex Steud., Nom. Bot. ed. 2. 1: 144. 1840, as synonym of Calamagrostis arenaria Roth.
(1) **Ammophila breviligulata** Fernald, Rhodora 22: 71. 1920. Milford, Conn., Bissell in 1902.

Ampelodesmos mauritanicus (Poir.) Dur. and Schinz, Consp. Fl. Afr. 5: 874. 1894. Based on Arundo mauritanica Poir.
Arundo mauritanica Poir., Voy. Barb. 2: 104. 1789. Algeria.
Arundo tenax Vahl, Symb. Bot. 2: 25. 1791. Tunis.
Ampelodesmos tenax Link, Hort. Berol. 1: 136. 1827. Based on Arundo tenax Vahl.

(146) AMPHICARPUM Kunth

(2) **Amphicarpum muhlenbergianum** (Schult.) Hitchc., Bartonia 14: 34. 1932. Based on Milium muhlenbergianum Schult.
Milium ? muhlenbergianum Schult., Mantissa 2: 178. 1824. Based on Milium No. 3 of Muhlenberg's Descriptio Graminum. Muhlenberg's specimen is without locality.
Amphicarpon floridanum Chapm., Fl. South. U. S. 572. 1860. Apalachicola River, Fla.
(1) **Amphicarpum purshii** Kunth, Rév. Gram. 1: 28. 1829. Based on Milium amphicarpon Pursh.
Milium amphicarpon Pursh, Fl. Amer. Sept. 1: 62. pl. 2. 1814. Egg Harbor, N. J.
Milium ciliatum Muhl., Descr. Gram. 77. 1817, Not M. ciliatum Moench, 1802. New Jersey. Name only, Muhl., Cat. Pl. 10. 1813.
Amphicarpon amphicarpon Nash, Torrey Bot. Club Mem. 5: 352. 1894. Based on Milium amphicarpon Pursh.

(154) ANDROPOGON L.

(19) **Andropogon arctatus** Chapm., Bot. Gaz. 3: 20. 1878. West Florida, Chapman [in 1875].
Andropogon tetrastachyus var. distachyus Chapm., Fl. South. U. S. 581. 1860. No locality cited. [Type specimen of A. arctatus is also type of this.]
Sorghum arctatum Kuntze, Rev. Gen. Pl. 2: 791. 1891. Based on Andropogon arctatus Chapm.
(33) **Andropogon barbinodis** Lag., Gen. et Sp. Nov. 3. 1816. Mexico, Sessé.
Andropogon leucopogon Nees, Linnaea 19: 694. 1845. Mexico, Aschenborn 141.

Andropogon saccharoides var. *barbinodis*
Hack. in DC., Monogr. Phan. 6: 494.
1889. Based on *A. barbinodis* Lag.
Andropogon saccharoides var. *leucopogon*
Hack. in DC., Monogr. Phan. 6: 496.
1889. Based on *A. leucopogon* Nees.
Amphilophis barbinodis Nash in Small,
Fl. Southeast. U. S. 65. 1903. Based
on *Andropogon barbinodis* Lag.
Holcus saccharoides var. *barbinodis* Hack.
ex Stuck., An. Mus. Nac. Buenos Aires
11: 48. 1904. Presumably based on
Andropogon barbinodis Lag.
Amphilophis leucopogon Nash, N. Amer.
Fl. 17: 126. 1912. Based on *Andropogon leucopogon* Nees.
Bothriochloa barbinodis Herter, Sudamer.
Bot. Rev. 6: 135. 1940. Based on
Andropogon barbinodis Lag.
(24) **Andropogon brachystachyus** Chapm.,
Fl. South. U. S. ed. 2. 668. 1883.
[Jacksonville], Fla., *Curtis* [3632].
Sorghum brachystachyum Kuntze, Rev.
Gen. Pl. 2:791. 1891. Based on *Andropogon brachystachyus* Chapm.
(17) **Andropogon cabanisii** Hack., Flora 68:
133. 1885. "Pennsylvania" [erroneous]
and Florida, *Cabanis*.
Sorghum cabanisii Kuntze, Rev. Gen. Pl.
2: 791. 1891. Based on *Andropogon cabanisii* Hack.
Andropogon ternarius var. *cabanisii* Fern.
and Grisc., Rhodora 37: 138. 1935.
Based on *A. cabanisii* Hack.
(30) **Andropogon campyloracheus** Nash,
N. Y. Bot. Gard. Bul. 1: 431. 1900.
Eustis, Fla., *Nash* 1738.
Andropogon elliottii var. *laxiflorus* Scribn.,
Torrey Bot. Club Bul. 23: 146. 1896
(Apr.). Eustis, Fla., *Nash* 1738. Published as new in Beal, Grasses N. Amer.
2: 51. 1896 (Nov.), *Nash* 1597 cited
as type.
(25) **Andropogon capillipes** Nash, N. Y.
Bot. Gard. Bul. 1: 431. 1900. Based
on *A. virginicus* var. *glaucus* Hack.
Andropogon glaucus Muhl., Descr. Gram.
278. 1817. Not *A. glaucus* Retz., 1789.
South Carolina.
Cymbopogon glaucus Schult., Mantissa 2:
459. 1824. Based on *Andropogon glaucus* Muhl.
Andropogon virginicus var. *glaucus* Hack.
in DC., Monogr. Phan. 6: 411. 1889.
[Jacksonville], Fla., *Curtis* 3638b.
(5) **Andropogon cirratus** Hack., Flora 68:
119. 1885. El Paso, Tex., *Wright* 804
[error for 805].
Sorghum cirratum Kuntze, Rev. Gen. Pl.
2: 791. 1891. Based on *Andropogon cirratus* Hack.
Schizachyrium cirratum Woot. and Standl.,
N. Mex. Col. Agr. Bul. 81: 30. 1912.
Based on *Andropogon cirratus* Hack.
(10) **Andropogon divergens** (Hack.) Anderss. ex Hitchc., Wash. Acad. Sci.
Jour. 23: 456. 1933. Based on *A.*

scoparius subsp. *maritimus* var. *divergens* Hack.
Andropogon scoparius subsp. *maritimus*
var. *divergens* Hack. in DC., Monogr.
Phan. 6: 385. 1889. Texas.
Andropogon divergens Anderss. ex Hack.,
in DC., Monogr. Phan. 6: 385. 1889,
as synonym of *A. scoparius* subsp.
maritimus var. *divergens* Hack.
(28) **Andropogon elliottii** Chapm.,Fl. South.
U.S. 581. 1860. Florida to North Carolina.
Chapman erroneously cites "*A. argenteus*
Ell., not of DC." but his description,
especially of the "dilated clustered
sheaths" shows that he did not know
Elliott's species (see synonymy under
A. ternarius Michx.), but was describing plants of his own collection, one of
which from Chapman's herbarium
named "*Andropogon Elliottii* S. Fl." in
his script is in the U. S. National
Herbarium.
Andropogon clandestinus Wood, Classbook ed. 1861, 809. 1861. Not *A.
clandestinus* Nees, 1854. Western Louisiana.
Andropogon elliottii var. *gracilior* Hack. in
DC., Monogr. Phan. 6: 415. 1889.
[Jacksonville], Fla., *Curtiss* 3636a.
Sorghum elliottii Kuntze, Rev. Gen. Pl. 2:
791. 1891. Based on *A. elliottii* Chapm.
?*Andropogon gyrans* Ashe, Elisha Mitchell Sci. Soc. Jour. 15: 113. 1898.
Durham County, N. C., *Ashe*.
Andropogon gracilior Nash in Small,
Fl. Southeast. U. S. 63. 1903. Based on
A. elliottii var. *gracilior* Hack.
Andropogon elliottii var. *projectus* Fern.
and Grisc., Rhodora 37: 139. 1935.
Biltmore, N. C., *Biltmore Herb.* No.
1421c.
(35) **Andropogon exaristatus** (Nash) Hitchc.,
Biol. Soc. Wash. Proc. 41: 163. 1928.
Based on *Amphilophis exaristatus* Nash.
Andropogon saccharoides var. *submuticus*
Vasey ex Hack., in DC., Monogr.
Phan. 6: 495. 1889. Not *A. submuticus* Steud., 1854. Texas, *Nealley*.
Amphilophis exaristatus Nash in Small,
Fl. Southeast. U. S. 65. 1903. Based
on *Andropogon saccharoides* var. *submuticus* Vasey.
Bothriochloa exaristata Henr., Blumea 4:
520. 1941. Based on *Amphilophis exaristatus* Nash.
(20) **Andropogon floridanus** Scribn., Torrey
Bot. Club Bul. 23: 145. 1896. [Eustis],
Fla., *Nash* 1572.
Andropogon bakeri Scribn. and Ball, U. S.
Dept. Agr., Div. Agrost. Bul. 24: 39.
1901. Grasmere, Fla., *C. H. Baker* 58.
(14) **Andropogon gerardi** Vitman, Summa
Pl. 6: 16. 1792. Based on the diagnosis
and figure in Gerard, Fl. Gallo-provincialis 107. f. 4. 1761. Provence,
France.
Andropogon provincialis Lam. Encycl. 1:

376. 1785. Not Retz., 1783. Based on Gerard's diagnosis and figure, and a plant in the Botanic Garden in Paris.
Andropogon furcatus Muhl. in Willd., Sp. Pl. 4: 919. 1806. North America [probably Pennsylvania].
?*Andropogon ternarius* [Michx. misapplied by] Bertol., Accad. Sci. Bologna Mem. 2: 600. 1850. Alabama.
Andropogon provincialis subvar. *furcatus* Hack. in DC., Monogr. Phan. 6: 442. 1889. Based on *A. furcatus* Muhl.
Andropogon provincialis subvar. *lindheimeri* Hack. in DC., Monogr. Phan. 6: 443. 1889. Texas, *Lindheimer 741*.
Andropogon provincialis subvar. *pycnanthus* Hack. in DC., Monogr. Phan. 6: 443. 1889. Texas, *Vinzent 69*.
Andropogon provincialis var. *tennesseensis* Scribn., Tenn. Agr. Expt. Sta. Bul. 7[2]: 23. 1894. Tennessee.
Andropogon hallii var. *grandiflorus* Scribn., U. S. Dept. Agr., Div. Agrost. Bul. 5: 21. 1897. Colorado, *Shear 747* [type], 605, 2366.
Andropogon tennesseensis Scribn., U. S. Dept. Agr., Div. Agrost. Cir. 16: 1. 1899. Based on *A. provincialis* var. *tennesseensis* Scribn.

(27) **Andropogon glomeratus** (Walt.) B. S. P., Prel. Cat. N. Y. 67. 1888. Based on *Cinna glomerata* Walt.
Cinna glomerata Walt., Fl. Carol. 59. 1788. South Carolina.
Andropogon macrourus Michx., Fl. Bor. Amer. 1: 56. 1803. Carolina to Florida, *Michaux*. [Type labeled "Virginia to Carolina."]
Andropogon spathaceus Trin., Fund. Agrost. 186. 1820, name only; Steud., Nom. Bot. ed. 2. 1: 93. 1840, as synonym of *A. macrourus* Michx.
Anatherum macrourum Griseb., Amer. Acad. Mem. (n.s.) 8: 534. 1863. Based on *Andropogon macrourus* Michx.
Andropogon macrourus var. *abbreviatus* Hack. in DC., Monogr. Phan. 6: 408. 1889. [Pleasant Bridge], N. J., *Gray*.
Andropogon macrourus var. *corymbosus* Chapm. ex Hack. in DC., Monogr. Phan. 6: 409. 1889. [Jacksonville], Fla., *Curtiss 3639c*.
Sorghum glomeratum Kuntze, Rev. Gen. Pl. 2: 790. 1891. Based on *Cinna glomerata* Walt.
Dimeiostemon macrurus Raf. ex Jacks., Ind. Kew. 1: 760. 1893, as synonym of *Andropogon macrourus* Michx.
Andropogon glomeratus var. *corymbosus* Scribn., U. S. Dept. Agr., Div. Agrost. Bul. 7 (ed. 3): 15. 1900. Based on *A. macrourus* var. *corymbosus* Chapm.
Andropogon glomeratus var. *abbreviatus* Scribn., U. S. Dept. Agr., Div. Agrost. Bul. 7 (ed. 3): 15. 1900. Based on *A. macrourus* var. *abbreviatus* Hack.
Andropogon corymbosus Nash in Britton,

Man. 69. 1901. Based on *A. macrourus* var. *corymbosus* Chapm.
Andropogon corymbosus abbreviatus Nash in Britton, Man. 70. 1901. Based on *A. macrourus* var. *abbreviatus* Hack.
Andropogon glomeratus tenuispatheus Nash in Small, Fl. Southeast. U. S. 61. 1903. Florida [type] to New Mexico.
Andropogon tenuispatheus Nash, N. Amer. Fl. 17: 113. 1912. Based on *A. glomeratus tenuispatheus* Nash.
Andropogon virginicus var. *corymbosus* Fern. and Grisc., Rhodora 37: 142. pl. 338. f. 2. 1935. Based on *A. macrourus* var. *corymbosus* Chapm.
Andropogon virginicus var. *abbreviatus* Fern. and Grisc., Rhodora 37: 142. pl. 338. f. 3. 1935. Based on *A. macrourus* var. *abbreviatus* Hack.
Andropogon virginicus var. *tenuispatheus* Fern. and Grisc., Rhodora 37: 142. pl. 338. f. 1. 1935. Based on *A. glomeratus tenuispatheus* Nash.
Andropogon virginicus var. *hirsutior* forma *tenuispatheus* Fernald, Rhodora 42: 416. 1940. Based on *A. glomeratus tenuispatheus* Nash.

(1) **Andropogon gracilis** Spreng., Syst. Veg. 1: 284. 1825. Hispaniola.
Andropogon juncifolius Desv. ex Hamilt., Prodr. Pl. Ind. Occ. 9. 1825. St. Croix, Virgin Islands.
Sorghum gracile Kuntze, Rev. Gen. Pl. 2: 791. 1891. Based on *Andropogon gracilis* Spreng.
Schizachyrium gracile Nash in Small, Fl. Southeast. U. S. 60. 1903. Based on *Andropogon gracilis* Spreng.

(15) **Andropogon hallii** Hack., Sitzungsb. Akad. Wiss. Math. Naturw. (Wien) 89[1]: 127. 1884. North America [Nebraska], *Hall* and *Harbour 651*.
Andropogon hallii var. *flaveolus* Hack., Sitzungsb. Akad. Wiss. Math. Naturw. (Wien) 89[1]: 128. 1884. [Nebraska] *Hall* and *Harbour 651*.
Andropogon hallii var. *incanescens* Hack., Sitzungsb. Akad. Wiss. Math. Naturw. (Wien) 89[1]: 128. 1884. [Nebraska. *Hall* and *Harbour*.
Andropogon hallii var. *muticus* Hack. in DC., Monogr. Phan. 6: 444. 1889. Brighton, Colo., *Vasey*.
Sorghum hallii Kuntze, Rev. Gen. Pl. 2: 791. 1891. Based on *Andropogon hallii* Hack.
Andropogon geminatus Hack. ex Beal, Grasses N. Amer. 2: 55. 1896. Texas, *Nealley*.
Andropogon hallii var. *bispicata* Vasey ex Beal, Grasses N. Amer. 2: 55. 1896, as synonym of *A. geminatus* Hack.
Andropogon chrysocomus Nash in Britton, Man. 70. 1901. Kansas [type, Stevens County, *Carleton 343*] and Texas.
Andropogon paucipilus Nash in Britton, Man. 70. 1901. Montana and Nebraska

[type, Whitman, *Rydberg* 1607].

Andropogon provincialis var. *paucipilus* Fern. and Grisc., Rhodora 37: 147. 1935. Based on *A. paucipilus* Nash.

Andropogon provincialis var. *chrysocomus* Fern. and Grisc., Rhodora 37: 147. 1935. Based on *A. chrysocomus* Nash.

Andropogon gerardi var. *chrysocomus* Fernald, Rhodora 45: 258. 1943. Based on *A. chrysocomus* Nash.

Andropogon gerardi var. *paucipilus* Fernald, Rhodora 45: 258. 1943. Based on *A. paucipilus* Nash.

(3) **Andropogon hirtiflorus** (Nees) Kunth, Rév. Gram. 1: Sup. 39. 1830. Based on *Schizachyrium hirtiflorum* Nees.

Streptachne domingensis Spreng. ex Schult., Mantissa 2: 188. 1824. Not *Andropogon domingensis* Steud., 1821. Santo Domingo, *Bertero*.

Schizachyrium hirtiflorum Nees, Agrost. Bras. 334. 1829. Brazil, *Sellow*.

Aristida domingensis Kunth, Rév. Gram. 1: 62. 1829. Based on *Streptachne domingensis* Spreng.

Andropogon oligostachyus Chapm., Fl. South. U. S. 581. 1860. Middle Florida, *Chapman*.

Andropogon hirtiflorus var. *oligostachyus* Hack. in DC., Monogr. Phan. 6: 372. 1889. Based on *A. oligostachyus* Chapm.

Sorghum hirtiflorum Kuntze, Rev. Gen. Pl. 2: 792. 1891. Based on *Schizachyrium hirtiflorum* Nees.

Schizachyrium oligostachyum Nash in Small, Fl. Southeast. U. S. 59. 1903. Based on *Andropogon oligostachyus* Chapm.

Schizachyrium domingense Nash, N. Amer. Fl. 17: 103. 1912. Based on *Streptachne domingensis* Spreng.

Andropogon domingensis Hubb., Amer. Acad. Sci. Proc. 49: 493. 1913. Not *A. domingensis* Steud., 1821. Based on *Streptachne domingensis* Spreng.

ANDROPOGON HIRTIFLORUS var. FEENSIS (Fourn.) Hack. in DC., Monogr. Phan. 6: 372. 1889. Based on *A. feensis* Fourn.

Andropogon feensis Fourn., Mex. Pl. 2: 62. 1886. Santa Fé, Mexico, *Bourgeau* 752.

Andropogon hirtiflorus var. *brevipedicellatus* Beal, Grasses N. Amer. 2: 44. 1896. Chihuahua, Mexico, *Pringle* 383.

Andropogon myosurus var. *feensis* Urbina, Pl. Mex. Cat. 379. 1897. Presumably based on *A. feensis* Fourn.

Schizachyrium feense A. Camus, Ann. Soc. Linn. Lyon 70: 89. 1923. Based on *Andropogon feensis* Fourn.

Andropogon ischaemum L., Sp. Pl. 1047. 1753. Southern Europe.

(9) **Andropogon littoralis** Nash in Britton, Man. 69. 1901. New York [type, Staten Island, *Nash* in 1894] and New Jersey.

Andropogon scoparius subsp. *euscoparius* Hack. ex Beal, Grasses N. Amer. 2: 46. 1896. Cape May, N. J., *Burk* in 1881 (misprinted as 1888).

Andropogon scoparius var. *littoralis* Hitchc., Rhodora 8: 205. 1906. Based on *A. littoralis* Nash.

Schizachyrium littorale Bicknell, Torrey Bot. Club Bul. 35: 182. 1908. Based on *Andropogon littoralis* Nash.

Andropogon scoparius var. *ducis* Fern. and Grisc., Rhodora 37: 145. pl. 340. f. 1. 2. 1935. West End Point, Naushon, Elizabeth Island, Mass., *Fogg* 2940.

(22) **Andropogon longiberbis** Hack., Flora 68: 131. 1885. Florida, *Garber* [in 1877].

Sorghum longiberbe Kuntze, Rev. Gen. Pl. 2: 792. 1891. Based on *Andropogon longiberbis* Hack.

(11) **Andropogon maritimus** Chapm., Fl. South. U. S. ed. 2: 668. 1883. West Florida, *Chapman*.

Andropogon scoparius subsp. *maritimus* Hack. in DC., Monogr. Phan. 6: 385. 1889. Based on *A. maritimus* Chapm.

Schizachyrium maritimum Nash in Small, Fl. Southeast. U. S. 59. 1903. Based on *Andropogon maritimus* Chapm.

(16) **Andropogon mohrii** (Hack.) Hack. ex Vasey, U. S. Natl. Herb. Contrib. 3: 11. 1892. Based on *A. liebmanni* subvar. *mohrii* Hack.

Andropogon liebmanni subvar. *mohrii* Hack. in DC., Monogr. Phan. 6: 413. 1889. Mobile, Ala., *Mohr* [in 1884].

Andropogon mohrii var. *pungens* Ashe, Elisha Mitchell Sci. Soc. Jour. 15: 114. 1898. Washington County, N. C., *Ashe*.

(6) **Andropogon niveus** Swallen, Wash. Acad. Sci. Jour. 31: 354. f. 7. 1941. Kissimmee, Fla., *Silveus* 6684.

Andropogon nodosus (Willem.) Nash, N. Amer. Fl. 17: 122. 1912. Based on *Dichanthium nodosum* Willem.

Dichanthium nodosum Willem., Ann. Bot. Usteri 18: 11. 1796. Mauritius.

Andropogon mollicomus Kunth, Rév. Gram. 1: 365. 1830. Mauritius.

Andropogon caricosus var. *mollicomus* Hack. in DC., Monogr. Phan. 6: 569. 1889. Based on *A. mollicomus* Kunth.

(23) **Andropogon perangustatus** Nash in Small, Fl. Southeast. U. S. 62. 1903. Based on *A. virginicus* var. [*viridis* sub-var.] *stenophyllus* Hack.

Andropogon virginicus var. *viridis* subvar. *stenophyllus* Hack. in DC., Monogr. Phan. 6: 411. 1889. Not *A. stenophyllus* Roem. and Schult., 1817. Florida, *Chapman* [in 1884].

Andropogon virginicus var. *stenophyllus* Fern. and Grisc., Rhodora 37: 142. 1935. Based on *A. virginicus* var. *viridis* subvar. *stenophyllus* Hack.

(32) **Andropogon perforatus** Trin. ex Fourn., Mex. Pl. 2: 59. 1886. [Mexico City] Mexico, *Berlandier* 641.

Andropogon emersus Fourn., Mex. Pl. 2: 58. 1886. Orizaba, Mexico, *Mueller* 2033.

Andropogon saccharoides var. *leucopogon* subvar. *perforatus* Hack. in DC., Monogr. Phan. 6: 496. 1889. Based on *A. perforatus* Trin.

Andropogon saccharoides var. *perforatus* Hack. ex L. H. Dewey, U. S. Natl. Herb. Contrib. 2: 497. 1894. Presumably based on *A. perforatus* Trin.

Amphilophis perforatus Nash in Small, Fl. Southeast. U. S. 66. 1903. Based on *Andropogon perforatus* Trin.

Holcus saccharoides var. *perforatus* Hack. ex Stuck., An. Mus. Nac. Buenos Aires 11: 48. 1904. Presumably based on *Andropogon perforatus* Trin.

Amphilophis emersus Nash, N. Am. Fl. 17: 126. 1912. Based on *Andropogon emersus* Fourn.

Bothriochloa perforata Herter, Rev. Sudamer. Bot. 6: 135. 1940. Based on *Andropogon perforatus* Trin.

Bothriochloa emersa Henr., Blumea 4: 520. 1941. Based on *Andropogon emersus* Fourn.

Andropogon pertusus (L.) Willd., Sp. Pl. 4: 922. 1806. Based on *Holcus pertusus* L.

Holcus pertusus L., Mant. Pl. 2: 301. 1771. East Indies.

Bothriochloa pertusa A. Camus, Ann. Soc. Linn. Lyon n. ser. 76 (1930): 164. 1931. Based on *Holcus pertusus* L.

(12) **Andropogon rhizomatus** Swallen, Wash. Acad. Sci. Jour. 31: 352. f. 6. 1941. Homestead, Fla., *Silveus* 6614.

(34) **Andropogon saccharoides** Swartz, Prodr. Veg. Ind. Occ. 26. 1788. Jamaica, *Swartz*.

Andropogon argenteus DC., Cat. Hort. Monsp. 77. 1813. Mexico, *Sessé*.

Andropogon laguroides DC., Cat. Hort. Monsp. 78. 1813. Grown from Mexican seed.

Andropogon glaucus Torr., Ann. Lyc. N. Y. 1: 153. 1824. Not *A. glaucus* Retz., 1789. Canadian River, Tex., *James*.

Trachypogon argenteus Nees, Agrost. Bras. 348. 1829. Based on *Andropogon argenteus* DC.

Trachypogon laguroides Nees, Agrost. Bras. 349. 1829. Based on *Andropogon laguroides* DC.

Andropogon torreyanus Steud., Nom. Bot. ed. 2. 1: 93. 1840. Based on *A. glaucus* Torr.

Andropogon jamesii Torr. in Marcy, Expl. Red. Riv. 302. 1853. Based on *A. glaucus* Torr.

Andropogon saccharoides var. *laguroides* Hack. in Mart., Fl. Bras. 2³: 293.

1883. Based on *A. laguroides* DC.

Andropogon tenuirachis Fourn., Mex. Pl. 2: 58. 1886. Mexico.

Andropogon saccharoides var. *torreyanus* Hack. in DC., Monogr. Phan. 6: 495. 1889. Based on *A. torreyanus* Steud.

Sorghum saccharoides Kuntze, Rev. Gen. Pl. 2: 792. 1891. Based on *Andropogon saccharoides* Swartz.

Andropogon saccharoides var. *glaucus* Scribn., Torrey Bot. Club Mem. 5: 28. 1894. Based on *A. glaucus* Torr.

Amphilophis torreyanus Nash in Britton, Man. 71. 1901. Based on *Andropogon torreyanus* Steud.

Holcus saccharoides Kuntze ex Stuck., An. Mus. Nac. Buenos Aires 11: 48. 1904. Presumably based on *Andropogon saccharoides* Swartz.

Holcus saccharoides var. *laguroides* Hack. ex Stuck., An. Mus. Nac. Buenos Aires 11: 48. 1904. Presumably based on *Andropogon laguroides* DC.

Amphilophis saccharoides Nash, N. Amer. Fl. 17: 125. 1912. Based on *Andropogon saccharoides* Swartz.

Bothriochloa saccharoides Rydb., Brittonia 1: 81. 1931. Based on *Andropogon saccharoides* Swartz.

Bothriochloa laguroides Herter, Rev. Sudamer. Bot. 6: 135. 1940. Based on *Andropogon laguroides* DC.

(8) **Andropogon scoparius** Michx., Fl. Bor. Amer. 1: 57. 1803. Carolina, *Michaux*.

Andropogon purpurascens Muhl. in Willd., Sp. Pl. 4: 913. 1806. North America [type received from Muhlenberg]. Listed by Muhlenberg in Amer. Phil. Soc. Trans. 4: 237. 1799. "Clayton 602" cited but without description.

Andropogon flexilis Bosc ex Poir. in Lam., Encyl. Sup. 1: 583. 1810. North America, *Bosc* [type, Carolina].

Pollinia scoparia Spreng., Pl. Pugill. 2: 13. 1815. Based on *Andropogon scoparius* Michx.

Andropogon halei Wood, Class-book ed. 1861. 809. 1861. [Louisiana, *Hale*.]

Andropogon scoparius subsp. *genuinus* Hack. in DC., Monogr. Phan. 6: 384. 1889. Based on *A. scoparius* Michx.

Andropogon scoparius subvar. *flexilis* Hack. in DC., Monogr. Phan. 6: 384. 1889. Based on *A. flexilis* Bosc.

Andropogon scoparius subvar. *caesia* Hack. in DC., Monogr. Phan. 6: 384. 1889. No locality cited. (Plants with pruinose sheaths.)

Andropogon scoparius subvar. *serpentinus* Hack. in DC., Monogr. Phan. 6: 384. 1889. No locality cited. (Plants with strongly flexuous rachis.)

Andropogon scoparius subvar. *simplicior* Hack. in DC., Monogr. Phan. 6: 384. 1889. No locality cited. (Sparingly branching plants.)

Sorghum scoparium Kuntze, Rev. Gen.

Pl. 2: 792.   1891. Based on *Andropogon scoparius* Michx.
Andropogon scoparius var. *polycladus* Scribn. and Ball, U. S. Dept. Agr., Div. Agrost. Bul. 24: 40. 1901. "Braidentown" (Bradenton), Fla., *Combs* 1298.
Andropogon scoparius var. *villosissimus* Kearney in Scribn. and Ball, U. S. Dept. Agr., Div. Agrost. Bul. 24: 41. 1901. Waynesboro, Miss., *Kearney* 136. (Foliage villous.)
Schizachyrium scoparium Nash in Small, Fl. Southeast. U. S. 59. 1903. Based on *Andropogon scoparius* Michx.
Schizachyrium villosissimum Nash in Small, Fl. Southeast. U. S. 59, 1326. 1903. Based on *Andropogon scoparius* var. *villosissimus* Kearney.
Schizachyrium acuminatum Nash in Small, Fl. Southeast. U. S. 59, 1326. 1903. Starkville, Miss., *Tracy* in 1890. (Sessile spikelets 10 mm. long.)
Andropogon scoparius var. *frequens* Hubb., Rhodora 19: 103. 1917. Block Island, R. I., *Fernald, Long,* and *Torrey* 8476.
Andropogon scoparius var. *glaucescens* House, N. Y. State Mus. Bul. 254: 68. 1924. West of Albany, N. Y. [*House* 3 in 1918].
Andropogon scoparius var. *genuinus* Fern. and Grisc., Rhodora 37: 143, 144. 1935. Based on *A. scoparius* Michx.
Andropogon scoparius var. *septentrionalis* Fern. and Grisc., Rhodora 37: 145. pl. 339, f. 1, 2. 1935. Canada, *Rolland* 19199.
Andropogon praematurus Fernald, Rhodora 42: 413. pl. 626. f. 1–3. 1940. Skipper's, Greenville County, Va., *Fernald* and *Long* 10092.
Andropogon praematurus forma *hirtivaginatus* Fernald, Rhodora 44: 383. 1942. Sussex County, Va., *Fernald* and *Long* 13248.
Andropogon scoparius var. *genuinus* forma *calvescens* Fernald, Rhodora 45: 390. 1943. Virginia, *Fernald* and *Lewis* 14474.
ANDROPOGON SCOPARIUS var. NEOMEXICANUS (Nash) Hitchc., Biol. Soc. Wash. Proc. 41: 163. 1928. Based on *A. neo-mexicanus* Nash.
Andropogon neo-mexicanus Nash, Torrey Bot. Club Bul. 25: 83. 1898. White Sands, Doña Ana County, N. Mex., *Wooton* [583] in 1897.
Schizachyrium neo-mexicanum Nash, N. Amer. Fl. 17: 107. 1912. Based on *Andropogon neo-mexicanus* Nash.
(4) **Andropogon semiberbis** (Nees) Kunth, Rév. Gram. 1: Sup. 39. 1830. Based on *Schizachyrium semiberbe* Nees.
Schizachyrium semiberbe Nees, Agrost. Bras. 336. 1829. Brazil, *Sellow.*
Andropogon vaginatus Presl, Rel. Haenk. 1: 336. 1830. Not *A. vaginatus* Ell.,

1816. Mexico, Haenke.
Andropogon velatus Kunth, Rév. Gram. 1: Sup. 39. 1830. Based on *A. vaginatus* Presl.
Andropogon semiberbis subvar. *pruinatus* Hack. in DC., Monogr. Phan. 6: 370. 1889. [Eau Gallie,] Fla., *Curtiss* 3633.
Andropogon tener Curtiss ex Hack. in DC., Monogr. Phan. 6: 370. 1889. Not *A. tener* Kunth, 1830. As synonym of *A. semiberbis* subvar. *pruinatus* Hack.
Sorghum semiberbe Kuntze, Rev. Gen. Pl. 2: 792. 1891. Based on *Schizachyrium semiberbe* Nees.
Andropogon hirtiflorus var. *semiberbis* Stapf in Dyer, Fl. Cap. 7: 337. 1898. Based on *A. semiberbis* Kunth.
(7) **Andropogon sericatus** Swallen, Wash. Acad. Sci. Jour. 31: 355. f. 8. 1941. Ramrod Key, Fla., *Silveus* 6633.
Andropogon sericeus R. Br., Prodr. Fl. Nov. Holl. 1: 201. 1810. Australia.
(13) **Andropogon stolonifer** (Nash) Hitchc., Amer. Jour. Bot. 2: 299. 1915. Based on *Schizachyrium stoloniferum* Nash.
Schizachyrium stoloniferum Nash in Small, Fl. Southeast. U. S. 59, 1326. 1903. Florida, *Chapman.*
Schizachyrium triaristatum Nash in Small, Fl. Southeast. U. S. 60, 1326. 1903. Florida, *Chapman.*
(29) **Andropogon subtenuis** Nash in Small, Fl. Southeast. U. S. 63. 1903. Biloxi, Miss., *Tracy* 2243.
(2) **Andropogon tener** (Nees) Kunth, Rév. Gram. 1: Sup. 39. 1830. Based on *Schizachyrium tenerum* Nees.
Schizachyrium tenerum Nees, Agrost. Bras. 336. 1829. Brazil, *Sellow.*
Andropogon gracilis Presl, Rel. Haenk. 1: 336. 1830. Not *A. gracilis* Spreng. 1825. Peru, *Haenke.*
Andropogon preslii Kunth, Rév. Gram. 1: Sup. 39. 1830. Based on *A. gracilis* Presl.
Andropogon leptophyllus Trin., Acad. St. Pétersb. Mém. VI. Math. Phys. Nat. 2: 264. 1832. Based on *Schizachyrium tenerum* Nees.
Sorghum tenerum Kuntze, Rev. Gen. Pl. 2: 792. 1891. Based on *Schizachyrium tenerum* Nees.
(18) **Andropogon ternarius** Michx., Fl. Bor. Amer. 1: 57. 1803. Carolina, *Michaux.*
Andropogon argenteus Ell., Bot. S. C. and Ga. 1: 148. 1816. Not *A. argenteus* DC., 1813. Presumably South Carolina.
Andropogon argyraeus Schult., Mantissa 2: 450. 1824. Based on *A. argenteus* Ell.
Andropogon muhlenbergianus Schult., Mantissa 2: 455. 1824. Based on Muhlenberg's *Andropogon* No. 4. North Carolina.
Andropogon belvisii Desv., Opusc. 67. 1831. No locality cited.
Sorghum argenteum Kuntze, Rev. Gen.

Pl. 2: 790. 1891. Based on *Andropogon argenteus* Ell.

Andropogon argyraeus var. *tenuis* Vasey, U. S. Natl. Herb. Contrib. 3: 12. 1892. Texas [Dallas, *Reverchon* 1161].

Andropogon argyraeus macrus Scribn., U. S. Dept. Agr., Div. Agrost. Bul. 1: 20. 1895. [Jacksonville,] Fla., *Curtiss* 4952. Published as new by Scribner and Ball (Hackel given as author), U. S. Dept. Agr., Div. Agrost. Bul. 24: 39. 1900, *Tracy* 3891 cited as type.

Andropogon elliottii var. *glaucescens* Scribn., Torrey Bot. Club Bul. 23: 145. 1896. Eustis, Fla., *Nash* 473.

Andropogon scribnerianus Nash, N. Y. Bot. Gard. Bul. 1: 432. 1900. Based on *A. elliottii* var. *glaucescens* Scribn.

Andropogon mississippiensis Scribn. and Ball, U. S. Dept. Agr., Div. Agrost. Bul. 24: 40. 1901. Biloxi, Miss., *Tracy* 3818.

Andropogon ternarius var. *glaucescens* Fern. and Grisc., Rhodora 37: 137. 1935. Based on *A. elliottii* var. *glaucescens* Scribn.

(21) **Andropogon tracyi** Nash, N. Y. Bot. Gard. Bul. 1: 433. 1900. Columbus, Miss., *Tracy* 3083.

(26) **Andropogon virginicus** L., Sp. Pl. 1046. 1753. America. The type specimen bears no data indicating origin. Linnaeus had also a specimen from Gronovius, *Clayton* 460 from Virginia.

Cinna lateralis Walt., Fl. Carol. 59. 1788. South Carolina.

Andropogon dissitiflorus Michx., Fl. Bor. Amer. 1: 57. 1803. Carolina to Florida, *Michaux*.

Anatherum virginicum Spreng., Pl. Pugill. 2: 16. 1815. Based on *Andropogon virginicus* L.

Andropogon vaginatus Ell., Bot. S. C. and Ga. 1: 148. 1816. Presumably South Carolina.

Andropogon tetrastachyus Ell., Bot. S. C. and Ga. 1: 150. pl. 8. f. 4. 1816. Charleston, S. C.

Holcus virginicus Muhl. ex Steud., Nom. Bot. ed. 2. 1: 773. 1840, as synonym of *Andropogon virginicus* L.

Andropogon eriophorus Scheele, Flora 27: 51. 1844. Not *A. eriophorus* Willd., 1806. Charles Town, W. Va.

?*Andropogon louisianae* Steud., Syn. Pl. Glum. 1: 383. 1854. Louisiana.

Andropogon curtisianus Steud., Syn. Pl. Glum. 1: 390. 1854. Carolina, *M. A. Curtis*. Referred by Hackel to *A. virginicus* var. *tetrastachyus*. Description does not well apply to any of our species.

Andropogon virginicus var. *vaginatus* Wood, Class-book ed. 1861. 808. 1861. Based on *A. vaginatus* Ell.

Andropogon virginicus subsp. *genuinus* Hack. in Mart., Fl. Bras. 2³: 285.

1883. Based on *A. virginicus* L.

Andropogon virginicus var. *viridis* Hack. in DC., Monogr. Phan. 6: 410. 1889. Group name for three subvarieties, 1. *genuinus* being *A. virginicus* L.

Andropogon virginicus var. *tetrastachyus* Hack. in DC., Monogr. Phan. 6: 411. 1889. Based on *A. tetrastachyus* Ell.

Sorghum virginicum Kuntze, Rev. Gen. Pl. 2: 792. 1891. Based on *Andropogon virginicus* L.

Dimeiostemon vaginatus Raf. ex Jacks., Ind. Kew. 1: 760. 1893, as synonym of *Andropogon virginicus* L.

Dimeiostemon tetrastachys Raf. ex Jacks., Ind. Kew. 1: 760. 1893, as synonym of *Andropogon virginicus* L.

Andropogon virginicus var. *genuinus* Fern. and Grisc., Rhodora 37: 142. 1935. Based on *A. virginicus* L.

ANDROPOGON VIRGINICUS var. GLAUCOPSIS (Ell.) Hitchc., Amer. Jour. Bot. 21:139. 1934. Based on *A. macrourus* var. *glaucopsis* Ell.

Andropogon macrourus var. *glaucopsis* Ell., Bot. S. C. and Ga. 1: 150. 1816. Presumably South Carolina.

Andropogon glaucopsis Steud., Nom. Bot. ed. 2. 1: 91. 1840. Not *A. glaucopsis* Steud., 1854. Based on *A. macrourus* var. *glaucopsis* Ell. Published as new by Nash, in Small, Fl. Southeast. U. S. 62. 1903, same basis.

Andropogon virginicus var. *dealbatus* Mohr ex Hack., in DC., Monogr. Phan. 6: 411. 1889. Mobile, Ala., *Mohr* [in 1894].

Andropogon glomeratus var. *glaucopsis* Mohr, Torrey Bot. Club Bul. 24: 21. 1897. Based on *A. macrourus* var. *glaucopsis* Ell.

ANDROPOGON VIRGINICUS var. HIRSUTIOR (Hack.) Hitchc., Wash. Acad. Sci. Jour. 23: 456. 1933. Based on *A. macrourus* var. *hirsutior* Hack.

Andropogon macrourus var. *hirsutior* Hack. in DC., Monogr. Phan. 6: 409. 1889. Mobile, Ala., *Mohr* [October 28, 1884].

Andropogon virginicus var. *viridis* subvar. *ditior* Hack. in DC., Monogr. Phan. 6: 411. 1889. [Jacksonville], Fla., *Curtiss* 3639d.

Andropogon macrourus var. *viridis* Curtiss ex Hack. in DC., Monogr. Phan. 6: 411. 1889, as synonym of *A. virginicus* var. *ditior* Hack. Florida, *Curtiss* N. Amer. Pl. 3639d.

Andropogon macrourus var. *pumilus* Vasey, Bot. Gaz. 16: 27. 1891. [Seminole Cave, Val Verde County], western Texas, *Nealley* [256 in 1890].

Andropogon macrourus var. *viridis* Chapm. ex Vasey, U. S. Natl. Herb. Contrib. 3: 11. 1892. Florida, *Chapman*.

Andropogon glomeratus var. *pumilus* Vasey ex L. H. Dewey, U. S. Natl. Herb.

818   MISC. PUBLICATION 200, U. S. DEPT. OF AGRICULTURE

Contrib. 2: 496. 1894. Presumably based on *A. macrourus* var. *pumilus* Vasey.

Andropogon glomeratus var. *hirsutior* Mohr, Torrey Bot. Club Bul. 24: 21. 1897. Based on *A. macrourus* var. *hirsutior* Hack.

Andropogon virginicus var. *tenuispatheus* forma *hirsutior* Fern. and Grisc., Rhodora 37: 142. 1935. Based on *A. macrourus* var. *hirsutior* Hack.

(31) **Andropogon wrightii** Hack., Flora 68: 139. 1885. [Silver City] N. Mex., *Wright* 2104.

Sorghum wrightii Kuntze, Rev. Gen. Pl. 2: 792. 1891. Based on *Andropogon wrightii* Hack.

Amphilophis wrightii Nash, N. Amer. Fl. 17: 124. 1912. Based on *Andropogon wrightii* Hack.

Bothriochloa wrightii Henr., Blumea 4: 520. 1941. Based on *Andropogon wrightii* Hack.

(127) **ANTHAENANTIA Beauv.**

(1) **Anthaenantia rufa** (Ell.) Schult., Mantissa 2: 258. 1824. Based on *Aulaxanthus rufus* Ell.

Aulaxanthus rufus Ell., Bot. S. C. and Ga. 1: 103. 1816. South Carolina.

Aulaxia rufa Nutt., Gen. Pl. 1: 47. 1818. Based on *Aulaxanthus rufus* Ell.

Panicum rufum Kunth, Rév. Gram. 1: 35. 1829. Based on *Aulaxanthus rufus* Ell.

Monachne rufa Bertol., Accad. Sci. Bologna Mem. 2: 596. pl. 41. f. 1. 1850. Based on *Panicum rufum* Kunth.

Leptocoryphium drummondii C. Muell., Bot. Ztg. 19: 314. 1861. Louisiana, *Drummond.*

Panicum ciliatiflorum var. *rufum* Wood, Amer. Bot. and Flor. pt. 2: 392. 1871. [Southern States.]

Panicum aulaxanthus Kuntze, Rev. Gen. Pl. 3²: 361. 1898. Based on *Aulaxanthus rufus* Ell.

Anthaenantia rufa scabra Nash in Small, Fl. Southeast. U. S. 79. 1903. South Carolina to Louisiana.

(2) **Anthaenantia villosa** (Michx.) Beauv., Ess. Agrost. 48, 151, pl. 10. f. 7. 1812. Based on *Phalaris villosa* Michx.

Phalaris villosa Michx., Fl. Bor. Amer. 1: 43. 1803. Carolina, *Michaux.*

Aulaxanthus ciliatus Ell., Bot. S. C. and Ga. 1: 102. 1816. South Carolina.

Panicum erianthum Poir., Encycl. Sup. 4: 284. 1816. Carolina, *Bosc.*

Panicum hirticalycinum Bosc ex Roem. and Schult., Syst. Veg. 2: 468. 1817, as synonym of *Anthaenantia villosa* Beauv.

Aulaxia ciliata Nutt., Gen. Pl. 1: 47. 1818. Based on *Aulaxanthus ciliatus* Ell.

Panicum hirticalycum Bosc ex Spreng.,

Syst. Veg. 1: 315. 1825, as synonym of *P. erianthum* Poir.

Oplismenus erianthos Kunth, Rév. Gram. 1: 45. 1829. Based on *Panicum erianthum* Poir.

Panicum ignoratum Kunth, Rév. Gram. 2: 217. pl. 20. 1830. Based on *Phalaris villosa* Michx.

Leptocoryphium obtusum Steud., Syn. Pl. Glum. 1: 34. 1854. Louisiana, *Riehl.*

Panicum ciliatiflorum Wood, Class-book pt. 2: 786. 1861. Not *P. ciliatiflorum* Kunth, 1829. Southern States.

Panicum anthaenantia Kuntze, Rev. Gen. Pl. 3²: 361. 1898. Based on *Anthaenantia villosa* Beauv.

ANTHEPHORA Schreb.

Anthephora hermaphrodita (L.) Kuntze, Rev. Gen. Pl. 2: 759. 1891. Based on *Tripsacum hermaphroditum* L.

Tripsacum hermaphroditum L., Syst. Nat. ed. 10. 2: 1261. 1759. Jamaica.

Anthephora elegans Schreb., Beschr. Gräs. 2: 105. pl. 44. 1810. Jamaica.

(117) **ANTHOXANTHUM L.**

(2) **Anthoxanthum aristatum** Boiss., Voy. Bot. Esp. 2: 638. 1845. Southern Europe.

Anthoxanthum puelii Lec. and Lam., Cat. Pl. France 385. 1847. France.

Anthoxanthum odoratum var. *puelii* Coss. and Dur., Expl. Sci. Alger. 2: 21. 1854. Based on *A. puelii* Lec. and Lam.

Anthoxanthum odoratum var. *aristatum* Coss. and Dur., Expl. Sci. Alger. 2: 22. 1854–55. Based on *A. aristatum* Boiss.

Anthoxanthum gracile Bivon., Stirp. Rar. Sic. 1: 13. pl. 1. f. 2. 1813. Italy.

(1) **Anthoxanthum odoratum** L., Sp. Pl. 28. 1753. Europe.

Anthoxanthum odoratum var. *altissimum* Eaton and Wright, Man. Bot. North. States 10. 1817. Probably Connecticut, *Ives.*

Xanthonanthos odoratum St. Lag., Ann. Soc. Bot. Lyon 7: 119. 1880. Based on *Anthoxanthum odoratum* L.

(70) **APERA Adans.**

(2) **Apera interrupta** (L.) Beauv., Ess. Agrost. 31, 151. 1812. Based on *Agrostis interrupta* L.

Agrostis interrupta L., Syst. Nat. ed. 10. 2: 872. 1759. Europe.

Anemagrostis interrupta Trin., Fund. Agrost. 129. 1820. Based on *Agrostis interrupta* L.

Muhlenbergia interrupta Steud., Syn. Pl. Glum. 1: 177. 1854. Based on *Agrostis interrupta* L.

Agrostis spica-venti var. *interrupta* Hook. f., Stud. Fl. 432. 1870. Based on *A. interrupta* L.

Agrostis anemagrostis subsp. *interrupta*
Syme in Sowerby, English Bot. ed. 3.
11:44. 1873. Based on *A. interrupta* L.
Apera spica-venti var. *interrupta* Beal,
Grasses N. Amer. 2: 357. 1896. Based
on *Agrostis interrupta* L.
Agrestis interrupta Bubani, Fl. Pyr. 4:
289. 1901. Based on *Agrostis inter-
rupta* L.
(1) **Apera spica-venti** (L.) Beauv., Ess.
Agrost. 151. 1812. Based on *Agrostis
spica-venti* L.
Agrostis spica-venti L., Sp. Pl. 61. 1753.
Europe.
Agrostis gracilis Salisb., Prodr. Stirp. 25.
1796. Based on *A. spica-venti* L.
Anemagrostis spica-venti Trin., Fund.
Agrost. 129. 1820. Based on *Agrostis
spica-venti* L.
Festuca spica-venti Raspail, Ann. Sci. Nat.,
Bot. 5: 445. 1825. Based on *Agrostis
spica-venti* L.
Muhlenbergia spica-venti Trin., Acad. St.
Pétersb. Mém. VI. Sci. Nat. 4¹: 285.
1841. Based on *Agrostis spica-venti* L.
Agrostis ventosa Dulac, Fl. Haut. Pyr. 74.
1867. Based on *Apera spica-venti*
Beauv.
Agrostis anemagrostis Syme in Sowerby,
English Bot. ed. 3. 11: 43. 1873.
Based on *Anemagrostis spica-venti* Trin.
Agrostis anemagrostis subsp. *spica-venti*
Syme in Sowerby, English Bot. ed. 3.
11: 43. 1873. Based on *A. spica-venti*
L.

(92) ARISTIDA L.

(14) **Aristida adscensionis** L., Sp. Pl. 82.
1753. Ascension Island.
Aristida interrupta Cav., Icon. Pl. 5:
45. pl. 471. f. 2. 1799. Mexico.
Chaetaria ascensionis Beauv., Ess. Agrost.
30, 151, 158. 1812. Based on *A.
adscensionis* L.
Aristida bromides H. B. K., Nov. Gen. et
Sp. 1: 122. 1815. Ecuador, *Humboldt*
and *Bonpland*.
Aristida coarctata H. B. K., Nov. Gen. et
Sp. 1: 122. 1815. Mexico, *Humboldt*
and *Bonpland*.
Chaetaria bromoides Roem. and Schult.,
Syst. Veg. 2: 396. 1817. Based on
Aristida bromoides H. B. K.
Chaetaria coarctata Roem. and Schult.,
Syst. Veg. 2: 396. 1817. Based on
Aristida coarctata H. B. K.
Aristida fasciculata Torr., Ann. Lyc. N. Y.
1: 154. 1824. Canadian River [Texas
or Oklahoma], *James*.
Chaetaria fasciculata Schult., Mantissa 3
(Add. 1): 578. 1827. Based on *Aristida
fasciculata* Torr.
Aristida nigrescens Presl, Rel. Haenk. 1:
223. 1830. Mexico, *Haenke*.
Aristida dispera Trin. and Rupr., Acad.
St. Pétersb. Mém. VI. Sci. Nat. 5¹: 129.
1842. Chile.

Aristida dispersa var. *bromoides* Trin. and
Rupr., Acad. St. Pétersb. Mém. VI. Sci.
Nat. 5¹: 130. 1842. Based on *A.
bromoides* H. B. K.
Aristida dispersa var. *coarctata* Trin. and
Rupr., Acad. St. Pétersb. Mém.
VI. Sci. Nat. 5¹: 130. 1842. Based on
A. coarctata H. B. K.
Aristida maritima Steud., Syn. Pl. Glum.
1: 137. 1854. Guadeloupe.
Aristida schaffneri Fourn., Mex. Pl. 2: 78.
1886. Mexico, *Schaffner*
Aristida grisebachiana Fourn., Mex. Pl.
2: 78. 1886. Mexico, *Schaffner* 175 in
part, 53.
Aristida grisebachiana var. *decolorata*
Fourn., Mex. Pl. 2: 78. 1886. Mexico,
Liebmann 663, 664.
Aristida adscensionis var. *coarctata*
Kuntze, Rev. Gen. Pl. 3: 340. 1898.
Based on *A. coarctata* H. B. K.
Aristida americana bromoides Scribn. and
Merr., U. S. Dept. Agr., Div. Agrost.
Cir. 32: 5. 1901. Based on *A. bromoides*
H. B. K.
Aristida debilis Mez, Repert. Sp. Nov.
Fedde 17: 151. 1921. Venezuela, *Mo-
ritz* [638]. [*Moritz* 1522 named *A. debilis*
by Mez is different. It has been named
A. moritzii Henr.] Jamaica, *MacNab*.
Aristida adscensionis var. *bromoides* Henr.,
Med. Rijks Herb. Leiden 54: 62. 1926.
Based on *A. bromoides* H. B. K.
Aristida adscensionis var. *mexicana* Hack.
ex Henr., Med. Rijks Herb. Leiden
54A: 265. 1927, as synonym of *A.
adscensionis*. Morelia, Mexico, *Arsène*.
(34) **Aristida affinis** (Schult.) Kunth, Rév.
Gram. 1: 61. 1829. Based on *Chaetaria
affinis* Schult.
Aristida racemosa Muhl., Descr. Gram.
172. 1817. Not *A. racemosa* Spreng.,
1807. Presumably Pennsylvania.
Chaetaria affinis Schult., Mantissa 2: 210.
1824. Based on *Aristida racemosa* Muhl.
Aristida purpurascens var. *alabamensis*
Trin. and Rupr., Acad. St. Pétersb.
Mém. VI. Sci. Nat. 5¹: 102. 1842.
Alabama.
Aristida virgata var. *palustris* Chapm., Fl.
South. U. S. 555. 1860. Western
Florida.
Aristida palustris Vasey, Grasses U. S.
Descr. Cat. 35. 1885. Based on *A.
virgata* var. *palustris* Chapm.
(29) **Aristida arizonica** Vasey, Torrey Bot.
Club Bul. 13: 27. 1886. Arizona
[*Rusby* 875; but the specimen bearing
the name and diagnosis in Vasey's
script was collected by G. R. Vasey at
Las Vegas, N. Mex.].
(16) **Aristida barbata** Fourn., Mex. Pl. 2:
78. 1886. Valley of Mexico, *Schaffner*
513.
Aristida havardii Vasey, Torrey Bot.
Club Bul. 13: 27. 1886. Western
Texas, *Havard* [28]. The date of publi-

cation is assumed to be subsequent to that of *A. barbata*.

(8) **Aristida basiramea** Engelm. ex Vasey, Bot. Gaz. 9: 76. 1884. Minneapolis, Minn., *Upham*.

(3) **Aristida californica** Thurb. in S. Wats., Bot. Calif. 2: 289. 1880. California, Colorado Desert, *Schott*; Fort ·Mohave, *Cooper*.

Aristida jonesii Vasey, U. S. Natl. Herb. Contrib. 3: 48. 1892, as synonym of *A. californica*. [The Needles, Calif., *Jones* 68a.]

Aristida californica var. *fugitiva* Vasey, U. S. Natl. Herb. Contrib. 3: 49. 1892. Colorado Desert, California, *Orcutt* [1486].

(39) **Aristida condensata** Chapm., Bot. Gaz. 3: 19. 1878. Florida [Apalachicola, *Chapman*].

Aristida stricta var. *condensata* Vasey, U. S. Natl. Herb. Contrib. 3: 45. 1892. Based on *A. condensata* Chapm.

Aristida combsii Scribn. and Ball, U. S. Dept. Agr., Div. Agrost. Bul. 24: 43. f. 17. 1901. Grasmere, Fla., *Combs* and *Baker* 1069.

Aristida condensata var. *combsii* Henr., Med. Rijks Herb. Leiden 54: 108. 1926. Based on *A. combsii* Scribn. and Ball.

(10) **Aristida curtissii** (A. Gray) Nash in Britton, Man. 94. 1901. Based on *A. dichotoma* var. *curtissii* A. Gray.

Aristida dichotoma var. *curtissii* A. Gray, Man. ed. 6. 640. 1890. [Bedford County, Va., *Curtiss*.]

Aristida basiramea var. *curtissii* Shinners, Amer. Midl. Nat. 23: 633. 1940. Based on *A. dichotoma* var. *curtissii* A. Gray.

(1) **Aristida desmantha** Trin. and Rupr., Acad. St. Pétersb. Mém. VI. Sci. Nat. 5¹: 109.· 1842. Texas, *Drummond* 285 [type], 333.

(9) **Aristida dichotoma** Michx., Fl. Bor. Amer. 1: 41. 1803. Lincoln, N. C., *Michaux*.

Curtopogon dichotomus Beauv., Ess. Agrost. 32, 159. pl. 8. f. 7. 1812. Based on *Aristida dichotoma* Michx.

Cyrtopogon dichotomus Spreng., Syst. Veg. 1: 266. 1825. Based on *Aristida dichotoma* Michx.

Avena setacea Muhl. ex Trin., Acad. St. Pétersb. Mém. VI. Math. Phys. Nat. 1: 87. 1830. Not *A. setacea* Vill., 1787. As synonym of *Aristida dichotoma* Michx.

Avena paradoxa Willd. ex Kunth, Enum. Pl. 1: 188. 1833, as synonym of *Aristida dichotoma* Michx.

Aristida dichotoma forma *major* Shinners, Amer. Midl. Nat. 23: 634. 1940. Starkville, Miss., *Kearney* in 1896.

(17) **Aristida divaricata** Humb. and Bonpl. ex Willd., Enum. Pl. 1: 99. 1809. Mexico, *Humboldt* and *Bonpland*.

Chaetaria divaricata Beauv., Ess. Agrost. 30, 158. 1812. Based on type of *Aristida divaricata* Humb. and Bonpl.

Aristida humboldtiana Trin. and Rupr., Acad. St. Pétersb. Mém. VI. Sci. Nat. 5¹: 118. 1842. Based on type of *A. divaricata* Humb. and Bonpl.

Aristida palmeri Vasey, Torrey Bot. Club Bul. 10: 42. 1883. Southern Arizona, *Palmer*.

Aristida lemmoni Scribn., N. Y. Acad. Sci. Trans. 14: 23. 1894. Arizona [Fort Huachuca, *Wilcox*].

(27) **Aristida fendleriana** Steud., Syn. Pl. Glum. 1: 420. 1855. New Mexico, *Fendler* 973.

Aristida purpurea var. *fendleri* Vasey in Rothr., Cat. Pl. Survey W. 100th Merid. 55. 1874. Name only, Denver [Wolf] 1110.

Aristida purpurea var. *fendleriana* Vasey, U. S. Natl. Herb. Contrib. 3: 46. 1892. Based on *A. fendleriana* Steud.

Aristida fasciculata var. *fendleriana* Vasey ex L. H. Dewey, U. S. Natl. Herb. Contrib. 2: 515. 1894. Based on *A. fendleriana* Steud.

Aristida longiseta fendleriana Merr., U. S. Dept. Agr., Div. Agrost. Cir. 34: 5. 1901. Based on *A. fendleriana* Steud.

Aristida subuniflora Nash in Small, Fl. Southeast. U. S. 116. 1903. New Mexico, *Vasey*.

(6) **Aristida floridana** (Chapm.) Vasey, Grasses U. S. Descr. Cat. 35. 1885. Based on *Streptachne floridana* Chapm.

Streptachne floridana Chapm., Fl. South. U. S. 554. 1860. South Florida, *Blodgett*.

Ortachne floridana Nash in Small, Fl. Southeast. U. S. 119. 1903. Based on *Streptachne floridana* Chapm.

(4) **Aristida glabrata** (Vasey) Hitchc., U. S. Natl. Herb. Contrib. 22: 522. 1924. Based on *A. californica* var. *glabrata* Vasey.

Aristida californica var. *major* Vasey, Calif. Acad. Sci. Proc. II. 2: 212. 1889. Name only [Magdalena Island, *Brandegee* in 1889].

Aristida californica var. *glabrata* Vasey, Calif. Acad. Sci. Proc. II. 3: 178. 1891. San José del Cabo, Baja California, [*Brandegee* 34 in 1890].

(22) **Aristida glauca** (Nees) Walp., Ann. Bot. [London] 1: 925. 1849. Based on *Chaetaria glauca* Nees.

Chaetaria glauca Nees, Linnaea 19: 688. 1847. Mexico, *Aschenborn* 251.

Aristida reverchoni Vasey, Torrey Bot. Club Bul. 13: 52. 1886. Crockett County, Tex., *Reverchon*.

Aristida stricta var. *nealleyi* Vasey, U. S. Natl. Herb. Contrib. 1: 55. 1890. Chenate Mountains, Tex., *Nealley* [709].

Aristida nealleyi Vasey, U. S. Natl. Herb. Contrib. 3: 45. 1892. Based on *A.*

stricta var. *nealleyi* Vasey.

Aristida reverchoni var. *augusta* [error for *angusta*] Vasey, U. S. Natl. Herb. Contrib. 3: 46. 1892. Comanche Peak, Tex., *Reverchon*.

Aristida vaseyi Woot. and Standl., N. Mex. Col. Agr. Bul. 81: 55. 1912. Based on *A. reverchoni* var. *augusta* Vasey.

(40) **Aristida gyrans** Chapm., Bot. Gaz. 3: 18. 1878. Roberts Key, Caximbas Bay, Fla. [*Chapman*].

(18) **Aristida hamulosa** Henr., Med. Rijks Herb. Leiden 54: 219. 1926. Tucson, Ariz., *Toumey*.

Aristida humboldtiana var. *minor* Vasey, U. S. Natl. Herb. Contrib. 3: 47. 1892. Texas [*Nealley*].

Aristida imbricata Henr., Med. Rijks Herb. Leiden 54A: 253. 1927. El Paso, Tex., *Griffiths* 7433.

Aristida gentilis var. *breviaristata* Henr., Med. Rijks Herb. Leiden 54A: 255. 1927. Santa Rita Mountains, Ariz., *Griffiths* 7270.

(15) **Aristida intermedia** Scribn. and Ball, U. S. Dept. Agr., Div. Agrost. Bul. 24: 44. f. 18. 1901. Biloxi, Miss., *Kearney* 204.

(28) **Aristida lanosa** Muhl. ex Ell., Bot. S. C. and Ga. 1: 143. 1816. South Carolina; name only, Muhl., Cat. Pl. 14. 1813.

Aristida lanata Poir. in Lam., Encycl. Sup. 1: 453. 1810. Not *A. lanata* Forsk., 1775. Carolina, *Bosc*.

Aristida gossypina Bosc ex Beauv., Ess. Agrost. 30, 152. 1812. Name only.

Chaetaria gossypina Bosc ex Beauv., Ess. Agrost. 30, 152, 158. 1812. Name only; Roem. and Schult., Syst. Veg. 2: 391. 1817. Based on *Aristida lanata* Poir.

Aristida lanuginosa Bosc ex Trin., Acad. St. Pétersb. Mém. VI. Sci. Nat. 2¹: 46. 1836, name only; Clarion in Trin. and Rupr., Acad. St. Pétersb. Mém. VI. Sci. Nat. 5¹: 103. 1842. North America, *Bosc*.

Moulinsia lanosa Raf. ex Jacks., Ind. Kew. 2: 267. 1894, as synonym of *Aristida lanosa* Muhl.

Aristida lanosa var. *macera* Fern. and Grisc., Rhodora 37: 135, pl. 335. 1935. Cape Henry, Va., *Fernald* and *Griscom* 2719.

(13) **Aristida longespica** Poir. in Lam., Encycl. Sup. 1: 452. 1810. Carolina, *Bosc*.

Aristida gracilis Ell., Bot. S. C. and Ga. 1: 142. pl. 8. f. 3. 1816. Charleston, S. C.

Aristida geniculata Raf., Amer. Monthly Mag. 2: 119. 1817. Long Island, N. Y.

Curtopogon gracilis Nees ex Trin. and Rupr., Acad. St. Pétersb. Mém. VI. Sci. Nat. 5¹: 101. 1842, as synonym of *Aristida gracilis* Ell.

Aristida gracilis var. *depauperata* A. Gray, Man. ed. 5. 618. 1867. Philadelphia, *Smith*.

Aristida simplicifolia [error for *simplici-flora*] var. *texana* Vasey, U. S. Natl. Herb. Contrib. 3: 44. 1892. Texas, [Marshall, *Riggs* 79].

Trixostis gracilis Raf. ex Jacks., Ind. Kew. 2: 1131. 1895, as synonym of *Aristida gracilis* Ell.

Aristida longespica var. *geniculata* Fernald, Rhodora 35: 318. 1933. Based on *A. geniculata* Raf.

(26) **Aristida longiseta** Steud., Syn. Pl. Glum. 1: 420. 1855. New Mexico, *Fendler* 978.

Aristida curtiseta Buckl., Acad. Nat. Sci. Phila. Proc. 1862: 92. 1862. Northern Texas [*Buckley*. Spikelets of type aborted by smut]. (Erroneously given in Index Kewensis as *A. breviseta*.)

Aristida purpurea var. *longiseta* Vasey in Rothr., in Wheeler, U. S. Survey W. 100th Merid. Rpt. 6: 286. 1878. Based on *A. longiseta* Steud.

Aristida fasciculata var. *nuttallii* Thurb. ex Beal, Grasses N. Amer. 2: 208. 1896. Based on *A. longiseta* Steud., though Thurber's name probably referred to *A. pallens* as used by Nuttall.

ARISTIDA LONGISETA var. RARIFLORA Hitchc., U. S. Natl. Herb. Contrib. 22: 565. 1924. Tom Green County, Tex., *Tweedy*. (Published as *A. longiseta rari-flora*.)

Aristida rariflora Henr., Med. Rijks Herb. Leiden 54A: 314. 1927. Based on *A. longiseta rariflora* Hitchc.

ARISTIDA LONGISETA var. ROBUSTA Merr., U. S. Dept. Agr., Div. Agrost. Cir. 34: 5. 1901. Indian Creek, Mont., *Scribner* 336.

Aristida purpurea robusta Piper, U. S. Natl. Herb. Contrib. 11: 107. 1906. Based on *A. longiseta* var. *robusta* Merr.

(37) **Aristida mohrii** Nash, N. Y. Bot. Gard. Bul. 1: 436. 1900. Spring Hill, near Mobile, Ala., *Mohr*.

(11) **Aristida oligantha** Michx., Fl. Bor. Amer. 1: 41. 1803. Illinois, *Michaux*.

?*Aristida adscensionis* [L. misapplied by] Walt., Fl. Carol. 74. 1788. South Carolina.

Chaetaria olygantha Beauv., Ess. Agrost. 30, 158. 1812. Based on *Aristida oligantha* Michx.

Aristida pallens [Cav. misapplied by] Nutt., Gen. Pl. 1: 57. 1818. Fort Mandan, N. Dak. [*Nuttall*].

Aristida micropoda Trin. and Rupr., Acad. St. Pétersb. Mém. VI. Sci. Nat. 5¹: 107. 1842. Arkansas, *Beyrich*.

Aristida macrochaeta Steud., Syn. Pl. Glum. 1: 134. 1854. Virginia, *M. A. Curtis*.

Aristida pauciflora Buckl., Acad. Nat. Sci. Phila. Proc. 1862: 92. 1862. Northern Texas [*Buckley*].

Aristida oligantha var. *nervata* Beal, Grasses N. Amer. 2: 202. 1896. Grants Pass, Oreg., *Howell*.

(7) **Aristida orcuttiana** Vasey, Torrey Bot. Club Bul. 13: 27. 1886. Hansen's Ranch, Baja California, *Orcutt* [507].

Aristida hypomegas Mez, Repert. Sp. Nov. Fedde 17: 146. 1921. New Mexico, *Bigelow* [34].

This species has been referred to *A. schiediana* Trin. and Rupr., a Mexican species not known from the United States.

(20) **Aristida pansa** Woot. and Standl., U. S. Natl. Herb. Contrib. 16: 112. 1913. Tortugas Mountain, N. Mex., *Wooton*.

(33) **Aristida parishii** Hitchc., in Jepson, Fl. Calif. 1: 101. 1912. Agua Caliente, Calif., *Parish Brothers* 1029a.

(19) **Aristida patula** Chapm. ex Nash, Torrey Bot. Club Bul. 23: 98. 1896. Based on *A. scabra* as described by Chapman (Fl. South. U. S. ed. 2. 663. 1883), not Kunth, 1829. Florida, *Chapman*.

(32) **Aristida purpurascens** Poir. in Lam., Encycl. Sup. 1: 452. 1810. South Carolina, *Bosc*.

Chaetaria purpurascens Beauv., Ess. Agrost. 30, 152, 158. 1812. Based on *Aristida purpurascens* Poir.

Aristida elliottiana Steud., Syn. Pl. Glum. 1: 133. 1854. Based on *A. stricta* as described by Elliott, not Michx., 1803.

Aristida geyeriana Steud., Syn. Pl. Glum. 1: 133. 1854. Illinois, *Geyer*.

Aristida stricta Steud., Syn. Pl. Glum. 1: 133. 1854. Not *A. stricta* Michx., 1803. As synonym of *A. geyeriana* Steud. Illinois.

Aristida purpurascens var. *minor* Vasey, U. S. Natl. Herb. Contrib. 1: 46. 1892. [Horn Island, Miss., *Tracy* 1564.]

Aristida purpurascens var. *glaucissima* Kearney ex Scribn. and Ball, U. S. Dept. Agr., Div. Agrost. Bull. 24: 45. 1901. Biloxi, Miss., *Kearney* 321.

(23) **Aristida purpurea** Nutt., Amer. Phil. Soc. Trans. (n. s.) 5: 145. 1837. Red River, Ark. [*Nuttall*].

Aristida purpurea var. *hookeri* Trin. and Rupr., Acad. St. Pétersb. Mém. VI. Sci. Nat. 5¹: 107. 1842. Texas, *Drummond* 293.

Aristida purpurea var. *berlandieri* Trin. and Rupr., Acad. St. Pétersb. Mém. VI. Sci. Nat. 5¹: 107. 1842. Bejar [Bexar], Tex., *Berlandier* 1777.

Aristida aequiramea Scheele, Linnaea 22: 343. 1849. New Braunfels, Tex., *Lindheimer* [562].

Aristida filipendula Buckl., Acad. Nat. Sci., Phila. Proc. 1862: 93. 1862. Western Texas [*Buckley*, the locality being northern Texas].

Aristida purpurea var. *californi*[*c*]*a* Vasey, U. S. Natl. Herb. Contrib. 3: 47. 1892. California [Capay Valley, *Lemmon* 5474].

Aristida fasciculata var. *californica* Vasey ex L. H. Dewey, U. S. Natl. Herb. Contrib. 2: 515. 1894. Presumably based on *A. purpurea* var. *californica* Vasey.

Aristida fasciculata var. *hookeri* L. H. Dewey, U. S. Natl. Herb. Contrib. 2: 515. 1894. Presumably based on *A. purpurea* var. *hookeri* Trin. and Rupr.

Aristida longiseta hookeri Merr., U. S. Dept. Agr., Div. Agrost. Cir. 34: 5. 1901. Based on *A. purpurea* var. *hookeri* Trin. and Rupr.

Aristida purpurea aequiramea Merr., U. S. Dept. Agr., Div. Agrost. Cir. 34: 7. 1901. Based on *A. aequiramea* Scheele.

Aristida purpurea capillarifolia Merr., U. S. Dept. Agr., Div. Agrost. Cir. 34: 8. 1901. Texas, *Nealley*.

Aristida berlandieri Hitchc., U. S. Natl. Herb. Contrib. 17: 280. 1913. Based on *A. purpurea* var. *berlandieri* Trin. and Rupr.

ARISTIDA PURPUREA var. LAXIFLORA Merr., U. S. Dept. Agr., Div. Agrost. Cir. 34: 8. 1901. Texas, *Reverchon* 12.

(12) **Aristida ramosissima** Engelm. ex A. Gray, Man. ed. 2. 550. 1856. Illinois, *Engelmann* [type] and Kentucky.

Aristida ramosissima var. *uniaristata* A. Gray, Man. ed. 5. 618. 1867. Odin, Ill., *Vasey*.

Aristida ramosissima var. *chaseana* Henr., Med. Rijks Herb. Leiden. 54B: 498. 1928. Lake Charles, La., *Chase* 4411.

(31) **Aristida rhizomophora** Swallen, Wash. Acad. Sci. Jour. 19: 196. f. 1. 1929. North of Lake Okeechobee, Fla., *Weatherwax* 1081.

(24) **Aristida roemeriana** Scheele, Linnaea 22: 343. 1849. New Braunfels, Tex., *Römer*.

Aristida muhlenbergioides Fourn., Mex. Pl. 2: 79. 1886. Mexico, *Virlet* 1424, *Karwinsky* 1008.

Aristida purpurea var. *micrantha* Vasey, U. S. Natl. Herb. Contrib. 3: 47. 1892. Western Texas [*Nealley*].

Aristida fasciculata var. *micrantha* Vasey ex L. H. Dewey, U. S. Natl. Herb. Contrib. 2: 515. 1894. Presumably based on *A. purpurea* var. *micrantha* Vasey.

Aristida micrantha Nash in Small, Fl. Southeast. U. S. 117. 1903. Based on *A. purpurea* var. *micrantha* Vasey.

(36) **Aristida simpliciflora** Chapm., Bot. Gaz. 3: 18. 1878. West Florida [*Chapman*].

(21) **Aristida spiciformis** Ell., Bot. S. C. and Ga. 1: 141. 1816. Presumably South Carolina.

Aristida stricta Muhl., Descr. Gram. 174.

1817. Not *A. stricta* Michx., 1803. Georgia.
Aristida squarrosa Trin. in Spreng., Neu. Entd. 2: 62. 1821. North America.
Chaetaria squarrosa Schult., Mantissa 3 (Add. 1): 577. 1827. Based on *Aristida squarrosa* Trin.

(30) **Aristida stricta** Michx., Fl. Bor. Amer. 1: 41. 1803. South Carolina, *Michaux.*
Chaetaria stricta Beauv., Ess. Agrost. 30, 152, 158. 1812. Based on *Aristida stricta* Michx.
Aristida beyrichiana Trin. and Rupr. Acad. St. Pétersb. Mém. VI. Sci. Nat. 5¹: 104. 1842. Georgia and Arkansas, *Beyrich.*

(38) **Aristida tenuispica** Hitchc., U. S. Natl. Herb. Contrib. 22: 581. 1924. Hillsboro, Fla., *Combs* 1384.

(5) **Aristida ternipes** Cav., Icon. Pl. 5: 46. 1799. Panama, *Née.*
Streptachne scabra H. B. K., Nov. Gen. et Sp. 1: 124. pl. 40. 1815. Near Toluca, Mex., *Humboldt* and *Bonpland.*
Streptachne tenuis H. B. K., Nov. Gen. et Sp. 1: 124. 1815. Venezuela, *Humboldt* and *Bonpland.*
Aristida scabra Kunth, Rév. Gram. 1: 62. 1829. Based on *Streptachne scabra* H. B. K.
Aristida tenuis Kunth, Rév. Gram. 1: 62. 1829. Based on *Streptachne tenuis* H. B. K.
Stipa tenuis Willd. ex Steud., Nom. Bot. ed. 2. 2: 643. 1841, as synonym of *Aristida tenuis.*
Muhlenbergia scabra Trin. and Rupr., Acad. St. Pétersb. Mém. VI. Sci. Nat. 5¹: 183. 1842. Based on *Aristida scabra* Kunth.
Streptachne cubensis A. Rich. in Sagra, Hist. Cuba 11: 311. 1850. Cuba, *Sagra.*
Ortachne scabra Fourn., Soc. Bot. France Bul. 27: 295. 1880. Based on *Streptachne scabra* H. B. K.
Ortachne tenuis Fourn., Soc. Bot. France Bul. 27: 295. 1880. Based on *Streptachne tenuis* H. B. K.
ARISTIDA TERNIPES var. MINOR (Vasey) Hitchc., Wash. Acad. Sci. Jour. 23: 453. 1933. Based on *A. schiedeana* var. *minor* Vasey.
Aristida schiedeana var. *minor* Vasey, Torrey Bot. Club Bul. 13: 28. 1886. Arizona, *Pringle* [type]; Bowie, *Jones.*
Aristida divergens Vasey, U. S. Natl. Herb. Contrib. 3: 48. 1892. Based on *A. schiedeana* var. *minor* Vasey.
Aristida ternipes divergens Hitchc., U. S. Natl. Herb. Contrib. 22: 525. 1924. Based on *A. divergens* Vasey.

(2) **Aristida tuberculosa** Nutt., Gen. Pl. 1: 57. 1818. Near Augusta, Ga.
Chaetaria tuberculosa Schult., Mantissa 2: 211. 1824. Based on *Aristida tuberculosa* Nutt.

(35) **Aristida virgata** Trin. in Spreng., Neu. Entd. 2: 60. 1821. North America [Philadelphia, Pa.].
Aristida stricta Steud., Nom. Bot. ed. 2. 1: 132. 1840. Not *A. stricta* Michx., 1803. As synonym of *A. virgata* Trin.
Aristida perennis Panz. in Trin. and Rupr., Acad. St. Pétersb. Mém. VI. Sci. Nat. 5¹: 104. 1842. South Carolina. (Fide Henrard, Med. Rijks Herb. Leiden 54A: 439. 1927. (Critical Revis. Aristida.)
Aristida gracilis var. *virgata* Wood, Amer. Bot. and Flor. pt. 2: 389. 1871. Presumably based on *A. virgata* Trin.
Aristida purpurascens var. *depauperata* Vasey ex Beal, Grasses N. Amer. 2: 201. 1896 [Ocean Springs], Miss., *Tracy* [107].
Aristida chapmaniana Nash in Small, Fl. Southeast. U. S. 118, 1327. 1903. Apalachicola, Fla., *Chapman.*

(25) **Aristida wrightii** Nash in Small, Fl. Southeast. U. S. 116. 1903. Dallas, Tex., *Reverchon* 1061.

(63) ARRHENATHERUM Beauv.

(1) **Arrhenatherum elatius** (L.) Presl, Fl. Cech. 17. 1819. Based on *Avena elatior* L.
Avena elatior L., Sp. Pl. 79. 1753. Europe.
Holcus avenaceus Scop., Fl. Carn. ed. 2. 2: 276. 1772. Based on *Avena elatior* L.
Avena elata Salisb., Prodr. Stirp. 23. 1796. Not *A. elata* Forsk., 1775. Based on *A. elatior* L.
Arrhenatherum avenaceum Beauv., Ess. Agrost. 55, 152, 164. pl. 11. f. 5. 1812. Based on *Holcus avenaceus* Scop.
Arrhenatherum americanum Beauv., Ess. Agrost. 56, 152, 1812. Name only.
Hordeum avenaceum Wigg. ex Beauv., Ess. Agrost. 165. 1812. Name only, referred to *Arrhenatherum;* Steud, Nom. Bot. 413. 1821, as synonym of *Holcus avenaceus* Scop.
ARRHENATHERUM ELATIUS var. BIARISTATUM (Peterm.) Peterm., Flora 27: 229. 1844. Based on *A. biaristatum* Peterm.
Arrhenatherum biaristatum Peterm., Fl. Lips. 106. 1838. Germany.
ARRHENATHERUM ELATIUS var. BULBOSUM (Willd.) Spenner, Fl. Friburg. 1: 113. 1825. Based on *Avena bulbosa* Willd.
Avena tuberosa Gilib., Exerc. Phyt. 2: 538. 1792. France.
Avena bulbosa Willd., Ges. Naturf. Freund. Berlin Neue Schrift. 2: 116. 1799. Switzerland.
Holcus bulbosus Schrad., Fl. Germ. 1: 248. 1806. Based on *Avena bulbosa* Willd.
Holcus avenaceus var. *bulbosus* Gaudin, Agrost. Helv. 1: 136. 1811. Based on *H. bulbosus* Schrad.
Avena elatior var. *bulbosa* St. Amans, Fl.

Agen. 47. 1821. Based on *A. bulbosa* Willd.

Arrhenatherum avenaceum var. *nodosum* Reichenb., Fl. Germ. 1: 53. 1830. Germany.

Arrhenatherum tuberosum Schultz, Pollichia 20–21: 272. 1863. Based on *Avena tuberosa* Gilib.

Avena elatior var. *tuberosa* Aschers., Fl. Brand. 1: 826. 1864. Based on *A. tuberosa* Gilib.

Arrhenatherum elatius var. *tuberosum* Thiel., Soc. Bot. Belg. Bul. 12: 184. 1873. Based on *Avena tuberosa* Gilib.

Arrhenatherum bulbosum variegatum Hitchc. in Bailey, Stand. Cycl. Hort. 1: 397. 1914. Cult.

Arrhenatherum elatius var. *nodosum* Hubb., Rhodora 18: 234. 1916. Not *A. elatius* var. *nodosum* Parl., 1848. Based on *A. avenaceum* var. *nodosum* Reichenb.

Arrhenatherum elatius var. *nodosum* forma *striatum* Hubb., Rhodora 18: 235. 1916. Based on *A. bulbosum variegatum* Hitchc.

Arrhenatherum elatius var. *bulbosum* forma *striatum* L. B. Smith, Rhodora 49: 267. 1947. Based on *A. elatius* var. *nodosum* forma *striatum* Hubb.

(153) ARTHRAXON Beauv.

(1) **Arthraxon hispidus** (Thunb.) Makino, Bot. Mag. [Tokyo] 26: 214. 1912. Based on *Phalaris hispida* Thunb.

Phalaris hispida Thunb., Fl. Japon. 44. 1784. Japan.

ARTHRAXON HISPIDUS var. CRYPTATHERUS (Hack.) Honda, Bot. Mag. [Tokyo] 39: 277. 1925. Based on *A. ciliaris* subsp. *langsdorffii* var. *cryptatherus* Hack.

Arthraxon ciliaris subsp. *langsdorffii* var. *cryptatherus* Hack., in DC., Monogr. Phan. 6: 355. 1889. Japan.

Arthraxon cryptatherus Koidz., Bot. Mag. [Tokyo] 39: 301. 1925. Based on *A. ciliaris* subsp. *langsdorffii* var. *cryptatherus* Hack.

(1) ARUNDINARIA Michx.

(1) **Arundinaria gigantea** (Walt.) Muhl., Cat. Pl. 14. 1813. Presumably based on *Arundo gigantea* Walt., the name published as "*Arundinaria gigantea* Walt." Carolina, Mississippi. The combination also made by Chapman, Fl. South. U. S. 561. 1860, Walter's name not cited, but *A. macrosperma* Michx. cited as synonym.

Arundo gigantea Walt., Fl. Carol. 81. 1788. South Carolina.

Arundinaria macrosperma Michx., Fl. Bor. Amer. 1: 74. 1803. Banks of Mississippi, Carolina, Florida, *Michaux*.

Miegia macrosperma Pers., Syn. Pl. 1: 102.

1805. Based on *Arundinaria macrosperma* Michx.

Ludolfia macrosperma Willd., Ges. Naturf. Freund. Berlin Mag. 2: 320. 1808. Based on *Arundinaria macrosperma* Michx.

Miegia gigantea Nutt., Gen. Pl. 1: 39. 1818. "Alluvions of the Mississippi." Based (through Elliott) on *Arundo gigantea* Walt.

Arundinaria gigantea Nutt., Gen. Pl. 1: 39. 1818, as synonym of *Miegia gigantea* Nutt.

Miegia arundinaria Raf., West. Rev. Misc. Mag. 1: 93. 1819. Name only. Kentucky.

Miegia arundinaria Raf., First Cat. Gard. Bot. Transylv. Univ. 14. 1824. Name only. Kentucky.

Nastus macrospermus Raspail, Ann. Sci. Nat., Bot. 5: 442, 458. pl. 8. f. 1. 1825. Based on *Arundinaria macrosperma* Michx.

Miegia pumila Nutt., Amer. Phil. Soc. Trans. (n. s.) 5: 149. 1837. Junction of Red and Kiamichi Rivers [Okla.]. A flowering basal shoot.

Arundinaria tecta var. *distachya* Rupr., Acad. St. Pétersb. Mém. VI. Sci. Nat. 3¹: 112. pl. 2. f. 1. γ. 1839. "Philadelphia." [Probably received from Philadelphia.]

Arundinaria tecta var. *pumila* Rupr., Acad. St. Pétersb. Mém. VI. Sci. Nat. 3¹: 112. 1839. Based on *Miegia pumila* Nutt.

Arundinaria macrosperma var. *arborescens* Munro, Linn. Soc. Trans. 26: 15. 1868. Based on *A. macrosperma* Michx.

Miegia arundinacea Torr. ex Munro, Linn. Soc. Trans. 26: 15. 1868, as synonym of *Arundinaria macrosperma* var. *arborescens* Munro.

Bambusa hermanni E. G. Camus, Bamb. Monogr. 36. 1913, horticultural name as synonym of *Arundinaria macrosperma* Michx.

(2) **Arundinaria tecta** (Walt.) Muhl., Cat. Pl. 14. 1813; Descr. Gram. 191. 1817. Based on *Arundo tecta* Walt.

Arundo tecta Walt., Fl. Carol. 81. 1788. South Carolina.

Ludolfia tecta A. Dietr., Sp. Pl. 2: 24. 1833. Based on *Arundo tecta* Walt.

Festuca grandiflora Lam., Tabl. Encycl. 1: 191. 1791. "Carolina, Fraser."

Arundinaria tecta var. *colorata* Rupr., Acad. St. Pétersb. Mém. VI. Sci. Nat. 3¹: 112. pl. 2. f. 1. δ. 1839. North America.

Arundinaria macrosperma var. *suffruticosa* Munro, Linn. Soc. Trans. 26: 15. 1868. Based on *A. tecta* Muhl.

Arundinaria macrosperma var. *tecta* Wood, Amer. Bot. and Flor. pt. 2: 404. 1871. Presumably based on *Arundo tecta* Walt. Published as new by Beal,

Grasses N. Amer. 2: 659. 1896, same basis.
Arundinaria gigantea tecta Scribn., Torrey Bot. Club Bul. 20: 478. 1893. Based on *Arundo tecta* Walt.

(26) ARUNDO L.

(1) **Arundo donax** L., Sp. Pl. 81. 1753. Southern Europe.
Arundo sativa Lam., Fl. Franç. 3: 616. 1778. France.
Arundo latifolia Salisb., Prodr. Stirp. 24. 1796. Based on *A. donax* L.
Donax arundinaceus Beauv., Ess. Agrost. 78, 152, 161. 1812. Based on *Arundo donax* L.
Scolochloa arundinacea Mert. and Koch ex Roehl., Deut. Fl. ed. 3. 1²: 530. 1823. Based on *Arundo donax* L.
Cynodon donax Raspail, Ann. Sci. Nat., Bot. 5: 302. 1825. Based on *Arundo donax* L.
Scolochloa donax Gaudin, Fl. Helv. 1: 202. 1828. Based on *Arundo donax* L.
Donax donax Aschers. and Graebn., Fl. Nordostd. Flachl. 101. 1898. Based on *Arundo donax* L.
Arundo glauca Bubani, Fl. Pyr. 4: 303. 1901. Not *A. glauca* Bieb., 1808. Based on *Arundo donax* L.
ARUNDO DONAX var. VERSICOLOR Stokes, Bot. Mat. Med. 1: 160. 1812. Presumably based on *Arundo versicolor* Mill.
Arundo versicolor Mill., Gard. Dict. ed. 8. No. 3. 1768. Cultivated from India.
Arundo donax var. *variegata* Vilm., Fl. Pl. Terre 90. 1863. France.

(61) AVENA L.

(3) **Avena barbata** Brot., Fl. Lusit. 1: 108. 1804. Europe. [*Avena barbata* Pott. ex Link, Jour. Bot. Schrad. 2: 315. 1799, inadequately described from garden plants and said to be wild about Lisbon, may be the same species.]
Avena brevis Roth, Bot. Abh. 42. 1787. Europe.
Avena byzantina C. Koch, Linnaea 21: 392. 1848. Constantinople.
(1) **Avena fatua** L., Sp. Pl. 80. 1753. Europe.
Avena fatua var. *glabrata* Peterm., Fl. Bienitz 13. 1841. Europe.
Avena nuda L., Amoen. Acad. 3: 401. 1756. Europe.
(2) **Avena sativa** L., Sp. Pl. 79. 1753. Europe.
Avena sativa var. *nigra* Wood, Class-book ed. 2. 610. 1847. Not *A. sativa* var. *nigra* Schrank as to name but probably the same form. Cultivated.
Avena sativa var. *secunda* Wood, Class-book ed. 2. 610. 1847. *A. sativa* var. *secunda* Provancher, Fl. Canad. 2: 689. 1862, is probably the same form. Cultivated.

Avena fatua var. *sativa* Hausskn., Mitt. Geogr. Ges. Thüringen 3: 238. 1885. Presumably based on *Avena sativa* L.
Avena fatua subsp. *sativa* Thell., Vierteljahrs. Nat. Ges. Zürich 56: 325. 1911. Based on *A. sativa* L.
Avena sterilis L., Sp. Pl. ed. 2. 118. 1762. Spain.
Avena algeriensis Trab., Bul. Agr. Alger. Tunis. 16: 354. 1910. Cult.
Avena sterilis algeriensis Trab., Jour. Hered. 5: 77. 1914. Presumably based on *A. algeriensis* Trab.
Avena strigosa Schreb., Spic. Fl. Lips. 52. 1771. Europe.

(134) AXONOPUS Beauv.

(3) **Axonopus affinis** Chase, Wash. Acad. Sci. Jour. 28: 180. f. 1, 2. 1938. Waynesboro, Miss., *Kearney* in 1896.
(2) **Axonopus compressus** (Swartz) Beauv., Ess. Agrost. 12. 1812. Based on *Milium compressum* Swartz.
Milium compressum Swartz, Prodr. Veg. Ind. Occ. 24. 1788. Jamaica.
Paspalum tristachyon Lam., Tabl. Encycl. 1: 176. 1791. South America, *Richard*.
Paspalum platicaulon Poir. in Lam., Encycl. Sup. 5: 34. 1804. Puerto Rico, *Ledru*.
Agrostis compressa Poir. in Lam., Encycl. Sup. 1: 259. 1810. Not *A. compressa* Poir., op. cit. 258, nor Willd., 1790. Based on *Milium compressum* Swartz.
Paspalum compressum Raspail, Ann. Sci. Nat., Bot. 5: 301. 1825. Based on *Axonopus compressus* Beauv.
Paspalum laticulmum Spreng., Syst. Veg. 1: 245. 1825. West Indies.
Digitaria platicaulis Desv., Opusc. 62. 1831. Based on *Paspalum platicaulon* Poir.
Digitaria domingensis Desv. ex Kunth, Enum. Pl. 1: 49. 1833, as synonym of *Paspalum platicaulon* Poir.
Paspalum platycaule Willd. ex Steud., Nom. Bot. ed. 2. 2: 272. 1840, erroneously cited as synonym of *P. furcatum* Flügge. Ecuador, *Humboldt*.
Paspalum guadaloupense Steud., Syn. Pl. Glum. 1: 18. 1854. Guadeloupe, *Duchaissing*.
Paspalum depressum Steud., Syn. Pl. Glum. 1: 20. 1854. Louisiana, *Hartmann 51*.
Paspalum filostachyum A. Rich. ex Steud., Syn. Pl. Glum. 1: 20. 1854. West Indies, *Sieber* [365].
Anastrophus compressus Schlecht. ex Doell, in Mart., Fl. Bras. 2²: 102. 1877. Presumably based on *Milium compressum* Swartz.
Paspalum furcatum var. *parviflorum* Doell in Mart., Fl. Bras. 2²: 104. 1877. [West Indies] *Sieber* 365; [Louisiana], *Hartmann 51*.

Anastrophus platycaulis Schlecht. ex Jacks., Ind. Kew. 1: 118. 1893, as synonym of *Paspalum platicaulon*.
Panicum platycaulon Kuntze, Rev. Gen. Pl. 3²: 363. 1898. Based on *Paspalum platicaulon* Poir.
Paspalum raunkiaerii Mez, Repert. Sp. Nov. Fedde 15: 60. 1917. St. Jan, West Indies, *Raunkiaer* 1313.

(1) **Axonopus furcatus** (Flügge) Hitchc., Rhodora 8: 205. 1906. Based on *Paspalum furcatum* Flügge.
Paspalum furcatum Flügge, Monogr. Pasp. 114. 1810. Carolina, *Bosc.*
Paspalum digitaria C. Muell., Bot. Ztg. 19: 324. 1861. Not *P. digitaria* Poir., 1816. Texas, *Drummond* 276.
Paspalum michauxianum var. *villosum* Vasey, Torrey Bot. Club Bul. 13: 163. 1886. No locality cited. [Type, Orange County, Fla., *Curtiss E.*]
Paspalum furcatum var. *villosum* Vasey, U. S. Natl. Herb. Contrib. 3: 16. 1892. Presumably based on *Paspalum michauxianum* var. *villosum* Vasey.
Paspalum paspaloïdes var. *villosum* Scribn. and Ball, U. S. Dept. Agr., Div. Agrost. Bul. 24: 42. 1901. Based on *P. furcatum* var. *villosum* Vasey.
Anastrophus furcatus Nash, N. Amer. Fl. 17: 162. 1912. Based on *Paspalum furcatum* Flügge.
This species was called *Paspalum paspaloides* by Scribner (Torrey Bot. Club Mem. 5: 29. 1894) and *Anastrophus paspaloides* by Nash (in Britton, Man. 75. 1901), but *Digitaria paspalodes* Michx., upon which these names are based, is *Paspalum distichum* L.

BAMBUSA Schreb.

Bambusa bambos (L.) Voss in Vilmorin,[21] Blumengartnerei 1: 1189. 1896. Based on *Arundo bambos* L.
Arundo bambos L., Sp. Pl. 81. 1753. India.
This is the thorny bamboo described by Gamble, Ann. Bot. Gard. Calcutta 7: 51. 1896, under "*Bambusa arundinacea* Willd.," and figured (op. cit. pl. 48) over the name "*Bambusa arundinacea* Retz."
Bambusa multiplex (Lour.) Raeusch., Nomenclature ed. 3. 103. 1797. Name only; Raeusch. ex Schult., Syst. Veg. 7: 1350. 1830. Based on *Arundo multiplex* Lour.
Arundo multiplex Lour. Fl. Cochinch. 58. 1790. Cochinchina.
Bambusa vulgaris Schrad. ex Wendl., Coll. Pl. 2: 26. pl. 47. 1810; (more fully described and illustrated by Gamble, Ann. Bot. Gard. Calcutta 7: 43. pl. 40. 1896). India.

[21] Contributed by F. A. McClure; see McClure, F. A., Blumea Sup. 3 (Henrard Jubilee vol.): 95. 1946.

Bambusa thouarsii Kunth, Rév. Gram. 2: 323. pl. 73, 74. 1830. Madagascar and Bourbon.
Bambusa surinamensis Rupr., Acad. St. Pétersb. Mém. VI. Sci. Nat. 3¹: 139. pl. 11, f. 49. 1839. Surinam, *Weigelt.*

(106) BECKMANNIA Host

(1) **Beckmannia syzigachne** (Steud.) Fernald, Rhodora 30: 27. 1928. Based on *Panicum syzigachne* Steud.
Panicum syzigachne Steud., Flora 29: 19. 1846. Japan.
Beckmannia erucaeformis var. *uniflora* Scribn. ex A. Gray, Man. ed. 6. 628. 1890. Iowa to Minnesota and westward.
Beckmannia erucaeformis var. *baicalensis* Kuznezow, Angew. Bot. Bul. 6: 584. 1913. Siberia.
Beckmannia baicalensis Hultén, Svensk. Vet. Akad. Handl. III. 5: 119. 1927. Based on *B. erucaeformis* var. *baicalensis* Kuznezow.
In most American botanical works, until recently, this is referred to *B. erucaeformis* (L.) Host, a European species. Nuttall (Gen. Pl. 1: 48. 1818) misspells the name *Bruchmannia*.

(37) BLEPHARIDACHNE Hack.

(2) **Blepharidachne bigelovii** (S. Wats.) Hack. in DC., Monogr. Phan. 6: 261. 1889. Based on *Eremochloë bigelovii* S. Wats.
Eremochloë bigelovii S. Wats. in King, Geol. Expl. 40th Par. 5: 382. pl. 40. f. 1–9. 1871. [Frontera, near El Paso, Tex.], *Wright* 2028.
Eremochloë thurberi S. Wats. in King, Geol. Expl. 40th Par. 5: pl. 40. f. 1–9. 1871. Name inadvertently given on the plate illustrating *E. bigelovii*.
(1) **Blepharidachne kingii** (S. Wats.) Hack. in DC., Monogr. Phan. 6: 261. 1889. Based on *Eremochloë kingii* S. Wats.
Eremochloë kingii S. Wats. in King, Geol. Expl. 40th Par. 5: 382. pl. 40. f. 10–16. 1871. Trinity Mountains, Nev., *Watson.*

(84) BLEPHARONEURON Nash

(1) **Blepharoneuron tricholepis** (Torr.) Nash, Torrey Bot. Club Bul. 25: 88. 1898. Based on *Vilfa tricholepis* Torr.
Vilfa tricholepis Torr., U. S. Expl. Miss. Pacif. Rpt. 4: 155. 1857. Sandia Mountains, N. Mex. [*Bigelow*].
Sporobolus tricholepis Coulter, Man. Rocky Mount. 411. 1885. Based on *Vilfa tricholepis* Torr.

(112) BOUTELOUA Lag.

(1) **Bouteloua aristidoides** (H. B. K.) Griseb., Fl. Brit. W. Ind. 537. 1864.

Based on *Dinebra artistidoides* H. B. K.
Dinebra aristidoides H. B. K., Nov. Gen.
et Sp. 1: 171. 1816. Mexico, *Humboldt
and Bonpland*.
Atheropogon aristidoides Roem. and
Schult., Syst. Veg. 2: 415. 1817. Based
on *Dinebra aristidoides* H. B. K.
Eutriana aristidoides Trin., Gram. Unifl.
242. 1824. Based on *Atheropogon aris-
tidoides* Roem. and Schult.
Dineba hirsuta Presl, Rel. Haenk. 1: 292.
1830. Peru, *Haenke*.
Eutriana hirsuta Kunth, Rév. Gram. 1:
Sup. 23. 1830. Based on *Dineba hirsuta*
Presl.
Aristida unilateralis Willd. ex Steud.,
Nom. Bot. ed. 2. 1: 132. 1840, as
synonym of *Eutriana aristidoides* Trin.
Bouteloua gracilis "Hook." ex Vasey in
Rothr., in Wheeler, U. S. Survey W
100th Merid. Rpt. 6: 287. 1878. Not
B. gracilis Lag., 1840. Arizona, *Roth-
rock* 701.
Bouteloua ciliata Griseb., Abh. Ges. Wiss.
Göttingen 24: 302. 1879. Juramento,
Argentina, *Lorenz* and *Hieronymus* 352.
Triathera aristidoides Nash in Small, Fl.
Southeast. U. S. 137. 1903. Based on
Dinebra aristidoides H. B. K.
BOUTELOUA ARISTIDOIDES var. ARIZONICA
Jones, West. Bot. Contrib. 14: 13.
1912. Tucson, Ariz., *Thornber* 177.
(10) **Bouteloua barbata** Lag., Var. Cienc. 4:
141. 1805. Mexico.
Actinochloa barbata Roem. and Schult.,
Syst. Veg. 2: 420. 1817. Based on
Bouteloua barbata Lag.
Eutriana barbata Kunth, Rév. Gram. 1:
96. 1829. Based on *Bouteloua barbata*
Lag.
Chondrosium polystachyum Benth., Bot.
Voy. Sulph. 56. 1844. Magdalena
Bay, Baja California, *Barclay*.
Chondrosium subscorpiodes C. Muell., Bot.
Ztg. 14: 347. 1856. Baja California,
Barclay.
Bouteloua polystachya Torr., U. S. Expl.
Miss. Pacif. Rpt. 5²: 366. pl. 10. 1857.
Based on *Chondrosium polystachyum*
Benth.
Bouteloua pumila Buckl., Acad. Nat. Sci.
Phila. Proc. 1862: 93. 1862. Texas,
Wright 754.
Bouteloua polystachya var. *major* Vasey in
Rothr., in Wheeler, U. S. Survey W.
100th Merid. Rpt. 6: 287. 1878.
Sonoyta Valley, Ariz., *Rothrock* 691.
Chondrosium exile Fourn., Mex. Pl. 2: 137.
1886. Mexico, *Berlandier* 842.
Chondrosium microstachyum Fourn., Mex.
Pl. 2: 138. 1886. Guadalupe, Mexico,
Bourgeau 667.
Bouteloua arenosa Vasey in S. Wats.,
Amer. Acad. Sci. Proc. 24: 81. 1889,
name only; U. S. Dept. Agr., Div. Bot.
Bul. 12¹: pl. 34. 1890. Guaymas,
Mexico, *Palmer* 189.

Bouteloua microstachya L. H. Dewey, U. S.
Natl. Herb. Contrib. 2: 531. 1894.
Based on *Chondrosium microstachyum*
Fourn.
Bouteloua micrantha Scribn. and Merr.,
U. S. Dept. Agr., Div. Agrost. Cir. 32: 8.
1901. Fort Lowell, Ariz., *Griffiths* 1556.
(16) **Bouteloua breviseta** Vasey, U. S. Natl.
Herb. Contrib. 1: 58. 1890. (July 18.)
Screw Bean, Presidio County, Tex.,
Nealley [669].
Bouteloua ramosa Scribn. ex Vasey, U. S.
Dept. Agr., Div. Bot. Bul. 12¹: pl. 44.
1890. (Oct. 13.) Mexico to Arizona and
western Texas, [type, *Nealley*].
Bouteloua oligostachya var. *ramosa* Scribn.
ex Beal, Grasses N. Amer. 2: 418.
1896. Based on *B. ramosa* Scribn.
(6) **Bouteloua chondrosioides** (H. B. K.)
Benth. ex S. Wats., Amer. Acad. Sci.
Proc. 18: 179. 1883. Based on *Dinebra
chondrosioides* H. B. K.
Dinebra chondrosioides H. B. K., Nov.
Gen. et Sp. 1: 173. pl. 53. 1816.
Michoácan, Mexico, *Humboldt* and
Bonpland.
Bouteloua ovata Lag., Gen. et Sp. Nov. 5.
1816. Mexico.
Atheropogon chondrosioides Roem. and
Schult., Syst. Veg. 2: 416. 1817. Based
on *Dinebra chondrosioides* H. B. K.
Actinochloa ovata Roem. and Schult.,
Syst. Veg. 2: 420. 1817. Based on
Bouteloua ovata Lag.
Eutriana cristata Trin., Gram. Unifl. 241.
1824. Based on *Atheropogon chondro-
sioides* Roem. and Schult.
Chondrosium humboldtianum Kunth, Rév.
Gram. 1: 93. 1829. Based on *Dinebra
chondrosioides* H. B. K.
Bouteloua havardii Vasey ex S. Wats.,
Amer. Acad. Sci. Proc. 18: 179. 1883.
Limpio Mountains, Tex., *Havard* in
1881.
(3) **Bouteloua curtipendula** (Michx.) Torr.
in Emory, Notes Mil. Reconn. 154.
1848. Based on *Chloris curtipendula*
Michx.
Chloris curtipendula Michx., Fl. Bor.
Amer. 1: 59. 1803. Illinois, *Michaux*.
Bouteloua racemosa Lag., Var. Cienc. 4:
141. 1805. Mexico.
Bouteloua pendula Lag., Var. Cienc. 4:
141. 1805, as synonym of *B. racemosa*.
Atheropogon apludoides Muhl. in Willd.,
Sp. Pl. 4: 937. 1806. North America.
Bouteloua melicaeformis Brouss. ex Hor-
nem., Enum. Pl. Hort. Hafn. 7. 1807.
Name only; Roem. and Schult., Syst.
Veg. 2: 414. 1817, as synonym of
Atheropogon apludoides Muhl.
Bouteloua melicoides Beauv., Ess. Agrost.
40, 155. pl. 9. f. 6. 1812. Based on *B.
melicoides* Hornem., doubtless error for
melicaeformis Brouss.
Dineba curtipendula Beauv., Ess. Agrost.
98, 158, 160. pl. 16. f. 1. 1812. Pre-

sumably based on *Chloris curtipendula*
Michx.

Dineba melicoides Beauv., Ess. Agrost.
160. 1812, name only, probably same
as *Bouteloua melicoides* Beauv.

Cynosurus secundus Pursh, Fl. Amer. Sept.
2: 728. 1814. "Upper Louisiana"
[northern Middle Western States],
Bradbury.

Atheropogon racemosus Roem. and Schult.,
Syst. Veg. 2: 414. 1817. Based on
Bouteloua racemosa Lag.

Dineba secunda Roem. and Schult., Syst.
Veg. 2: 711. 1817. Based on *Cynosurus
secundus* Pursh.

Aristida secunda Rud. ex Roem. and
Schult., Syst. Veg. 2: 711. 1817, as
synonym of *Dineba secunda* Roem. and
Schult.

Eutriana curtipendula Trin., Fund. Agrost.
161. 1820. Based on *Chloris curtipendula* Michx.

Melica curtipendula Michx. ex Steud.,
Nom. Bot. 1: 91, 519. 1821, as synonym of *Atheropogon apludoides* Muhl.

Cynodon curtipendula Raspail, Ann. Sci.
Nat., Bot. 5: 303. 1825. Based on
Dineba curtipendula Beauv.

Cynodon melicoides Raspail, Ann. Sci.
Nat., Bot. 5: 303. 1825. Based on
Bouteloua melicoides Beauv.

Chloris secundus Eaton, Man. ed. 5. 173.
1829. Based on *Cynosurus secundus*
Pursh.

Andropogon curtipendulus Spreng. ex
Steud., Nom. Bot. ed. 2. 1: 90. 1840,
as synonym of *Eutriana curtipendula*
Trin.

Eutriana affinis Hook. f., Linn. Soc.
Trans. 20: 174. 1847. St. Louis, Mo.;
Texas, *Drummond.*

Heterostegon curtipendula Schwein. in
Hook. f., Linn. Soc. Trans. 20: 175.
1851, as synonym of *Eutriana affinis.*

Bouteloua curtipendula var. *aristosa* A.
Gray, Man. ed. 2. 553. 1856. Illinois,
Geyer.

Atheropogon curtipendulus Fourn., Mex.
Pl. 2: 138. 1886. Based on *Bouteloua
curtipendula* A. Gray [error for Torrey].

Atheropogon medius Fourn., Mex. Pl. 2:
139. 1886. Mexico, *Liebmann* 581.

Atheropogon affinis Fourn., Mex. Pl. 2:
141. 1886. Based on *Eutriana affinis*
Hook. f.

Bouteloua racemosa var. *aristosa* Wats. and
Coult. ex Gray, Man. ed. 6. 656. 1890.
Illinois, *Geyer.*

(5) **Bouteloua eludens** Griffiths, U. S. Natl.
Herb. Contrib. 14: 401. 1912. Santa
Rita Mountains, Ariz., *Griffiths* 7269.

(17) **Bouteloua eriopoda** (Torr.) Torr., U. S.
Expl. Miss. Pacif. Rpt. 4: 155. 1856.
Based on *Chondrosium eriopodum* Torr.

Chondrosium eriopodum Torr. in Emory,
Notes Mil. Reconn. 154. 1848. Del

Norte [Rio Grande] River, N. Mex.
[*Bigelow*].

Bouteloua brevifolia Buckl., Acad. Nat.
Sci. Phila. Proc. 1862: 93. 1862.
Northwestern Texas [*Wright* 748,
Fendler 950].

(8) **Bouteloua filiformis** (Fourn.) Griffiths,
U. S. Natl. Herb. Contrib. 14: 413.
1912. Based on *Atheropogon filiformis*
Fourn.

Bouteloua juncifolia Vasey, Descr. Cat.
Grasses U. S. 62. 1885. Name only,
Texas [*Havard* 89] to Arizona. (*B. hum-
boldtiana* Griseb., doubtfully cited, is *B.
heterostega* (Trin.) Griffiths of the West
Indies.)

Atheropogon filiformis Fourn., Mex. Pl. 2:
140. 1886. Mexico, *Karwinsky* 991b.

(13) **Bouteloua glandulosa** (Cervant.) Swallen, N. Amer. Fl. 17: 621. 1939. Based
on *Erucaria glandulosa* Cervant.

Erucaria glandulosa Cervant., Naturaleza
1: 347. 1870. "Guadalupe et Mocte-
zuma," Mexico.

Bouteloua hirticulmis Scribn., U. S. Dept.
Agr., Div. Agrost. Cir. 30: 4. 1901.
Sierra de San Francisquito Mountains,
Baja California, Mexico, *Brandegee* 11.

(15) **Bouteloua gracilis** (H. B. K.) Lag. ex
Steud., Nom. Bot. ed. 2. 1: 219. 1840.
Based on *Chondrosium gracile* H. B. K.

Chondrosium gracile H. B. K., Nov. Gen.
et Sp. 1: 176. pl. 58. 1816. Mexico,
Humboldt and *Bonpland.*

Actinochloa gracilis Willd. ex Roem. and
Schult., Syst. Veg. 2: 418. 1817. Based
on *Chrondrosium gracile* H. B. K.

Atheropogon oligostachyus Nutt., Gen. Pl.
1: 78. 1818. Plains of the upper Missouri [*Nuttall*].

Eutriana gracilis Trin., Gram. Unifl. 240.
1824. Based on *Actinochloa gracilis*
Willd.

Atheropogon gracilis Spreng., Syst. Veg. 1:
293. 1825. Based on *Chondrosium
gracile* H. B. K.

Eutriana oligostachya Kunth, Rév. Gram.
1: 96. 1829. Based on *Atheropogon
oligostachyus* Nutt.

Chondrosium gracile var. *polystachyum*
Nees, Linnaea 19: 692. 1847. Mexico,
Aschenborn 153. [Spikes 2 or 3.]

Chondrosium oligostachyum Torr. in
Marcy, Expl. Red Riv. 300. 1852.
Based on *Atheropogon oligostachyum*
Nutt.

Bouteloua oligostachya Torr. ex A. Gray,
Man. ed. 2. 553. 1856. Based on
Atheropogon oligostachyus Nutt.

Bouteloua oligostachya var. *intermedia*
Vasey, Grasses U. S. 33. 1883. Name
only. Texas to Arizona.

Bouteloua major Vasey, Torrey Bot. Club
Bul. 14: 9. 1887. Name only, for a
plant grown from seed collected in
Mexico by Palmer.

Bouteloua oligostachya var. *major* Vasey ex

L. H. Dewey, U. S. Natl. Herb. Contrib. 2: 531. 1894. Texas to Arizona [type, *Lemmon 427*].
Bouteloua oligostachya var. *pallida* Scribn. ex Beal, Grasses N. Amer. 2: 418. 1896. Mexico, *Pringle 407*.
BOUTELOUA GRACILIS var. STRICTA (Vasey) Hitchc., Wash. Acad. Sci. Jour. 23: 454. 1933. Based on *B. stricta* Vasey.
Bouteloua stricta Vasey, Torrey Bot. Club Bul. 15: 49. 1888. Western Texas, *Nealley*, scarcely described; U. S. Dept. Agr., Div. Bot. Bul. 12¹: pl. 45. 1890.
(14) **Bouteloua hirsuta** Lag., Var. Cienc. 4: 141. 1805. Mexico.
Bouteloua hirta Lag., Var. Cienc. 4: 141. 1805, as synonym of *B. hirsuta* Lag.
Chondrosium hirtum H. B. K., Nov. Gen. et Sp. 1: 176. pl. 59. 1816. Mexico, *Humboldt* and *Bonpland*.
Actinochloa hirsuta Roem. and Schult., Syst. Veg. 2: 419. 1817. Based on *Bouteloua hirsuta* Lag.
Eutriana hirta Trin., Gram. Unifl. 240. 1824. Based on *Actinochloa hirsuta* Roem. and Schult.
Atheropogon hirtus Spreng., Syst. Veg. 1: 293. 1825. Based on *Chondrosium hirtum* H. B. K.
Chondrosium hirsutum Sweet, Hort. Brit. 1: 455. 1826. Presumably based on *Actinochloa hirsuta* Roem. and Schult.
Atheropogon papillosus Engelm., Amer. Jour. Sci. 46: 104. 1843. Beardstown, Ill., *Geyer*.
Chondrosium aschenbornianum Nees, Linnaea 19: 692. 1847. Mexico, *Aschenborn 331*.
Chondrosium foeneum Torr. in Emory, Notes Mil. Reconn. 154. pl. 12. 1848. Valley of the Del Norte [N. Mex., *Emory Exped.*].
Chondrosium papillosum Torr. in Marcy, Expl. Red Riv. 300. 1852. Based on *Atheropogon papillosus* Engelm.
Bouteloua foenea Torr. in S. Wats. and Rothr., Cat. Pl. Survey W. 100th Merid. 18. 1874. Based on *Chondrosium foeneum* Torr.
Bouteloua aschenborniana Griseb. ex Fourn., Mex. Pl. 2: 137. 1886, as synonym of *Chondrosium aschenbornianum* Nees.
Chondrosium drummondii Fourn., Mex. Pl. 2: 137. 1886. Texas, *Drummond 323*.
Bouteloua palmeri Vasey, Torrey Bot. Club Bul. 14: 9. 1887. Name only, later described as *B. hirsuta* var. *palmeri* Vasey ex Beal.
Bouteloua hirsuta var. *minor* Vasey, U. S. Dept. Agr., Div. Bot. Bul. 12¹: pl. 39. f. 2. 1890, nomen seminudum. [Texas, *Reverchon 1153*.]
Bouteloua hirsuta var. *major* Vasey, U. S. Dept. Agr., Div. Bot. Bul. 12¹: pl. 39.

f. 3. 1890. Without description. [Austin, Tex., *Stiles* in 1884.]
Bouteloua hirta Scribn., U. S. Natl. Herb. Contrib. 2: 531. 1894. Based on *Chondrosium hirtum* H. B. K.
Bouteloua hirta var. *major* Vasey ex L. H. Dewey, U. S. Natl. Herb. Contrib. 2: 531. 1894. Western Texas to Mexico.
Bouteloua hirta var. *minor* Vasey ex L. H. Dewey, U. S. Natl. Herb. Contrib. 2: 531. 1894. Central Texas.
Bouteloua hirsuta var. *palmeri* Vasey ex Beal, Grasses N. Amer. 2: 417. 1896. Cultivated, seed collected by Palmer in Mexico.
Bouteloua bolanderi Vasey ex Beal, Grasses N. Amer. 2: 417. 1896, as synonym of *B. hirsuta* var. *palmeri* Vasey.
Bouteloua pectinata Featherly, Bot. Gaz. 91: 103. f. 1–4. 1931. Oklahoma, *English 71*.
Bouteloua hirsuta var. *pectinata* Cory, Rhodora 38: 405. 1936. Based on *B. pectinata* Featherly.
(11) **Bouteloua parryi** (Fourn.) Griffiths, U. S. Natl. Herb. Contrib. 14: 381. 1912. Based on *Chondrosium parryi* Fourn.
Bouteloua polystachya var. *vestita* S. Wats., Amer. Acad. Sci. Proc. 18: 177. 1883. Sierra Madre south of Saltillo, Mexico, *Palmer 1357* in 1880.
Chondrosium parryi Fourn., Mex. Pl. 2: 150. 1886. San Luis Potosí, *Parry* and *Palmer 923½* [error for 943½].
Bouteloua vestita Scribn. ex L. H. Dewey, U. S. Natl. Herb. Contrib. 2: 531. 1894. Based on *B. polystachya* var. *vestita* S. Wats.
(7) **Bouteloua radicosa** (Fourn.) Griffiths, U. S. Natl. Herb. Contrib. 14: 411. 1912. Based on *Atheropogon radicosus* Fourn.
Dinebra bromoides H. B. K., Nov. Gen. et Sp. 1: 172. pl. 51. 1816. Not *Bouteloua bromoides* Lag., 1816. Mexico, *Humboldt* and *Bonpland*.
Atheropogon bromoides Roem. and Schult., Syst. Veg. 2: 415. 1817. Based on *Dinebra bromoides* H. B. K.
Eutriana bromoides Trin., Fund. Agrost. 161. 1820. Based on *Dinebra bromoides* H. B. K.
Nestlera festucaeformis Willd. ex Steud., Nom. Bot. ed. 2. 2: 192. 1841, as synonym of *Eutriana bromoides* Trin.
Atheropogon radicosus Fourn., Mex. Pl. 2: 140. 1886. Mexico City, *Bourgeau 450*.
Bouteloua bromoides var. *radicosa* Vasey ex L. H. Dewey, U. S. Natl. Herb. Contrib. 2: 533. 1894. Based on *Atheropogon radicosus* Fourn.
(4) **Bouteloua rigidiseta** (Steud.) Hitchc., Wash. Acad. Sci. Jour. 23: 453. 1933. Based on *Aegopogon rigidisetus* Steud.
Aegopogon rigidisetus Steud., Syn. Pl.

Glum. 1: 146. 1854. Texas, *Drummond.*

Bouteloua texana S. Wats., Amer. Acad. Sci. Proc. 18: 196. 1883. Texas, *Berlandier* 1535, *Drummond* 340, 374.

Polyodon texanus Nash in Small, Fl. Southeast. U. S. 138, 1327. 1903. Based on *Bouteloua texana* S. Wats.

(12) **Bouteloua rothrockii** Vasey, U. S. Natl. Herb. Contrib. 1: 268. 1893. Cottonwood, Ariz., *Rothrock* 347.

(9) **Bouteloua simplex** Lag., Var. Cienc. 4: 141. 1805. Peru.

Chloris procumbens Durand, Chlor. Sp. 16. 1808. Grown at Madrid, seed said to come from the Philippine Islands (collected by Née) where the species is not known to occur. Probably from South America or Mexico, which regions Née visited.

Chloris filiformis Poir. in Lam., Encycl. Sup. 2: 237. 1811. Grown at Paris, the source unknown.

Chondrosium procumbens Desv. ex Beauv., Ess. Agrost. 41, 158. pl. 9. f. 7. 1812. Based on *Chloris procumbens* Durand.

Chondrosium humile Beauv., Ess. Agrost. 41, 158. 1812. Name only.

Chondrosium tenue Beauv., Ess. Agrost. 41, 158. 1812. Name only.

Atheropogon procumbens Jacq., Eclog. Gram. 2: 16. pl. 12. 1813. Based on *Chloris procumbens* Durand.

Bouteloua prostrata Lag., Gen. et Sp. Nov. 5. 1816. Mexico.

Chondrosium humile H. B. K., Nov. Gen. et Sp. 1: 175. pl. 56. 1816. Ecuador, *Humboldt* and *Bonpland.*

Chondrosium tenue Beauv. ex H. B. K., Nov. Gen. et Sp. 1: 176. pl. 57. 1816. Mexico, *Humboldt* and *Bonpland.*

Chloris tenuis Poir. in Lam., Encycl. Sup. 5: 614. 1817. Based on *C. filiformis* Poir., p. 237, not *C. filiformis* Poir., op. cit. p. 238.

Actinochloa procumbens Roem. and Schult., Syst. Veg. 2: 417. 1817. Based on *Chloris procumbens* Durand.

Actinochloa humilis Willd. ex Roem. and Schult., Syst. Veg. 2: 417. 1817. Based on *Chondrosium humile* H. B. K.

Actinochloa simplex Roem. and Schult., Syst. Veg. 2: 418. 1817. Based on *Bouteloua simplex* Lag.

Actinochloa tenuis Willd. ex Roem. and Schult., Syst. Veg. 2: 418. 1817. Based on *Chondrosium tenue* H. B. K.

Actinochloa prostrata Roem. and Schult., Syst. Veg. 2: 419. 1817. Based on *Bouteloua prostrata* Lag.

Eutriana humilis Trin., Gram. Unifl. 239. 1824. Based on *Actinochloa humilis* Willd.

Eutriana tenuis Trin., Gram. Unifl. 240. 1824. Based on *Actinochloa tenuis* Willd.

Atheropogon humilis Spreng., Syst. Veg. 1: 293. 1825. Based on *Chondrosium*

humile H. B. K.

Cynodon procumbens Raspail, Ann. Sci. Nat., Bot. 5: 303. 1825. Based on *Chondrosium procumbens* Desv.

Chondrosium prostratum Sweet, Hort. Brit. 1: 455. 1826. Based on *Bouteloua prostrata* Lag.

Chondrosium simplex Kunth, Rév. Gram. 1: 94. 1829. Based on *Bouteloua simplex* Lag.

Bouteloua tenuis Griseb., Abh. Ges. Wiss. Göttingen 19: 259. 1874. Based on *Chondrosium tenue* Beauv.

Bouteloua humilis Hieron., Bol. Acad. Cienc. Córdoba 4: 495. 1882. Based on *Chondrosium humile* Beauv.

Bouteloua pusilla Vasey, Torrey Bot. Club Bul. 11: 6. 1884. Kingman, N. Mex., *Vasey.*

Bouteloua brachyathera Phil., An. Mus. Nac. Chile Bot. 8: 85. 1891. Tarapacá, Chile.

Bouteloua rahmeri Phil., An. Mus. Nac. Chile Bot. 8: 85. 1891. Tarapacá, Chile.

Bouteloua procumbens Griffiths, U. S. Natl. Herb. Contrib. 14: 364. 1912. Based on *Chloris procumbens* Durand.

Bouteloua simplex var. *rahmeri* Henr., Med. Rijks Herb. Leiden No. 40: 66. 1921. Based on *B. rahmeri* Phil.

(18) **Bouteloua trifida** Thurb. in S. Wats., Amer. Acad. Sci. Proc. 18: 177. 1883. Monclova, Coahuila, *Palmer* 1355 in 1880.

Bouteloua burkii Scribn. in S. Wats., Amer. Acad. Sci. Proc. 18: 179. 1883. Western Texas and New Mexico, *Berlandier* 167 and 1427.

Chondrosium trinii Fourn., Mex. Pl. 2: 136. 1886. Laredo, Tex., *Berlandier* 1427.

Chondrosium polystachyum Trin. ex Fourn., Mex. Pl. 2: 136. 1886, as synonym of *C. trinii* Fourn.

Chondrosium virletii Fourn., Mex. Pl. 2: 136. 1886. San Luis Potosí, Mexico, *Virlet* 1373.

Bouteloua trifida var. *burkii* Vasey ex L. H. Dewey, U. S. Natl. Herb. Contrib. 2: 532. 1894. Based on *B. burkii* Scribn.

Bouteloua trinii Griffiths, U. S. Natl. Herb. Contrib. 14: 387. 1912. Based on *Chondrosium trinii* Fourn. Griffiths accepts 1881 as the date for Fournier's work.

(2) **Bouteloua uniflora** Vasey, Bot. Gaz. 16: 26. 1891. Crockett County, Tex., *Nealley* [222].

(133) BRACHIARIA (Trin.) Griseb.

(1) **Brachiaria ciliatissima** (Buckl.) Chase in Hitchc., U. S. Dept. Agr. Bul. 772: 221. 1920. Based on *Panicum ciliatissimum* Buckl.

Panicum ciliatissimum Buckl., Tex. Geol. Agr. Survey Prel. Rpt. App. 4. 1866. Northern Texas [*Buckley*].

Brachiaria erucaeformis (J. E. Smith) Griseb. in Ledeb., Fl. Ross. 4: 469. 1853. Based on *Panicum erucaeforme* J. E. Smith.

Panicum erucaeforme J. E. Smith in Sibth., Fl. Graec. 1: 44. pl. 59. 1806. Greece.

Panicum isachne Roth in Roem. and Schult., Syst. Veg. 2: 458. 1819. East Indies.

Echinochloa eruciformis Koch, Linnaea 21: 437. 1848. Based on *Panicum erucaeforme* J. E. Smith.

Panicum isachne var. *mexicana* Beal, Grasses N. Amer. 2: 114. 1896. Grown from seed said to come from Mexico.

Brachiaria isachne Stapf in Prain, Fl. Trop. Afr. 9: 552. 1917. Based on *Panicum isachne* Roth.

(3) **Brachiaria plantaginea** (Link) Hitchc., U. S. Natl. Herb. Contrib. 12: 212. 1909. Based on *Panicum plantagineum* Link.

Panicum plantagineum Link, Hort. Berol. 1: 206. 1827. Grown in Berlin, origin unknown.

Panicum leandri Trin., Gram. Icon. 3: pl. 335. 1836. *Brazil*.

Panicum distans Salzm. ex Steud., Syn. Pl. Glum. 1: 61. 1854. Not *P. distans* Trin., 1829. Bahia, Brazil [*Salzmann*].

Panicum disciferum Fourn., Mex. Pl. 2: 19. 1886. San Luis Potosí, Mexico, *Virlet* 1292.

(2) **Brachiaria platyphylla** (Griseb.) Nash in Small, Fl. Southeast. U. S. 81. 1327. 1903. Based on *Paspalum platyphyllum* Griseb.

Paspalum platyphyllum Griseb., Pl. Cub. Cat. 230. 1866. Not *P. platyphyllum* Schult., 1827. Zarabanda, Cuba, *Wright* 3441.

Panicum platyphyllum Munro ex Wright, An. Acad. Cienc. Habana 8: 206. 1871. Based on *Paspalum platyphyllum* Griseb.

Brachiaria extensa Chase, U. S. Natl. Herb. Contrib. 28: 240. 1929. Based on *Paspalum platyphyllum* Griseb. Not *P. platyphyllum* Schult.

Brachiaria subquadripara (Trin.) Hitchc., Lingnan Sci. Jour. 7: 214. 1931. Based on *Panicum subquadriparum* Trin.

Panicum subquadriparum Trin., Gram. Pan. 145. 1826. Marianas Islands.

(87) BRACHYELYTRUM Beauv.

(1) **Brachyelytrum erectum** (Schreb.) Beauv., Ess. Agrost. 155. 1812. Based on *Muhlenbergia erecta* Schreb.

Muhlenbergia erecta Schreb. in Roth, Neue Beyträge Bot. 1: 97. 1802. "America boreali." It is stated (p. 96) that a fuller description will be furnished by Schreber

in a new part of his Beschreibung der Gräser, which, however, did not appear until 1810. (Beschreib. Gräs. 2: 139. pl. 50. 1810.) Name only, Muhl., Amer. Phil. Soc. Trans. 3: 160. 1793.

Dilepyrum aristosum Michx., Fl. Bor. Amer. 1: 40. 1803. Georgia and Carolina, *Michaux*.

Muhlenbergia aristata Pers., Syn. Pl. 1: 73. 1805. Based on *Dilepyrum aristosum* Michx.

Brachyelytrum aristatum Roem. and Schult., Syst. Veg. 2: 413. 1817. Based on *Dilepyrum aristosum* Michx.

Muhlenbergia brachyelytrum Trin., Gram. Unifl. 188. 1824. Based on *Brachyelytrum erectum* Beauv.

Agrostis erecta Spreng., Syst. Veg. 1: 264. 1825. Based on *Muhlenbergia erecta* Schreb.

Brachyelytrum aristatum var. *engelmanni* A. Gray, Man. ed. 5. 614. 1867. "A western form."

Brachyelytrum aristosum Trel., Brann. and Coville, in Branner, Ark. Geol. Survey Rpt. 4: 235. 1891. Based on *Dilepyrum aristosum* Michx.

Brachyelytrum aristosum var. *glabratum* Vasey in Millsp., W. Va. Agr. Expt. Sta. Bul. 24: 469. 1892. Fayette near Nuttallburg, W. Va., *Nuttall*.

Dilepyrum erectum Farwell, Amer. Midl. Nat. 8: 33. 1922. Based on *Muhlenbergia erecta* Schreb.

Brachyelytrum erectum var. *septentrionale* Babel, Rhodora 45: 260. 1943. Durham, N. H., *Babel* 46.

(3) BRACHYPODIUM Beauv.

Brachypodium caespitosum (Host) Roem, and Schult., Syst. Veg. 2: 737. 1817, Based on *Bromus caespitosus* Host.

Bromus caespitosus Host, Icon. Gram. Austr. 4: 11. pl. 18. 1809. Tyrol. (Spelled *"cespitosus."*)

(1) **Brachypodium distachyon** (L.) Beauv., Ess. Agrost. 101, 155. 1812. Based on *Bromus distachyos* L.

Bromus distachyos L., Cent. Pl. 2: 8. 1756; Amoen. Acad. 4: 304. 1759. Europe and the Orient.

Festuca distachyos Roth, Cat. Bot. Fasc. 1: 11. 1797. Based on *Bromus distachyos* L.

Trachynia distachya Link, Hort. Berol. 1: 43. 1827. Based on *Bromus distachyos* L.

Zerna distachyos Panz. ex Jacks., Ind. Kew. 2: 1249. 1895. Based on *Festuca distachyos* Roth.

Brachypodium pinnatum (L.) Beauv., Ess. Agrost. 101, 155. pl. 19. f. 3. 1812. Based on *Bromus pinnatus* L.

Bromus pinnatus L. Sp. Pl. 78. 1753. Europe.

Brachypodium sylvaticum (Huds.) Beauv., Ess. Agrost. 101, 155. 1810. Based

ultimately on *Festuca sylvatica* Huds.
Festuca sylvatica Huds., Fl. Angl. 1: 38.
1762. England.

(13) BRIZA L.

(1) **Briza maxima** L., Sp. Pl. 70. 1753.
Europe.
(3) **Briza media** L., Sp. Pl. 70. 1753.
Europe.
(2) **Briza minor** L., Sp. Pl. 70. 1753.
Europe.

(2) BROMUS L.

(3) **Bromus aleutensis** Trin. ex Griseb., in
Ledeb., Fl. Ross. 4: 361. 1853. Una-
laska, *Eschscholz.*
Bromus alopecuros Poir., Voy. Barb. 2: 100.
1789. Algeria, *Poiret.*
Bromus contortus Desf., Fl. Atlant. 1: 95.
pl. 25. 1800. Algeria.
Bromus alopecurus Pers., Syn. Pl. 1: 95.
1805. "*B. contortus* Desf." and "Poiret,
iter" [Voyage Barbarie] both cited.
(24) **Bromus anomalus** Rupr. ex Fourn.,
Acad. Sci. Brux. Bul. 9²: 236. 1840.
Name only; Mex. Pl. 2: 126. 1886.
Mexico, *Galeotti* 5757, 5815.
Bromus kalmii var. *porteri* Coult., Man.
Rocky Mount. 425. 1885. Twin Lakes,
Colo., *Porter.*
Bromus ciliatus var. *minor* Munro ex L.
H. Dewey, U. S. Natl. Herb. Contrib.
2: 548. 1894. West Texas [Chisos
Mountains, *Havard* 20].
Bromus ciliatus porteri Rydb., U. S. Natl.
Herb. Contrib. 3: 192. 1895. Based on
B. kalmii var. *porteri* Coult.
Bromus porteri Nash, Torrey Bot. Club
Bul. 22: 512. 1895. Based on *B. kalmii*
var. *porteri* Coult.
Bromus ciliatus var. *montanus* Vasey ex
Beal, Grasses N. Amer. 2: 619. 1896.
Colorado, *Patterson* 264.
Bromus kalmii var. *occidentalis* Vasey ex
Beal, Grasses N. Amer. 2: 624. 1896.
Montana [type, *Canby* and *Scribner*
384].
Bromus scabratus Scribn., U. S. Dept.
Agr., Div. Agrost. Bul. 13: 46. 1898.
Not *B. scabratus* Link, 1843. Vermilion
Creek, Wyo., *A. Nelson* 3800.
Bromus kalmii var. *major* Vasey ex Shear,
U. S. Dept. Agr., Div. Agrost. Bul. 23:
35. 1900, as synonym of *B. porteri*
Nash.
Bromus porteri havardii Shear, U. S. Dept.
Agr., Div. Agrost. Bul. 23: 37. 1900.
Based on *B. ciliatus* var. *minor* Munro.
Zerna anomala Henr., Blumea 4: 499.
1941. Based on *Bromus anomalus* Rupr.
BROMUS ANOMALUS var. LANATIPES (Shear)
Hitchc., Wash. Acad. Sci. Jour. 23: 449.
1933. Based on *B. porteri* var. *lanatipes*
Shear.
Bromus porteri var. *lanatipes* Shear, U S..
Dept. Agr., Div. Agrost. Bul. 23:. 37

1900. Idaho Springs, Colo., *Shear* 739.
Bromus lanatipes Rydb., Colo. Agr. Expt.
Sta. Bul. 100: 52. 1906. Based on *B.
porteri* var. *lanatipes* Shear.
(36) **Bromus arenarius** Labill., Nov. Holl.
Pl. 1: 23. pl. 28. 1804. Australia.
(6) **Bromus arizonicus** (Shear) Stebbins,
Calif. Acad. Sci. IV. Proc. 25: 309.
1944. Based on *Bromus carinatus* var.
arizonicus Shear.
Bromus carinatus var. *arizonicus* Shear,
U. S. Dept. Agr., Div. Agrost. Bul. 23:
62. 1900. Santa Cruz Valley, Tucson,
Ariz., *Pringle* in 1884.
(35) **Bromus arvensis** L., Sp. Pl. 77. 1753.
Europe.
Bromus erectus var. *arvensis* Huds., Fl.
Angl. ed. 2. 50. 1778. Based on *B.
arvensis* L.
Serrafalcus arvensis Godr., Fl. Lorr. 3:
185. 1844. Based on *Bromus arvensis*
L.
Forasaccus arvensis Bubani, Fl. Pyr. 4:
385. 1901. Based on *Bromus arvensis*
L.
(4) **Bromus breviaristatus** Buckl., Acad.
Nat. Sci. Phila. Proc. 1862: 98. 1862.
Rocky Mountains, *Nuttall.*
Bromus parviflorus Nutt. ex A. Gray,
Acad. Nat. Sci. Phila. Proc. 1862: 336.
1862, as synonym of *B. breviaristatus*
Buckl.
Bromus subvelutinus Shear, U. S. Dept.
Agr., Div. Agrost. Bul. 23: 52. 1900.
Reno, Nev., *Tracy* 249.
Bromus pauciflorus Nutt. ex Shear, U. S.
Dept. Agr., Div. Agrost. Bul. 23: 53.
1900. This name, on Nuttall's ticket on
the type of *B. breviaristatus* Buckl., was
misread as "*parviflorus*" by Gray.
Bromus carinatus var. *linearis* Shear, U. S.
Dept. Agr., Div. Agrost. Bul. 23: 61.
1900. California, *Vasey* in 1875.
(27) **Bromus brizaeformis** Fisch. and Mey.,
Ind. Sem. Hort. Petrop. 3: 30. 1837.
Europe.
(5) **Bromus carinatus** Hook. and Arn., Bot.,
Beechey Voy. 403. 1840. California.
Ceratochloa grandiflora Hook., Fl. Bor.
Amer. 2: 253. 1840. Not *Bromus
grandiflorus* Weigel, 1772. Plains of the
Columbia [Oregon], *Scouler, Douglas.*
Bromus oregonus Nutt. ex Hook. f., Jour.
Bot. Kew Misc. 8: 18. 1856. Name
only for *Geyer* 244, "Upper Missouri and
Oregon territories." Nutt. ex Shear,
U. S. Dept. Agr., Div. Agrost. Bul. 23:
59. 1900, as synonym of *B. carinatus.*
Bromus virens Buckl., Acad. Nat. Sci.
Phila. Proc. 1862: 98. 1862. Rocky
Mountains and Columbia River, *Nut-
tall.* The specimen in the herbarium of
the Philadelphia Academy is the Pacific
coast form with long awns, and probably
came from the Columbia River.
Bromus californicus Nutt. ex Buckl.,
Acad. Nat. Sci. Phila. Proc. 1862: 336.

1862, as synonym of *B. virens* Buckl. [California, *Nuttall*.]

Bromus nitens Nutt. ex A. Gray, Acad. Nat. Sci. Phila. Proc. 1862: 336. 1862, as synonym of *B. virens* Buckl. [Columbia woods, *Nuttall*.]

Bromus hookerianus Thurb. in Wilkes, U. S. Expl. Exped. Bot. 17: 493. 1874. Based on *Ceratochloa grandiflora* Hook.

Bromus hookerianus var. *minor* Scribn. ex Vasey, Descr. Cat. Grasses U. S. 92. 1885, name only, Oregon; Macoun, Cat. Can. Pl. 2⁴: 238. 1888, without description, *B. virens* Buckl., cited as synonym.

Bromus virens var. *minor* Scribn. in Beal, Grasses N. Amer. 2: 614. 1896. Arizona and Oregon.

Bromus carinatus californicus Shear, U. S. Dept. Agr., Div. Agrost. Bul. 23: 60. 1900. [California, *Nuttall*.]

Bromus carinatus hookerianus Shear, U. S. Dept. Agr., Div. Agrost. Bul. 23: 60. 1900. Based on *B. hookerianus* Thurb.

Bromus carinatus var. *densus* Shear, U. S. Dept. Agr., Div. Agrost. Bul. 23: 61. 1900. San Nicolas Island, Calif., *Trask* [12].

(1) **Bromus catharticus** Vahl, Symb. Bot. 2: 22. 1791. Lima, Peru.

Festuca unioloides Willd., Hort. Berol. 3. pl. 3. 1803. Described from a plant grown at Berlin from seed from "Carolina," where it must have been cultivated.

Ceratochloa unioloides Beauv., Ess. Agrost. 75. pl. 15. f. 7. 1812. Based on *Festuca unioloides* Willd.

Bromus unioloides H. B. K., Nov. Gen. et Sp., 1: 151. 1815. Quito, Ecuador, *Humboldt* and *Bonpland*.

Schedonorus unioloides Roem. and Schult., Syst. Veg. 2: 708. 1817. Based on *Bromus unioloides* H. B. K.

Bromus unioloides Raspail, Ann. Sci. Nat., Bot. 5: 439. 1825. Based on *Ceratochloa unioloides* Beauv.

Bromus willdenovii Kunth, Rév. Gram. 1: 134. 1829. Based on *Festuca unioloides* Willd.

Ceratochloa pendula Schrad., Linnaea 6: Litt. 72. 1831. Grown at Göttingen from seed from Carolina.

Bromus schraderi Kunth, Enum. Pl. 1: 416. 1833. Based on *Ceratochloa pendula* Schrad.

Bromus mucronatus Willd. ex Steud., Nom. Bot. ed. 2. 1: 228. 1840, as synonym of *B. unioloides* H. B. K.

Ceratochloa breviaristata Hook., Fl. Bor. Amer. 2: 253. pl. 234. 1840. Lewis and Clark River and near the sources of the Columbia. *Douglas* [in 1826].

Bromus breviaristatus Thurb. in Wilkes, U. S. Expl. Exped. Bot. 17²: 493. 1874. Not *B. breviaristatus* Buckl., 1862. Based on *Ceratochloa breviaristata* Hook.

Tragus unioloides Panz. ex Jacks., Ind.

Kew. 2: 1099. 1895, as synonym of *Festuca unioloides* Willd.

Forasaccus brebiaristatus [error for *breviaristatus*] Lunell, Amer. Midl. Nat. 4: 225. 1915. Based on *Ceratochloa breviaristata* Hook.

Zerna unioloides Lindm., Svensk Fanerogamfl. 101. 1918. Based on *Bromus unioloides* H. B. K.

Ceratochloa cathartica Herter, Rev. Sudamer. Bot. 6: 144. 1940. Based on *Bromus catharticus* Vahl.

The form described by Shear (U. S. Dept. Agr., Div. Agrost. Bul. 23: 52. 1900) as *Bromus unioloides haenkeanus* (Presl) Shear is a form of rescue grass, but *Ceratochloa haenkeana* Presl, upon which the name is based, is a different species with purplish, awned spikelets, as shown by examination of the type, from Chile, at the herbarium of the German University at Prague.

(19) **Bromus ciliatus** L., Sp. Pl. 1: 76. 1753. Grown at Uppsala from seed collected by Kalm in Canada.

Bromus canadensis Michx., Fl. Bor. Amer. 1: 65. 1803. Canada, Lac St. Jean, *Michaux*.

Bromus richardsoni Link, Hort. Berol. 2: 281. 1833. Grown at Berlin from seed sent by Richardson from northwestern North America.

Bromus purgans var. *longispicatus* Hook., Fl. Bor. Amer. 2: 252. 1840. Rocky Mountains, *Drummond*.

Bromus purgans var. *pallidus* Hook., Fl. Bor. Amer. 2: 252. 1840. Saskatchewan to Rocky Mountains, *Drummond*.

Bromus inermis var. *ciliatus* Traut., Act. Hort. Petrop. 5: 135. 1877. Based on *B. ciliatus* L.

Bromus hookeri var. *canadensis* Fourn., Mex. Pl. 2: 128. 1886. Based on *B. canadensis* Michx.

Bromus hookeri var. *ciliatus* Fourn., Mex. Pl. 2: 128. 1886. Based on *B. ciliatus* L.

Bromus ciliatus scariosus Scribn., U. S. Dept. Agr., Div. Agrost. Bul. 13: 46. 1898. Sheep Mountain, Wyo., *A. Nelson* 3305.

Bromus richardsoni var. *pallidus* Shear, U. S. Dept. Agr., Div. Agrost. Bul. 23: 34. 1900. Based on *B. purgans* var. *pallidus* Hook.

Forasaccus ciliatus Lunell, Amer. Midl. Nat. 4: 225. 1915. Based on *Bromus ciliatus* L.

Bromus ciliatus forma *denudatus* Wiegand, Rhodora 24: 91. 1922. Ashfield, Mass., *Williams* in 1909.

Bromus ciliatus var. *denudatus* Fernald, Rhodora 28: 20. 1926. Based on *B. ciliatus* forma *denudatus* Wiegand.

Bromus dudleyi Fernald, Rhodora 32: 63. pl. 196. f. 1–3. 1930. Deer Brook, Bonne Bay, Newfoundland, *Fernald*, *Long*, and *Fogg* 1223.

Bromus ciliatus var. *intonsus* Fernald, Rhodora 32: 70. 1930. Ashfield, Mass., *Williams*, August 4, 1909. The form with more densely pilose sheaths. According to Fernald (Rhodora 32: 70. 1930) this, as shown by specimens so named in the Gray Herbarium, is the form described as *B. asper* Murray in Gray's Manual, eds. 5 and 6, and in Britton and Brown's Illustrated Flora. Shear in his revision of *Bromus* (U. S. Dept. Agr., Div. Agrost. Bul. 23: 30. 1900) uses the earlier name *B. ramosus* Huds., but says he had seen no American speci.nens.

Zerna richardsoni Nevski, Act. Univ. Asiae Med. VIII b. Bot. 17: 17. 1934. Based on *Bromus richardsoni* Link.

Zerna ciliata Henr., Blumea 4: 498. 1941. Based on *Bromus ciliatus* L.

(29) **Bromus commutatus** Schrad., Fl. Germ. 353. 1806. Germany.

Bromus pratensis Ehrh., Beiträge 6: 84. 1791. Name only; Hoffm. Deut. Fl. ed. 2. 2: 52. 1800. Not *B. pratensis* Lam., 1785. Europe.

Brachypodium commutatum Beauv., Ess. Agrost. 101, 155. 1812. Based on *Bromus commutatus* Lam. (error for Schrad.).

Serrafalcus commutatus Bab., Man. Brit. Bot. ed. 1. 374. 1843. Based on *Bromus commutatus* Schrad.

Bromus mutabilis var. *commutatus* Schultz, Flora 32: 234. 1849. Based on *B. commutatus* Schrad.

Bromus racemosus var. *commutatus* Coss. and Dur., Expl. Sci. Alger. 2: 165. 1855. Based on *B. commutatus* Schrad.

Bromus mollis var. *commutatus* Sanio, Verh. Bot. Ver. Brand. 23: Abh. 31. 1882. Based on *B. commutatus* Schrad.

Serrafalcus racemosus var. *commutatus* Husnot, Gram. Fr. Belg. 72. 1899. Based on *Bromus commutatus* Schrad.

Forasaccus commutatus Bubani, Fl. Pyr. 4: 387. 1901. Based on *Bromus commutatus* Schrad.

Bromus secalinus var. *gladewitzii* Farwell, Amer. Midl. Nat. 10: 24. 1926. Michigan, *Farwell* and *Gladewitz* 7434.

BROMUS COMMUTATUS var. APRICORUM Simonkai, Enum., Fl. Transsilv. 583. 1886. Europe.

(12) **Bromus erectus** Huds., Fl. Angl. 39. 1762. England.

Festuca erecta Wallr., Sched. Crit. 35. 1822. Based on *Bromus erectus* Smith (error for Huds.).

Bromus macounii Vasey, Torrey Bot. Club Bul. 15: 48. 1888. Vancouver Island, *Macoun* in 1887.

Zerna erecta Panz. ex Jacks., Ind. Kew. 2: 1249. 1895. Based on *Bromus erectus* Huds.

Forasaccus erectus Bubani, Fl. Pyr. 4: 384. 1901. Based on *Bromus erectus* Huds.

(26) **Bromus frondosus** (Shear) Woot. and Standl., N. Mex. Col. Agr. Bul. 81: 144. 1912. Based on *B. porteri* var. *frondosus* Shear.

Bromus porteri var. *frondosus* Shear, U. S. Dept. Agr., Div. Agrost. Bul. 23: 37. 1900. Mangas, N. Mex., *J. G. Smith* in 1897.

(15) **Bromus grandis** (Shear) Hitchc. in Jepson, Fl. Calif. 1: 175. 1912. Based on *B. orcuttianus* var. *grandis* Shear.

Bromus orcuttianus var. *grandis* Shear, U. S. Dept. Agr., Div. Agrost. Bul. 23: 43. 1900. San Diego, Calif., *Orcutt* 472.

Bromus porteri var. *assimilis* Davy, Calif. Univ. Pubs., Bot. 1: 55. 1902. San Jacinto Mountains, *Hall* 2228.

(10) **Bromus inermis** Leyss., Fl. Hal. 16. 1761. Europe.

Festuca inermis DC. and Lam., Fl. Franc. 3: 49. 1805. Based on *Bromus inermis* Leyss.

Schedonorus inermis Beauv., Ess. Agrost. 99, 177. 1812. Based on *Festuca inermis* DC.

Festuca inermis var. *villosa* Mert. and Koch, Deutschl. Fl. 1: 675. 1823. Germany.

Bromus inermis var. *aristatus* Schur, Enum. Pl. Transsilv. 805. 1866. Europe.

Bromus inopinatus Brues, Trans. Wis. Acad. Sci., Arts, and Letters 17: 73. 1911. Milwaukee, Wis. [*Brues* 78].

Forasaccus inermis Lunell, Amer. Midl. Nat. 4: 225. 1915. Based on *Bromus inermis* Leyss.

Zerna inermis Lindm., Svensk Fanerogamfl. 101. 1918. Based on *Bromus inermis* Leyss.

Bromus inermis forma *villosus* Fernald, Rhodora 35: 316. 1933. Based on *Festuca inermis* var. *villosa* Mert. and Koch.

Bromus inermis forma *aristatus* Fernald, Rhodora 35: 316. 1933. Based on *B. inermis* var. *aristatus* Schur.

Bromus inermis forma *bulbiferus* Moore, Rhodora 43: 76. 1941. Ramsey County, Minn., *Kaufman* in 1938.

(34) **Bromus japonicus** Thunb., Fl. Japon. 52. 1784. Japan.

Bromus patulus Mert. and Koch, Deut. Fl. 1: 685. 1823. Europe.

Bromus arvensis var. *patulus* Mutel, Fl. Franç. 4: 134. 1837. Based on *B. patulus* Mert. and Koch.

Serrafalcus patulus Parl., Fl. Ital. 1: 394. 1848. Based on *Bromus patulus* Mert. and Koch.

Bromus squarrosus var. *patulus* Regel, Act. Hort. Petrop. 7: 602. 1881. Based on *B. patulus* Mert. and Koch.

Forasaccus patulus Bubani, Fl. Pyr. 4: 387. 1901. Based on *Bromus patulus* Mert. and Koch.

Bromus japonicus var. *porrectus* Hack.

Magyar Bot. Lapok (Ungar. Bot. Bl.)
2: 58. 1903. Eurasia.
Bromus japonicus var. *subsquarrosus*
(Borb.) Savul. and Rays., (Rumania)
Min. Agr. Bul. 4 (Sup. 2): 39. 1924,
As synonym of *B. japonicus* var. *porrectus* Hack.
(25) **Bromus kalmii** A. Gray, Man. 600.
1848. Canada or northeastern United
States, *Kalm.*
Bromus laciniatus Beal, Grasses N. Amer.
2: 615. 1896. Mexico.
Bromus pendulinus Sessé ex Lag., Gen.
and Sp. Nov. 4. 1816. Not *B. pendulinus* Schrad. 1810. Mexico.
(17) **Bromus laevipes** Shear, U. S. Dept.
Agr., Div. Agrost. Bul. 23: 45. 1900.
West Klickitat County, Wash., *Suksdorf 178.*
(21) **Bromus latiglumis** (Shear) Hitchc.,
Rhodora 8: 211. 1906. Based on *B.
purgans* var. *latiglumis* Shear.
Bromus altissimus Pursh, Fl. Amer.
Sept. 2: 728. 1814. Not *B. altissimus*
Gilib., 1790. On the banks of the
Missouri [*Nuttall*].
Bromus purgans var. *latiglumis* Shear,
U. S. Dept. Agr., Div. Agrost. Bul. 23:
40. 1900. Dakota City, Iowa, *Pammel
222.*
Bromus ciliatus latiglumis Scribn. ex
Shear, U. S. Dept. Agr., Div. Agrost.
Bul. 23: 40. 1900, as synonym of *B.
purgans* var. *latiglumis* Shear.
Bromus purgans var. *incanus* Shear,
U. S. Dept. Agr., Div. Agrost. Bul. 23:
41. 1900. Canton, Ill., *Wolf 3.*
Bromus incanus Hitchc., Rhodora 8: 212.
1906. Based on *B. purgans* var. *incanus*
Shear.
Forasaccus latiglumis Lunell, Amer. Midl.
Nat. 4: 225. 1915. Based on *Bromus
latiglumis* Hitchc.
Bromus ciliatus var. *incanus* Farwell,
Amer. Midl. Nat. 10: 204. 1927.
Based on *B. purgans* var. *incanus* Shear.
Bromus ciliatus var. *incanus* subvar. *latiglumis* Farwell, Amer. Midl. Nat. 10:
204. 1927. Based on *B. purgans* var.
latiglumis Shear.
Bromus latiglumis forma *incanus* Fernald,
Rhodora 35: 316. 1933. Based on *B.
purgans* var. *incanus* Shear.
Zerna latiglumis Henr., Blumea 4: 498.
1941. Based on *Bromus purgans* var.
latiglumis Shear.
Bromus macrostachys Desf., Fl. Atlant. 1:
96. pl. 19. f. 2. 1798. Algeria.
Serrafalcus macrostachys Parl., Fl. Ital. 1:
397. 1848. Based on *Bromus macrostachys* Desf.
Zerna macrostachys Panz. ex Jacks., Ind.
Kew. 2: 1249. 1895. Based on *Bromus
macrostachys* Desf.
(40) **Bromus madritensis** L., Cent. Pl. 1:
5. 1755; Amoen. Acad. 4: 265. 1759.
Spain. (The name is spelled *matritensis*

in Roem. and Schult., Syst. Veg. 2:
651. 1817.)
Festuca madritensis Desf., Fl. Atlant. 1:
91. 1798. Based on *Bromus madritensis* L.
Zerna madritensis Panz. ex Jacks., Ind.
Kew. 2: 1249. 1895, as synonym of
Bromus madritensis L.
Anisantha madritensis Nevski, Act. Univ.
Asiae Med. VIII b. Bot. 17: 21. 1934.
Based on *Bromus madritensis* L.
(7) **Bromus marginatus** Nees in Steud., Syn.
Pl. Glum. 1: 322. 1854. Columbia
River, *Douglas.*
Bromus hookeri var. *marginatus* Fourn.,
Mex. Pl. 2: 127. 1886. Based on *B.
marginatus* Nees. [*B. hookeri* Fourn.
(not *B. hookerianus* Thurb.) is based on
"*B. purgans* Hook. f., Bot. of Capt.
Beech. Voy. 119," name only.]
Ceratochloa marginata Nees ex Steud. ex
Jacks., Ind. Kew. 1: 487. 1893, presumably referring to *Bromus marginatus*
Nees.
Bromus marginatus var. *seminudus* Shear,
U. S. Dept. Agr., Div. Agrost. Bul. 23:
55. 1900. Wallowa Lake, Oreg., *Shear*
1811.
Bromus marginatus var. *latior* Shear, U. S.
Dept. Agr., Div. Agrost. Bul. 23: 55.
1900. Walla Walla, Wash., *Shear* 1615.
Bromus flodmanii Rydb., Torrey Bot.
Club Bul. 36: 538. 1909. Sheep Creek,
Mont., *Flodman* 187.
Forasaccus marginatus Lunell, Amer.
Midl. Nat. 4: 225. 1915. Based on
Bromus marginatus Nees.
Bromus latior Rydb., Fl. Rocky Mount.
89. 1917. Based on *B. marginatus* var.
latior Shear.
(8) **Bromus maritimus** (Piper) Hitchc. in
Jepson, Fl. Calif. 1: 177. 1912. Based
on *B. marginatus maritimus* Piper.
Bromus marginatus maritimus Piper, Biol.
Soc. Wash. Proc. 18: 148. 1905. Point
Reyes, Calif., *Davy* 6798.
(31) **Bromus molliformis** Lloyd, Fl. Loire-
Inf. 315. 1844. France.
(30) **Bromus mollis** L., Sp. Pl. ed. 2. 1: 112.
1762. Europe.
Serrafalcus mollis Parl., Pl. Rar. Sic. 2:
11. 1840. Based on *Bromus mollis* L.
Forasaccus mollis Bubani, Fl. Pyr. 4: 386.
1901. Based on *Bromus mollis* L.
Bromus hordeaceus L. subsp. *mollis* Hylander, Uppsala Univ. Årsskr: 7: 84.
1945. Based on *B. mollis* L.
This is the species referred to *B. hordeaceus*
L. in recent American works. The specimen
referred by Shear (U. S. Dept. Agr., Div.
Agrost. Bul. 23: 19. 1900) to *B. hordeaceus*
var. *intermedius* (Guss.) Shear belongs to
B. mollis.
(22) **Bromus nottowayanus** Fernald, Rhodora 43: 530. pl. 670. f. 1–7. 1941.
Sussex County, Va., *Fernald* and *Long*
12239.

(14) **Bromus orcuttianus** Vasey, Bot. Gaz. 10: 223. 1885. San Diego, Calif., *Orcutt* in 1884.
Bromus brachyphyllus Merr., Rhodora 4: 146. 1902. Crook County Oreg., *Cusick* 2677.
BROMUS ORCUTTIANUS var. HALLII Hitchc. in Jepson, Fl. Calif. 1: 175. 1912. San Jacinto Mountains, *Hall* 2301.
(16) **Bromus pacificus** Shear, U. S. Dept. Agr., Div. Agrost. Bul. 23: 38. 1900. Seaside, Oreg., *Scribner* and *Shear* 1703.
Bromus magnificus Elmer, Bot. Gaz. 36: 53. 1903. Port Angeles, Wash., *Elmer* 1957.
(9) **Bromus polyanthus** Scribn. in Shear, U. S. Dept. Agr., Div. Agrost. Bul. 23: 56. 1900. Based on *B. multiflorus* Scribn.
Bromus multiflorus Scribn., U. S. Dept. Agr., Div. Agrost. Bul. 13: 46. 1898. Not *B. multiflorus* Weigel, 1772. Battle Lake, Wyo., *A. Nelson* 4021.
Bromus polyanthus var. *paniculatus* Shear, U. S. Dept. Agr., Div. Agrost. Bul. 23: 57. 1900. West Mancos Canyon, Colo., *Tracy,* Earle, and *Baker* 333.
Bromus paniculatus Rydb., Fl. Rocky Mount. 90. 1917. Based on *B. polyanthus* var. *paniculatus* Shear.
(11) **Bromus pumpellianus** Scribn., Torrey Bot. Club Bul. 15: 9. 1888. Belt Mountains, Mont., *Scribner* 418.
Bromus purgans var. *purpurascens* Hook., Fl. Bor. Amer. 2: 252. 1840. Bear Lake to Arctic seacoast, *Richardson.*
Bromus ciliatus var. *coloradensis* Vasey, Torrey Bot. Club Bul. 15: 10. 1888, name only; Beal, Grasses N. Amer. 2: 619. 1896. [Colo. Expl. 100th Merid. *Wolf* 1158.]
Bromus pumpellianus var. *melicoides* Shear, U. S. Dept. Agr., Div. Agrost. Bul. 23: 50. 1900. Beaver Creek Camp, Colo., *Pammel* in 1896.
Forasaccus pumpellianus Lunell, Amer. Midl. Nat. 4: 225. 1915. Based on *Bromus pumpellianus* Scribn.
BROMUS PUMPELLIANUS var. TWEEDYI Scribn. in Beal, Grasses N. Amer. 2: 622. 1896. Yellowstone Park, *Tweedy* 587.
(20) **Bromus purgans** L., Sp. Pl. 1: 76. 1753. Canada, *Kalm.*
Bromus pubescens Muhl. in Willd., Enum. Pl. 120. 1809. Pennsylvania, *Muhlenberg.*
Bromus imperialis Steud., Nom. Bot. ed. 2. 1: 229. 1840, as synonym of *B. purgans* L.
Bromus steudelii Frank ex Steud., Nom. Bot. ed. 2. 1: 229. 1840, as synonym of *B. purgans* L.
Bromus ciliatus var. *purgans* A. Gray, Man. 600. 1848. Based on *B. purgans* L.
Bromus hookeri var. *pubescens* Fourn.,

Mex. Pl. 2: 127. 1886. Based on *B. pubescens* Muhl.
Forasaccus purgans Lunell, Amer. Midl. Nat. 4: 225. 1915. Based on *Bromus purgans* L.
Bromus purgans forma *laevivaginatus* Wiegand, Rhodora 24: 92. 1922. Ithaca, N. Y., *Metcalf* 5821.
Bromus ciliatus var. *purgans* subvar. *laevivaginatus* Farwell, Amer. Midl. Nat. 10: 204. 1927. Presumably based on *B. purgans* forma *laevivaginatus* Wiegand.
Zerna purgans Henr., Blumea 4: 498. 1941. Based on *Bromus purgans* L.
BROMUS PURGANS var. LAEVIGLUMIS (Scribn.) Swallen, Biol. Soc. Wash. Proc. 54: 45. 1941. Based on *B. ciliatus* var. *laeviglumis* Scribn.
Bromus ciliatus var. *laeviglumis* Scribn. in Shear, U. S. Dept. Agr., Div. Agrost. Bul. 23: 32. 1900. Galt, Ontario, *Herriot* in 1898.
Forasaccus ciliatus var. *laeviglumis* Lunell, Amer. Midl. Nat. 4: 225. 1915. Based on *Bromus ciliatus* var. *laeviglumis* Scribn.
Bromus purgans forma *glabriflorus* Wiegand, Rhodora 24: 92. 1922. Ithaca, N. Y., *Metcalf* 5813.
Bromus laeviglumis Hitchc., Biol. Soc. Wash. Proc. 41: 157. 1928. Based on *B. ciliatus* var. *laeviglumis* Scribn.
(32) **Bromus racemosus** L., Sp. Pl. ed. 2. 1: 114. 1762. Europe.
Bromus mollis var. *leiostachys* Hartm., Skand. Fl. Handb. ed. 2: 33. 1832. Sweden.
Serrafalcus racemosus Parl., Rar. Pl. Sic. 2: 14. 1840. Based on *Bromus racemosus* L.
Bromus arvensis var. *racemosus* Neilreich, Fl. Nieder-Oesterr. 81. 1859. Based on *B. racemosus* L.
Bromus squarrosus var. *racemosus* Regel, Act. Hort. Petrop. 7: 602. 1881. Based on *B. racemosus* L.
Forasaccus racemosus Bubani, Fl. Pyr. 4: 387. 1901. Based on *Bromus racemosus* L.
Bromus mollis forma *leiostachys* Fernald, Rhodora 35: 316. 1933. Based on *B. mollis* var. *leiostachys* Hartm.
The specimens referred by Shear (U. S. Dept. Agr., Div. Agrost. Bul. 23: 20. 1900) to *B. hordeaceus* var. *glabrescens* (Coss.) Shear belong to *B. racemosus.*
Bromus ramosus Huds., Fl. Angl. 40. 1762. England.
Zerna ramosa Lindm., Svensk. Fanerogamfl. 101. 1918. Based on *Bromus ramosus* Huds.
(37) **Bromus rigidus** Roth, Mag. Bot. Roem. and Ust. 10: 21. 1790. Europe.
Bromus villosus Forsk., Fl. Aegypt. Arab. 23. 1775. Not *B. villosus* Scop., 1772. Egypt.

Bromus maximus Desf., Fl. Atlant. 1: 95. pl. 26. 1798. Not *B. maximus* Gilib., 1790. North Africa.

Bromus madritensis var. *maximus* St. Amans, Fl. Agen. 45. 1821. Based on *B. maximus* Desf.

Bromus rubens var. *rigidus* Mutel, Fl. Franç. 4: 133. 1837. Based on *B. rigidus* Roth.

Bromus madritensis var. *rigidus* Bab. ex Syme in Sowerby, English Bot. ed. 3. 11: 161. 1873. Based on *B. rigidus* Roth.

Bromus villosus var. *maximus* Aschers. and Graebn., Syn. Mitteleur. Fl. 2: 595. 1901. Based on *B. maximus* Desf.

Bromus villosus var. *rigidus* Aschers. and Graebn., Syn. Mitteleur. Fl. 2: 596. 1901. Based on *B. rigidus* Roth.

Forasaccus maximus Bubani, Fl. Pyr. 4: 382. 1901. Based on *Bromus maximus* Desf.

Anisantha rigida Hylander, Uppsala Univ. Årskr. 7: 32. 1945. Based on *Bromus rigidus* Roth.

BROMUS RIGIDUS var. GUSSONEI (Parl.) Coss. and Dur., Expl. Sci. Alger. 2: 159. 1855. Based on *B. gussonii* Parl.

Bromus gussonii Parl., Rar. Pl. Sic. 2: 8. 1840. Europe.

Bromus maximus var. *gussonii* Parl., Fl. Ital. 1: 407. 1848. Based on *B. gussonii* Parl.

Bromus villosus var. *gussonii* Aschers. and Graebn., Syn. Mitteleur. Fl. 2: 595. 1901. Based on *B. gussonii* Parl.

Zerna gussonii Grossh., Akad. Nauk. S. S. S. R. Bot. Inst. Trudy Azerbaidzh. Fil. 8: 305. 1939. Based on *Bromus gussonii* Parl.

(39) **Bromus rubens** L., Cent. Pl. 1: 5. 1755; Amoen. Acad. 4: 265. 1759. Spain.

Festuca rubens Pers., Syn. Pl. 1: 94. 1805. Based on *Bromus rubens* L.

Bromus scoparius var. *rubens* St. Amans, Fl. Agen. 45. 1821. Based on *B. rubens* L.

Bromus madritensis subsp. *rubens* Husnot, Gram. Fr. Belg. 71. 1899. Based on *B. rubens* L.

Anisantha rubens Nevski, Act. Univ. Asiae Med. VIII b. Bot. 17: 19. 1934. Based on *Bromus rubens* L.

Zerna rubens Grossh., Akad. Nauk S. S. S. R. Bot. Inst. Trudy Azerbaidzh. Fil. 8: 306. 1939. Based on *Bromus rubens* L.

Bromus scoparius L., Cent. Pl. 1: 6. 1755; Amoen. Acad. 4: 266. 1759. Spain.

Serrafalcus scoparius Parl., Fl. Palerm. 1: 174. 1845. Based on *Bromus scoparius* L.

(28) **Bromus secalinus** L., Sp. Pl. 76. 1753. Europe.

Bromus mollis var. *secalinus* Huds., Fl. Angl. ed. 2. 49. 1778. Based on *B. secalinus* L.

Avena secalinus Salisb., Prodr. Stirp. 22. 1796. Based on *Bromus secalinus* L.

Serrafalcus secalinus Bab., Man. Brit. Bot. ed. 1. 374. 1843. Based on *Bromus secalinus* L.

?Bromus submuticus Steud., Syn. Pl. Glum. 1: 321. 1854. St. Louis, Mo.

Forasaccus secalinus Bubani, Fl. Pyr. 4: 388. 1901. Based on *Bromus secalinus* L.

BROMUS SECALINUS var. VELUTINUS Koch, Syn. Fl. Germ. Helv. 819. 1837. Based on *B. velutinus* Schrad.

Bromus velutinus Schrad., Fl. Germ. 1: 349. pl. 6. f. 3. 1806. Germany.

(2) **Bromus sitchensis** Trin., Acad. St. Pétersb. Mém. VI. Math. Phys. Nat. 2: 173. 1832. Sitka, Alaska [*Mertens*].

(33) **Bromus squarrosus** L., Sp. Pl. 1: 76. 1753. France, Switzerland, Siberia.

(38) **Bromus sterilis** L., Sp. Pl. 77. 1753. Europe.

Schedonorus sterilis Fries, Bot. Not. 131. 1843. Based on *Bromus sterilis* L.

Zerna sterilis Panz. ex. Jacks., Ind. Kew. 2: 1249. 1895, as synonym of *Bromus sterilis* L.

Anisantha sterilis Nevski, Act. Univ. Asiae Med. VIII b. Bot. 17: 20. 1934. Based on *Bromus sterilis* L.

(13) **Bromus suksdorfii** Vasey, Bot. Gaz. 10: 23. 1885. Mount Adams, Wash., *Suksdorf* [74 in 1883].

(41) **Bromus tectorum** L., Sp. Pl. 77. 1753. Europe.

Schedonorus tectorum Fries, Bot. Not. 131. 1843. Based on *Bromus tectorum* L.

Bromus setaceus Buckl., Acad. Nat. Sci. Phila. Proc. 1862; 98. 1862. Northern Texas, *Buckley*.

Zerna tectorum Panz. ex Jacks., Ind. Kew. 2: 1249. 1895, as synonym of *Bromus tectorum* L.

BROMUS TECTORUM var. GLABRATUS Spenner, Fl. Friburg. 1: 152. 1825. Germany.

Bromus tectorum var. *nudus* Klett. and Richt., Fl. Leipzig 109. 1830. Germany.

Anisantha tectorum Nevski, Act. Univ. Asiae Med. VIII b. Bot. 17: 20, 22. 1934. Based on *Bromus tectorum* L.

Bromus tectorum forma *nudus* St. John, Fl. Southeast. Wash. and Adj. Idaho 36. 1937. Based on *B. tectorum* var. *nudus* Klett. and Richt.

(23) **Bromus texensis** (Shear) Hitchc., U. S. Natl. Herb. Contrib. 17: 381. 1913. Based on *B. purgans* var. *texensis* Shear.

Bromus purgans var. *texensis* Shear, U. S. Dept. Agr., Div. Agrost. Bul. 23: 41. 1900. Bexar County, Tex., *Jermy* 230.

(42) **Bromus trinii** Desv. in Gay, Fl. Chil. 6: 441. 1853. Based on *Trisetum hirtum* Trin.

Trisetum hirtum Trin., Linnaea 10: 300. 1836. Not *Bromus hirtus* Lichtst., 1817. Chile.

Bromus trinii var. *pallidiflorus* Desv. in
Gay, Fl. Chil. 6: 441. 1853. Chile.
Trisetum barbatum Steud., Syn. Pl. Glum.
1: 229. 1854. Not *T. barbatum* Nees,
1841. Chile, *Bertero* 806.
Danthonia pseudo-spicata C. Muell., Bot.
Ztg. 14: 348. 1856. Valparaiso, Chile,
Cuming 466.
Trisetum barbatum var. *major* Vasey,
U. S. Dept. Agr., Div. Bot. Bul 13²:
60. 1893. Mexico, *Palmer* 667.
Bromus barbatoides Beal, Grasses N.
Amer. 2: 614. 1896. Based on *Trisetum
barbatum* Steud.
Bromus barbatoides var. *sulcatus* Beal,
Grasses N. Amer. 2: 615. 1896. Mex-
ico, *Palmer* 667.
Trisetum trinii Louis-Marie, Rhodora 30:
243. 1928. Based on *Bromus trinii*
Desv.
Trisetum trinii var. *pallidiflorus* Louis-
Marie, Rhodora 30: 243. 1928. Based
on *Bromus trinii* var. *pallidiflorus*
Desv.
Trisetum trinii var. *majus* Louis-Marie,
Rhodora 30: 243. 1928. Based on
T. barbatum var. *major* Vasey.
Trisetobromus hirtus Nevski, Acta Univ.
Asiae Med. VIII b. Bot. 17: 15. 1934.
Based on *Trisetum hirtum* Trin.
BROMUS TRINII var. EXCELSUS Shear, U. S.
Dept. Agr., Div. Agrost. Bul. 23: 25.
1900. Panamint Mountains, Calif.,
Coville and *Funston* 522.
(18) **Bromus vulgaris** (Hook.) Shear, U. S.
Dept. Agr., Div. Agrost. Bul. 23: 43.
1900. Based on *B. purgans* var. *vulgaris*
Hook.
Bromus purgans var. *vulgaris* Hook., Fl.
Bor. Amer. 2: 252. 1840. Canada,
Goldie, Richardson; Red River, *Douglas;*
Columbia River, *Scouler.*
Bromus ciliatus var. *ligulatus* Vasey ex
Macoun, Can. Pl. Cat. 2⁴: 238. 1888.
Name only, Vancouver Island, *Macoun*
in 1887.
Bromus ciliatus var. *pauciflorus* Vasey ex
Macoun, Can. Pl. Cat. 2⁴: 238. 1888,
name only; Beal, Grasses N. Amer. 2:
619. 1896. Oregon, *Howell.*
Bromus debilis Nutt. ex Shear, U. S.
Dept. Agr., Div. Agrost. Bul. 23: 43.
1900, as synonym of *B. vulgaris.*
[Columbia River, *Scouler.*]
Bromus vulgaris var. *eximius* Shear, U. S.
Dept. Agr., Div. Agrost. Bul. 23: 44.
1900. Near Wallowa Lake, Oreg., *Shear*
1791.
Bromus vulgaris var. *robustus* Shear, U. S.
Dept. Agr., Div. Agrost. Bul. 23: 44.
1900. Seaside, Oreg., *Shear* 1710.
Bromus ciliatus var. *glaberrimus* Suksdorf,
Deut. Bot. Monatsschr. 19: 93. 1901.
Skamania County, Wash., *Suksdorf* in
1894 [2335].
Bromus eximius Piper, U. S. Natl. Herb.
Contrib. 11: 143. 1906. Based on *B.*

vulgaris var. *eximius* Shear.
Bromus eximius robustus Piper, U. S.
Natl. Herb. Contrib. 11: 143. 1906.
Based on *B. vulgaris* var. *robustus*
Shear.
Bromus eximius umbraticus Piper, U. S.
Natl. Herb. Contrib. 11: 144. 1906.
Based on *Bromus vulgaris* Shear, not
Bromus purgans var. *vulgaris* Hook.,
Piper considering the specimens re-
ferred by Shear to this species to be
distinct from the form described by
Hooker.
Zerna vulgaris Henr., Blumea 4: 498.
1941. Based on *Bromus purgans* var.
vulgaris Hook.

(115) BUCHLOË Engelm.

(1) **Buchloë dactyloides** (Nutt.) Engelm.,
Acad. Sci. St. Louis, Trans. 1: 432. pl.
12, 14, f. 1–17. 1859. Based on *Sesleria
dactyloides* Nutt.
Sesleria dactyloides Nutt., Gen. Pl. 1:
65. 1818. Grassy plains of the Mis-
souri [*Nuttall*, type a staminate plant].
Anthephora axilliflora Steud., Syn. Pl.
Glum. 1: 111. 1854. [Misspelled *Ante-
phora.*] Texas, *Drummond* [pistillate
plant].
Calanthera dactyloides Kunth ex Hook.,
Jour. Bot. Kew Misc. 8: 18. 1856.
Based on *Sesleria dactyloides* Nutt.
Lasiostega humilis Rupr. ex Munro in
Benth., Pl. Hartw. 347. 1857. Name
only (error for *Casiostega*). Aguas
Calientes, Mexico, *Hartweg* 250.
Casiostega dactyloides Fourn., Soc. Bot.
Belg. Bul. 15: 470. 1876. Based on
Sesleria dactyloides Nutt.
Casiostega hookeri Rupr. ex Fourn., Soc.
Bot. Belg. Bul. 15: 471. 1876, as
synonym of *Buchloë dactyloides* Engelm.
Bouteloua mutica Griseb. ex Fourn.,
Soc. Bot. Belg. Bul. 15: 471. 1876, as
synonym of *Buchloë dactyloides* Engelm.
Mexico, *Schaffner* 134 [staminate plant].
Melica mexicana Link ex Fourn., Soc.
Bot. Belg. Bul. 15: 471. 1876, as syno-
nym of *Buchloë dactyloides* Engelm.
Bulbilis dactyloides Raf. ex Kuntze, Rev.
Gen. Pl. 2: 763. 1891. Based on
Sesleria dactyloides Nutt.

(67) CALAMAGROSTIS Adans.

(1) **Calamagrostis bolanderi** Thurb. in S.
Wats., Bot. Calif. 2: 280. 1880. Mendocino County, Calif., *Bolander* 6471
in part.
Calamagrostis varia Boland. ex Thurb. in
S. Wats., Bot. Calif. 2: 280. 1880.
Not *C. varia* Host, 1809. As synonym
of *C. bolanderi* Thurb.
Deyeuxia bolanderi Vasey, Grasses U. S.
28. 1883. Based on *Calamagrostis
bolanderi* Thurb.

(3) **Calamagrostis breweri** Thurb. in S. Wats., Bot. Calif. 2: 280. 1880. Carson Pass, Calif., *Brewer 2128*.

Deyeuxia breweri Vasey, Grasses U. S. 28. 1883. Based on *Calamagrostis breweri* Thurb.

Calamagrostis lemmoni Kearney, U. S. Dept. Agr., Div. Agrost. Bul. 11: 16. 1898. California, *Lemmon* in 1875.

(8) **Calamagrostis cainii** Hitchc., Wash. Acad. Sci. Jour. 24: 480. 1934. Mount LeConte, Tenn., *Cain 48*.

(26) **Calamagrostis californica** Kearney, U. S. Dept. Agr., Div. Agrost. Bul. 11: 37. 1898. Sierra Nevada, Calif., *Lemmon 444* in 1875.

(21) **Calamagrostis canadensis** (Michx.) Beauv., Ess. Agrost. 15, 152, 157. 1812. Based on *Arundo canadensis* Michx.

Arundo canadensis Michx., Fl. Bor. Amer. 1: 73. 1803. Canada, *Michaux*.

Arundo agrostoides Pursh, Fl. Amer. Sept. 1: 86. 1814. New Jersey and Pennsylvania.

Calamagrostis mexicana Nutt., Gen. Pl. 1: 46. 1818. North America. "*Agrostis mexicana?* Persoon, *Arundo agrostoides* Pursh" are cited. *Agrostis mexicana* L., in Persoon's work is a species of *Muhlenbergia*, but Nuttall's description agrees with Pursh's.

Calamagrostis agrostoides Pursh ex Spreng., Syst. Veg. 1: 252. 1825. Presumably based on *Arundo agrostoides* Pursh.

Cinna purshii Kunth, Rév. Gram. 1: 67. 1829. Based on *Arundo agrostoides* Pursh.

Arundo fissa Willd. ex Steud., Nom. Bot. ed. 2. 1: 144. 1840, as synonym of *Calamagrostis michauxii* Trin.

Calamagrostis michauxii Trin. ex Steud., Nom. Bot. ed. 2. 1: 250. 1840. Based on *Arundo canadensis* Michx.

Calamagrostis hirtigluma Steud., Syn. Pl. Glum. 1: 188. 1854. Labrador.

Deyeuxia canadensis Munro ex Hook. f., Linn. Soc. Trans. 23: 345. 1861. Presumably based on *Arundo canadensis* Michx., indirect citations given. See also, Vasey, Grasses U. S. 28. 1883; Agr. Grasses U. S. 69. pl. 59. 1884; Cassidy, Colo. Agr. Expt. Sta. Bul. 12: 48, with plate. 1890.

Calamagrostis oregonensis Buckl., Acad. Nat. Sci. Phila. Proc. 1862: 92. 1862. Columbia River, *Nuttall*.

Calamagrostis columbiensis Nutt. ex A. Gray, Acad. Nat. Sci. Phila. Proc. 1862: 334. 1862. Name only [Columbia River, *Nuttall*].

Calamagrostis canadensis var. *robusta* Vasey in Rothr. in Wheeler, U. S. Survey W. 100th Merid. Rpt. 6: 285. 1878. Twin Lakes, Colo., Expl. 100th Merid. [*Wolf*] 1093.

Calamagrostis pallida Vasey and Scribn. ex Vasey, U. S. Natl. Herb. Contrib. 3: 79. 1892. Not *C. pallida* C. Muell., 1861. Washington, *Suksdorf* in 1883.

Calamagrostis blanda Beal, Grasses N. Amer. 2: 349. 1896. Based on *C. pallida* Vasey and Scribn.

Calamagrostis canadensis acuminata Vasey ex Shear and Rydb., U. S. Dept. Agr., Div. Agrost. Bul. 5: 26. 1897. Georgetown, Colo., *Shear 615* [type]; Montana, Idaho.

Calamagrostis canadensis var. *campestris* Kearney, U. S. Dept. Agr., Div. Agrost. Bul. 11: 31. 1898. Louis Plain, Assiniboia, *Macoun 56*.

Calamagrostis alaskana Kearney, U. S. Dept. Agr., Div. Agrost. Bul. 11: 32. 1898. Yukon River, Alaska, *Funston 157*.

Calamagrostis atropurpurea Nash, N. Y. Bot. Gard. Bul. 2: 153. 1901. Dawson, Yukon Territory, *R. S. Williams* in 1899.

Calamagrostis anomala Suksdorf, Allg. Bot. Ztschr. 12: 43. 1906. Mount Paddo [Adams], Wash., *Suksdorf 2824*.

Calamagrostis langsdorfi var. *acuminata* Litw., Trav. Mus. Bot. Acad. Sci. Petrograd 18: 52. 1920. Based on *C. canadensis* var. *acuminata* Vasey.

Calamagrostis canadensis var. *pallida* Stebbins, Rhodora 32: 45. 1930. Based on *C. pallida* Vasey and Scribn.

Calamagrostis scribneri var. *imberbis* Stebbins, Rhodora 32: 46. 1930. Based on *C. anomala* Suksdorf "not Steud., in Lechl., Berb. Am. Aust. 56. (1857)," a name only.

CALAMAGROSTIS CANADENSIS var. MACOUNIANA (Vasey) Stebbins, Rhodora 32: 41. 1930. Based on *Deyeuxia macouniana* Vasey.

Deyeuxia macouniana Vasey, Bot. Gaz. 10: 297. 1885. Northwest Territory, *Macoun*.

Calamagrostis macouniana Vasey, U. S. Natl. Herb. Contrib. 3: 81. 1892. Based on *Deyeuxia macouniana* Vasey.

CALAMAGROSTIS CANADENSIS var. SCABRA (Presl) Hitchc., Amer. Jour. Bot. 21: 135. 1934. Based on *C. scabra* Presl.

Calamagrostis scabra Presl, Rel. Haenk. 1: 234. 1830. Nootka Sound, Vancouver Island, *Haenke*.

Deyeuxia preslii Kunth, Rév. Gram. 1: Sup. 20. 1830. Based on *Calamagrostis scabra* Presl.

This variety has been referred to *Calamagrostis langsdorfi* (Link) Trin. by many American authors. A fragment of the type of *Arundo langsdorfi* Link, sent by Dr. Pilger from the Berlin Herbarium, shows that it is not an American species. The rachilla is very minute or wanting, the spikelets are smaller than in *C. scabra*, the glumes are thinner, showing the nerves distinctly, and the blades are narrower. The following names, typonyms of *C. langsdorfi*, found in American works, belong to the Old World species:

Arundo langsdorfi Link, Enum. Pl. 1: 74. 1821. Described from a garden specimen.

Calamagrostis langsdorfi Trin., Gram. Unifl. 225. pl. 4. f. 10. 1824. Based on *Arundo langsdorfi* Link.

Deyeuxia langsdorfi Kunth, Rév. Gram. 1: 77. 1829. Based on *Arundo langsdorfi* Link.

Calamagrostis canadensis var. *langsdorfi* Inman, Rhodora 24: 143. 1922. Based on *Arundo langsdorfi* Link.

(23) **Calamagrostis cinnoides** (Muhl.) Barton, Compend. Fl. Phila. 1: 45. 1818. Based on *Arundo cinnoides* Muhl.

Agrostis glauca Muhl., Descr. Gram. 76. 1817. Not *Calamagrostis glauca* Reichenb., 1830. Pennsylvania, New Jersey, Carolina. Name only, Muhl., Cat. Pl. 10. 1813.

Arundo cinnoides Muhl., Descr. Gram. 187. 1817. Pennsylvania, Massachusetts. Name only, Muhl., Cat. Pl. 13. 1813. "*A. confinis* Willd." cited as synonym.

Arundo conoides Eaton, Man. ed. 2. 174. 1818. Error for *A. cinnoides* Muhl.

Arundo coarctata Torr., Fl. North. and Mid. U. S. 1: 94. 1823. New Jersey.

Calamagrostis langsdorfi var. *marylandica* Trin., Gram. Unifl. 225. 1824. Based on *Arundo cinnoides* Muhl.

Calamagrostis coarctata Torr. ex Eaton, Man. ed. 5. 144. 1829. Presumably based on *Arundo coarctata* Torr.; *Calamagrostis coarctata* Torr. in A. Gray, N. Amer. Gram. and Cyp. 1: 19. 1834. Based on *Arundo coarctata* Torr. Published as new in Torr., Fl. N. Y. 2: 444. pl. 151. 1843. Based on *A. coarctata* Torr.

Arundo canadensis Nutt. ex Steud., Nom. Bot. ed. 2. 1: 144. 1840, as synonym of *Calamagrostis nuttalliana* Steud. [Philadelphia, *Nuttall*.]

Calamagrostis nuttalliana Steud., Nom. Bot. ed. 2. 1: 251. 1840. Based on the species described by Nuttall [from specimen from Philadelphia] as *C. canadensis* (Nutt., Gen. Pl. 1: 46. 1818).

Deyeuxia nuttalliana Vasey, Grasses U. S. 28. 1883. Based on *Calamagrostis nuttalliana* Steud.

(28) **Calamagrostis crassiglumis** Thurb. in S. Wats., Bot. Calif. 2: 281. 1880. Mendocino County, Calif., *Bolander* 4766, 4787.

Deyeuxia crassiglumis Vasey, Grasses U. S. 28. 1883. Based on *Calamagrostis crassiglumis* Thurb.

Calamagrostis neglecta var. *crassiglumis* Beal, Grasses N. Amer. 2: 353. 1896. Based on *C. crassiglumis* Thurb.

(18) **Calamagrostis densa** Vasey, Bot. Gaz. 16: 147. 1891. Julian, San Diego County, Calif., *Orcutt*.

Calamagrostis koelerioides var. *densa* Beal, Grasses N. Amer. 2: 345. 1896. Based on *C. densa* Vasey.

Calamagrostis vilfaeformis Kearney, U. S. Dept. Agr., Div. Agrost. Bul. 11: 20. 1898. Based on *C. densa* Vasey.

(29) **Calamagrostis epigeios** (L.) Roth, Tent. Fl. Germ. 1: 34. 1788. Based on *Arundo epigeios* L.

Arundo epigeios L., Sp. Pl. 81. 1753. Europe.

Calamagrostis georgica C. Koch, Linnaea 21: 387. 1848. Georgia (Russia) near Tiflis.

Calamagrostis epigejos var. *georgica* Ledeb., Fl. Ross. 4: 433. 1853. Based on *C. georgica* C. Koch.

Calamagrostis arenicola Fernald, Rhodora 30: 203. 1928. Barnstable County, Mass., *Fernald* 757.

(14) **Calamagrostis fernaldii** Louis-Marie, Rhodora 46: 290. pl. 836, f. 4. 1944. Boarstone Mountain, Piscataquis County, Maine, *Fernald* 427.

(5) **Calamagrostis foliosa** Kearney, U. S. Dept. Agr., Div. Agrost. Bul. 11: 17. 1898. Based on *C. sylvatica* var. *longifolia* Vasey.

Calamagrostis sylvatica var. *longifolia* Vasey, U. S. Natl. Herb. Contrib. 3: 83. 1892. Not *C. longifolia* Hook., 1840. [Humboldt County] Calif., *Bolander* 6470.

(2) **Calamagrostis howellii** Vasey, Bot. Gaz. 6: 271. 1881. Oregon, *Howell*.

Deyeuxia howellii Vasey, Grasses U. S. 28. 1883. Based on *Calamagrostis howellii* Vasey.

(25) **Calamagrostis inexpansa** A. Gray, N. Amer. Gram. and Cyp. 1. No. 20. 1834. Penn Yan, N. Y. *Sartwell*.

Calamagrostis stricta var. *brevior* Vasey in Rothr. in Wheeler, U. S. Survey W. 100th Merid. Rpt. 6: 285. 1878. Mosquito, Colo., [*Wolf*] 1098.

Calamagrostis stricta var. *robusta* Vasey in Rothr. in Wheeler, U. S. Survey W. 100th Merid. Rpt. 6: 285. 1878. Twin Lakes, Colo., [*Wolf*] 1099.

Deyeuxia neglecta var. *americana* Vasey in Macoun, Can. Pl. Cat. 2⁴: 206. 1888. Donald, Columbia Valley, *Macoun* in 1885.

Deyeuxia neglecta var. *robusta* Vasey in Macoun, Can. Pl. Cat. 2⁴: 206. 1888. Alberta, *Macoun*.

Deyeuxia glomerata Vasey ex Macoun, Bot. Gaz. 16: 288. 1891. Name only. Rocky Mountains, British Columbia, *J.* and *J. M. Macoun* in 1890.

Calamagrostis robusta Vasey, U. S. Natl. Herb. Contrib. 3: 82. 1892. Not *C. robusta* Muell., 1861. Presumably based

on *C. stricta* var. *robusta* Vasey, the description being an amplification of that.
Calamagrostis americana Scribn., U. S. Dept. Agr., Div. Agrost. Bul. 5: 27. 1897. Based on *Deyeuxia neglecta* var. *americana* Vasey.
Calamagrostis inexpansa var. *cuprea* Kearney, U. S. Dept. Agr., Div. Agrost. Bul. 11: 37. 1898. Falcon Valley, Wash., *Suksdorf* 910.
Calamagrostis hyperborea var. *stenodes* Kearney, U. S. Dept. Agr., Div. Agrost. Bul. 11: 39. 1898. Marshall Pass, Colo., *Clements* 206.
Calamagrostis hyperborea elongata Kearney, U. S. Dept. Agr., Div. Agrost. Bul. 11: 40. 1898. Plummer County, Nebr., *Rydberg* 1494.
Calamagrostis hyperborea americana Kearney, U. S. Dept. Agr., Div. Agrost. Bul. 11: 41. 1898. Based on *Deyeuxia neglecta* var. *americana* Vasey.
Calamagrostis micrantha var. *sierrae* Jones, West. Bot. Contrib. 14: 9. 1912. Prattville and Susanville, Calif. [*Jones.*]
Calamagrostis neglecta var. *inexpansa* Jones, West. Bot. Contrib. 14: 9. 1912. Based on *C. inexpansa* A. Gray.
Deyeuxia hyperborea elongata Lunell, Amer. Midl. Nat. 4: 218. 1915. Based on *Calamagrostis hyperborea elongata* Kearney.
Deyeuxia hyperborea stenodes Lunell, Amer. Midl. Nat. 4: 218. 1915. Based on *Calamagrostis hyperborea stenodes* Kearney.
Calamagrostis elongata Rydb., Fl. Rocky Mount. 58. 1917. Based on *C. hyperborea elongata* Kearney.
Calamagrostis wyomingensis Gandog., Soc. Bot. France Bul. 66⁷: 299. 1920. Granger, Wyo., [A.] *Nelson* 3884.
Calamagrostis scopulorum var. *bakeri* Stebbins, Rhodora 32: 47. 1930. Pagosa Peak, Colo., *Baker* 162.
Calamagrostis inexpansa var. *robusta* Stebbins, Rhodora 32: 48. 1930. Based on *C. stricta* var. *robusta* Vasey.
Calamagrostis inexpansa var. *brevior* Stebbins, Rhodora 32: 50. 1930. Based on *C. stricta* var. *brevior* Vasey.
Calamagrostis expansa Rickett and Gilly, Torrey Bot. Club Bul. 69: 464. 1942. Error for *C. inexpansa* A. Gray.
This species has been referred by American authors to *C. hyperborea* Lange (*C. neglecta* var. *hyperborea* Jones, *Deyeuxia hyperborea* Lunell); and to *C. stricta* Trin.
CALAMAGROSTIS INEXPANSA var. BARBULATA Kearney, U. S. Dept. Agr., Div. Agrost. Bul. 11: 37. 1898. Mason County, Wash., *Piper* 947.
CALAMAGROSTIS INEXPANSA var. NOVAE-ANGLIAE Stebbins, Rhodora 32: 51. 1930. Mount Desert, Maine, *Williams* and *Rand* in 1899.

(12) **Calamagrostis insperata** Swallen, Wash. Acad. Sci. Jour. 25: 413. 1935. Jackson County, Ohio, *Bartley* and *Pontius* in 1934.
(19) **Calamagrostis koelerioides** Vasey, Bot. Gaz. 16: 147. 1891. Julian, San Diego County, Calif., *Orcutt.*
(22) **Calamagrostis lactea** Beal, Grasses N. Amer. 2: 346. 1896. Washington, *Suksdorf* 1022.
Deyeuxia lactea Beal, Grasses N. Amer. 2: 346. 1896, as synonym of *Calamagrostis lactea* Beal; Suksdorf, Deut. Bot. Monatsschr. 19: 92. 1901. Based on *C. lactea* Beal.
Calamagrostis langsdorfi lactea Kearney, U. S. Dept. Agr., Div. Agrost. Bul. 11: 28. 1898. Based on *C. lactea* Beal.
(13) **Calamagrostis lacustris** (Kearney) Nash, in Britt. and Brown, Illus. Fl. ed. 2. 1: 208. 1913. Based on *C. breviseta* var. *lacustris* Kearney.
Calamagrostia breviseta var. *lacustris* Kearney, U. S. Dept. Agr., Div. Agrost. Bul. 11: 25. 1898. Fond du Lac, Minn., *Wood* in 1889.
Calamagrostis pickingeri var. *lacustris* Hitchc., Rhodora 8: 210. 1906. Based on *C. breviseta* var. *lacustris* Kearney.
(7) **Calamagrostis montanensis** Scribn. in Vasey, U. S. Natl. Herb. Contrib. 3: 82. 1892. Montana, *Scribner.* Type is type specimen of *Deyeuxia montanensis* Scribn.
Deyeuxia montanensis Scribn., Soc. Prom. Agr. Sci. Proc. 2: 52. 1885. Helena, Mont., *Scribner.*
Calamagrostis neglecta var. *candidula* Kearney, U. S. Dept. Agr., Div. Agrost. Bul. 11: 35. 1898. Cypress Hills, Assiniboia, *Macoun* 7483.
(27) **Calamagrostis neglecta** (Ehrh.) Gaertn., Mey., and Scherb., Fl. Wett. 1: 94. 1799. Based on *Arundo neglecta* Ehrh.
Arundo neglecta Ehrh., Beiträge 6: 137. 1791. Europe.
Deyeuxia neglecta Kunth, Rév. Gram. 1: 76. 1829. Based on *Arundo neglecta* Ehrh.
Deyeuxia neglecta var. *gracilis* Scribn., Bot. Gaz. 11: 175. 1886. Yellowstone Park, *Tweedy* 582.
Deyeuxia vancouverensis Vasey, Torrey Bot. Club Bul. 15: 48. 1888. Locality erroneously given as "Vancouver Island," *Macoun* in 1887. Correction made in Macoun, Cat. Can. Pl. 2⁴: 207. 1888. Fort George, James Bay, Quebec.
Deyeuxia neglecta var. *brevifolia* Vasey in Macoun, Can. Pl. Cat. 2⁴: 206. 1888. Pelly Banks, Northwest Territory, *Dawson.*
Deyeuxia borealis Macoun, Can. Pl. Cat. 2⁴: 207. 1888. Change of name for *D. vancouverensis* Vasey, erroneously ascribed to Vancouver Island; collected at

Fort George, James Bay, Quebec, *J. M. Macoun.*

Calamagrostis laxiflora Kearney, U. S. Dept. Agr., Div. Agrost. Bul. 11: 34. 1898. Not *C. laxiflora* Phil., 1896. Based on "*C. neglecta gracilis* Scribn.," error for *Deyeuxia neglecta gracilis* Scribn.

Calamagrostis neglecta gracilis Scribn. ex Kearney, U. S. Dept. Agr., Div. Agrost. Bul. 11: 34. 1898, as synonym of *C. laxiflora* Kearney.

Calamagrostis micrantha Kearney, U. S. Dept. Agr., Div. Agrost. Bul. 11: 36. 1898. Prince Albert, Saskatchewan, *Macoun* 13111.

Calamagrostis lucida Scribn., U. S. Dept. Agr., Div. Agrost. Cir. 30: 8. 1901. Not *C. laxiflora* Phil. Based on *C. laxiflora* Kearney.

Calamagrostis neglecta var. *micrantha* Stebbins, Rhodora 32: 55. 1930. Based on *C. micrantha* Kearney.

(15) **Calamagrostis nubila** Louis-Marie, Rhodora 46: 296. pl. 836. f. 1–4. 1944. Lake of the Clouds, Mount Washington, N. H., *Boott.*

(17) **Calamagrostis nutkaensis** (Presl) Steud., Syn. Pl. Glum. 1: 190. 1854. Based on *Deyeuxia nutkaensis* Presl.

Deyeuxia nutkaensis Presl, Rel. Haenk. 1: 250. 1830. Nootka Sound, Vancouver Island, *Haenke.*

Calamagrostis aleutica Trin. in Bong., Acad. St. Pétersb. Mém. VI. Math. Phys. Nat. 2: 171. 1832. Unalaska Island, Alaska.

Deyeuxia aleutica Munro ex Hook. f., Linn. Soc. Trans. 23: 345. 1862. Based on *Calamagrostis aleutica* Trin.

Calamagrostis albicans Buckl., Acad. Nat. Sci. Phila. Proc. 1862: 92. 1862. Columbia Plains, Oreg., *Nuttall.*

Calamagrostis pallida Nutt. ex A. Gray, Acad. Nat. Sci. Phila. Proc. 1862: 334. 1862, as synonym of *C. albicans* Buckl. ["Columbia alluvions," *Nuttall*].

Calamagrostis albescens Buckl. ex A. Gray, Acad. Nat. Sci. Phila. Proc. 1862: 334. 1862, herbarium name, as synonym of *C. albicans* Buckl.

Deyeuxia breviaristata Vasey, Torrey Bot. Club Bul. 15: 48. 1888. Vancouver Island, *Macoun* in 1887.

Calamagrostis aleutica var. *patens* Kearney, U. S. Dept. Agr., Div. Agrost. Bul. 11: 20. 1898. Mendocino, Calif. [probably collected by *Bolander*].

(11) **Calamagrostis perplexa** Scribn., U. S. Dept. Agr., Div. Agrost. Cir. 30: 7. 1901. Based on *C. nemoralis* Kearney.

Calamagrostis nemoralis Kearney, U. S. Dept. Agr., Div. Agrost. Bul. 11: 26. 1898. Not *C. nemoralis* Phil., 1896. Ithaca, N. Y., *Dudley* in 1884.

(16) **Calamagrostis pickeringii** A. Gray, Man. ed. 2. 547. 1856. White Mountains, N. H., *Pickering.*

Calamagrostis sylvatica var. *breviseta* A. Gray, Man. 582. 1848. White Mountains, N. H.

Deyeuxia pickeringii Vasey, Grasses U. S. 28. 1883. Based on *Calamagrostis pickeringii* A. Gray.

Calamagrostis breviseta Scribn., Torrey Bot. Club Mem. 5: 41. 1894. Based on *C. sylvatica* var. *breviseta* A. Gray.

Calamagrostis breviseta var. *debilis* Kearney, U. S. Dept. Agr., Div. Agrost. Bul. 11: 25. 1898. Newfoundland, *Robinson* and *Schrenk* 205.

Calamagrostis pickeringii var. *debilis* Fern. and Wieg., Rhodora 15: 135. 1913. Based on *C. breviseta* var. *debilis* Kearney.

Calamagrostis pickeringii forma *vivipara* Louis-Marie, Rhodora 46: 296. 1944. Digby County, Nova Scotia, *Fernald* and *Long* 19924.

(10) **Calamagrostis porteri** A. Gray, Amer. Acad. Sci. Proc. 6: 79. 1862. Huntingdon County, Pa., *Porter* in 1862.

Deyeuxia porteri Vasey, Grasses U. S. 28. 1883. Based on *Calamagrostis porteri* A. Gray.

(6) **Calamagrostis purpurascens** R. Br. in Richards., Bot. App. Franklin Jour. 731. 1823. Northern British America.

Arundo purpurascens Schult., Mantissa 3 (Add. 1): 603. 1827. Based on *Calamagrostis purpurascens* R. Br.

Deyeuxia purpurascens Kunth, Rév. Gram. 1: 77. 1829. Based on *Calamagrostis purpurascens* R. Br.

Trisetum sesquiflorum Trin., Acad. St. Pétersb. Mém. VI. Sci. Nat. 2¹: 14. 1836. Unalaska.

Calamagrostis sylvatica var. *purpurascens* Thurb. ex Vasey, U. S. Natl. Herb. Contrib. 3: 83. 1892. [Mount Dana,] Calif., *Bolander* 5071.

Calamagrostis sylvatica var. *americana* Vasey, U. S. Natl. Herb. Contrib. 3: 83. 1892. British America to Colorado. [Type, Pen Gulch, Colo., *Vasey* in 1884.]

Calamagrostis arctica Vasey, U. S. Dept. Agr., Div. Bot. Bul. 13²: pl. 55. 1893. St. Paul Island, Bering Sea, *J. M. Macoun.*

Calamagrostis vaseyi Beal, Grasses N. Amer. 2: 344. 1896. Cascade Mountains, Wash., *Vasey.*

Calamagrostis purpurascens arctica Kearney, U. S. Dept. Agr., Div. Agrost. Bul. 11: 19. 1898. Based on *C. arctica* Vasey.

Calamagrostis yukonensis Nash, N. Y. Bot. Gard. Bul. 2: 154. 1901. Dawson, Yukon Territory, *R. S. Williams.*

Calamagrostis purpurascens var. *vaseyi* Jones, West Bot. Contrib. 14: 9. 1912. Based on *C. vaseyi* Beal.

Calamagrostis purpurascens var. *ophitidis* J. T. Howell, West. Bot. Leaflets 4:

246. 1946. Mount Tamalpais, Calif., *J. T. Howell 16334.*
This species has been referred to *Deyeuxia sylvatica* (DC.) Kunth by American authors.
(9) **Calamagrostis rubescens** Buckl., Acad. Nat. Sci. Phila. Proc. 1862: 92. 1862. Oregon, *Nuttall.*
Deyeuxia rubescens Vasey, Grasses U. S. 28. 1883. Based on *Calamagrostis rubescens* Buckl.
Deyeuxia cusickii Vasey, Bot. Gaz. 10: 224. 1885. Eagle Mountains, Oreg., *Cusick 1159.*
Deyeuxia suksdorfii Scribn., Torrey Bot. Club Bul. 15: 9. pl. 76. 1888. Washington, *Suksdorf 26.*
Calamagrostis aleutica var. *angusta* Vasey, U. S. Natl. Herb. Contrib. 3: 80. 1892. Santa Cruz, Calif., *Anderson.*
Calamagrostis cusickii Vasey, U. S. Natl. Herb. Contrib. 3: 81. 1892. Based on *Deyeuxia cusickii* Vasey.
Calamagrostis suksdorfii Scribn. in Vasey, U. S. Natl. Herb. Contrib. 3: 82. 1892. Based on *Deyeuxia suksdorfii* Scribn.
Calamagrostis angusta Kearney, U. S. Dept. Agr., Div. Agrost. Bul. 11: 21. 1898. Based on *C. aleutica* var. *angusta* Vasey.
Calamagrostis subflexuosa Kearney, U. S. Dept. Agr., Div. Agrost. Bul. 11: 22. 1898. Oakland, Calif., *Bolander 2274.*
Calamagrostis fasciculata Kearney, U. S. Dept. Agr., Div. Agrost. Bul. 11: 23. 1898. Mendocino County, Calif., *Pringle* in 1882.
Calamagrostis suksdorfii var. *luxurians* Kearney, U. S. Dept. Agr., Div. Agrost. Bul. 11: 24. 1898. Lake Coeur d'Alene, Idaho, *Sandberg, Heller,* and *McDougal* 630.
Calamagrostis luxurians Rydb., Fl. Rocky Mount. 57. 1917. Based on *Calamagrostis suksdorfii* var. *luxurians* Kearney.
This species has been referred by some American authors to *Calamagrostis sylvatica* DC., and to *Deyeuxia varia* Kunth.
(24) **Calamagrostis scopulorum** Jones, Calif. Acad. Sci. Proc. II. 5: 722. 1895. Springdale, Utah, *Jones 6075.*
Calamagrostis scopulorum var. *lucidula* Kearney, U. S. Dept. Agr., Div. Agrost. Bul. 11: 33. 1898. Wasatch Mountains, Utah, *Jones 1145.*
(20) **Calamagrostis scribneri** Beal, Grasses N. Amer. 2: 343. 1896. Based on *Deyeuxia dubia* Scribn. and Tweedy.
Deyeuxia dubia Scribn. and Tweedy, Bot. Gaz. 11: 174. 1886. Not *Calamagrostis dubia* Bunge, 1854. Yellowstone Park, *Tweedy.*
Calamagrostis dubia Scribn. in Vasey, U. S. Natl. Herb. Contrib. 3: 80. 1892. Based on *Deyeuxia dubia* Scribn. and Tweedy.
Calamagrostis canadensis var. *dubia* Vasey, U. S. Natl. Herb. Contrib. 3: 80. 1892.

Based on *C. dubia* Scribn. and Tweedy.
Calamagrostis langsdorfii var. *scribneri* Jones, West. Bot. Contrib. 14: 9. 1912. Based on *C. scribneri* Beal.
(4) **Calamagrostis tweedyi** (Scribn.) Scribn. in Vasey, U. S. Natl. Herb. Contrib. 3: 83. 1892. Based on *Deyeuxia tweedyi* Scribn.
Deyeuxia tweedyi Scribn., Torrey Bot. Club Bul. 10: 64. 1883. Cascade Mountains, Wash., *Tweedy.*

(69) **CALAMOVILFA Hack.**

(2) **Calamovilfa brevipilis** (Torr.) Scribn. in Hack., True Grasses 113. 1890. Based on *Arundo brevipilis* Torr.
Arundo brevipilis Torr., Fl. North. and Mid. U. S. 1: 95. 1823. Quaker Bridge, N. J.
Calamagrostis brevipilis L. C. Beck, Bot. North. and Mid. States 401. 1833. Based on *Arundo brevipilis* Torr.
Ammophila brevipilis Benth. ex Vasey, Grasses U. S. 29. 1883. Based on *Calamagrostis brevipilis* Beck.
Calamovilfa brevipilis var. *typica* Fernald, Rhodora 41: 502. 1939. Based on *Arundo brevipilis* Torr.
CALAMOVILFA BREVIPILIS var. CALVIPES Fernald, Rhodora 41: 501. pl. 573. f. 1. 2. 1939. Greensville County, Va., *Fernald* and *Long 8548.*
CALAMOVILFA BREVIPILIS var. HETEROLEPIS Fernald, Rhodora 41: 502. pl. 573. f. 4. 1939. Harnett County, N. C., *Correll* and *Blomquist 2539.*
(1) **Calamovilfa curtissii** (Vasey) Scribn., U. S. Dept. Agr., Div. Agrost. Bul. 17: 199. f. 495. 1899. Based on *Ammophila curtissii* Vasey.
Ammophila curtissii Vasey, Torrey Bot. Club Bul. 11: 7. 1884. Indian River, Fla., *Curtiss.*
Calamagrostis curtissii Vasey, Bot. Gaz. 15: 269. 1890. Based on *Ammophila curtissii* Vasey.
(4) **Calamovilfa gigantea** (Nutt.) Scribn. and Merr., U. S. Dept. Agr., Div. Agrost. Cir. 35: 2. 1901. Based on *Calamagrostis gigantea* Nutt.
Calamagrostis gigantea Nutt., Amer. Phil. Soc. Trans. (n. s.) 5: 143. 1837. Great Salt River of the Arkansas.
Toxeumia gigantea Nutt. ex Scribn. and Merr., U. S. Dept. Agr., Div. Agrost. Cir. 35: 2. 1901, as synonym of *Calamovilfa gigantea.* Salt River, Ark., *Nuttall.*
(3) **Calamovilfa longifolia** (Hook.) Scribn. in Hack., True Grasses 113. 1890. Based on *Calamagrostis longifolia* Hook.
Calamagrostis longifolia Hook., Fl. Bor. Amer. 2: 241. 1840. Saskatchewan, *Drummond.*
Vilfa rigida Buckl., Acad. Nat. Sci. Phila.

Proc. 1862: 89. 1862. "Oregon?" the locality probably erroneous.
Ammophila longifolia Benth. ex Vasey, Grasses U. S. 29. 1883. Based on *Calamagrostis longifolia* Gray [error for Hook].
Athernotus longifolius Lunell, Amer. Midl. Nat. 4: 218. 1915. Based on *Calamagrostis longifolia* Hook.
CALAMOVILFA LONGIFOLIA var. MAGNA Scribn. and Merr., U. S. Dept. Agr., Div. Agrost. Cir. 35: 3. 1901. Mouth of Kalamazoo River, Mich., *Taylor* in 1894.

(15) CATABROSA Beauv.

(1) **Catabrosa aquatica** (L.) Beauv., Ess. Agrost. 97, 149, 157. pl. 19. f. 8. 1812. Based on *Aira aquatica* L.
Aira aquatica L., Sp. Pl. 64. 1753. Europe.
Molinia aquatica Wib., Prim. Fl. Werthem. 116. 1799. Based on *Aira aquatica* L.
Poa airoides Koel., Descr. Gram. 194. 1802. Based on *Aira aquatica* L.
Glyceria aquatica Presl, Fl. Cech. 25. 1819. Based on *Aira aquatica* L.
Hydrochloa airoides Hartm., Gen. Gram. Skand. 8. 1819. Based on *Aira aquatica* L.
Catabrosa aquatica var. *uniflora* S. F. Gray, Nat. Arr. Brit. Pl. 2: 133. 1821. Great Britain.
Diarrhena aquatica Raspail, Ann. Sci. Nat., Bot. 5: 447. 1825. Based on *Catabrosa aquatica* Beauv.
Melica aquatica Loisel., Fl. Gall. ed. 2. 1: 59. 1828. Based on *Aira aquatica* L.
Glyceria airoides Reichenb. in Moessl., Handb. Gewächsk. ed. 2. 3: 1827-1829. Based on *Poa airoides* Koel.
Colpodium aquaticum Trin., Acad. St. Pétersb. Mém. VI. Math. Phys. Nat. 1: 395. 1830. Based on *Aira aquatica* L.
Glyceria catabrosa Klett and Richt., Fl. Leipzig 96. 1830. Based on *Catabrosa aquatica* Beauv.
Catapodium aquaticum Trin. ex Willk. and Lange, Prodr. Fl. Hisp. 1: 77. 1861, as synonym of *Catabrosa aquatica* Beauv.

(113) CATHESTECUM Presl

(1) **Cathestecum erectum** Vasey and Hack., Torrey Bot. Club Bul. 11: 37. pl. 45. 1884. Presidio, Tex., *Havard*.
This is the species described and figured by Scribner (U. S. Dept. Agr., Div. Agrost. Bul. 7: 242. f. 224. 1897) under the name *Cathestecum prostratum* Presl.

(145) CENCHRUS L.

Cenchrus biflorus Roxb., Hort. Beng. 81. 1814. Name only; Fl. Ind. 1: 238. 1820. Coromandel coast, India.
Cenchrus barbatus Schum., Beskr. Guin. Pl. 63. 1827. Guinea, Africa.

Cenchrus catharticus Delile, Cat. Hort. Monsp. 1838: 4. 1839. Grown from seed from Nubia, Africa.
(2) **Cenchrus brownii** Roem. and Schult., Syst. Veg. 2: 258. 1817. Based on *C. inflexus* R. Br.
Cenchrus inflexus R. Br., Prodr. Fl. Nov. Holl. 1: 195. 1810. Not *C. inflexus* Poir., 1804. Australia.
Cenchrus viridis Spreng., Syst. Veg. 1: 301. 1825. Guadeloupe, [*Bertero*].
Cenchrus echinatus var. *viridis* Spreng. ex Griseb., Fl. Brit. W. Ind. 556. 1864. Based on *C. viridis* Spreng.
(3) **Cenchrus echinatus** L., Sp. Pl. 1050. 1753. Jamaica, Curaçao.
Cenchrus pungens H. B. K., Nov. Gen. et Sp. 1: 115. 1815. Guayaquil, Ecuador, *Humboldt* and *Bonpland*.
Cenchrus brevisetus Fourn., Mex. Pl. 2: 50. 1886. Orizaba, Mexico, *Schaffner* 198; *Bourgeau* 3140; *Botteri* 133.
Cenchrus echinatus brevisetus Scribn. in Millsp., Field Mus. Bot. 2: 26. 1900. Based on *Cenchrus brevisetus* Fourn.
(4) **Cenchrus gracillimus** Nash, Torrey Bot. Club Bul. 22: 299. 1895. Eustis, Fla., *Nash* 188 [type], 288.
(5) **Cenchrus incertus** M. A. Curtis, Boston Soc. Nat. Hist. Jour. 1: 135. 1837. Smithville, N. C., *Curtis*.
?*Cenchrus carolinianus* Walt., Fl. Carol. 79. 1788. South Carolina.
Cenchrus strictus Chapm., Bot. Gaz. 3: 20. 1878. West Florida, [*Chapman*].
? *Nastus carolinianus* Lunell, Amer. Midl. Nat. 4: 214. 1915. Based on *Cenchrus carolinianus* Walt.
(1) **Cenchrus myosuroides** H. B. K., Nov. Gen. et Sp. 1: 115. pl. 35. 1815. Flamingo Key, Cuba, *Humboldt* and *Bonpland*.
Panicum cenchroides Ell., Bot. S. C. and Ga. 1: 111. 1816. Not *P. cenchroides* L. Rich., 1792. Jekyl Island, Ga., *Baldwin*.
Pennisetum pungens Nutt., Gen. Pl. 1: 54. 1818. Based on *Panicum cenchroides* Ell.
Setaria elliottiana Schult., Mantissa 2: 279. 1824. Based on *Panicum cenchroides* Ell.
Pennisetum myosuroides Spreng., Syst. Veg. 1: 303. 1825. Based on *Cenchrus myosuroides* H. B. K.
Cenchrus elliottii Kunth, Rév. Gram. 1: 51. 1829. Based on *Panicum cenchroides* Ell.
Cenchrus alopecuroides Presl, Rel. Haenk. 1: 317. 1830. Not *C. alopecuroides* Thunb., 1794. Original locality unknown, probably Peru.
Cenchrus setoides Buckl., Tex. Geol. Agr. Survey Prel. Rpt. App. 2. 1866. Northern Texas [*Linscum* and *Buckley*].
Cenchropsis myosuroides Nash in Small,

Fl. Southeast. U. S. 109, 1327. 1903.
Based on *Cenchrus myosuroides* H. B. K.
(6) **Cenchrus pauciflorus** Benth., Bot. Voy.
Sulph. 56. 1840. Magdalena Bay,
Baja California, [*Barclay*].
Cenchrus roseus Fourn., Mex. Pl. 2: 50.
1886. Vera Cruz, Mexico, *Gouin* 42 in
part, 43.
Cenchrus echinatus forma *longispina* Hack.
in Kneucker, Allg. Bot. Ztschr. 9:
169. 1903. Oxford, Conn., *Harger*,
Gram. Exs. *Kneucker* 426.
Cenchrus albertsonii Runyon, Amer. Jour.
Bot. 26: 485. f. 1, 2. 1939. Woodward,
Okla., *Runyon* 200.
Cenchrus longispinus Fernald, Rhodora
45: 388. 1943. Based on *C. echinatus*
forma *longispina* Hack.
Cenchrus pauciflorus var. *longispinus*
Jensen and Wachter, Nederl. Kruid.
Archief 56: 246. 1949. Based on *C.
echinatus* forma *longispinus* Hack.
(7) **Cenchrus tribuloides** L., Sp. Pl. 1050.
1753. Seacoast of Virginia, [*Clayton*].
Cenchrus echinatus var. *tribuloides* Torr.,
Fl. North. and Mid. U. S. 1: 69. 1823.
Based on *C. tribuloides* L.
Cenchrus vaginatus Steud., Syn. Pl. Glum.
1: 110. 1854. Cultivated in the botani-
cal garden, Paris.
Cenchrus tribuloides var. *macrocephalus*
Doell, in Mart., Fl. Bras. 2²: 312. 1877.
Brazil, *Martius*.
Cenchrus macrocephalus Scribn., U. S.
Dept. Agr., Div. Agrost. Bul. 17: 110. f.
406. 1899. Based on *C. tribuloides* var.
macrocephalus Doell.

(110) CHLORIS Swartz

(7) **Chloris andropogonoides** Fourn., Mex.
Pl. 2: 143. 1886. San Luis Potosí,
Mexico, *Virlet* 1462.
Chloris tenuispica Nash, Torrey Bot. Club
Bul. 25: 436. 1898. Texas, *Nealley* in
1889.
Chloris argentina (Hack.) Lillo and Parodi,
Physis 4: 180. 1918. Based on *C.
distichophylla* var. *argentina* Hack.
Chloris distichophylla var. *argentina* Hack.
ex Stuck., An. Mus. Nac. Buenos Aires
11: 113. 1904. Argentina, *Stuckert*.
Chloris berroi Arech., Anal. Mus. Nac.
Montevideo 1: 388. pl. 44. 1896.
Uruguay, *Berro*.
Chloris cantérai Arech., Anal. Mus. Nac.
Montevideo 1: 385. 1896. Paysandú,
Uruguay.
Chloris capensis (Houtt.) Thell., Repert.
Sp. Nov. Fedde 10: 289. 1912. Based
on *Andropogon capense* Houtt.
Andropogon capense Houtt. Nat. Hist. II.
13: Aanwyz. Plaat. [2]. pl. 103. f. 3.
1782; Panzer, Pflanzensyst. 12: Ver-
zeich. Kuppertaf. [4]. pl. 93. f. 3.
1785. Cape of Good Hope, Africa.
(6) **Chloris chloridea** (Presl) Hitchc., Biol.

Soc. Wash. Proc. 41: 162. 1928. Based
on *Dineba chloridea* Presl.
Dineba chloridea Presl, Rel. Haenk. 1:
291. 1830. Mexico, *Haenke*.
Eutriana chloridea Kunth, Rév. Gram. 1:
Sup. 23. 1830. Based on *Dineba
chloridea* Presl.
Gymnopogon longifolius Fourn., Mex. Pl.
2: 144. 1886. Vera Cruz, Mexico,
Gouin 52.
Gymnopogon virletii Fourn., Mex. Pl. 2:
144. 1886. San Luis Potosí, Mexico,
Virlet 1441.
Chloris longifolia Vasey, U. S. Natl.
Herb. Contrib. 1: 284. pl. 19. 1893.
Not *C. longifolia* Steud., 1854. Based on
Gymnopogon longifolius Fourn.
Chloris clandestina Scribn. and Merr.,
U. S. Dept. Agr., Div. Agrost. Bul. 24:
25. 1901. Based on *Gymnopogon longi-
folius* Fourn.
(11) **Chloris ciliata** Swartz, Prodr. Veg. Ind.
Occ. 25. 1788. Jamaica, *Swartz*.
Cynodon ciliatus Raspail, Ann. Sci. Nat.,
Bot. 5: 303. 1825. Based on *Chloris
ciliata* Swartz.
Chloris propinqua Steud., Syn. Pl. Glum.
1: 204. 1854. Guadeloupe, *Duchaissing*.
Chloris ciliata var. *texana* Vasey, U. S.
Dept. Agr., Div. Bot. Bul. 12¹: pl. 30.
1890. Brownsville, Tex. [*Nealley*].
Chloris texana Nash, Torrey Bot. Club
Bul. 25: 441. 1898. Based on *C. ciliata*
var. *texana* Vasey.
Chloris nashii Heller, Muhlenbergia 5:
120. 1909. Based on *C. texana* Nash.
(15) **Chloris cucullata** Bisch., Ann. Sci.
Nat., Bot. III. 19: 357. 1853. Culti-
vated, seed from Matamoros, Mexico.
Chloris distichophylla Lag., Gen. et Sp.
Nov. 4. 1816. Argentina and Chile.
Eustachys distichophylla Nees, Agrost.
Bras. 418. 1829. Based on *Chloris
distichophylla* Lag.
(3) **Chloris floridana** (Chapm.) Wood, Amer.
Bot. and Flor. pt. 2: 407. 1871. Based
on *Eustachys floridana* Chapm.
Eustachys floridana Chapm., Fl. South.
U. S. 557. 1860. Middle Florida.
(5) **Chloris gayana** Kunth, Rév. Gram. 1:
89. 1829. Senegal, Africa.
(1) **Chloris glauca** (Chapm.) Wood, Amer.
Bot. and Flor. pt. 2: 407. 1871. Based
on *Eustachys glauca* Chapm.
Eustachys glauca Chapm., Fl. South.
U. S. 557. 1860. West Florida.
(14) **Chloris latisquamea** Nash, Torrey Bot.
Club Bul. 25: 439. 1898. Kerrville,
Tex., *Heller* 1767.
Chloris verticillata var. *intermedia* Vasey,
in Coult., U. S. Nat. Herb. Contrib. 2:
528. 1894. Texas, [Houston, *Hall*
773].
(4) **Chloris neglecta** Nash, Torrey Bot.
Club Bul. 22: 423. 1895. Orange Bend,
Fla., *Nash* 2149.
Eustachys neglecta Nash, Torrey Bot.

Club Bul. 25: 450. 1898. Based on *Chloris neglecta* Nash.

(2) **Chloris petraea** Swartz, Prodr. Veg. Ind. Occ. 25. 1788. Jamaica, *Swartz.*

?*Aira aegilopsoides* Walt., Fl. Carol. 78. 1788. South Carolina.

Agrostis complanata Ait., Hort. Kew. 1: 96. 1789. Grown in England, seed from Jamaica.

Eustachys petraea Desv., Nouv. Bul. Soc. Philom. Paris 2: 189. 1810. Based on *Chloris petraea* Swartz.

Schultesia petraea Spreng., Pl. Pugill. 2: 17. 1815. Based on *Chloris petraea* Swartz.

Aira complanata Steud., Nom. Bot. ed. 2. 1: 44. 1840, as synonym of *Chloris petraea* Swartz.

Chloris swartzii C. Muell., Bot. Ztg. 19: 341. 1861. Based on *C. petraea* Swartz.

Chloris septentrionalis C. Muell., Bot. Ztg. 19: 340. 1861. Rio Brazos, Tex., *Drummond.*

Chloris swartziana Doell in Mart., Fl. Bras. 2³: 68. 1878. Based on *C. petraea* Swartz.

(10) **Chloris polydactyla** (L.) Swartz, Prodr. Veg. Ind. Occ. 26. 1788. Based on *Andropogon polydactylon* L.

Andropogon barbatus L., Syst. Nat. ed. 10. 2: 1305. 1759. Jamaica. Not *Chloris barbata* Swartz, 1797, based on *A. barbatus* L., 1771, from the East Indies, which is *C. inflata* Link (*C. paraguayensis* Steud.).

Andropogon polydactylon L., Sp. Pl. ed. 2. 2: 1483. 1763. Jamaica. Diagnosis of *A. barbatus* L. (1759) copied.

Saccharum ¦polydactylum Thunb., Fl. Jap. 42. 1784. Based on *Andropogon polydactylon* L.

Chloris barbata Nash, Torrey Bot. Club Bul. 25: 443. 1898. Not *C. barbata* Swartz, 1797. Based on *Andropogon barbatus* L. (1759).

Chloris prieurii Kunth, Rév. Gram. 1. 89. 1829. Senegambia, Africa.

Chloris radiata (L.) Swartz, Prodr. Veg. Ind. Occ. 26. 1788. Based on *Agrostis radiata* L.

Agrostis radiata L., Syst. Nat. ed. 10. 2: 873. 1759. Jamaica.

Chloris glaucescens Steud., Syn. Pl. Glum. 1: 206. 1854. Guadeloupe, *Duchaissing.*

(13) **Chloris subdolichostachya** C. Muell., Bot. Ztg. 19: 341. 1861. Texas, *Drummond 372.*

Chloris verticillata var. *aristulata* Torr. and Gray, U. S. Expl. Miss. Pacif. Rpt. 2: 176. 1855. Lower Rio Grande, *Gregg.*

Chloris brevispica Nash, Torrey Bot. Club Bul. 25: 438. 1898. Nueces County, Tex., *Heller 1471.*

Chloris submutica H. B. K., Nov. Gen. et Sp. 1: 167. pl. 50. 1816. Mexico, *Humboldt* and *Bonpland.*

(8) **Chloris texensis** Nash, Torrey Bot. Club Bul. 23: 151. 1896. Texas, *Thurow; Nealley.*

Chloris nealleyi Nash, Torrey Bot. Club Bul. 25: 435. 1898. Based on *C. texensis* Nash, not *C. texana* (Vasey) Nash, 1890.

Chloris truncata R. Br., Prodr. Fl. Nov. Holl. 186. 1810. Australia.

Chloris ventricosa R. Br., Prodr. Fl. Nov. Holl. 186. 1810. Australia.

(12) **Chloris verticillata** Nutt., Amer. Phil. Soc. Trans. (n.s.) 5: 150. 1837. Fort Smith, Ark., [*Nuttall*].

(9) **Chloris virgata** Swartz, Fl. Ind. Occ. 1. 203. 1797. Antigua, *Swartz.*

Chloris pubescens Lag., Var. Cien. 4: 143. 1805. [Peru.]

Rabdochloa virgata Beauv., Ess. Agrost. 84, 158. 1812. Presumably based on *Chloris virgata* Swartz.

Chloris compressa DC., Cat. Hort. Monsp. 94. 1813. Cultivated at Montpellier.

Chloris elegans H. B. K., Nov. Gen. et Sp. 1: 166. pl. 49. 1816. Mexico, *Humboldt* and *Bonpland.*

Chloris alba Presl, Rel. Haenk. 1: 289. 1830. Mexico, *Haenke.*

Chloris penicillata Willd. ex Steud., Nom. Bot. ed. 2. 1: 353. 1840, as synonym of *C. elegans* H. B. K.

Chloris alba var. *aristulata* Torr., U. S. Expl. Miss. Pacif. Rpt. 4: 155. 1857. Banks of the upper Rio Grande [*Emory* Exped.]; Tex., *Drummond 395* also mentioned.

Agrostomia barbata Cervant., Naturaleza 1: 346. 1870. Cuernavaca, Mexico.

(158) CHRYSOPOGON Trin.

(1) **Chrysopogon pauciflorus** (Chapm.) Benth. ex Vasey, Descr. Cat. Grasses U. S. 1883. Based on *Sorghum pauciflorum* Chapm.

Sorghum pauciflorum Chapm., Bot. Gaz. 3: 20. 1878. Jacksonville, Fla., *Chapman.*

Chrysopogon wrightii Munro ex Vasey, Descr. Cat. Grasses U. S. 29. 1885. Based on *Sorghum pauciflorum* Chapm.

Andropogon pauciflorus Hack. in DC., Monogr. Phan. 6: 548. 1889. Based on *Sorghum pauciflorum* Chapm.

Rhaphis pauciflorus Nash in Small, Fl. Southeast. U. S. 67. 1903. Based on *Sorghum pauciflorum* Chapm.

(74) CINNA L.

(1) **Cinna arundinacea** L., Sp. Pl. 5. 1753. Canada, *Kalm.*

Agrostis cinna Retz., Obs. Bot. 5: 18. 1789. Based on *Cinna arundinacea* L. but, according to Obs. Bot. 6: 12. 1791, misapplied to a species of *Muhlenbergia.*

Agrostis cinna Lam., Tabl. Encycl. 1: 162. 1791. Based on *Cinna arundinacea* L.

Agrostis cinna Pursh, Fl. Amer. Sept. 1: 64. 1814. Based on *Cinna arundinacea* Willd. (error for L.).
Cinna agrostoides Beauv. ex Steud., Nom. Bot. 1: 20, 198. 1821, as synonym of *Agrostis cinna* Lam.
Muhlenbergia cinna Trin., Gram. Unifl. 191. 1824. Based on *Agrostis cinna* Lam.
CINNA ARUNDINACEA var. INEXPANSA Fern. and Grisc., Rhodora 37: 135. pl. 334. f. 1, 2. 1935. Virginia Beach, Va., *Fernald* and *Long* 3648.
(2) **Cinna latifolia** (Trevir.) Griseb. in Ledeb., Fl. Ross. 4: 435. 1853. Based on *Agrostis latifolia* Trevir.
Agrostis latifolia Trevir. ex Göpp., Beschr. Bot. Gart. Breslau 82. 1830. Europe.
Muhlenbergia pendula Trin., Acad. St. Pétersb. Mém. VI. Math. Phys. Nat. 2: 172. 1832. Sitka.
Cinna expansa Link, Hort. Berol. 2: 236. 1833. Western North America, *Richardson*.
Cinna pendula Trin., Acad. St. Pétersb. Mém. VI. Sci. Nat. 4¹: 280. 1841. Norway, Sitka, Baikal. The earlier *Muhlenbergia pendula* Trin., not mentioned.
Cinna arundinacea var. *pendula* A. Gray, Man. ed. 2. 545. 1856. Based on *C. pendula* Trin.
Cinna pendula var. *glomerula* Scribn., Acad. Nat. Sci. Phila. Proc. 1884: 290. 1884. Washington, *Tweedy*.
Cinna bolanderi Scribn., Acad. Nat. Sci. Phila. Proc. 1884: 290. 1884. California, *Bolander* 6090.
Cinna pendula var. *acutiflora* Vasey ex Macoun, Can. Pl. Cat. 2⁴: 203. 1888, name only, Vancouver Island; 2⁵: 393. 1890, as synonym of *C. pendula* var. *glomerula* Scribn. [error for var. *glomerula*].
Cinna pendula var. *mutica* Vasey in Macoun, Can. Pl. Cat. 2⁴: 202. 1888. Name only for collection at Pelly Banks, Northwest Territory, *Dawson* in 1887; Vasey, U. S. Natl. Herb. Contrib. 3: 57. 1892. Oregon, [*Cusick*].
Cinna pendula var. *bolanderi* Vasey, U. S. Natl. Herb. Contrib. 3: 57. 1892. Based on *C. bolanderi* Scribn.

(165) COIX L.

1) **Coix lacryma-jobi** L., Sp. Pl. 972. 1753. India.
Coix lacryma L., Syst. Nat. ed. 10. 1261. 1759. Based on *C. lacryma-jobi* L.
Lithagrostis lacryma-jobi Gaertn., Fruct. et Sem. 1: 7. 1788. Based on *Coix lacryma-jobi* L.
Sphaerium lacryma Kuntze, Rev. Gen. Pl. 2: 793. 1891. Based on *Coix lacryma* L.

(73) COLEANTHUS Seidel

(1) **Coleanthus subtilis** (Tratt.) Seidel in Roem. and Schult., Syst. Veg. 2: 276. 1817. Based on *Schmidtia subtilis* Tratt.
Schmidtia subtilis Tratt., Fl. Oesterr. 1: 12. 1816. Bohemia.
Zizania subtilis Raspail, Ann. Sci. Nat., Bot. 5: 452, 458. 1825. Based on *Coleanthus subtilis* [Seidel] Roem. and Schult.
Wilibaldia subtilis Roth, Enum. Pl. Phan. Germ. 1: 92. 1827. Based on *Schmidtia subtilis* Tratt.
Smidetia humilis Raf., Autikon Bot. 187. 1840. Based on *Schmidtia subtilis* Tratt.

CORIDOCHLOA Nees

Coridochloa cimicina (L.) Nees ex Jacks., Ind. Kew. 1: 618. 1893, as synonym of *Panicum cimicinum*; Chase, Biol. Soc. Wash. Proc. 24: 129. 1911. This name is usually credited to Nees, Edinb. New Phil. Jour. 15: 381. 1833, but though Nees adds, after briefly distinguishing the genus, that its type is *Panicum cimicinum* Retz., he does not transfer the name to *Coridochloa*.
Milium cimicinum L., Mant. Pl. 2: 184. 1771. Malabar, India.
Agrostis digitata Lam., Encycl. 1: 59. 1783. "*Milium cimicinum* L." cited and description from "L. Mant. 184" quoted. Malabar.
Panicum cimicinum Retz., Obs. Bot. 3: 9. 1783. Based on *Milium cimicinum* L.
Axonopus? cimicinus Beauv., Ess. Agrost. 12, 154. 1812. Based on *Milium cimicinum* L.

(27) CORTADERIA Stapf

Cortaderia rudiuscula Stapf, Gard. Chron. III. 22: 396. 1897. Argentina.
This is the species described by Stapf under *C. quila* (Nees) Stapf, but that name is ultimately based on *Arundo quila* Molina, a species of bamboo, *Chusquea quila* (Molina) Kunth.
(1) **Cortaderia selloana** (Schult.) Aschers. and Graebn., Syn. Mitteleur. Fl. 2: 325. 1900. Based on *Arundo selloana* Schult.
Arundo dioeca Spreng., Syst. Veg. 1: 361. 1825. Not *A. dioica* Lour., 1793. Monte Video, Uruguay, *Sello*.
Arundo selloana Schult., Mantissa 3 (Add. 1): 605. 1827. Based on *A. dioeca* Spreng. Schultes cites "*A. dioeca* Spreng., S.V. p. 361," hence the date is later than 1824, the title-page date.
Gynerium argenteum Nees, Agrost. Bras. 462. 1829. Brazil.
Cortaderia argentea Stapf, Gard. Chron. III. 22: 396. 1897. Based on *Gynerium argenteum* Nees.
Cortaderia dioica Speg., An. Mus. Nac.

Buenos Aires **7**: 194. 1902. Based on *Arundo dioica* Spreng.

(60) CORYNEPHORUS Beauv.

(1) **Corynephorus canescens** (L.) Beauv., Ess. Agrost. 90, 149, 159. 1812. Based on *Aira canescens* L.

Aira canescens L., Sp. Pl. 65. 1753. Europe.

Avena canescens Web. in Wigg., Prim. Fl. Hols. 9. 1780. Based on *Aira canescens* L.

Weingaertneria canescens Bernh., Syst. Verz. Pflanz. 51. 1800. Based on *Aira canescens* L.

(38) COTTEA Kunth

(1) **Cottea pappophoroides** Kunth, Rév. Gram. 1: 84. 1829. Peru.

(85) CRYPSIS Ait.

(1) **Crypsis niliaca** Fig. and De Not., Mem. Accad. Torino II. 14: 322. 1854. (Separate 1853.) Island in the Nile, lower Egypt.

Referred to *C. aculeata* (L.) Ait. in Manual, ed. 1. That species is not known from America.

(108) CTENIUM Panzer

(1) **Ctenium aromaticum** (Walt.) Wood, Class-book ed. 1861. 806. 1861. Based on *Aegilops aromatica* Walt.

?Nardus gangitis L., Sp. Pl. 53. 1753. Garden specimen, southern France, (probably Montpellier). The specimen under this name in the Linnaean Herbarium is from Montpellier and is said by Munro (Jour. Linn. Soc. Bot. 6: 35. 1862) to be *Lepturus incurvatus* Trin. (*Parapholis incurvus* (L.) C. E. Hubb.). The Linnaean citations refer to *Andropogon* and to *Rottboellia* according to Trinius (Clav. Agrost. 346. 1822), except that to Morison (Pl. Hist. 3: Sect. 8, tab. 3, last figure) which is a species of *Ctenium*. Linnaeus gives as the origin of his plant "Habitat in G. Narbonensi" (Gallia Narbonensis is southern France). The application of the name *N. gangitis* is too uncertain to be accepted for *Ctenium aromaticum*, as proposed by Druce.

Aegilops aromatica Walt., Fl. Carol. 249. 1788. South Carolina.

Nardus scorpioides Lam., Tabl. Encycl. 1: 152. 1791. America.

Chloris monostachya Michx., Fl. Bor. Amer. 1: 59. 1803. South Carolina, *Michaux*.

Campulosus gracilior Desv., Nouv. Bul. Soc. Philom. (Paris) 2: 189. 1810. Based on *Chloris monostachya* Michx.

Campulosus monostachyus Beauv., Ess.

Agrost. 64, 157, 158. pl. 13. f. 1. 1812. Based on *Chloris monostachya* Michx.

Ctenium carolinianum Panz., Denkschr. Bayer. Akad. Wiss. 4: 311. pl. 13. f. 1, 2. 1813. South Carolina.

Campuloa gracilis Desv., Jour. Bot. 1: 69. 1813. Based on *Chloris monostachya* Michx.

Monocera aromatica Ell., Bot. S. C. and Ga. 1: 177. pl. 11. f. 3. 1816. Based on *Aegilops aromatica* Walt.

Campuloa monostachya Roem. and Schult., Syst. Veg. 2: 516. 1817. Based on *Chloris monostachya* Michx.

?Monerma ? gangitis Roem. and Schult., Syst. Veg. 2: 800. 1817. Based on *Nardus gangitis* L.

Triatherus aromaticus Raf., Amer. Month. Mag. 3: 99. 1818. Based on *Monocera aromatica* Ell.

Cynodon monostachyos Raspail, Ann. Sci. Nat., Bot. 5: 303. 1825. Based on *Campulosus monostachyus* Desv. [error for Beauv.].

Ctenium americanum Spreng., Syst. Veg. 1: 274. 1825. North America, *Chloris monostachya* Michx., cited as synonym.

Aplocera maritima Raf., Med. Fl. 2: 193. 1830. *Aplocera* proposed as change of name for *Monocera* Ell., no basis given for the specific name.

Campulosus aromaticus Trin. ex Steud., Nom. Bot. ed. 2. 1: 272. 1840, as synonym of *C. monostachyus* Beauv.

Chloris piperita Michx. ex Steud., Nom. Bot. ed. 2. 1: 353. 1840, as synonym of *Campulosus monostachyus* Beauv.

Rottboellia scorpioides Poir. ex Steud. Nom. Bot. ed. 2. 2: 474. 1841, as synonym of *Ctenium americanum* Spreng.

Campulosus gracilis Bertol., Accad. Sci. Bologna Mem. 2: 602. pl. 43. f. a.b.c. 1850. Alabama.

?Campulosus gangitis Kuntze, Rev. Gen. Pl. 2: 764. 1891. Based on *Nardus gangitis* L., taken up for *Ctenium aromaticum*.

Campulosus aromaticus Scribn., Torrey Bot. Club Mem. 5: 45. 1894. Based on *Aegilops aromaticus* Walt.

?Ctenium gangitum Druce, Bot. Exch. Club Brit. Isles Rpt. 3: 416. 1914. Based on *Nardus gangitis* L., taken up for *C. aromaticum*.

(2) **Ctenium floridanum** (Hitchc.) Hitchc., Biol. Soc. Wash. Proc. 41: 162. 1928. Based on *Campulosus floridanus* Hitchc.

Campulosus floridanus Hitchc., Amer. Jour. Bot. 2: 306. 1915. East Florida, *Curtiss* in 1875.

This is the species described by Scribner (U. S. Dept. Agr., Div. Agrost. Bul. 7: 197. f. 179. 1897) and by Nash (Small, Fl. Southeast. U. S. 133. 1903) under *Campulosus chapadensis* Trin. That is a Brazilian species not known from North America.

Cutandia memphitica (Spreng.) Richt. Pl. Eur. 1: 77. 1890. Based on *Dactylis memphitica* Spreng.
Dactylis memphitica Spreng., Nachtr. Bot. Gart. Halle 20. 1801. Egypt.

CYMBOPOGON Spreng.

Cymbopogon citratus (DC.) Stapf, Kew Bul. Misc. Inf. 1906: 322. 1906. Based on *Andropogon citratus* DC.
Andropogon citratus DC., Cat. Hort. Monsp. 78. 1813, without description. DC. ex Nees, Allg. Gartenz. 3: 267. 1835. Garden plant.
Cymbopogon nardus (L.) Rendle, Cat. Afr. Pl. Welw. 2: 155. 1899. Based on *Andropogon nardus* L.
Andropogon nardus L., Sp. Pl. 1046. 1753. India.
Sorghum nardus Kuntze, Rev. Gen. Pl. 2: 792. 1891. Based on *Andropogon nardus* L.

(103) CYNODON L. Rich.

(1) **Cynodon dactylon** (L.) Pers., Syn. Pl. 1: 85. 1805. Based on *Panicum dactylon* L.
Panicum dactylon L., Sp. Pl. 58. 1753. Southern Europe.
Digitaria dactylon Scop., Fl. Carn. ed. 2. 1: 52. 1772. Based on *Panicum dactylon* L.
Dactilon officinale Vill., Hist. Pl. Dauph. 2: 69. 1787. Based on *Panicum dactylon* L.
?*Cynosurus uniflorus* Walt., Fl. Carol. 82. 1788. South Carolina.
Paspalum dactylon Lam., Tabl. Encycl. 1: 176. 1791. Based on *Panicum dactylon* L.
Digitaria littoralis Salisb., Prodr. Stirp. 19. 1796. Based on *Panicum dactylon* L.
Milium dactylon Moench, Meth. Pl. Sup. 67. 1802. Based on *Panicum dactylon* L.
Fibichia umbellata Koel., Descr. Gram. 308. 1802. Based on *Panicum dactylon* L.
Digitaria stolonifera Schrad., Fl. Germ. 1: 165. 1806. Based on *Panicum dactylon* L.
Cynodon maritimus H. B. K., Nov. Gen. et Sp. 1: 170. 1816. Peru, *Humboldt and Bonpland.*
Cynodon tenuis Trin. in Spreng., Neu. Entd. 2: 63. 1821. North America.
Chloris cynodon Trin., Gram. Unifl. 229. 1824. Based on *Cynodon dactylon* Pers.
Digitaria maritima Spreng., Syst. Veg. 1: 272. 1825. Based on *Cynodon maritimus* H. B. K.
Cynodon erectus Presl, Rel. Haenk. 1: 290. 1830. Mexico [type, *Haenke*] and Peru.
Agrostis bermudiana Tussac ex Kunth,

Enum. Pl. 1: 259. 1833, as synonym of *Cynodon dactylon* Pers.
Cynodon occidentalis Willd. ex Steud., Nom. Bot. ed. 2. 1: 463. 1840, as synonym of *C. dactylon* Pers.
Cynodon portoricensis Willd. ex Steud., Nom. Bot. ed. 2. 1: 463. 1840, as synonym of *C. dactylon* Pers.
Capriola dactylon Kuntze, Rev. Gen. Pl. 2: 764. 1891. Based on *Panicum dactylon* L.
Fibichia dactylon Beck, Wiss. Mitt. Bosn. Herzeg. 9: 436. 1904. Based on *Panicum dactylon* L.
Cynodon dactylon var. *maritimus* Hack. in Fries, Arkiv Bot. 8: 40. 1909. Based on *C. maritimus* H. B. K.
Capriola dactylon maritima Hitchc., U. S. Dept. Agr. Bul. 772: 179. 1920. Based on *Cynodon maritimus* H. B. K.
Cynodon transvaalensis Burtt-Davy,[22] Kew Bul. Misc. Inf. 1921: 281. 1921. Transvaal, South Africa, *Burtt-Davy* 18156.

(24) CYNOSURUS L.

(1) **Cynosurus cristatus** L., Sp. Pl. 72. 1753. Europe.
(2) **Cynosurus echinatus** L., Sp. Pl. 72. 1753. Europe.
Phalona echinata Dum., Obs. Gram. Belg. 114. 1823. Based on *Cynosurus echinatus* L.

(23) DACTYLIS L.

(1) **Dactylis glomerata** L., Sp. Pl. 71. 1753. Europe.
Bromus glomeratus Scop., Fl. Carn. ed. 2. 1: 76. 1772. Based on *Dactylis glomerata* L.
Festuca glomerata All., Fl. Pedem. 2: 252. 1785. Based on *Dactylis glomerata* L.
Limnetis glomerata Eaton, Man. 14. 1817. Based on *Dactylis glomerata* L.
Trachypoa vulgaris Bubani. Fl. Pyr. 4: 359. 1901. Based on *Dactylis glomerata* L.

(101) DACTYLOCTENIUM Willd.

(1) **Dactyloctenium aegyptium** (L.) Beauv., Ess. Agrost. Expl. Pl. 15, 159. 1812. Based on *Cynosurus aegyptius* L. The same combination made by Richt. Pl. Eur. 1: 68. 1890, based on the same species.
Cynosurus aegyptius L., Sp. Pl. 72. 1753. Africa, Asia, America.
Aegilops saccharinum Walt., Fl. Carol. 249. 1788. South Carolina.
Eleusine aegyptiaca Desf., Fl. Atlant. 1: 85. 1798. Based on *Cynosurus aegyptius* L.

---

[22] J. Burtt-Davy in earlier papers on American grasses used Davy as author, as in *Elymus divergens* and others, but in later papers used Burtt-Davy.

Eleusine pectinata Moench, Meth. Pl. Sup. 68. 1802. Based on *Cynosurus aegyptius* L.

Chloris mucronata Michx., Fl. Bor. Amer. 1: 59. 1803. Carolina, *Michaux*.

Eleusine aegyptia Pers., Syn. Pl. 1: 87. 1805. Based on *Cynosurus aegyptius* L.

Dactyloctenium aegyptiacum Willd., Enum. Pl. 1029. 1809. Based on *Cynosurus aegyptius* L.

Dactyloctenium mucronatum Willd., Enum. Pl. 1029. 1809. Based on *Chloris mucronata* Michx.

Eleusine mucronata Stokes, Bot. Mat. Med. 1: 150. 1812. Not *E. mucronata* Michx., 1803. Jamaica, *Broughton*.

Rabdochloa mucronata Beauv., Ess. Agrost. 84, 158, 176. 1812. Presumably based on *Chloris mucronata* Michx.

Cenchrus aegyptius L. ex Beauv., Ess. Agrost. 157. 1812, as synonym of *Dactyloctenium aegyptium*, doubtless error for *Cynosurus*.

Eleusine egyptia Raf., Précis Décour. Somiol. 45. 1814.

Eleusine aegyptia Raf., Chloris Aetn. 7. 1815.

Eleusine cruciata Ell., Bot. S. C. and Ga. 1: 176. 1816. Presumably South Carolina.

Eleusine mucronata Hornem., Hort. Hafn. Sup. 116. 1819. Not *E. mucronata* Michx., 1803. Based on *Dactyloctenium mucronatum* Willd.

Dactyloctenium meridionale Hamilt., Prodr. Pl. Ind. Occ. 6. 1825. West Indies and tropical America.

Cynosurus carolinianus Willd. ex Steud., Nom. Bot. ed. 2. 1: 465. 1840. Name only, referred to *Dactyloctenium*.

Dactyloctenium mucronatum var. *erectum* Fourn., Mex. Pl. 2: 144. 1886. Mexico, *Gouin 68*; *Karwinsky 989, 989b*.

(66) DANTHONIA Lam. and DC.

(6) **Danthonia californica** Boland., Calif. Acad. Sci. Proc. 2: 182. 1863. Oakland and San Francisco, Calif., *Bolander*.

Merathrepta californica Piper, U. S. Natl. Herb. Contrib. 11: 122. 1906. Based on *Danthonia californica* Boland.

Pentameris californica Nels. and Macbr., Bot. Gaz. 56: 469. 1913. Based on *Danthonia californica* Boland.

DANTHONIA CALIFORNICA var. AMERICANA (Scribn.) Hitchc., Biol. Soc. Wash. Proc. 41: 160. 1928. Based on *D. americana* Scribn. (Published as *D. californica americana*.)

Danthonia grandiflora Phil., An. Univ. Chile 48: 568. 1873. Not *D. grandiflora* Hochst., 1851. Province Nuble, Chile.

Danthonia americana Scribn., U. S. Dept. Agr., Div. Agrost. Cir. 30: 5. 1901. Based on *D. grandiflora* Phil.

Merathrepta americana Piper, U. S. Natl. Herb. Contrib. 11: 123. 1906.

Based on *Danthonia americana* Scribn.

Pentameris americana Nels. and Macbr., Bot. Gaz. 56: 469. 1913. Based on *Danthonia americana* Scribn.

Danthonia macounii Hitchc., Amer. Jour. Bot. 2: 305. 1915. Nanaimo, Vancouver Island, *Macoun 78825*.

Danthonia californica var. *palousensis* St. John, Fl. Southeast. Wash. and Adj. Idaho 38. 1937. Potlatch, Idaho, *Beattie 4061*.

Danthonia californica var. *piperi* St. John, Fl. Southeast. Wash. and Adj. Idaho 38. 1937. Pullman, Wash., *Piper 1744*.

(2) **Danthonia compressa** Austin in Peck, N. Y. State Mus. Ann. Rpt. 22: 54. 1869. Herkimer County, N. Y., *Austin* in 1868.

Danthonia spicata var. *compressa* Wood, Amer. Bot. and Flor. pt. 2: 396. 1871. Based on *D. compressa* Austin.

Danthonia alleni Austin, Torrey Bot. Club Bul. 3: 21. 1872. Rockaway, Long Island, *Allen*.

Danthonia faxoni Austin, Torrey Bot. Club Bul. 6: 190. 1877. White Mountains, N. H., *Faxon* in 1877.

Merathrepta compressa Heller, Muhlenbergia 5: 120. 1909. Based on *Danthonia compressa* Austin.

Pentameris compressa Nels. and Macbr., Bot. Gaz. 56: 469. 1913. Based on *Danthonia compressa* Austin.

(4) **Danthonia intermedia** Vasey, Torrey Bot. Club Bul. 10: 52. 1883. California; Rocky Mountains; Plains of British America to Mount Albert, Quebec, *Allen* [in 1881, type].

Danthonia intermedia var. *cusickii* Williams, U. S. Dept. Agr., Div. Agrost. Cir. 30: 7. 1901. Oregon, *Cusick 2427*.

Merathrepta intermedia Piper, U. S. Natl. Herb. Contrib. 11: 122. 1906. Based on *Danthonia intermedia* Vasey.

Merathrepta intermedia cusickii Piper, U. S. Natl. Herb. Contrib. 11: 122. 1906. Based on *Danthonia intermedia cusickii* Williams.

Pentameris intermedia Nels. and Macbr., Bot. Gaz. 56:-470. 1913. Based on *Danthonia intermedia* Vasey.

Danthonia cusickii Hitchc., Amer. Jour. Bot. 2: 305. 1915. Based on *D. intermedia* var. *cusickii* Williams.

(5) **Danthonia parryi** Scribn., Bot. Gaz. 21: 133. 1896. Colorado, *Parry*.

Danthonia parryi var. *longifolia* Scribn., Bot. Gaz. 21: 134. 1896. Twin Lakes, Colo., *Wolf 1170*.

Merathrepta parryi Heller, Muhlenbergia 5: 120. 1909. Based on *Danthonia parryi* Scribn.

Danthonia pilosa R. Br., Prodr. Fl. Nov. Holl. 177. 1810. Australia.

Danthonia semiannularis (Labill.) R. Br., Prodr. Fl. Nov. Holl. 177. 1810. Based on *Arundo semiannularis* Labill.

Arundo semiannularis Labill., Nov. Holl. Pl. 1: 26. pl. 33. 1804. Australia.
(3) **Danthonia sericea** Nutt., Gen. Pl. 1: 71. 1818. Carolina to Florida.
Danthonia glabra Nash, Torrey Bot. Club Bul. 24: 43. 1897. Not *D. glabra* Phil., 1896. Little Stone Mountain, Ga., *Small* in 1895.
Danthonia epilis Scribn., U. S. Dept. Agr., Div. Agrost. Cir. 30: 7. 1901. Based on *D. glabra* Nash.
Merathrepta sericea Heller, Muhlenbergia 5: 120. 1909. Based on *Danthonia sericea* Nutt.
Pentameris epilis Nels. and Macbr., Bot. Gaz. 56: 469. 1913. Based on *Danthonia epilis* Scribn.
Pentameris sericea Nels. and Macbr., Bot. Gaz. 56: 470. 1913. Based on *Danthonia sericea* Nutt.
This is the species described by Elliott (Bot. S. C. and Ga. 1: 174. 1816) under the name *Avena spicata* L.
(1) **Danthonia spicata** (L.) Beauv. ex Roem. and Schult., Syst. Veg. 2: 690. 1817. Based on *Avena spicata* L.
Avena spicata L., Sp. Pl. 80. 1753. Pennsylvania;
Avena glumosa Michx., Fl. Bor. Amer. 1: 72. 1803. Pennsylvania; Carolina, *Michaux*. (In Index Kewensis this name is erroneously credited to Ell. Elliott cited *A. glumosa* Michx. as synonym of *A. spicata* L.)
Danthonia glumosa Beauv., Ess. Agrost. 92, 153, 160. 1812. Based on *Avena glumosa* Michx.
Avena spicaeformis Beauv., Ess. Agrost. 154. 1812, name only; Roem. and Schult., Syst. Veg. 2: 690. 1817, as synonym of *Danthonia spicata* L.
Triodia glumosa Beauv., Ess. Agrost. Atlas 12. pl. 18. f. 7. 1812. Evidently an error for *Danthonia glumosa* Beauv.
Merathrepta spicata Raf. ex Jacks., Ind. Kew. 2: 211. 1894, as synonym of *Danthonia spicata*.
Danthonia spicata var. *villosa* Peck, N. Y. State Mus. Ann. Rpt. 47: 168. 1894. Brownville [Peck] and Taberg, N. Y.
Danthonia spicata pinetorum Piper, Erythea 7: 103. 1899. Mason County, Wash., *Piper* 943.
Danthonia thermale Scribn., U. S. Dept. Agr., Div. Agrost. Cir. 30: 5. 1901. Yellowstone Park, Wyo., *A. Nelson* and *E. Nelson* 6140.
Danthonia spicata var. *longipila* Scribn. and Merr., U. S. Dept. Agr., Div. Agrost. Cir. 30: 7. 1901. Benton County, Ark., *Plank* 38.
Merathrepta pinetorum Piper, U. S. Natl. Herb. Contrib. 11: 122. 1906. Based on *Danthonia spicata pinetorum* Piper.
Merathrepta thermale Heller, Muhlenbergia 5: 120. 1909. Based on *Danthonia thermale* Scribn.

Merathrepta thermale var. *pinetorum* Piper ex Fedde and Schust., in Just's Bot. Jahresber. 37: 128. 1911 (erroneously ascribed to Heller, Muhlenbergia 5: 120. 1909).
Pentameris spicata Nels. and Macbr., Bot. Gaz. 56: 470. 1913. Based on *Avena spicata* L.
Pentameris thermale Nels. and Macbr., Bot. Gaz. 56: 470. 1913. Based on *Danthonia thermale* Scribn.
Danthonia pinetorum Piper in Piper and Beattie, Fl. Northw. Coast 46. 1915. Based on *D. spicata pinetorum* Piper.
Danthonia spicata var. *typica* Fernald, Rhodora 45: 242. 1943. Based on *Avena spicata* L.
(7) **Danthonia unispicata** (Thurb.) Munro ex Macoun, Can. Pl. Cat. 2⁴: 215. 1888. Based on *D. californica* var. *unispicata* Thurb. The name was earlier listed without description as follows: Thurb. in A. Gray, Proc. Acad. Phila., 1863: 78. 1863, name only, for *Geyer* 189. Thurb. in S. Wats., Bot. Calif. 2: 294. 1880, as synonym of *D. californica* var. *unispicata* Thurb. Munro; Vasey, Descr. Cat. Grasses U. S. 59. 1885. Name only.
Danthonia californica var. *unispicata* Thurb. in S. Wats., Bot. Calif. 2: 294. 1880. San Diego to San Francisco, Calif., *Bolander, Parry, Lemmon*.
Merathrepta unispicata Piper, U. S. Natl. Herb. Contrib. 11: 123. 1906. Based on *Danthonia unispicata* Munro.
Pentameris unispicata Nels. and Macbr., Bot. Gaz. 56: 470. 1913. Based on *Danthonia unispicata* Munro.

(58) **DESCHAMPSIA Beauv.**

(4) **Deschampsia atropurpurea** (Wahl.) Scheele, Flora 27: 56. 1844. Based on *Aira atropurpurea* Wahl.
Aira atropurpurea Wahl., Fl. Lapp. 37. 1812. Lapland.
Holcus atropurpureus Wahl., Svensk Bot. pl. 677. 1826–29. Based on *Aira atropurpurea* Wahl.
Avena atropurpurea Link, Hort. Berol. 1: 119. 1827. Based on *Aira atropurpurea* Wahl.
Aira latifolia Hook., Fl. Bor. Amer. 2: 243. pl. 227. 1840. Rocky Mountains, *Drummond*.
Vahlodea atropurpurea Fries, Bot. Not. 178. 1842. Presumably based on *Aira atropurpurea* Wahl.
Deschampsia latifolia Vasey, Grasses U. S. 29. 1883. Not *D. latifolia* Hochst., 1851. Based on *Aira latifolia* Hook.
Deschampsia hookeriana Scribn., Bot. Gaz. 11: 97. 1886. Based on *Aira latifolia* Hook.
Deschampsia atropurpurea var. *minor* Vasey, Torrey Bot. Club Bul. 15: 48.

1888. Vancouver Island, *Macoun* in 1887.
Deschampsia atropurpurea var. *latifolia* Scribn. ex Macoun, Cat. Can. Pl. 2⁴: 209. 1888. Based on *Aira latifolia* Hook.
Vahlodea latifolia Hultén, Fl. Aleut. Isl. 83. 1937. Based on *Aira latifolia* Hook.

(6) **Deschampsia caespitosa** (L.) Beauv., Ess. Agrost. 91, 149, 160. pl. 18. f. 3. 1812. Based on *Aira caespitosa* L.
Aira caespitosa L., Sp. Pl. 64. 1753. Europe.
Agrostis caespitosa Salisb., Prodr. Stirp. 25. 1796. Based on *Aira caespitosa* L.
Aira ambigua Michx., Fl. Bor. Amer. 1: 61. 1803. Canada, *Michaux*.
Aira caespitosa var. *ambigua* Pursh, Fl. Amer. Sept. 1: 77. 1814. Based on *A. ambigua* Michx.
Aira cespitosa Muhl., Descr. Gram. 85. 1817. Pennsylvania; New England.
Aira aristulata Torr., Fl. North. and Mid. U. S. 1: 132. 1823. New York, *Cooper*.
Campbella caespitosa Link, Hort. Berol. 1: 122. 1827. Based on *Aira caespitosa* L.
Aira caespitosa var. *genuina* Reichenb., Icon. 1: pl. 96. f. 1682. 1834. Based on *A. caespitosa* L.
Podionapus caespitosus Dulac, Fl. Haut. Pyr. 82. 1867. Based on *Deschampsia caespitosa* Beauv.
Avena caespitosa Kuntze, Taschenfl. Leipzig 45. 1867. Based on *Aira caespitosa* L.
Aira major subsp. *caespitosa* Syme in Sowerby, English Bot. ed. 3. 11: 64. 1873. Based on *A. caespitosa* L.
Aira caespitosa var. *montana* Vasey in Rothr. in Wheeler, U. S. Survey, W. 100th Merid. Rpt. 6: 294. 1878. Not *A. caespitosa* var. *montana* Reichenb., 1850. Utah, Colorado, and Arizona.
Deschampsia caespitosa var. *maritima* Vasey, Torrey Bot. Club Bul. 15: 48. 1888. Vancouver Island, *Macoun* in 1887.
Deschampsia ambigua Beauv. ex Jacks., Ind. Kew. 1: 735. 1893. Name only, presumably referring to *Aira ambigua* Michx.
Deschampsia caespitosa var. *alpina* Vasey ex Beal, Grasses N. Amer. 2: 369. 1896. Not *D. caespitosa* var. *alpina* Ducomm., 1869. Alaska, *Elliott;* Colorado, *Letterman*.
Deschampsia caespitosa var. *confinis* Vasey ex Beal, Grasses N. Amer. 2: 369. 1896. Southern California, *Palmer 231* in 1888.
Deschampsia caespitosa var. *longiflora* Beal, Grasses N. Amer. 2: 369. 1896. Vancouver Island, *Macoun* in 1887.
Deschampsia alpicola Rydb., Torrey Bot. Club Bul. 32: 601. 1905. Based on *D. caespitosa* var. *alpina* Vasey.

Deschampsia confinis Rydb., Torrey Bot. Club Bul. 36: 533. 1909. Based on *D. caespitosa* var. *confinis* Vasey.
Deschampsia pungens Rydb., Torrey Bot. Club Bul. 39: 103. 1912. Banff, Alberta, *McCalla 2309.*
Deschampsia caespitosa var. *genuina* Volk, Bot. Jahrb. 47: 312. 1912. Based on *Aira caespitosa* var. *genuina* Reichenb.
Aira alpicola Rydb., Fl. Rocky Mount. ed. 2. 1112. 1922. Based on *Deschampsia alpicola* Rydb.
Deschampsia caespitosa subsp. *genuina* W. E. Lawr., Amer. Jour. Bot. 32: 302. 1945. Based on *Aira caespitosa* var. *genuina* Reichenb.

The following names based on *Deschampsia brevifolia* R. Br. (Sup. App. Parry's Voy. 191. 1821) described from Melville Island, Arctic America, and not known from the United States, have been misapplied to *D. caespitosa* by various American authors:
Aira arctica Spreng., Syst. Veg. 4: Cur. Post. 32. 1827. Based on *Deschampsia brevifolia* R. Br.
Aira caespitosa var. *arctica* Thurb. ex A. Gray, Acad. Nat. Sci. Phila. Proc. 1863: 78. 1863. Based on *Deschampsia brevifolia* R. Br.
Deschampsia brachyphylla Nash in Rydb., N. Y. Bot. Gard. Mem. 1: 37. 1900. Not *D. brachyphylla* Phil., 1896. Based on *D. brevifolia* R. Br.
Deschampsia curtifolia Scribn., U. S. Dept. Agr., Div. Agrost. Cir. 30: 7. 1901. Based on *D. brachyphylla* Nash.
Deschampsia arctica Merr., Rhodora 4: 143. 1902. Based on *Aira arctica* Spreng.
Aira curtifolia Rydb., Fl. Rocky Mount. ed. 2. 1112. 1922. Based on *Deschampsia curtifolia* Scribn.

Other names based on Old World species were misapplied to *Deschampsia caespitosa* by Beal:
Deschampsia caespitosa var. *bottnica* Vasey ex Beal, Grasses N. Amer. 2: 369. 1896. Based on *Aira bottnica* Wahl.
Deschampsia caespitosa var. *brevifolia* Vasey ex Beal, Grasses N. Amer. 2: 369. 1896. Based on *Aira brevifolia* Bieb.
Deschampsia caespitosa var. *montana* Vasey ex Beal, Grasses N. Amer. 2: 369. 1896. Based on *D. montana* Schur.

The following names applied to various collections of *D. caespitosa* are based on European types. Some of these collections agree fairly well, some do not, with European specimens distributed under these names:
Deschampsia caespitosa var. *glauca* (Hartm.) Lindm., Svensk. Fanerogampl. 81. 1918. Not *D. caespitosa* var. *glauca* Regel, 1881. Based on *D. glauca* Hartm.
Deschampsia glauca Hartm., Handb. Skand. Fl. 448. 1820.

Deschampsia caespitosa var. *littoralis* (Gaudin) Richt., Pl. Eur. 1: 56. 1890. Based on *Aira caespitosa* var. *littoralis* Gaudin, 1828.

DESCHAMPSIA CAESPITOSA var. PARVIFLORA (Thuill.) Coss. and Germ., Fl. Env. Paris ed. 2. 806. 1861. Based on *Aira parviflora* Thuill.

Aira parviflora Thuill., Fl. Env. Paris ed. 2. 1: 38. 1799. Paris.

(3) **Deschampsia congestiformis** Booth, Rhodora 45: 414. 1943. Gallatin Valley, Bozeman, Mont., *Hawkins* in 1903.

(1) **Deschampsia danthonioides** (Trin.) Munro ex Benth., Pl. Hartw. 342. 1857. Based on *Aira danthonioides* Trin.

Aira danthonioides Trin., Mém. Acad. St. Pétersb. VI. Math. Phys. Nat. 1: 57. 1830. Western North America.

Deschampsia calycina Presl, Rel. Haenk. 1: 251. 1830. "Peru" is the published locality, but the type specimen is labeled Monterey, Calif., *Haenke.*

Aira calycina Steud., Syn. Pl. Glum. 1: 220. 1854. Based on *Deschampsia calycina* Presl.

Trisetum glabrum Buckl., Acad. Nat. Sci. Phila. Proc. 1862: 100. 1862. "Texas *Dr. Linsecum.*" [Locality probably erroneous, the plants bearing this ticket in the herbarium of the Academy of Sciences, Philadelphia, being very like two on the same sheet labeled "Rocky Mountains of Columbia, *Nuttall.*" The species is not otherwise known east of Arizona.]

Deschampsia gracilis Vasey, Bot. Gaz. 10: 224. 1885. San Diego, Calif., *Orcutt* [1072].

Deschampsia danthonioides var. *gracilis* Munz, Man. South Calif. Bot. 45: 597. 1935. Based on *Deschampsia gracilis* Vasey.

(2) **Deschampsia elongata** (Hook.) Munro ex Benth., Pl. Hartw. 342. 1857. Based on *Aira elongata* Hook.

Aira elongata Hook., Fl. Bor. Amer. 2: 243. pl. 228. 1840. Columbia River, *Douglas.*

Deschampsia elongata var. *ciliata* Vasey ex Beal, Grasses N. Amer. 2: 371. 1896. Oregon, *Howell;* California, *Anderson* [Santa Cruz, type].

Deschampsia elongata var. *tenuis* Vasey ex Beal, Grasses N. Amer. 2: 372. 1896. Santa Cruz, Calif., *Jones* 2201. Published as new in Jepson, Fl. West. Mid. Calif. 51. 1901. Evergreen, Santa Clara County, Calif.

Deschampsia ciliata Rydb., Fl. Rocky Mount. 60. 1917. Based on *D. elongata* var. *ciliata* Vasey.

Aira vaseyana Rydb., Fl. Rocky Mount. ed. 2. 1112. 1922. Based on *Deschampsia elongata* var. *ciliata* Vasey.

(5) **Deschampsia flexuosa** (L.) Trin., Acad.

St. Pétersb. Mém. VI. Sci. Nat. 2¹: 9. 1836. Based on *Aira flexuosa* L.

Aira flexuosa L., Sp. Pl. 65. 1753. Europe.

Avena flexuosa Mert. and Koch in Roehl, Deut. Fl. ed. 3. 1²: 570. 1823. Based on *Aira flexuosa* L.

Avenella flexuosa Parl., Fl. Ital. 1: 246. 1848. Based on *Aira flexuosa* L.

Lerchenfeldia flexuosa Schur, Enum. Pl. Transsilv. 753. 1866. Based on *Aira flexuosa* L.

Podionapus flexuosus Dulac, Fl. Haut. Pyr. 83. 1867. Based on *Deschampsia flexuosa* Trin.

Salmasia flexuosa Bubani, Fl. Pyr. 4: 319. 1901. Based on *Aira flexuosa* L.

(7) **Deschampsia holciformis** Presl, Rel. Haenk. 1: 251. 1830. Monterey, Calif., *Haenke.*

Aira holciformis Steud., Syn. Pl. Glum. 1: 221. 1854. Based on *Deschampsia holciformis* Presl.

Deschampsia caespitosa subsp. *holciformis* W. E. Lawr., Amer. Jour. Bot. 32: 302. 1945. Based on *D. holciformis* Presl.

Desmazeria sicula (Jacq.) Dum., Comm. Bot. 27. 1822. Based on *Cynosurus siculus* Jacq. The generic name spelled *Demazeria;* later (Obs. Gram. Belg. 46. 1823) corrected to *Desmazeria* by Dumortier.

Cynosurus siculus Jacq., Obs. Bot. 2: 22. pl. 43. 1767. Europe.

(17) DIARRHENA Beauv.

(1) **Diarrhena americana** Beauv., Ess. Agrost. 142. pl. 25. f. 2. 1812. Based on *Festuca diandra* Michx.

Festuca diandra Michx., Fl. Bor. Amer. 1: 67. 1803. Not *F. diandra* Moench., 1794. "Kentucky, Tennessee, etc." *Michaux.*

Diarina festucoides Raf., Med. Repos. N. Y. 5: 352. 1808. Not *Diarrhena festucoides* Raspail, 1825. Based on *Festuca diandra* Michx.

Festuca americana Michx. ex Beauv., Ess. Agrost. 162. 1812. Name only.

Korycarpus arundinaceus Zea ex Lag., Gen. et Sp. Nov. 4. 1816. America.

Roemeria zeae Roem. and Schult., Syst. Veg. 1: 61, 287. 1817. Source unknown.

Diarina sylvatica Raf., Jour. Phys. Chym. 89: 104. 1819. Based on *Festuca diandra* Michx.

Diarrhena diandra Wood, Class-book ed. 2. 612. 1847. Based on *Festuca diandra* Michx.

Corycarpus diandrus Kuntze, Rev. Gen. Pl. 2: 772. 1891. Based on *Festuca diandra* Michx.

Diarrhena festucoides Fernald, Rhodora 34: 204. 1932. Not *D. festucoides* Raspail, 1825. Based on *Diarina festucoides* Raf.

Diarrhena arundinacea Rydb., Fl. Prairie and Plains Centr. N. Amer. 114. 1932. Based on *Korycarpus arundinaceus* Zea.

(129) DIGITARIA Heister

(17) **Digitaria albicoma** Swallen, Wash. Acad. Sci. Jour. 30: 214. f. 3. 1940. Brooksville, Fla., *Swallen* 5644.

Digitaria decumbens Stent, Bothalia 3: 150. 1930. Transvaal, South Africa, *Pentz*, Nat. Herb. Pretoria 8485.

(11) **Digitaria dolichophylla** Henr., Blumea 1: 94. 1934. Florida, *A. A. Eaton* 459.

(9) **Digitaria filiformis** (L.) Koel., Descr. Gram. 26. 1802. Based on *Panicum filiforme* L.

Panicum filiforme L., Sp. Pl. 57. 1753. North America, *Kalm*.

Paspalum filiforme Flügge, Monogr. Pasp. 139. 1810. Not *P. filiforme* Swartz, 1788. Based on *Panicum filiforme* L.

Paspalum furcatum var. *filiforme* Doell in Mart., Fl. Bras. 2²: 104. 1877. Based on *Digitaria filiformis* Muhl. (the same as Flügge) but misapplied to a species of *Axonopus*.

Syntherisma filiformis Nash, Torrey Bot. Club Bul. 22: 420. 1895. Based on *Panicum filiforme* L.

Digitaria laeviglumis Fernald, Rhodora 22: 102. 1920. Manchester, N. H., *Batchelder*.

(4) **Digitaria floridana** Hitchc., Biol. Soc. Wash. Proc. 41: 163. 1928. Hernando County, Fla., *Hitchcock* Fla. Pl. 2517.

Syntherisma floridanum Hitchc. in Small, Man. Southeast. Fl. 51. 1933. Hernando County, Fla.

(12) **Digitaria gracillima** (Scribn.) Fernald, Rhodora 22: 101. 1920. Based on *Panicum gracillimum* Scribn.

Panicum gracillimum Scribn., Torrey Bot. Club Bul. 23: 146. 1896. Eustis, Fla., *Nash* 1192.

Syntherisma gracillima Nash, Torrey Bot. Club Bul. 25: 295. 1898. Based on *Panicum gracillimum* Scribn.

Syntherisma bakeri Nash, Torrey Bot. Club Bul. 25: 296. 1898. Grasmere, Fla., *C. H. Baker* 47.

Digitaria bakeri Fernald, Rhodora 22: 102. 1920. Based on *Syntherisma bakeri* Nash.

(2) **Digitaria horizontalis** Willd., Enum. Pl. 92. 1809. Dominican Republic.

Milium digitatum Swartz, Prodr. Veg. Ind. Occ. 24. 1788. Not *Digitaria digitata* Buse, 1854. Jamaica.

Agrostis digitata Poir. in Lam., Encycl. Sup. 1: 258. 1810. Not *A. digitata* Lam., 1783. Based on *Milium digitatum* Swartz.

Axonopus digitatus Beauv., Ess. Agrost. 12, 154. 1812. Based on *Milium digi-*

tatum Swartz.

Panicum horizontale G. Meyer, Prim. Fl. Esseq. 54. 1818. Based on *Digitaria horizontalis* Willd.

Digitaria jamaicensis Spreng., Syst. Veg. 1: 272. 1825. Jamaica.

Digitaria setosa Desv. ex Hamilt., Prodr. Pl. Ind. Occ. 6. 1825. West Indies.

Paspalum digitatum Kunth, Rév. Gram. 1: 24. 1829. Based on *Milium digitatum* Swartz.

Panicum hamiltonii Kunth, Rév. Gram. 1: Sup. 9. 1830. Based on *Digitaria setosa* Desv.

Syntherisma setosum Nash, Torrey Bot. Club Bul. 25: 300. 1898. Based on *Digitaria setosa* Desv.

Digitaria sanguinalis var. *horizontalis* Rendle, Cat. Afr. Pl. 2: 163. 1899. Based on *D. horizontalis* Willd.

Panicum sanguinale var. *digitatum* Hack. ex Urban, Symb. Antill. 4: 86. 1903. Based on *Milium digitatum* Swartz.

Panicum sanguinale subsp. *horizontale* Hack., Ergeb. Bot. Exped. Akad. Wiss. Südbras. 8. 1906; Denkschr. Akad. Wiss. Math. Naturw. (Wien) 79: 69. 1908. Based on *Digitaria horizontalis* Willd.

Syntherisma digitatum Hitchc., U. S. Natl. Herb. Contrib. 12: 142. 1908. Based on *Milium digitatum* Swartz.

Digitaria digitata Urban, Symb. Antill. 8: 24: 1920. Not *D. digitata* Buse, 1854. Based on *Milium digitatum* Swartz.

(3) **Digitaria ischaemum** (Schreb.) Schreb. ex Muhl., Descr. Gram. 131. 1817. Presumably based on *Panicum ischaemum* Schreb. Name only, Muhl., Cat. Pl. 9. 1813.

Panicum ischaemum Schreb. in Schweigger, Spec. Fl. Erland. 16. 1804. Germany.

Digitaria humifusa Pers., Syn. Pl. 1: 85. 1805. France.

Syntherisma glabrum Schrad., Fl. Germ. 1: 163. pl. 3. f. 6. 1806. Germany.

Panicum glabrum Gaudin, Agrost. Helv. 1: 22. 1811. Based on *Syntherisma glabrum* Schrad. (In Index Kewensis "Ell." is erroneously given as author of *P. glabrum*.)

Digitaria glabra Beauv., Ess. Agrost. 51. 1812. Presumably based on *Syntherisma glabrum* Schrad.

Paspalum humifusum Poir. in Lam., Encycl. Sup. 4: 316. 1816. Based on *Digitaria humifusa* Pers.

Panicum humifusum Kunth, Rév. Gram. 1: 33. 1829. Based on *Digitaria humifusa* Pers.

Panicum phaeocarpum var. *drummondianum* Nees, Fl. Afr. Austr. 22. 1841. St. Louis, Mo., *Drummond*.

Paspalum glabrum Wood, Amer. Bot. and Flor. pt. 2: 390. 1871. Not *P. glabrum* Poir., 1804. "(Gaud.)," given in parentheses by Wood, doubtless refers to *Panicum glabrum* Gaudin.

Paspalum glabrum Cassidy, Colo. Agr. Expt. Sta. Bul. 12: 91. 1890. Not *P. glabrum* Poir., 1804. Colorado.

Syntherisma humifusum Rydb., N. Y. Bot. Gard. Mem. 1: 469. 1900. Based on *Digitaria humifusa* Pers.

Syntherisma ischaemum Nash, N. Amer. Fl. 17: 151. 1912. Based on *Panicum ischaemum* Schreb.

The name *Panicum lineare* L. (*Syntherisma lineare* Nash) has been used for *Digitaria ischaemum*, but the description does not apply (e.g. "calycis squama exterior brevior, patens, rachi adhaerens"). It is probably *Cynodon dactylon*.

DIGITARIA ISCHAEMUM var. MISSISSIPPIENSIS (Gattinger) Fernald, Rhodora 22: 103. 1920. Based on *Panicum glabrum* var. *mississippiense* Gattinger.

Panicum glabrum var. *mississippiense* Gattinger, Tenn. Fl. 95. 1887, name only, Nashville. Scribn., Tenn. Agr. Expt. Sta. Bul. 7: 39. 1894. Knoxville, Tenn.

Panicum lineare var. *mississippiense* Gattinger ex Beal, Grasses N. Amer. 2: 111. 1896. Presumably based on *P. glabrum* var. *mississippiense* Gattinger.

Syntherisma linearis mississippiensis Nash, Torrey Bot. Club. Bul. 25: 300. 1898. Based on *Panicum glabrum* var. *mississippiense* Gattinger.

(7) **Digitaria longiflora** (Retz.) Pers., Syn. Pl. 1: 85. 1805. Based on *Paspalum longiflorum* Retz.

Paspalum longiflorum Retz., Obs. Bot. 4: 15. 1786. India.

Panicum longiflorum Gmel., Syst. Nat. 2: 158. 1791. Presumably based on *Paspalum longiflorum* Retz.

Syntherisma longiflora Skeels, U. S. Dept. Agr., Bur. Plant Indus. Bul. 261: 30. 1912. Based on *Paspalum longiflorum* Retz.

(15) **Digitaria pauciflora** Hitchc., Biol. Soc. Wash. Proc. 41: 162. 1928. Southern Florida, *Eaton* 207.

Syntherisma pauciflorum Hitchc. in Small, Man. Southeast. Fl. 51. 1933. Southern Florida. [*Eaton* 207.]

Digitaria pentzii Stent, Bothalia 3: 147. 1930. Cape Province, South Africa. *Pentz* 8510.

(14) **Digitaria runyoni** Hitchc., Wash. Acad. Sci. Jour. 23: 455. 1933. Mouth of Rio Grande, near Brownsville, Tex., *Runyon* 188.

(1) **Digitaria sanguinalis** (L.) Scop., Fl. Carn. ed. 1: 52. 1772. Based on *Panicum sanguinale* L.

Panicum sanguinale L., Sp. Pl. 57. 1753. America and southern Europe.

Dactylon sanguinalis Vill., Hist. Pl. Dauph. 2: 69. 1787. Based on *Panicum sanguinale* L.

Syntherisma praecox Walt., Fl. Carol. 76. 1788. South Carolina.

Paspalum sanguinale Lam., Tabl. Encycl. 1: 176. 1791. Based on *Panicum sanguinale* L.

Digitaria praecox Willd., Enum. Pl. 91. 1809. Based on *Syntherisma praecox* Walt.

Panicum adscendens H. B. K., Nov. Gen. et Sp. 1: 97. 1815. Venezuela, Peru, and Mexico, *Humboldt* and *Bonpland*.

Cynodon praecox Roem. and Schult., Syst. Veg. 2: 412. 1817. Based on *Syntherisma praecox* Walt.

Digitaria marginata Link, Enum. Pl. 1: 102. 1821. Brazil.

Digitaria fimbriata Link, Hort. Berol. 1: 226. 1827. Brazil.

Panicum fimbriatum Kunth, Rév. Gram. 1: 33. 1829. Based on *Digitaria fimbriata* Link.

Panicum linkianum Kunth, Rév. Gram. 1: 33. 1829. Based on *Digitaria marginata* Link.

Syntherisma sanguinalis Dulac, Fl. Haut. Pyr. 77. 1867. Based on *Panicum sanguinale* L.

Syntherisma fimbriatum Nash, Torrey Bot. Club Bul. 25: 302. 1898. Based on *Digitaria fimbriata* Link.

Syntherisma marginatum Nash, N. Amer. Fl. 17: 154. 1912. Based on *Digitaria marginata* Link.

Digitaria marginata var. *fimbriata* Stapf in Prain, Fl. Trop. Afr. 9: 440. 1919. Based on *D. fimbriata* Link.

Panicum sanguinale subsp. *marginatum* Thell., Vierteljahrs. Nat. Ges. Zürich 64: 699. 1919. Based on *Digitaria marginata* Link.

Digitaria sanguinalis var. *marginata* Fernald, Rhodora 22: 103. 1920. Based on *D. marginata* Link.

Digitaria adscendens Henr., Blumea 1: 92. 1934. Based on *Panicum adscendens* H. B. K.

Digitaria nealleyi Henr., Blumea 1: 94. 1934. Texas, *Nealley* in 1884. A duplicate of the type in the National Herbarium is distorted by a fungus. Described as *Syntherisma barbata* (Willd.) Nash in Small's Flora.

DIGITARIA SANGUINALIS var. CILIARIS (Retz.) Parl., Fl. Ital. 1: 126. 1848. Based on *Panicum ciliare* Retz.

Panicum ciliare Retz. Obs. Bot. 4: 16. 1786. Asia.

Digitaria sanguinalis subsp. *ciliaris* Domin, Preslia 13/15: 47. 1935. Based on *Panicum ciliare* Retz.

(6) **Digitaria serotina** (Walt.) Michx., Fl. Bor. Amer. 1: 46. 1803. Based on *Syntherisma serotinum* Walt.

Syntherisma serotinum Walt., Fl. Carol. 76. 1788. South Carolina.

Paspalum serotinum Flügge, Monogr. Pasp. 145. 1810. Based on *Digitaria serotina* Michx.

(8) **Digitaria simpsoni** (Vasey) Fernald, Rhodora 22: 103. 1920. Based on *Panicum sanguinale* var. *simpsoni* Vasey.

Panicum sanguinale var. *simpsoni* Vasey, U. S. Natl. Herb. Contrib. 3: 25. 1892. Manatee, Fla., *Simpson.*

Panicum simpsoni Beal, Grasses N. Amer. 2: 109. 1896. Based on *Panicum sanguinale* var. *simpsoni* Vasey.

Syntherisma simpsoni Nash, Torrey Bot. Club Bul. 25: 297. 1898. Based on *Panicum sanguinale* var. *simpsoni* Vasey.

Digitaria swazilandensis Stent, Bothalia 3: 156. 1930. Swaziland, Africa.

(16) **Digitaria subcalva** Hitchc., Amer. Jour. Bot. 21: 138. f. 4. 1934. Plant City, Fla., *C. P. Wright.*

(13) **Digitaria texana** Hitchc., Biol. Soc. Wash. Proc. 41: 162. 1928. Sarita, Tex., *Hitchcock 5479.*

(10) **Digitaria villosa** (Walt.) Pers., Syn. Pl. 1: 85. 1805. Based on *Syntherisma villosa* Walt.

Syntherisma villosa Walt., Fl. Carol. 77. 1788. South Carolina.

Digitaria pilosa Michx., Fl. Bor. Amer. 1: 45. 1803. Carolina and Georgia, *Michaux.* Willdenow (Enum. Pl. 1: 91. 1809) uses this name, doubtfully citing *D. pilosa* Michx. The description suggests that Willdenow's plant, from Carolina, is also *D. villosa.*

Paspalum carolinianum Poir. in Lam., Encycl. Sup. 4: 311. 1816. Carolina and Georgia, *Bosc.*

Syntherisma leucocoma Nash, Torrey Bot. Club Bul. 25: 295. 1898. Lake Ella, Fla., *Nash 1155.*

Panicum leucocomum Scribn., U. S. Dept. Agr., Div. Agrost. Bul. 7. (ed. 2): 58. 1898. Based on *Syntherisma leucocoma* Nash.

Digitaria leucocoma Urban, Symb. Antill. 8: 24. 1920. Based on *Syntherisma leucocoma* Nash.

Digitaria filiformis var. *villosa* Fernald, Rhodora 36: 19. 1934. Based on *Syntherisma villosa* Walt.

(5) **Digitaria violascens** Link, Hort. Berol. 1: 229. 1827. Brazil.

Panicum violascens Kunth, Rév. Gram. 1: 33. 1829. Based on *Digitaria violascens* Link.

Paspalum chinense Nees in Hook. and Arn., Bot. Beechey Voy. 231. 1836. Macao, China.

Syntherisma chinensis Hitchc., U. S. Natl. Herb. Contrib. 22: 468. 1922. Based on *Paspalum chinense* Nees.

Digitaria chinensis A. Camus, Not. Syst. Lecomte 4: 48. 1923. Not *D. chinensis* Hornem., 1819. Based on *Paspalum chinense* Nees.

(18) **DISSANTHELIUM Trin.**

(1) **Dissanthelium californicum** (Nutt.) Benth. in Hook. f., Icon. Pl. III. 4: 56. pl. 1375. 1881. Based on *Stenochloa californica* Nutt.

Stenochloa californica Nutt., Jour. Acad. Sci. Phila. II. 1: 189. 1848. Santa Catalina Island, Calif., *Gambel.*

(21) **DISTICHLIS Raf.**

(1) **Distichlis spicata** (L.) Greene, Calif. Acad. Sci. Bul. 2: 415. 1887. Based on *Uniola spicata* L.

Uniola spicata L., Sp. Pl. 71. 1753. Atlantic coast of North America.

Briza spicata Lam., Encycl. 1: 465. 1785. Based on *Uniola spicata* L.

?Festuca multiflora Walt., Fl. Carol. 81. 1788. South Carolina.

Festuca triticoides Lam., Tabl. Encycl. 1: 191. 1791. Carolina, *Fraser.*

Festuca distichophylla Michx., Fl. Bor. Amer. 1: 67. 1803. Carolina, *Michaux.*

Uniola distichophylla Roem. and Schult., Syst. Veg. 2: 596. 1817. Based on *Festuca distichophylla* Michx.

Distichlis maritima Raf., Jour. Phys. Chym. 89: 104. 1819. Based on *Uniola spicata* L.

Distichlis nodosa Raf., Jour. Phys. Chym. 89: 104. 1819. Based on *Festuca distichophylla* Michx.

Brizopyrum americanum Link, Hort. Berol. 1: 160. 1827. Based on *Uniola spicata* L.

Poa michauxii Kunth, Rév. Gram. 1: 111. 1829. Based on *Festuca distichophylla* Michx.

Brizopyrum boreale Presl, Rel. Haenk. 1: 280. 1830. Nootka Sound, Vancouver Island, *Haenke.*

Poa borealis Kunth, Rév. Gram. 1: Sup. 28. 1830. Based on *Brizopyrum boreale* Presl.

Festuca triticea Lam. ex Kunth, Enum. Pl. 1: 325. 1833, as synonym of *Poa michauxii* Kunth. (Probably error for *F. triticoides* Lam.)

Brizopyrum spicatum Hook. and Arn., Bot. Beechey Voy. 403. 1841. Based on *Uniola spicata* L.

Distichlis spicata var. *borealis* Beetle, Torrey Bot. Club Bul. 70: 643. f. 1. 1943. Based on *Brizopyrum boreale* Presl.

Distichlis spicata var. *stolonifera* Beetle, Torrey Bot. Club Bul. 70: 644. f. 4, 7, 12. 1943. Ferndale, Humboldt County, Calif., *Davy* and *Blasdale 6202* (pistillate); Arcata, Humboldt County, Calif., *Davy* and *Blasdale 5604* (staminate).

Distichlis spicata var. *divaricata* Beetle, Torrey Bot. Club Bul. 70:,647. f. 10. 1943. Salton, Calif., *Davy* in 1902.

Distichlis spicata var. *distichophylla* Beetle, Rhodora 47: 148. 1945. Based on

Uniola distichophylla Roem. and Schult., this based on *Festuca distichophylla* Michx.

DISTICHLIS SPICATA var. NANA Beetle, Torrey Bot. Club Bul. 70: 647. f. 3, 9. 1943. Whitaker Forest, Tulare County, Calif., *Kennedy* in 1928.

(2) **Distichlis stricta** (Torr.) Rydb., Torrey Bot. Club Bul. 32: 602. 1905. Based on *Uniola stricta* Torr.

Festuca spicata Nutt., Gen. Pl. 1: 72. 1818. Not *F. spicata* Pursh, 1814. "On the banks of the Missouri."

Uniola stricta Torr., Ann. Lyc. N. Y. 1: 155. 1824. Canadian River [Okla.].

Uniola multiflora Nutt., Amer. Phil. Soc. Trans. (n.s.) 5: 148. 1837. Arkansas River, *Nuttall*.

Uniola flexuosa Buckl., Acad. Nat. Sci. Phila. Proc. 1862: 99. 1862. Fort Belknap, Tex., *Buckley*.

Brizopyrum spicatum var. *strictum* A. Gray ex S. Wats., in King, Geol. Expl. 40th Par. 5: 385. 1871. Based on *Uniola stricta* Torr.

Distichlis maritima var. *stricta* Thurb., in S. Wats., Bot. Calif. 2: 306. 1880. Based on *Uniola stricta* Torr.

Distichlis spicata stricta Scribn., Torrey Bot. Club Mem. 5: 51. 1894. Based on *Uniola stricta* Torr.

Distichlis spicata var. *laxa* Vasey ex Beal, Grasses N. Amer. 2: 519. 1896. Lake Park, Utah, *Tracy* in 1887.

Distichlis dentata Rydb., Torrey Bot. Club Bul. 36: 536. 1909. Washington, *Sandberg* and *Leiberg* 463. (Pistillate plant.)

Distichlis stricta var. *laxa* Fawcett and West ex Munz, Man. South. Calif. Bot. 52, 597. 1935. Based on *D. spicata* var. *laxa* Vasey.

Distichlis spicata var. *stricta* Beetle, Torrey Bot. Club Bul. 70: 645. f. 2, 6, 11, 13. 1943. Based on *Brizopyrum spicatum* var. *strictum* A. Gray; S. Wats.," this based on *Uniola stricta* Torr.

(3) **Distichlis texana** (Vasey) Scribn., U. S. Dept. Agr., Div. Agrost. Cir. 16: 2. 1899. Based on *Poa texana* Vasey.

Poa texana Vasey, U. S. Natl. Herb. Contrib. 1: 60. 1890. Region of Rio Grande, Tex., *Nealley*.

Sieglingia wrightii Vasey, U. S. Natl. Herb. Contrib. 1: 269. 1893. Valley of the Limpio, Tex., *Wright* 2038.

(141) ECHINOCHLOA Beauv.

(2) **Echinochloa colonum** (L.) Link, Hort. Berol. 2: 209. 1833. Based on *Panicum colonum* L.

Panicum colonum L., Syst. Nat. ed. 10. 2: 870. 1759. Jamaica, *Browne*.

Milium colonum Moench, Meth. Pl. 202. 1794. Based on *Panicum colonum* L.

Oplismenus colonum H. B. K., Nov. Gen. et Sp. 1: 108. 1815. Based on *Panicum colonum* L.

Panicum zonale Guss., Fl. Sic. Prodr. 1: 62. 1827. Sicily.

Oplismenus repens Presl, Rel. Haenk. 1: 321. 1830. Mexico, *Haenke*.

Oplismenus colonum var. *zonalis* Schrad., Linnaea 12: 429. 1838. Based on *Panicum zonale* Guss.

Panicum incertum Bosc ex Steud., Nom. Bot. ed. 2. 2: 258. 1841. Name only. Carolina.

Echinochloa zonalis Parl., Fl. Panorm. 1: 119. 1845. Based on *Panicum zonale* Guss.

Panicum prorepens Steud., Syn. Pl. Glum. 1: 46. 1854. Based on *Oplismenus repens* Presl.

Oplismenus crusgalli var. *colonum* Coss. and Dur., Expl. Sci. Alger. 2: 28. 1854. Based on *Panicum colonum* L.

Panicum colonum var. *zonale* L. H. Dewey, U. S. Natl. Herb. Contrib. 2: 502. 1894. Based on *P. zonale* Guss.

Echinochloa colonum var. *zonalis* Woot. and Standl., N. Mex. Col. Agr. Bul. 81: 45. 1912. Based on *Panicum zonale* Guss.

Echinochloa crusgalli subsp. *colonum* Honda, Bot. Mag. [Tokyo] 37: 122. 1923. Based on *Panicum colonum* L.

Panicum crusgalli subsp. *colonum* Makino and Nemoto, Fl. Jap. 1470. 1925. Based on *P. colonum* L.

(3) **Echinochloa crusgalli** (L.) Beauv., Ess. Agrost. 53, 161. 1812. Based on *Panicum crusgalli* L.

Panicum crusgalli L., Sp. Pl. 56. 1753. Europe; Virginia.

Milium crusgalli Moench, Meth. Pl. 202. 1794. Based on *Panicum crusgalli* L.

Panicum grossum Salisb., Prod. Stirp. 18. 1796. Based on *P. crusgalli* L.

Panicum muricatum Michx., Fl. Bor. Amer. 1: 47. 1803. Not *P. muricatum* Retz., 1786. Canada, Lake Champlain [type] and Lake Ontario, *Michaux*.

?*Panicum echinatum* Willd., Enum. Pl. 1032. 1809. "America meridionali." Wiegand (Rhodora 23: 60. 1921) takes up this name for *Echinochloa crus-pavonis*. The specimen in the Willdenow Herbarium named *P. echinatum* (Magdalena, Colombia, *Humboldt*) is *Pseudechinolaena polystachya* (H. B. K.) Stapf. The brief description does not apply to the specimen so named nor to *E. crus-pavonis*. Willdenow differentiates the species from *P. crusgalli* (with "glumis aristatis hispidis") by "glumis aristatus muricato-echinatus," whereas in *E. crus-pavonis* the glumes are less muricate than in *E. crusgalli*.

Setaria muricata Beauv., Ess. Agrost. 51, 170, 178. 1812. Based on *Panicum muricatum* Michx.

?*Echinochloa echinata* Beauv., Ess. Agrost.
53, 161, 169. 1812. Based on *Panicum echinatum* Willd.

Panicum crusgalli var. *aristatum* Pursh, Fl. Amer. Sept. 66. 1814. North America.

Panicum pungens Poir. in Lam., Encycl. Sup. 4: 273. 1816. Based on *P. muricatum* Michx.

Pennisetum crusgalli Baumg., Enum. Stirp. Transsilv. 3: 277. 1816. Based on *Panicum crusgalli* L.

Echinochloa crusgalli var. *aristata* S. F. Gray, Nat. Arr. Brit. Pl. 2: 158. 1821. Great Britain.

Oplismenus crusgalli Dum., Obs. Gram. Belg. 138. 1823. Based on *Panicum crusgalli* L.

?*Orthopogon echinatus* Spreng., Syst. Veg. 1: 307. 1825. Based on *Panicum echinatum* Willd.

Orthopogon crusgalli Spreng., Syst. Veg. 1: 307. 1825. Based on *Panicum crusgalli* L.

Oplismenus muricatus Kunth, Rév. Gram. 1: 44. 1829. Based on *Panicum muricatum* Michx.

?*Oplismenus echinatus* Kunth, Rév. Gram. 1: 45. 1829. Based on *Panicum echinatum* Willd.

?*Panicum crusgalli* var. *echinatum* Doell in Mart., Fl. Bras. 2²: 143. 1877. Based on *P. echinatum* Willd.

Echinochloa muricata Fernald, Rhodora 17: 106. 1915. Based on *Panicum muricatum* Michx.

Echinochloa crusgalli forma *vittata* Hubb., Rhodora 18: 232. 1916. New Brunswick, *Hubbard* 763.

Echinochloa crusgalli var. *muricata* Farwell, Mich. Acad. Sci. Rpt. 21: 350. 1920. Based on *Panicum muricatum* Michx.

Echinochloa crusgalli var. *michauxii* House, N. Y. State Mus. Bul. 243–244: 42. 1923. Based on *Panicum muricatum* Michx.

Echinochloa pungens Rydb., Brittonia 1: 81. 1931. Based on *Panicum pungens* Poir.

Echinochloa pungens var. *coarctata* Fern. and Grisc., Rhodora 37: 136. pl. 336. f. 1, 2. 1935. Pungo Ferry, Va., *Fernald and Griscom* 2760.

ECHINOCHLOA CRUSGALLI var. FRUMENTACEA (Roxb.) W. F. Wight, Cent. Dict. Sup. 810. 1909. Presumably based on *Panicum frumentaceum* Roxb. (Published as *E. crusgalli frumentacea*.)

Panicum frumentaceum Roxb., Fl. Ind. 1: 307. 1820. Not *P. frumentaceum* Salisb., 1796. India.

Echinochloa frumentacea Link, Hort. Berol. 1: 204. 1827. Based on *Panicum frumentaceum* Roxb.

Oplismenus frumentaceus Kunth, Rév.

Gram. 1: 445. 1829. Based on *Panicum frumentaceum* Roxb.

Panicum crusgalli var. *frumentaceum* Trimen, Syst. Cat. Ceylon Pl. 104. 1885. Based on *P. frumentaceum* Roxb.

Echinochloa crusgalli edulis Hitchc., U. S. Dept. Agr. Bul. 772: 238. 1920. Based on *Panicum frumentaceum* Roxb.

Echinochloa crusgalli subsp. *colonum* var. *edulis* Honda, Bot. Mag. [Tokyo] 37:123. 1923. Based on *E. crusgalli* var. *edulis* Hitchc.

Echinochloa colonum var. *frumentacea* Ridl., Fl. Malay Pen. 5: 223. 1925. Presumably based on *Panicum frumentaceum* Roxb.

Panicum crusgalli subsp. *colonum* var. *edulis* Makino and Nemoto, Fl. Jap. 1470. 1925. Based on *P. frumentaceum* Roxb.

ECHINOCHLOA CRUSGALLI var. MITIS (Pursh) Peterm., Fl. Lips. 82. 1838. Based on *Panicum crusgalli* var. *mite* Pursh.

Panicum crusgalli var. *mite* Pursh, Fl. Amer. Sept. 66. 1814. North America.

Panicum crusgalli var. *purpureum* Pursh, Fl. Amer. Sept. 66. 1814. North America.

Panicum crusgalli var. *muticum* Ell., Bot. S. C. and Ga. 1: 114. 1816. Probably South Carolina.

Panicum scindens Nees ex Steud., Syn. Pl. Glum. 1: 47. 1854. St. Louis [*Drummond*].

Oplismenus crusgalli var. *muticus* Wood, Amer. Bot. and Flor. pt. 2: 393. 1871. Eastern States.

Panicum crusgalli α *normale* var. *mite* forma *hispidum* Kuntze, Rev. Gen. Pl. 2: 783. 1891. Pennsylvania.

Echinochloa crusgalli var. *mutica* Rydb., Colo. Agr. Col. Bul. 100: 21. 1906. Presumably based on *Panicum crusgalli* var. *muticum* Ell.

Echinochloa crusgalli forma *purpurea* Farwell, Mich. Acad. Sci. Rpt. 21: 349. 1920. Based on *Panicum crusgalli* var. *purpureum* Pursh.

Echinochloa zelayensis var. *macera* Wiegand, Rhodora 23: 54. 1921. Matamoros, Mexico, *Berlandier* 890.

Echinochloa muricata var. *ludoviciana* Wiegand, Rhodora 23: 58. 1921. Baton Rouge, La., *Billings* 14.

Echinochloa muricata var. *occidentalis* Wiegand, Rhodora 23: 58. 1921. Grand Tower, Ill., *Gleason* 1720. (See *E. pungens* var. *wiegandii* below.)

Echinochloa muricata var. *microstachya* Wiegand, Rhodora 23: 58. 1921. Cayuga Lake Basin, N. Y., *Palmer* 97.

Echinochloa muricata var. *multiflora* Wiegand, Rhodora 23: 59. 1921. Lincoln County, Okla., *Blankenship*.

Echinochloa microstachya Rydb., Brittonia 1: 82. 1931. Based on *E. muricata* var. *microstachya* Wiegand.

Echinochloa occidentalis Rydb., Brittonia 1: 82. 1931. Based on *E. muricata* var. *occidentalis* Wiegand.

Echinochloa pungens var. *ludoviciana* Fern. and Grisc., Rhodora 37: 137. 1935. Based on *E. muricata* var. *ludoviciana* Wiegand.

Echinochloa pungens var. *microstachya* Fern. and Grisc., Rhodora 37: 137. 1935. Based on *E. muricata* var. *microstachya* Wiegand.

Echinochloa pungens var. *multiflora* Fern. and Grisc., Rhodora 37: 137. 1935. Based on *E. muricata* var. *multiflora* Wiegand.

Echinochloa pungens var. *occidentalis* Fern. and Grisc., Rhodora 37: 137. 1935. Based on *E. muricata* var. *occidentalis* Wiegand.

Echinochloa pungens var. *wiegandii* Fassett, Rhodora 51: 2. 1949. Based on the description of *E. muricata* var. *occidentalis* Wieg., but excluding the cited type, *Gleason* 1720.

ECHINOCHLOA CRUSGALLI var. ZELAYENSIS (H. B. K.) Hitchc., U. S. Dept. Agr. Bul. 772: 238. 1920. Based on *Oplismenus zelayensis* H. B. K. (Published as *E. crusgalli zelayensis*.)

Oplismenus zelayensis H. B. K., Nov. Gen. et Sp. 1: 108. 1815. Zelaya, Mexico, *Humboldt* and *Bonpland*.

Echinochloa zelayensis Schult., Mantissa 2: 269. 1824. Based on *Oplismenus zelayensis* H. B. K.

Panicum zelayense Steud., Nom. Bot. ed. 2. 2: 265. 1841. Based on *Oplismenus zelayensis* H. B. K.

Panicum crus-pici Willd. ex Doell, in Mart., Fl. Bras. 2²: 143. 1877. Name only. South America.

?*Panicum crusgalli α normale* var. *pygmaeum* Kuntze, Rev. Gen. Pl. 2: 783. 1891. Colorado.

Echinochloa crusgalli forma *zelayensis* Farwell, Mich. Acad. Sci. Papers 26: 4. 1941. Based on *Oplismenus zelayensis* H. B. K.

(4) **Echinochloa crus-pavonis** (H. B. K.) Schult., Mantissa 2: 269. 1824. Based on *Oplismenus crus-pavonis* H. B. K.

Oplismenus crus-pavonis H. B. K., Nov. Gen. et Sp. 1: 108. 1815. Cumaná, Venezuela, *Humboldt* and *Bonpland*.

Panicum crusgalli var. *sabulicolum* Trin., Gram. Icon. 2: pl. 163. 1828. Brasil.

Panicum sabulicola Nees, Agrost. Bras. 258. 1829. Pará, Brazil, *Sieber;* Uruguay and Paraguay.

Panicum crus-pavonis Nees, Agrost. Bras. 259. 1829. Based on *Oplismenus crus-pavonis* H. B. K.

Echinochloa composita Presl ex Nees, Agrost. Bras. 259. 1829, as synonym of *Panicum crus-pavonis* Nees. Acapulco, Mexico, *Haenke*.

Oplismenus sabulicola Kunth, Rev. Gram.

1: Sup. 11, 1830. Based on *Panicum sabulicola* Nees.

Panicum aristatum Macfad., Bot. Misc. Hook. 2: 115. 1831. Jamaica, [*Macfadden*].

Oplismenus jamaicensis Kunth, Enum. Pl. 1: 147. 1833. Based on *Panicum aristatum* Macfad.

Panicum jamaicense Steud., Nom. Bot. ed. 2. 2: 257. 1841. Based on *Oplismenus jamaicensis* Kunth.

Panicum crusgalli var. *sabulicola* Doell in Mart., Fl. Bras. 2²: 142. 1877. Based on *P. sabulicola* Nees.

Oplismenus angustifolius Fourn., Mex. Pl. 2: 40. 1886. Vera Cruz, Mexico, *Gouin* 54 [error for 50].

Echinochloa sabulicola Hitchc., U. S. Natl. Herb. Contrib. 17: 257. 1913. Based on *Panicum sabulicola* Nees.

Echinochloa crusgalli crus-pavonis Hitchc., U. S. Natl. Herb. Contrib. 22: 148. 1920. Based on *Oplismenus crus-pavonis* H. B. K.

Echinochloa crusgalli forma *sabulonum* Farwell, Mich. Acad. Sci. Rpt. 21: 349. 1920. Based on "*Panicum crusgalli* var. *sabulonum* Trin.," error for var. *sabulicolum* Trin.

Echinochloa zelayensis var. *subaristata* Wiegand, Rhodora 23: 54. 1921. Pierce, Texas, *Tracy* 7743.

(5) **Echinochloa paludigena** Wiegand, Rhodora 23: 64. 1921. Hillsborough County, Fla., *Fredholm* 6390.

Echinochloa paludigena var. *soluta* Wiegand, Rhodora 23: 64. 1921. Manatee, Fla., *Tracy* 7754.

(1) **Echinochloa polystachya** (H. B. K.) Hitchc., U. S. Natl. Herb. Contrib. 22: 135. 1920. Based on *Oplismenus polystachyus* H. B. K.

Oplismenus polystachyus H. B. K., Nov. Gen. et Sp. 1: 107. 1815. Colombia, *Humboldt* and *Bonpland*.

(6) **Echinochloa walteri** (Pursh) Heller, Cat. N. Amer. Pl. ed. 2. 21. 1900. Based on *Panicum walteri* Pursh.

Panicum hirtellum Walt., Fl. Carol. 72. 1788. Not *P. hirtellum* L., 1759. South Carolina.

Panicum walteri Pursh, Fl. Amer. Sept. 66. 1814. Based on *P. hirtellum* Walter.

Panicum crusgalli var. *hispidum* Ell., Bot. S. C. and Ga. 1: 114. 1816. Based on *P. hispidum* Muhl., in manuscript.

Panicum hispidum Muhl., Descr. Gram. 107. 1817. Not *P. hispidum* Forst., 1786. New York to Carolina. Name only, Muhl., Cat. Pl. 9. 1813.

Orthopogon hispidus Spreng., Syst. Veg. 1: 307. 1825. Based on *Panicum hispidum* Muhl.

Oplismenus hispidus Wood, Class-book ed. 2. 604. 1847. Based on *Panicum hispidum* Muhl.

Oplismenus crusgalli var. *hispidus* Wood, Amer. Bot. and Flor. pt. 2: 393. 1871. Presumably based on *Panicum hispidum* Muhl.

Echinochloa longearistata Nash in Small, Fl. Southeast. U. S. 84. 1903. Louisiana, *Hale.*

Panicum crusgalli var. *walteri* Farwell, Mich. Acad. Sci. Rpt. 6: 202. 1904. Based on *P. walteri* Pursh.

Echinochloa crusgalli var. *hispida* Farwell, Amer. Midl. Nat. 9: 4. 1925. Based on *Panicum hispidum* Muhl.

Echinochloa walteri forma *breviseta* Fern. and Grisc., Rhodora 37: 137. 1935. North Landing, Norfolk County, Va., *Fernald* and *Griscom* 2761.

ECHINOCHLOA WALTERI forma LAEVIGATA Wiegand, Rhodora 23: 62. 1921. Based on *Panicum longisetum* Torr.

Panicum longisetum Torr., Amer. Jour. Sci. 4: 58. 1822. Not *P. longisetum* Poir., 1816. Fox River, Wis. [*Douglas* in 1820].

Oplismenus longisetus Kunth, Rév. Gram. 1: 45. 1829. Based on *Panicum longisetum* Torr.

Echinochloa crusgalli var. *hispida* subvar. *laevigata* Farwell, Amer. Midl. Nat. 9: 4. 1925. Based on *E. walteri* forma *laevigata* Wiegand.

(32A) ECTOSPERMA Swallen

(1) **Ectosperma alexandrae** Swallen, Wash. Acad. Sci. Jour. 40: 19 f. 1. 1950. Inyo County, Calif., *Alexander* and *Kellogg* 5655.

EHRHARTA Thunb.

Ehrharta calycina J. E. Smith, Pl. Icon. Ined. pl. 33. 1790. Cape of Good Hope, Africa.

Ehrharta capensis Thunb., Svensk. Akad. Handl. 40: 217. pl. 8. 1779. Cape of Good Hope, Africa.

Ehrharta erecta Lam., Encycl. 2: 347. 1786. South Africa.

(100) ELEUSINE Gaertn.

Eleusine coracana (L.) Gaertn., Fruct. et Sem. 1: 8. pl. 1. 1788. Based on *Cynosurus coracanus* L.

Cynosurus coracanus L., Syst. Nat. ed. 10. 2: 875. 1759. East Indies.

(1) **Eleusine indica** (L.) Gaertn., Fruct. et Sem. 1: 8. 1788. Based on *Cynosurus indicus* L.

Cynosurus indicus L., Sp. Pl. 72. 1753. India.

Eleusine gracilis Salisb., Prodr. Stirp. 19. 1796. Based on *Cynosurus indicus* L.

Eleusine domingensis Sieber ex Schult., Mantissa 2: 323. 1824. Not *E. domingensis* Pers., 1805. As synonym of *E. indica* Lam. (error for Gaertn.).

Cynodon indicus Raspail, Ann. Sci. Nat., Bot. 5: 303. 1825. Based on *Eleusine*

indica Lam. (error for Gaertn.).

Chloris repens Steud., Nom. Bot. ed. 2. 1: 353. 1840, as synonym of *Eleusine indica* Pers. (error for Gaertn.).

Eleusine scabra Fourn. ex Hemsl., Biol. Centr. Amer. Bot. 3: 565. 1885, name only; Fourn., Mex. Pl. 2: 145. 1886. Mexico, *Bourgeau* 1030, 2378 in part, 2634, 2743; *Virlet* 1435; *Bilimek* 454; *Müller* 1392; *Gouin* 67.

Eleusine indica var. *major* Fourn., Mex. Pl. 2: 145. 1886. Mexico, *Liebmann* 222, 223, 227; *Karwinsky* 955.

Eleusine tristachya (Lam.) Lam., Tabl. Encyl. 1: 203. 1791. Based on *Cynosurus tristachyus* Lam.

Cynosurus tristachyus Lam., Tabl. Encycl. 2: 188. 1786. Uruguay, *Commerson.*

(46) ELYMUS L.

(11) **Elymus ambiguus** Vasey and Scribn., U. S. Natl. Herb. Contrib. 1: 280. 1893. Pen Gulch, Colo., *Vasey* in 1884.

ELYMUS AMBIGUUS var. STRIGOSUS (Rydb.) Hitchc., Amer. Jour. Bot. 21: 133. 1934. Based on *E. strigosus* Rydb.

Elymus strigosus Rydb., Torrey Bot. Club Bul. 32: 609. 1905. Boulder, Colo., *Letterman* 553 [type]; Wyoming, *A. Nelson* 7151.

Elymus villiflorus Rydb., Torrey Bot. Club Bul. 32: 609. 1905. Boulder, Colo., *Tweedy* 4818.

(5) **Elymus arenicola** Scribn. and Smith, U. S. Dept. Agr., Div. Agrost. Cir. 9: 7. 1899. Suferts, Oreg., *Leckenby* in 1898.

Leymus arenicola Pilger, Bot. Jahrb. 74: 6. 1945. Based on *Elymus arenicola* Scribn. and Smith.

(19) **Elymus aristatus** Merr., Rhodora 4: 147. 1902. Harney County, Oreg., *Cusick* 2712.

Elymus glaucus var. *aristatus* Hitchc. in Abrams, Illustr. Fl. 1: 252. 1923. Based on *E. aristatus* Merr.

(22) **Elymus canadensis** L., Sp. Pl. 83. 1753. Canada, *Kalm.*

Elymus philadelphicus L., Cent. Pl. 1: 6. 1755; Amoen. Acad. 4: 266. 1759. Pennsylvania, *Kalm.*

Hordeum patulum Moench, Meth. Pl. 199. 1794. Garden plant, *Elymus canadensis* L., cited as synonym.

Elymus glaucifolius Muhl. in Willd., Enum. Pl. 1: 131. 1809. Pennsylvania, *Muhlenberg.*

Elymus canadensis var. *glaucifolius* Torr., Fl. North. and Mid. U. S. 1: 137. 1823. Based on *E. glaucifolius* Muhl.

Elymus canadensis var. *pendulus* Eaton and Wright, N. Amer. Bot. ed. 8. 232. 1840. No locality cited.

Sitanion brodiei Piper, Erythea 7: 100. 1899. Bishop's Bar, Snake River, Wash., *Brodie* in 1895.

Hordeum canadense Aschers. and Graebn., Syn. Mitteleur. Fl. 2: 745. 1902. Based on *Elymus canadensis* L.

Terrellia canadensis Lunell, Amer. Midl. Nat. 4: 228. 1915. Based on *Elymus canadensis* L.

Terrellia canadensis var. *glaucifolia* Lunell, Amer. Midl. Nat. 4: 228. 1915. Based on *Elymus glaucifolius* Muhl.

Elymus robustus var. *vestitus* Wiegand, Rhodora 20: 90. 1918. Cedar Point, Ohio, *MacDaniels* 106.

Elymus canadensis var. *philadelphicus* Farwell, Mich. Acad. Sci. Rpt. 21: 357. 1920. Based on *E. philadelphicus* L.

Elymus philadelphicus var. *hirsutus* Farwell, Amer. Midl. Nat. 10: 314. 1927. Name proposed for *E. canadensis* as described by Wiegand (Rhodora 20: 87. 1918) "in large part."

Elymus philadelphicus var. *pendulus* Farwell, Amer. Midl. Nat. 10: 314. 1927. Based on *E. canadensis* var. *pendulus* Eaton and Wright.

Clinelymus canadensis Nevski, Jard. Bot. Acad. Sci. U. R. S. S. Bul. 30: 650. 1932. Based on *Elymus canadensis* L.

Elymus canadensis forma *glaucifolius* Fernald, Rhodora 35: 191. 1933. Based on *E. glaucifolius* Muhl.

Elymus wiegandii Fernald, Rhodora 35: 192. 1933. St. Francis, Maine, *Fernald* 197.

Elymus wiegandii forma *calvescens* Fernald, Rhodora 35: 192. 1933. Dead River, Maine, *Fernald* and *Strong* in 1896.

ELYMUS CANADENSIS var. BRACHYSTACHYS (Scribn. and Ball) Farwell, Mich. Acad. Sci. Rpt. 21: 357. 1920. Based on *E. brachystachys* Scribn. and Ball.

Elymus brachystachys Scribn. and Ball, U. S. Dept. Agr., Div. Agrost. Bul. 24: 47. f. 21. 1901. Indian Territory [Oklahoma], *Palmer* 420.

Elymus philadelphicus var. *brachystachys* Farwell, Amer. Midl. Nat. 10: 314. 1927. Based on *E. brachystachys* Scribn. and Ball.

ELYMUS CANADENSIS var. ROBUSTUS (Scribn. and Smith) Mackenz. and Bush, Man. Fl. Jackson County 38. 1902. Based on *E. robustus* Scribn. and Smith.

Elymus canadensis forma *crescendus* Ramaley, Minn. Bot. Studies 1: 114. 1894. Springfield, Minn., *Sheldon* 1120.

Elymus robustus Scribn. and Smith, U. S. Dept. Agr., Div. Agrost. Bul. 4: 37. 1897. Illinois [type, *Wolf*], Iowa, Kansas, and Montana.

Elymus crescendus Wheeler, Minn. Bot. Studies 3: 106. 1903. Based on *E. canadensis* forma *crescendus* Ramaley.

Elymus canadensis villosus Bates, Amer. Bot. 20: 17. 1914. Loup City and Arcadia, Nebr., *Bates* in 1911.

Elymus glaucifolius crescendus Bush,

Amer. Midl. Nat. 10: 83. 1926. Based on *E. canadensis* forma *crescendus* Ramaley.

Elymus glaucifolius robustus Bush, Amer. Midl. Nat. 10: 87. 1926. Based on *E. robustus* Scribn. and Smith.

Elymus philadelphicus var. *robustus* Farwell, Amer. Midl. Nat. 10: 314. 1927. Based on *E. robustus* Scribn. and Smith.

(1) **Elymus caput-medusae** L., Sp. Pl. 84. 1753. Southern Europe.

Hordeum caput-medusae Coss. and Dur., Expl. Sci. Alger. 2: 198. 1855. Based on *Elymus caput-medusae* L.

Taeniatherum caput-medusae Nevski, Act. Univ. Asiae Med. VIII b. Bot. 17: 38. 1934. Based on *Elymus caput-medusae* L.

(14) **Elymus cinereus** Scribn. and Merr., Torrey Bot. Club Bul. 29: 467. 1902. Pahrump Valley, Nev., *Purpus* 6050.

Elymus condensatus var. *pubens* Piper, Erythea 7: 101. 1899. Yakima City, Wash., *Piper* 2591. (Published as *E. condensatus pubens*.)

Elymus condensatus forma *pubens* St. John, Fl. Southeast. Wash. and Adj. Idaho 42. 1937. Based on *E. condensatus* var. *pubens* Piper.

(13) **Elymus condensatus** Presl, Rel. Haenk. 1: 265. 1830. Monterey, Calif. *Haenke*.

Aneurolepidium condensatum Nevski, Akad. Nauk. S. S. R. Bot. Inst. Trudy I. (Acad. Sci. U. R. S. S. Inst. Bot. Acta I, Flora et Syst. Plant. Vasc.) 1: 14. 1933. Based on *Elymus condensatus* Presl.

(4) **Elymus flavescens** Scribn. and Smith, U. S. Dept. Agr., Div. Agrost. Bul. 8: 8. f. 1. 1897. Columbus, Wash., *Suksdorf* 916.

Leymus flavescens Pilger, Bot. Jahrb. 74: 6. 1945. Based on *Elymus flavescens* Scribn. and Smith.

Elymus giganteus Vahl, Symb. Bot. 3: 10. 1794. Old World.

Leymus giganteus Pilger, Bot. Jahrb. 74: 6. 1945. Based on *Elymus giganteus* Vahl.

(15) **Elymus glaucus** Buckl., Acad. Nat. Sci. Phila. Proc. 1862: 99. 1862. Columbia River, Oreg., *Nuttall*.

Elymus villosus var. *glabriusculus* Torr., U. S. Expl. Miss. Pacif. Rpt. 4: 157. 1857. Napa Valley, Calif.

Elymus nitidus Vasey, Torrey Bot. Club Bul. 13: 120. 1886. Eagle Mountains, Oreg., *Cusick* [1130].

Elymus americanus Vasey and Scribn. ex Macoun, Can. Pl. Cat. 2⁴: 245. 1888, name only; Cassidy, Colo. Agr. Expt. Sta. Bul. 12: 57. 1890. Arapahoe Pass, Colo.

Elymus sibiricus var. *americanus* Wats. and Coult. in A. Gray, Man. ed. 6. 673. 1890. Michigan and westward.

Elymus sibiricus var. *glaucus* Ramaley,

Minn. Bot. Studies 9: 112. 1894. Based on *E. glaucus* Buckl.

Elymus glaucus var. *breviaristatus* Davy in Jepson, Fl. West. Mid. Calif. 79. 1901. Point Reyes, Calif, *Davy.*

Elymus glaucus var. *maximus* Davy in Jepson, Fl. West. Mid. Calif. 79. 1901. Napa Valley, Calif., *Jepson.*

Elymus hispidulus Davy in Jepson, Fl. West. Mid. Calif. 79. 1901. Olema, Calif., *Davy* 4306b.

Elymus angustifolius Davy in Jepson, Fl. West. Mid. Calif. 80. 1901. San Francisco, Calif., *Davy.*

Elymus angustifolius var. *caespitosus* Davy in Jepson, Fl. West. Mid. Calif. 81. 1901. Berkeley Hills, Calif., *Davy* 4255.

Elymus marginalis Rydb., Torrey Bot. Club Bul. 36: 539. 1909. Lower Arrow Lake, British Columbia, *Macoun* 44.

Terrellia glauca Lunell, Amer. Midl. Nat. 4: 228. 1915. Based on *Elymus glaucus* Buckl.

Elymus mackenzii Bush, Amer. Midl. Nat. 10: 53. 1926. Eagle Rock, Mo., *Bush* 77.

Clinelymus glaucus Nevski, Jard. Bot. Acad. Sci. U. R. S. S. Bul. 30: 648. 1932. Based on *Elymus glaucus* Buckl.

Clinelymus glaucus subsp. *californicus* Nevski, Jard. Bot. Acad. Sci. U. R. S. S. Bul. 30: 649. 1932. California, *Heller* 5714-a, first of several cited from California.

Clinelymus glaucus subsp. *coloratus* Nevski, Jard. Bot. Acad. Sci. U. R. S. S. Bul. 30: 648. 1932. Washington, *Heller* 3965.

ELYMUS GLAUCUS var. JEPSONI Davy in Jepson, Fl. West. Mid. Calif. 79. 1901. Napa Valley, Calif., *Jepson.*

Elymus divergens Davy in Jepson, Fl. West. Mid. Calif. 80. 1901. Petaluma, Calif., *Davy* 4037.

Elymus velutinus Scribn. and Merr., Torrey Bot. Club Bul. 29: 466. 1902. San Bernardino Mountains, Calif., *Abrams* 2056.

Elymus parishii Davy and Merr., Calif. Univ. Pubs., Bot. 1: 58. 1902. San Jacinto Mountains, Calif., *Hall* 2097.

Elymus edentatus Suksdorf, Werdenda 1²: 4. 1923. Bingen, Wash., *Suksdorf* 10057.

Clinelymus glaucus subsp. *californicus* var. *pubescens* Nevski, Jard. Bot. Acad. Sci. U. R. S. S. Bul. 30: 649. 1932. California, *Tiling* 8822; *Palmer* 417.

Clinelymus velutinus Nevski, Jard. Bot. Acad. Sci. U. R. S. S. Bul. 30: 649. 1932. Based on *Elymus velutinus* Scribn. and Merr.

Elymus glaucus forma *jepsoni* St. John, Fl. Southeast. Wash. and Adj. Idaho 42. 1937. Based on *E. glaucus* var. *jepsoni* Davy.

Elymus glaucus subsp. *jepsoni* Gould, Madroño 9: 126. 1947. Based on *E. glaucus* var. *jepsoni* Davy.

ELYMUS GLAUCUS var. TENUIS Vasey, U. S. Natl. Herb. Contrib. 1: 280. 1893. "Type specimen collected by John Macoun on Vancouver Island in 1887 (No. 3)" comprises two forms, all Macoun's No. 3. One specimen is a small form of *E. glaucus* var. *jepsoni;* the others have spikes with fragile rachises, spikelets with 5- to 6-nerved glumes and lemmas with divergent awns and apparently represent a form not found in the United States. There is another Macoun specimen upon which Vasey has written the varietal name *tenuis,* this specimen having glabrous sheaths and divergent awns. The description states that the sheaths are glabrous or pubescent and that the awns are divergent. Hence the plant of number 3 with divergent awns is selected as the type of *E. glaucus* var. *tenuis,* and the name is excluded from our flora.

(17) **Elymus hirsutus** Presl, Rel. Haenk. 1: 264. 1830. Nootka Sound, Vancouver Island, *Haenke.*

Elymus ciliatus Scribn., U. S. Dept. Agr., Div. Agrost. Bul. 11: 57. pl. 16. 1898. Not *E. ciliatus* Muhl., 1817. Sitka, Alaska, *Evans* 210.

Elymus borealis Scribn., U. S. Dept. Agr., Div. Agrost. Cir. 27: 9. 1900. Based on *E. ciliatus* Scribn.

Clinelymus borealis Nevski, Jard. Bot. Acad. Sci. U. R. S. S. Bul. 30: 645. 1932. Based on *Elymus borealis* Scribn.

(7) **Elymus hirtiflorus** Hitchc., Amer. Jour. Bot. 21: 132. f. 2. 1934. Green River, Wyo., *Shear* 284.

(6) **Elymus innovatus** Beal, Grasses N. Amer. 2: 650. 1896. North Fork Sims River, Mont., *Williams* in 1887.

Elymus mollis R. Br. in Richards., Bot. App. Franklin Jour. 732. 1823. Not *E. mollis* Trin., 1821. Canada [*Richardson*].

Elymus brownii Scribn. and Smith, U. S. Dept. Agr., Div. Agrost. Bul. 8: 7. pl. 4. 1897. Banff, Alberta, *Canby* 24 in 1895.

Leymus innovatus Pilger, Bot. Jahrb. 74: 6. 1945. Based on *Elymus innovatus* Beal.

(21) **Elymus interruptus** Buckl., Acad. Nat. Sci. Phila. Proc. 1862: 99. 1862. Llano County, Tex., *Buckley.*

Elymus occidentalis Scribn., U. S. Dept. Agr., Div. Agrost. Bul. 13: 49. 1898. Laramie River, Wyo., *Nelson* 4470.

Elymus diversiglumis Scribn. and Ball, U. S. Dept. Agr., Div. Agrost. Bul. 24: 48. f. 22. 1901. Bear Lodge Mountains, Wyo., *Williams* 2653.

Terrellia diversiglumis Lunell, Amer. Midl. Nat. 4: 228. 1915. Based on *Elymus diversiglumis* Scribn. and Ball.

(18) **Elymus macounii** Vasey, Torrey Bot. Club Bul. 13: 119. 1886. Great Plains of British Columbia, *Macoun.*

Terrellia macounii Lunell, Amer. Midl.
Nat. 4: 228. 1915. Based on *Elymus
macounii* Vasey.
(2) **Elymus mollis** Trin. in Spreng., Neu.
Entd. 2: 72. 1821. Kamchatka and the
Aleutian Islands.
Elymus dives Presl, Rel. Haenk. 1: 265.
1830. Nootka Sound, Vancouver Island,
Haenke.
Elymus arenarius var. *villosus* E. Meyer,
Pl. Labrad. 20. 1830. Labrador.
Elymus ampliculmis Provancher, Fl.
Canad. 2: 706. 1862. Canada.
Elymus capitatus Scribn., U. S. Dept.
Agr., Div. Agrost. Bul. 11: 55. pl. 14.
1898. Homer, Alaska, *Evans* 471. Ab-
normal form.
Elymus mollis brevispicus Scribn. and
Smith, U. S. Dept. Agr., Div. Agrost.
Bul. 11: 56. 1898. St. Lawrence Bay,
Siberia.
Elymus villosissimus Scribn., U. S. Dept.
Agr., Div. Agrost. Bul. 17: 326. f. 622.
1899. St. Paul Island, *Macoun* 16226.
Elymus arenarius forma *compositus* Abro-
meit, Bibl. Bot. 8: heft 42: 96. 1899.
Greenland.
Elymus arenarius var. *mollis* Koidzumi,
Tokyo Imp. Univ., Col. Sci. Jour. 27:
24. 1910. Based on *E. mollis* Trin.
Elymus arenarius var. *compositus* St.
John, Rhodora 17: 102. 1915. Based on
E. arenarius forma *compositus* Abro-
meit.
Leymus mollis Pilger, Bot. Jahrb. 74: 6.
1945. Based on *Elymus mollis* Trin.
(9) **Elymus pacificus** Gould, Madroño 9:
127. 1947. Based on *Agropyron areni-
cola* Davy, not *Elymus arenicola*
Scribn. and Smith.
Agropyron arenicola Davy in Jepson, Fl.
West. Mid. Calif. 76. 1901. Point
Reyes, Calif., *Davy* 6879.
(23) **Elymus riparius** Wiegand, Rhodora
20: 84. 1918. Ithaca, N. Y., *Eames* and
MacDaniels 3567.
(12) **Elymus salinus** Jones, Calif. Acad.
Sci. Proc. II. 5: 725. 1895. Salina
Pass, Utah, *Jones* 5447. The name is
spelled "salina" in the text, but
"salinus" in Jones' index; salina was
doubtless a slip of the pen.
(10) **Elymus simplex** Scribn. and Williams,
U. S. Dept. Agr., Div. Agrost. Bul.
11: 57. pl. 17. 1898. Green River,
Wyo., *Williams* 2334.
Elymus triticoides var. *simplex* Hitchc.,
Amer. Jour. Bot. 21: 132. 1934. Based
on *E. simplex* Scribn. and Williams.
(8) **Elymus triticoides** Buckl., Acad. Nat.
Sci. Phila. Proc. 1862: 99. 1862.
"Rocky Mountains," *Nuttall.*
Elymus condensatus var. *triticoides* Thurb.
in S. Wats., Bot. Calif. 2: 326. 1880.
Based on *E. triticoides* Buckl.
Elymus orcuttianus Vasey, Bot. Gaz. 10:
258. 1885. San Diego, Calif., *Orcutt.*

Elymus simplex var. *luxurians* Scribn.
and Williams, U. S. Dept. Agr., Div.
Agrost. Bul. 11: 58. 1898. Green River,
Wyo., *Williams* 2338.
Elymus acicularis Suksdorf, Werdenda
1^2: 3. 1923. Bingen, Wash., *Suksdorf*
7861.
Leymus triticoides Pilger, Bot. Jahrb. 74:
6. 1945. Based on *Elymus triticoides*
Buckl.
ELYMUS TRITICOIDES subsp. MULTIFLORUS
Gould, Madroño 8: 46. 1945. Contra
Costa County, Calif., *Gould* 1304.
ELYMUS TRITICOIDES var. PUBESCENS Hitchc.
in Jepson, Fl. Calif. 1: 186. 1912.
Griffin, Calif., *Elmer* 3748.
(3) **Elymus vancouverensis** Vasey, Torrey
Bot. Club Bul. 15: 48. 1888. Van-
couver Island, *Macoun* in 1887.
Leymus vancouverensis Pilger, Bot. Jahrb.
74: 6. 1945. Based on *Elymus van-
couverensis* Vasey.
(20) **Elymus villosus** Muhl. in Willd.,
Enum. Pl. 1: 131. 1809. Pennsylvania,
Muhlenberg.
Elymus ciliatus Muhl., Descr. Gram. 179.
1817. North Carolina. Name only,
Muhl., Cat. Pl. 14. 1813.
Elymus hirsutus Schreb. ex Roem. and
Schult., Syst. Veg. 2: 776. 1817, as
synonym of *E. villosus* Muhl.
Elymus striatus var. *villosus* A. Gray,
Man. 603. 1848. Based on *E. villosus*
Muhl.
Elymus propinquus Fresen. ex Steud., Syn.
Pl. Glum. 1: 349. 1854. Illinois.
Elymus striatus var. *ballii* Pammel, Iowa
Geol. Survey Sup. Rpt. 1903: 347.
f. 246. 1905. Iowa [type, from which
figure was drawn, Johnson County,
Fitzpatrick].
Hordeum villosum Schenck, Bot. Jahrb.
40: 109. 1907. Based on *Elymus
villosus* Muhl.
ELYMUS VILLOSUS forma ARKANSANUS
(Scribn. and Ball) Fernald, Rhodora
35: 195. 1933. Based on *E. arkansanus*
Scribn. and Ball.
Elymus arkansanus Scribn. and Ball, U. S.
Dept. Agr., Div. Agrost. Bul. 24: 45. f.
19. 1901. Arkansas, *Harvey.*
Elymus striatus var. *arkansanus* Hitchc.,
Rhodora 8: 212. 1906. Based on *E.
arkansanus* Scribn. and Ball.
(16) **Elymus virescens** Piper, Erythea 7:
101. 1899. Olympic Mountains, Wash.,
Piper 1988.
Elymus pubescens Davy in Jepson, Fl.
West. Mid. Calif. 78. 1901. Point
Reyes, Calif.
Elymus howellii Scribn. and Merr., U. S.
Natl. Herb. Contrib. 13: 88. 1910.
Revillagigedo Island, British Columbia,
Howell 1723.
Elymus strigatus St. John, Rhodora 17:
102. 1915. Westport, Mendocino Coun-
ty, Calif., *Congdon* in 1902.

864 MISC. PUBLICATION 200, U. S. DEPT. OF AGRICULTURE

Elymus glaucus var. virescens Gould, Madroño 9: 126. 1947. Based on E. virescens Piper.

(24) **Elymus virginicus** L., Sp. Pl. 84. 1753. Virginia.

Elymus carolinianus Walt., Fl. Carol. 82. 1788. South Carolina.

Hordeum cartilagineum Moench, Meth. 199. 1794. Grown in botanic garden, Marburg, Germany.

Elymus striatus Willd., Sp. Pl. 1: 470. 1797. North America. Name only in Muhl., Amer. Phil. Soc. 2: 161. 1793.

Elymus hordeiformis Desf., Tabl. Ecol. Bot. Mus. 15. 1804, name only; Cat. Pl. Paris. ed. 3. 18, 387. 1829. Grown in botanical garden, Paris. "E. striatus Willd." cited as synonym.

Elymus virginicus Hedw. ex Steud., Nom. Bot. ed. 2. 1: 550. 1840, as synonym of E. virginicus L.

Elymus virginicus var. minor Vasey ex L. H. Dewey, U. S. Natl. Herb. Contrib. 2: 550. 1892. Northern Texas, [Buckley].

Elymus virginicus forma jejunus Ramaley, Minn. Bot. Stud. 9: 114. 1894. Lake Benton, Minn., Sheldon 1735 (error for 1375).

Hordeum virginicum Schenck, Bot. Jahrb. 40: 109. 1907. Based on Elymus virginicus L.

Hordeum striatum Schenck, Bot. Jahrb. 40: 109. 1907. Based on Elymus striatus Willd.

Elymus jejunus Rydb., Torrey Bot. Club Bul. 36: 539. 1909. Based on E. virginicus forma jejunus Ramaley.

Terrellia virginica Lunell, Amer. Midl. Nat. 4: 228. 1915. Based on Elymus virginicus L.

Terrellia striata Lunell, Amer. Midl. Nat. 4: 228. 1915. Based on Elymus striatus Willd.

Elymus virginicus var. jejunus Bush, Amer. Midl. Nat. 10: 65. 1926. Based on E. virginicus forma jejunus Ramaley.

Terrella jejuna Nevski, Jard. Bot. Acad. Sci. U. R. S. S. Bul. 30: 639. 1932. Based on Elymus virginicus forma jejunus Ramaley.

Terrella virginica Nevski, Jard. Bot. Acad. Sci. U. R. S. S. Bul. 30: 639. 1932. Based on Elymus virginicus L.

Elymus virginicus var. typicus Fernald, Rhodora 35: 198. 1933. Based on E. virginicus L.

Elymus virginicus var. micromeris Schmoll, Rhodora 39: 416. 1937. Leeds, N. Dak., Lunell in 1900.

ELYMUS VIRGINICUS var. AUSTRALIS (Scribn. and Ball) Hitchc. in Deam, Ind. Dept. Conserv. Pub. 82: 113. 1929. Based on E. australis Scribn. and Ball.

Elymus australis Scribn. and Ball, U. S. Dept. Agr., Div. Agrost. Bul. 24: 46.

f. 20. 1901. Biltmore, N. C., Biltmore Herbarium 411b.

Elymus virginicus var. glabriflorus forma australis Fernald, Rhodora 35: 198. 1933. Based on E. australis Scribn. and Ball.

ELYMUS VIRGINICUS var. GLABRIFLORUS (Vasey) Bush, Amer. Midl. Nat. 10: 62. 1926. Based on E. canadensis var. glabriflorus Vasey.

Elymus canadensis var. glabriflorus Vasey ex L. H. Dewey, U. S. Natl. Herb. Contrib. 2: 550. 1894. Texas to Georgia [Louisiana, Langlois].

?Elymus virginicus var. glaucus Beal, Grasses N. Amer. 2: 653. 1896. Agricultural College, Michigan, Beal 164, 165.

Elymus glabriflorus Scribn. and Ball, U. S. Dept. Agr., Div. Agrost. Bul. 24: 49. f. 23. 1901. Based on E. canadensis var. glabriflorus Vasey.

Elymus australis var. glabriflorus Wiegand, Rhodora 20: 84. 1918. Based on E. canadensis var. glabriflorus Vasey.

ELYMUS VIRGINICUS var. HALOPHILUS (Bicknell) Wiegand, Rhodora 20: 83. 1918. Based on E. halophilus Bicknell.

Elymus halophilus Bicknell, Torrey Bot. Club Bul. 35: 201. 1908. Nantucket Island, Mass., Bicknell.

Terrella halophila Nevski, Jard. Bot. Acad. Sci. U. R. S. S. Bul. 30: 639. 1932. Based on Elymus halophilus Bicknell.

Elymus virginicus var. halophilus forma lasiolepis Fernald, Rhodora 35: 198. 1933. Nova Scotia, Fernald, Long, and Linder 20113.

ELYMUS VIRGINICUS var. INTERMEDIUS (Vasey) Bush, Amer. Midl. Nat. 10: 60. 1926. Based on E. canadensis var. intermedius Vasey.

Elymus canadensis var. intermedius Vasey ex A. Gray, Man. ed. 6. 673. 1890. Northeastern United States. [Type, Lansingburg, N. Y., Howe in 1886.]

Elymus intermedius Scribn. and Smith, U. S. Dept. Agr., Div. Agrost. Bul. 4: 38. 1897. Not E. intermedius Bieb., 1808. Maine to Virginia, west to Illinois and Nebraska. [Herbarium evidence shows this to be based on E. canadensis var. intermedius Vasey.]

Elymus hirsutiglumis Scribn., U. S. Dept. Agr., Div. Agrost. Bul. 11: 58. 1898. Based on E. intermedius Scribn. and Smith.

Elymus virginicus var. hirsutiglumis Hitchc., Rhodora 10: 65. 1908. Based on E. hirsutiglumis Scribn.

Terrella hirsutiglumis Nevski, Jard. Bot. Acad. Sci. U. R. S. S. Bul. 30: 639. 1932. Based on Elymus hirsutiglumis Scribn.

Elymus virginicus var. typicus forma hir-

sutiglumis Fernald, Rhodora 35: 198. 1933. Based on *E. hirsutiglumis* Scribn.
ELYMUS VIRGINICUS var. SUBMUTICUS Hook., Fl. Bor. Amer. 2: 255. 1840. Cumberland House Fort, Saskatchewan, *Drummond*.
?*Elymus virginicus* var. *arcuatus* Wood, Amer. Bot. and Flor. pt. 2: 405. 1871. Southern States.
Elymus curvatus Piper, Torrey Bot. Club Bul. 30: 233. 1903. Stevens County, Wash., *Kreager 375*.
Elymus submuticus Smyth, Kans. Acad. Sci. Trans. 25: 99. 1913. Based on *E. virginicus* var. *submuticus* Hook.
Terrellia virginica var. *submutica* Lunell, Amer. Midl. Nat. 4: 228. 1915. Based on *Elymus virginicus* var. *submuticus* Hook.
Terrella curvata Nevski, Jard. Bot. Acad. Sci. U. R. S. S. Bul. 30: 639. 1932. Based on *Elymus curvatus* Piper.
Elymus virginicus forma *submutica* Pohl, Amer. Midl. Nat. 38: 549. 1947. Based on *E. virginicus* var. *submuticus* Hook.

(161) ELYONURUS Humb. and Bonpl. ex Willd.

(1) Elyonurus barbiculmis Hack. in DC., Monogr. Phan. 6: 339. 1889. Texas, *Wright 804*; New Mexico, *Wright 2106*; Arizona, *Lemmon 2926* [type]; *Rothrock 638*.
Elyonurus barbiculmis var. *parviflorus* Scribn., U. S. Dept. Agr., Div. Agrost. Cir. 32: 1. 1901. Arizona, *Griffiths 1849*.

(2) Elyonurus tripsacoides Humb. and Bonpl. ex Willd., Sp. Pl. 4: 941. 1806. Caracas, Venezuela, *Humboldt and Bonpland*.
Rottboellia ciliata Nutt., Gen. Pl. 1: 83. 1818. Georgia, *Baldwin*.
Anatherum tripsacoides Spreng., Syst. Veg. 1: 290. 1825. Based on *Elyonurus tripsacoides* Humb. and Bonpl.
Andropogon tripsacoides Steud., Syn. Pl. Glum. 1: 364. 1854. Based on *Elyonurus tripsacoides* Humb. and Bonpl.
Andropogon nuttallii Chapm., Fl. South. U. S. 580. 1860. Based on *Rottboellia ciliata* Nutt.
Elyonurus nuttallianus Benth. ex Vasey, Grasses U. S. 17. 1883. Based on *Andropogon nuttallianus* [error for *nuttallii* Chapm.].
Elyonurus nuttallii Vasey, Grasses U. S. Descr. Cat. 25. 1885. Based on *Andropogon nuttallii* Chapm.

(40) ENNEAPOGON Desv. ex Beauv.

(1) Enneapogon desvauxii Beauv., Ess. Agrost. 82, 161. pl. 16. f. 11. 1812; ex Desv., Opusc. 98. 1831. Locality erroneously given as "Manilia," probably Argentina.

Enneapogon phleoides Roem. and Schult., Syst. Veg. 2: 616. 1817. South America.
Pappophorum wrightii S. Wats., Amer. Acad. Sci. Proc. 18: 178. 1883. [Devils River, Tex.], *Wright 751* and *2029*.
Pappophorum mexicanum Griseb. ex Fourn., Mex. Pl. 2: 133. 1886. Mexico, Guadalupe, *Bourgeau*; valley of Mexico, *Schaffner 184*.
Enneapogon wrightii C. E. Hubb. in Hook. f., Icon. Pl. (pl. 3337). 2. 1937. Based on *Pappophorum wrightii* S. Wats.

(14) ERAGROSTIS Beauv.

(43) Eragrostis acuta Hitchc., Biol. Soc. Wash. Proc. 41: 159. 1928. Punta Rassa, Fla., *Hitchcock 263*.
Eragrostis alba Presl, Rel. Haenk. 1: 279. 1830. "Hab. ad Monte-Rey, Californiae. ♃" The label with the type specimen bears "Regio montana," indicating that the plant came from Peru. The species is not known from the United States.
(8) Eragrostis amabilis (L.) Wight and Arn. ex Nees in Hook. and Arn., Bot. Beechey Voy. 251. 1838. Based on *Poa amabilis* L.
Poa amabilis L., Sp. Pl. 68. 1753. India.
Poa plumosa Retz., Obs. Bot. 4: 20. 1786. East Indies.
Megastachya amabilis Beauv., Ess. Agrost. 74, 167, 173. 1812. Based on *Poa amabilis* L.
Cynodon amabilis Raspail, Ann. Sci. Nat., Bot. 5: 302. 1825. Based on *Megastachya amabilis* Beauv.
Eragrostis plumosa Link, Hort. Berol. 1: 192. 1827. Based on *Poa plumosa* Retz.
Erochloë amabilis Raf. ex Jacks., Ind. Kew. 1: 886. 1893, as synonym of *Eragrostis amabilis*.
Erochloë spectabilis Raf. ex Jacks., Ind. Kew. 1: 886. 1893, as synonym of *Eragrostis amabilis*.
Eragrostis ciliaris var. *patens* Chapm. ex Beal, Grasses N. Amer. 2: 479. 1896. Jesup, Ga., *Curtiss 3493**.
Eragrostis tenella var. *plumosa* Stapf in Hook. f., Fl. Brit. Ind. 7: 315. 1896. Based on *Poa plumosa* Retz.
Eragrostis amabilis var. *plumosa* E. G. and A. Camus in Lecomte, Fl. Gen. Ind.-Chin. 7: 557. 1923. Based on *Poa plumosa* Retz.
(29) Eragrostis arida Hitchc., Wash. Acad. Sci. Jour. 23: 449. 1933. Del Rio, Tex., *Hitchcock 13650*.
(46) Eragrostis bahiensis Schrad. in Schult., Mantissa 2: 318. 1824. Brazil.
(26) Eragrostis barrelieri Daveau in Morot., Jour. Bot. 8: 289. 1894. Southern Europe.

Eragrostis vulgaris subsp. *barrelieri* Douin in Bonn., Fl. Compl. 12: 32. 1927–32. Based on *E. barrelieri* Daveau.
(5) **Eragrostis beyrichii** J. G. Smith, Mo. Bot. Gard. Rpt. 6: 117. pl. 56. 1895. "Arkansas," Beyrich in 1834, but there is no recent record from that State. In 1834 the boundaries were as at present, but earlier included parts of Texas.
(14) **Eragrostis capillaris** (L.) Nees, Agrost. Bras. 505. 1829. Based on *Poa capillaris* L.
Poa capillaris L., Sp. Pl. 68. 1753. Canada, *Kalm.*
Aira capillacea Lam., Tabl. Encycl. 1. 177. 1791. Carolina, *Fraser.*
Poa tenuis Ell., Bot. S. C. and Ga. 1: 156. 1816. South Carolina.
Eragrostis tenuis Steud., Syn. Pl. Glum. 1: 273. 1854. Based on *Poa tenuis* Ell.
(45) **Eragrostis chariis** (Schult.) Hitchc., Lingnan Sci. Jour. 7: 193. 1931. Based on *Poa chariis* Schult.
Poa elegans Roxb., Hort. Beng. 82. 1814. Fl. Ind. 1: 339. 1820. Not *P. elegans* Poir., 1804. India.
Poa chariis Schult., Mantissa 2: 314. 1824. Based on *P. elegans* Roxb.
Poa elegantula Kunth, Rév. Gram. 1: 114. 1829. Based on *P. elegans* Roxb.
Eragrostis elegantula Nees ex Steud., Syn. Pl. Glum. 1: 266. 1854. Not *E. elegantula* Nees, 1851. Based on *Poa elegantula* Kunth.
Eragrostis chloromelas Steud., Syn. Pl. Glum. 1: 271. 1854. Based on the species described under *E. atrovirens* by Nees, that name based on *Poa atrovirens* Desf., a different species.
(24) **Eragrostis cilianensis** (All.) Lutati, Malpighia 18: 386. 1904. Based on *Poa cilianensis* All.
Briza eragrostis L., Sp. Pl. 70. 1753. Europe.
Poa cilianensis All., Fl. Pedem. 2: 246. 1785. Italy.
?Briza caroliniana Walt., Fl. Carol. 79. 1788. Not *B. caroliniana* Lam. South Carolina.
Poa megastachya Koel., Descr. Gram. 181. 1802. Based on *Briza eragrostis* L.
Eragrostis major Host, Icon. Gram. Austr. 4: 14. pl. 24. 1809; Fl. Austr. 1: 135. 1827. Austria.
Megastachya eragrostis Beauv. ex Roem. and Schult., Syst. Veg. 2: 575, in obs.; 584. 1817. Based on *Briza eragrostis* L.
Briza purpurascens Muhl., Descr. Gram. 154. 1817. Carolina.
Poa obtusa Nutt., Gen. Pl. 1: 67. 1818. Not *P. obtusa* Muhl., 1817. Philadelphia, *Barton.*
Poa pennsylvanica Nutt., Gen. Pl. 2: errata. 1818. Based on *P. obtusa* Nutt.
Poa philadelphica Barton, Compend. Fl. Phila. 1: 62. 1818. Based on *P. obtusa* Nutt.

Megastachya obtusa Schult., Mantissa 2: 326. 1824. Based on *Poa obtusa* Nutt.
Megastachya purpurascens Schult., Mantissa 2: 326. 1824. Based on *Briza purpurascens* Muhl.
Poa nuttallii Spreng., Syst. Veg. 1: 344. 1825. Based on *P. obtusa* Nutt.
Calotheca purpurascens Spreng., Syst. Veg. 1: 348. 1825. Based on *Briza purpurascens* Muhl.
Eragrostis megastachya Link, Hort. Berol. 1: 187. 1827. Based on *Poa megastachya* Koel.
Briza megastachya Steud., Nom. Bot. ed. 2. 1: 225. 1840, as synonym of *Poa megastachya* Koel.
Eragrostis vulgaris var. *megastachya* Coss. and Germ., Fl. Env. Paris 2: 641. 1845. Based on *Poa megastachya* Koel.
Eragrostis poaeoides var. *megastachya* A. Gray, Man. ed. 2. 563. 1856. Based on *E. megastachya* Link.
Eragrostis virletii Fourn., Mex. Pl. 2: 116. 1886. San Luis Potosí, Mexico, *Virlet* 1391.
Eragrostis eragrostis MacM., Met. Minn. Vall. 75. 1892. Not *E. eragrostis* Beauv., 1812. Based on *Briza eragrostis* L.
Eragrostis megastachya var. *cilianensis* Aschers. and Graebn., Syn. Mitteleur. Fl. 2: 371. 1900. Based on *Poa cilianensis* All.
Eragrostis minor var. *megastachya* Davy in Jepson, Fl. West Mid. Calif. 60. 1901. Based on *E. megastachya* Link.
Erosion ciliare Lunell, Amer. Midl. Nat. 4: 221. 1915. Lunell cites "*Eragrostis ciliaris* (All.) Link" as basis. Reference to "Hubbard, Philippine Jour. Sci. Bot. 8: 159–161. 1913" and the fact that this name is included in a list of plants of North Dakota indicate that Lunell meant *Eragrostis cilianensis* (All.) Lutati, rather than *E. ciliaris* (L.) Link.
Eragrostis eragrostis var. *megastachya* Farwell, Mich. Acad. Sci. Rpt. 17: 182. 1916. Based on *Poa megastachya* Koel.
?Eragrostis eragrostis subvar. *leersioides* Farwell, Amer. Midl. Nat. 10: 306. 1927. Based on *E. multiflora* var. *leersioides* Richt., this based on *Megastachya leersioides* Presl, described from Sicily, the description not applying to American forms.
(7) **Eragrostis ciliaris** (L.) R. Br. in Tuckey, Narr. Exp. Congo App. 478. 1818. Based on *Poa ciliaris* L.
Poa ciliaris L., Syst. Nat. ed. 10. 2: 875. 1759. Jamaica.
Megastachya ciliaris Beauv., Ess. Agrost. 74, 167, 174, 1812. Based on *Poa ciliaris* L.
Eragrostis villosa Trin., Fund. Agrost. 137. 1820. Based on *Poa ciliaris* L.
Cynodon ciliaris Raspail, Ann. Sci. Nat.,

Bot. 5: 302. 1825. Based on *Megastachya ciliaris* Beauv.

Macroblepharus contractus Phil., Linnaea 19: 101. 1858. Chile, *Gay* 129.

Eragrostis ciliaris var. *laxa* Kuntze, Rev. Gen. Pl. 2: 774. 1891. West Indies.

(3) **Eragrostis curtipedicellata** Buckl., Acad. Nat. Sci. Phila. Proc. 1862: 97. 1862. Northern Texas, *Buckley*.

Eragrostis brevipedicellata A. Gray, Acad. Nat. Sci. Phila. Proc. 1862: 336. 1862, as synonym of *E. curtipedicellata* Buckl.

Eragrostis viscosa Scribn., U. S. Dept. Agr., Div. Agrost. Bul. 11: 51. pl. 7. 1898. Not *E. viscosa* Trin., 1830. Midland, Tex., *J. G. Smith*.

(47) **Eragrostis curvula** (Schrad.) Nees, Fl. Afr. Austr. 397. 1841. Based on *Poa curvula* Schrad.

Poa curvula Schrad., Gött. Anz. Ges. Wiss. 3: 2073. 1821. Cape of Good Hope.

Eragrostis cyperoides (Thunb.) Beauv., Ess. Agrost. 71, 162, 174. 1812. Based on *Poa cyperoides* Thunb.

Poa cyperoides Thunb., Prodr. Pl. Cap. 22. 1794. South Africa, *Thunberg*.

(19) **Eragrostis diffusa** Buckl., Acad. Nat. Sci. Phila. Proc. 1862: 97. 1862. Northern Texas, *Buckley*.

Eragrostis purshii var. *delicatula* Munro ex Scribn., Torrey Bot. Club Bul. 10: 30. 1883. Name only. Arizona, *Pringle*.

Eragrostis purshii var. *diffusa* Vasey, U. S. Natl. Herb. Contrib. 1: 59. 1890. Based on *E. diffusa* Buckl.

(42) **Eragrostis elliottii** S. Wats., Amer. Acad. Sci. Proc. 25: 140. 1890. Based on *Poa nitida* Ell.

Poa nitida Ell., Bot. S. C. and Ga. 1: 162. 1816. Not *P. nitida* Lam., 1791. South Carolina.

Eragrostis nitida Chapm., Fl. South. U. S. 564. 1860. Not *E. nitida* Link, 1827. Based on *Poa nitida* Ell.

Eragrostis macropoda Pilger in Urban, Symb. Antill. 4: 106. 1903. Puerto Rico, *Sintenis* 1233.

(33) **Eragrostis erosa** Scribn. in Beal, Grasses N. Amer. 2: 483. 1896. Chihuahua, Mexico, *Pringle* 415.

(15) **Eragrostis frankii** C. A. Meyer ex Steud., Syn. Pl. Glum. 1: 273. 1854. Ohio, *Frank*.

Poa parviflora Nutt., Gen. Pl. 1: 67. 1818. Not *P. parviflora* R. Br. [United States].

Poa micrantha Schult., Mantissa 2: 305. 1824. Not *Eragrostis micrantha* Hack. 1895. Based on *P. parviflora* Nutt.

Eragrostis erythrogona Nees in Steud., Syn. Pl. Glum. 1: 273. 1854. St. Louis, *Drummond*.

Eragrostis capillaris var. *frankii* Farwell, Mich. Acad. Sci. Rpt. 17: 182. 1916. Based on *E. frankii* "Steud."

ERAGROSTIS FRANKII var. BREVIPES Fassett,

Rhodora 34: 95. 1932. Glenhaven, Wis., *Fassett* 12899.

(9) **Eragrostis glomerata** (Walt.) L. H. Dewey, U. S. Natl. Herb. Contrib. 2: 543. 1894. Based on *Poa glomerata* Walt.

Poa glomerata Walt., Fl. Carol. 80. 1788. South Carolina.

Poa conferta Ell., Bot. S. C. and Ga. 1: 158. 1816. South Carolina.

Megastachya glomerata Schult., Mantissa 2: 327. 1824. Based on *Poa glomerata* Walt.

Poa walteri Kunth, Rév. Gram. 1: 116. 1829. Based on *P. glomerata* Walt.

Eragrostis conferta Trin., Acad. St. Pétersb. Mém. VI. Math. Phys. Nat. 1: 409. 1830. Based on *Poa conferta* Ell.

Eragrostis pallida Vasey, U. S. Natl. Herb. Contrib. 1: 285. 1893. Colima, Mexico, *Palmer* 1268.

(30) **Eragrostis hirsuta** (Michx.) Nees, Agrost. Bras. 508. 1829. Based on *Poa hirsuta* Michx.

?*Poa simplex* Walt., Fl. Carol. 79. 1788. Not *Eragrostis simplex* Scribn., 1900. South Carolina.

Poa hirsuta Michx., Fl. Bor. Amer. 1: 68. 1803. South Carolina, *Michaux*.

Eragrostis sporoboloides Smith and Bush, Mo. Bot. Gard. Rpt. 6: 116. pl. 54. 1895. Sapulpa, Indian Territory [Okla.], *Bush* [766].

Eragrostis hirsuta var. *laevivaginata* Fernald, Rhodora 41: 500. 1939. Southampton County, Va., *Fernald* and *Long* 9273.

(11) **Eragrostis hypnoides** (Lam.) B. S. P., Prel. Cat. N. Y. 69. 1888. Based on *Poa hypnoides* Lam.

Poa hypnoides Lam., Tabl. Encycl. 1: 185. 1791. Tropical America.

Megastachya hypnoides Beauv., Ess. Agrost. 74, 167, 175. 1812. Based on *Poa hypnoides* Lam.

Poa reptans var. *caespitosa* Torr., Fl. North. and Mid. U. S. 1: 115. 1823. New Jersey.

Neeragrostis hypnoides Bush, Acad. Sci. St. Louis, Trans. 13: 180. 1903. Based on *Poa hypnoides* Lam.

Erosion hypnoides Lunell, Amer. Midl. Nat. 4: 221. 1915. Based on *Poa hypnoides* Lam.

(35) **Eragrostis intermedia** Hitchc., Wash. Acad. Sci. Jour. 23: 450. 1933. San Antonio, Tex., *Hitchcock* 5491.

Eragrostis lehmanniana Nees, Fl. Afr. Austr. 402. 1841. South Africa, *Drege*.

(31) **Eragrostis lugens** Nees, Agrost. Bras. 505. 1829. Brazil.

Poa lugens Kunth, Rév. Gram. 1: Sup. 28. 1830. Based on *Eragrostis lugens* Nees.

(23) **Eragrostis lutescens** Scribn., U. S. Dept. Agr., Div. Agrost. Cir. 9: 7. 1899. Almota, Wash., *Piper* 2624.

(28) **Eragrostis mexicana** (Hornem.) Link, Hort. Berol. 1: 190. 1827. Based on *"Poa mexicana* Lag. Hornem."
Poa mexicana Hornem., Hort. Hafn. 2: 953. 1815. Garden specimen from Mexican seed.
Poa mexicana Lag., Gen. et Sp. Nov. 3. 1816. Grown in Madrid from Mexican seed.
Small specimens of this species have been referred to *Eragrostis limbata* Fourn., a Mexican species, not known from the United States.

(21) **Eragrostis multicaulis** Steud., Syn. Pl. Glum. 1: 426. 1855. Japan.
Eragrostis pilosa var. *damiensiana* Bonnet, Naturaliste 3: 412. 1881. France.
Eragrostis pilosa var. *condensata* Hack., Allg. Bot. Ztschr. 7: 13. 1901. Karlsruhe, Germany, *Kneucker Gram. Exs.* 115.
Eragrostis peregrina Wiegand, Rhodora 19: 95. 1917. Based on *E. pilosa* var. *condensata* Hack.
Eragrostis damiensiana Thell., Repert. Sp. Nov. Fedde 24: 323. 1928. Based on *E. pilosa* var. *damiensiana* Bonnet.
Eragrostis damiensiana var. *condensata* Thell., Repert. Sp. Nov. Fedde 24: 328. 1928. Based on *E. pilosa* var. *condensata* Hack.

(27) **Eragrostis neomexicana** Vasey, U. S. Natl. Herb. Contrib. 2: 542. 1894. New Mexico, *Vasey.*
Eragrostis obtusa Munro ex Stapf, in Dyer, Fl. Cap. 7: 625. 1898. South Africa.

(1) **Eragrostis obtusiflora** (Fourn.) Scribn., U. S. Dept. Agr., Div. Agrost. Bul. 8: 10. pl. 5. 1897. Based on *Brizopyrum obtusiflorum* Fourn.
Brizopyrum obtusiflorum Fourn., Mex. Pl. 2: 120. 1886. Orizaba, Mexico, *Émy.*

(22) **Eragrostis orcuttiana** Vasey, U. S. Natl. Herb. Contrib. 1: 269. 1893. San Diego, Calif., *Orcutt* 1313.

(4) **Eragrostis oxylepis** (Torr.) Torr., U. S. Expl. Miss. Pacif. Rpt. 4: 156. 1857. Based on *Poa oxylepis* Torr.
Poa interrupta Nutt., Amer. Phil. Soc. Trans. (n.s.) 5: 146. 1837. Not *P. interrupta* Lam., 1791. Banks of the Arkansas [*Nuttall*].
Poa oxylepis Torr. in Marcy, Expl. Red Riv. 301. 1853. Based on *Poa interrupta* Nutt.
Eragrostis veraecrucis Rupr., Acad. Sci. Brux. Bul. 9²: 235. 1842, name only; Fourn., Mex. Pl. 2: 118. 1886, as synonym of *Megastachya oxylepis* var. *capitata* Fourn.
Megastachya oxylepis Fourn., Mex. Pl. 2: 118. 1886. Based on *Poa oxylepis* Torr.
Megastachya oxylepis var. *capitata* Fourn., Mex. Pl. 2: 118. 1886. Vera Cruz, Mexico.
Eragrostis interrupta Trel. in Branner and Coville, Ann. Rpt. Geol. Survey Ark. 4:

237. 1891. Not *E. interrupta* Beauv., 1812. Based on *Poa interrupta* Nutt.
Referred to *Eragrostis secundiflora* Presl in Manual ed. 1. That is a rare Mexican species not known from the United States.

(34) **Eragrostis palmeri** S. Wats., Amer. Acad. Sci. Proc. 18: 182. 1883. Juarez, Coahuila, *Palmer* 1368.
Eragrostis caudata Fourn., Mex. Pl. 2: 115. 1886. Not *E. caudata* Nees ex Steud. 1854. Mexico, *Berlandier* 2345.

(18) **Eragrostis pectinacea** (Michx.) Nees, Fl. Afr. Austr. 406. 1841. Based on *Poa pectinacea* Michx., the name given as *"Er. pectinacea* Michx."
Poa pectinacea Michx., Fl. Bor. Amer. 1: 69. 1803. Illinois, *Michaux.*
Poa eragrostis Ell., Bot. S. C. and Ga. 1: 161. 1816. Not *P. eragrostis* L., 1753. South Carolina and Georgia.
Poa tenella Nutt., Gen. Pl. 1: 67. 1818. Not *P. tenella* L., 1753. North America.
Eragrostis brizoides Schult., Mantissa 2: 319. 1824. Based on *Poa tenella* Nutt.
Poa nuttallii Kunth, Rév. Gram. 1: 116. 1829. Not *P. nuttallii* Spreng., 1825. Based on *Poa tenella* Nutt.
Eragrostis purshii Schrad., Linnaea 12: 451. 1838. North America; description inadequate; Gray, Man. ed. 2. 564. 1856.
Poa diandra Schrad., Linnaea 12: 451. 1838, as synonym of *Eragrostis purshii* Schrad.
Eragrostis nuttalliana Steud., Nom. Bot. ed. 2. 1: 563. 1840. Based on *Poa tenella* Nutt.
Eragrostis pennsylvanica Scheele, Flora 27: 58. 1844. Pennsylvania.
Eragrostis unionis Steud., Syn. Pl. Glum. 1: 273. 1854. Miami, Ohio.
Eragrostis cognata Steud., Syn. Pl. Glum. 1: 273. 1854. Ohio.

(17) **Eragrostis perplexa** L. H. Harvey, Univ. Microfilms, Publ. 967: 194. 1948. Mellette County, S. Dak., *W. L. Tolstead,* Aug. 30, 1935.

(40) **Eragrostis pilifera** Scheele, Linnaea 22: 344. 1849. New Braunfels, Tex., *Lindheimer.*
Eragrostis grandiflora Smith and Bush, Mo. Bot. Gard. Rpt. 6: 117. pl. 55. 1895. Sapulpa, Indian Territory [Okla.], *Bush* [808].
Eragrostis trichodes var. *pilifera* Fernald, Rhodora 40: 331. 1938. Based on *E. pilifera* Scheele.

(16) **Eragrostis pilosa** (L.) Beauv., Ess. Agrost. 71, 162, 175. 1812. Based on *Poa pilosa* L.
Poa pilosa L., Sp. Pl. 68. 1753. Italy.
Poa eragrostis Walt., Fl. Carol. 80. 1788. Not *P. eragrostis* L., 1753. South Carolina.
?*Poa tenella* [L. misapplied by] Pursh, Fl. Amer. Sept. 1: 80. 1814. New Jersey to Carolina. Elliott (Bot. S. C. and Ga. 1:

160. 1816) follows Pursh. According to
Merrill (U. S. Dept. Agr., Div. Agrost.
Cir. 29: 11. 1901) Elliott's plant is *E.
pilosa.*
Eragrostis filiformis Link, Hort. Berol. 1:
191. 1827. North America.
Poa linkii Kunth, Rév. Gram. 1: 113.
1829. Based on *Eragrostis filiformis*
Link.
Eragrostis linkii Steud., Syn. Pl. Glum. 1:
273. 1854. Based on *Poa linkii* Kunth.
(25) **Eragrostis poaeoides** Beauv., Ess.
Agrost. 162. 1812, name only; ex
Roem. and Schult., Syst. Veg. 2: 574.
1817. Based on *Poa eragrostis* L.
(Spelled *pooides* by Hylander, Uppsala
Univ. Årskr. 7: 71. 1945.)
Poa eragrostis L., Sp. Pl. 68. 1753. Italy.
Eragrostis minor Host, Icon. Gram.
Austr. 4: 15. 1809 [name untenable
because the genus was not validly
published until 1812]; Fl. Austr. 1:
135. 1827. Based on *Poa eragrostis* L.
Eragrostis eragrostis Beauv., Ess. Agrost.
71, 174. pl. 14. f. 11. 1812. Based on
Poa eragrostis L.
Eragrostis poaeformis Link, Hort. Berol.
1: 188. 1827. Based on *Poa eragrostis*
L.
Eragrostis vulgaris Presl ex Steud., Nom.
Bot. ed. 2. 1: 564. 1840, as synonym
of *E. poaeformis* Link.
Eragrostis vulgaris Coss. and Germ., Fl.
Env. Paris 2: 641. 1845. Based on
"*Poa eragrostis* et *Briza eragrostis* L.,"
the two species named var. *micro-
stachya* and *megastachya*, respectively.
Eragrostis vulgaris var. *microstachya* Coss.
and Germ., Fl. Env. Paris 2: 641.
1845. Based on *Poa eragrostis* L.
Eragrostis eragrostis var. *microstachya*
Farwell, Amer. Midl. Natl. 10: 306.
1927. Based on *E. vulgaris* var. *micro-
stachya* Coss. and Germ.
Eragrostis vulgaris subsp. *poaeoides* Douin
in Bonn., Fl. Compl. 12: 32. 1927–32.
Based on *E. poaeoides* Beauv.
(44) **Eragrostis refracta** (Muhl.) Scribn.,
Torrey Bot. Club Mem. 5: 49. 1894.
Based on *Poa refracta* Muhl.
?*Poa virginica* Zucc. ex Roemer, Col. Bot.
1: 124. 1809. Virginia.
Poa refracta Muhl. ex Ell., Bot. S. C. and
Ga. 1: 162. 1816. South Carolina.
Name only, Muhl., Cat. Pl. 12. 1813.
Eragrostis campestris Trin., Acad. St.
Pétersb. Mém. VI. Sci. Nat. 2¹: 72.
1836. North America.
Eragrostis longeradiata Steud., Syn. Pl.
Glum. 1: 272. 1854. Carolina, *Curtis.*
?*Eragrostis virginica* Steud., Syn. Pl.
Glum. 1: 273. 1854. Based on *Poa
virginica* Zucc.
Eragrostis pectinacea var. *refracta* Chapm.,
Fl. South. U. S. 564. 1860. Based on
Poa refracta Muhl.
Eragrostis campestris var. *refracta* Chapm.,

Fl. South. U. S. ed. 3. 617. 1897.
Based on *Poa refracta* Muhl.
Poa reflexa Ell. ex Scribn. and Merr.,
U. S. Dept. Agr., Div. Agrost. Cir. 27:
5. 1900, as synonym of *Eragrostis
refracta* Scribn.
This species was described under the
name *Poa capillaris* L., in Michx., Fl. Bor.
Amer. 1: 67. 1803.
(10) **Eragrostis reptans** (Michx.) Nees,
Agrost. Bras. 514. 1829. Based on
Poa reptans Michx.
Poa reptans Michx., Fl. Bor. Amer. 1:
69. pl. 11. 1803. Illinois, *Michaux.*
Poa dioica Michx. ex Poir. in Lam.,
Encycl. 5: 87. 1804, erroneously cited
as synonym of *P. hypnoides* Lam.
Kaskaskia River, Ill., *Michaux.*
Megastachya reptans Beauv., Ess. Agrost.
74, 167, 175. 1812. Based on *Poa
reptans* Michx.
Poa weigeltiana Reichenb. ex Trin., Acad.
St. Pétersb. Mém. VI. Math. Phys.
Nat. 1: 410. 1830, as synonym of
Eragrostis reptans Nees. Dutch Guiana,
Weigelt.
Poa dioica Vent. ex Kunth, Enum. Pl. 1:
336. 1833, as synonym of *P. reptans*
Michx.
Poa capitata Nutt., Amer. Phil. Soc.
Trans. (n.s.) 5: 146. 1837. Arkansas
River, *Nuttall.*
Eragrostis capitata Nash in Britton, Man.
1042. 1901. Based on *Poa capitata*
Nutt.
Neeragrostis weigeltiana Bush, Acad. Sci.
St. Louis, Trans. 13: 178. 1903.
Based on *Poa weigeltiana* Reichenb.
Eragrostis weigeltiana Bush, Acad. Sci.
St. Louis, Trans. 13: 180. 1903. Based
on *Poa weigeltiana* Reichenb.
(2) **Eragrostis sessilispica** Buckl., Acad.
Nat. Sci. Phila. Proc. 1862: 97. 1862.
Austin, Tex., *Buckley.*
Diplachne rigida Vasey, U. S. Dept. Agr.,
Div. Bot. Bul. 12²: pl. 44. 1891.
Texas [type, *Reverchon* in 1879], and
New Mexico, northward to Kansas.
Leptochloa rigida Munro ex Vasey, U. S.
Dept. Agr., Div. Bot. Bul. 12²: pl. 44.
1891, as synonym of *Diplachne rigida*
Vasey.
Eragrostis rigida Scribn., Acad. Nat. Sci.
Phila. Proc. 1891: 304. 1891. Based
on *Diplachne rigida* Vasey.
Acamptoclados sessilispicus Nash in Small,
Fl. Southeast. U. S. 140. 1903. Based
on *Eragrostis sessilispica* Buckl.
(38) **Eragrostis silveana** Swallen, Amer.
Jour. Bot. 19: 438. f. 3. 1932. Taft,
Tex., *Silveus* 360.
(12) **Eragrostis simplex** Scribn., U. S. Dept.
Agr., Div. Agrost. Bul. 7 (ed. 3):
250. f. 244. 1900. Florida, *Curtiss* 6073.
Eragrostis brownei Kunth ex Chapm., Fl.
South. U. S. ed. 2. 664. 1883. Not *E.
brownei* Nees, 1841. East Florida,

Garber. (Chapman probably had *E. brownei* (Kunth) Nees, an Australian species, in mind, but he cites nothing that can connect his publication with that. The name *E. brownei* Nees is used for *E. simplex* by Scribner (U. S. Dept. Agr., Div. Agrost. Bul. 7: 262. 1897.)

(41) **Eragrostis spectabilis** (Pursh) Steud., Nom. Bot. ed. 2. 1: 564. 1840. Based on *Poa spectabilis* Pursh.

?Poa amabilis Walt., Fl. Carol. 80. 1788. Not *P. amabilis* L., 1753. South Carolina.

Poa spectabilis Pursh, Fl. Amer. Sept. 1: 81. 1814. New York to Carolina.

Megastachya spectabilis Roem. and Schult., Syst. Veg. 2: 589. 1817. Based on *Poa spectabilis* Pursh.

Poa hirsuta var. *spectabilis* Torr., Fl. North. and Mid. U. S. 1: 114. 1823. Based on *Poa spectabilis* Pursh.

?Eragrostis velutina Schrad., Linnaea 12: 451. 1838. Carolina.

?Poa villosa Beyr. ex Schrad., Linnaea 12: 451. 1838, as synonym of *E. velutina* Schrad.

Eragrostis geyeri Steud., Syn. Pl. Glum 1: 272. 1854. Illinois, *Geyer.*

Poa pectinacea Geyer ex Steud., Syn. Pl. Glum. 1: 272. 1854. Not *P. pectinacea* Michx., 1803. As synonym of *Eragrostis geyeri* Steud.

Eragrostis pectinacea var. *spectabilis* A. Gray, Man. ed. 2. 565. 1856. Based on *Poa spectabilis* Pursh.

Erochloë spectabilis Raf. ex Jacks., Ind. Kew. 1: 886. 1893.

Eragrostis spectabilis var. *sparsihirsuta* Farwell, Amer. Midl. Nat. 10: 306. 1927. Michigan.

This is the species called *Poa pectinacea* Michx. and *Eragrostis pectinacea* Nees by American authors, not Michaux's species.

(6) **Eragrostis spicata** Vasey, Bot. Gaz. 16: 146. 1891. Baja California, *Brandegee.*

Sporobolus tenuispica Hack., Repert. Sp. Nov. Fedde 6: 344. 1909. Pilcomayo River, Paraguay, *Rojas 258.*

Eragrostis stenophylla Hochst. ex Miquel, An. Bot. Ind. 2: 27. 1851. Asia.

(36) **Eragrostis swalleni** Hitchc., Wash. Acad. Sci. Jour. 23: 451. 1933. Riviera, Tex., *Swallen 1847.*

Eragrostis tef (Zuccagni) Trotter, Soc. Bot. Ital. Bul. 1918: 62. 1918. Based on *Poa tef* Zuccagni.

Poa tef Zuccagni, Diss. Concern. l'Ist. Pianta Paniz. Abiss. 1775. Abyssinia (no page).

Poa abyssinica Jacq., Misc. Austr. 2: 364. 1781. Abyssinia.

Eragrostis abyssinica Link, Hort. Berol. 1: 192. 1827. Based on *Poa abyssinica* Jacq.

(20) **Eragrostis tephrosanthos** Schult., Mantissa 2: 316. 1824. Martinique, *Sieber.*

Poa tephrosanthos Spreng. ex Schult., Mantissa 2: 316. 1824, as synonym of *Eragrostis tephrosanthos* Schult.

Eragrostis delicatula Trin., Acad. St. Pétersb. Mém. VI. Sci. Nat. 2^1: 73. 1836. Brazil.

Eragrostis pilosa var. *delicatula* Hack. in Stuck., An. Mus. Nac. Buenos Aires 11: 133. 1904. Based on *E. delicatula* Trin.

(37) **Eragrostis tracyi** Hitchc., Amer. Jour. Bot. 21: 130. f. 1. 1934. Sanibel Island, Fla., *Tracy 7168.*

(32) **Eragrostis trichocolea** Hack. and Arech., An. Mus. Nac. Montevideo 1: 444. 1896. Uruguay.

Eragrostis floridana Hitchc., Amer. Jour. Bot. 2: 308. 1915. Tampa, Fla., *Curtiss 3494*.*

(39) **Eragrostis trichodes** (Nutt.) Wood, Class-book ed. 1861. 796. 1861. Based on *Poa trichodes* Nutt.

Poa trichodes Nutt., Amer. Phil. Soc. Trans. (n.s.) 5: 146. 1837. Arkansas, *Nuttall.*

Eragrostis tenuis var. *texensis* Vasey, U. S. Natl. Herb. Contrib. 1: 59. 1890. Texas, *Nealley.*

Eragrostis tenuis A. Gray, Man. ed. 6. 661. 1890. Not *E. tenuis* Steud., 1854. Ohio to Illinois, Kansas and southward.

Eragrostis capillacea Jedw., Bot. Archiv Mez 5: 196. 1924. Nebraska, *Rydberg 1832.*

(13) **Eragrostis unioloides** (Retz.) Nees in Steud., Syn. Pl. Glum. 1: 264. 1854. Based on *Poa unioloides* Retz.

Poa unioloides Retz., Obs. Bot. 5: 19. 1789. East Indies.

Eragrostis virescens Presl, Rel. Haenk. 1: 276. 1830. Chile, *Haenke.*

Eremochloa ciliaris (L.) Merr., Philippine Jour. Sci. 1 (Sup. 5): 331. 1906. Based on *Nardus ciliaris* L.

Nardus ciliaris L. Sp. Pl. 53. 1753. India.

Eremochloa ophiuroides (Munro) Hack. in DC., Monogr. Phan. 6: 261. 1889. Based on *Ischaemum ophiuroides* Munro.

Ischaemum ophiuroides Munro, Amer. Acad. Sci. Proc. 4: 363. 1860. Southern China.

(151) ERIANTHUS Michx.

(3) **Erianthus alopecuroides** (L.) Ell., Bot. S. C. and Ga. 1: 38. 1816. Based on *Andropogon alopecuroides* L.

Andropogon divaricatus L., Sp. Pl. 1045. 1753. Virginia [*Clayton 70*].

Andropogon alopecuroides L., Sp. Pl. 1045. 1753. Virginia [*Clayton 601*].

Saccharum alopecuroideum Nutt., Gen. Pl. 1: 60. 1818. Based inferentially on *Erianthus alopecuroides* Ell.

Erianthus divaricatus Hitchc., U. S. Natl.

Herb. Contrib. 12: 125. 1908. Based on *Andropogon divaricatus* L.

ERIANTHUS ALOPECUROIDES var. HIRSUTIS Nash in Small, Fl. Southeast. U. S. 55. 1903. Florida [*Chapman*]. (Published as *E. alopecuroides hirsutus.*)

(4) **Erianthus brevibarbis** Michx., Fl. Bor. Amer. 1: 55. 1803. "Tennessee and Carolina," *Michaux.* The only specimen in the Michaux Herbarium bearing this name is from dry hills 5 days distant from the Wabash River toward the mouth of the Missouri, that is, in southern Illinois, where it has not since been found.

Saccharum brevibarbe Pers., Syn. Pl. 1: 103. 1805. Based on *Erianthus brevibarbis* Michx.

Calamagrostis rubra Bosc ex Kunth, Enum. Pl. 1: 478. 1833, as synonym of *Erianthus brevibarbis* Michx.

Erianthus alopecuroides var. *brevibarbis* Chapm., Fl. South U. S. 583. 1860. Based on *E. brevibarbis* Michx.

Erianthus saccharoides subsp. *brevibarbis* Hack. in DC., Monogr. Phan. 6: 131. 1889. Based on *E. brevibarbis* Michx.

(5) **Erianthus coarctatus** Fernald, Rhodora 45: 246. pl. 758. 1943. Homeville, Sussex County, Va., *Fernald* and *Long* 7301.

ERIANTHUS COARCTATUS var. ELLIOTTIANUS Fernald, Rhodora 45: 246. 1943. Live Oak, Fla., *Curtiss 6940.*

(2) **Erianthus contortus** Baldw. ex Ell., Bot. S. C. and Ga. 1: 40. 1816. Savannah, Ga., *Baldwin.*

Saccharum contortum Nutt., Gen. Pl. 1: 60. 1818. Based on *Erianthus contortus* Ell.

Erianthus alopecuroides var. *contortus* Chapm., Fl. South. U. S. 582. 1860. Based on *E. contortus* Ell.

Erianthus saccharoides subsp. *contortus* Hack. in DC., Monogr. Phan. 6: 131. 1889. Based on *E. contortus* Ell.

Erianthus smallii Nash, N. Y. Bot. Gard. Bul. 1: 429. 1900. Stone Mountain, Ga., *Small* in 1894.

(6) **Erianthus giganteus** (Walt.) Muhl., Cat. Pl. 4. 1813. Based on *Anthoxanthum giganteum* Walt. Later (Descr. Gram. 192. 1817) Muhlenberg uses the name for both *E. saccharoides* and *E. alopecuroides* (his herbarium specimen under this name including both species), but the description (awn twisted) applies better to *E. alopecuroides. Erianthus giganteus* was published as new by Hubbard (Rhodora 14: 166. 1912) based on *Anthoxanthum giganteum* Walt.

Anthoxanthum giganteum Walt., Fl. Carol. 65. 1788. South Carolina.

Erianthus saccharoides Michx., Fl. Bor. Amer. 1: 55. 1803. Carolina to Florida, *Michaux.*

Saccharum giganteum Pers., Syn. Pl. 1: 103. 1805. Based on *Anthoxanthum giganteum* Walt.

Saccharum erianthoides Raspail, Ann. Sci. Nat., Bot. 5: 308. 1825. Based on *Erianthus saccharoides* Rich. [same as Michx.].

Andropogon erianthus Link, Hort. Berol. 1: 243. 1827. Based on *Erianthus saccharoides* Michx.

Erianthus saccharoides var. *michauxii* Hack. in Mart., Fl. Bras. 2³: 257. 1883. Based on *E. saccharoides* Michx.

Erianthus compactus Nash, Torrey Bot. Club Bul. 22: 419. 1895. New Jersey to North Carolina and Tennessee [type, Washington, D. C., *Nash* in 1895].

Erianthus laxus Nash, Torrey Bot. Club Bul. 24: 344. 1897. Near Paola, Fla., *Swingle* 1732a.

Erianthus tracyi Nash, Torrey Bot. Club Bul. 24: 37. 1897. Starkville, Miss., *Tracy* in 1896.

Erianthus saccharoides var. *compactus* Fernald, Rhodora 45: 252. 1943. Based on *E. compactus* Nash.

Erianthus ravennae (L.) Beauv., Ess. Agrost. 14, 162, 177. 1812. Based on *Saccharum ravennae* Murr., this based on *Andropogon ravennae* L.

Andropogon ravennae L., Sp. Pl. ed. 2. 2: 1481. 1763. Italy.

Saccharum ravennae Murr. in L., Syst. Veg. ed., 13. 88. 1774. Based on *Andropogon ravennae* L.

Ripidium ravennae Trin., Fund. Agrost. 169. 1820. Based on *Saccharum ravennae* Murr.

ERIANTHUS RAVENNAE var. PURPURASCENS (Anderss.) Hack. in DC., Monogr. Phan. 6: 140. 1889. Based on *E. purpurascens* Anderss.

Erianthus purpurascens Anderss., Svenska Vetensk. Akad. Öfversigt af Förhandl. 12: 161. 1855. India, *Hugel.*

(1) **Erianthus strictus** Baldw. in Ell., Bot. S. C. and Ga. 1: 39. 1816. Savannah, Ga., *Baldwin.*

Saccharum strictum Nutt., Gen. Pl. 1: 60. 1818. Based on *Erianthus strictus* Baldw.

Saccharum baldwinii Spreng., Syst. Veg. 1: 282. 1825. Based on *Erianthus strictus* Baldw.

Pollinia dura Trin., Acad. St. Pétersb. Mém. VI. Sci. Nat. 2¹: 91. 1836. Carolina.

Andropogon durus Steud., Nom. Bot. ed. 2. 1: 91. 1840. Based on *Pollinia dura* Trin.

(132) **ERIOCHLOA** H. B. K.

(1) **Eriochloa aristata** Vasey, Torrey Bot. Club Bul. 13: 229. 1886. Southwest Chihuahua, *Palmer* in 1885 [110e].

Eriochloa punctata var. *aristata* Jones, West. Bot. Contrib. 14: 11. 1912. Based on *E. arıstata* Vasey.

(6) **Eriochloa contracta** Hitchc., Biol. Soc. Wash. Proc. 41: 163. 1928. Based on *Helopus mollis* C. Muell.

Helopus mollis C. Muell., Bot. Ztg. 19: 314. 1861. Not *Eriochloa mollis* Kunth, 1829. Texas, *Drummond* 370.

(5) **Eriochloa gracilis** (Fourn.) Hitchc., Wash. Acad. Sci. Jour. 23: 455. 1933. Based on *Helopus gracilis* Fourn.

Helopus gracilis Fourn., Mex. Pl. 2: 13. 1886. Oaxaca, Mexico, *Liebmann* 436.

ERIOCHLOA GRACILIS var. MINOR (Vasey) Hitchc., Wash. Acad. Sci. Jour. 23: 456. 1933. Based on *E. punctata* var. *minor* Vasey.

Eriochloa punctata var. *minor* Vasey, U. S. Natl. Herb. Contrib. 3: 21. 1892. Texas, *Wright* 2087, *Nealley.*

Eriochloa texana Mez, Bot. Jahrb. 56: Beibl. 125: 12. 1921. [El Paso] Tex., *Jones* 4177.

(3) **Eriochloa lemmoni** Vasey and Scribn., Bot. Gaz. 9: 185. pl. 2. 1884. [Huachuca Mountains], Ariz., *Lemmon* 2910.

(8) **Eriochloa michauxii** (Poir.) Hitchc., U. S. Natl. Herb. Contrib. 12: 147. 1908. Based on *Panicum michauxii* Poir.

Panicum molle Michx., Fl. Bor. Amer. 1: 47. 1803. Not *P. molle* Swartz, 1788. Florida, *Michaux.*

Panicum michauxii Poir. in Lam., Encycl. Sup. 4: 278. 1816. Based on *P. molle* Michx.

Panicum michauxianum Schult., Mantissa 2: 227. 1824. Based on *P. molle* Michx.

Panicum georgicum Spreng., Syst. Veg. 1: 308. 1825. Based on *P. molle* Michx.

Eriochloa mollis Kunth, Rév. Gram. 1: 30. 1829. Based on *Panicum molle* Michx.

Eriochloa mollis var. *longifolia* Vasey, Torrey Bot. Club Bul. 13: 25. 1886. Key West, Fla., *Curtiss.*

Eriochloa longifolia Vasey, U. S. Natl. Herb. Contrib. 3: 21. 1892. Based on *E. mollis* var. *longifolia* Vasey.

Eriochloa debilis Mez, Bot. Jahrb. 56: Beibl. 125: 12. 1921. [No-name Key], Fla., *Curtiss* 3600. The same form as *E. longifolia* Vasey.

ERIOCHLOA MICHAUXII var. SIMPSONI Hitchc., Biol. Soc. Wash. Proc. 41: 163. 1928. Cape Romano, Fla., *Simpson* 262. (Published as *E. michauxii simpsoni*).

(4) **Eriochloa procera** (Retz.) C. E. Hubb., Kew Bul. Misc. Inf. 1930: 256. 1930. Based on *Agrostis procera* Retz.

Agrostis procera Retz., Obs. Bot. 4: 19. 1786. India.

Milium ramosum Retz., Obs. Bot. 6: 22. 1791. Asia.

Paspalum annulatum Flügge, Monogr. Pasp. 133. 1810. Asia.

Agrostis ramosa Poir. in Lam., Encycl. Sup. 1: 257. 1810. Based on *Milium ramosum* Retz.

Eriochloa annulata Kunth, Rév. Gram. 1: 30. 1829. Based on *Paspalum annulatum* Flügge.

Helopus annulatus Nees, Agrost. Bras. 17. 1829. Based on *Paspalum annulatum* Flügge.

Eriochloa ramosa Kuntze, Rev. Gen. Pl. 2: 775. 1891. Based on *Milium ramosum* Retz.

Eriochloa polystachya var. *annulata* Maid. and Betche, Cens. N. S. Wales Pl. 16. 1916. Based on *E. annulata* Kunth.

Thysanolaena procera Mez in Janow., Bot. Archiv Mez 1: 27. 1922. Based on *Agrostis procera* Retz., but misapplied to *T. maxima* (Roxb.) Kuntze.

(7) **Eriochloa punctata** (L.) Desv. ex Hamilt., Prodr. Pl. Ind. Occ. 5. 1825. Based on *Milium punctatum* L.

Milium punctatum L., Syst. Nat. ed. 10. 2: 872. 1759. Jamaica.

Agrostis punctata Lam., Encycl. 1: 58. 1783. Based on *Milium punctatum* L.

Paspalum punctatum Flügge, Mongr. Pasp. 127. 1810. Based on *Milium punctatum* L.

Piptatherum punctatum Beauv., Ess. Agrost. 18, 173. 1812. Based on *Milium punctatum* L.

Eriochloa kunthii G. Meyer, Prim. Fl. Esseq. 47. 1818. British Guiana.

Oedipachne punctata Link, Hort. Berol. 1: 51. 1827. Based on *Milium punctatum* L.

Helopus punctatus Nees, Agrost. Bras. 16. 1829. Based on *Milium punctatum* L.

Helopus kunthii Trin. ex Steud., Nom. Bot. ed. 2. 1: 747. 1840. Based on *Eriochloa kunthii* G. Meyer.

Monachne punctata Nash, Torrey Bot. Club Bul. 30: 374. 1903. Based on *Milium punctatum* L.

Eriochloa polystachya var. *punctata* Maid. and Betche, Cens. N. S. Wales Pl. 16. 1916. Based on *E. punctata* Desv.

(2) **Eriochloa sericea** (Scheele) Munro in Vasey, U. S. Dept. Agr., Div. Bot. Bul. 12¹: pl. 1. 1890. Based on *Paspalum sericeum* Scheele, as shown by Munro manuscript in Kew Herbarium.

Paspalum racemosum Nutt., Amer. Phil. Soc. Trans. (n. s.) 5: 145. 1837. Not *P. racemosum* Lam., 1791. Red River, Ark. [**Nuttall**].

Paspalum sericeum Scheele, Linnaea 22: 341. 1849. New Braunfels, Tex., *Lindheimer.*

Panicum sericatum Scheele ex Steud., Syn. Pl. Glum. 1: 58. 1854. Based on *Paspalum sericeum* Scheele.

Helopus junceus C. Muell., Bot. Ztg. 19: 314. 1861. Texas, *Drummond* 305 and 368.

Eriochloa villosa (Thunb.) Kunth, Rév.

Gram. 1: 30. 1829. Based on *Paspalum villosum* Thunb.

Paspalum villosum Thunb., Fl. Jap. 45. 1784. Japan.

(167) EUCHLAENA Schrad.

(1) Euchlaena mexicana Schrad., Ind. Sem. Hort. Goettingen 1832; reprinted in Linnaea 8: Litt. 25. 1833. Mexico, *Muhlenfordt.*

Reana luxurians Durieu, Soc. Acclim. Bul. II. 9: 581. 1872. This and the following are names only. They have, however, come into frequent use for teosinte.

Euchlaena luxurians Durieu and Aschers., Soc. Linn. Paris Bul. 1: 107. 1877. Based on *Reana luxurians* Durieu.

Euchlaena mexicana var. *luxurians* Haines, Bot. Bihar and Orissa pt. 6: 1065. 1924. Based on *Reana luxurians* "Brogn." (error for Durieu).

Zea mexicana Reeves and Mangelsd., Amer. Jour. Bot. 29: 817. 1942. Based on *Euchlaena mexicana* Schrad.

(2) Euchlaena perennis Hitchc., Wash. Acad. Sci. Jour. 12: 207. 1922. Zapotlan, Jalisco, Mexico, *Hitchcock 7146.*

(4) FESTUCA L.

Festuca amethystina L., Sp. Pl. 74. 1753. Europe.

(9) Festuca arida Elmer, Bot. Gaz. 36: 52. 1903. North Yakima, Wash., *Henderson 2196.*

This species was referred by Piper to *Festuca eriolepis* Desv., a South American species not known from North America.

(33) Festuca arizonica Vasey, U. S. Natl. Herb. Contrib. 1: 277. 1893. Flagstaff, Ariz., *Tracy 118.*

Festuca ovina var. *arizonica* Hack. ex Beal, Grasses N. Amer. 2: 598. 1896. Based on *F. arizonica* Vasey.

Festuca vaseyana Hack. ex Beal., Grasses N. Amer. 2: 601. 1896. Veta Pass, Colo., *Vasey.*

Festuca scabrella var. *vaseyana* Hack. ex Beal, Grasses N. Amer. 2: 605. 1896. Veta Pass, Colo., *Vasey.*

Festuca altaica subsp. *arizonica* St. Yves, Candollea 2: 267. 1925. Based on *F. arizonica* Vasey.

Festuca arundinacea Schreb., Spic. Fl. Lips. 57. 1771. Germany.

Bromus arundinaceus Roth, Tent. Fl. Germ. 2: 141. 1789. Based on *Festuca arundinacea* Schreb.

Festuca elatior var. *arundinacea* Wimm., Fl. Schles. ed. 3. 59. 1857. Based on *F. arundinacea* Schreb.

(25) Festuca californica Vasey, U. S. Natl. Herb. Contrib. 1: 277. 1893. Oakland, Calif., *Bolander 1505.*

Bromus kalmii var. *aristulatus* Torr., U. S.

Expl. Miss. Pacif. Rpt. 4: 157. 1856. Mark West Creek, Calif., *Bigelow.*

Festuca aristulata Shear ex Piper, U. S. Natl. Herb. Contrib. 10: 32. 1906. Based on *Bromus kalmii* var. *aristulatus* Torr.

Festuca aristulata parishii Piper, U. S. Natl. Herb. Contrib. 10: 33. 1906. Mill Creek Falls, San Bernardino Mountains, Calif., *Parish 5036.*

Festuca parishii Hitchc. in Jepson, Fl. Calif. 1: 169. 1912. Based on *F. aristulata parishii* Piper.

Festuca californica parishii Hitchc. in Abrams, Illustr. Fl. 1: 222. 1923. Based on *F. aristulata parishii* Piper.

Festuca altaica var. *aristulata* St. Yves, Candollea 2: 273. 1925. Based on *Bromus kalmii* var. *aristulatus* Torr.

(31) Festuca capillata Lam., Fl. Franç. 3: 597. 1778. France.

Festuca ovina var. *capillata* Alefeld, Landw. Fl. 354. 1866. Based on *F. capillata* Lam.

(7) Festuca confusa Piper, U. S. Natl. Herb. Contrib. 10: 13. pl. 1. 1906. Western Klickitat County, Wash., *Suksdorf 1140.*

Festuca microstachya var. *ciliata* A. Gray, Amer. Acad. Sci. Proc. 8: 410. 1872. Name only, for *Hall 639* in 1871, Silver Creek, Oreg.

Festuca suksdorfii Piper in Suksdorf, Werdenda 1²: 2. 1923. Bingen, Wash., *Suksdorf 5604.*

(26) Festuca dasyclada Hack. ex Beal, Grasses N. Amer. 2: 602. 1896. Utah, *Parry* in 1875.

(4) Festuca dertonensis (All.) Aschers. and Graebn., Syn. Mitteleur. Fl. 2: 588. 1900. Based on *Bromus dertonensis* All.

Bromus dertonensis All., Fl. Pedem. 2: 249. 1785. Italy.

Vulpia dertonensis Volk. in Schinz and Keller, Fl. Schweiz ed. 2. 57 (not in Washington); Dur. and Barr., Fl. Lib. Prodr. 269. 1910. Based on *Festuca dertonensis* Aschers. and Graebn.

This is the species referred by American authors to *F. bromoides* L. That seems to be a mixture; the name is referred to *F. myuros* by European authors.

(12) Festuca eastwoodae Piper, U. S. Natl. Herb. Contrib. 10: 16. 1906. Santa Lucia Mountains, Monterey County, Calif., *Eastwood.*

(17) Festuca elatior L., Sp. Pl. 75. 1753. Europe.

Festuca pratensis Huds., Fl. Angl. 37. 1762. England.

Festuca fluitans var. *pratensis* Huds., Fl. Angl. ed. 2. 47. 1778. Based on *F. pratensis* Huds.

Avena secunda Salisb., Prodr. Stirp. 22. 1796. Based on *Festuca elatior* L.

Bromus elatior Koel., Descr. Gram. 214. 1802. Based on *Festuca elatior* L.

Festuca poaeoides Michx., Fl. Bor. Amer. 1: 67. 1803. St. Lawrence River, *Michaux*.

Festuca poaeoides americana Pers., Syn. Pl. 1: 94. 1805. Based on *F. poaeoides* Michx.

Schedonorus elatior Beauv., Ess. Agrost. 99, 156, 177. 1812. Based on *Bromus elatior* Koel.

Schedonorus pratensis Beauv., Ess. Agrost. 99, 163, 177. 1812. Based on *Festuca pratensis* Huds.

Festuca americana F. G. Dietr., Vollst. Lex. Gärtn. Bot. Nachtr. 3: 332. 1817. Based on *F. poaeoides americana* Pers.

Schenodorus americanus Roem. and Schult., Syst. Veg. 2: 706. 1817. (Error for *Schedonorus*). Based on *Festuca poaeoides americana* Pers.

Bromus pratensis Spreng., Syst. Veg. 1: 359. 1825. Not *B. pratensis* Lam., 1785. Based on *Festuca pratensis* Huds.

Bucetum pratense Parnell, Grasses Scotl. 105. pl. 46. 1842. Based on *Festuca pratensis* Huds.

Bucetum elatius Parnell, Grasses Scotl. 107. pl. 46. 1842. Based on *Festuca elatior* L.

Festuca elatior var. *pratensis* A. Gray, Man. ed. 5. 634. 1867. Based on *F. pratensis* Huds.

Tragus elatior Panz. ex Jacks., Ind. Kew. 2: 1098. 1895, as synonym of *Festuca elatior* L.

Gnomonia elatior Lunell, Amer. Midl. Nat. 4: 224. 1915. Based on *Festuca elatior* L.

(16) **Festuca elmeri** Scribn. and Merr., Torrey Bot. Club Bul. 29: 468. 1902. Stanford University, Calif., *Elmer* 2101.

Festuca elmeri var. conferta (Hack.) Hitchc., Amer. Jour. Bot. 21: 128. 1934. Based on *F. jonesii* var. *conferta* Hack.

Festuca jonesii var. *conferta* Hack. ex Beal, Grasses N. Amer. 2: 593. 1896. San Jose Normal School, California.

Festuca elmeri luxurians Piper, U. S. Natl. Herb. Contrib. 10: 38. 1906. Based on *F. jonesii* var. *conferta* Hack.

Festuca geniculata (L.) Cav., An. Cienc. Nat. Madrid 6: 150. 1803. Based on *Bromus geniculatus* L.

Bromus geniculatus L., Mant. Pl. 33. 1767. Portugal.

Festuca gigantea (L.) Vill., Hist. Pl. Dauph. 2: 110. 1787. Based on *Bromus giganteus* L.

Bromus giganteus L., Sp. Pl. 77. 1753. Europe.

Zerna gigantea Panz. ex Jacks., Ind. Kew. 2: 1249. 1895. Based on *Bromus giganteus* L.

Forasaccus giganteus Bubani, Fl. Pyr. 4: 383. 1901. Based on *Bromus giganteus* L.

(8) **Festuca grayi** (Abrams) Piper, U. S.

Natl. Herb. Contrib. 10: 14. pl. 3. 1906. Based on *F. microstachys grayi* Abrams.

Festuca microstachys var. *ciliata* A. Gray ex Beal, Grasses N. Amer. 2: 585. 1896. Not *F. ciliata* Gouan, 1762. Grants Pass, Oreg., *Howell*. Beal's specimen is a mixture of *F. grayi* and *F. confusa*, but the description applies to *F. grayi*.

Festuca microstachys grayi Abrams, Fl. Los Angeles 52. 1904. Based on *F. microstachys* var. *ciliata* A. Gray ex Beal.

Festuca pacifica var. *ciliata* Hoover, Madroño 3: 227. 1936. Based on *F. microstachys* var. *ciliata* A. Gray.

(32) **Festuca idahoensis** Elmer, Bot. Gaz. 36: 53. 1903. Smiths Valley, Shoshone County, Idaho, *Abrams* 688.

Festuca ovina var. *ingrata* Hack. ex Beal, Grasses N. Amer. 2: 598. 1896. Oregon, *Howell*.

Festuca ovina var. *columbiana* Beal, Grasses N. Amer. 2: 599. 1896. [Blue Mountains], Wash., *Lake*.

Festuca ovina var. *oregona* Hack. ex Beal, Grasses N. Amer. 2: 599. 1896. Oregon, *Cusick* 753.

Festuca ingrata Rydb., Torrey Bot. Club Bul. 32: 608. 1905. Based on *F. ovina* var. *ingrata* Hack.

Fsetuca ingrata nudata Rydb., Colo. Agr. Expt. Sta. Bul. 100: 50. 1906. "*F. ovina* var. *nudata* Vasey," (herbarium name only), Colorado [*Beardslee* in 1892].

Festuca amethystina var. *asperrima* subvar. *idahoensis* St.-Yves, Candollea 2: 260. 1925. Based on *F. idahoensis* Elmer.

Festuca amethystina var. *asperrima* subvar. *robusta* St. Yves, Candollea 2: 264. 1925. Walla Walla, Wash., *Piper* 2410.

(23) **Festuca ligulata** Swallen, Amer. Jour. Bot. 19: 436. f. 1. 1932. Guadalupe Mountains, Tex., *Moore* and *Steyermark* 3576.

(3) **Festuca megalura** Nutt., Jour. Acad. Phila. II. 1: 188. 1848. Santa Barbara, Calif., *Gambel*.

Vulpia megalura Rydb., Torrey Bot. Club Bul. 36: 538. 1909. Based on *Festuca megalura* Nutt.

(11) **Festuca microstachys** Nutt., Jour. Acad. Phila. II. 1: 187. 1848. Los Angeles, Calif., *Gambel*.

Vulpia microstachya Munro ex Benth., Pl. Hartw. 342. 1857. Based on *Festuca microstachys* Nutt.

?*Vulpia microstachya* var. *ciliata* Munro ex Benth., Pl. Hartw. 342. 1857. Name only, for *Hartweg* 281, Sacramento, Calif.

Festuca microstachys var. *subappressa* Suksdorf, Werdenda 1²: 3. 1923. Bingen, Wash., *Suksdorf* 6236.

(5) **Festuca myuros** L., Sp. Pl. 74. 1753.
Europe.
Avena muralis Salisb., Prodr. Stirp. 22.
1796. Based on *Festuca myuros* L.
Vulpia myuros K. Gmel., Fl. Badens.
1: 8. 1805. Based on *Festuca myuros* L.
Festuca myuros Muhl., Descr. Gram. 160.
1817. Maryland; Georgia. Probably *F. myuros* L. is referred to, Muhlenberg's
specimen being a mixture of this and
F. sciurea Nutt.
Distomomischus myuros Dulac, Fl. Haut.
Pyr. 91. 1867. Based on *Vulpia
myuros* K. Gmel.
Zerna myuros Panz. ex Jacks., Ind. Kew.
2: 1249. 1895, as synonym of *Festuca
myuros* L.
(20) **Festuca obtusa** Bieler, Pl. Nov. Herb.
Spreng. Cent. 11. 1807. Pennsylvania,
Muhlenberg. Name only, Muhl., Amer.
Phil. Soc. Trans. 3: 161. 1793. Pennsylvania.
Poa laxa Lam., Tabl. Encycl. 1: 183.
1791. Not *P. laxa* Haenke, 1791.
Virginia.
Panicum divaricatum Michx., Fl. Bor.
Amer. 1: 50. 1803. Not *P. divaricatum*
L., 1753. Carolina. (Michaux's plant an
old specimen with all but the lowest
floret fallen from the spikelets.)
Poa subverticillata Pers., Syn. Pl. 1: 92.
1805. Based on *Poa laxa* Lam.
Panicum gracilentum Poir. in Lam.,
Encycl. Sup. 4: 276. 1816. Cultivated
in Paris botanic garden.
Panicum debile Poir. in Lam., Encycl.
Sup. 4: 283. 1816. Not *P. debile* Desf.,
1798. Based on *P. divaricatum* Michx.
Panicum patentissimum Roem. and
Schult., Syst. Veg. 2: 448. 1817. Not
P. patentissimum Desv., 1816. Based
on *P. divaricatum* Michx.
Schedonorus obtusus Bieler ex Roem. and
Schult., Syst. Veg. 2: 710. 1817. Based
on *Festuca obtusa* Bieler.
Poa festucoides LeConte ex Torr. in
Eaton, Man. Fl. ed. 2. 367. 1818.
New York, *LeConte*.
Poa brachiata Desv., Opusc. 100. 1831.
Based on *Panicum divaricatum* Michx.
Festuca pseudoduriuscula Steud., Syn. Pl.
Glum. 1: 312. 1854. Texas, *Drummond* 398.
Steinchisma divaricatum Raf. ex Jacks.,
Ind. Kew. 2: 982. 1895, as doubtful
synonym of *Panicum debile*. Rafinesque
(Bul. Bot. Seringe 1: 220. 1830) cites
Panicum divaricatum [Michx.] under
Steinchisma, but does not transfer the
name.
Festuca nutans palustris Muhl. ex Piper,
U. S. Natl. Herb. Contrib. 10: 34.
1906, as synonym of *F. obtusa* "Spreng."
Festuca obtusa var. *sprengeliana* St.-Yves,
Candollea 2: 276. 1925. Based on *F.
obtusa* Bieler.
(29) **Festuca occidentalis** Hook., Fl. Bor.

Amer. 2: 249. 1840. Mouth of Columbia River, *Scouler, Douglas*.
Festuca ovina var. *polyphylla* Vasey ex
Beal, Grasses N. Amer. 2: 597. 1896.
Cascade Mountains, Oreg., *Howell*.
(1) **Festuca octoflora** Walt., Fl. Carol. 81.
1788. South Carolina.
Festuca setacea Poir. in Lam., Encyl. Sup.
2: 638. 1811. Grown in Jardin du Val
de Grace, France, source unknown.
[?Carolina, Bosc.]
Festuca parviflora Ell., Bot. S. C. and Ga.
1: 170. 1816. Orangeburg, S. C.
Diarrhena setacea Roem. and Schult.,
Syst. Veg. 1: 289. 1817. Based on
Festuca setacea Poir.
Festuca octoflora var. *aristulata* Torr. ex
L. H. Dewey, U. S. Natl. Herb. Contrib.
2: 547. 1894. Texas, *Nealley*.
Vulpia octoflora Rydb., Torrey Bot. Club
Bul. 36: 538. 1909. Based on *Festuca
octoflora* Walt.
Gnomomia octoflora Lunell, Amer. Mid.
Nat. 4: 224. 1915. Based on *Festuca
octoflora* Walt.
FESTUCA OCTOFLORA var. GLAUCA (Nutt.)
Fernald, Rhodora 34: 209. 1932.
Based on *F. tenella* var. *glauca* Nutt.
Festuca tenella var. *glauca* Nutt., Amer.
Phil. Soc. Trans. (n.s.) 5: 147. 1837.
Fort Smith, Ark., *Nuttall*.
Vulpia octoflora var. *glauca* Fernald,
Rhodora 47: 107. 1945. Based on
Festuca tenella var. *glauca* Nutt.
FESTUCA OCTOFLORA var. HIRTELLA Piper,
U. S. Natl. Herb. Contrib. 10: 12.
1906. Santa Catalina Mountains, Ariz.,
Shear 1962. (Published as *F. octoflora*
subsp. *hirtella*.)
Festuca pusilla Buckl., Acad. Nat. Sci.
Phila. Proc. 1862: 98. 1862. California, *Nuttall*.
Vulpia octoflora var. *hirtella* Henr.,
Blumea 2: 320. 1937. Based on *Festuca
octoflora* subsp. *hirtella* Piper.
FESTUCA OCTOFLORA var. TENELLA (Willd.)
Fernald, Rhodora 34: 209. 1932.
Based on *F. tenella* Willd.
Festuca tenella Willd., Sp. Pl. 1: 419.
1797. North America [Pennsylvania].
Name only, Muhl., Amer. Phil. Soc.
Trans. 3: 161. 1793.
Schedonorus tenellus Beauv., Ess. Agrost.
99, 163, 177. 1812. Based on *Festuca
tenella* Willd.
Brachypodium festucoides Link, Enum. Pl.
1: 95. 1821. Based on *Festuca tenella*
L. (error for Willd.)
Vulpia tenella Heynh., Nom. 1: 854.
1840. Based on *Festuca tenella* Willd.
Festuca tenella var. *aristulata* Torr., U. S.
Expl. Miss. Pacif. Rpt. 4: 156. 1856.
Name only. Napa Valley, Calif.,
Bigelow.
Festuca gracilenta Buckl., Acad. Nat. Sci.
Phila. Proc. 1862: 97. 1862. Northern
Texas, *Buckley*.

Vulpia octoflora var. *tenella* Fernald, Rhodora 47: 107. 1945. Based on *Festuca tenella* Willd.

(30) **Festuca ovina** L., Sp. Pl. 73. 1753. Europe.

Festuca ovina var. *vivipara* L., Sp. Pl. ed. 2. 1: 108. 1762. Sweden.

Bromus ovinus Scop., Fl. Carn. 1: 77. 1772. Based on *Festuca ovina* L.

Avena ovina Salisb., Prodr. Stirp. 22. 1796. Based on *Festuca ovina* L.

Festuca ovina var. *duriuscula* A. Gray ex Port. and Coult., Syn. Fl. Colo. 150. 1874. Not *F. ovina* var. *duriuscula* Koch, 1837. Name only, for alpine specimens from Colorado [*Hall* and *Harbour* 665]. No reference to *F. duriuscula* L.

Festuca amethystina var. *asperrima* Hack. ex Beal, Grasses N. Amer. 2: 601. 1896. Arizona, *Rusby* 901.

Festuca minutiflora Rydb., Torrey Bot. Club Bul. 32: 608. 1905. Cameron Pass, Colo., *Baker.*

Festuca ovina calligera Piper, U. S. Natl. Herb. Contrib. 10: 27. 1906. Based on *F. amethystina* var. *asperrima* Hack.

Festuca saximontana Rydb., Torrey Bot. Club Bul. 36: 536. 1909. Banff, Alberta, *MacCalla* 2331.

Festuca calligera Rydb., Torrey Bot. Club Bul. 36: 537. 1909. Based on *F. ovina calligera* Piper.

Gnomonia ovina Lunell, Amer. Midl. Nat. 4: 224. 1915. Based on *Festuca ovina* L.

Festuca ovina subsp. *saximontana* St.-Yves, Candollea 2: 245. 1925. Based on *F. saximontana* Rydb.

Festuca ovina subsp. *saximontana* var. *rydbergii* St.-Yves, Candollea 2: 245. 1925. Based on *F. saximontana* Rydb.

Festuca brevifolia var. *utahensis* St.-Yves, Candollea 2: 257. 1925. Wasatch Mountains, Utah; Colorado, *Baker* 175.

F ESTUCA OVINA var. BRACHYPHYLLA (Schult.) Piper, U. S. Natl. Herb. Contrib. 10: 27. 1906. Based on *F. ovina brachyphylla* Schult. (Published as *F. ovina brachyphylla.*)

Festuca brevifolia R. Br., Sup. App. Parry's Voy. 289. 1824. Not *F. brevifolia* Muhl., 1817. Melville Island, Arctic America.

Festuca brachyphylla Schult., Mantissa 3 (Add. 1): 646. 1827. Based on *F. brevifolia* R. Br.

Festuca ovina var. *brevifolia* S. Wats. in King, Geol. Expl. 40th Par. 5: 389. 1871. Based on *F. brevifolia* R. Br.

Festuca ovina subsp. *saximontana* var. *purpusiana* St.-Yves, Candollea 2: 247. 1925. Farewell Gap, Calif., *Purpus* 3076, 5117.

F ESTUCA OVINA var. DURIUSCULA (L.) Koch, Syn. Fl. Germ. Helv. 812. 1837. Based on *F. duriuscula* L.

Festuca duriuscula L., Sp. Pl. 74. 1753. Europe.

F ESTUCA OVINA var. GLAUCA (Lam.) Koch, Syn. Fl. Germ. Helv. 812. 1837. Based on *F. glauca* Lam.

Festuca glauca Lam., Encycl. 2: 459. 1788. France.

The following varieties of *F. ovina*, recognized by Piper (North American Species of Festuca, U. S. Natl. Herb. Contrib. 10: 26–28. 1906), are based on European types. The specimens cited by him are in this Manual referred as follows:

F. ovina amethystina (Schur) Aschers. and Graebn., to *F. ovina.*

F. ovina supina (Schur) Hack., to *F. ovina* var. *brachyphylla.*

F. ovina pseudovina Hack., to *F. ovina.*

(6) **Festuca pacifica** Piper, U. S. Natl. Herb. Contrib. 10: 12. 1906. Pullman, Wash., *Elmer* 262.

Vulpia pacifica Rydb., Torrey Bot. Club Bul. 36: 538. 1909. Based on *Festuca pacifica* Piper.

Festuca subbiflora Suksdorf, Werdenda 1²: 2. 1923. Bingen, Wash., *Suksdorf* 6144.

Festuca dives Suksdorf, Werdenda 1²: 3. 1923. Not *F. dives* Muell., 1863. Bingen, Wash., *Suksdorf* 6153.

F ESTUCA PACIFICA var. SIMULANS Hoover, Madroño 3: 228. 1936. Kern County, Calif., *Hoover* 451.

(21) **Festuca paradoxa** Desv., Opusc. 105. 1831. Habitat unknown [United States].

Festuca nutans Bieler, Pl. Nov. Herb. Spreng. Cent. 10. 1807. Not *F. nutans* Moench, 1794. Pennsylvania, *Muhlenberg.*

Poa nutans Link, Enum. Pl. 1: 86. 1821. Based on *Festuca nutans* Bieler.

Festuca shortii Kunth ex Wood, Classbook ed. 1861. 794. 1861; A. Gray, Man. ed. 6. 669. 1890.

?*Festuca nutans* var. *palustris* Wood, Amer. Bot. and Flor. pt. 2: 399. 1871. Eastern States.

Festuca nutans var. *major* Vasey, U. S. Dept. Agr. Spec. Rpt. 63: 43. 1883. Name only; Beal, Grasses N. Amer. 2: 589. 1896, as synonym of *F. nutans* var. *shortii* Beal.

Festuca nutans var. *shortii* Beal, Grasses N. Amer. 2: 589. 1896. Based on *F. shortii* Kunth.

Gnomonia nutans Lunell, Amer. Midl. Nat. 4: 224. 1915. Based on "*Festuca nutans* Willd.''

(10) **Festuca reflexa** Buckl., Acad. Nat. Sci. Phila. Proc. 1862: 98. 1862. California.

Festuca microstachys var. *pauciflora* Scribn. ex Beal, Grasses N. Amer. 2: 586. 1896. Oregon, *Howell.*

Vulpia reflexa Rydb., Torrey Bot. Club Bul. 36: 538. 1909. Based on *Festuca reflexa* Buckl.

Festuca rigescens (Presl) Kunth, Rév. Gram. 1: Sup. 31. 1830. Based on *Diplachne rigescens* Presl.

Diplachne rigescens Presl, Rel. Haenk. 1: 260. 1830. Peru, *Haenke.*

(28) **Festuca rubra** L., Sp. Pl. 74. 1753. Europe.

Festuca ovina var. *rubra* Smith, English Fl. 1: 139. 1824. Based on *F. rubra* L.

Festuca duriuscula var. *rubra* Wood, Amer. Bot. and Flor. pt. 2: 399. 1871. Presumably based on *F. rubra* L.

Festuca oregona Vasey, Bot. Gaz. 2: 126. 1877. Oregon.

Festuca ovina subsp. *rubra* Hook. f., Stud. Fl. ed. 3. 497. 1884. Based on *F. rubra* L.

Festuca rubra var. *littoralis* Vasey ex Beal, Grasses N. Amer. 2: 607. 1896. Tillamook Bay, Oreg., *Howell* in 1882.

Festuca vallicola Rydb., N. Y. Bot. Gard. Mem. 1: 57. 1900. Silver Bow, Mont., *Rydberg* 2108.

Festuca earlei Rydb., Torrey Bot. Club Bul. 32: 608. 1905. La Plata Canyon, Colo., *Baker, Earle,* and *Tracy* 920.

Festuca rubra prolifera Piper, U. S. Natl. Herb. Contrib. 10: 21. 1906. Mount Washington, N. H., *Pringle* in 1877.

Festuca rubra var. *densiuscula* Hack. ex Piper, U. S. Natl. Herb. Contrib. 10: 22. 1906. Crescent City, Calif., *Davy* and *Blasdale* 5931.

Festuca rubra var. *prolifera* Piper in Robinson, Rhodora 10: 65. 1908. Based on *F. rubra prolifera* Piper.

Festuca prolifera Fernald, Rhodora 35: 133. 1933. Based on *F. rubra prolifera* Piper.

Festuca rubra var. *mutica* Hartm. forma *prolifera* Hylander, Uppsala Univ. Årskr. 7: 83. 1945. Based on *F. rubra* var. *prolifera* Piper.

FESTUCA RUBRA var. COMMUTATA Gaud., Fl. Helv. 1: 287. 1828. Switzerland.

Festuca fallax Thuill., Fl. Env. Paris n. ed. 50. 1799. France.

Festuca rubra var. *fallax* Hack., Bot. Centralbl. 8: 407. 1881. Based on *F. fallax* Thuill.

Festuca rubra subsp. *eurubra* var. *commutata* subvar. *eu-commutata* St.-Yves, Ann. Cons. Jard. Genève 17: 129. 1913. Based on *F. commutata* Gaud.

FESTUCA RUBRA var. HETEROPHYLLA Mutel, Fl. Franç. 4: 103. 1837. Based on *F. heterophylla* Lam.

Festuca heterophylla Lam., Fl. Franç. 3: 600. 1778. France.

FESTUCA RUBRA var. LANUGINOSA Mert. and Koch, Deut. Fl. ed. 3. 1: 654. 1823. Prussia.

Festuca arenaria Osbeck in Retz., Sup. Prodr. Fl. Scand. 1: 4. 1805. Not *F. arenaria* Lam., 1791. Scandinavia.

Festuca rubra var. *arenaria* Fries, Fl. Halland. 28. 1818. Based on *F. arenaria* Osbeck.

Bromus secundus Presl, Rel. Haenk. 1: 263. 1830. Nootka Sound, Vancouver Island, *Haenke.*

Festuca richardsoni Hook., Fl. Bor. Amer. 2: 250. 1840. Arctic seacoast of North America, *Richardson.*

Festuca rubra var. *villosa* Vasey ex Macoun, Can. Pl. Cat. 2⁴: 236. 1888. Name only, for specimen collected by Macoun at Dawson, Yukon Territory.

Festuca rubra var. *pubescens* Vasey ex Beal, Grasses N. Amer. 2: 607. 1896. Not *F. rubra* var. *pubescens* Spenner, 1825. Oregon, *Howell.*

Festuca rubra secunda Scribn., Mo. Bot. Gard. Rpt. 10: 39. 1899. Based on *Bromus secundus* Presl.

Festuca rubra var. *subvillosa* forma *vivipara* Eames, Rhodora 11: 89. 1909. Newfoundland, Governors Island, *Eames* and *Godfrey.*

Festuca rubra subsp. *richardsoni* Hultén, Acta Univ. Lund. n. ser. 38: 246. map 178c. 1942. Based on *F. richardsoni* Hook.

The following varieties of *Festuca rubra,* recognized by Piper (North American Species of Festuca, U. S. Natl. Herb. Contrib. 10: 21–23. 1906), are based on European types. The specimens cited by him are in this Manual referred as follows:

F. rubra megastachya Gaud., to *F. rubra.*

F. rubra glaucodea Piper (based on *F. glaucescens* Hegetschw.), to *F. rubra.*

F. rubra multiflora (Hoffm.) Aschers. and Graebn., to *F. rubra.*

F. rubra pruinosa Hack., to *F. rubra.*

F. rubra lanuginosa Mert. and Koch, to *F. rubra* var. *lanuginosa.*

F. rubra kitaibeliana (Schult.) Piper, to *F. rubra* var. *lanuginosa.*

(24) **Festuca scabrella** Torr. in Hook., Fl. Bor. Amer. 2: 252. 1840. Rocky Mountains, *Drummond.*

Melica hallii Vasey, Bot. Gaz. 6: 296. 1881. Rocky Mountains, latitude 39° to 41° [north half of Colorado], *Hall* and *Harbour* 621.

Festuca hallii Piper, U. S. Natl. Herb. Contrib. 10: 31. 1906. Based on *Melica hallii* Vasey.

Festuca confinis subsp. *rabiosa* Piper, U. S. Natl. Herb. Contrib. 10: 41. 1906. Crazy Womans Creek, Wyo., *Williams* and *Griffiths* 25.

Daluca hallii Lunell, Amer. Midl. Nat. 4: 221. 1915. Based on *Melica hallii* Vasey.

Festuca altaica subsp. *arizonica* subvar. *hallii* St.-Yves, Candollea 2: 271. 1925. Based on *Melica hallii* Vasey.

Festuca kingii var. *rabiosa* Hitchc., Amer. Jour. Bot. 21: 128. 1934. Based on *F. confinis* subsp. *rabiosa* Piper.

Hesperochloa kingii var. *rabiosa* Swallen, Biol. Soc. Wash. Proc. 54: 45. 1941.

Based on *F. confinis* subsp. *rabiosa* Piper.

FESTUCA SCABRELLA var. MAJOR Vasey, U. S. Natl. Herb. Contrib. 1: 278. 1893. Spokane County, Wash., *Suksdorf* 118.

Festuca campestris Rydb., N. Y. Bot. Gard. Mem. 1: 57. 1900. Based on *F. scabrella* var. *major* Vasey.

(2) **Festuca sciurea** Nutt., Amer. Phil. Soc. Trans. (n.s.) 5: 147. 1837. Arkansas, *Nuttall*.

?*Festuca quadriflora* Walt., Fl. Carol. 81. 1788. Not *F. quadriflora* Honck., 1782. South Carolina.

Festuca monandra Ell., Bot. S. C. and Ga. 1: 170. 1816, as synonym of *F. myuros* L., as misapplied by Elliott.

Dasiola elliotea Raf., Neogenyt. 4. 1825. Not *Festuca elliotii* Hack., 1906. Based on *Festuca monandra* Ell.

Vulpia quadriflora Trin. ex Steud., Nom. Bot. ed. 2. 2: 780. 1841. Based on *Festuca quadriflora* Walt.

Vulpia sciurea Henr., Blumea 2: 323. 1937. Presumably based on *Festuca sciurea* Nutt.

Vulpia elliotea Fernald, Rhodora 47: 106. 1945. Based on *Dasiola elliotea* Raf.

(18) **Festuca sororia** Piper, U. S. Natl. Herb. Contrib., 16: 197. 1913. Rincon Mountains, Ariz., *Nealley* 177.

Festuca subulata var. *sororia* St.-Yves, Candollea 2: 285. 1925. Based on *F. sororia* Piper.

(15) **Festuca subulata** Trin. in Bong., Acad. St. Pétersb. Mém. VI. Math. Phys. Nat. 2: 173. 1832. Sitka, Alaska, *Mertens*.

Festuca jonesii Vasey, U. S. Natl. Herb. Contrib. 1: 278. 1893. Utah, *Jones* in 1880.

Festuca subulata var. *jonesii* St.-Yves, Candollea 2: 284. 1925. Based on *F. jonesii* Vasey.

(14) **Festuca subuliflora** Scribn. in Macoun, Can. Pl. Cat. 2⁵: 396. 1890. Goldstream, Vancouver Island, *Macoun* 7. (By a slip of the pen the name is given as "*subulifolia*" in a note following.)

Festuca ambigua Vasey, U. S. Natl. Herb. Contrib. 1: 277. 1893. Not *F. ambigua* Le Gal., 1852. Oregon, *Howell* 19 in 1881.

Festuca denticulata Beal, Grasses N. Amer. 2: 589. 1896. Based on *F. ambigua* Vasey.

(22) **Festuca thurberi** Vasey in Rothr., Cat. Pl. Survey W. 100th Merid. 56. 1874. South Park, Colo., *Wolf* 1154.

Poa festucoides Jones, Calif. Acad. Sci. Proc. II. 5: 723. 1895. Not *P. festucoides* Lam., 1791. Mount Ellen, Henry Mountains, Utah, *Jones* 5671.

Poa kaibensis Jones, Erythea 4: 36. 1896. Based on *P. festucoides* Jones.

Festuca tolucensis subsp. *thurberi* St.-

Yves, Candollea 2: 304. 1925. Based on *F. thurberi* Vasey.

(13) **Festuca tracyi** Hitchc. in Abrams, Illustr. Fl. 1: 220. 1923. Howell Mountain, Napa County, Calif., *J. P. Tracy* 1479.

Festuca valesiaca Schleich. ex Gaud., Agrost. Helv. 1: 242. 1811. Switzerland.

Festuca ovina var. *valesiaca* Link, Hort. Berol. 2: 267. 1833. Based on "*F. valesiaca* Gaud.*" the name spelled "*vallesiaca.*"

(19) **Festuca versuta** Beal, Grasses N. Amer. 2: 589. 1896. Based on *F. texana* Vasey.

Festuca texana Vasey, Torrey Bot. Club Bul. 13: 119. 1886. Not *F. texana* Steud., 1854. Upper Llano, Tex., *Reverchon* 1618.

Festuca nutans var. *johnsoni* Vasey, U. S. Natl. Herb. Contrib. 2: 548. 1894. Harrison City, Tex., *Johnson.*

Festuca johnsoni Piper, U. S. Natl. Herb. Contrib. 10: 35. 1906. Based on *F. nutans* var. *johnsoni* Vasey.

Festuca obtusa subsp. *versuta* St.-Yves, Candollea 2: 280. 1925. Based on *F. versuta* Beal.

(27) **Festuca viridula** Vasey, U. S. Dept. Agr., Div. Bot. Bul. 13²: pl. 93. 1893. California (probably Summit Station), *Bolander.*

Festuca howellii Hack. ex Beal, Grasses N. Amer. 2: 591. 1896. Oregon, *Howell* [248].

Gnomonia viridula Lunell, Amer. Midl. Nat. 4: 224. 1915. Based on *Festuca viridula* Vasey.

Festuca viridula var. *vaseyana* St.-Yves, Candollea 2: 265. 1925. Based on *F. viridula* Vasey.

Festuca viridula var. *howellii* St.-Yves, Candollea 2: 266. 1925. Based on *F. howellii* Hack.

(80) GASTRIDIUM Beauv.

(1) **Gastridium ventricosum** (Gouan) Schinz and Thell., Vierteljahrs. Nat. Ges. Zürich 58: 39. 1913. Based on *Agrostis ventricosa* Gouan.

Agrostis ventricosa Gouan, Hort. Monsp. 39. pl. 1. f. 2. 1762. France.

Milium lendigerum L., Sp. Pl. ed. 2. 91. 1762. Europe.

Agrostis australis L., Mant. Pl. 1: 30. 1767. Portugal.

Alopecurus ventricosus Huds., Fl. Angl. ed. 2. 1: 28. 1778. Based on *Agrostis ventricosa* Gouan.

Agrostis lendigera Neck., Elem. Bot. 3: 219. 1791. Based on *Milium lendigerum* L.

Avena lendigera Salisb., Prodr. Stirp. 23. 1796. Based on *Milium lendigerum* L.

Gastridium australe Beauv., Ess. Agrost. 21, 164. pl. 6. f. 6. 1812. Europe.

Gastridium lendigerum Desv., Obs. Angers 48. 1818. Based on *Milium lendigerum* L.

Chilochloa ventricosa Beauv. ex Steud., Nom. Bot. ed. 2. 1: 350. 1840, as synonym of *Alopecurus ventricosus* Huds.

Lachnagrostis phleoides Nees and Meyen in Nees, Nov. Act. Acad. Caes. Leop. Carol. 19: Sup. 1: 14. 1841; 146, 1843. Valparaiso, Chile.

(7) GLYCERIA R. Br.

(1) **Glyceria acutiflora** Torr., Fl. North. and Mid. U. S. 1: 104. 1823. New York, New Jersey, and Massachusetts. *Festuca brevifolia* Muhl. erroneously cited as synonym.

Festuca acutiflora Bigel., Fl. Bost. ed. 3. 39. 1840. Based on *Glyceria acutiflora* Torr.

Panicularia acutiflora Kuntze, Rev. Gen. Pl. 2: 783. 1891. Based on *Glyceria acutiflora* Torr.

(4) **Glyceria arkansana** Fernald, Rhodora 31: 49. 1929. Varner, Ark., *Bush 9 in 1898.*

(2) **Glyceria borealis** (Nash) Batchelder, Manchester Inst. Proc. 1: 74. 1900. Based on *Panicularia borealis* Nash.

Glyceria fluitans var. *angustata* Vasey ex Fernald, Portland Soc. Nat. Hist. Proc. 2: 91. 1895. Maine, *Fernald* [193].

Panicularia borealis Nash, Torrey Bot. Club Bul. 24: 348. 1897. Maine, *Fernald.*

(13) **Glyceria canadensis** (Michx.) Trin., Acad. St. Pétersb. Mém. VI. Math. Phys. Nat. 1: 366. 1830. Based on *Briza canadensis* Michx.

Briza canadensis Michx., Fl. Bor. Amer. 1: 71. 1803. Canada, *Michaux.*

Megastachya canadensis Michx. ex Roem. and Schult., Syst. Veg. 2: 593. 1817. Based on *Briza canadensis* Michx.

?Briza canadensis Nutt., Gen. Pl. 1: erratum. 1818. Not op. cit. 69. New Jersey, near Philadelphia.

Nevroloma canadensis Raf., Jour. Phys. Chym. 89: 106. 1819. Based on *Briza canadensis* Michx.

Poa canadensis Torr., Fl. North. and Mid. U. S. 1: 112. 1823. Based on *Briza canadensis* Michx.

Panicularia canadensis Kuntze, Rev. Gen. Pl. 2: 783. 1891. Based on *Briza canadensis* Michx.

GLYCERIA CANADENSIS var. LAXA (Scribn.) Hitchc., Amer. Jour. Bot. 21: 128. 1934. Based on *Panicularia laxa* Scribn.

Panicularia laxa Scribn., Torrey Bot. Club Bul. 21: 37. 1894. Mount Desert, Maine, *Redfield* and *Rand.*

Glyceria laxa Scribn. in Rand and Redfield, Fl. Mt. Desert 180. 1894. Based on *Panicularia laxa* Scribn.

Glyceria canadensis var. *parviflora* Fernald, Portland Soc. Nat. Hist. Proc. 2: 91. 1895, as synonym of *G. laxa* Scribn.

(8) **Glyceria declinata** Brébiss. Fl. Normandie 354. 1859. Orne River, France.

Glyceria plicata var. *declinata* Druce, List Brit. Pl. 83. 1908. Presumably based on *G. declinata* Brébiss.

Glyceria cookei Swallen, Wash. Acad. Sci. Jour. 31: 348. f. 1. 1941. Mount Shasta City, Calif., *Cooke 15312.*

(15) **Glyceria elata** (Nash) Hitchc. in Jepson, Fl. Calif. 1: 162. 1912. Based on *Panicularia elata* Nash.

Panicularia elata Nash in Rydb., N. Y. Bot. Gard. Mem. 1: 54. 1900. Montana, *Flodman 176.*

Glyceria latifolia Cotton, Torrey Bot. Club Bul. 29: 573. 1902. Washington, *Elmer 721.*

Panicularia nervata elata Piper, U. S. Natl. Herb. Contrib. 11: 140. 1906. Based on *P. elata* Nash.

(16) **Glyceria erecta** Hitchc. in Jepson, Fl. Calif. 1: 161. 1912. Yosemite, Calif., *Hitchcock 3250½.*

Panicularia erecta Hitchc., Amer. Jour. Bot. 2: 309. 1915. Based on *Glyceria erecta* Hitchc.

Glyceria californica Beetle, Madroño 8: 161. 1946. Farwell, Tulare County, Calif., *Purpus 2057.*

Torreyochloa erecta Church, Amer. Jour. Bot. 36: 163. 1949. Based on *Glyceria erecta* Hitchc.

Torreyochloa californica Church, Amer. Jour. Bot. 36: 163. 1949. Based on *Glyceria californica* Beetle.

(20) **Glyceria fernaldii** (Hitchc.) St. John, Rhodora 19: 76. 1917. Based on *Glyceria pallida* var. *fernaldii* Hitchc.

Glyceria pallida var. *fernaldii* Hitchc., Rhodora 8: 211. 1906. Maine, *Fernald 191.*

Panicularia fernaldii Hitchc. in House, N. Y. State Mus. Bul. 233–234: 11. 1921. Based on *Glyceria pallida* var. *fernaldii* Hitchc.

Torreyochloa fernaldii Church, Amer. Jour. Bot. 36: 164. 1949. Based on *Glyceria pallida* var. *fernaldii* Hitchc.

(6) **Glyceria fluitans** (L.) R. Br., Prodr. Fl. Nov. Holl. 1: 179. 1810. Based on *Festuca fluitans* L.

Festuca fluitans L., Sp. Pl. 75. 1753. Europe.

Poa fluitans Scop., Fl. Carn. ed. 2. 73. 1772. Based on "Gramen aquaticum fluitans" Bauhin, cited by Linnaeus sub *Festuca fluitans.*

Hydrochloa fluitans Hartm., Gen. Gram. Scand. 8. 1819. Presumably based on *Festuca fluitans* L.

Melica fluitans Raspail, Ann. Sci. Nat., Bot. 5: 443. 1825. Based on *Festuca fluitans* L.

Devauxia fluitans Beauv. ex Kunth, Enum. Pl. 1: 367. 1833, as synonym of *Glyceria fluitans* R. Br.

Panicularia fluitans Kuntze, Rev. Gen. Pl. 2: 782. 1891. Based on *Festuca fluitans* L.

Panicularia brachyphylla Nash, Torrey Bot. Club Bul. 24: 349. 1897. Near New York City, *Nash.*

(9) **Glyceria grandis** S. Wats. ex A. Gray, Man. ed. 6. 667. 1890. [Type from Quebec, *Munro* in 1858.] New England to western New York, Michigan, Minnesota, and westward.

Poa aquatica var. *americana* Torr., Fl. North. and Mid. U. S. 1: 108. 1823. Massachusetts, *Cooley.*

Panicularia americana MacM., Met. Minn. Vall. 81. 1892. Based on *Poa aquatica* var. *americana* Torr.

Glyceria americana Pammel, Iowa Geol. Survey Sup. Rpt. 1903: 271. 1905. Based on *Poa aquatica* var. *americana* Torr.

Glyceria flavescens Jones, Mont. Univ. Bul. Biol. Ser. 15: 17. pl. 2. 1910. Swan Lake, Mont., *Jones* [9697].

Panicularia grandis Nash in Britt. and Brown, Illus. Fl. ed. 2. 1: 265. 1913. Based on *Glyceria grandis* S. Wats.

Glyceria grandis forma *pallescens* Fernald, Rhodora 23: 231. 1921. Nova Scotia, *Bissell, Pease, Long,* and *Linder* 20,026.

Glyceria maxima subsp. *grandis* Hultén, Acta Univ. (n.s.) 38: 229. 1942. Based on *G. grandis* S. Wats.

(3) **Glyceria leptostachya** Buckl., Acad. Nat. Sci. Phila. Proc. 1862: 95. 1862. Oregon, *Nuttall.*

Panicularia davyi Merr., Rhodora 4: 145. 1902. Sonoma County, Calif., *Davy* 6005.

Panicularia leptostachya Piper in Piper and Beattie, Fl. Northw. Coast 59. 1915. Not *P. leptostachya* Maclosk., 1904. Based on *Glyceria leptostachya* Buckl.

(12) **Glyceria melicaria** (Michx.) Hubb., Rhodora 14: 186. 1912. Based on *Panicum melicarium* Michx.

Panicum melicarium Michx., Fl. Bor. Amer. 1: 50. 1803. Carolina, *Michaux.* [Michaux's specimen overmature, all the florets but the lowermost fallen.]

Poa torreyana Spreng., Neu. Entd. 2: 104. 1821. Massachusetts.

Poa elongata Torr. ex Spreng., Neu. Entd. 2: 104. 1821. Not *P. elongata* Willd., 1809. As synonym of *P. torreyana* Spreng.

Poa elongata Torr., Fl. North. and Mid. U. S. 1: 112. 1823. Not *P. elongata* Willd., 1809. Massachusetts, *Cooley.*

Glyceria elongata Trin., Acad. St. Pétersb. Mém. VI. Sci. Nat. 2¹: 58. 1836. Based on *Poa elongata* Torr.

Panicularia elongata Kuntze, Rev. Gen.

Pl. 2: 783. 1891. Based on *Poa elongata* Torr.

Panicularia torreyana Merr., Rhodora 4: 146. 1902. Based on *Poa torreyana* Spreng.

Glyceria torreyana Hitchc., Rhodora 8: 211. 1906. Based on *Poa torreyana* Spreng.

Panicularia melicaria Hitchc., U. S. Natl. Herb. Contrib. 12: 149. 1908. Based on *Panicum melicarium* Michx.

(10) **Glyceria nubigena** W. A. Anders., Rhodora 35: 321. f. B. 1933. Clingmans Dome, Great Smoky Mountains, Tenn., *Anderson* and *Jennison* 1418.

(11) **Glyceria obtusa** (Muhl.) Trin., Acad. St. Pétersb. Mém. VI. Math. Phys. Nat. 1: 366. 1830. Based on *Poa obtusa* Muhl.

Poa obtusa Muhl., Descr. Gram. 147. 1817. Pennsylvania, *Muhlenberg.* Name only, Muhl., Cat. Pl. 11. 1813.

Panicularia obtusa Kuntze, Rev. Gen. Pl. 2: 783. 1891. Based on *Poa obtusa* Muhl.

(7) **Glyceria occidentalis** (Piper) J. C. Nels., Torreya 19: 224. 1919. Based on *Panicularia occidentalis* Piper.

Panicularia occidentalis Piper in Piper and Beattie, Fl. Northw. Coast 59. 1915. Vancouver, Wash., *Piper* 4905.

(18) **Glyceria otisii** Hitchc., Amer. Jour. Bot. 21: 128. 1934. Jefferson County, Wash., *Otis* 1548.

Torreyochloa otisii Church, Amer. Jour. Bot. 36: 163. 1949. Based on *Glyceria otisii* Hitchc.

(19) **Glyceria pallida** (Torr.) Trin., Acad. St. Pétersb. Mém. VI. Sci. Nat. 2¹: 57. 1836. Based on *Windsoria pallida* Torr.

Windsoria pallida Torr., Cat. Pl. N. Y. 91. 1819. New York.

Triodia pallida Spreng., Neu. Entd. 1: 246. 1820. New York, "*Windsoria pallida* Eddy in litt"; Spreng., Syst. Veg. 1: 330. 1825. Based on *Windsoria pallida* Torr.

Poa dentata Torr., Fl. North. and Mid. U. S. 1: 107. 1823. Based on *Windsoria pallida* Torr.

Uralepis pallida Kunth, Rév. Gram. 1: 108. 1829. Based on *Windsoria pallida* Torr.

Panicularia pallida Kuntze, Rev. Gen. Pl. 2: 783. 1891. Based on *Windsoria pallida* Torr.

Panicularia pallida var. *flava* Farwell, Mich. Acad. Sci. Rpt. 6: 203. 1904. "*Glyceria flava* Scribn." ined. Keweenaw County, Mich.

Glyceria flava Scribn. ex Farwell, Mich. Acad. Sci. Rpt. 6: 203. 1904, as synonym of *Panicularia pallida* var. *flava* Farwell.

Torreyochloa pallida Church, Amer. Jour. Bot. 36: 164. 1949. Based on *Windsoria pallida* Torr.

(17) **Glyceria pauciflora** Presl, Rel. Haenk. 1: 257. 1830. Nootka Sound, Vancouver Island, *Haenke.*

Glyceria microtheca Buckl., Acad. Nat. Sci. Phila. Proc. 1862: 96. 1862. Oregon, *Nuttall.*

Glyceria spectabilis var. *flaccida* Trin. ex A. Gray, Acad. Nat. Sci. Phila. Proc. 1862: 336. 1863, as synonym of *G. microtheca* Buckl., *G. leptostachya* Buckl. confused with it.

Panicularia pauciflora Kuntze, Rev. Gen. Pl. 2: 783. 1891. Based on *Glyceria pauciflora* Presl.

Panicularia holmii Beal, Torreya 1: 43. 1901. Longs Peak, Colo., *Holm 249.*

Panicularia multifolia Elmer, Bot. Gaz. 36: 54. 1903. Olympic Mountains, Wash., *Elmer 1939.*

Panicularia flaccida Elmer, Bot. Gaz. 36: 55. 1903. Olympic Mountains, Wash., *Elmer 1940.*

Torreyochloa pauciflora Church, Amer. Jour. Bot. 36: 163. 1936. Based on *Glyceria pauciflora* Presl.

(5) **Glyceria septentrionalis** Hitchc., Rhodora 8: 211. 1906. New Jersey, *Van Sickle.*

Panicularia septentrionalis Bicknell, Torrey Bot. Club Bul. 35: 196. 1908. Based on *Glyceria septentrionalis* Hitchc.

Panicularia fluitans var. *septentrionalis* Farwell, Mich. Acad. Sci. Rpt. 21: 353. 1920. Based on *Glyceria septentrionalis* Hitchc.

(14) **Glyceria striata** (Lam.) Hitchc., Biol. Soc. Wash. Proc. 41: 157. 1928. Based on *Poa striata* Lam.

Poa striata Lam., Tabl. Encycl. 1: 183. 1791. Virginia; Carolina.

Poa nervata Willd., Sp. Pl. 1: 389. 1797. North America.

Poa striata Michx., Fl. Bor. Amer. 1: 69. 1803. Pennsylvania, *Michaux.*

Poa lineata Pers., Syn. Pl. 1: 89. 1805. Based on *P. striata* Michx.

Poa parviflora Pursh, Fl. Amer. Sept. 1: 80. 1814. Not *P. parviflora* R. Br., 1810. New York to Virginia.

Poa sulcata Roem. and Schult., Syst. Veg. 2: 550. 1817. Not *P. sulcata* Lag., 1816. Based on *P. striata* Lam.

Briza canadensis Nutt., Gen. Pl. 1: 69. 1818. Not *B. canadensis* Michx., 1803. Canada and Pennsylvania. (Canada refers to Michaux's species, Nuttall misunderstanding it.)

Glyceria michauxii Kunth, Rév. Gram. 1: 118. 1829. Based on *Poa striata* Michx.

Glyceria nervata Trin., Acad. St. Pétersb. Mém. VI. Math. Phys. Nat. 1: 365. 1830. Based on *Poa nervata* Willd.

Poa lamarckii Kunth, Enum. Pl. 1: 362. 1833. Based on *P. striata* Lam.

Glyceria neogaea Steud., Syn. Pl. Glum. 1: 285. 1854. Newfoundland.

Panicularia nervata Kuntze, Rev. Gen. Pl.

2: 783. 1891. Based on *Poa nervata* Willd.

Panicularia nervata forma *major* Millsp., Fl. W. Va. 473. 1892. Monongalia, W. Va.

Panicularia nervata stricta Scribn., U. S. Dept. Agr., Div. Agrost. Bul. 13: 44. 1898. Colorado-Wyoming State line, *A. Nelson 3818.*

Panicularia nervata rigida Nash in Rydb., N. Y. Bot. Gard. Mem. 1: 54. 1900. Montana, *Rydberg 2068.*

Panicularia nervata var. *parviglumis* Scribn. and Merr., U. S. Dept. Agr., Div. Agrost. Cir. 30: 8. 1901. Racine, Wis. *Wadmond 36.*

Glyceria nervata var. *stricta* Scribn. ex Hitchc. in A. Gray, Man. ed. 7. 159. 1908. Based on *Panicularia nervata stricta* Scribn.

Glyceria nervata var. *rigida* Lunell, Amer. Midl. Nat. 4: 223. 1915. Based on *Panicularia nervata rigida* Nash.

Panicularia rigida Rydb., Fl. Rocky Mount. 83. 1917. Based on *P. nervata rigida* Nash.

Panicularia nervata var. *filiformis* Farwell, Mich. Acad. Sci. Rpt. 20: 168. 1919. Michigan, *Farwell 4514½.*

Panicularia nervata var. *purpurascens* Farwell, Mich. Acad. Sci. Rpt. 20: 168. 1919. Michigan, *Farwell 4495½* (first of several specimens cited).

Panicularia nervata var. *viridis* Farwell, Mich. Acad. Sci. Rpt. 22: 180. 1921. Michigan, *Farwell 5234.*

Glyceria striata var. *stricta* Fernald, Rhodora 31: 47. 1929. Based on *Panicularia nervata stricta* Scribn.

Glyceria rigida Rydb., Fl. Prairies and Plains Cent. N. Amer. 122. 1932. Not *G. rigida* Smith, 1824. Based on *Panicularia nervata rigida* Nash.

Panicularia striata Hitchc. in Small, Man. Southeast. Fl. 132. 1933. Based on *Poa striata* Lam.

(109) GYMNOPOGON Beauv.

(1) **Gymnopogon ambiguus** (Michx.) B. S. P., Prel. Cat. N. Y. 69. 1888. Presumably based on *Andropogon ambiguus* Michx.

Andropogon ambiguus Michx., Fl. Bor. Amer. 1: 58. 1803. Carolina, *Michaux.*

Gymnopogon racemosus Beauv., Ess. Agrost. 41, 164. pl. 9. f. 3. 1812. Based on *Andropogon ambiguus* Michx.

Andropogon ambiguus sive *latifolius* Muhl. Cat. Pl. 94. 1813. Suggested change of name.

Anthopogon lepturoides Nutt., Gen. Pl. 82. 1818. Banks of the Potomac, near Harpers Ferry, Va.

Gymnopogon scoparius Trin., Gram. Unifl. 237. 1824. New Jersey.

Alloiatheros lepturoides Steud., Nom. Bot.

ed. 2. 1: 55. 1840, as synonym of *Gymnopogon racemosus* Beauv.
Stipa expansa Willd. ex Steud., Nom. Bot. ed. 2. 2: 643. 1841, as synonym of *Gymnopogon racemosus* Beauv.
Gymnopogon distichophyllus Steud., Syn. Pl. Glum. 1: 218. 1854. Texas, Seubert Herb. [coll. *Vinzent*] 128; Louisiana, *Hartmann* 57.
Sciadonardus distichophyllus Steud., Flora 33: 229. 1850; Syn. Pl. Glum. 1: 218. 1854, as synonym of *Gymnopogon distichophyllus*. Louisiana, *Hartmann* 57.
Agrostis boeckeleri Seubert ex Steud., Syn. Pl. Glum. 1: 218. 1854, as synonym of *Gymnopogon distichophyllus*. Texas [*Vinzent* 128].
Alloiatheros ambiguus Ell. ex Jacks., Ind. Kew. 1: 83. 1893, as synonym of *Gymnopogon racemosus*.
Alloiatheros aristatus Raf. ex Jacks., Ind. Kew. 1: 83. 1893, as synonym of *Gymnopogon racemosus*.
(2) **Gymnopogon brevifolius** Trin., Gram. Unifl. 238. 1824. Delaware.
Anthopogon brevifolius Nutt. ex. Trin., Gram. Unifl. 238. 1824, as synonym of *Gymnopogon brevifolius* Trin.
Anthopogon filiforme Nutt., Amer. Phil. Soc. Trans. (n. s.) 5: 152. 1837. Banks of the Arkansas and in Delaware.
(3) **Gymnopogon chapmanianus** Hitchc., Amer. Jour. Bot. 2: 306. 1915. Sanford, Fla., *Chase* 4135.
(4) **Gymnopogon floridanus** Swallen, N. Amer. Fl. 17: 607. 1939. Clay County, Fla., *Swallen* 5596.

GYNERIUM Willd. ex Beauv.

Gynerium sagittatum (Aubl.) Beauv., Ess. Agrost. 138, 153. 1812. Based on *Saccharum sagittatum* Aubl.
Saccharum sagittatum Aubl., Pl. Guian. 1: 50. 1775. French Guiana.
Arundo sagittata Pers., Syn. Pl. 1: 102. 1805. Based on *Saccharum sagittatum* Aubl.
Gynerium procerum Beauv., Ess. Agrost. Atlas, pl. 24. f. 6. 1812. Based on *Saccharum sagittatum* Aubl.
Arundo sagittata Aubl. ex Beauv., Ess. Agrost. 153. 1812. Error for *Saccharum sagittatum* Aubl.
Gynerium saccharoides Humb. and Bonpl., Pl. Aequin. 2: 105. pl. 115. 1813. Venezuela, *Humboldt* and *Bonpland*.
Arundo saccharoides Poir. in Lam., Encycl. Sup. 4: 703. 1816. Based on *Gynerium saccharoides* Humb. and Bonpl.

(164) HACKELOCHLOA Kuntze

(1) **Hackelochloa granularis** (L.) Kuntze, Rev. Gen. Pl. 2: 776. 1891. Based on *Cenchrus granularis* L.
Cenchrus granularis L., Mant. Pl. 2: 575. 1771. East Indies.

Manisuris granularis Swartz, Prodr. Veg. Ind. Occ. 25. 1788. Based on *Cenchrus granularis* L. The name was earlier given (L. f. Nov. Gram. Gen. 40. pl. 1. f. 4–7. 1779) without description or basis. *Manisuris*, based on this species, has been credited to Swartz (not *Manisuris* L.), but Swartz does not propose the genus as new. He includes the original *M. myuros* L. and adds *M. granularis*.
Manisuris polystachya Beauv., Fl. Oware et Benin 1: 24. pl. 14. 1804. Oware and Benin, West Africa.
Tripsacum granulare Raspail, Ann. Sci. Nat., Bot. 5: 306. 1825. Based on *Manisuris granularis* Swartz.
Rytilix glandulosa Raf., Bul. Bot. Seringe 1: 219. 1830. Change of name or slip of the pen for "granularis," "*Manisuris granularis*" being cited.
Rytilix granularis Skeels, U. S. Dept. Agr., Bur. Plant Indus. Bul. 282: 20. 1913. Based on *Cenchrus granularis* L.

(86) HELEOCHLOA Host ex Roemer

Heleochloa alopecuroides (Pill. and Mitterp.) Host, Icon. Gram. Austr. 1: 23. pl. 29. 1801; ex Roemer, Collect. Rem. Bot. 233. 1809. Based on *Phleum alopecuroides* Pill. and Mitterp.
Phleum alopecuroides Pill. and Mitterp., Iter Posegan. 147. pl. 16. 1783. Europe.
Crypsis alopecuroides Schrad., Fl. Germ. 1: 167. 1806. Based on *Heleochloa alopecuroides* Host.
(1) **Heleochloa schoenoides** (L.) Host, Icon. Gram. Austr. 1: 23. pl. 30. 1801; ex Roemer, Collect. Rem. Bot. 233. 1809. Based on *Phleum schoenoides* L.
Phleum schoenoides L., Sp. Pl. 60. 1753. Southern Europe.
Crypsis schoenoides Lam., Tabl. Encycl. 1: 166. pl. 42. 1791. Based on *Phleum schoenoides* L. This name is spelled *C. schenoides* by Beauv., Ess. Agrost. 23. 1812.

(62) HELICTOTRICHON Besser

(2) **Helictotrichon hookeri** (Scribn.) Henr., Blumea 3: 429. 1940. Based on *Avena hookeri* Scribn.
Avena pratensis var. *americana* Scribn., Bot. Gaz. 11: 177. 1886. Based on *A. versicolor* as described by Hooker (Fl. Bor. Amer. 2: 244. 1840), not *A. versicolor* Vill., 1779. Rocky Mountains, *Drummond* [209].
Avena hookeri Scribn. in Hack., True Grasses 123. 1890. Based on *A. versicolor* as described by Hooker.
Avena americana Scribn., U. S. Dept. Agr., Div. Agrost. Bul. 7: 183. f. 165. 1897. Based on *A. pratensis* var. *americana* Scribn.

(3) **Helictotrichon mortonianum** (Scribn.)
Henr., Blumea 3: 429. 1940. Based on
Avena mortoniana Scribn.
Avena mortoniana Scribn., Bot. Gaz. 21:
133. pl. 11. 1896. Silver Plume, Colo.,
Shear 697 [type]; *Rydberg* 2439.

(1) **Helictotrichon pubescens** (Huds.) Pil-
ger, Repert. Sp. Nov. Fedde 45: 6.
1938. Based on *Avena pubescens* Huds.
Avena pubescens Huds., Fl. Angl. 42.
1762. England.
Heuffelia pubescens Schur, Enum. Pl.
Transsilv. 760. 1866. Based on *Avena
pubescens* L. (error for Huds.).
Avenula pubescens Dum., Soc. Bot. Belg.
Bul. 7¹: 68. 1868. Based on *Avena
pubescens* Huds.
Avenastrum pubescens Jess. ex Dalla
Torre, Alpenfl. 44. 1899. Based on
Avena pubescens Huds.

(11) HESPEROCHLOA (Piper) Rydb.

(1) **Hesperochloa kingii** (S. Wats.) Rydb.,
Torrey Bot. Club Bul. 39: 106. 1912.
Based on *Poa kingii* S. Wats.
Poa kingii S. Wats. in King, Geol. Expl.
40th Par. 5: 387. 1871. East Hum-
boldt Mountains, *Watson 1317.*
Festuca confinis Vasey, Torrey Bot. Club
Bul. 11: 126. 1884. Pen Gulch, Colo.,
Vasey.
Festuca kingii Cassidy, Colo. Agr. Expt.
Sta. Bul. 12: 36. 1890. On the North
Poudre, Colo. It may be based on *Poa
kingii* S. Wats., though that is not cited;
there is a description. Proposed as new
by Scribner, U. S. Dept. Agr., Div.
Agrost. Bul. 5: 36. 1897. Based on *Poa
kingii* S. Wats.
Festuca watsoni Nash in Britt., Man. 148.
1901. Based on *Festuca kingii* Scribn.
Wasatchia kingii Jones, West. Bot.
Contrib. 14:16. 1912. Based on *Poa
kingii* S. Wats.

(159) HETEROPOGON Pers.

(1) **Heteropogon contortus** (L.) Beauv. ex
Roem. and Schult., Syst. Veg. 2: 836.
1817. Based on *Andropogon contortus* L.
Andropogon contortus L., Sp. Pl. 1045.
1753. India.
Heteropogon glaber Pers., Syn. Pl. 2: 533.
1807. Europe.
Heteropogon hirtus Pers., Syn. Pl. 2: 533.
1807. Based on *Andropogon contortus* L.
Andropogon glaber Raspail, Ann. Sci.
Nat., Bot. 5: 307. 1825. Not *A. glaber*
Roxb., 1820. Based on *Heteropogon
glaber* Pers.
Andropogon secundus Willd. ex Nees,
Agrost. Bras. 364. 1829, as synonym
of *Heteropogon contortus.* Described in
Griseb., Fl. Brit. W. Ind. 558. 1864.
Not *A. secundus* Ell., 1821. Antigua,
Wullschlaegel.
Heteropogon firmus Presl, Rel. Haenk. 1:

334. 1830. Mexico, *Haenke.*
Andropogon firmus Kunth, Rév. Gram. 1:
Sup. 39. 1830. Based on *Heteropogon
firmus* Presl.
Heteropogon contortus var. *hirtus* Fenzl ex
Hack. in Mart. Fl. Bras. 2³: 267. 1883.
Based on *H. hirtus* Pers.
Heteropogon contortus var. *glaber* Hack. in
Mart., Fl. Bras. 2³: 268. 1883. Based
on *H. glaber* Pers.
Andropogon contortus subvar. *secundus*
Hack. in DC., Monogr. Phan. 6: 587.
1889. Based on *A. secundus* Willd.
Andropogon contortus subvar. *glaber* Hack.
in DC., Monogr. Phan. 6: 587. 1889.
Based on *Heteropogon glaber* Pers.
Sorghum contortum Kuntze, Rev. Gen. Pl.
2: 791. 1891. Based on *Andropogon
contortus* L.
Holcus contortus Kuntze ex Stuck., An.
Mus. Nac. Buenos Aires 11: 48. 1904.
Based on *Andropogon contortus* L.
Heteropogon contortus subvar. *secundus*
Domin, Bibl. Bot. 85: 276. 1915. Based
on *Andropogon contortus* var. *secundus*
Hack.

(2) **Heteropogon melanocarpus** (Ell.)
Benth., Linn. Soc. Jour. Bot. 19: 71.
1881. Based on *Andropogon melano-
carpus* Ell.
Andropogon melanocarpus Ell., Bot. S. C.
and Ga. 1: 146. 1816. Between Alta-
maha and Jefferson, Ga.
Stipa melanocarpa Muhl., Descr. Gram.
183. 1817. Georgia. Name only, Muhl.
Cat. Pl. 13. 1813.
Cymbopogon melanocarpus Spreng., Syst.
Veg. 1: 289. 1825. Based on *Andro-
pogon melanocarpus* Ell.
Trachypogon scrobiculatus Nees, Agrost.
Bras. 347. 1829. Piauhy, Brazil,
[Martius].
Andropogon scrobiculatus Kunth, Rév.
Gram. 1: Sup. 40. 1830. Based on
Trachypogon scrobiculatus Nees.
Heteropogon acuminatus Trin., Acad. St.
Pétersb. Mém. VI. Math. Phys. Nat. 2:
254. 1832. Brazil.
Heteropogon scrobiculatus Fourn., Mex.
Pl. 2: 64. 1886. Based on *Trachypogon
scrobiculatus* Nees.
Sorghum melanocarpum Kuntze, Rev.
Gen. Pl. 2: 792. 1891. Based on *Andro-
pogon melanocarpus* Ell.
Heteropogon melanocarpus Coult., U. S.
Natl. Herb. Contrib. 2: 493. 1894.
Based on *Stipa melanocarpa* Muhl.
Spirotheros melanocarpus Raf. ex Jacks.,
Ind. Kew. 2: 967. 1895, as synonym of
Heteropogon acuminatus Trin.

(116) HIEROCHLOË R. Br.

(1) **Hierochloë alpina** (Swartz) Roem. and
Schult., Syst. Veg. 2: 515. 1817. Based
on *Holcus alpinus* Swartz.
Aira alpina Liljebl., Utk. Svensk Fl. 49,

1792. Not *A. alpina* L., 1753. Sweden.
Holcus alpinus Swartz in Willd., Sp. Pl. 4:
937. 1806. Lapland.
Holcus monticola Bigel., New England
Jour. Med. and Surg. 5: 334. 1816;
Eaton, Man. Bot. ed. 2. 273. 1818.
White Hills, N. H., *Bigelow*.
Dimesia montwola Raf., Amer. Month.
Mag. 1: 442. 1817. Based on *Holcus
monticola* Bigel.
Hierochloë alpina var. *aristata* Raspail in
Saig. and Rasp., Ann. Sci. Obs. 2: 85.
1829. Based on "*H. alpina* R. Br."
(probably in Parry's Voyage), same as
Roem. and Schult.
Dimesia monticola Raf. ex Jacks., Ind.
Kew. 1: 760. 1893, as synonym of
Holcus monticola Bigel.
Savastana alpina Scribn., Torrey Bot.
Club Mem. 5: 34. 1894. Based on
Holcus alpinus Swartz.
Torresia alpina Hitchc., Amer. Jour. Bot.
2: 300. 1915. Based on *Holcus alpinus*
Swartz.
(3) **Hierochloë occidentalis** Buckl., Acad.
Nat. Sci. Phila. Proc. 1862: 100. 1862.
Columbia woods, [Oregon], *Nuttall*.
Hierochloë macrophylla Thurb. ex Boland.,
Calif. Agr. Soc. Trans. 1864–65: 132.
1866; S. Wats., Bot. Calif. 2: 265.
1880. Coast Range, Calif., *Bolander*
2279.
Savastana macrophylla Beal, Grasses N.
Amer. 2: 187. 1896. Based on *Hiero-
chloë macrophylla* Thurb.
Torresia macrophylla Hitchc., Amer. Jour.
Bot. 2: 300. 1915. Based on *Hierochloë
macrophylla* Thurb.
(2) **Hierochloë odorata** (L.) Beauv., Ess.
Agrost. 62, 164. pl. 12. f. 5. 1812.
Based on *Holcus odoratus* L.
Holcus odoratus L., Sp. Pl. 1048. 1753.
Europe.
Avena odorata Koel., Descr. Gram. 299.
1802. Based on *Holcus odoratus* L.
Holcus fragrans Willd., Sp. Pl. 4: 936.
1806. Hudson Bay, Canada.
Holcus borealis Schrad., Fl. Germ. 1:
252. 1806. Germany.
Hierochloë borealis Roem. and Schult.,
Syst. Veg. 2: 513. 1817. Based on
Holcus borealis Schrad.
Hierochloa fragrans Roem. and Schult.,
Syst. Veg. 2: 514. 1817. Based on
Holcus fragrans Willd.
Dimesia fragrans Raf., Amer. Month.
Mag. 1: 442. 1817. Based on "*Holcus
fragrans* of Mx. and Pursh." In Mich-
aux the name is *Holcus odoratus* L.
Hierochloë arctica Presl, Rel. Haenk. 1:
252. 1830. Nootka Sound, Vancouver
Island, *Haenke*.
Hierochloë odorata var. *fragrans* Richt.,
Pl. Eur. 1: 31. 1890. Based on *Holcus
fragrans* Willd.
Savastana odorata Scribn., Torrey Bot.
Club Mem. 5: 34. 1894. Based on

Holcus odoratus L.
Savastana nashii Bicknell, Torrey Bot.
Club Bul. 25: 104. pl. 328. 1898.
Van Cortlandt Park, New York City
[*Bicknell* in 1897].
Hierochloë nashii Kaczmarek, Amer.
Midl. Nat. 3: 198. 1914. Based on
Savastana nashii Bicknell.
Torresia odorata Hitchc., Amer. Jour.
Bot. 2: 301. 1915. Based on *Holcus
odoratus* L.
Hierochloa odorata var. *fragrans* forma
eamesii Fernald, Rhodora 19: 152.
1917. Connecticut, *Eames* 8734.
Savastana odorata var. *fragrans* Farwell,
Mich. Acad. Sci. Rpt. 21: 350. 1920.
Based on *Holcus fragrans* Willd.
Torresia nashii House, N. Y. State Mus.
Bul. 243–244: 58. 1923. Based on
Savastana nashii Bicknell.

(95) **HILARIA** H. B. K.

(1) **Hilaria belangeri** (Steud.) Nash, N.
Amer. Fl. 17: 135. 1912. Based on
Anthephora belangeri Steud.
Anthephora belangeri Steud., Syn. Pl.
Glum. 1: 111. 1854. "Mexico, *Belanger*
1428." Belanger is evidently an error
for Berlandier, since *Berlandier* 1428,
collected between Laredo and Bejar
[Bexar], now Texas, agrees with the
description. Belanger collected in India.
Schleropelta stolonifera Buckl., Prel. Rpt.
Geol. Agr. Survey Tex. App. 1. 1866.
Northwestern Texas.
Hilaria cenchroides var. *texana* Vasey,
U. S. Natl. Herb. Contrib. 1: 53. 1890.
Pena, Duval County, Tex., *Nealley*
[600].
Hilaria texana Nash in Small, Fl. South-
east. U. S. 68. 1903. Based on *Hilaria
cenchroides* var. *texana* Vasey.
HILARIA BELANGERI var. LONGIFOLIA
(Vasey) Hitchc., Biol. Soc. Wash. Proc.
41: 162. 1928. Based on *H. cenchroides*
var. *longifolia* Vasey. (Published as
H. belangeri longifolia.)
Hilaria cenchroides var. *longifolia* Vasey,
Amer. Acad. Sci. Proc. 24: 80. 1889,
name only; Beal, Grasses N. Amer. 2:
69. 1896. Islands in Guaymas harbor,
Mexico, *Palmer* 347 in 1887.
(4) **Hilaria jamesii** (Torr.) Benth., Linn.
Soc. Jour., Bot. 19: 62. 1881. Based
on *Pleuraphis jamesii* Torr.
Pleuraphis jamesii Torr., Ann. Lyc. N. Y.
1: 148. pl. 10. 1824. Sources of the
Canadian River [Texas or New Mex-
ico], *James*.
Hilaria sericea Benth., Linn. Soc. Jour.,
Bot. 19: 62. 1881. Name only.
Pleuraphis sericea Nutt. ex Benth., Linn.
Soc. Jour., Bot. 19: 62. 1881, as syno-
nym of *Hilaria sericea* Benth. [Harris
Fork of the Colorado, *Nuttall*.]
(3) **Hilaria mutica** (Buckl.) Benth., Linn.
Soc. Jour., Bot. 19: 62. 1881. Based

on *Pleuraphis mutica* Buckl.
Pleuraphis mutica Buckl., Acad. Nat.
Sci. Phila. Proc. 1862: 95. 1862.
Northern Texas [*Wright* 760–2108].
(5) **Hilaria rigida** (Thurb.) Benth. ex
Scribn., Torrey Bot. Club Bul. 9:
86. 1882. Based on *Pleuraphis rigida*
Thurb.
Pleuraphis rigida Thurb. in S. Wats., Bot.
Calif. 2: 293. 1880. California, Fort
Mojave and Providence Mountains,
Cooper [2230, the type]; Fort Yuma,
Thomas; Colorado Desert, *Schott*.
(2) **Hilaria swalleni** Cory, Wrightia 1: 215.
1948. Musquiz Canyon, Texas, July 28,
1938. *Sperry* T778, type.

(64) HOLCUS L.

(1) **Holcus lanatus** L., Sp. Pl. 1048. 1753.
Europe.
Aira holcus-lanata Vill., Hist. Pl. Dauph.
2: 87. 1787. Based on *Holcus lanatus* L.
Avena pallida Salisb., Prodr. Stirp. 24.
1796. Not *A. pallida* Thunb., 1794.
Based on *Holcus lanatus* L.
Avena lanata Koel., Descr. Gram. 300.
1802. Based on *Holcus lanatus* L.
Same published by Cav., Descr. Pl. 308.
1802.
Ginannia pubescens Bubani, Fl. Pyr. 4:
321. 1901. Based on *Holcus lanatus* L.
Notholcus lanatus Nash ex Hitchc., in
Jepson, Fl. Calif. 1: 126. 1912. Based
on *Holcus lanatus* L.
Nothoholcus lanatus Nash in Britt. and
Brown, Illustr. Fl. ed. 2. 1: 214. 1913.
Based on *Holcus lanatus* L.
Ginannia lanata Hubb., Rhodora 18: 234.
1916. Based on *Holcus lanatus* L.
(2) **Holcus mollis** L. Syst., Nat. ed. 10. 2:
1305. 1759. Europe.
Aira mollis Schreb., Spic. Fl. Lips. 51.
1771. Based on *Holcus mollis* L.
Aira holcus-mollis Vill., Hist. Pl. Dauph.
2: 88. 1787. Based on *Holcus mollis* L.
Avena sylvatica Salisb., Prodr. Stirp. 24.
1796. Based on *Holcus mollis* L.
Avena mollis Koel., Descr. Gram. 300.
1802. Not *A. mollis* Salisb., 1796.
Based on *Holcus mollis* L.
Ginannia mollis Bubani, Fl. Pyr. 4: 321.
1901. Based on *Holcus mollis* L.
Notholcus mollis Hitchc., Amer. Jour. Bot.
2: 304. 1915. Based on *Holcus mollis* L.

(49) HORDEUM L.

(6) **Hordeum arizonicum** Covas, Madroño
10: 16. 1949. Fort Lowell, Arizona,
J. J. Thornber 536. Referred to *H.
adscendens* H. B. K. in Manual ed. 1.
(3) **Hordeum brachyantherum** Nevski, Acta
Inst. Bot. Acad. Sci. U. R. S. S. I. 2:
61. 1936. Based on *H. boreale* Scribn.
and Smith. Not *H. boreale* Gandog.,
1881.
Hordeum boreale Scribn. and Smith, U. S.

Dept. Agr., Div. Agrost. Bul. 4: 24.
1897. Aleutian Islands [type, Atka
Island, *Turner* 1193] and Alaska to
California.
Hordeum nodosum var. *boreale* Hitchc.,
Amer. Jour. Bot. 21: 134. 1934. Based
on *H. boreale* Scribn. and Smith.
This is the species to which the name
H. nodosum L. has generally been mis-
applied in this country.
(4) **Hordeum californicum** Covas and Steb-
bins, Madroño 10: 5. 1949. Hastings
Reservation, Jamesburg, Monterey
County, Calif. *Stebbins* 3944.
(7) **Hordeum depressum** (Scribn. and
Smith) Rydb., Torrey Bot. Club Bul.
36: 539. 1909. Based on *H. nodosum*
var. *depressum* Scribn. and Smith.
Hordeum nodosum var. *depressum* Scribn.
and Smith, U. S. Dept. Agr., Div.
Agrost. Bul. 4: 24. 1897. Type, Lexing-
ton, Oreg., *Leiberg* 39.
Hordeum distichon L., Sp. Pl. 85. 1753.
Cultivated.
Hordeum hexastichon L., Sp. Pl. 85.
1753. Cultivated.
(8) **Hordeum hystrix** Roth, Cat. Bot. 1:
23. 1797. Spain.
Hordeum gussonianum Parl., Fl. Palerm.
1: 246. 1845. Italy.
Hordeum maritimum var. *gussonianum*
Richt., Pl. Eur. 1: 131. 1890. Based on
H. gussonianum Parl.
Hordeum marinum var. *gussonianum*
Thell., Vierteljahrs. Nat. Ges. Zürich
52: 441. 1908. Based on *H. gussoni-
anum* Parl.
(2) **Hordeum jubatum** L., Sp. Pl. 85. 1753.
Canada, *Kalm*.
?*Critesion geniculatum* Raf., Jour. Phys.
Chym. 89: 103. 1819. Illinois.
?*Elymus jubatus* Link, Hort. Berol. 1: 19.
1827. Garden specimen, *Hordeum ju-
batum* L., doubtfully cited as synonym.
Critesion jubatum Nevski in Komorov, Fl.
U. R. S. S. 2: 721. 1934. Based on
Hordeum jubatum L.
HORDEUM JUBATUM var. CAESPITOSUM
(Scribn.) Hitchc., Biol. Soc. Wash.
Proc. 41: 160. 1928. Based on *H.
jubatum caespitosum*.)
Hordeum caespitosum Scribn., Davenport
Acad. Sci. Proc. 7: 245. 1899. Edge-
mont, S. Dak., *Pammel* 143; Geranium
Park, Wyo., *Pammel* 157 type.
(9) **Hordeum leporinum** Link, Linnaea 9:
133. 1835. Greece.
Hordeum murinum var. *leporinum* Arc-
ang. Comp. Fl. Ital. 805. 1882. Based
on *H. leporinum* Link. This and *H.
stebbinsii* have been referred to *Hor-
deum murinum* L. by American authors.
Hordeum marinum Huds., Fl. Angl. ed. 2.
57. 1778. England.
Hordeum maritimum With., Bot. Arr.
Veg. Brit. ed. 2. 1: 127. 1787. Based

886   MISC. PUBLICATION 200, U. S. DEPT. OF AGRICULTURE

on *H. marinum* Huds.
(1) **Hordeum montanense** Scribn. in Beal,
Grasses N. Amer. 2: 644. 1896.
Montana, *Scribner* 429, 430.
Hordeum pammeli Scribn. and Ball, Iowa
Geol. Survey Sup. Rpt. 1903: 335.
1905. Dakota City, Iowa, *Pammel* 3824.
(5) **Hordeum pusillum** Nutt., Gen. Pl. 1:
87. 1818. Plains of the Missouri [*Nuttall*].
Hordeum riehlii Steud., Syn. Pl. Glum. 1:
353. 1854. St. Louis, Mo., *Riehl* 181.
HORDEUM PUSILLUM var. PUBENS Hitchc.,
Wash. Acad. Sci. Jour. 23: 453. 1933.
La Verkin, Utah, *Jones* 5196W.
Hordeum spontaneum C. Koch, Linnaea 21:
430. 1848. Caucasus.
(10) **Hordeum stebbinsii** Covas, Madroño
10: 17. 1949. Middletown, Lake County, Calif., *G. L. Stebbins* and *G. Covas*
3927.
(11) **Hordeum vulgare** L., Sp. Pl. 84. 1753.
Cultivated in Europe.
Hordeum sativum Pers., Syn. Pl. 1: 108.
1805, as synonym of *H. vulgare* L.
Hordeum polystichum var. *vulgare* Doell,
Rhein. Fl. 67. 1843. Based on *H. vulgare* L.
Hordeum sativum var. *vulgare* Richt., Pl.
Eur. 1: 130. 1890. Based on *H. vulgare* L.
HORDEUM VULGARE var. TRIFURCATUM
(Schlecht.) Alefeld, Landw. Fl. 341.
1866. Based on *H. trifurcatum* Jess.
(probably error for Wender.).
Hordeum coeleste var. *trifurcatum*
Schlecht., Linnaea 11: 543. 1837.
Cultivated at Halle, seed from Montpellier.
Hordeum trifurcatum Wender, Flora 26:
233. 1843. Cultivated in Marburg,
Germany.

(124) **HYDROCHLOA** Beauv.

(1) **Hydrochloa caroliniensis** Beauv., Ess.
Agrost. 135, 165, 182. pl. 3. f. 18;
pl. 24. f. 4. 1812. No specific description except explanation of figures.
"*Zizania natans* Mich." (an unpublished
name) is cited under the genus, and
Z. fluitans Michx. is referred in the
index to *Hydrochloa*. The name for
pl. 3. f. 18 is given as *H. caroliniana*.
Zizania fluitans Michx., Fl. Bor. Amer. 1:
75. 1803. Not *Hydrochloa fluitans*
Hartm., 1819. The published locality,
Lake Champlain, is an error. The
type specimen indicates Charleston,
S. C., *Michaux*.
Zizania natans Michx. ex Beauv., Ess.
Agrost. 136. 1812, name only; Bosc in
Trin., Acad. St. Pétersb. Mém. VI.
Sci. Nat. 3¹: 186. 1840, as synonym of
Hydrochloa caroliniensis Beauv. The
name is misspelled *Zizania nutans* in
Steud., Nom. Bot. ed. 2. 2: 799. 1841.
Luziola caroliniensis Raspail, Ann. Sci.

Nat., Bot. 5: 304. 1825. Based on
Hydrochloa caroliniensis Beauv.
Hydrochloa fluitans Torr., Comp. Fl.
North. Mid. States 354, 403. 1826.
Not *H. fluitans* Hartm., 1819. Based on
Zizania fluitans Michx.
Hydropyrum fluitans Kunth, Rév. Gram.
1: 7. 1829. Based on *Zizania fluitans*
Michx.
Luziola caroliniana Trin. ex Steud.,
Nom. Bot. ed. 2. 2: 79. 1841. Based on
Zizania natans "Bosc in Kunth hrb.
(ex Trin. mpt.)".

(155) **HYPARRHENIA** Anderss. ex Stapf

Hyparrhenia hirta (L.) Stapf in Prain, Fl.
Trop. Afr. 9: 315. 1918. Based on
Andropogon hirtus L.
Andropogon hirtus L., Sp. Pl. 1046. 1753.
Southern Europe and Asia Minor.
Trachypogon hirtus Nees, Agrost. Bras.
346. 1829. Based on *Andropogon hirtus*
L.
Sorghum hirtum Kuntze, Rev. Gen. Pl. 2:
792. 1891. Based on *Andropogon hirtus*
L.
(1) **Hyparrhenia rufa** (Nees) Stapf in Prain,
Fl. Trop. Afr. 9: 304. 1918. Based on
Trachypogon rufus Nees.
Trachypogon rufus Nees, Agrost. Bras.
345. 1829. Piauhy, Brazil, *Martius*.
Andropogon rufus Kunth, Rév. Gram. 1:
Sup. 39. 1830. Based on *Trachypogon
rufus* Nees.
Sorghum rufum Kuntze, Rev. Gen. Pl. 2:
792. 1891. Based on *Andropogon rufus*
Kunth.
Cymbopogon rufus Rendle, Cat. Afr. Pl.
Welw. 2: 155. 1899. Based on *Andropogon rufus* Kunth.

(48) **HYSTRIX** Moench

(2) **Hystrix californica** (Boland.) Kuntze,
Rev. Gen. Pl. 2: 778. 1891. Based on
Gymnostichum californicum Boland.
Gymnostichum californicum Boland. ex
Thurb., in S. Wats. Bot. Calif. 2: 327.
1880. Near San Francisco, *Bolander*;
Sausalito, *Kellogg* and *Harford* 1107.
Asperella californica Beal, Grasses N.
Amer. 2: 657. 1896. Based on *Gymnostichum californicum* Boland.
Asprella californica Benth. ex Beal,
Grasses N. Amer. 2: 657. 1896, as
synonym of *Asperella californica*.
(1) **Hystrix patula** Moench, Meth. Pl. 295.
1794. Based on *Elymus hystrix* L.
Elymus hystrix L., Sp. Pl. 560. 1753.
[Virginia, *Clayton*.]
Asperella hystrix Humb., Mag. Bot.
Roem. et Ust. 7: 5. 1790. Based on
Elymus hystrix L.
Asprella hystrix Willd., Enum. Pl. 132.
1809. Based on *Elymus hystrix* L.
Gymnostichum hystrix Schreb., Beschr.
Gräs. 2: 127. pl. 47. 1810. Based on

Elymus hystrix L.
Zeocriton hystrix Beauv., Ess. Agrost. 115, 182. 1812. Presumably based on *Elymus hystrix* L.
Asperella echidnea Raf., Amer. Monthly Mag. 4: 190. 1819. Based on *Elymus hystrix* L.
Elymus pseudohystrix Schult., Mantissa 2: 427. 1824. Based on *"Elymus hystrix* Nutt." (error for L., Nuttall applying the Linnaean name correctly).
Asprella americana Nutt., Amer. Phil. Soc. Trans. (n.s.) 5: 151. 1837. Arkansas, *Nuttall.*
Asprella angustifolia Nutt., Amer. Phil. Soc. Trans. (n.s.) 5: 151. 1837. Arkansas, *Nuttall.*
Asprella major Fres. ex Steud., Nom. Bot. ed. 2. 1: 152. 1840, as synonym of *Elymus hystrix* L.
Hystrix hystrix Millsp., Fl. W. Va. 474. 1892. Based on *Elymus hystrix* L.
Hystrix elymoides Mackenz. and Bush, Man. Fl. Jackson County 39. 1902. Based on *Elymus hystrix* L.
Hordeum hystrix Schenck, Bot. Jahrb. 40: 109. 1907. Not *H. hystrix* Roth, 1797. Based on *Elymus hystrix* L.
Gymnostichum patulum Lunell, Amer. Midl. Nat. 4: 228. 1915. Based on *Hystrix patula* Moench.
Asperella hystrix var. *bigeloviana* Fernald, Rhodora 24: 230. 1922. Hanover, Conn., *Williams* in 1910.
Hystrix patula var. *bigeloviana* Deam, Ind. Dept. Conserv. Pub. 82: 117. 1929. Based on *Asperella hystrix* var. *bigeloviana* Fernald.
Elymus californicus Gould, Madroño 9: 127. 1947. Based on *Gymnostichum californicum* Boland.

(148) IMPERATA Cyrillo

(1) **Imperata brasiliensis** Trin., Acad. St. Pétersb. Mém. VI. Math. Phys. Nat. 2: 331. 1832. Brazil.
Imperata brasiliensis var. *mexicana* Rupr., Acad. Sci. Brux. Bul. 9²: 245. 1842. Name only. Mexico, *Galeotti* 5678.
Imperata arundinacea var. *americana* Anderss., Öfv. Svensk. Vet. Akad. Förh. 12: 160. 1855. British Guiana, *Schomburgk* 665; Mexico, *Galeotti* 5678; Chile, *d' Urville.*
This is the species described as *Imperata caudata* Cyrillo in Chapm., Fl. South. U.S. ed. 2. 668. 1883.
(2) **Imperata brevifolia** Vasey, Torrey Bot. Club Bul. 13: 26. 1886. Southern California, *Parish* 1031.
Imperata arundinacea subsp. *hookeri* Rupr. ex Anderss., Öfv. Svensk. Vet. Akad. Förh. 12: 160. 1855. Texas, *Drummond* II. 283.
Imperata hookeri Rupr. ex Hack. in DC. Monogr. Phan. 6: 97. 1889. Based on

Imperata arundinacea subsp. *hookeri* Rupr. ex Anderss.
Imperata cylindrica (L.) Beauv., Ess. Agrost. 8, 165, 166, 177. pl. 5. f. 1. 1812. Based on *Lagurus cylindricus* L.
Lagurus cylindricus L., Syst. Nat. ed. 10. 2: 878. 1759. Europe.
Saccharum cylindricum Lam., Encycl. 1: 594. 1783. Based on *Lagurus cylindricus* L.
Imperata arundinacea Cyrillo, Pl. Rar. Neap. 2: 27. pl. 11. 1788. Italy.

(55) KOELERIA Pers.

(1) **Koeleria cristata** (L.) Pers., Syn. Pl. 1: 97. 1805. "*Poa cristata* auctorum," presumably *Poa cristata* L., used by Willd. (Sp. Pl. 1: 402. 1797), Lamarck (Tabl. Encycl. 1: 182. 1791), and others.
Aira cristata L., Sp. Pl. 63. 1753. Europe.
Poa cristata L., Syst. Nat. ed. 12. 94. 1767. Based on *Aira cristata* L.
Festuca cristata Vill., Hist. Pl Dauph. 1: 250. 1786. Not *F. cristata* L., 1753. Based on *Aira cristata* L.
Koeleria gracilis Pers., Syn. Pl. 1: 97. 1805. Europe.
Koeleria nitida Nutt., Gen. Pl. 1: 74. 1818. Plains of the Missouri.
Aira gracilis Trin., Fund. Agrost. 144. 1820. Based on *Koeleria gracilis* Pers.
Airochloa cristata Link, Hort. Berol. 1: 127. 1827. Based on *Aira cristata* L. The specific name was misspelled "*aristata*" in Link, Handb. Gewächs. 1: 64. 1829.
Airochloa gracilis Link, Hort. Berol. 2: 276. 1827. Based on *Koeleria gracilis* Pers.
Koeleria cristata var. *nuttalii* Wood, Class-book ed. 2. 613. 1847. Presumably based on *K. nitida* Nutt.
Koeleria cristata var. *gracilis* A. Gray, Man. 591. 1848. No definite locality cited. Presumably based on *K. gracilis* Pers.
Brachystylus cristatus Dulac, Fl. Haut. Pyr. 85. 1867. Based on *Koeleria cristata* Pers.
Koeleria nitida var. *arkansana* Scribn., Kans. Acad. Sci. Trans. 9: 118. 1885. [Arkansas.]
Koeleria arkansana Nutt. ex Scribn., Kans. Acad. Sci. Trans. 9: 118. 1885, [Arkansas, *Nuttall*] as synonym of *K. nitida* var. *arkansana.*
Achaeta geniculata Fourn., Mex. Pl. 2: 109. 1886. Mexico, *Liebmann* 609.
Koeleria cristata var. *major* Vasey in Macoun, Can. Pl. Cat. 2⁴: 218. 1888. Not *K. cristata* var. *major* Koch, 1837. Name only, for *Macoun*, Vancouver Island.
Koeleria cristata var. *pubescens* Vasey ex Davy in Jepson, Fl. West. Mid. Calif.

62. 1901. Not *K. cristata* var. *pubescens* Mutel, 1837. San Francisco, Calif., *Michener* and *Bioletti*.
Koeleria cristata var. *longifolia* Vasey ex Davy in Jepson, Fl. West. Mid. Calif. 62. 1901. Santa Cruz County, Calif., *Anderson*.
Koeleria cristata pinetorum Abrams, Fl. Los Angeles 46. 1904. Based on *K. cristata* var. *pubescens* Vasey.
Koeleria pseudocristata var. *californica* Domin, Magyar Bot. Lapok 3: 264. 1904. San Diego, Calif., *Pringle* in 1882.
Koeleria elegantula Domin, Bibl. Bot. 65: 172. 1907. Gunnison, Colo., *Baker* 578.
Koeleria robinsoniana Domin, Bibl. Bot. 65: 172. 1907. Wenatchee, Wash., *Whited* 1131.
Koeleria robinsoniana var. *australis* Domin, Bibl. Bot. 65: 173. 1907. Blalocks, Oreg., *Leckenby* 28 in 1900.
Koeleria gracilis var. *dasyclada* Domin, Bibl. Bot. 65: 211. 1907. California, *Lemmon* in 1882.
Koeleria pseudocristata Domin, Bibl. Bot. 65: 222. 1907. With two American forms: *densevestita*, California, *Hall* 2206; *laxa*, California, *Heller* 7443.
Koeleria pseudocristata var. *longifolia* Domin, Bibl. Bot. 65: 224. 1907. California, *Nuttall*.
Koeleria pseudocristata var. *oregona* Domin, Bibl. Bot. 65: 224. 1907. Oregon, *Nuttall*.
Koeleria pseudocristata var. *pseudonitida* Domin, Bibl. Bot. 65: 224. 1907. Wyoming, *Nelson* 273.
Koeleria polyantha var. *californiensis* Domin, Bibl. Bot. 65: 226. 1907. San Jacinto Mountains, Calif., *Hall* 2131.
Koeleria nitida var. *missouriana* Domin, Bibl. Bot. 65: 233. 1907. St. Louis, *Riehl* 44; Courtney, Mo., *Bush* 773.
Koeleria nitida var. *californica* Domin, Bibl. Bot. 65: 233. 1907. Based on *K. pseudocristata* var. *californica* Domin. With three subvarieties from California: *transiens*, *Brandegee* 3678; *multiflora*, *Parish Brothers* 855; *vestita*, *Palmer* 405.
Koeleria nitida var. *sublanuginosa* Domin, Bibl. Bot. 65: 234. 1907. Miranda, S. Dak., *Griffiths* 235. With subvar. *pubiflora*, Washington, *Lyall* in 1860.
Koeleria nitida var. *laxa* Domin, Bibl. Bot. 65: 235. 1907. Arizona, *Palmer* in 1890; New Mexico, *Metcalfe*.
Koeleria nitida var. *subrepens* Domin, Bibl. Bot. 65: 235. 1907. Arboles, Colo., *Baker* 185.
Koeleria nitida var. *munita* Domin, Bibl. Bot. 65: 235. 1907. Montana, *Rydberg* 3294.
Koeleria nitida var. *latifrons* Domin, Bibl. Bot. 65: 236. 1907. Nebraska, *Rydberg*.
Koeleria nitida var. *breviculmis* Domin,

Bibl. Bot. 65: 236. 1907. Colorado, *Baker, Earle,* and *Tracy* 114.
Koeleria nitida var. *caudata* Domin, Bibl. Bot. 65: 236. 1907. Wisconsin, *Kumlien* 99.
Koeleria idahoensis Domin, Bibl. Bot. 65: 237. 1907. Lewiston, Idaho, *Heller* 309 (error for 3091).
Koeleria idahoensis var. *pseudocristatoides* Domin, Bibl. Bot. 65: 238. 1907. Nez Perce County, Idaho, *Heller* 3291.
Koeleria macrura Domin, Bibl. Bot. 65: 238. 1907. With three forms: *quadriflora*, Arizona, *Nealley* in 1891; *triflora*, Organ Mountains, N. Mex., *Wooton* 110; *biflora*, Chiricahua Mountains, Ariz., *Toumey* in 1896.
Koeleria latifrons Rydb., Brittonia 1: 84. 1931. Based on *K. nitida* var. *latifrons* Domin.

(2) **Koeleria phleoides** (Vill.) Pers., Syn. Pl. 1: 97. 1805. Based on *Festuca phleoides* Vill.
Festuca phleoides Vill., Fl. Delph. 7. 1785. Europe.
Koeleria brachystachys DC., Cat. Hort. Monsp. 120. 1813. Europe.
Lophochloa phleoides Reichenb., Fl. Germ. 42. 1830. Based on *Festuca phleoides* Vill.

(81) LAGURUS L.

(1) **Lagurus ovatus** L., Sp. Pl. 81. 1753. Southern Europe.

(25) LAMARCKIA Moench

(1) **Lamarckia aurea** (L.) Moench, Meth. Pl. 201. 1794. Based on *Cynosurus aureus* L.
Cynosurus aureus L., Sp. Pl. 73. 1753. Europe.
Chrysurus cynosuroides Pers., Syn. Pl. 1: 80. 1805. Based on *Cynosurus aureus* L.
Chrysurus aureus Beauv. ex Spreng., Syst. Veg. 1: 296. 1825. Based on *Cynosurus aureus* L.
Achyrodes aureum Kuntze, Rev. Gen. Pl. 2: 758. 1891. Based on *Cynosurus aureus* L.

(138) LASIACIS (Griseb.) Hitchc.

(1) **Lasiacis divaricata** (L.) Hitchc., U. S. Natl. Herb. Contrib. 15: 16. 1910. Based on *Panicum divaricatum* L.
Panicum divaricatum L., Syst. Nat. ed. 10. 2: 871. 1759. Jamaica, *Browne*.
Panicum bambusioides Desv. ex Hamilt., Prodr. Pl. Ind. Occ. 10. 1825. Puerto Rico.
Panicum chauvinii Steud., Syn. Pl. Glum. 1: 68. 1854. Guadeloupe, *Duchaissing*.
Panicum divaricatum var. *stenostachyum* Griseb., Fl. Brit. W. Ind. 551. 1864.

Jamaica, Alexander, *Wilson*, *March* [type].

(120) LEERSIA Swartz

(4) **Leersia hexandra** Swartz, Prodr. Veg. Ind. Occ. 21. 1788. Jamaica, *Swartz.*
Asprella hexandra Beauv., Ess. Agrost. 2, 153. 1812. Based on *Leersia hexandra* Swartz.
Leersia mexicana H. B. K., Nov. Gen. et Sp. 1: 195. 1816. Mexico, *Humboldt* and *Bonpland.*
Asprella mexicana Roem. and Schult., Syst. Veg. 2: 267. 1817. Based on *Leersia mexicana* H. B. K.
Leersia contracta Nees, Agrost. Bras. 516. 1829. Brazil, *Sellow.*
Leersia elongata Willd. ex Trin., Acad. St. Pétersb. Mém. VI. Sci. Nat. 3¹: 172. 1840, as synonym of *L. mexicana* H. B. K.
Oryza hexandra Doell in Mart., Fl. Bras. 2²: 10. 1871. Based on *Leersia hexandra* Swartz.
Oryza mexicana Doell in Mart., Fl. Bras. 2²: 10. 1871. Based on *Leersia mexicana* H. B. K.
Leersia gouinii Fourn., Mex. Pl. 2: 2. 1886. Vera Cruz, Mexico, *Gouin.*
Homalocenchrus gouinii Kuntze, Rev. Gen. Pl. 2: 777. 1891. Based on *Leersia gouinii* Fourn.
Homalocenchrus hexandrus Kuntze, Rev. Gen. Pl. 2: 777. 1891. Based on *Leersia hexandra* Swartz.
Leersia dubia Areschoug, Svensk Freg. Eugenies Resa 1910: 115. 1910. Ecuador, *Andersson.*

(1) **Leersia lenticularis** Michx., Fl. Bor. Amer. 1: 39. 1803. Illinois, *Michaux.*
Asprella lenticularis Beauv., Ess. Agrost. 2, 153. 1812. Based on *Leersia lenticularis* Michx.
Zizania lenticularis Michx. ex Beauv., Ess. Agrost. 182. 1812. Name only, doubtless error for *Leersia lenticularis* Michx.
Homalocenchrus lenticularis Kuntze, Rev. Gen. Pl. 2: 777. 1891. Based on *Leersia lenticularis* Michx.
Endodia lenticularis Raf. ex Jacks., Ind. Kew. 1: 840. 1893, as synonym of *Leersia lenticularis* Michx.

(5) **Leersia monandra** Swartz, Prodr. Veg. Ind. Occ. 21. 1788. Jamaica, *Swartz.*
Asprella monandra Beauv., Ess. Agrost. 2, 153. 1812. Based on *Leersia monandra* Swartz.
Paspalum cubense Spreng., Neu. Entd. 3: 12. 1822. Cuba and neighboring islands.
Oryza monandra Doell in Mart., Fl. Bras. 2²: 9. 1871. Based on *Leersia monandra* Swartz.
Homalocenchrus monandrus Kuntze, Rev. Gen. Pl. 2: 777. 1891. Based on *Leersia monandra* Swartz.

(2) **Leersia oryzoides** (L.) Swartz, Prodr. Veg. Ind. Occ. 21. 1788. Based on *Phalaris oryzoides* L.
Phalaris oryzoides L., Sp. Pl. 55. 1753. Virginia.
Homalocenchrus oryzoides Poll., Hist. Pl. Palat. 1: 52. 1776. Based on *Phalaris oryzoides* L.
Ehrhartia clandestina Web., Prim. Fl. Hols. 64. 1780. Based on *Phalaris oryzoides* L.
Asperella oryzoides Lam., Tabl. Encycl. 1: 167. 1791. Based on *Phalaris oryzoides* L.
Asprella oryzoides Beauv., Ess. Agrost. 2, 153. pl. 4. f. 2. 1812. Based on *Phalaris oryzoides* L.
Leersia asperrima Willd. ex Trin., Acad. St. Pétersb. Mém. VI. Sci. Nat. 3¹: 171. 1840, as synonym of *L. oryzoides* Swartz.
Oryza clandestina A. Br. in Aschers., Fl. Brand. 799. 1864. Based on *Ehrhartia clandestina* Web.
Laertia oryzoides Gromow. in Trautv., Act. Hort. Petrop. 9: 354. 1884. Error for *Leersia oryzoides* Swartz.
Oryza clandestina forma *inclusa* Wiesb. in Baenitz., Deut. Bot. Monatschr. 15: 19. 1897. Hungary.
Leersia oryzoides forma *glabra* A. A. Eaton, Rhodora 5: 118. 1903. Newburyport, Mass.
Oryza oryzoides Dalla Torre and Sarnth., Fl. Tirol 6: 142. 1906. Based on *Phalaris oryzoides* L.
Leersia oryzoides forma *inclusa* Dörfl., Herb. Norm. Sched. Cent. 55–56. 164. 1915. Based on *Oryza clandestina* forma *inclusa* Wiesb. (Published as new, Fogg, Rhodora 30: 84. 1928, same basis.)

(3) **Leersia virginica** Willd., Sp. Pl. 1: 325. 1797. North America.
Asprella virginica Beauv., Ess. Agrost. 2, 153. 1812. Based on *Leersia virginica* Willd.
Leersia imbricata Poir. in Lam., Encycl. Sup. 3: 329. 1813. Carolina, *Bosc.*
Leersia ovata Poir. in Lam., Encycl. Sup. 3: 329. 1813. North America.
Asprella ovata Roem. and Schult., Syst. Veg. 2: 267. 1817. Based on *Leersia ovata* Poir.
Asprella imbricata Roem. and Schult., Syst. Veg. 2: 268. 1817. Based on *Leersia imbricata* Poir.
Leersia virgata Raf., Bot. Seringe Bul. 1: 220. 1830 [probably error for *L. virginica*]. Cited as type of the genus *Aplexia*, but the name not transferred.
Homalocenchrus virginicus Britton, N. Y. Acad. Sci. Trans. 9: 14. 1889. Based on *Leersia virginica* Willd.
Homalocenchrus ovata Kuntze, Rev. Gen. Pl. 2: 777. 1891. Based on *Leersia ovata* Poir.

Aplexia virgata Raf. ex Jacks., Ind. Kew. 1: 162. 1893, as synonym of *Leersia virginica*.

Aplexia virginica Raf. ex Jacks., Ind. Kew. 1: 162. 1893, as synonym of *Leersia virginica*.

Leersia virginica var. *ovata* Fernald, Rhodora 38: 386. pl. 440. f. 9–13. 1936. Based on *L. ovata* Poir.

(97) LEPTOCHLOA Beauv.

(2) **Leptochloa chloridiformis** (Hack.) Parodi, Physis 4: 184. 1918. Based on *Diplachne chloridiformis* Hack.

Diplachne chloridiformis Hack. in Stuck., An. Mus. Nac. Buenos Aires 13: 498. 1906. Prov. Córdoba, Argentina, *Stuckert* 2329.

Baldomiria chloridiformis Herter, Rev. Sudamer. Bot. 6: 145. 1940. Based on *Diplachne chloridiformis* Hack.

(4) **Leptochloa domingensis** (Jacq.) Trin., Fund. Agrost. 133. 1820. Based on *Cynosurus domingensis* Jacq.

Cynosurus domingensis Jacq., Misc. Austr. 2: 363. 1781. Dominican Republic.

Festuca domingensis Lam., Tabl. Encycl. 1: 189. 1791. Based on *Cynosurus domingensis* Jacq.

Eleusine domingensis Pers., Syn. Pl. 1: 87. 1805. Based on *Cynosurus domingensis* Jacq.

Rabdochloa domingensis Beauv., Ess. Agrost. 84, 176. 1812. Based on *Cynosurus domingensis* Jacq.

Leptostachys domingensis G. Meyer, Prim. Fl. Esseq. 74. 1818. Based on *Eleusine domingensis* Pers.

Cynodon domingense Raspail, Ann. Sci. Nat., Bot. 5: 302. 1825. Based on *Rabdochloa domingensis* Beauv.

Leptochloa virgata var. *domingensis* Link ex Griseb., Fl. Brit. W. Ind. 538. 1864. Based on *L. domingensis* Link (same as *L. domingensis* Trin.).

Diplachne domingensis Chapm., Fl. South. U. S. ed. 3. 609. 1897. Based on *Leptochloa domingensis* Link (same as *L. domingensis* Trin.).

(1) **Leptochloa dubia** (H. B. K.) Nees, Syll. Pl. Ratisb. 1: 4. 1824. Based on *Chloris dubia* H. B. K.

Chloris dubia H. B. K., Nov. Gen. et Sp. 1: 169. 1816. Mexico, *Humboldt and Bonpland.*

Leptostachys dubia G. Meyer, Prim. Fl. Esseq. 74. 1818. Based on *Chloris dubia* H. B. K.

Festuca obtusiflora Willd. ex Spreng., Syst. Veg. 1: 356. 1825. Mexico.

Schismus patens Presl, Rel. Haenk. 1: 269. 1830. Chile, *Haenke*.

Leptochloa patens Kunth, Rév. Gram. 1: Sup. 22. 1830. Based on *Schismus patens* Presl.

Leptochloa obtusiflora Trin. ex Steud.,

Nom. Bot. ed. 2. 2: 30. 1841, as synonym of *L. dubia* Nees.

Diplachne patens Desv. in Gay, Fl. Chil. 6: 371. 1853. Based on *Schismus patens* Presl.

Uralepis brevicuspidata Buckl., Acad. Nat. Sci. Phila. Proc. 1862: 93. 1862. Texas [*Wright* 767].

Ipnum mendocinum R. A. Phil., An. Univ. Chile 36: 211. 1870. Mendoza, Argentina.

Diplachne dubia Scribn., Torrey Bot. Club Bul. 10: 30. 1883. Based on *Leptochloa dubia* Nees.

Molinia retusa Griseb. ex Fourn., Mex. Pl. 2: 147. 1886, as synonym of *Leptochloa dubia* Nees.

Diplachne dubia var. *aristata* Vasey, Calif. Acad. Sci. Proc. II. 2: 213. 1889. Name only. Baja California, *Brandegee*.

Leptochloa pringlei Beal, Grasses N. Amer. 2: 436. 1896. Arizona, *Pringle* in 1884.

Diplachne pringlei Vasey ex Beal, Grasses N. Amer. 2: 436. 1896, as synonym of *Leptochloa pringlei*.

Diplachne mendocina Kurtz, Bol. Acad. Cienc. Córdoba 15: 521. 1897. Based on *Ipnum mendocinum* R. A. Phil.

Diplachne dubia var. *pringleana* Kuntze, Rev. Gen. Pl. 3²: 349. 1898. Chihuahua, Mexico, *Pringle* 422.

Diplachne dubia var. *humboldtiana* Kuntze, Rev. Gen. Pl. 3²: 349. 1898. Presumably the original form collected by *Humboldt* and *Bonpland*.

Leptochloa dubia pringleana Scribn. and Merr., U. S. Dept. Agr., Div. Agrost. Bul. 24: 27. 1901. Based on *Diplachne dubia* var. *pringleana* Kuntze.

Rabdochloa dubia Kuntze ex Stuck., An. Mus. Nac. Buenos Aires 11: 121. 1904. Based on *Leptochloa dubia* Nees.

Sieglingia dubia Kuntze ex Stuck., An. Mus. Nac. Buenos Aires 11: 128. 1904. Based on *Chloris dubia* H. B. K.

Eragrostis mendozina Jedw., Bot. Archiv Mez 5: 192. 1924. Based on *Ipnum mendocinum* Phil.

(7) **Leptochloa fascicularis** (Lam.) A. Gray, Man. 588. 1848. Based on *Festuca fascicularis* Lam.

Festuca fascicularis Lam., Tabl. Encycl. 1: 189. 1791. South America.

Festuca polystachya Michx., Fl. Bor. Amer. 1: 66. 1803. Illinois, *Michaux*.

Diplachne fascicularis Beauv., Ess. Agrost. 81, 160. pl. 16. f. 9. 1812. Based on *Festuca fascicularis* Lam.

Festuca procumbens Muhl., Descr. Gram. 160. 1817. Carolina. Name only, Muhl., Cat. Pl. 13. 1813.

Festuca clandestina Muhl., Descr. Gram. 162. 1817. New York. Name only, Muhl., Cat. Pl. 13. 1813.

Festuca aquatica Bosc ex Roem. and Schult., Syst. Veg. 2: 615. 1817, as synonym of *Diplachne fascicularis* Beauv.

Cynodon fascicularis Raspail, ˙Ann. Sci. Nat. Bot. 5: 303. 1825. Based on *Diplachne fascicularis* Beauv.

Leptochloa polystachya Kunth, Rév. Gram. 1: 91. 1829. Based on *Festuca polystachya* Michx.

Diachroa procumbens Nutt., Amer. Phil. Soc. Trans. (n.s.) 5: 147. 1837. Based on *Festuca procumbens* Muhl.

Festuca texana Steud., Syn. Pl. Glum. 1: 310. 1854. Texas, *Drummond* 387.

Uralepsis composita Buckl., Acad. Nat. Sci. Phila. Proc. 1862: 94. 1862. New Mexico, *Woodhouse.*

Diplachne patens Fourn. ex Hemsl., Biol. Centr. Amer. Bot. 3: 570. 1885, name only; Mex. Pl. 2: 148. 1886. Not *D. patens* Desv., 1853. Vera Cruz, Mexico, *Gouin* 93.

Diplachne tracyi Vasey, Torrey Bot. Club Bul. 15: 40. 1888. Reno, Nev., *Tracy* [216].

Leptochloa tracyi Beal, Grasses N. Amer. 2: 436. 1896. Based on *Diplachne tracyi* Vasey.

Festuca prostrata Muhl. ex Scribn. and Merr., U. S. Dept. Agr., Div. Agrost. Cir. 27: 5. 1900, as synonym of *F. procumbens* Muhl.

Diplachne procumbens Nash in Britton, Man. 128. 1901. Not *D. procumbens* Arech., 1896. Based on *Festuca procumbens* Muhl.

Diplachne acuminata Nash in Britton, Man. 128. 1901. Arkansas to Nebraska and Colorado. [Type, Kansas, *Thompson.*]

Diplachne maritima Bicknell, Torrey Bot. Club Bul. 35: 195. 1908. Based on *D. procumbens* Nash.

(5) **Leptochloa filiformis** (Lam.) Beauv., Ess. Agrost. 71, 161, 166. 1812. Based on *Festuca filiformis* Lam.

Festuca filiformis Lam., Tabl. Encycl. 1: 191. 1791. South America.

Eleusine mucronata Michx., Fl. Bor. Amer. 1: 65. 1803. Illinois, *Michaux.*

Eleusine filiformis Pers., Syn. Pl. 1: 87. 1805. South America.

Eleusine sparsa Muhl., Cat. Pl. 12. 1813. Based on *E. filiformis* Pers.; Descr. Gram. 135. 1817. Carolina and Georgia.

Oxydenia attenuata Nutt., Gen. Pl. 1: 76. 1818. New Orleans, La. [*Nuttall*].

Leptostachys filiformis G. Meyer, Prim. Fl. Esseq. 74. 1818. Based on *Eleusine filiformis* Pers.

Leptochloa mucronata Kunth, Rév. Gram. 1: 91. 1829. Based on *Eleusine mucronata* Michx.

Aira panicea Willd. ex Steud., Nom. Bot. ed. 2. 1: 45. 1840, as synonym of *Leptochloa filiformis* Beauv.

Eleusine stricta Willd. ex Steud. Nom.

Bot. ed. 2. 1: 549. 1840. Not *E. stricta* Roxb., 1820. As synonym of *Leptochloa filiformis* Beauv.

Eleusine elongata Willd. ex Steud., Nom. Bot. ed. 2. 1: 549. 1840, as synonym of *Leptochloa filiformis* Beauv.

Leptochloa brachiata Steud., Syn. Pl. Glum. 1: 209. 1854. Guadeloupe, *Duchaissing.*

Leptochloa attenuata Steud., Syn. Pl. Glum. 1: 209. 1854. Based on *Oxydenia attenuata* Nutt.

Leptochloa pellucidula Steud., Syn. Pl. Glum. 1: 209. 1854. Panama, *Duchaissing.*

Leptochloa paniculata Fourn., Soc. Bot. France Bul. II. 27: 296. 1880. Nicaragua, *Levy* 1079.

Leptochloa mucronata var. *pulchella* Scribn., Torrey Bot. Club Bul. 9: 147. 1882. Santa Cruz Valley, Ariz., *Pringle* in 1881.

Oxydenia filiformis Nutt. ex Jacks., Ind. Kew. 2: 392. 1894, as synonym of *Leptochloa filiformis* Beauv.

Leptochloa pilosa Scribn., U. S. Dept. Agr., Div. Agrost. Cir. 32: 9. 1901. Travis County, Tex., *Bodin* 294 in 1891.

Leptochloa filiformis forma *attenuata* Gates, Kans. State Col. Agr., Dept. Bot. Contrib. 391: 130. 1940. Based on *Oxydenia attenuata* Nutt.

(9) **Leptochloa nealleyi** Vasey, Torrey Bot. Club Bul. 12: 7. 1885. Texas, *Nealley.*

Leptochloa stricta Fourn., Mex. Pl. 2: 147. 1886. Vera Cruz, Mexico, *Gouin* 73.

(11) **Leptochloa panicoides** (Presl) Hitchc., Amer. Jour. Bot. 21: 137. 1934. Based on *Megastachya panicoides* Presl. (Not invalidated by *L. panicoides* Wight, 1854; listed as a synonym only.)

Megastachya panicoides Presl, Rel. Haenk. 1: 283. 1830. Acapulco, Mexico, *Haenke.*

Poa panicoides Kunth, Rév. Gram. 1: Sup. 28. 1830. Based on *Megastachya panicoides* Presl.

Eragrostis panicoides Steud., Syn. Pl. Glum. 1: 278. 1854. Based on *Megastachya panicoides* Presl.

Leptochloa floribunda Doell in Mart., Fl. Bras. 2³: 89. 1878. Amazon River, Brazil.

Diplachne halei Nash, N. Y. Bot. Gard. Bul. 1: 292. 1899. Louisiana, *Hale.*

Leptochloa halei Scribn. and Merr., U. S. Dept. Agr., Div. Agrost. Bul. 24: 27. 1901. Based on *Diplachne halei* Nash.

(10) **Leptochloa scabra** Nees, Agrost. Bras. 435. 1829. Amazon River, Brazil.

Leptochloa langloisii Vasey, Torrey Bot. Club Bul. 12: 7. 1885. Louisiana, *Langlois.*

Leptochloa liebmanni Fourn., Mex. Pl. 2: 147. 1886. Antigua, Mexico, *Liebmann* 244, 248.

(8) **Leptochloa uninervia** (Presl) Hitchc. and Chase, U. S. Natl. Herb. Contrib. 18: 383. 1917. Based on *Megastachya uninervia* Presl.

Megastachya uninervia Presl, Rel. Haenk. 1: 283. 1830. Mexico, *Haenke.*

Poa uninervia Kunth, Rév. Gram. 1: Sup. XXVIII. 1830. Based on *Megastachya uninervia* Presl.

Diplachne verticillata Nees and Mey., Nov. Act. Acad. Caes. Leop. Carol. 19: Sup. 1: 27. 1841; 159. 1843. Chile and Peru, *Meyen.*

Uralepis verticillata Steud., Syn. Pl. Glum. 1: 248. 1854. Based on *Diplachne verticillata* Nees and Mey.

Eragrostis uninervia Steud., Syn. Pl. Glum. 1: 278. 1854. Based on *Megastachya uninervia* Presl.

Atropis carinata Griseb., Abh. Ges. Wiss. Göttingen 24: 291. 1879. Argentina.

Leptochloa imbricata Thurb. in S. Wats., Bot. Calif. 2: 293. 1880. California, Larken's Station, San Diego County, *Palmer 404;* Fort Yuma, *Thomas;* Gila Valley to Rio Grande.

Diplachne imbricata Scribn., Torrey Bot. Club Bul. 10: 30. 1883. Based on *Leptochloa imbricata* Thurb.

Brizopyrum uninervium Fourn., Mex. Pl. 2: 121. 1886. Based on *Megastachya uninervia* Presl.

Leptochloa virletii Fourn., Mex. Pl. 2: 147. 1886. San Luis Potosí, Mexico, *Virlet 1404.*

Diplachne tarapacana Phil., An. Mus. Nac. Chile. Bot. 8: 88. 1891. Tarapacá, Chile.

Rabdochloa imbricata Kuntze, Rev. Gen. Pl. 2: 788. 1891. Based on *Leptochloa imbricata* Thurb.

Diplachne carinata Hack., Bol. Acad. Cienc. Córdoba 16: 253. 1900. Based on *Atropis carinata* Griseb.

Diplachne uninervia Parodi, Univ. Nac. Buenos Aires Rev. Céntr. Estud. 18: 147. 1925. Based on *Megastachya uninervia* Presl.

(3) **Leptochloa virgata** (L.) Beauv., Ess. Agrost. 71, 161, 166. pl. 15. f. 1. 1812. Based on *Eleusine virgata* Pers., which is based on *Cynosurus virgatus* L.

Cynosurus virgatus L., Syst. Nat. ed. 10. 2: 876. 1759. Jamaica.

Festuca virgata Lam., Tabl. Encycl. 1: 189. 1791. Based on *Cynosurus virgatus* L.

Eleusine virgata Pers., Syn. Pl. 1: 87. 1805. Based on *Cynosurus virgatus* L.

Chloris poaeformis H. B. K., Nov. Gen. et Sp. 1: 169. 1816. Colombia and Ecuador, *Humboldt* and *Bonpland.*

Leptostachys virgata G. Meyer, Prim. Fl. Esseq. 74. 1818. Based on *Cynosurus virgatus* Willd. [error for L.].

Cynodon virgatus Raspail, Ann. Sci. Nat.,

Bot. 5: 302. 1825. Based on *Leptochloa virgata* Beauv.

Eleusine unioloides Willd. ex Steud., Nom. Bot. ed. 2. 1: 549. 1840, as synonym of *Leptochloa virgata* Pers.

Leptochloa mutica Steud., Syn. Pl. Glum. 1: 208. 1854. Surinam [Dutch Guiana], *Kappler 1553.*

Leptochloa virgata var. *mutica* Doell in Mart., Fl. Bras. 2³: 91. 1878. Based on *L. mutica* Steud.

Leptochloa virgata var. *aristata* Fourn., Mex. Pl. 2: 146. 1886. Mexico.

Leptochloa virgata var. *intermedia* Fourn., Mex. Pl. 2: 146. 1886. Mexico, *Liebmann 243, 251.*

Oxydenia virgata Nutt. ex Jacks., Ind. Kew. 2: 392. 1894, as synonym of *Leptochloa virgata.*

Leptochloa perennis Hack., Inf. Est. Centr. Agron. Cuba 1: 411. 1906. Cuba, *Baker 4617.*

(6) **Leptochloa viscida** (Scribn.) Beal, Grasses N. Amer. 2: 434. 1896. Based on *Diplachne viscida* Scribn.

Diplachne viscida Scribn., Torrey Bot. Club Bul. 10: 30. 1883. Santa Cruz Valley, Tucson, Ariz., *Pringle* in 1881.

(130) LEPTOLOMA Chase

(2) **Leptoloma arenicola** Swallen, Tex. Res. Found. Contrib. 1: 1. 1950. Kennedy County, Tex., *Swallen.*

(1) **Leptoloma cognatum** (Schult.) Chase, Biol. Soc. Wash. Proc. 19: 192. 1906. Based on *Panicum cognatum* Schult.

?*Panicum nudum* Walt., Fl. Carol. 73. 1788. South Carolina. Description inadequate, no specimen in the Walter Herbarium in the British Museum.

Panicum divergens Muhl. ex Ell., Bot. S. C. and Ga. 1: 130. 1816. Not *P. divergens* H. B. K., 1815. South Carolina. Name only, Muhl., Cat. Pl. 9. 1813.

Panicum cognatum Schult., Mantissa 2: 235. 1824. Based on *P. divergens* Muhl.

Panicum autumnale Bosc. ex Spreng., Syst. Veg. 1: 320. 1825. Origin unknown.

Panicum fragile Kunth, Rév. Gram. 1: 36. 1829. Based on *P. divergens* Muhl.

Panicum autumnale var. *pubiflorum* Vasey ex L. H. Dewey, U. S. Natl. Herb. Contrib. 2: 508. 1894. Texas.

Digitaria cognata Pilger in Engl. and Prantl, Pflanzenfam. ed. 2. 14e: 50. 1940. Based on *Panicum cognatum* Schult.

(75) LIMNODEA L. H. Dewey

(1) **Limnodea arkansana** (Nutt.) L. H. Dewey, U. S. Natl. Herb. Contrib. 2: 518. 1894. Based on *Greenia arkansana* Nutt.

Greenia arkansana Nutt., Amer. Phil. Soc. Trans. (n.s.) 5: 142. 1837. Red River, Ark.

Sclerachne arkansana Torr. ex Trin., Acad. St. Pétersb. Mém. VI. Sci. Nat. 4¹: 274. 1841. Based on *Greenia arkansana* Nutt.

Sclerachne pilosa Trin., Acad. St. Pétersb. Mém. VI. Sci. Nat. 4¹: 275. 1841. Texas, *Drummond.*

Limnas arkansana Trin. ex Steud., Nom. Bot. ed. 2. 2: 45. 1841. Based on *Greenia arkansana* Nutt.

Stipa demissa Steud., Syn. Pl. Glum. 1: 130. 1854. New Orleans, La., *Drummond* 465.

Muhlenbergia hirtula Steud., Syn. Pl. Glum. 1: 180. 1854. Texas, *Drummond.*

Limnas pilosa Steud., Syn. Pl. Glum. 1: 421. 1854. Based on *Sclerachne pilosa* Trin.

Thurberia arkansana Benth. ex Vasey, U. S. Dept. Agr. Spec. Rpt. 63: 16. 1883. Based on *Greenia arkansana* Nutt.

Thurberia pilosa Vasey, U. S. Dept. Agr. Spec. Rpt. 63: 16. 1883. Based on *Sclerachne pilosa* Trin.

Limnodea arkansana pilosa Scribn., U. S. Dept. Agr., Div. Agrost. Bul. 7 (ed. 3): 139. 1900. Based on *Sclerachne pilosa* Trin.

(50) LOLIUM L.

(2) **Lolium multiflorum** Lam., Fl. Franç. 3: 621. 1778. France.

Lolium scabrum Presl, Rel. Haenk. 1: 267. 1830. Peru, *Haenke.*

Lolium italicum A. Br., Flora 17: 241. 1834. Europe.

Lolium perenne var. *italicum* Parnell, Grasses Scotl. 1¹: 142. pl. 65. 1842. Presumably based on *L. italicum* A. Br.

Lolium perenne var. *multiflorum* Parnell, Grasses Brit. 302. pl. 140. 1845. Presumably based on *L. multiflorum* Lam.

Lolium multiflorum forma *microstachyum* Uechtritz, Jahresb. Schles. Ges. Vaterl. Cult. 1876: 334. 1880. Germany.

Lolium temulentum var. *multiflorum* Kuntze, Rev. Gen. Pl. 2: 779. 1891. Based on *L. multiflorum* Lam.

Lolium multiflorum var. *italicum* Beck., Wiss. Mitt. Bosn. Herzeg. 9: 459. 1904. Based on *L. italicum* A. Br.

Lolium multiflorum var. *diminutum* Mutel, as used by Harger et al. (Conn. State Geol. Nat. Hist. Survey Bul. 48: 25. 1930) appears to be *L. multiflorum.* Mutel's variety, described from France, is uncertain.

LOLIUM MULTIFLORUM var. RAMOSUM Guss. ex Arcang. Comp. Fl. Ital. 799. 1882. Sicily and Corsica.

(1) **Lolium perenne** L., Sp. Pl. 83. 1753. Europe.

Lolium brasilianum Nees, Agrost. Bras. 443. 1829. Montevideo, *Sellow.*

Lolium canadense Bernh. in Rouv., Monogr. Lolium 27. 1853. Not *L. canadense* Michx., 1817. As synonym of *L. perenne* L.

Lolium perenne var. *pacyi* Sturtev., N. Y. State Agr. Expt. Sta. Rpt. 1882¹: 77. 1883. Name only, Experiment Station, Geneva, N. Y.

LOLIUM PERENNE var. CRISTATUM Pers., Syn. Pl. 1: 110. 1805. Europe.

(4) **Lolium persicum** Boiss. and Hohen. in Boiss., Diagn. Pl. Orient. Nov. I. 2¹³: 66. 1853. Northern Persia, *Kotschy* 278.

Lolium remotum Schrank, Baier. Fl. 1: 382. 1789 (description inadequate); Schrank ex Hoffm., Deutschl. Fl. ed. 2. 1: 63. 1800. Germany.

Lolium strictum Presl, Cyp. Gram. Sicul. 49. 1820. Sicily.

(5) **Lolium subulatum** Vis. Fl. Dalm. 1: 90. pl. 3. 1842. Europe.

(3) **Lolium temulentum** L., Sp. Pl. 83. 1753. Europe.

Craepalia temulenta Schrank, Baier. Fl. 1: 382. 1789. Based on *Lolium temulentum* L.

LOLIUM TEMULENTUM var. LEPTOCHAETON A. Br., Flora 1: 252. 1834. Germany.

Lolium arvense With., Bot. Arr. Veg. Brit. ed. 3. 2: 168. 1796. Great Britain.

Lolium temulentum var. *arvense* Bab., Man. Brit. Bot. 377. 1843. Based on *L. arvense* With.

(123) LUZIOLA Juss.

(2) **Luziola bahiensis** (Steud.) Hitchc., U. S. Natl. Herb. Contrib. 12: 234. 1909. Based on *Caryochloa bahiensis* Steud.

Caryochloa bahiensis Steud., Syn. Pl. Glum. 1: 5. 1854. Bahia, Brazil.

Luziola alabamensis Chapm., Fl. South. U. S. 584. 1860. Brooklyn, Conecuh County, Ala., *Beaumont.*

Luziola longivalvula Doell in Mart., Fl. Bras. 2²: 17. 1871. Bahia, Brazil, *Salzmann* [type]; Minas Geraes, *Widgren, Regnell* III. 1376. (Misspelled *longivalvula* but correct in index.)

Luziola striata Bal. and Poitr., Soc. Hist. Nat. Toulouse Bul. 12: 231. pl. 4. f. 2. 1878. Paraguay, *Balansa* 181.

Luziola pusilla S. Moore, Linn. Soc. Bot. Trans. II. 4: 507. pl. 37. f. 1–8. 1895. Santa Cruz, Mato Grosso, Brazil, *Moore* 760.

Luziola bahiensis var. *alabamensis* Prodoehl, Bot. Archiv Mez 1: 242. 1922. Based on *Luziola alabamensis* Chapm.

(1) **Luziola peruviana** Gmel., Syst. Nat. 2: 637. 1791. Based on a species described but not named by Jussieu, Gen. Pl. 33. 1789. Peru, *Dombey.*

(78) LYCURUS H. B. K.

(1) **Lycurus phleoides** H. B. K., Nov. Gen. et Sp. 1: 142. pl. 45. 1815. Mexico, *Humboldt* and *Bonpland.*
Pleopogon setosum Nutt., Acad. Nat. Sci. Phila. Jour. II. 1: 189. 1848. Santa Fe, N. Mex., *Gambel.*
Lycurus phleoides var. *glaucifolius* Beal, Grasses N. Amer. 2: 271. 1896. Mexico, *Pringle 426*; Texas, *Havard, Nealley.*

(163) MANISURIS L.

(1) **Manisuris altissima** (Poir.) Hitchc., Wash. Acad. Sci. Jour. 24: 292. 1934. Based on *Rottboellia altissima* Poir.
Rottboellia altissima Poir., Voy. Barb. 2: 105. 1789. North Africa.
Rottboellia fasciculata Lam., Tabl. Encycl. 1: 204. 1791. North Africa.
Hemarthria fasciculata Kunth, Rév. Gram. 1: 153. 1829. Based on *Rottboellia fasciculata* Lam.
Rottboellia compressa var. *fasciculata* Hack. in DC., Monogr. Phan. 6: 286. 1889. Based on *R. fasciculata* Lam.
Manisuris fasciculata Hitchc., Amer. Jour. Bot. 2: 299. 1915. Based on *Rottboellia fasciculata* Lam.
(2) **Manisuris cylindrica** (Michx.) Kuntze, Rev. Gen. Pl. 2: 779. 1891. Based on *Tripsacum cylindricum* Michx.
?*Ischaemum scariosum* Walt., Fl. Carol. 249. 1788. South Carolina.
Tripsacum cylindricum Michx., Fl. Bor. Amer. 1: 60. 1803. Florida, *Michaux.*
Rottboellia campestris Nutt., Amer. Phil. Soc. Trans. (n.s.) 5: 151. 1837. Arkansas [*Nuttall*].
Rottboellia cylindrica Torr., U. S. Expl. Miss. Pacif. Rpt. 4: 159. 1857. Not *R. cylindrica* Willd., 1797. Based on *Tripsacum cylindricum* Michx.
Coelorachis cylindrica Nash, N. Amer. Fl. 17: 85. 1909. Based on *Tripsacum cylindricum* Michx.
Manisuris campestris Hitchc. in Small, Man. Southeast. Fl. 41. 1933. Based on *Rottboellia campestris* Nutt.
(4) **Manisuris rugosa** (Nutt.) Kuntze, Rev. Gen. Pl. 2: 780. 1891. Based on *Rottboellia rugosa* Nutt.
Rottboellia rugosa Nutt., Gen. Pl. 1: 84. 1818. Florida, *Baldwin.*
Rottboellia corrugata Baldw., Amer. Jour. Sci. 1: 355. 1819. Camden County, Ga., *Baldwin.*
Hemarthria rugosa Kunth, Rév. Gram. 1: 153. 1829. Based on *Rottboellia rugosa* Nutt.

Rottboellia rugosa var. *chapmani* Hack. in DC., Monogr. Phan. 6: 308. 1889. Florida, *Chapman.*
Manisuris corrugata Kuntze, Rev. Gen. Pl. 2: 779. 1891. Based on *Rottboellia corrugata* Baldw.
Manisuris rugosa var. *chapmani* Scribn., Torrey Bot. Club Mem. 5: 28. 1894. Based on *Rottboellia rugosa* var. *chapmani* Hack.
Manisuris chapmani Nash in Small, Fl. Southeast. U. S. 56. 1903. Based on *Rottboellia rugosa* var. *chapmani* Hack.
Coelorachis rugosa Nash, N. Amer. Fl. 17: 86. 1909. Based on *Rottboellia rugosa* Nutt.
Coelorachis corrugata A. Camus, Ann. Soc. Linn. Lyon 68: 197. 1921. Based on *Rottboellia corrugata* Baldw.
(3) **Manisuris tessellata** (Steud.) Scribn., U. S. Dept. Agr., Div. Agrost. Bul. 20: 20. f. 9. 1900. Based on *Rottboellia tessellata* Steud.
Rottboellia tessellata Steud., Syn. Pl. Glum. 1: 362. 1854. Louisiana, *Riehl* 60.
Rottboellia corrugata var. *areolata* Hack. in DC., Monogr. Phan. 6: 309. 1889. Mobile, Ala., *Mohr* in 1884.
Manisuris corrugata var. *areolata* Mohr, Torrey Bot. Club Bul. 24: 21. 1897. Based on *Rottboellia corrugata* var. *areolata* Hack.
Manisuris tessellata var. *areolata* Scribn., U. S. Dept. Agr., Div. Agrost. Bul. 17 (ed. 2): 9. 1901. Presumably based on *Rottboellia corrugata* var. *areolata* Hack.
Coelorachis tessellata Nash, N. Amer. Fl. 17: 86. 1909. Based on *Rottboellia tessellata* Steud.
(5) **Manisuris tuberculosa** Nash, N. Y. Bot. Gard. Bul. 1: 430. 1900. Eustis, Fla., *Nash* 1074.
Coelorachis tuberculosa Nash, N. Amer. Fl. 17: 86. 1909. Based on *Manisuris tuberculosa* Nash.
Rottboellia tuberculosa Hitchc., Biol. Soc. Wash. Proc. 41: 163. 1928. Based on *Manisuris tuberculosa* Nash.

(30) MELICA L.

Melica altissima L., Sp. Pl. 66. 1753. Siberia.
(2) **Melica aristata** Thurb. ex Boland., Calif. Acad. Sci. Proc. 4: 103. 1870. "Number 4861 [*Bolander*] Catalogue, 1867," Clarks (now Wawona) [type]; Yosemite Valley; Shady Canyon to Summit; Bear Valley to Eureka, Calif.
Bromelica aristata Farwell, Rhodora 21: 77. 1919. Based on *Melica aristata* Thurb.
(7) **Melica bulbosa** Geyer. ex Port. and Coult., Syn. Fl. Colo. 149. 1874. Porter

and Coulter cite Gray, Amer. Acad. Sci. Proc. 8: 409. 1872. Gray gives no description but cites *M. bulbosa* Geyer, Jour. Bot. Kew Misc. (Pl. Geyer.) 8: 19. 1856. In the latter work "Geyer no. 11, Upper Platte," is listed without description. The description by Porter and Coulter applies to this collection as represented in the Gray Herbarium.

Melica bella Piper, U. S. Dept. Agr., Div. Agrost. Cir. 27: 10. 1900. Upper Platte, *Geyer* [11]. A new name for "*M. bulbosa* Geyer, in U. S. Dept. Agr., Div. Bot. Bul. 13: 63. pl. 63. 1893, not *M. bulbosa* Geyer, in S. Wats., Bot. Calif. 2: 304. 1880," the description by Porter and Coulter having been overlooked. The Thurber publication refers to *M. californica* (No. 18 of this work).

Melica bella subsp. *intonsa* Piper, U. S. Natl. Herb. Contrib. 11: 128. 1906. Wenas, Wash., *Griffiths* and *Cotton* 103.

Melica bulbosa var. *typica* Cronquist, Madroño 7: 77. 1943. Based on *M. bulbosa* Geyer.

Melica bulbosa var. *caespitosa* Cronquist, Madroño 7: 77. 1943. West side of Alturas Lake, Blaine County, Idaho, *Cronquist* and *Cronquist* 2603.

(18) **Melica californica** Scribn., Acad. Nat. Sci. Phila. Proc. 1885: 46. pl. 1. f. 6. 1885. Based on *M. poaeoides* as described by Torrey in Pacific Railroad Report (see below), the specimen cited by Torrey, in N. Y. Bot. Gard., being named "*M. californica* Scribn." in Scribner's script.

Melica poaeoides Nutt. [misapplied by] Torr., U. S. Rpt. Expl. Miss. Pacif. 4: 157. 1857. Not *M. poaeoides* Nutt., ·1848. Corte Madera, Calif., [*Bigelow*].

Melica bulbosa Geyer ex Thurb. in S. Wats., Bot. Calif. 2: 304. 1880. Not *M. bulbosa* Geyer ex Port. and Coult., 1874. Santa Inez, Calif., Brewer 569.

Melica longiligula Scribn. and Kearn., U. S. Dept. Agr., Div. Agrost. Bul. 17: 225. f. 521. 1899. Southern California, *Parish Brothers* 865.

MELICA CALIFORNICA var. NEVADENSIS Boyle, Madroño 8: 17. 1945. Calaveras County, Calif., *Rutter* 163.

Melica ciliata L., Sp. Pl. 66. 1753. Europe.

(17) **Melica frutescens** Scribn., Acad. Nat. Sci. Phila. Proc. 1885: 45. pl. 1. f. 15, 16. 1885. Southern California, *Parry* and *Lemmon* 401 [type, labeled by Scribner].

(8) **Melica fugax** Boland., Calif. Acad. Proc. 4: 104. 1870. Donner Lake, Calif. [*Bolander* and *Kellogg*].

Melica geyeri [Munro misapplied by] Thurb. in Wilkes, U. S. Expl. Exped. Bot. 17: 492. 1874. Cascade Mountains, Oreg.

Melica fugax subsp. *madophylla* Piper, U. S. Natl. Herb. Contrib. 11: 128. 1906. Falcon Valley, Wash., *Suksdorf* 61.

Melica macbridei Rowland in Nels., Bot. Gaz. 54: 404. 1912. Silver City, Idaho, *Macbride* 948.

Melica fugax var. *inexpansa* Suksdorf, Werdenda 1²: 1. 1923. Falcon Valley, Wash., *Suksdorf* 6989.

Melica fugax var. *macbridei* Beetle, West. Bot. Leaflets 4: 286. 1946. Based on *M. macbridei* Rowland.

(5) **Melica geyeri** Munro in Boland., Calif. Acad. Sci. Proc. 4: 103. 1870. [Ukiah] Calif., *Bolander* 7, the specimen examined by Munro (in U. S. Natl. Herb.). The same collection was later distributed as 6119.

Glyceria bulbosa Buckl., Acad. Nat. Sci. Phila. Proc. 1862: 95. 1862. Columbia woods, *Nuttall*.

Bromus muticus Nutt. ex A. Gray, Acad. Nat. Sci. Phila. Proc. 1862: 335. 1862, as synonym of *Glyceria bulbosa* Buckl.

Melica poaeoides var. *bromoides* Boland., Calif. Acad. Sci. Proc. 4: 103. 1870, as synonym of *M. geyeri* Munro. *Bolander* 40 and 6119.

Melica bromoides Boland. ex A. Gray, Amer. Acad. Sci. Proc. 8: 409. 1872. Based on *M. poaeoides* var. *bromoides* Boland. [*Bolander* 6119].

Melica poaeoides Boland. ex Scribn., Acad. Nat. Sci. Phila. Proc. 1885: 47. 1885, as synonym of "*M. bromoides* Gray."

Melica bromoides var. *howellii* Scribn., Acad. Nat. Sci. Phila. Proc. 1885: 47. 1885. Near Waldo, Oreg., *Howell* 335 in 1884.

Melica pammeli Scribn., Davenport Acad. Sci. Proc. 7: 240. 1899. Geranium Park, Wyo., *Pammel* 159.

Bromelica geyeri Farwell, Rhodora 21: 78. 1919. Based on *Melica geyeri* Munro.

Bromelica geyeri var. *howellii* Farwell, Rhodora 21: 78. 1919. Based on *Melica bromoides* var. *howellii* Scribn.

MELICA GEYERI var. ARISTULATA J. T. Howell, West. Bot. Leaflets 4: 245. 1946. Marin County, Calif., *J. T. Howell* 17906.

(3) **Melica harfordii** Boland., Calif. Acad. Sci. Proc. 4: 102. 1870. Santa Cruz, *Bolander* 53 [type]; Redwood, *Bolander* 6464; Yosemite Valley and Bear Valley, both *Bolander*.

Melica harfordii var. *minor* Vasey, Torrey Bot. Club Bul. 15: 48. 1888. Siskiyou Mountains, Oreg., *Howell* in 1887.

Melica harfordii tenuior Piper, U. S. Natl. Herb. Contrib. 11: 127. 1906. Based on *M. harfordii* var. *minor* Vasey.

Bromelica harfordii Farwell, Rhodora 21:

78. 1919. Based on *Melica harfordii* Boland.

Bromelica harfordii var. *minor* Farwell, Rhodora 21: 78. 1919. Based on *Melica harfordii* var. *minor* Vasey.

Melica harfordii var. *tenuis* Suksdorf, Werdenda 1: 17. 1927. Bingen, Wash., *Suksdorf* 12018.

Melica harfordii var. *viridifolia* Suksdorf, Werdenda 1: 17. 1927. Bingen, Wash., *Suksdorf* 11686, 11777.

(16) **Melica imperfecta** Trin., Acad. St. Pétersb. Mém. VI. Sci. Nat. 2^1: 59. 1836. California.

Melica colpodioides Nees, Ann. Nat. Hist. 1: 283. 1838. California, *Douglas.*

Melica panicoides Nutt., Acad. Nat. Sci. Phila. Jour. II. 1: 188. 1848. Santa Barbara, Calif., *Gambel.*

Melica poaeoides Nutt., Acad. Nat. Sci. Phila. Jour. II. 1: 188. 1848. Santa Catalina Island, Calif., *Gambel.* [The type at the British Museum is labeled San Diego.]

Melica parishii Vasey ex Beal, Grasses N. Amer. 2: 500. 1896. Southern California, *Parish* 1997.

Melica imperfecta var. *pubens* Scribn., U. S. Dept. Agr., Div. Agrost. Cir. 30: 8. 1901. Santa Cruz Island, Calif., *Brandegee* 64.

The name is erroneously given as *Melica imperforata* Nees in Hook. and Arn., Bot. Beechey Voy. 403. 1840. This is the species described and figured by Vasey (U. S. Dept. Agr., Div. Bot. Bul. 13^2: pl. 84. 1893) as *Poa thurberiana* Vasey, but the name is based on *Panicularia thurberiana* Kuntze.

MELICA IMPERFECTA var. FLEXUOSA Boland., Calif. Acad. Sci. Proc. 4: 101. 1870. "Mariposa to Clark's" [Yosemite Valley region], Calif., *Bolander* in 1866.

MELICA IMPERFECTA var. MINOR Scribn., Acad. Nat. Sci. Phila. Proc. 1885: 42. 1885. San Bernardino Mountains, *Parish Brothers* 856.

MELICA IMPERFECTA var. REFRACTA Thurb. in S. Wats., Bot. Calif. 2: 303. 1880. San Bernardino, Calif., *Lemmon.*

(9) **Melica inflata** (Boland.) Vasey, U. S. Natl. Herb. Contrib. 1: 269. 1893. Based on *M. poaeoides* var. *inflata* Boland.

Melica poaeoides var. *inflata* Boland., Calif. Acad. Sci. Proc. 4: 101. 1870. Yosemite Valley, Calif., *Bolander* 6121.

Melica bulbosa var. *inflata* Boyle, Madroño 8: 19. 1945. Based on *M. poaeoides* var. *inflata* Boland.

(14) **Melica montezumae** Piper, Biol. Soc. Wash. Proc. 18: 144. 1905. Chihuahua, Mex., *Pringle* 430.

Melica alba Hitchc., U. S. Natl. Herb. Contrib. 17: 367. 1913. Chihuahua, Mex., *Pringle* 430.

(12) **Melica mutica** Walt., Fl. Carol. 78. 1788. South Carolina.

Melica glabra Michx., Fl. Bor. Amer. 1: 62. 1803. Virginia to Florida, *Michaux.*

Melica rariflora Schreb., Beschr. Gräs. 2: 157. 1810. Based on *M. glabra* Michx.

Melica diffusa Pursh, Fl. Amer. Sept. 1: 77. 1814. Virginia and Carolina.

Melica speciosa Muhl., Descr. Gram. 87. 1817. Pennsylvania.

Melica racemosa Muhl., Descr. Gram. 88. 1817. Not *M. racemosa* Thunb., 1794. Carolina; Georgia. Name only, Muhl., Cat. Pl. 11. 1813.

Melica muhlenbergiana Schult., Mantissa 2: 294. 1824. Based on *M. racemosa* Muhl.

Melica mutica var. *glabra* A. Gray, Man. ed. 5. 626. 1867. Based on *M. glabra* Pursh (error for Michx.).

Melica mutica var. *diffusa* A. Gray, Man. ed. 5. 626. 1867. Based on *M. diffusa* Pursh.

Melica mutica forma *diffusa* Fernald, Rhodora 41: 501. 1939. Based on *M. diffusa* Pursh.

(13) **Melica nitens** (Scribn.) Nutt. ex Piper, Torrey Bot. Club Bul. 32: 387. 1905. Based on *M. diffusa* var. *nitens* Scribn.

Melica scabra Nutt., Amer. Phil. Soc. Trans. (n.s.) 5: 148. 1837. Not *M. scabra* H. B. K., 1816. Fort Smith, Ark., *Nuttall.*

Melica diffusa var. *nitens* Scribn., Acad. Nat. Sci. Phila. Proc. 1885: 44. 1885. Arkansas, *Nuttall.* [The type in the Academy of Natural Sciences, Philadelphia, is labeled *M. nitens* Nutt.]

Melica nitens Nutt. ex Scribn., Acad. Nat. Sci. Phila. Proc. 1885: 44. 1885, as synonym of *M. diffusa* var. *nitens.*

(11) **Melica porteri** Scribn., Acad. Nat. Sci. Phila. Proc. 1885: 44. pl. 1. f. 17, 18. 1885. Based on *M. mutica* var. *parviflora* Porter.

Melica mutica var. *parviflora* Porter in Port. and Coult., Syn. Fl. Colo. 149. 1874. Glen Eyrie, Colo., *Porter* [type]; *Meehan*; Sierra Madre Range, *Coulter.*

Melica parviflora Scribn., Torrey Bot. Club Mem. 5: 50. 1894. Based on *M. mutica* var. *parviflora* Porter.

MELICA PORTERI var. LAXA Boyle, Madroño 8: 25. 1945. White Mountains, N. Mex., *Wooton* 680.

(1) **Melica smithii** (Porter) Vasey, Torrey Bot. Club Bul. 15: 294. 1888. Based on *Avena smithii* Porter.

Avena smithii Porter ex A. Gray, Man. ed. 5. 640. 1867. Sault Sainte Marie, Mich., *C. E. Smith.*

Melica retrofracta Suksdorf, Deut. Bot. Monatsschr. 19: 92. 1901. Skamania County, Wash. [*Suksdorf* 2334].

Bromelica smithii Farwell, Rhodora 21: 77. 1919. Based on *Avena smithii* Porter.

Schizachne smithii Wiegand ex Muenscher, Fl. Whatcom County, Washington 66.

1941. Presumably based on *Avena smithii* Porter.

(6) **Melica spectabilis** Scribn., Acad. Nat. Sci. Phila. Proc. 1885: 45. pl. 1. f. 11, 12, 13. 1885. Montana, Crow Mountains, *Scribner* 385 [type]; Boseman Pass, *Canby* 368. Colorado, *Porter* in 1872. Yellowstone Park, *Parry* 295. Nevada (erroneously given as Utah), *Watson* 1303. Idaho, *Watson* 455. The synonyms cited by Scribner are erroneous, "*M. bulbosa* S. Wats., Bot. King Exp. 383" being an error for *M. poaeoides* Nutt., Bot. King Exp. 383; "Porter and Coulter Fl. Colorado 149." refers to the valid *M. bulbosa*.
Melica scabrata Scribn. in Piper, Fl. Palouse 25. 1901. Pullman, Wash., *Piper* 1745.

(10) **Melica stricta** Boland., Calif. Acad. Sci. Proc. 3: 4. 1863. Silver City, Nev., *Dunn*.
Melica stricta var. *albicaulis* Boyle, Madroño 8: 24. 1945. San Antonio Mountains, Calif., *I. M. Johnston* 1516.

(4) **Melica subulata** (Griseb.) Scribn., Acad. Nat. Sci. Phila. Proc. 1885: 47. 1885. Based on *Bromus subulatus* Griseb.
Bromus subulatus Griseb. in Ledeb., Fl. Ross. 4: 358. 1853. Unalaska, *Eschscholtz*.
Melica acuminata Boland., Calif. Acad. Sci. Proc. 4: 104. 1870. Mendocino County, Calif., *Bolander* 4698.
Festuca acerosa Trin. ex A. Gray, Amer. Acad. Sci. Proc. 8: 410. 1872, as synonym of *Bromus subulatus* Griseb.
Melica poaeoides var. *acuminata* Boland. ex Scribn., Acad. Nat. Sci. Phila. Proc. 1885: 47. 1885, as synonym of *M. subulata* Scribn. California, *Bolander* 4698.
Bromelica subulata Farwell, Rhodora 21: 78. 1919. The name is based on *Festuca subulata* Bong., doubtless an error for *Bromus subulatus* Griseb., since *Melica acuminata* Boland is also cited.
This is the species to which Scribner (U.S. Dept. Agr., Div. Agrost. Cir. 30: 8. 1901.) applied the name *Melica cepacea* Scribn., based on *Festuca cepacea* Phil., a Chilean species of *Melica*.

(15) **Melica torreyana** Scribn., Acad. Nat. Sci. Phila. Proc. 1885: 43. pl. 1. f. 3, 4. 1885. California, *Bigelow* in 1853–4.
Melica imperfecta var. *sesquiflora* Torr. ex Scribn., Acad. Nat. Sci. Phila. Proc. 1885: 43. 1885, as synonym of *M. torreyana*, a herbarium name given to a specimen collected by Bigelow in California in 1853–4.

(126) **MELINIS Beauv.**

(1) **Melinis minutiflora** Beauv., Ess. Agrost. 54. pl. 11. f. 4. 1812. Rio de Janeiro, Brazil.

Tristegis glutinosa Nees, Horae Phys. Berol. 47, 54. pl. 7. 1820. Brazil.
Panicum minutiflorum Raspail, Ann. Sci. Nat., Bot. 5: 299. 1825. Based on *Melinis minutiflora* Beauv.
Panicum melinis Trin., Acad. St. Pétersb. Mém. VI. Sci. Nat. 1: 291. 1834. Based on *Melinis minutiflora* Beauv.
Muhlenbergia brasiliensis Steud., Syn. Pl. Glum. 1: 177. 1854. Bahia, Brazil, *Salzmann* [652].
Agrostis polypogon Salzm. ex Steud., Syn. Pl. Glum. 1: 177. 1854, as synonym of *Muhlenbergia brasiliensis*.

Mibora minima (L.) Desv., Obs. Angers 45. 1818. Based on *Agrostis minima* L.
Agrostis minima L., Sp. Pl. 63. 1753. France.

(102) **MICROCHLOA R. Br.**

(1) **Microchloa kunthii** Desv., Opusc. 75. 1831. Mexico, *Humboldt* and *Bonpland*.
Paspalum tenuissimum Jones, West. Bot. Contrib. 18: 24. 1935. Baja California, *Jones* 27584.

(152) **MICROSTEGIUM Nees**

(1) **Microstegium vimineum** (Trin.) A. Camus, Ann. Soc. Linn. Lyon 68: 201. 1921. Based on *Andropogon vimineus* Trin.
Andropogon vimineus Trin., Acad. St. Pétersb. Mém. VI. Math. Phys. Nat. 2: 268. 1832. Nepal.
Microstegium willdenovianum Nees in Lindl., Nat. Syst. Bot. (ed. 2 of his Introd. Bot.) 447. 1836. Nepal.
Pollinia willdenoviana Benth., Linn. Soc. Jour., Bot. 19: 67. 1881. Based on *Microstegium willdenovianum* Nees.
Pollinia imberbis var. *willdenoviana* Hack. in DC., Monogr. Phan. 6: 178. 1889. Based on *Microstegium willdenovianum* Nees.
Eulalia viminea Kuntze, Rev. Gen. Pl. 2: 775. 1891. Based on *Andropogon vimineus* Trin.
Pollinia viminea Merr., Enum. Philipp. Pl. 1: 35. 1922. Based on *Andropogon vimineus* Trin.
MICROSTEGIUM VIMINEUM var. IMBERBE (Nees) Honda, Tokyo Univ. Facult. Sci. Jour. Sec. 3. Bot. 3: 408. 1930. Based on *Pollinia imberbis* Nees.
Pollinia imberbis Nees in Steud., Syn. Pl. Glum. 1: 410. 1855. Nepal.
Eulalia viminea var. *variabilis* Kuntze, Rev. Gen. Pl. 2: 775. 1891. Sikkim.

(88) **MILIUM L.**

(1) **Milium effusum** L., Sp. Pl. 61. 1753. Europe.
Miliarium effusum Moench, Meth. Pl.

204. 1794. Based on *Milium effusum* L.
Melica effusa Salisb., Prodr. Stirp. 20. 1796. Based on *Milium effusum* L.
Decandolia effusa Bast., Fl. Maine-et-Loire 28. 1808. Based on *Milium effusum* L.
Paspalum effusum Raspail, Ann. Sci. Nat. Bot. 5: 301. 1825. Based on *Milium effusum* L.

(149) MISCANTHUS Anderss.

Miscanthus nepalensis (Trin.) Hack. in DC., Monogr. Phan. 6: 104. 1889. Based on *Eulalia nepalensis* Trin.
Eulalia nepalensis Trin., Acad. St. Pétersb. Mém. VI. Math. Phys. Nat. 2: 333. 1832. Nepal, India.
Miscanthus sacchariflorus (Maxim.) Hack. in Engl. and Prantl, Pflanzenfam. 2: 23, 102. 1887. Based on *Imperata sacchariflora* Maxim.
Imperata sacchariflora Maxim. (Prim. Fl. Amur. 331.) Acad. St. Pétersb. Sav. Étrang. Mém. 9: 331. 1859. East Siberia.
(1) **Miscanthus sinensis** Anderss., Öfv. Svensk, Vet. Akad. Förh. 12: 166. 1856. China.
Saccharum japonicum Thunb., Linn. Soc. Trans. 2: 328. 1794. Not *Miscanthus japonicus* Anderss., 1855. Japan.
Eulalia japonica Trin., Acad. St. Pétersb. Mém. VI. Math. Phys. Nat. 2: 333. 1832. Based on *Saccharum japonicum* Thunb.
Miscanthus sinensis var. *variegatus* Beal, Grasses N. Amer. 2: 25. 1896. Cultivated.
Miscanthus sinensis var. *zebrinus* Beal, Grasses N. Amer. 2: 25. 1896. Cultivated.
Xiphagrostis japonica Coville, U. S. Natl. Herb. Contrib. 9: 400. 1905. Based on *Saccharum japonicum* Thunb.
Miscanthus sinensis var. *gracillimus* Hitchc. in Bailey, Cycl. Amer. Hort. 1021. f. 1408. 1901. Cultivated under the garden name *Eulalia japonica* var. *gracillima*.
Eulalia japonica var. *gracillima* Grier, Amer. Midl. Nat. 11: 331. 1929. Based on *Miscanthus sinensis* var. *gracillimus* Hitchc.

(16) MOLINIA Schrank

(1) **Molinia caerulea** (L.) Moench, Meth. Pl. 183. 1794. Based on *Aira caerulea* L.
Aira caerulea L., Sp. Pl. 63. 1753. Europe.
Festuca caerulea Lam. and DC., Fl. Franç. ed. 3. 3: 46. 1805. Based on *Aira caerulea* L.
Enodium caeruleum Gaudin, Agrost. Helv.

1: 145. 1811. Based on *Aira caerulea* L.
Cynodon caeruleus Raspail, Ann. Sci. Nat., Bot. 5: 302. 1825. Based on *Molinia caerulea* Koel. (error for Moench).
Amblytes caerulea Dulac, Fl. Haut. Pyr. 80. 1867. Based on *Molinia caerulea* Moench.

(20) MONANTHOCHLOË Engelm.

(1) **Monanthochloë littoralis** Engelm., St. Louis Acad. Sci. Trans. 1: 437. pl. 13. 14. 1859. Texas, *Drummond, Berlandier* 3227 (Matamoros region), Galveston Island, *Lindheimer;* Florida, Key West, *Blodgett.*

(51) MONERMA Beauv.

(1) **Monerma cylindrica** (Willd.) Coss. and Dur., Expl. Sci. Alger. 2: 214. 1855. Based on *Rottboellia cylindrica* Willd.
Rottboellia cylindrica Willd., Sp. Pl. 1: 464. 1797. Europe.
Ophiurus cylindricus Beauv., Ess. Agrost. 116, 168, 176. 1812. Based on *Rottboellia cylindrica* Willd.
Monerma monandra Beauv., Ess. Agrost. 117, 168, 177. pl. 20. f. 10. 1812. "*Rottboellia monandra* Lin." (p. 117) and "*R. monandra* Roth" (p. 177) are referred to *Monandra monerma*, but neither Linnaeus nor Roth published the name. Pl. 10, fig. 9, and the generic description unquestionably indicate *Rottboellia cylindrica* Willd. No locality is cited for *M. monandra* Beauv.
Lepturus cylindricus Trin., Fund. Agrost. 123. 1820. Based on *Rottboellia cylindrica* Willd.
Lolium cylindricum Aschers. and Graebn., Syn. Mitteleur. Fl. 2: 761. 1902. Based on *Rottboellia cylindrica* Willd.

(82) MUHLENBERGIA Schreb.

(31) **Muhlenbergia andina** (Nutt.) Hitchc., U. S. Dept. Agr. Bul. 772: 145. 1920. Based on *Calamagrostis andina* Nutt.
Calamagrostis andina Nutt., Acad. Nat. Sci. Phila. Jour. II. 1: 187. 1848. California, on the Colorado of the West, *Gambel.*
Vaseya comata Thurb. in A. Gray, Acad. Nat. Sci. Phila. Proc. 1863: 79. 1863. Nebraska [probably Wyoming, *Hall* and *Harbour* 685].
Muhlenbergia comata Thurb. ex Benth., Linn. Soc. Jour., Bot. 19: 83. 1881. Based on *Vaseya comata* Thurb.
(8) **Muhlenbergia appressa** C. O. Goodding, Wash. Acad. Sci. Jour. 31: 504. 1941. Pinal or Gila County, Ariz., *Harrison* and *Kearney* 1493.
(26) **Muhlenbergia arenacea** (Buckl.) Hitchc., Biol. Soc. Wash. Proc. 41: 161.

1928. Based on *Sporobolus arenaceus* Buckl.

Sporobolus arenaceus Buckl., Acad. Nat. Sci. Phila. Proc. 1862: 89. 1862. Western Texas [*Wright* 737].

Sporobolus asperifolius var. *brevifolius* Vasey, U. S. Natl. Herb. Contrib. 1: 56. 1890, name only, Pena, Duval County, Tex., *Nealley;* U. S. Natl. Herb. Contrib. 3: 64. 1892, as synonym of *S. auriculatus* Vasey.

Sporobolus auriculatus Vasey, Contrib. U. S. Natl. Herb. 3: 64. 1892. Pena, Tex., *Nealley.*

(54) **Muhlenbergia arenicola** Buckl., Acad. Nat. Sci. Phila. Proc. 1862: 91. 1862. Western Texas [*Wright* 735].

Podosaemum arenicola Bush, Amer. Midl. Nat. 7: 40. 1921. Based on *Muhlenbergia arenicola* Buckl.

(52) **Muhlenbergia arizonica** Scribn., Torrey Bot. Club Bul. 15: 8. pl. 76. f. A. 1888. Near Mexican Boundary, Arizona, *Pringle* in 1884.

(23) **Muhlenbergia arsenei** Hitchc., Biol. Soc. Wash. Proc. 41: 161. 1928. Sulphur Springs, N. Mex., *Arsène* and *Benedict* 16405.

(27) **Muhlenbergia asperifolia** (Nees and Mey.) Parodi, Univ. Nac. Buenos Aires Rev. Agron. 6: 117. f. 1. 1928. Based on *Sporobolus asperifolius* Nees and Mey.

Vilfa asperifolia Meyen, Reis. Erd. 1: 408. 1834, name only; Nees and Mey., Acad. St. Pétersb. Mém. VI. Sci. Nat. 4¹: 95. 1840. Chile, *Meyen.*

Sporobolus asperifolius Nees, Nov. Act. Acad. Caes. Leop. Carol. 19: Sup. 1: 9. 1841; 141. 1843. Based on *Vilfa asperifolia* Nees and Mey.

Agrostis distichophylla R. A. Phil., Fl. Atac. 54. 1860. Not *A. distichophylla* Roem. and Schult., 1817. Chile. (Fide Parodi.)

Sporobolus sarmentosus Griseb., Abhandl. Gesell. Wiss. Göttingen 24: 295. 1879. Argentina.

Sporobolus deserticolus Phil., An. Mus. Nac. Chile Bot. 8: 82. 1891. Chile. (Fide Parodi.)

Sporobolus asperifolius var. *major* Vasey, U. S. Natl. Herb. Contrib. 3: 64. 1892. [Marfa, Tex., *Havard* 10 in 1883.]

Sporobolus distichophyllus Phil., An. Univ. Chile 94: 7. 1896. Based on *Agrostis distichophylla* Phil.

Agrostis eremophila Speg., An. Mus. Nac. Buenos Aires 7: 190. 1902. Based on *A. distichophylla* Phil.

(37) **Muhlenbergia brachyphylla** Bush, Amer. Midl. Nat. 6: 41. 1919. Webb City, Mo., *Palmer* 2734. (Not invalidated by *M. brachyphylla* Nees ex Jacks., Ind. Kew. 2: 269. 1894, a clerical error for *Podosaemum brachyphyllum* Nees.)

(12) **Muhlenbergia brevis** C. O. Goodding, Wash. Acad. Sci. Jour. 31: 505. 1941. Socorro County, N. Mex., *Metcalfe* 671.

(34) **Muhlenbergia californica** Vasey, Torrey Bot. Club Bul. 13: 53. 1886. Based on *M. glomerata* var. *brevifolia* Vasey.

Muhlenbergia glomerata var. *brevifolia* Vasey, Bot. Gaz. 7: 92. 1882. [San Bernardino Mountains], Calif., *Parish* [1028].

Muhlenbergia sylvatica var. *californica* Vasey, Bot. Gaz. 7: 93. 1882. San Bernardino Mountains, Calif., *Parish* [1076].

Muhlenbergia parishii Vasey, Torrey Bot. Club Bul. 13: 53. 1886. Based on *M. sylvatica* var. *californica* Vasey.

Muhlenbergia racemosa var. *brevifolia* Vasey ex Beal, Grasses N. Amer. 2: 253. 1896. Based on *M. glomerata* var. *brevifolia* Vasey.

Muhlenbergia californica Abrams, Fl. Los Angeles 32. 1904. Based on *M. sylvatica* var. *californica* Vasey.

(62) **Muhlenbergia capillaris** (Lam.) Trin., Gram. Unifl. 191. 1824. Based on *Trichochloa capillaris* DC., this based on *Stipa capillaris* Lam.

Stipa diffusa Walt., Fl. Carol. 78. 1788. Not *Muhlenbergia diffusa* Willd., 1798. South Carolina.

Stipa capillaris Lam., Tabl. Encycl. 1: 158. 1791. Carolina, *Fraser.*

Podosaemum capillare Desv., Nouv. Bul. Soc. Philom. (Paris) 2: 188. 1810. Based on *Stipa capillaris* Lam.

Tosagris agrostidea Beauv., Ess. Agrost. 29. pl. 8. f. 3. 1812. United States.

Podosaemum agrostideum Beauv., Ess. Agrost. 176, 179. 1812. Based on *Tosagris agrostidea* Beauv.

Trichochloa capillaris DC., Cat. Hort. Monsp. 152. 1813. Based on *Stipa capillaris* Lam.

Trichochloa polypogon DC., Cat. Hort. Monsp. 152. 1813. Carolina, *Fraser.*

Muhlenbergia polypogon Kunth, Rév. Gram. 1: 64. 1829. Based on *Trichochloa polypogon* DC.

Agrostis setosa Willd. ex Trin., Acad. St. Pétersb. Mém. VI. Sci. Nat. 4¹: 300. 1841, as synonym of *Muhlenbergia capillaris.* "Willd. hb. 1682," received from Muhlenberg.

MUHLENBERGIA CAPILLARIS var. FILIPES (M. A. Curtis) Chapm. ex Beal, Grasses N. Amer. 2: 256. 1896. Based on *M. filipes* M. A. Curtis.

Stipa sericea Michx., Fl. Bor. Amer. 1: 54. 1803. South Carolina, *Michaux.*

Agrostis sericea Ell., Bot. S. C. and Ga. 1: 135. 1816. Based on *Stipa sericea* Michx. In Muhl., Descr. Gram. 64. 1817, the name is misapplied to *M. capillaris* (Lam.) Trin.

Polypogon sericeus Spreng., Syst. Veg. 1:

243. 1825. Based on *Stipa sericea* Michx.

Stipa cericea Michx. ex Raf., Neogen 4. 1825. Error for *S. sericea*.

Muhlenbergia filipes M. A. Curtis, Amer. Jour. Sci. 44: 83. 1843. Sea Islands of North Carolina; Florida [*M. A. Curtis*].

Podosaemum filipes Bush, Amer. Midl. Nat. 7: 29. 1921. Based on *Muhlenbergia filipes* M. A. Curtis.

(19) **Muhlenbergia curtifolia** Scribn., Torrey Bot. Club Bul. 38: 328. 1911. Between Kanab and Carmel, Utah, *Jones* 6047j.

Muhlenbergia curtifolia subsp. *griffithsii* Scribn., Torrey Bot. Club Bul. 38: 328. 1911. Du Chelly Canyon, Ariz., *Griffiths* 5837.

(43) **Muhlenbergia curtisetosa** (Scribn.) Bush, Amer. Midl. Nat. 6: 35. 1919. Based on *M. schreberi* subsp. *curtisetosa* Scribn.

Muhlenbergia schreberi subsp. *curtisetosa* Scribn., Rhodora 9: 17. 1907. Illinois, *Wolf* in 1881.

(24) **Muhlenbergia cuspidata** (Torr.) Rydb., Torrey Bot. Club Bul. 32: 599. 1905. Based on *Vilfa cuspidata* Torr.

Agrostis brevifolia Nutt., Gen. Pl. 1: 44. 1818. Fort Mandan [N. Dak.].

Vilfa cuspidata Torr. in Hook., Fl. Bor. Amer. 2: 238. 1840. Saskatchewan River, Rocky Mountains, *Drummond.*

Vilfa gracilis Trin., Acad. St. Pétersb. Mém. VI. Sci. Nat. 4¹: 104. 1840. Not *V. gracilis* Trin., op. cit. 74. North America, received from Hooker.

Sporobolus cuspidatus Wood, Amer. Bot. and Flor. pt. 2: 385. 1871. Based on *Vilfa cuspidata* Torr.

Sporobolus brevifolius Scribn., Torrey Bot. Club Mem. 5: 39. 1894. Not *S. brevifolius* Nees, 1841. Based on *Agrostis brevifolia* Nutt. As new, Nash, in Britton, Man. 105. 1901, same basis.

Muhlenbergia brevifolia Jones, West. Bot. Contrib. 14: 12. 1912. Not *M. brevifolia* Scribn., 1896. Based on *Agrostis brevifolia* Nutt.

(11) **Muhlenbergia depauperata** Scribn., Bot. Gaz. 9: 187. 1884. Arizona, *Pringle*, in 1884.

Muhlenbergia schaffneri Fourn., Mex. Pl. 2: 85. 1886. Mexico, Tacubaya, *Schaffner* 50, 514; Mirador, *Schaffner* 142.

Lycurus schaffneri Mez, Repert. Sp. Nov. Fedde 17: 212. 1921. Based on *Muhlenbergia schaffneri* Fourn.

(58) **Muhlenbergia dubia** Fourn. in Hemsl., Biol. Centr. Amer. Bot. 3: 540. 1885. Chinantla, Mexico, *Liebmann* [688].

Muhlenbergia acuminata Vasey, Bot. Gaz. 11: 337. 1886. New Mexico, *Wright* 1993.

Sporobolus ligulatus Vasey and Dewey, U. S. Natl. Herb. Contrib. 1: 268.

1893. Presidio County, Tex., *Nealley*, 127.

Sporobolus inflatus Vasey and Dewey ex Beal, Grasses N. Amer. 2: 289. 1896. Error for *S. ligulatus* Vasey and Dewey.

Crypsinna breviglumis Jones, West. Bot. Contrib. 14: 8. 1912. Chihuahua, Mexico [*Jones* in 1903].

(59) **Muhlenbergia dubioides** C. O. Goodding, Wash. Acad. Sci. Jour. 30: 20. 1940. Box Canyon, Ariz., *Silveus* 3490.

(30) **Muhlenbergia dumosa** Scribn. in Vasey, U. S. Natl. Herb. Contrib. 3: 71. 1892. Santa Catalina Mountains, Ariz., *Pringle* [in 1884], *Lemmon; Mexico, Pringle;* southern California, *Orcutt.*

Muhlenbergia dumosa var. *minor* Scribn. in Beal, Grasses N. Amer. 2: 261. 1896. Mexico, *Pringle* 2355.

(6) **Muhlenbergia eludens** C. G. Reeder, Wash. Acad. Sci. Jour. 39: 365. f. 1, *B*. 1949. Minaca, Chihuahua, Mex., *Hitchcock* 7768.

(67) **Muhlenbergia emersleyi** Vasey, U. S. Natl. Herb. Contrib. 3: 66. 1892. Southern Arizona, *Emersley.*

Muhlenbergia vaseyana Scribn., Mo. Bot. Gard. Rpt. 10: 52. 1899. Based on *M. distichophylla* as described by Vasey (Rothr. in Wheeler, U. S. Survey W. 100th Merid. Rpt. 6: 283. 1878, Arizona, *Rothrock* 282, type).

Epicampes emerslyi Hitchc., U. S. Dept. Agr. Bul. 772: 144. 1920. Based on *Muhlenbergia emersleyi* Vasey.

Epicampes subpatens Hitchc., U. S. Dept. Agr. Bul. 772: 144. 1920. Guadalupe Mountains, N. Mex., *Hitchcock* 13541.

(60) **Muhlenbergia expansa** (DC.) Trin., Gram. Pan. 26. 1826. Based on *Trichochloa expansa* DC.

Agrostis arachnoidea Poir., in Lam., Encycl. Sup. 1: 249. 1810. Carolina, *Bosc.*

Trichochloa purpurea Beauv., Ess. Agrost. 29. pl. 8. f. 2. 1812. United States.

Vilfa arachnoidea Beauv., Ess. Agrost. 147, 181. 1812. Presumably based on *Agrostis arachnoidea* Poir.

Podosaemum purpureum Beauv., Ess. Agrost. 176, 179. pl. 8. f. 2. 1812. Based on *Trichochloa purpurea* Beauv.

Trichochloa expansa DC., Cat. Hort. Monsp. 151. 1813. Carolina, *Bosc.* *Stipa expansa* Poir. doubtfully cited.

Agrostis rubicunda Bosc. ex DC., Cat. Hort. Monsp. 151. 1813, as synonym of *Trichochloa expansa* DC.

Agrostis trichopodes Ell., Bot. S. C. and Ga. 1: 135. pl. 8. f. 1. 1816. Chatham County, Ga., *Baldwin.*

Cinna arachnoidea Kunth, Rév. Gram. 1: 67. 1829. Based on *Agrostis arachnoidea* Poir.

Muhlenbergia arachnoidea Trin. ex Kunth, Enum Pl. 1: 207. 1833, as synonym of *Cinna arachnoidea* Kunth.

Agrostis expansa Poir. ex Steud., Nom. Bot. ed. 2. 1: 40. 1840, as synonym of *Cinna arachnoidea* Kunth.
Agrostis longiflora Willd. ex Steud., Nom. Bot. ed. 2. 1: 41. 1840, as synonym of *Cinna arachnoidea* Kunth.
Muhlenbergia trichopodes Chapm., Fl. South. U. S. 553. 1860. Based on *Agrostis trichopodes* Ell.
Muhlenbergia caespitosa Chapm., Bot. Gaz. 3: 18. 1878. Apalachicola, Fla., *Chapman*.
Muhlenbergia capillaris var. *trichopodes* Vasey, U. S. Natl. Herb. Contrib. 3: 66. 1892. Based on *Agrostis trichopodes* Ell.
Podosaemum trichopodes Bush, Amer. Midl. Nat. 7: 30. 1921. Based on *Agrostis trichopodes* Ell.

(46) **Muhlenbergia filiculmis** Vasey, U. S. Natl. Herb. Contrib. 1: 267. 1893. Green Mountain Falls, Colo., *Sheldon* [321].

(13) **Muhlenbergia filiformis** (Thurb.) Rydb., Torrey Bot. Club Bul. 32: 600. 1905. Based on *Vilfa depauperata* var. *filiformis* Thurb.
Vilfa depauperata var. *filiformis* Thurb. in S. Wats. in King, Geol. Expl. 40th Par. 5: 376. 1871. Yosemite Valley, Calif., *Bolander* 6091; Donner Lake, *Torrey* 565; East Humboldt Mountains, Nev., *Watson* 1280; Uinta Mountains, Utah, *Watson* 1281.
Vilfa gracillima Thurb. in S. Wats., Bot. Calif. 2: 268. 1880. Not *Muhlenbergia gracillima* Torr. 1856. California, Sierra Nevada, *Brewer* [2827]; Yosemite Valley, *Bolander* [6091].
Sporobolus gracillimus Vasey, Grasses U. S. Descr. Cat. 44. 1885. Based on *Vilfa gracillima* Thurb.
Sporobolus filiformis Rydb., U. S. Natl. Herb. Contrib. 3: 189. 1895. Based on *Vilfa depauperata* var. *filiformis* Thurb.
Sporobolus depauperatus var. *filiformis* Beal, Grasses N. Amer. 2: 296. 1896. Montana, *Williams*; Utah, *Jones*.
Sporobolus simplex Scribn., U. S. Dept. Agr., Div. Agrost. Bul. 11: 48. 1898. Georgetown, Colo., *Rydberg* 2411.
Sporobolus aristatus Rydb., Torrey Bot. Club Bul. 28: 266. 1901. Not *Muhlenbergia aristata* Pers. 1805. Big Horn Mountains, Wyo., *Tweedy* 2196.
Sporobolus simplex var. *thermale* Merr., Rhodora 4: 48. 1902. Lolo Hot Springs Mont., *Griffiths* 302a.
Muhlenbergia simplex Rydb., Torrey Bot. Club Bul. 32: 600. 1905. Not *M. simplex* Kunth, 1829. Based on *Sporobolus simplex* Scribn.
Muhlenbergia aristulata Rydb., Torrey Bot. Club Bul. 32: 600. 1905. Based on *Sporobolus aristatus* Rydb.
Muhlenbergia filiformis var. *fortis* E. H.

Kelso, Rhodora 38: 298. 1936. Based on *Sporobolus simplex* Scribn.
Muhlenbergia idahoensis St. John, Fl. Southeast. Wash. and Adj. Idaho 50. 1937. Zaza, Nez Perce County, Idaho, *St. John* 9085.

(3) **Muhlenbergia fragilis** Swallen, U. S. Natl. Herb. Contrib. 29: 206. 1947. Alpine, Brewster County, Tex., *Warnock* 517.

(38) **Muhlenbergia frondosa** (Poir.) Fernald, Rhodora 45: 235. pl. 750. 1943. Based on *Agrostis frondosa* Poir.
Agrostis frondosa Poir. in Lam., Encycl. Sup. 1: 252. 1790. Described from a cultivated or adventive specimen grown in Germany.
Agrostis lateriflora Michx. Fl. Bor. Amer. 1: 53. 1803. Mississippi River [Illinois], *Michaux*.
Vilfa lateriflora Beauv., Ess. Agrost. 16, 147, 181. 1812. Based on *Agrostis lateriflora* Michx. (Appears erroneously as *laterifolia* on pages 16 and 147, but correctly on page 181.)
Cinna lateriflora Kunth, Rév. Gram. 1: 67. 1829. Based on *Agrostis lateriflora* Michx.
Muhlenbergia lateriflora Trin. ex Kunth, Enum. Pl. 1: 207. 1833, as synonym of *Cinna lateriflora* Kunth.
Calamagrostis compressa Doell in Mart., Fl. Bras. 2³: 56. 1878. "E seminibus a cl. Glaziou prope Rio de Janeiro lectis in horto bot. Monacensi anno 1869 cultura enata." A specimen named in Doell's script and bearing the above data was examined in Doell's herbarium in the Botanical Institute at Freiburg. This agrees perfectly with Doell's description. It is a characteristic specimen of *Muhlenbergia frondosa* except that the rachilla is minutely produced beyond the palea, a very rare occurrence in *Muhlenbergia*. Presumably the seed from Rio de Janeiro failed to germinate, and this species, probably in a neighboring plot, intruded.

MUHLENBERGIA FRONDOSA forma COMMUTATA (Scribn.) Fernald, Rhodora 45: 235. 1943. Based on *M. mexicana* subsp. *commutata* Scribn.
Muhlenbergia mexicana subsp. *commutata* Scribn., Rhodora 9: 18. 1907. Maine, *Fernald* 522 in 1896.
Muhlenbergia mexicana var. *commutata* Farwell, Mich. Acad. Sci. Rpt. 17: 181. 1916. Based on *M. mexicana* subsp. *commutata* Scribn.
Muhlenbergia commutata Bush, Amer. Midl. Nat. 6: 61. 1919. Based on *M. mexicana* subsp. *commutata* Scribn.
Muhlenbergia mexicana forma *commutata* Wiegand, Rhodora 26: 1. 1924. Based on *M. mexicana* subsp. *commutata* Scribn.

(39) **Muhlenbergia glabriflora** Scribn., Rhodora 9: 22. 1907. Texas, *Reverchon* 5.

(20) **Muhlenbergia glauca** (Nees) Mez, Repert. Sp. Nov. Fedde 17: 214. 1921. Based on *Podosaemum glaucum* Nees. In Index Kewensis this name is credited to Nees in Linnaea 19: 689. 1847, but the name there is *Podosaemum glaucum*.

Podosaemum glaucum Nees, Linnaea 19: 689. 1847. Mexico, *Aschenborn* 335.

Agrostis glauca Steud., Syn. Pl. Glum. 1: 175. 1854. Not *A. glauca* Muhl., 1817. Based on *Podosaemum glaucum* Nees.

Muhlenbergia lemmoni Scribn. in Coulter, U. S. Natl. Herb. Contrib. 1: 56. 1890. Ballinger, Tex., *Nealley;* New Mexico; Arizona. [*Lemmon* 2918, type, the species being named for Lemmon] Mexico.

Muhlenbergia huachucana Vasey, U. S. Natl. Herb. Contrib. 3: 69. 1892. Huachuca Mountains, Ariz., *Lemmon* [2915].

(32) **Muhlenbergia glomerata** (Willd.) Trin., Gram. Unifl. 191. pl. 5. f. 10. 1824. Based on *Polypogon glomeratus* Willd.

Polypogon setosus Bieler, Pl. Nov. Herb. Spreng. Cent. 7. 1807. Pennsylvania, *Muhlenberg.* Not *Muhlenbergia setosa* Kunth, 1829.

Polypogon glomeratus Willd., Enum. Pl. 87. 1809. North America [Pennsylvania].

Agrostis setosa Muhl., Cat. Pl. 10. 1813. "Polypogon W." cited.

Alopecurus glomeratus Poir., in Lam., Encycl. 5: 495. 1817. Based on *Polypogon glomeratus* Willd.

Agrostis glomerata Muhl., Descr. Gram. 68. 1817. Pennsylvania. "*Polypogon setosus* C. Sprengel, *glomeratus* Willd." cited.

Agrostis festucoides Muhl. ex. Roem and Schult., Syst. Veg. 1: 326. 1817, as synonym of *Polypogon glomeratus* Willd.

Trichochloa glomerata Trin., Fund. Agrost. 117. 1820. Based on *Polypogon glomeratus* Willd.

Trichochloa calycina Trin., Fund. Agrost. 117. 1820. "*Agrostis setosa* Spreng." (ined.) cited; no description.

Agrostis setosa Spreng. ex Trin., Fund. Agrost. 117. 1820. As synonym of *Trichochloa calycina* Trin., not *A. setosa* Spreng. himself, 1824 (see synonymy under *Muhlenbergia microsperma*).

Muhlenbergia calycina Trin., Gram. Unifl. 193. 1824. Based on *Trichochloa calycina* Trin., and cited as synonym of "*Polypogon setosus* Spreng."

Podosaemum glomeratum Link, Hort. Berol. 1: 84. 1827. Based on *Polypogon glomeratus* Willd.

Cinna glomerata Link, Hort. Berol. 2: 237. 1833. Not *C. glomerata* Walt., 1788. Based on *Podosaemum glomeratum* Link.

Dactylogramma cinnoides Link, Hort. Berol. 2: 248. 1833. Grown in Berlin, seed from *Richardson,* western North America.

Muhlenbergia setosa Trin. ex Jacks., Ind. Kew. 2: 269. 1894. Not *Muhlenbergia setosa* Kunth, 1829. Based on "*Polypogon setosus* Spreng.*"

Muhlenbergia racemosa subsp. *violacea* Scribn., Rhodora 9: 22. 1907. North Hannibal, N. Y., *Pearce* in 1883.

Muhlenbergia setosa var. *cinnoides* Fernald, Rhodora 45: 238. 1943. Based on *Dactylogramma cinnoides* Link.

Muhlenbergia glomerata var. *cinnoides* Hermann, Rhodora 48: 64. 1946. Based on *Dactylogramma cinnoides* Link.

(66) **Muhlenbergia involuta** Swallen, Amer. Jour. Bot. 19: 436. f. 2. 1932. San Antonio, Tex., *Silveus* 358.

(44) **Muhlenbergia jonesii** (Vasey) Hitchc., in Jepson, Fl. Calif. 1: 111. 1912. Based on *Sporobolus jonesii* Vasey.

Sporobolus jonesii Vasey, Bot. Gaz. 6: 297. 1881. Soda Springs, Calif., *Jones* [303] in 1881.

(65) **Muhlenbergia lindheimeri** Hitchc., Wash. Acad. Sci. Jour. 24: 291. 1934. Texas, *Lindheimer* 725. (This species has been referred to *Epicampes berlandieri* Fourn., and to *Muhlenbergia fournieriana* Hitchc., based upon it, but that species is confined to Mexico.)

Epicampes gracilis Trin., Acad. St. Pétersb. Mém. VI. Sci. Nat. 4[1]: 271. 1841. Not *Muhlenbergia gracilis* Kunth, 1829. Mexico [eastern Texas, probably *Berlandier*].

(64) **Muhlenbergia longiligula** Hitchc., Amer. Jour. Bot. 21: 136. 1934. Based on *Epicampes ligulata* Scribn.

Epicampes ligulata Scribn. in Vasey, U. S. Natl. Herb. Contrib. 3: 58. 1892. Not *Muhlenbergia ligulata* Scribn. and Merr. Texas to Arizona [type, Santa Rita Mountains, *Pringle* in 1884] and Mexico.

Epicampes distichophylla var. *mutica* Scribn. in Beal, Grasses N. Amer. 2: 308. 1896. Arizona, *Toumey* 740 [type]; Mexico, *Pringle* 1427. The other specimens cited do not agree with the description.

Epicampes anomala Scribn. in Beal, Grasses N. Amer. 2: 311. 1896. Not *Muhlenbergia anomalis* Fourn., 1886. Chihuahua, Mexico, *Pringle* 1423.

Melica anomala Scribn., in Beal, Grasses N. Amer. 2: 311. 1896, as synonym of *Epicampes anomala.*

Epicampes stricta var. *mutica* Jones, West. Bot. Contrib. 14: 6. 1912. Based on *E. distichophylla* var. *mutica* Scribn.

(70) **Muhlenbergia marshii** I. M. Johnston, Arnold Arboretum Jour. 24: 392. 1943. Coahuila, Mex., *Marsh* 746.

(57) **Muhlenbergia metcalfei** Jones, West.

Bot. Contrib. 14: 12. 1912. Santa Rita Mountains, N. Mex., *Metcalfe* 1485. The name was published as *"Metcalfi."*

(41) **Muhlenbergia mexicana** (L.) Trin., Gram. Unifl. 189. 1824.

Agrostis mexicana L., Mant. Pl. 1: 31. 1767. Grown at Upsala, tropical America erroneously given as the source, received from Jacquin.

Vilfa mexicana Beauv., Ess. Agrost. 16, 148, 181. 1812. Based on *Agrostis mexicana* L.

Cinna? mexicana Beauv., Ess. Agrost. 32, 148, 158. 1812. Based on *Agrostis mexicana* L.

Trichochloa mexicana Trin., Fund. Agrost. 117. 1820. Based on *Agrostis mexicana* L.

Podosaemum mexicanum Link, Hort. Berol. 1: 84. 1827. Based on *Agrostis mexicana* L.

Cinna arundinacea Retz. ex Steud., Nom. Bot. ed. 2. 1: 365. 1840. Not *C. arundinacea* L., 1753. As synonym of *C. mexicana* Beauv.

Muhlenbergia mexicana var. *purpurea* Wood, Amer. Bot. and Flor. pt. 2: 386. 1871. Illinois, *Wolf.*

Polypogon canadensis Fourn., Mex. Pl. 2: 92. 1886. Based on *Agrostis mexicana* L.

Muhlenbergia polystachya Mackenz. and Bush, Man. Fl. Jackson County 23. 1902. Sibley, Mo., *Mackenzie* 637.

Lepyroxis canadensis Beauv. ex Jacks., Ind. Kew Suppl. 1: 244. 1906, as synonym of *Agrostis mexicana* L.

MUHLENBERGIA MEXICANA forma AMBIGUA (Torr.) Fernald, Rhodora 45: 236. 1943. Based on *M. ambigua* Torr.

Agrostis filiformis Willd., Enum. Pl. 1: 95. 1809. Not *A. filiformis* Vill., 1787, nor *Muhlenbergia filiformis* Rydb., 1905. [Pennsylvania] North America.

Agrostis foliosa "Hortul." Roem. and Schult., Syst. Veg. 2: 373. 1817. Garden specimen; seed from North America.

Trichochloa foliosa Trin., Fund. Agrost. 117. 1820. Based on *Agrostis filiformis* Willd.

Cinna filiformis Link, Enum. Pl. 1: 70. 1821. Based on *Agrostis filiformis* Willd.

Agrostis lateriflora var. *filiformis* Torr., Fl. North. and Mid. U. S. 1: 86. 1823. Based on *A. filiformis* Muhl. (error for Willd.).

Trichochloa filiformis Trin. ex Torr., Fl. North. and Mid. U. S. 1: 86. 1823, as synonym of *Agrostis lateriflora* var. *filiformis* Torr.

Podosaemum foliosum Link, Hort. Berol. 1: 83. 1827. Based on *Agrostis foliosa* Roem. and Schult.

Muhlenbergia ambigua Torr. in Nicoll.,

Rpt. Miss. 164. 1843. "Okaman Lake, Sioux Country," *Geyer.*

Muhlenbergia mexicana var. *filiformis* Vasey, Grasses U. S. 23. 1883. Name only.

Muhlenbergia mexicana filiformis Scribn., Torrey Bot. Club Mem. 5: 36. 1894. Based on *A. filiformis* Muhl. (error for Willd.).

Muhlenbergia foliosa ambigua Scribn., Rhodora 9: 20. 1907. Based on *M. ambigua* Torr.

Muhlenbergia ambigua var. *filiformis* Farwell, Mich. Acad. Sci. Rpt. 20: 168. 1919. Based on *Agrostis filiformis* Muhl. [error for Willd.].

Muhlenbergia foliosa forma *ambigua* Wiegand, Rhodora 26: 1. 1924. Based on *M. ambigua* Torr.

MUHLENBERGIA MEXICANA forma SETIGLUMIS (S. Wats.) Fernald, Rhodora 45: 236. 1943. Based on *M. sylvatica* var. *setiglumis* S. Wats.

Muhlenbergia sylvatica var. *setiglumis* S. Wats. in King, Geol. Expl. 40th Par. 5: 378. 1871. Humboldt Pass, Nev., *Watson* 1288.

Muhlenbergia foliosa setiglumis Scribn., Rhodora 9: 20. 1907. Based on *M. sylvatica* var. *setiglumis* S. Wats.

Muhlenbergia setiglumis Nels. and Macbr., Bot. Gaz. 61: 30. 1916. Based on *M. sylvatica* var. *setiglumis* S. Wats.

(7) **Muhlenbergia microsperma** (DC.) Kunth, Rév. Gram. 1: 64. 1829. Based on *Trichochloa microsperma* DC.

Trichochloa microsperma DC., Cat. Hort. Monsp. 151. 1813. Mexico.

Podosaemum setosum H. B. K., Nov. Gen. et Sp. 1: 129. 1815. Mexico, *Humboldt* and *Bonpland.*

Podosaemum debile H. B. K., Nov. Gen. et Sp. 1: 128. 1815. Ecuador, *Humboldt* and *Bonpland.*

Agrostis microsperma Lag., Gen. et Sp. Nov. 2. 1816. Mexico, *Sessé.*

Trichochloa debilis Roem. and Schult., Syst. Veg. 2: 385. 1817. Based on *Podosaemum debile* H. B. K.

Trichochloa setosa Roem. and Schult., Syst. Veg. 2: 386. 1817. Based on *Podosaemum setosum* H. B. K.

Muhlenbergia fasciculata Trin., Gram. Unifl. 192. 1824. North America.

Agrostis setosa Spreng., Syst. Veg. 1: 262. 1825. Based on *Podosaemum setosum* H. B. K.

Agrostis debilis Spreng., Syst. Veg. 1: 262. 1825. Not *A. debilis* Poir., 1810. Based on *Podosaemum debile* H. B. K.

Muhlenbergia setosa Kunth, Rév. Gram. 1: 63. 1829. Based on *Podosaemum setosum* H. B. K.

Muhlenbergia debilis Kunth, Rév. Gram. 1: 63. 1829. Based on *Podosaemum debile* H. B. K.

Agrostis microcarpa Steud., Nom. Bot. ed.

2. 1: 41. 1840; 2: 164. 1841, as synonym of *Muhlenbergia microsperma* Kunth.

Muhlenbergia purpurea Nutt., Acad. Nat. Sci. Phila. Jour. II. 1: 186. 1848. Santa Barbara and Santa Catalina Island, Calif., *Gambel.*

Muhlenbergia ramosissima Vasey, Torrey Bot. Club Bul. 13: 231. 1886. Chihuahua, Mexico, *Palmer* [158] in 1885.

(2) **Muhlenbergia minutissima** (Steud.) Swallen, U. S. Natl. Herb. Contrib. 29: 207. 1947. Based on *Agrostis minutissima* Steud.

Agrostis minutissima Steud., Syn. Pl. Glum. 1: 171. 1854. New Mexico, *Fendler* 986.

Sporobolus minutissimus Hitchc., Biol. Soc. Wash. Proc. 41: 161. 1928. Based on *Agrostis minutissima* Steud.

This is the species described in the Manual, ed. 1, under *Sporobolus microspermus* (Lag.) Hitchc. That is the same as *Muhlenbergia confusa* (Fourn.) Swallen, known only from Mexico and Guatemala.

(45) **Muhlenbergia montana** (Nutt.) Hitchc., U. S. Dept. Agr. Bul. 772: 145, 147. 1920. Based on *Calycodon montanum* Nutt.

Calycodon montanum Nutt., Acad. Nat. Sci. Phila. Jour. II. 1: 186. 1848. Santa Fe, [New] Mexico, *Gambel.*

Muhlenbergia gracilis var. *breviaristata* Vasey in Rothr., Cat. Pl. Survey W. 100th Merid. 54. 1874. Twin Lakes, Colo., [*Wolf*] 1090 in 1873.

Muhlenbergia gracilis var. *major* Vasey in Rothr., in Wheeler, U. S. Survey W. 100th Merid. Rpt. 6: 284. 1878. Mount Graham, Ariz., Wheeler Exped. [*Rothrock*] 744.

Muhlenbergia subalpina Vasey, Grasses U. S. Descr. Cat. 40. 1885. Based on *M. gracilis* var. *breviaristata* Vasey.

Muhlenbergia trifida Hack., Repert. Sp. Nov. Fedde 8: 518. 1910. Michoacan, Mexico, *Arsène* 3217.

This is the species referred to *Muhlenbergia gracilis* by American authors, not *M. gracilis* (H. B. K.) Kunth.

(48) **Muhlenbergia monticola** Buckl., Acad. Nat. Sci. Phila. Proc. 1862: 91. 1862. "Northwestern Texas," [*Wright* 731].

Muhlenbergia sylvatica var. *flexuosa* Vasey in Rothr., in Wheeler, U. S. Survey W. 100th Merid. Rpt. 6: 284. 1878. New Mexico, *Wright* 731; Camp Crittenden, Ariz., *Rothrock* 681.

(69) **Muhlenbergia mundula** I. M. Johnston, Arnold Arboretum Jour. 24: 392. 1943. Chihuahua, Mexico, *Pringle* 417.

(49) **Muhlenbergia parviglumis** Vasey, U. S. Natl. Herb. Contrib. 3: 71. 1892. Texas, *Nealley.*

(22) **Muhlenbergia pauciflora** Buckl., Acad. Nat. Sci. Phila. Proc. 1862: 91. 1862. Western Texas [*Wright* 732].

Muhlenbergia sylvatica var. *pringlei* Scribn., Torrey Bot. Club Bul. 9: 89. 1882. Santa Rita Mountains, N. Mex., *Pringle* 480.

Muhlenbergia neo-mexicana Vasey, Bot. Gaz. 11: 337. 1886. New Mexico [type, *G. R. Vasey*] and Arizona.

Muhlenbergia pringlei Scribn. in Vasey, U. S. Natl. Herb. Contrib. 3: 71. 1892. Santa Rita Mountains, Ariz., *Pringle* 480.

(10) **Muhlenbergia pectinata** C. O. Goodding, Wash. Acad. Sci. Jour. 31: 505. 1941. Guadalajara, Mexico, *Pringle* 1745.

(21) **Muhlenbergia polycaulis** Scribn., Torrey Bot. Club Bul. 38: 327. 1911. Chihuahua, Mexico, *Pringle* 1414.

(51) **Muhlenbergia porteri** Scribn. in Beal, Grasses N. Amer. 2: 259. 1896. Based on *M. texana* Thurb.

Muhlenbergia texana Thurb.; Port. and Coult., Syn. Fl. Colo. 144. 1874. Not *M. texana* Buckl., 1863. Texas, *Bigelow; Parry; Wright* 734.

Podosaemum porteri Bush, Amer. Midl. Nat. 7: 36. 1921. Based on *Muhlenbergia porteri* Scribn.

(9) **Muhlenbergia pulcherrima** Scribn. in Beal, Grasses N. Amer. 2: 240. 1896. Sierra Madre, Mexico, *Pringle* 1416.

(50) **Muhlenbergia pungens** Thurb. in A. Gray, Acad. Nat. Sci. Phila. Proc. 1863: 78. 1863. Rocky Mountains, Colo., *Hall* and *Harbour* 632.

Podosaemum pungens Bush, Amer. Midl. Nat. 7: 32. 1921. Not *P. pungens* Link, 1827. Based on *Muhlenbergia pungens* Thurb.

(33) **Muhlenbergia racemosa** (Michx.) B. S. P., Prel. Cat. N. Y. 67. 1888. Presumably based on *Agrostis racemosa* Michx.

Agrostis racemosa Michx., Fl. Bor. Amer. 1: 53. 1803. Mississippi River [Ill.], *Michaux.*

Vilfa racemosa Beauv., Ess. Agrost. 16, 148, 182. 1812. Based on *Agrostis racemosa* Michx.

Polypogon racemosus Nutt., Gen. Pl. 1: 51. 1818. Based on *Agrostis racemosa* Michx.

Cinna racemosa Kunth, Rév. Gram. 1: 67. 1829. Based on *Agrostis racemosa* Michx.

Muhlenbergia glomerata var. *ramosa* Vasey, Grasses U. S. Descr. Cat. 40. 1885. Illinois to Colorado and Montana. [Type, collected by *Vasey*, marked "Dakota and Wisconsin."]

Muhlenbergia racemosa var. *ramosa* Vasey ex Beal, Grasses N. Amer. 2: 253. 1896. Presumably based on *M. glomerata* var. *ramosa* Vasey.

(14) **Muhlenbergia repens** (Presl) Hitchc. in Jepson, Fl. Calif. 1: 111. 1912. Based on *Sporobolus repens* Presl.

Sporobolus repens Presl, Rel. Haenk. 1:
241. 1830. Mexico, *Haenke.*
Vilfa repens Trin., Acad. St. Pétersb.
Mém. VI. Sci. Nat. 4¹: 102. 1840.
Based on *Sporobolus repens* Presl.
Muhlenbergia subtilis Nees, Linnaea 19:
689. 1847. Mexico, *Aschenborn* 206.
Muhlenbergia abata I. M. Johnston,
Arnold Arboretum Jour. 24: 387. 1943.
Valley of the Rio Grande, New Mexico,
Wright 1982.
(61) **Muhlenbergia reverchoni** Vasey and
Scribn., U. S. Natl. Herb. Contrib. 3:
66. 1892. Texas, *Reverchon* [73].
Podosaemum reverchoni Bush, Amer. Midl.
Nat. 7: 38. 1921. Based on *Muhlenbergia reverchoni* Vasey and Scribn.
(16) **Muhlenbergia richardsonis** (Trin.)
Rydb., Torrey Bot. Club Bul. 32: 600.
1905. Based on *Vilfa richardsonis* Trin.
Vilfa squarrosa Trin., Acad. St. Pétersb.
Mém. VI. Sci. Nat. 4¹: 100. 1840.
Menzies Island [Columbia River,
Wash.].
Vilfa richardsonis Trin., Acad. St. Pétersb.
Mém. VI. Sci. Nat. 4¹: 103. 1840.
North America, *Richardson.*
Muhlenbergia aspericaulis Nees ex Trin.,
Acad. St. Pétersb. Mém. VI. Sci. Nat.
4¹: 103. 1840, as synonym of *Vilfa
richardsonis* Trin.
Vilfa depauperata Torr. in Hook., Fl. Bor.
Amer. 2: 257. pl. 2. 36. 1840. Columbia River, from Menzies Island upward,
Douglas.
Sporobolus depauperatus Scribn., Torrey
Bot. Club Bul. 9: 103. 1882. Based on
Vilfa depauperata Torr.
Sporobolus aspericaulis Scribn., Bot. Gaz.
21: 15. 1896. Based on *Muhlenbergia
aspericaulis* Nees.
Sporobolus richardsonii Merr., Rhodora 4:
46. 1902. Based on *Vilfa richardsonis*
Trin.
Muhlenbergia squarrosa Rydb., Torrey
Bot. Club Bul. 36: 531. 1909. Based
on *Vilfa squarrosa* Trin.
Muhlenbergia brevifolia var. *richardsonis*
Jones, West. Bot. Contrib. 14: 12.
1912. Based on *Vilfa richardsonis* Trin.
This is the species which Nash (Britton
Man. 105. 1901) called *Sporobolus brevifolius*, but that name is based on *Agrostis
brevifolius* Nutt., which is *Muhlenbergia
cuspidata* (which see).
(68) **Muhlenbergia rigens** (Benth.) Hitchc.,
Wash. Acad. Sci. Jour. 23: 453. 1933.
Based on *Epicampes rigens* Benth.
Cinna macroura (Kunth misapplied by)
Thurb. in S. Wats., Bot. Calif. 2: 276.
1880. Not *C. macroura* (H. B. K.)
Kunth, 1835. California.
Vilfa rigens Thurb. ex S. Wats., Bot.
Calif. 2: 276. 1880. Not *V. rigens*
Trin., 1830. As synonym of *C. macroura*
Kunth. "Sonora" [probably error for
Sonoma] California, *Bolander* [6124].

Epicampes rigens Benth., Linn. Soc. Jour.,
Bot. 19: 88. 1881. Based on the species
Thurber described as *Cinna macroura*,
not that of (H. B. K.) Kunth.
Crypsinna rigens Jones, West. Bot.
Contrib. 14: 8. 1912. Based on *Epicampes rigens* Benth.
(63) **Muhlenbergia rigida** (H. B. K.) Kunth,
Rév. Gram. 1: 63. 1829. Based on
Podosaemum rigidum H. B. K.
Podosaemum rigidum H. B. K., Nov. Gen.
et Sp. 1: 129. 1815. Mexico, *Humboldt*
and *Bonpland.*
Trichochloa rigida Roem. and Schult.,
Syst. Veg. 2: 386. 1817. Based on
Podosaemum rigidum H. B. K.
Muhlenbergia berlandieri Trin., Acad. St.
Pétersb. Mém. VI. Sci. Nat. 4¹: 299.
1841. Mexico, *Berlandier.*
Muhlenbergia affinis Trin., Acad. St.
Pétersb. Mém. VI. Sci. Nat. 4¹: 301.
1841. "Toluco," *Berlandier.*
(42) **Muhlenbergia schreberi** Gmel., Syst.
Nat. 2: 171. 1791. Based on the species
described by Schreber (Gen. Pl. 1: 44.
1789) under *Muhlenbergia* with no
specific name [Pennsylvania].
Muhlenbergia diffusa Willd., Sp. Pl. 1:
320. 1797. North America. Name only,
Muhl., Amer. Phil. Soc. Trans. 3: 160.
1793.
Dilepyrum minutiflorum Michx., Fl. Bor.
Amer. 1: 40. 1803. Kentucky and
Illinois, *Michaux.* Listed as *Dylepyrum
multiflorum* by Beauv., Ess. Agrost. 160.
1812.
Dylepyrum diffusum Beauv., Ess. Agrost.
160. 1812. Name only, referred to
Muhlenbergia. Probably the same as *M.
diffusa* Willd.
Anthipsimus gonopodus Raf., Jour. Phys.
Chym. 89: 105. 1819. Dry hills of the
Ohio.
Cynodon diffusus Raspail, Ann. Sci. Nat.,
Bot. 5: 303. 1825. Based on "*Muhlenbergia* Schr." (error for Willd.).
Agrostis apetala Bosc. ex Trin., Acad. St.
Pétersb. Mém. VI. Sci. Nat. 4¹: 287.
1841, as synonym of *Muhlenbergia
diffusa* Schreb.
Muhlenbergia botteri Fourn., Mex. Pl. 2:
85. 1886. Orizaba, Mexico, *Botteri* 87.
Muhlenbergia minutiflora Hitchc., Kans.
Acad. Sci. Trans. 14: 140. 1896. Based
on *Dilepyrum minutiflorum* Michx.
MUHLENBERGIA SCHREBERI var. PALUSTRIS
(Scribn.) Scribn., Rhodora 9: 17. 1907.
Based on *M. palustris* Scribn. (Published
as *M. schreberi palustris.*)
Muhlenbergia palustris Scribn., U. S.
Dept. Agr., Div. Agrost. Bul. 11: 47.
1898. District of Columbia, *Steele* in
1896.
Muhlenbergia schreberi var. *palustris*
Scribn. ex Robinson, Rhodora 10: 65.
1908. Based on *M. palustris* Scribn.
(55) **Muhlenbergia setifolia** Vasey, Bot.

Gaz. 7: 92. 1882. Guadalupe Mountains, Tex., *Havard*.

(4) **Muhlenbergia sinuosa** Swallen, U. S. Natl. Herb. Contrib. 29: 204. 1947. San Luis Mountains, New Mexico, *Mearns* 2457.

Sporobolus confusus var. *aberrans* Jones, West. Bot. Contrib. 14: 10. 1912. Bowie, Ariz., *Jones*.

(35) **Muhlenbergia sobolifera** (Muhl.) Trin., Gram. Unifl. 189. pl. 5. f. 4. 1824. Based on *Agrostis sobolifera* Muhl.

Agrostis sobolifera Muhl. in Willd., Enum. Pl. 95. 1809. Pennsylvania. Name only, Muhl., Amer. Phil. Soc. Trans. 4: 236. 1799.

Achnatherum soboliferum Beauv., Ess. Agrost. 20, 146. 1812. Based on *Agrostis sobolifera* Muhl.

Trichochloa sobolifera Trin., Fund. Agrost. 117. 1820. Based on *Agrostis sobolifera* Muhl.

Cinna sobolifera Link, Enum. Pl. 1: 71. 1821. Based on *Agrostis sobolifera* Willd.

Podosaemum soboliferum Link, Hort. Berol. 1: 83. 1827. Based on *Agrostis sobolifera* Muhl.

MUHLENBERGIA SOBOLIFERA var. SETIGERA Scribn., Rhodora 9: 18. 1907. Texas, *Reverchon* 70. (Published as *M. sobolifera* (subsp.) *setigera*.)

Muhlenbergia sobolifera forma *setigera* Deam, Ind. Dept. Conserv. Pub. 82: 163. 1929. Based on *M. sobolifera setigera* Scribn.

(40) **Muhlenbergia sylvatica** (Torr.) Torr. in A. Gray, N. Amer. Gram. et Cyp. 1: 13. 1834. Based on *Agrostis sylvatica* Torr.

Agrostis diffusa Muhl., Descr. Gram. 64. 1817. Not *A. diffusa* Host, 1809. Pennsylvania.

Agrostis sylvatica Torr., Fl. North. and Mid. U. S. 1: 87. 1823. Not *A. sylvatica* Huds., 1762. Mountains of New Jersey.

Muhlenbergia sylvatica var. *gracilis* Scribn., Kans. Acad. Sci. Trans. 9: 116. 1885. Topeka, Kans., *Popenoe*.

Muhlenbergia umbrosa Scribn., Rhodora 9: 20. 1907. Based on *Agrostis sylvatica* Torr.

Muhlenbergia diffusa Farwell, Mich. Acad. Sci. Rpt. 20: 168. 1919. Not *M. diffusa* Willd., 1797. Based on *Agrostis diffusa* Muhl.

MUHLENBERGIA SYLVATICA forma ATTENUATA (Scribn.) Palmer and Steyermark, Mo. Bot. Gard. Ann. 22: 467. 1935. Based on *M. umbrosa* subsp. *attenuata* Scribn.

Muhlenbergia umbrosa subsp. *attenuata* Scribn., Rhodora 9: 21. 1907. Aurora County, S. Dak., *Wilcox* 25.

Muhlenbergia umbrosa forma *attenuata* Deam, Ind. Dept. Conserv. Pub. 82:

171. 1929. Based on *M. umbrosa* subsp. *attenuata* Scribn.

Muhlenbergia diffusa var. *attenuata* Farwell, Mich. Acad. Sci. Papers 23: 125. 1938. Presumably based on *M. umbrosa* subsp. *attenuata* Scribn.

MUHLENBERGIA SYLVATICA var. ROBUSTA Fernald, Rhodora 45: 236. 1943. Sydney, Maine, *Fernald* and *Long* 12597.

(36) **Muhlenbergia tenuiflora** (Willd.) B. S. P., Prel. Cat. N. Y. 67. 1888. Based on *Agrostis tenuiflora* Willd.

Agrostis tenuiflora Willd., Sp. Pl. 1: 364. 1797. North America.

Apera tenuiflora Beauv., Ess. Agrost. 151. 1812. Based on *Agrostis tenuiflora* Willd.

Trichochloa longiseta Trin., Fund. Agrost. 117. 1820. Based on *Agrostis tenuiflora* Willd. Erroneously given as *T. longiflora* Trin., in Kunth, Enum. Pl. 1: 601. 1833.

Cinna tenuiflora Link, Enum. Pl. 1: 71. 1821. Based on *Agrostis tenuiflora* Willd.

Muhlenbergia wildenowii Trin., Gram. Unifl. 188. pl. 5. f. 3. 1824. Based on *Agrostis tenuiflora* Willd.

Trichochloa tenuiflora Sweet, Hort. Brit. 443. 1826. Based on *Agrostis tenuiflora* Willd.

Podosaemum tenuiflorum Link, Hort. Berol. 1: 82. 1827. Based on *Agrostis tenuiflora* Willd.

Muhlenbergia tenuiflora subsp. *variabilis* Scribn., Rhodora 9: 18. 1907. Chimney Mountain, N. C., *Biltmore Herbarium* 654a.

(5) **Muhlenbergia texana** Buckl., Acad. Nat. Sci. Phila. Proc. 1862: 91. 1862. Northern Texas.

Agrostis barbata Buckl. ex A. Gray, Acad. Nat. Sci. Phila. Proc. 1862: 334. 1862. Not *A. barbata* Pers., 1805. As synonym of *Muhlenbergia texana* Buckl.

Muhlenbergia buckleyana Scribn., U. S. Natl. Herb. Contrib. 1: 56. 1890. Based on *M. texana* Buckl.

Podosaemum texanum Bush, Amer. Midl. Nat. 7: 41. 1921. Based on *Muhlenbergia texana* Buckl.

(18) **Muhlenbergia thurberi** Rydb., Torrey Bot. Club Bul. 32: 601. 1905. Based on *Sporobolus filiculmis* Vasey ex Beal. *Vilfa filiculmis* Thurb., also cited, is a name only, and no reference is made to *Sporobolus thurberi* Scribn.

Sporobolus filiculmis Vasey, Descr. Cat. Grasses U. S. 44. 1885, name only; Vasey ex Beal, Grasses N. Amer. 2: 288. 1896. Not *S. filiculmis* L. H. Dewey, 1894. New Mexico, Whipple Exped. [Plaza Larga, *Bigelow* 778].

Vilfa filiculmis Thurb. ex Vasey, Descr. Cat. Grasses U. S. 44. 1885, as synonym of *Sporobolus filiculmis* Vasey.

Sporobolus thurberi Scribn., U. S. Dept.

Agr., Div. Agrost. Bul. 11: 48. f. 5. 1898. "*Vilfa filiculmis* Thurb." Plaza Larga, N. Mex., *Bigelow*.
Vilfa filiculmis Thurb. ex Scribn., U. S. Dept. Agr., Div. Agrost. Bul. 11: 48. 1898, as synonym of *Sporobolus thurberi* Scribn.
Muhlenbergia filiculmis Jones, West. Bot. Contrib. 14: 12. 1912. Not *M. filiculmis* Vasey, 1893. Based on *Vilfa filiculmis* Thurb., name only.

(28) **Muhlenbergia torreyana** (Schult.) Hitchc., Amer. Jour. Bot. 21: 136. 1934. Based on *Agrostis torreyana* Schult.
Agrostis compressa Torr., Cat. Pl. N. Y. 91. 1819. Not *A. compressa* Willd., 1790. New Jersey, *Goldy*.
Vilfa compressa Trin. in Spreng., Neu. Entd. 2: 58. 1821. Not *V. compressa* Beauv., 1812. North America.
Colpodium compressum Trin. ex Spreng., Neu. Entd. 2: 58. 1821, as synonym of *Vilfa compressa* Trin.
Agrostis torreyana Schult., Mantissa 2: 203. 1824. Based on *Agrostis compressa* Torr.
Sporobolus compressus Kunth, Enum. Pl. 1: 217. 1833. Based on *Agrostis compressa* Torr.
Sporobolus torreyanus Nash in Britton, Man. 107. 1901. Based on *Agrostis torreyana* Schult.

(53) **Muhlenbergia torreyi** (Kunth) Hitchc. ex Bush, Amer. Midl. Nat. 6: 84. 1919. Based on *Agrostis torreyi* Kunth.
Agrostis caespitosa Torr., Ann. Lyc. N. Y. 1: 152. 1824. Not *A. caespitosa* Salisb., 1796, nor *Muhlenbergia caespitosa* Chapm., 1878. Prairies of Missouri and Platte River.
Agrostis torreyi Kunth, Rév. Gram. 1: Sup. 17. 1830. Based on *A. caespitosa* Torr.
Muhlenbergia gracillima Torr., U. S. Expl. Miss. Pacif. Rpt. 4: 155. 1857. Llano Estacado and near Antelope Hills, Canadian River, Tex. [*Bigelow*.]
Muhlenbergia nardifolia Griseb., Abh. Ges. Wiss. Göttingen 24: 294. 1879. Argentina.
Podosaemum gracillimum Bush, Amer. Midl. Nat. 7: 33. 1921. Based on *Muhlenbergia gracillima* Torr.

(29) **Muhlenbergia uniflora** (Muhl.) Fernald, Rhodora 29: 10. 1927. Based on *Poa uniflora* Muhl.
Poa? uniflora Muhl., Descr. Gram. 151. 1817. New England. Name only, Muhl., Cat. Pl. 11. 1813.
Agrostis serotina Torr., Fl. North. and Mid. U. S. 1: 88. 1823. Not *A. serotina* Lam., 1767. New Jersey.
Vilfa serotina Trin., Gram. Icon. 3: pl. 251. 1830. North America, "*Agrostis serotina* Nutt. ms."
Vilfa serotina Torr. in A. Gray, N. Amer.

Gram. and Cyp. 1: 2. 1834. Based on *Agrostis serotina* Torr.
Vilfa tenera Trin., Acad. St. Pétersb. Mém. VI. Sci. Nat. 4[1]: 87. 1804. Boston, *Boott*.
Poa modesta Tuckerm., Amer. Jour. Sci. 45: 45. 1843. Cambridge, Mass. [*Tuckerman*.]
Sporobolus serotinus A. Gray, Man. 577. 1848. Based on *Agrostis serotina* Torr.
Sporobolus uniflorus Scribn. and Merr., U. S. Dept. Agr., Div. Agrost. Cir. 27: 5. 1900. Based on *Poa uniflora* Muhl.
Poa stricta uniflora Muhl. ex Scribn. and Merr., U. S. Dept. Agr., Div. Agrost. Cir. 27: 5. 1900, as synonym of *Sporobolus uniflorus* Muhl.
Muhlenbergia uniflora var. *terrae-novae* Fernald, Rhodora 29: 11. 1927. Newfoundland, *Fernald*, *Long*, and *Dunbar* 26244.

(15) **Muhlenbergia utilis** (Torr.) Hitchc., Wash. Acad. Sci. Jour. 23: 453. 1933. Based on *Vilfa utilis* Torr.
Vilfa utilis Torr., U. S. Expl. Miss. Pacif. Rpt. 5[2]: 365. 1858. Between Tejon Pass and Lost Hills, Calif. [*Blake*].
Vilfa sacatilla Fourn., Mex. Pl. 2: 101. 886. 1881. Chapultepec, Mexico, *Schaffner;* San Luis de Potosí, *Virlet* 1455; Texas, *Wright*.
Sporobolus sacatilla Griseb. ex Fourn., Mex. Pl. 2: 101. 1886, as synonym of *Vilfa sacatilla* Fourn.
Sporobolus utilis Scribn., U. S. Dept. Agr., Div. Agrost. Bul. 17: 171. f. 467. 1899. Based on *Vilfa utilis* Torr.

(17) **Muhlenbergia villosa** Swallen, Wash. Acad. Sci. Jour. 31: 350. f. 2. 1941. Stanton, Tex., *Tharp* 5048.

(47) **Muhlenbergia virescens** (H. B. K.) Kunth, Rév. Gram. 1: 64. 1829. Based on *Podosaemum virescens* H. B. K.
Podosaemum virescens H. B. K., Nov. Gen. et Sp. 1: 132. 1815. Mexico, *Humboldt* and *Bonpland*.
Trichochloa virescens Roem. and Schult., Syst. Veg. 2: 389. 1817. Based on *Podosaemum virescens* H. B. K.
Muhlenbergia straminea Hitchc., U. S. Natl. Herb. Contrib. 17: 302. 1913. Chihuahua, Mexico, *Endlich* 1210.

(1) **Muhlenbergia wolfii** (Vasey) Rydb., Torrey Bot. Club Bul. 32: 600. 1905. Based on *Sporobolus wolfii* Vasey.
Vilfa minima Vasey, U. S. Dept. Agr. Monthly Rpt. 1874: 155. 1874. Not *V. minima* Trin. ex Steud., 1854. Twin Lakes, Colo., *Wolf* 1077.
Sporobolus wolfii Vasey, Torrey Bot. Club Bul. 10: 52. 1883. Twin Lakes, Colo., *Wolf* [1077].
Sporobolus racemosus Vasey, Torrey Bot. Club Bul. 14: 9. 1887. Chihuahua, Mexico, *Palmer* [4 B in 1885].
This species was described under *Sporobolus ramulosus* in the Manual, ed. 1. That

species is *Muhlenbergia ramulosus* (H. B. K.) Swallen, known only from Mexico and Central America.

(25) **Muhlenbergia wrightii** Vasey in Coulter, Man. Rocky Mount. 409. 1885. Colorado and New Mexico [type *Wright* 1986].

Muhlenbergia wrightii var. *annulata* Vasey, Grasses U. S. Descr. Cat. 41. 1885. Name only. [Arizona, *Lemmon*, 3179.]

Muhlenbergia coloradensis Mez, Repert. Sp. Nov. Fedde 17: 213. 1921. "Chiann [Cheyenne] Canyon," Colo., *Jones* [806].

(56) **Muhlenbergia xerophila** C. O. Goodding, Wash. Acad. Sci. Jour. 30: 19. 1940. Sycamore Canyon, Ariz., *Goodding* M. 262.

(114) MUNROA Torr.

(1) **Munroa squarrosa** (Nutt.) Torr., U. S. Expl. Miss. Pacif. Rpt. 4⁵: 158. 1857. Based on *Crypsis squarrosa* Nutt.

Crypsis squarrosa Nutt., Gen. Pl. 1: 49. 1818. Grand detour of the Missouri River [S. Dak., *Nuttall*].

Munroa squarrosa var. *floccuosa* Vasey ex Beal, Grasses N. Amer. 2: 456. 1896. Arizona, [Peach Springs], *Jones*. (See p. 545.)
_____

Nardus stricta L., Sp. Pl. 53. 1753. Europe.
_____

Nassella chilensis (Trin. and Rupr.) E. Desv. in Gay, Fl. Chil. 6: 267. 1853. Based on *Urachne chilensis* Trin.

Urachne chilensis Trin., Acad. St. Pétersb. Mém. VI. Sci. Nat. 1: 123. 1834. Chile.

Urachne major Trin. and Rupr., Acad. St. Pétersb. Mém. VI. Sci. Nat. 5¹: 21. 1842. Chile.

Nassella major E. Desv. in Gay, Fl. Chil. 6: 265. 1853. Based on *Urachne chilensis* Trin. and Rupr.

(35) NEOSTAPFIA Davy

(1) **Neostapfia colusana** (Davy) Davy, Erythea 7: 43. 1899. Based on *Stapfia colusana* Davy.

Stapfia colusana Davy, Erythea 6: 110. pl. 3. 1898. Colusa County, Calif., *Davy*.

Davyella colusana Hack., Oesterr. Bot. Ztschr. 49: 134. 1899. Based on *Stapfia colusana* Davy.

Anthochloa colusana Scribn., U. S. Dept. Agr., Div. Agrost. Bul. 17: 221. f. 517. 1899.

(29) NEYRAUDIA Hook. f.

(1) **Neyraudia reynaudiana** (Kunth) Keng, Amer. Jour. Bot. 21: 131. 1934. Based on *Arundo reynaudiana* Kunth.

Arundo reynaudiana Kunth, Rév. Gram. 2: 275. pl. 49. 1830. Burma, *Reynaud*.

(147) OLYRA L.

(1) **Olyra latifolia** L., Syst. Nat. ed. 10. 2: 1261. 1759. Jamaica, *Sloane*.

Olyra paniculata Swartz, Prodr. Veg. Ind. Occ. 21. 1788. Jamaica, *Swartz*.

Olyra arundinacea H. B. K., Nov. Gen. et Sp. 1: 197. 1816. Colombia, *Humboldt* and *Bonpland*.

Stipa latifolia Raspail, Ann. Sci. Nat., Bot. 5: 449. 1825. Based on *Olyra latifolia* L.

Olyra latifolia var. *arundinacea* Griseb., Fl. Brit. W. Ind. 535. 1864. Presumably based on *O. arundinacea* H. B. K.

(140) OPLISMENUS Beauv.

(2) **Oplismenus hirtellus** (L.) Beauv., Ess. Agrost. 54, 168. 1812. Based on *Panicum hirtellum* L.

Panicum hirtellum L., Syst. Nat. ed. 10. 2: 870. 1759. Jamaica [*Browne*].

Orthopogon hirtellus Nutt., Gen. Pl. 1: 55. 1818. Based on *Panicum hirtellum* L.

Orthopogon cubensis Spreng., Syst. Veg. 1: 307. 1825. Cuba.

Echinochloa cubensis Schult., Mantissa 3 (Add. 1): 596. 1827. Based on *Orthopogon cubensis* Spreng.

Oplismenus cubensis Kunth, Rév. Gram. 1: 45. 1829. Based on *Orthopogon cubensis* Spreng.

Panicum cubense Steud., Nom. Bot. ed. 2. 2: 255. 1841. Based on *Orthopogon cubensis* Spreng.

Oplismenus chondrosioides Fourn., Mex. Pl. 2: 39. 1886. Mexico, *Liebmann* 367. This species is cultivated under the name *Panicum variegatum* Hort. (see Gard. Chron. 458. 1867.)

(1) **Oplismenus setarius** (Lam.) Roem. and Schult., Syst. Veg. 2: 481. 1817. Based on *Panicum setarium* Lam.

Panicum setarium Lam., Tabl. Encycl. 1: 170. 1791. South America, *Commerson*.

Panicum velutinum G. Meyer, Prim. Fl. Esseq. 51. 1818. British Guiana [*Meyer*].

Orthopogon parvifolium Nutt., Gen. Pl. 1: 55, errata. 1818. Florida and South Carolina. On page 55 this is described under *Orthopogon hirtellus* Nutt., the name based on *Panicum hirtellum* L., but misapplied.

Setaria hirtella Schult., Mantissa 2: 276. 1824. Based on the species described by Muhlenberg (Descr. Gram. 103. 1817) under the name *Panicum hirtellum*.

Orthopogon setarius Spreng., Syst. Veg. 1: 306. 1825. Based on *Panicum setarium* Lam.

Oplismenus parvifolius Kunth, Rév. Gram. 1: 45. 1829. Based on *Orthopogon parvifolium* Nutt.

Orthopogon hirtellus Eaton and Wright, N.

Amer. Bot. ed. 8. 336. 1840. Southern States. No reference to Nuttall, nor synonym cited.
Panicum nuttallianum Steud., Nom. Bot. ed. 2. 2: 260. 1841. Based on *Orthopogon parvifolius* Nutt.
Oplismenus compositus var. *setarius* F. M. Bailey, Queensl. Grasses 19. 1888. Based on *Panicum setarium* Lam.
Hippagrostis setarius Kuntze, Rev. Gen. Pl. 2: 777. 1891. Based on *Panicum setarium* Lam.
Oplismenus hirtellus subsp. *setarius* Mez ex Ekman, Arkiv Bot. 11⁴: 26. 1912. Based on *Panicum setarium* Lam.

(36) ORCUTTIA Vasey

(2) **Orcuttia californica** Vasey, Torrey Bot. Club Bul. 13: 219. pl. 60. 1886. San Quentin Bay, Baja California, *Orcutt.*
ORCUTTIA CALIFORNICA var. INAEQUALIS (Hoover) Hoover, Torrey Bot. Club Bul. 68: 154. 1941. Based on *O. inaequalis* Hoover.
Orcuttia inaequalis Hoover, Madroño 3: 229. 1936. Montpellier, Calif., *Hoover 582.*
ORCUTTIA CALIFORNICA var. VISCIDA Hoover, Torrey Bot. Club Bul. 68: 155. 1941. Folsom, Calif., *Hoover 3709.*
(1) **Orcuttia greenei** Vasey, Bot. Gaz. 16: 146. 1891. Chico, Calif., *Greene.*
(4) **Orcuttia pilosa** Hoover, Torrey Bot. Club Bul. 68: 155. 1941. Waterford, Calif., *Hoover 3624.*
(3) **Orcuttia tenuis** Hitchc., Amer. Jour. Bot. 21: 131. 1934. Goose Valley, Shasta County, Calif., *Eastwood 1013* (distributed in *Amer. Gr. Natl. Herb.* No. 686 as *Orcuttia californica*). .

(119) ORYZA L.

(1) **Oryza sativa** L., Sp. Pl. 333. 1753. Africa and India.
Oryza sativa var. *rubribarbis* Desv., Jour. Bot. Desv. 1: 76. 1813. Cultivated in North America.
Oryza rubribarbis Steud., Nom. Bot. 577. 1821. Based on *O. sativa* var. *rubribarbis* Desv.
Oryza sativa var. *savannae* Koern. in Koern. and Wern., Handb. Getreidebau. 1: 233, 236. 1885. Cultivated. Savannah, Ga.

(89) ORYZOPSIS Michx.

(8) **Oryzopsis asperifolia** Michx., Fl. Bor. Amer. 1: 51. pl. 9. 1803. Hudson Bay to Quebec, *Michaux.*
Oryzopsis aspera "Mx." ex Muhl., Cat. Pl. 11. 1813., error for *O. asperifolia.*
Oryzopsis mutica Link, Enum. Pl. 1: 41. 1821. North America.
Urachne asperifolia Trin., Gram. Unifl. 174. 1824. Based on *Oryzopsis asperifolia* Michx.

Urachne leucosperma Link, Hort. Berol. 1: 94. 1827. Albany, N. Y.
Urachne mutica Steud., Nom. Bot. ed. 2. 2: 731. 1841. Based on *Oryzopsis mutica* Link.
Oryzopsis leucosperma Link ex Walp., Ann. Bot. [London] 3: 728. 1853, as synonym of *Urachne asperifolia* Trin.
(10) **Oryzopsis bloomeri** (Boland.) Ricker in Piper, U. S. Natl. Herb. Contrib. 11: 109. 1906. Based on *Stipa bloomeri* Boland.
Stipa bloomeri Boland., Calif. Acad. Sci. Proc. 4: 168. 1872. Bloody Canyon, near Mono Lake, Calif., *Bolander* [6116].
Oryzopsis caduca Beal, Bot. Gaz. 15: 111. 1890. Belt Mountains, Mont., *Scribner.*
Stipa caduca Scribn., U. S. Natl. Herb. Contrib. 3: 54. 1892. Based on *Oryzopsis caduca* Beal.
Eriocoma caduca Rydb., N. Y. Bot. Gard. Mem. 1: 25. 1900. Based on *Stipa caduca* Scribn.
× *Stiporyzopsis caduca* B. L. Johnson and Rogler, Amer. Jour. Bot. 30: 55. f. 10, 14, 28–33. 1943. Based on *Oryzopsis caduca* Beal. "*Oryzopsis hymenoides* × *Stipa viridula.*"
× *Stiporyzopsis bloomeri* B. L. Johnson, Amer. Jour. Bot. 32: 602. f. 14–18. 1945. Based on *Stipa bloomeri* Bolander. "*Oryzopsis hymenoides* × *Stipa occidentalis.*"
This is the species described by Beal (Grasses N. Amer. 2: 226. 1896) under the name *Oryzopsis sibirica* Beal, but the name is based on *Stipa sibirica* Lam., not known from America.
(6) **Oryzopsis canadensis** (Poir.) Torr., Fl. N. Y. 2: 433. 1843. Based on *Stipa canadensis* Poir.
Stipa juncea Michx., Fl. Bor. Amer. 1: 54. 1803. Not *S. juncea* L., 1753. Hudson Bay, Canada, *Michaux.*
Stipa canadensis Poir. in Lam., Encycl. 7: 452. 1806. Based on *S. juncea* Michx.
Urachne canadensis Torr. in A. Gray, N. Amer. Gram. and Cyp. 2: 114. 1835. Based on *Stipa canadensis* Poir.
Oryzopsis juncea B. S. P., Prel. Cat. N. Y. 67. 1888. Based on *Stipa juncea* Michx.
Stipa macounii Scribn. in Macoun, Can. Pl. Cat. 2⁵: 390. 1890. New Brunswick.
Oryzopsis macounii Beal, Grasses N. Amer. 2: 229. 1896. Based on *Stipa macounii* Scribn.
This is the species to which the name *Stipa richardsonii* Link was applied by A. Gray in the earlier editions of the Manual.
(4) **Oryzopsis exigua** Thurb. in Wilkes, U. S. Expl. Exped. Bot. 17: 481. 1874. Cascade Mountains, Oreg., *Wilkes Expl. Exped.*

(3) **Oryzopsis hendersoni** Vasey, U. S. Natl. Herb. Contrib. 1: 267. 1893. [Clements Mountain, near North Yakima] *Henderson 2249.*
Oryzopsis exigua var. *hendersoni* Jones, West. Bot. Contrib. 14: 11. 1912. Based on *O. hendersoni* Vasey.

(12) **Oryzopsis hymenoides** (Roem. and Schult.) Ricker in Piper, U. S. Natl. Herb. Contrib. 11: 109. 1906. Based on *Stipa hymenoides* Roem. and Schult.
Stipa membranacea Pursh, Fl. Amer. Sept. 2: 728. 1814. Not *S. membranacea* L., 1753. Banks of the Missouri River, *Bradbury.*
Stipa hymenoides Roem. and Schult., Syst. Veg. 2: 339. 1817. Based on *Stipa membranacea* Pursh.
Eriocoma cuspidata Nutt., Gen. Pl. 1: 40, 1818. Grassy plains of the Missouri [type from "Platte Plains," *Nuttall*].
Milium cuspidatum Spreng., Syst. Veg. 1: 251. 1825. Based on *Eriocoma cuspidata* Nutt.
Urachne lanata Trin. and Rupr., Acad. St. Pétersb. Mém. VI. Sci. Nat. 1: 126. 1834. North America.
Eriocoma membranacea Steud., Nom. Bot. ed. 2. 1: 586. 1840, as synonym of *Urachne lanata* Trin.
Fendleria rhynchelytroides Steud., Syn. Pl. Glum. 1: 420. 1854. New Mexico, *Fendler 979.*
Oryzopsis cuspidata Benth. ex Vasey, Grasses U. S. 23. 1883. Based on *Eriocoma cuspidata* Nutt.
Oryzopsis membranacea Vasey, U. S. Dept. Agr., Div. Bot. Bul. 12²: pl. 10. 1891. Based on *Stipa membranacea* Pursh.
Eriocoma membranacea Beal, Grasses N. Amer. 2: 232. 1896. Based on *Stipa membranacea* Pursh.
Eriocoma hymenoides Rydb., Torrey Bot. Club Bul. 39: 102. 1912. Based on *Stipa hymenoides* Roem. and Schult.
ORYZOPSIS HYMENOIDES var. CONTRACTA B. L. Johnson, Bot. Gaz. 107: 24. 1945. Wyoming, *Elias Nelson 4850.*

(7) **Oryzopsis kingii** (Boland.) Beal, Grasses N. Amer. 2: 229. 1896. Based on *Stipa kingii* Boland.
Stipa kingii Boland., Calif. Acad. Sci. Proc. 4: 170. 1872. Mount Dana, Calif., *Bolander 6076* [error for 6097].

(2) **Oryzopsis micrantha** (Trin. and Rupr.) Thurb., Acad. Nat. Sci. Phila. Proc. 1863: 78. 1863. Based on *Urachne micrantha* Trin. and Rupr.
Urachne micrantha Trin. and Rupr., Acad. St. Pétersb. Mém. VI. Sci. Nat. 5¹: 16. 1842. North America [type from Saskatchewan].

(1) **Oryzopsis miliacea** (L.) Benth. and Hook. ex Aschers. and Schweinf., Mém. Inst. Egypte 2: 169. 1887. Presumably based on *Agrostis miliacea* L.

Agrostis miliacea L., Sp. Pl. 61. 1753. Europe.
Achnatherum miliaceum Beauv., Ess. Agrost. 20, 146, 148. 1812. Based on *Agrostis miliacea* L.
Piptatherum miliaceum Coss., Notes Crit. 129. 1851. Based on *Agrostis miliacea* L.

(5) **Oryzopsis pungens** (Torr.) Hitchc., U. S. Natl. Herb. Contrib. 12: 151. 1908. Based on *Milium pungens* Torr.
Milium pungens Torr. in Spreng., Neu. Entd. 2: 102. 1821. "Schenectady in Massachusetana." [Error for New York.]
Oryzopsis parviflora Nutt., Acad. Nat. Sci. Phila. Jour. 3: 125. 1823. Bellows Falls, Vt.
Panicum firmum Kunth, Rév. Gram. 1: 37. 1829. Based on *Milium pungens* Torr.
Urachne brevicaudata Trin., Acad. St. Pétersb. Mém. VI. Sci. Nat. 1: 127. 1834. Lake Winnipeg, Canada.
Urachne canadensis Torr. and Gray ex Trin. and Rupr., Acad. St. Pétersb. Mém. VI. Sci. Nat. 5¹: 17. 1842, as synonym of *Urachne brevicaudata* Trin.

(9) **Oryzopsis racemosa** (J. E. Smith) Ricker in Hitchc., Rhodora 8: 210. 1906. Based on *Milium racemosum* J. E. Smith.
Milium racemosum J. E. Smith, in Rees's Cycl. 23: Milium No. 15. 1813. Lancaster, Pa., *Muhlenberg.*
Oryzopsis melanocarpa Muhl., Descr. Gram. 79. 1817. Pennsylvania, *Muhlenberg.* Name only, Muhl., Cat. Pl. 11. 1813.
Piptatherum nigrum Torr., Fl. North. and Mid. U. S. 1: 79. 1823. Williamstown and Deerfield, Mass.; Kingston and Fishkill Mountains, N. Y.; Pennsylvania, *Muhlenberg.*
Urachne racemosa Trin., Gram. Unifl. 174. 1824. Based on *Milium racemosum* J. E. Smith.
Urachne melanosperma Link, Hort. Berol. 1: 94. 1827. Based on *Oryzopsis melanocarpa* Muhl.
Piptatherum racemosum Eaton, Man. ed. 5. 331. 1829. Presumably based on *Milium racemosum* J. E. Smith.

(11) **Oryzopsis webberi** (Thurb.) Benth. ex Vasey, Grasses U. S. 23. 1883. Based on *Eriocoma webberi* Thurb.
Eriocoma webberi Thurb. in S. Wats., Bot. Calif. 2: 283. 1880. Sierra Valley, Calif., *Bolander.*
Stipa webberi B. L. Johnson, Bot. Gaz. 107: 25. 1945. Based on *Eriocoma webberi* Thurb.

(137) PANICUM L.

(156) **Panicum abscissum** Swallen, Wash. Acad. Sci. Jour. 30: 215. f. 4. 1940. Sebring, Fla., *Weatherwax* in 1925.

(14) **Panicum aciculare** Desv. ex Poir. in Lam., Encycl. Sup. 4: 274. 1816. "Indes orientales," erroneous; probably from southeastern United States.

Panicum setaceum Muhl., Descr. Gram. 99. 1817. Georgia. Name only, Muhl., Cat. Pl. 9. 1813.

Panicum subuniflorum Bosc ex Spreng., Syst. Veg. 1: 312. 1825. Carolina, *Bosc.*

Panicum arenicola Ashe, Elisha Mitchell Sci. Soc. Jour. 15: 56. 1898. Chapel Hill, N. C., *Ashe.*

Panicum pungens Muhl. ex Scribn. and Merr., U. S. Dept. Agr., Div. Agrost. Bul. 27: 2. 1900. Not *P. pungens* Poir., 1816. As synonym of *P. setaceum* Muhl.

Panicum filirameum Ashe, Elisha Mitchell Sci. Soc. Jour. 16: 88. 1900. New Hanover County, N. C., *Ashe.*

This is the species described in Britton's Manual and in Small's Flora (ed. 1) under the name *Panicum neuranthum* Griseb.

(103) **Panicum aculeatum** Hitchc. and Chase, Rhodora 8: 209. 1906. District of Columbia, *Chase 2520.*

(68) **Panicum addisonii** Nash, Torrey Bot. Club Bul. 25: 83. 1898. Wildwood, N. J., *Bicknell* in 1897.

Panicum owenae Bicknell, Torrey Bot. Club Bul. 35: 185. 1908. Nantucket, Mass., *Bicknell* in 1907.

Panicum commonsianum subsp. *addisonii* Stone, N. J. State Mus. Ann. Rpt. 1910: 205. 1911. Based on *P. addisonii* Nash.

Panicum commonsianum var. *addisonii* Pohl, Amer. Midl. Nat. 38: 582. 1947. Based on *P. addisonii* Nash.

(121) **Panicum adspersum** Trin., Gram. Pan. 146. 1826. Dominican Republic.

Panicum thomasianum Steud ex Doell, in Mart., Fl. Bras. 2²: 188. 1877, as synonym of *P. adspersum*. St. Thomas, *Duchaissing.*

Panicum keyense Mez, Notizbl. Bot. Gart. Berlin 7: 61. 1917. Sand Key, Fla., *Curtiss 3606**, 5431, 6705.*

This is the species described as *Panicum striatum* Lam. by Chapman (Fl. South. U. S. ed. 2. 666. 1883).

(157) **Panicum agrostoides** Spreng., Pl. Pugill. 2: 4. 1815. Pennsylvania, *Muhlenberg.* Name only, Muhl., Amer. Phil. Soc. Trans. 4: 236. 1799.

Panicum rigidulum Bosc ex Spreng., Syst. Veg. 1: 320. 1825; Nees, Agrost. Bras. 163. 1829. [South Carolina? *Bosc.*]

Agrostis polystachya Bosc ex Steud., Nom. Bot. ed. 2. 1: 40. 1840, erroneously cited as synonym of *A. composita* Poir. [Carolina, *Bosc.*]

Panicum elongatum var. *ramosior* Mohr, U. S. Natl. Herb. Contrib. 6: 357. 1901. Near Mobile, Ala. [*Mohr*].

PANICUM AGROSTOIDES var. RAMOSIUS (Mohr) Fernald, Rhodora 38: 390. 1936. Based on *P. elongatum* var. *ramosior* Mohr.

(46) **Panicum albemarlense** Ashe, Elisha Mitchell Sci. Soc. Jour. 16: 84. 1900. Scranton, Hyde County, N. C., *Ashe* in 1899.

Panicum velutinum Bosc ex Spreng., Syst. Veg. 1: 315. 1825. Not *P. velutinum* Meyer, 1818. Name only. [*Bosc.*]

Panicum meridionale var. *albermarlense* Fernald, Rhodora 36: 76. 1934. Based on *P. albermarlense* Ashe.

(77) **Panicum albomarginatum** Nash, Torrey Bot. Club Bul. 24: 40. 1897. Eustis, Fla., *Nash 925.*

(154) **Panicum amarulum** Hitchc. and Chase, U. S. Natl. Herb. Contrib. 15: 96. f. 87. 1910. Virginia Beach, Va., *Williams 3090.*

(153) **Panicum amarum** Ell., Bot. S. C. and Ga. 1: 121. 1816. Presumably South Carolina.

Panicum amarum var. *minus* Vasey and Scribn., U. S. Dept. Agr., Div. Bot. Bul. 8: 38. 1889. Fortress Monroe, Va., *Vasey.*

Panicum amaroides Scribn. and Merr., U. S. Dept. Agr., Div. Agrost. Cir. 29: 5. f. 1. 1901. Based on *P. amarum* var. *minus* Vasey and Scribn.

Chasea amara Nieuwl., Amer. Midl. Nat. 2: 64. 1911. Based on *Panicum amarum* Ell.

(162) **Panicum anceps** Michx., Fl. Bor. Amer. 1: 48. 1803. Carolina, *Michaux.*

Panicum rostratum Muhl. in Willd., Enum. Pl. 1032. 1809. Pennsylvania [type, *Muhlenberg*] and Carolina. Name only, Muhl., Amer. Phil. Soc. Trans. 4: 236. 1799.

Agrostis nutans Poir. in Lam., Encycl. Sup. 1: 255. 1810. Carolina, *Bosc.*

Vilfa nutans Beauv., Ess. Agrost. 16, 148, 181. 1812. Based on *Agrostis nutans* Poir.

Panicum nutans Desv., Opusc. 93. 1831. Based on *Agrostis nutans* Poir.

Panicum anceps var. *angustum* Vasey, U. S. Dept. Agr., Div. Bot. Bul. 8: 37. 1889. Texas, *Nealley.*

Panicum anceps var. *densiflorum* Vasey, U. S. Dept. Agr., Div. Bot. Bul. 8: 37. 1889. [Marshall] Tex., *Riggs* [91].

(18) **Panicum angustifolium** Ell., Bot. S. C. and Ga. 1: 129. 1816. Presumably South Carolina.

?*Panicum ramulosum* Michx., Fl. Bor. Amer. 1: 50. 1803. Carolina, *Michaux.*

Panicum nitidum var. *angustifolium* A. Gray, N. Amer. Gram. and Cyp. 2: 112. 1835. Based on *P. angustifolium* Ell.

Panicum curtisii Steud., Syn. Pl. Glum. 1: 66. 1854. South Carolina, *M. A. Curtis.*

Chasea angustifolia Nieuwl., Amer. Midl.

Nat. 2: 64. 1911. Based on *Panicum angustifolium* Ell.
(29) **Panicum annulum** Ashe, Elisha Mitchell Sci. Soc. Jour. 15: 58. 1898. Maryland to North Carolina and Georgia, Washington, D. C., *Ward* in 1892 [type].
Panicum bogueanum Ashe, Elisha Mitchell Sci. Soc. Jour. 16: 85. 1900. Based on *P. annulum* Ashe.
Panicum antidotale Retz., Obs. Bot. 4: 17. 1786. Botanic garden, India.
(21) **Panicum arenicoloides** Ashe, Elisha Mitchell Sci. Soc. Jour. 16: 89. 1900. Wilmington, N. C., *Ashe* in 1899.
Panicum orthophyllum Ashe, Elisha Mitchell Sci. Soc. Jour. 16: 90. 1900. New Hanover County, N. C., *Ashe* in 1899.
(123) **Panicum arizonicum** Scribn. and Merr., U. S. Dept. Agr., Div. Agrost. Cir. 32: 2. 1901. Camp Lowell, Ariz., *Pringle 465.*
Panicum fuscum var. *majus* Vasey, U. S. Dept. Agr., Div. Bot. Bul. 8: 26. 1889. Mexico [southwestern Chihuahua, *Palmer* 1b in 1885].
Panicum dissitiflorum Vasey in S. Wats., Amer. Acad. Sci. Proc. 24: 80. 1889. Name only. Guaymas, Mexico, *Palmer* 159 in part, 190.
Panicum fasiculatum var. *majus* Beal, Grasses N. Amer. 2: 117. 1896. Based on *P. fuscum* var. *majus* Vasey.
Panicum fasiculatum dissitiflorum Vasey ex Scribn. and Merr., U. S. Dept. Agr., Div. Agrost. Cir. 32: 2. 1901, as synonym of *P. arizonicum.*
Panicum arizonicum var. *tenue* Scribn. and Merr., U. S. Dept. Agr., Div. Agrost. Cir. 32: 3. 1901. Fort Huachuca, Ariz., *Wilcox* in 1894.
Panicum arizonicum var. *laeviglume* Scribn. and Merr., U. S. Dept. Agr., Div. Agrost. Cir. 32: 3. 1901. Mescal, Ariz., *Griffiths* 1810.
Panicum arizonicum var. *majus* Scribn. and Merr., U. S. Dept. Agr., Div. Agrost. Cir. 32: 3. 1901. Based on *P. fuscum* var. *majus* Vasey.
(108) **Panicum ashei** Pearson in Ashe, Elisha Mitchell Sci. Soc. Jour. 15: 35. 1898. Ithaca, N. Y., *Ashe* in 1898.
Panicum umbrosum LeConte ex Torr. in Eaton, Man. Bot. 342. 1818. Not *P. umbrosum* Retz., 1786. New York.
Panicum commutatum var. *ashei* Fernald, Rhodora 36: 83. 1934. Based on *P. ashei* Pearson.
(51) **Panicum auburne** Ashe, N. C. Agr. Expt. Sta. Bul. 175: 115. 1900. Auburn, Ala., *Earle* and *Baker* 1527.
(34) **Panicum barbulatum** Michx., Fl. Bor. Amer. 1: 49. 1803. "Carolina" [but type from Canada].
Panicum dichotomum var. *barbulatum* Wood, Class-book ed. 3. 786. 1861.

Presumably based on *P. barbulatum* Michx.
Panicum pubescens var. *barbulatum* Britton, Cat. Pl. N. J. 280. 1889. Presumably based on *P. barbulatum* Michx.
Panicum nitidum var. *barbulatum* Chapm., Fl. South. U. S. ed. 3. 586. 1897. Based on *P. barbulatum* Michx.
Panicum gravius Hitchc. and Chase, Rhodora 8: 205. 1906. Between Centreville and Mount Cuba, Del., *Chase* 3620.
(126) **Panicum bartowense** Scribn. and Merr., U. S. Dept. Agr., Div. Agrost. Cir. 35: 3. 1901. Bartow, Fla., *Combs* 1220.
Panicum dichotomiflorum var. *bartowense* Fernald, Rhodora 38: 387. 1936. Based on *P. bartowense* Scribn. and Merr.
(60) **Panicum benneri** Fernald, Rhodora 46: 2. pl. 807. 1944. New Jersey, along the Delaware River, Hunterdon County, *Benner* 9635.
(15) **Panicum bennettense** M. V. Brown, Torrey Bot. Club Bul. 69: 539. f. 1. 1942. North Carolina, at Bennett Memorial, Durham County, *M. V. Brown* 2492.
Panicum bergii Arech., An. Mus. Nac. Montevideo 1: 147. 1894. Uruguay.
(24) **Panicum bicknellii** Nash, Torrey Bot. Club Bul. 24: 193. 1897. Bronx Park, N. Y., *Bicknell* in 1895.
Panicum nemopanthum Ashe, Elisha Mitchell Sci. Soc. Jour. 15: 42. 1898. Raleigh, N. C., *Ashe* in 1895.
Panicum bushii Nash, Torrey Bot. Club Bul. 26: 568. 1899. McDonald County, Mo., *Bush* 413.
Panicum bicknellii var. *bushii* Farwell, Mich. Acad. Sci. Papers 1: 85. 1921. Based on *P. bushii* Nash.
(32) **Panicum boreale** Nash, Torrey Bot. Club Bul. 22: 421. 1895. Cairo, N. Y., *Nash* in 1893.
Panicum boreale var. *michiganense* Farwell, Rhodora 42: 306. 1940. Detroit, Mich., *Farwell* 1425.
(115) **Panicum boscii** Poir. in Lam., Encycl. Sup. 4: 278. 1816. Carolina, *Bosc.*
Panicum waltheri Poir. in Lam., Encycl. Sup. 4: 282. 1816. Not *P. walteri* Pursh, 1814. Based on *P. latifolium* as described by Michaux.
Panicum latifolium var. *australe* Vasey, U. S. Dept. Agr., Div. Bot. Bul. 8: 34. 1889. Alabama [type, Thomasville, *Mohr*] to Texas.
Panicum porterianum Nash, Torrey Bot. Club Bul. 22: 420. 1895. Based on *P. waltheri* Poir.
PANICUM BOSCII var. MOLLE (Vasey) Hitchc. and Chase in Robinson, Rhodora 10:

64. 1908. Based on *P. latifolium* var. *molle* Vasey.

Panicum latifolium var. *molle* Vasey ex Ward, Fl. Washington 135. 1881. District of Columbia, [*Ward*].

Panicum walteri var. *molle* Porter, Torrey Bot. Club Bul. 20: 194. 1893. Presumably based on *P. latifolium* var. *molle* Vasey.

Panicum pubifolium Nash, Torrey Bot. Club Bul. 26: 577. 1899. Based on *P. latifolium* var. *molle* Vasey.

(166) **Panicum brachyanthum** Steud., Syn. Pl. Glum. 1: 67. 1854. [Rusk County] Tex., *Vinzent* 124.

Panicum sparsiflorum Vasey, U. S. Dept. Agr., Div. Bot. Bul. 8: 36. 1889. Not *P. sparsiflorum* Doell, 1877. South Carolina to Texas [type, San Bernardino, *Ridell* 20].

This species was described as *Panicum angustifolium* Ell. by Chapman (Fl. South. U. S. 574. 1860).

(86) **Panicum breve** Hitchc. and Chase, U. S. Natl. Herb. Contrib. 15: 271. f. 301. 1910. Jensen, Fla., *Hitchcock* 734.

(148) **Panicum bulbosum** H. B. K., Nov. Gen. et Sp. 1: 99. 1815. Guanajuato, Mexico, *Humboldt* and *Bonpland*.

Panicum avenaceum H. B. K., Nov. Gen. et Sp. 1: 99. 1815. Ecuador, *Humboldt* and *Bonpland*.

Panicum gongylodes Jacq., Eclog. Gram. 30. pl. 21. 1815–1820. Cultivated at Vienna.

Panicum nodosum Willd. ex Steud., Nom. Bot. ed. 2. 2: 260. 1841, as synonym of *P. bulbosum*.

Panicum maximum var. *gongylodes* Doell in Mart., Fl. Bras. 2²: 203. 1877. Based on *P. gongylodes* Jacq.

Panicum maximum var. *bulbosum* Vasey in Rothr., in Wheeler, U. S. Survey W. 100th Merid. Rpt. 6: 295. 1878. Presumably based on *P. bulbosum* H. B. K.

Panicum polygamum var. *gongylodes* Fourn., Mex. Pl. 2: 28. 1886. Based on *P. gongylodes* Jacq.

Panicum bulbosum subvar. *violaceum* Fourn., Mex. Pl. 2: 27. 1886. Chinantla, Mexico, *Liebmann* 451.

Panicum bulbosum var. *avenaceum* Beal, Grasses N. Amer. 2: 132. 1896. Based on *P. avenaceum* H. B. K.

PANICUM BULBOSUM var. MINUS Vasey, U. S. Dept. Agr., Div. Bot. Bul. 8: 38. 1889. Texas, New Mexico, and Arizona [type New Mexico, *Rusby* in 1880].

Panicum sciaphilum Rupr. in Fourn., Mex. Pl. 2: 19. 1886. Yavesia, Mexico, *Galeotti* 5759.

Panicum bulbosum sciaphilum Hitchc. and Chase, U. S. Natl. Herb. Contrib. 15: 83. f. 73. 1910. Based on *P. sciaphilum* Rupr.

(37) **Panicum caerulescens** Hack. ex

Hitchc., U. S. Natl. Herb. Contrib. 12: 219. 1909. Miami, Fla., *Hitchcock* 706.

(25) **Panicum calliphyllum** Ashe, Elisha Mitchell Sci. Soc. Jour. 15: 31. 1898. Watkins, N. Y., *Ashe* in 1898.

(133) **Panicum capillare** L., Sp. Pl. 58. 1753. Virginia, [*Clayton* 454].

Milium capillare Moench, Meth. Pl. 203. 1794. Based on *P. capillare* L.

Panicum bobarti Lam., Encycl. 4: 748. 1798. [Virginia, *Bobart*.]

Panicum capillare var. *agreste* Gattinger, Tenn. Fl. 94. 1887. Tennessee [Ridgetop, *Gattinger*].

Panicum capillare var. *vulgare* Scribn., Tenn. Agr. Expt. Sta. Bul. 7: 44. 1894. Presumably Knoxville, Tenn.

Chasea capillaris Nieuwl., Amer. Midl. Nat. 2: 64. 1911. Based on *Panicum capillare* L.

Leptoloma capillaris Smyth, Kans. Acad. Sci. Trans. 25: 86. 1913. Based on *Panicum capillare* L.

PANICUM CAPILLARE var. OCCIDENTALE Rydb., U. S. Natl. Herb. Contrib. 3: 186. 1895. Whitman, Nebr., *Rydberg* 1788.

Panicum capillare brevifolium Vasey ex Rydb. and Shear, U. S. Dept. Agr., Div. Agrost. Bul. 5: 21. 1897. Manhattan, Mont., *Shear* 436.

Panicum barbipulvinatum Nash in Rydb., N. Y. Bot. Gard. Mem. 1: 21. 1900. Based on *P. capillare* var. *brevifolium* Vasey.

Leptoloma barbipulvinata Smyth, Kans. Acad. Sci. Trans. 25: 86. 1913. Based on *Panicum barbipulvinatum* Nash.

Milium barbipulvinatum Lunell, Amer. Midl. Nat. 4: 212. 1915. Based on *Panicum barbipulvinatum* Nash.

Panicum barbipulvinatum var. *hirsutipes* Suksdorf, Werdenda 1: 17. 1927. Spokane, Wash., *Suksdorf* 9068.

Panicum elegantulum Suksdorf, Werdenda 1: 16. 1927. Not *P. elegantulum* Mez, 1917. Spokane, Wash., *Suksdorf* 9069. (No. 11792, also cited, is *P. capillare*.)

(139) **Panicum capillarioides** Vasey in Coulter, U. S. Natl. Herb. Contrib. 1: 54. 1890. Point Isabel, Tex., *Nealley* [634].

(84) **Panicum chamaelonche** Trin., Gram. Pan. 242. 1826. North America, *Enslin*.

Panicum nitidum var. *minus* Vasey, U. S. Natl. Herb. Contrib. 3: 30. 1892. Florida, [type, St. Augustine, *Canby*].

Panicum baldwinii Nutt. ex Kearney, U. S. Dept. Agr., Div. Agrost. Bul. 1: 21. 1895, name only; Chapm. Fl. South. U. S. ed. 3. 586. 1897. Florida, *Baldwin*.

Panicum dichotomum var. *nitidum* Chapm. ex. Scribn., U. S. Dept. Agr., Div. Agrost. Bul. 11: 43. 1898, as synonym of *P. baldwinii*.

(1) **Panicum chapmani** Vasey, Torrey Bot. Club Bul. 11: 61. 1884. Southern Florida, *Chapman*.
Setaria chapmani Pilger in Engl. and Prantl, Pflanzenfam. ed. 2. 14e: 72. 1940. Based on *Panicum chapmani* Vasey.
This is the species described as *Panicum tenuiculmum* Meyer by Chapman (Fl. South. U. S. 572. 1860).

(16) **Panicum chrysopsidifolium** Nash in Small, Fl. Southeast. U. S. 100, 1327. 1903. Leon County, Fla., *Curtiss* (No. D).

(11) **Panicum ciliatum** Ell., Bot. S. C. and Ga. 1: 126. 1816. Presumably South Carolina.
Panicum leucoblepharis Trin., Clav. Agrost. 234. 1822. North America [type, *Enslin*].
Panicum ciliatifolium Kunth, Rév. Gram. 1: 36. 1829. Based on *P. ciliatum* Ell.
Panicum ciliatifolium Desv., Opusc. 88. 1831. North America.

(113) **Panicum clandestinum** L., Sp. Pl. 58. 1753. Pennsylvania, *Kalm*.
Milium clandestinum Moench, Meth. Pl. 204. 1794. Based on *Panicum clandestinum* L.
Panicum latifolium var. *clandestinum* Pursh, Fl. Amer. Sept. 1: 68. 1814. Based on *P. clandestinum* L.
Panicum pedunculatum Torr., Fl. North. and Mid. U. S. 141. 1823. "Island of New York."
Panicum clandestinum var. *pedunculatum* Torr., Fl. N. Y. 2: 426. 1843. Based on *P. pedunculatum* Torr.
Panicum decoloratum Nash, Torrey Bot. Club Bul. 26: 570. 1899. Tullytown, Pa., *Bicknell* in 1899.
Chasea clandestina Nieuwl., Amer. Midl. Nat. 2: 64. 1911. Based on *Panicum clandestinum* L.

(31) **Panicum clutei** Nash, Torrey Bot. Club Bul. 26: 569. 1899. Between Tuckerton and Atsion, N. J., *Clute*.
Panicum mattamuskeetense var. *clutei* Fernald, Rhodora 39: 386. 1937. Based on *P. clutei* Nash.

(71) **Panicum columbianum** Scribn., U. S. Dept. Agr., Div. Agrost. Bul. 7: 78. f. 60. 1897. District of Columbia, *Scribner* in 1894.
Panicum heterophyllum Bosc ex Nees, Agrost. Bras. 227. 1829. Not *P. heterophyllum* Spreng., 1822. North America, *Bosc*.
Panicum psammophilum Nash, Torrey Bot. Club Bul. 26: 576. 1899. Not *P. psammophilum* Welw., 1899. Toms River, N. J., *Clute* 175.
PANICUM COLUMBIANUM var. THINIUM Hitchc. and Chase in Robinson, Rhodora 10: 64. 1908. Based on *P. unciphyllum thinium* Hitchc. and Chase.
Panicum unciphyllum thinium Hitchc. and

Chase, Rhodora 8: 209. 1906. Toms River, N. J., *Chase* 3577.
Panicum heterophyllum var. *thinium* Hubb., Rhodora 14: 172. 1912. Based on *P. unciphyllum thinium* Hitchc. and Chase.

(161) **Panicum combsii** Scribn. and Ball, U. S. Dept. Agr., Div. Agrost. Bul. 24: 42. f. 16. 1901. Chipley, Fla., *Combs* 583.
Panicum longifolium var. *combsii* Fernald, Rhodora 36: 69. 1934. Based on *P. combsii* Scribn. and Ball.

(67) **Panicum commonsianum** Ashe, Elisha Mitchell Sci. Soc. Jour. 15: 55. 1898. Cape May, N. J., *Commons* 341.

(109) **Panicum commutatum** Schult., Mantissa 2: 242. 1824. Based on *P. nervosum* Muhl.
Panicum nitidum var. *majus* Pursh, Fl. Amer. Sept. 1: 67. 1814. North America.
Panicum nervosum Muhl. ex Ell., Bot. S. C. and Ga. 1: 122. 1816. Not *P. nervosum* Lam., 1797. Carolina and Georgia.
Panicum enslini Trin., Gram. Pan. 230. 1826. North America, *Enslin*.
Panicum polyneuron Steud., Syn. Pl. Glum. 1: 91. 1854. Based on *P. nervosum* Muhl.
Panicum commutatum var. *minus* Vasey, U. S. Dept. Agr., Div. Bot. Bul. 8: 34. 1889. Southern States [type, Aiken, S. C., *Ravenel*].
Panicum commutatum var. *latifolium* Scribn. in Kearney, Torrey Bot. Club Bul. 20: 476. 1893. Pine Mountain, Ky., *Kearney* 299.
Panicum commelinaefolium Ashe, Elisha Mitchell Sci. Soc. Jour. 15: 29. 1898. Not *P. commelinaefolium* Rudge, 1805. Stone Mountain, Ga., *Small* in 1895.
Panicum currani Ashe, Elisha Mitchell Sci. Soc. Jour. 15: 113. 1899. Based on *P. commelinaefolium* Ashe.
Panicum subsimplex Ashe, N. C. Agr. Expt. Sta. Bul. 175: 115. 1900. Wilmington, Del., *Commons*.

(80) **Panicum concinnius** Hitchc. and Chase, U. S. Natl. Herb. Contrib. 15: 263. f. 289. 1910. Based on *P. gracilicaule* Nash.
Panicum gracilicaule Nash in Small, Fl. Southeast. U. S. 98. 1903. Not *P. gracilicaule* Rendle, 1899. Sand Mountain, Ala., *Harbison* 2415.

(158) **Panicum condensum** Nash in Small, Fl. Southeast. U. S. 93. 1903. [Jacksonville], Fla., *Curtiss* 5576.
Agrostis purpurascens Bert. ex Steud., Nom. Bot. ed. 2. 1: 42. 1840. Name only. Dominican Republic, *Bertero, Balbis*.
Panicum contractum Trin. ex Steud., Nom. Bot. ed. 2. 2: 254. 1841. Name

only. Guadeloupe and Dominican Republic, *Balbis.*

Panicum agrostoides var. *condensum* Fernald, Rhodora 36: 74. 1934. Based on *P. condensum* Nash.

(17) **Panicum consanguineum** Kunth, Rév. Gram. 1: 36. 1829. Based on *P. villosum* Ell.

Panicum villosum Ell., Bot. S. C. and Ga. 1: 124. 1816. Not *P. villosum* Lam., 1791. Presumably South Carolina.

Panicum commutatum var. *consanguineum* Beal, Grasses N. Amer. 2: 141. 1896. Based on *P. consanguineum* Kunth.

Panicum georgianum Ashe, Elisha Mitchell Sci. Soc. Jour. 15: 36. 1898. Darien Junction, Ga., *Small* in 1895.

Panicum cahoonianum Ashe, Elisha Mitchell Sci. Soc. Jour. 15: 113. 1899. Based on *P. georgianum* Ashe.

(107) **Panicum cryptanthum** Ashe, N. C. Agr. Expt. Sta. Bul. 175: 115. 1900. Wilsons Mills, N. C., *Ashe* in 1897.

(83) **Panicum curtifolium** Nash, Torrey Bot. Club Bul. 26: 569. 1899. Ocean Springs, Miss., *Tracy 4598.*

Panicum earlei Nash, Torrey Bot. Club Bul. 26: 571. 1899. Auburn, Ala., *Earle* and *Baker 1532.*

Panicum austro-montanum Ashe, Elisha Mitchell Sci. Soc. Jour. 16: 85. 1900. Northern Alabama and adjacent parts of Tennessee, *Ashe.*

(66) **Panicum deamii** Hitchc. and Chase in Deam, Ind. Dept. Conserv. Pub. 82: 284. pl. 75. f. 18. 1929. Pine, Lake County, Ind., *Deam 43287.*

(5) **Panicum depauperatum** Muhl., Descr. Gram. 112. 1817. Pennsylvania, Carolina [type]. Name only, Muhl., Cat. Pl. 9. 1813.

Panicum strictum Pursh, Fl. Amer. Sept. 1: 69. 1814. Not *P. strictum* R. Br., 1810. Pennsylvania.

Panicum rectum Roem. and Schult., Syst. Veg. 2: 457. 1817. Based on *P. strictum* Pursh.

Panicum involutum Torr., Fl. North. and Mid. U. S. 144. 1823. Deerfield, Mass., *Cooley.*

Panicum muhlenbergii Spreng., Syst. Veg. 1: 314. 1825. North America. [Type, New Jersey, *Torrey.*]

Panicum junceum Trin., Gram. Pan. 220. 1826. North America.

Panicum sprengelii Kunth, Rév. Gram. 1: 39. 1829. Based on *P. muhlenbergii* Spreng.

Panicum depauperatum var. *involutum* Wood, Class-book ed. 1861. 786. 1861. Based on *P. involutum* Torr.

?*Panicum depauperatum* var. *laxum* Vasey, U. S. Dept. Agr., Div. Bot. Bul. 8: 29. 1889. "Virginia, Florida, Texas, Arkansas, Missouri."

Panicum depauperatum var. *psilophyllum*

Fernald, Rhodora 23: 193. 1921. Canton, Maine, *Parlin 1957.*

Panicum strictum var. *psilophyllum* Farwell, Mich. Acad. Sci. Papers 26: 5. 1941. Based on *P. depauperatum* var. *psilophyllum* Fernald.

(125) **Panicum dichotomiflorum** Michx., Fl. Bor. Amer. 1: 48. 1803. Western Allegheny Mountains, *Michaux.*

Panicum miliaceum Walt., Fl. Carol. 72. 1788. Not *P. miliaceum* L., 1753. South Carolina.

Panicum geniculatum Muhl., Cat. Pl. 9. 1813. Not Lam. 1798. Based on *P. dichotomiflorum* Michx. Name only, Muhl., Amer. Phil. Soc. Trans. 4: 235. 1799.

Panicum multiflorum Poir. in Lam., Encycl. Sup. 4: 282. 1816. Carolina, *Bosc.*

Panicum brachiatum Bosc ex Spreng., Syst. Veg. 1: 321. 1825. Not *P. brachiatum* Poir. Bermuda cited [but type probably from South Carolina, *Bosc*].

Panicum elliottii Trin. ex Nees, Agrost. Bras. 170. 1829, as synonym of *P. proliferum* Lam. [misapplied to *P. dichotomiflorum*].

Panicum retrofractum Delile ex Desv., Opusc. 96. 1831. North America. [Type from Carolina.]

Panicum proliferum var. *pilosum* Griseb., Cat. Pl. Cub. 232. 1866. Hanábana, Cuba, *Wright* [186].

Panicum proliferum var. *geniculatum* Wood, Amer. Bot. and Flor. pt. 2: 392. 1871. Eastern States.

Panicum amplectans Chapm., Bot. Gaz. 3: 20. 1878. South Florida [*Blodgett*].

Leptoloma dichotomiflora Smyth, Kans. Acad. Sci. Trans. 25: 86. 1913. Based on *Panicum dichotomiflorum* Michx.

Panicum dichotomiflorum var. *geniculatum* Fernald, Rhodora 38: 387. pl. 441. f. 2. 1936. Based on *P. proliferum* var. *geniculatum* Wood.

Panicum dichotomiflorum var. *imperiorum* Fernald, Rhodora 44: 380. 1942. Greensville County, Va., *Fernald* and *Long 13877.*

This species has been referred to *P. proliferum* Lam., an Old World species.

PANICUM DICHOTOMIFLORUM var. PURITANORUM Svenson, Rhodora 22: 154. f. 1–5. 1920. Barnstable, Mass., *Fernald* in 1919.

(33) **Panicum dichotomum** L., Sp. Pl. 58. 1753. Virginia, [*Clayton 458*].

Panicum angustifolium LeConte ex Torr. in Eaton, Man. Bot. ed. 2: 342. 1818. Not *P. angustifolium* Ell., 1816. New York.

Panicum tremulum Spreng., Neu. Entd. 2: 103. 1821. New Jersey [*Torrey*].

Panicum dichotomum var. *viride* Vasey, U. S. Dept. Agr., Div. Bot. Bul. 8: 30. 1889. No locality cited. [Type, Washington, D. C., *Ward* in 1881.]

Panicum dichotomum var. *divaricatum* Vasey, U. S. Dept. Agr., Div. Bot. Bul. 8: 30. 1889. No locality cited. [Type, Lake, Miss., *Tracy* 127.]

Panicum nitidum var. *pauciflorum* Britton, N. Y. Acad. Sci. Trans. 9: 14. 1889. Morris County, N. J., *Britton.*

Panicum nitidum var. *viride* Britton, N. Y. Acad. Sci. Trans. 9: 14. 1889. Based on *P. dichotomum* var. *viride* Vasey.

Panicum dichotomum var. *commune* Wats. and Coult. in A. Gray, Man. ed. 6: 633. 1890. No locality cited.

Panicum ramulosum var. *viride* Porter, Torrey Bot. Club Bul. 20: 194. 1893. Presumably based on *P. dichotomum* var. *viride* Vasey.

Chasea dichotoma Nieuwl., Amer. Midl. Nat. 2: 64. 1911. Based on *Panicum dichotomum* L.

(81) **Panicum ensifolium** Baldw. ex Ell., Bot. S. C. and Ga. 1: 126. 1816. Georgia, *Baldwin.*

Panicum nitidum var. *ensifolium* Vasey, U. S. Dept. Agr., Div. Bot. Bul 8: 29. 1889. Based on *P. ensifolium* Baldw.

Panicum brittoni Nash, Torrey Bot. Club Bul. 24: 194. 1897. Forked River, N. J., *Britton* in 1896.

Panicum cuthbertii Ashe, Elisha Mitchell Sci. Soc. Jour. 15: 48. 1898. St. Helena Island, S. C., *Cuthbert.*

Panicum glabrissimum Ashe, Elisha Mitchell Sci. Soc. Jour. 15: 62. 1898. Manteo, N. C., *Ashe* in 1898.

Panicum shallotte Ashe, Elisha Mitchell Sci. Soc. Jour. 16: 84. 1900. Based on *P. glabrissimum* Ashe.

Panicum parvipaniculatum Ashe, Elisha Mitchell Sci. Soc. Jour. 16: 87. 1900. Onslow County, N. C., *Ashe* in 1899.

(112) **Panicum equilaterale** Scribn., U. S. Dept. Agr., Div. Agrost. Bul. 11: 42. pl. 2. 1898. Eustis, Fla., *Nash* 1674.

Panicum epilifolium Nash, Torrey Bot. Club Bul. 26: 571. 1899. Eustis, Fla., *Nash 45.*

(75) **Panicum erectifolium** Nash, Torrey Bot. Club Bul. 23: 148. 1896. Based on *P. sphaerocarpon* var. *floridanum* Vasey.

Panicum sphaerocarpon var. *floridanum* Vasey, U. S. Dept. Agr., Div. Bot. Bul. 8: 33. 1889. Not *P. floridanum* Trin., 1834. Florida [type, Mosquito Inlet, *Curtiss* 3599].

Panicum floridanum Chapm., Fl. South. U. S. ed. 3. 585. 1897. Not *P. floridanum* Trin., 1834. Presumably based on *P. sphaerocarpon* var. *floridanum* Vasey.

(120) **Panicum fasciculatum** Swartz, Prodr. Veg. Ind. Occ. 22. 1788. Jamaica, *Swartz.*

Panicum chartaginense Swartz, Prodr. Veg. Ind. Occ. 22. 1788. Cartagena, Colombia.

Panicum fuscum Swartz, Prodr. Veg. Ind. Occ. 23. 1788. Jamaica, *Swartz.*

Panicum flavescens Swartz, Prodr. Veg. Ind. Occ. 23. 1788. Jamaica, *Swartz.*

Panicum fusco-rubens Lam., Tabl. Encycl. 1: 171. 1791. West Indies.

Panicum fastigiatum Poir. in Lam., Encycl. Sup. 4: 277. 1816. Based on *P. fasciculatum* Swartz.

Panicum spithamaeum Willd. ex Nees, Agrost. Bras. 152. 1829. Name only. South America, *Humboldt.*

Panicum illinoniense Desv., Opusc. 91. 1831. North America.

Panicum reticulatum Griseb., Abhandl. Gesell. Wiss. Göttingen 7: 264. 1857. Not *P. reticulatum* Torr. 1852. West Indies or Panama.

Panicum fuscum var. *fasciculatum* Griseb., Fl. Brit. W. Ind. 547. 1864. Based on *P. fasciculatum* Swartz.

Panicum fasciculatum var. *flavescens* Doell in Mart., Fl. Bras. 2²: 205. 1877. Based on *P. flavescens* Swartz.

Panicum fasciculatum var. *fuscum* Doell in Mart., Fl. Bras. 2²: 205. 1877. Based on *P. fuscum* Swartz.

Panicum fasciculatum var. *chartaginense* Doell in Mart., Fl. Bras. 2²: 205. 1877. Based on *P. chartaginense* Swartz.

PANICUM FASCICULATUM var. RETICULATUM (Torr.) Beal, Grasses N. Amer. 2: 117. 1896. Based on *P. reticulatum* Torr.

Panicum reticulatum Torr. in Marcy, Expl. Red Riv. 299. 1852. Red River, Tex.

Panicum fuscum reticulatum Scribn. and Merr., U. S. Dept. Agr., Div. Agrost. Cir. 32: 4. 1901. Based on *P. reticulatum* Torr.

(140) **Panicum filipes** Scribn. in Heller, Herb. Frankl. Marsh. Col. Contrib. 1: 13. 1895. Corpus Christi, Tex., *Heller* 1809.

(4) **Panicum firmulum** Hitchc. and Chase, U. S. Natl. Herb. Contrib. 15: 27. f. 9. 1910. Elsordo, Tex., *Griffiths* 6446.

Setaria firmula Pilger in Engl. and Prantl, Pflanzenfam. ed. 2. 14e: 72. 1940. Based on *Panicum firmulum* Hitchc. and Chase.

(79) **Panicum flavovirens** Nash, Torrey Bot. Club Bul. 26: 572. 1899. Lake County, Fla., *Nash* 2061.

(128) **Panicum flexile** (Gattinger) Scribn. in Kearney, Torrey Bot. Club Bul. 20: 476. 1893. Based on *P. capillare* var. *flexile* Gattinger.

Panicum capillare var. *flexile* Gattinger, Tenn. Fl. 94. 1887. [Nashville, Tenn., *Gattinger.*]

Chasea flexilis Nieuwl., Amer. Midl. Nat. 2: 65. 1911. Based on *Panicum flexile* Scribn.

(20) **Panicum fusiforme** Hitchc., U. S. Natl. Herb. Contrib. 12: 222. 1909.

Based on *P. neuranthum* var. *ramosum* Griseb.

Panicum neuranthum var. *ramosum* Griseb., Cat. Pl. Cub. 232. 1866. Not *P. ramosum* L., 1767. Western Cuba, *Wright* 3454.

(129) **Panicum gattingeri** Nash in Small, Fl. Southeast. U. S. 92, 1327. 1903. Based on *P. capillare* var. *campestre* Gattinger.

Panicum capillare var. *campestre* Gattinger, Tenn. Fl. 94. 1887. Not *P. campestre* Nees. [Nashville, Tenn., *Gattinger.*]

Panicum capillare var. *geniculatum* Scribn. in Kearney, Torrey Bot. Club Bul. 20: 477. 1893. Wasioto, Ky., [*Kearney* 378].

Panicum capillare gattingeri Nash in Britt. and Brown, Illustr. Fl. 1: 123. 1896. Based on *P. capillare* var. *campestre* Gattinger.

(116) **Panicum geminatum** Forsk., Fl. Aegypt. Arab. 18. 1775. Rosetta, Egypt.

Paspalum appressum Lam., Tabl. Encycl. 1: 176. 1791. South America.

Digitaria appressa Pers., Syn. Pl. 1: 85. 1805. Based on *Paspalum appressum* Lam.

Panicum beckmanniaeforme Mikan ex Trin. in Spreng., Neu. Entd. 2: 83. 1821. Brazil.

Panicum brizaeforme Presl, Rel. Haenk. 1: 302. 1830. Luzon.

Panicum glomeratum Buckl., Prel. Rpt. Geol. Agr. Survey Tex. App. 3. 1866. Not *P. glomeratum* Moench, 1794. Western Texas.

Panicum appressum Lam. ex Doell, in Mart., Fl. Bras. 2²: 184. 1877. Not *P. appressum* Forsk., 1775. Based on *Paspalum appressum* Lam.

Paspalidium geminatum Stapf in Prain, Fl. Trop. Afr. 9: 583. 1920. Based on *P. geminatum* Forsk.

This species has been referred to *Panicum paspalodes* Pers., an Old World species, probably a synonym of *P. punctatum* Burm.

(143) **Panicum ghiesbreghtii** Fourn., Mex. Pl. 2: 29. 1886. Mexico, *Ghiesbreght.*

Panicum hirtivaginum Hitchc., U. S. Natl. Herb. Contrib. 12: 223. 1909. Cuba, *Wright* 758.

(85) **Panicum glabrifolium** Nash, Torrey Bot. Club Bul. 24: 196. 1897. Tampa, Fla., *Nash* 2415a.

(150) **Panicum gouini** Fourn., Mex. Pl. 2: 28. 1886. Vera Cruz, Mexico, *Gouin* 4.

Panicum gouini var. *pumilum* Fourn., Mex. Pl. 2: 28. 1886. Mexico, Vera Cruz, *Virlet* 1300; Antigua, *Liebmann* 450.

Panicum repens var. *confertum* Vasey, Torrey Bot. Club Bul. 13: 25. 1886. "Louisiana" [erroneous, type from Bay St. Louis, Miss., *Langlois*].

Panicum halophilum Nash in Lloyd and Tracy, Torrey Bot. Club Bul. 28: 86. 1901. Based on *P. repens* var. *confertum* Vasey.

(170) **Panicum gymnocarpon** Ell., Bot. S. C. and Ga. 1: 117. 1816. Savannah, Ga., *Baldwin.*

Panicum monachnoides Desv., Opusc. 86. 1831. "Brazil" [locality erroneous].

Panicum drummondii Nees in Steud., Syn. Pl. Glum. 1: 63. 1854. New Orleans, La., *Drummond* [574].

Phanopyrum gymnocarpon Nash in Small, Fl. Southeast. U. S. 104. 1903. Based on *Panicum gymnocarpon* Ell.

(141) **Panicum hallii** Vasey, Torrey Bot. Club Bul. 11: 61. 1884. Austin, Tex., *Hall* 816 (in part).

Panicum virletii Fourn., Mex. Pl. 2: 29. 1886. San Luis Potosí, Mexico, *Virlet* 1305, 1371.

(152) **Panicum havardii** Vasey, Torrey Bot. Club Bul. 14: 95. 1887. Described from type of *P. virgatum* var. *macranthum* Vasey.

Panicum virgatum var. *macranthum* Vasey, Torrey Bot. Club Bul. 13: 26. 1886. Not *P. macranthum* Trin., 1826. Guadalupe Mountains, Tex., *Havard.*

(94) **Panicum helleri** Nash, Torrey Bot. Club Bul. 26: 572. 1899. Kerrville, Tex., *Heller* 1759.

Panicum pernervosum Nash, Torrey Bot. Club Bul. 26: 576. 1899. Houston, Tex., *Hall* 830.

Panicum oligosanthes var. *helleri* Fernald, Rhodora 36: 80. 1934. Based on *P. helleri* Nash.

(169) **Panicum hemitomon** Schult., Mantissa 2: 227. 1824. Based on *P. walteri* Muhl.

Panicum dimidiatum Walt., Fl. Carol. 72. 1788. Not *P. dimidiatum* L., 1753. South Carolina. Referred by Elliott to *P. walteri.*

Panicum walteri Ell., Bot. S. C. and Ga. 1: 115. 1816. Not *P. walteri* Pursh, 1814. Charleston, S. C.; Savannah, Ga., [type].

Panicum walteri Muhl., Descr. Gram. 108. 1817. Not *P. walteri* Pursh, 1814. No locality cited, probably Georgia.

Panicum carolinianum Spreng., Syst. Veg. 1: 310. 1825. Based on *P. walteri* Ell.

Oplismenus walteri Kunth, Rév. Gram. 1: 45. 1829. Based on *Panicum walteri* Muhl.

Panicum carinatum Torr. in Curtis, Bost. Jour. Nat. Hist. 1: 137. 1835. Not *P. carinatum* Presl, 1830. [Wilmington] N. C. [*M. A. Curtis*].

Panicum digitarioides Carpenter ex Curtis, Amer. Jour. Sci. (II) 7: 410. 1849, not *P. digitarioides* Raspail, 1833, as synonym of *P. carinatum* Torr.; Steud., Syn. Pl. Glum. 1: 75. 1854. North America [type, Louisiana, *Carpenter*].

Panicum curtisii Chapm., Fl. South. U. S.
573. 1860. Not *P. curtisii* Steud.,
1854. Based on *P. walteri* Ell.
Oplismenus colonum var. *walteri* Fourn.,
Mex. Pl. 2: 40. 1886. Based on *O.
walteri* Kunth.
Brachiaria digitarioides Nash in Britton,
Man. 77. 1901. Based on *P. digitari-
oides* Carpenter.
(164) **Panicum hians** Ell., Bot. S. C. and
Ga. 1: 118. 1816. Charleston, S. C.
Panicum oblongiflorum Desv., Opusc.
89. 1831. Carolina, *Bosc.*
Panicum jejunum Trin., Acad. St.
Pétersb. Mém. VI. Sci. Nat. 2¹: 103.
1836. Louisiana.
Aira incompleta Bosc ex Steud., Nom.
Bot. ed. 2. 1: 45. 1840. Name only.
[Carolina, *Bosc.*]
Steinchisma hians Nash in Small, Fl.
Southeast. U. S. 105. 1903. Based on
Panicum hians Ell. This name, credited
to Raf., is listed in Index Kewensis (2:
982. 1895.) as synonym of *Panicum
debile* [Poir.] which is *Festuca obtusa.*
(134) **Panicum hillmani** Chase, Wash. Acad.
Sci. Jour. 14: 345. f. 1. 1934. Amarillo,
Tex., *Hitchcock* 16206.
(144) **Panicum hirsutum** Swartz, Fl. Ind.
Occ. 1: 173. 1797. Jamaica, Hispani-
ola, *Swartz.*
Panicum elatum Willd. ex Steud., Nom.
Bot. ed. 2. 2: 256. 1841. Name only.
South America, *Humboldt.*
(135) **Panicum hirticaule** Presl, Rel. Haenk.
1: 308. 1830. Acapulco, Mexico,
Haenke.
Panicum flabellatum Fourn., Soc. Bot.
France Bul. II. 27: 293. 1880. Omo-
tepe Island, Nicaragua, *Lévy* 1166.
Panicum polygamum var. *hirticaule*
Fourn., Mex. Pl. 2: 28. 1886. Based on
P. hirticaule Presl, but misapplied to
P. maximum Jacq.
Panicum capillare var. *glabrum* Vasey ex
T. S. Brandeg., Proc. Calif. Acad. II. 2:
211. 1889. Name only. Baja Cali-
fornia, *Brandegee* in 1889.
Panicum capillare var. *hirticaule* Gould,
Madroño 10: 94. 1949. Based on *P.
hirticaule* Presl.
(48) **Panicum huachucae** Ashe, Elisha
Mitchell Sci. Soc. Jour. 15: 51. 1898.
Huachuca Mountains, Ariz., *Lemmon*
in 1882.
Panicum nitidum var. *pilosum* Torr., Fl.
North. and Mid. U. S. 146. 1824. Not
P. pilosum Swartz. New York.
Panicum languinosum var. *huachucae*
Hitchc., Rhodora 8: 208. 1906. Based
on *P. huachucae* Ashe.
Panicum lindheimeri var. *fasciculatum*
subvar. *pilosum* Farwell, Amer. Midl.
Nat. 11: 45. 1928. New York.
Panicum lanuginosum var. *fasciculatum*
subvar. *pilosum* Farwell, Mich. Acad.

Sci. Papers 26: 5. 1941. Based on *P.
nitidum* var. *pilosum* Torr.
PANICUM HUACHUCAE var. FASCICULATUM
(Torr.) Hubb., Rhodora 14: 171. 1912.
Based on *P. dichotomum* var. *fascicula-
tum* Torr.
Panicum dichotomum var. *fasciculatum*
Torr., Fl. North. and Mid. U. S. 145.
1824. New Jersey.
Panicum nitidum var. *ciliatum* Torr., Fl.
North. and Mid. U. S. 146. 1824.
New Jersey.
Panicum huachucae var. *silvicola* Hitchc.
and Chase in Robinson, Rhodora 10: 64.
1908. District of Columbia, *Chase* 2400.
Panicum lindheimeri var. *fasciculatum*
Fernald, Rhodora 23: 228. 1921.
Based on *P. dichotomum* var. *fascicu-
latum* Torr.
Panicum lanuginosum var. *fasciculatum*
Fernald, Rhodora 36: 77. 1934. Based
on *P. dichotomum* var. *fasciculatum*
Torr.
Panicum glutinoscabrum Fernald, Rho-
dora 49: 122. pl. 1059. 1947. Nanse-
mond County, Va., *Fernald, Long,* and
Clement 15186.
(47) **Panicum implicatum** Scribn. in Britt.
and Brown, Illustr. Fl. 3: 498. f. 267a.
1898. Cape Elizabeth, Maine, *Scribner*
in 1895.
Panicum unciphyllum implicatum Scribn.
and Merr., Rhodora 3: 123. 1901.
Based on *P. implicatum* Scribn.
Panicum lindheimeri var. *implicatum*
Fernald, Rhodora 23: 228. 1921.
Based on *P. implicatum* Scribn.
Panicum lanuginosum var. *implicatum*
Fernald, Rhodora 36: 77. 1934. Based
on *P. implicatum* Scribn.
(111) **Panicum joorii** Vasey, U. S. Dept.
Agr., Div. Bot. Bul. 8: 31. 1889.
Louisiana, *Joor.*
Panicum leiophyllum Fourn., Mex. Pl. 2:
20. 1886. Not *P. leiophyllum* Nees,
1829. Córdoba, Mexico, *Bourgeau.*
Panicum manatense Nash, Torrey Bot.
Club Bul. 24: 42. 1897. Manatee
County, Fla., *Nash* 2428a.
Panicum commutatum var. *joorii* Fernald,
Rhodora 39: 388. 1937. Based on *P.
joorii* Vasey.
(127) **Panicum lacustre** Hitchc. and Ekman,
U. S. Dept. Agr. Misc. Pub. 243: 253.
f. 205. 1936. Pinar del Rio, Cuba,
Ekman 17878.
(88) **Panicum lancearium** Trin., Gram. Pan.
223. 1826. North America, *Enslin.*
Panicum nashianum Scribn., U. S. Dept.
Agr., Div. Agrost. Bul. 7: 79. f. 61.
1897. Eustis, Fla., *Nash* 466.
(58) **Panicum languidum** Hitchc. and
Chase, U. S. Natl. Herb. Contrib. 15:
232. f. 245. 1910. Based on *P. unciphyllum* forma *prostratum* Scribn. and
Merr.
Panicum unciphyllum forma *prostratum*

Scribn. and Merr., Rhodora 3: 124. 1901. Not *P. prostratum* Lam., 1791. South Berwick, Maine, *Fernald* in 1897.

(50) **Panicum lanuginosum** Ell., Bot. S. C. and Ga. 1: 123. 1816. Georgia, *Baldwin.*

Panicum dichotomum var. *lanuginosum* Wood, Class-book ed. 3. 786. 1861. Presumably based on *P. lanuginosum* Ell.

Panicum orangense Ashe, Elisha Mitchell Sci. Soc. Jour. 15: 113. 1899. Orange County, N. C., *Ashe* in 1898.

Panicum ciliosum Nash, Torrey Bot. Club Bul. 26: 568. 1899. Biloxi, Miss., *Tracy* 4580.

(114) **Panicum latifolium** L., Sp. Pl. 58. 1753. America.

Milium latifolium Moench, Meth. Pl. 204. 1794. Based on *P. latifolium* L.

Panicum macrocarpon LeConte ex Torr. in Eaton, Man. Bot. ed. 2: 341. 1818. New York.

Panicum schnecki Ashe, N. C. Agr. Expt. Sta. Bul. 175: 116. 1900. Southern Indiana and Illinois [*Schneck*].

(9) **Panicum laxiflorum** Lam., Encycl. 4: 748. 1798. North America.

Panicum dichotomum var. *laxiflorum* Beal, Grasses N. Amer. 2: 139. 1896. Based on *Panicum laxiflorum* Lam.

Panicum pyriforme Nash, Torrey Bot. Club Bul. 26: 579. 1899. Orange Bend, Fla., *Nash* 239.

Panicum aureum Muhl. ex Scribn. and Merr., U. S. Dept. Agr., Div. Agrost. Cir. 27: 4. 1900, as synonym of *P. laxiflorum* Lam.

(98) **Panicum leibergii** (Vasey) Scribn. in Britt. and Brown, Illustr. Fl. 3: 497. 1898. Based on *P. scoparium* var. *leibergii* Vasey.

Panicum scoparium var. *leibergii* Vasey, U. S. Dept. Agr., Div. Bot. Bul. 8: 32. 1889. Plymouth County, Iowa, *Leiberg.*

Panicum scribnerianum var. *leibergii* Scribn., U. S. Dept. Agr., Div. Agrost. Bul. 6: 32. 1897. Presumably based on *P. scoparium* var. *leibergii* Vasey.

Milium leibergii Lunell, Amer. Midl. Nat. 4: 213. 1915. Based on *Panicum scoparium* var. *leibergii* Vasey.

(142) **Panicum lepidulum** Hitchc. and Chase, U. S. Natl. Herb. Contrib. 15: 75. f. 64. 1910. Chihuahua, Mexico, *Pringle* 497.

(42) **Panicum leucothrix** Nash, Torrey Bot. Club Bul. 24: 41. 1897. Eustis, Fla., *Nash* 1338.

Panicum parvispiculum Nash, Torrey Bot. Club Bul. 24: 347. 1897. Darien Junction, Ga., *Small* in 1895.

(41) **Panicum lindheimeri** Nash, Torrey Bot. Club Bul. 24: 196. 1897. [New Braunfels] Tex., *Lindheimer* 565.

Panicum funstoni Scribn. and Merr., U. S. Dept. Agr., Div. Agrost. Cir. 35:

4. 1901. Three Rivers, Calif., *Coville* and *Funston* 1286.

Panicum lindheimeri var. *typicum* Fernald, Rhodora 23: 227. 1921. Based on *P. lindheimeri* Nash.

Panicum lanuginosum var. *lindheimeri* Fernald, Rhodora 36: 77. 1934. Based on *P. lindheimeri* Nash.

(7) **Panicum linearifolium** Scribn. in Britt. and Brown, Illustr. Fl. 3: 500. f. 268a. 1898. New York and New Jersey to Missouri. [Type, Washington, D. C., *Vasey* in 1882.]

Panicum strictum var. *linearifolium* Farwell, Amer. Midl. Nat. 11: 44. 1928. Based on *P. linearifolium* Scribn.

(131) **Panicum lithophilum** Swallen, Biol. Soc. Wash. Proc. 54: 43. 1941. Stone Mountain, Ga., *Hitchcock* (Amer. Gr. Natl. Herb. No. 24) as "*Panicum philadelphicum.*"

(160) **Panicum longifolium** Torr., Fl. North. and Mid. U. S. 149. 1824. New Jersey, *Goldy.*

Panicum anceps var. *pubescens* Vasey, U. S. Dept. Agr., Div. Bot. Bul. 8: 37. 1889. Mobile, Ala., *Mohr.*

Panicum pseudanceps Nash, Torrey Bot. Club Bul. 25: 85. 1898. Florida, *Simpson* in 1889.

Panicum longifolium var. *pubescens* Fernald, Rhodora 36: 69. 1934. Based on *P. anceps* var. *pubescens* Vasey.

(43) **Panicum longiligulatum** Nash, Torrey Bot. Club Bul. 26: 574. 1899. Apalachicola, Fla., *Vasey* in 1892.

(38) **Panicum lucidum** Ashe, Elisha Mitchell Sci. Soc. Jour. 15: 47. 1898. Lake Mattamuskeet, N. C., *Ashe* in 1898.

Panicum taxodiorum Ashe, Elisha Mitchell Sci. Soc. Jour. 16: 91. 1900. Lake Charles, La., *Mackenzie* 460.

PANICUM LUCIDUM var. OPACUM Fernald, Rhodora 39: 386. 1937. Prince George County, Va., *Fernald* and *Long* 6484.

(65) **Panicum malacon** Nash, Torrey Bot. Club Bul. 24: 197. 1897. Eustis, Fla., *Nash* 628.

Panicum strictifolium Nash, Torrey Bot. Club Bul. 26: 579. 1899. Eustis, Fla., *Nash* 603.

(93) **Panicum malacophyllum** Nash, Torrey Bot. Club Bul. 24: 198. 1897. Sapulpa, Indian Territory [Okla.], *Bush* 1228.

Panicum scoparium var. *minus* Scribn., Tenn. Agr. Expt. Sta. Bul. 7: 48. 1894. Tennessee, *Gattinger.*

(30) **Panicum mattamuskeetense** Ashe, Elisha Mitchell Sci. Soc. Jour. 15: 45. 1898. Lake Mattamuskeet, N. C., *Ashe* and *Pearson* in 1898.

?*Panicum barbatum* LeConte ex Torr., in Eaton, Man. Bot. ed. 2. 342. 1818. Not *P. barbatum* Lam., 1791. New York.

?*Panicum nitidum* var. *barbatum* Torr.,

Fl. North. and Mid. U. S. 146. 1824. No locality cited.

Panicum flexuosum Muhl. ex Scribn. and Merr., U. S. Dept. Agr., Div. Agrost. Cir. 27: 3. 1900. Not *P. flexuosum* Retz., 1791. Name only for specimen in Muhlenberg Herb. (See "(174)" Hitchcock, Bartonia 14: 39. 1932.)

(146) **Panicum maximum** Jacq., Col. Bot. 1: 76. 1786. Guadeloupe.

Panicum polygamum Swartz, Prodr. Veg. Ind. Occ. 24. 1788. Not *P. polygamum* Forsk., 1775. [Jamaica, *Swartz.*]

Panicum laeve Lam., Tabl. Encyl. 1: 172. 1791. Dominican Republic.

Panicum jumentorum Pers., Syn. Pl. 1: 83. 1805. Based on *P. polygamum* Swartz.

Panicum scaberrimum Lag., Gen. et Sp. Nov. 2. 1816. Mexico, *Sessé.*

Panicum trichocondylum Steud., Syn. Pl. Glum. 1: 74. 1854. Guadeloupe, *Duchaissing.*

Panicum praticola Salzm. ex Doell in Mart., Fl. Bras. 2²: 203. 1877, as synonym of *P. maximum*. Bahia, Brazil, *Salzmann* 683.

(45) **Panicum meridionale** Ashe, Elisha Mitchell Sci. Soc. Jour. 15: 59. 1898. Chapel Hill and Burke County, N. C., *Ashe.*

Panicum filiculme Ashe, Elisha Mitchell Sci. Soc. Jour. 15: 59. 1898. Not *P. filiculme* Hack., 1895. Chapel Hill, N. C., *Ashe* in 1898; Stone Mountain, Ga., *Small* in 1895.

?*Panicum microphyllum* Ashe, Elisha Mitchell Sci. Soc. Jour. 15: 61. 1898. Chapel Hill, N. C., *Ashe* in 1898.

Panicum unciphyllum meridionale Scribn. and Merr., Rhodora 3: 123. 1901. Based on *P. meridionale* Ashe.

Panicum lindheimeri var. *implicatum* subvar. *meridionale* Farwell, Amer. Midl. Nat. 11: 45. 1928. Based on *P. meridionale* Ashe.

Panicum lanuginosum var. *implicatum* subvar. *meridionale* Farwell, Mich. Acad. Sci. Papers 26: 5. 1941. Based on *P. meridionale* Ashe.

(27) **Panicum microcarpon** Muhl. ex Ell., Bot. S. C. and Ga. 1: 127. 1816. [Georgia, *Baldwin.*]

Panicum heterophyllum Muhl., Amer. Phil. Soc. Trans. 3: 160. 1793. Name only.

Panicum nitidum var. *ramulosum* Torr., Fl. North. and Mid. U. S. 146. 1824. Quaker Bridge, N. J.

(138) **Panicum miliaceum** L., Sp. Pl. 58. 1753. India.

Milium panicum Mill., Gard. Dict. Milium No. 1. 1768. Based on *Panicum miliaceum* L.

Milium esculentum Moench, Meth. Pl. 203. 1794. Based on *Panicum miliaceum* L,

Panicum milium Pers., Syn. Pl. 1: 83. 1805. Based on *P. miliaceum* L.

Leptoloma miliacea Smyth, Kans. Acad. Sci. Trans. 25: 86. 1913. Based on *Panicum miliaceum* L.

(105) **Panicum mundum** Fernald, Rhodora 38: 392. pl. 443. f. 1–5. 1936. Homeville, Va., *Fernald* and *Long* 6499.

(110) **Panicum mutabile** Scribn. and Smith ex Nash in Small, Fl. Southeast. U. S. 103. 1903. Biloxi, Miss., *Tracy* 3074.

(23) **Panicum neuranthum** Griseb., Cat. Pl. Cub. 232. 1866. Eastern Cuba, *Wright* 3453.

(28) **Panicum nitidum** Lam., Tabl. Encycl. 1: 172. 1791. Carolina, *Fraser.*

Panicum nodiflorum Lam., Encycl. 4: 744. 1798. Carolina, *Fraser;* South Carolina, *Michaux.*

Panicum dichotomum var. *nitidum* Wood, Class-book ed. 1861. 786. 1861. Presumably based on *P. nitidum* Lam.

Panicum dichotomum var. *nodiflorum* Griseb., Cat. Pl. Cub. 234. 1866. Based on *P. nodiflorum* Lam.

Panicum subbarbulatum Scrib. and Merr., U. S. Dept. Agr., Div. Agrost. Cir. 29: 9. 1901. Based on *P. barbulatum* Michx. as described by Elliott, not Michaux's species. Presumably South Carolina.

(101) **Panicum nodatum** Hitchc. and Chase, U. S. Natl. Herb. Contrib. 15: 293. 1910. Sarita, Tex., *Hitchcock* 3865.

(26) **Panicum nudicaule** Vasey, U. S. Dept. Agr., Div. Bot. Bul. 8: 31. 1889. Santa Rosa County, Fla., *Curtiss* [3583*].

(168) **Panicum obtusum** H. B. K., Nov. Gen. et Sp. 1: 98. 1815. Near Guanajuato, Mexico, *Humboldt* and *Bonpland.*

Panicum polygonoides C. Muell., Bot. Ztg. 19: 323. 1861. Not *P. polygonoides* Lam., 1798. Texas, *Drummond* 371.

Panicum repens Buckl., Prel. Rpt. Geol. Agr. Survey Tex. App. 3. 1866. Texas [*Buckley*].

Brachiaria obtusa Nash in Britton, Man. 77. 1901. Based on *Panicum obtusum* H. B. K.

(55) **Panicum occidentale** Scribn., Mo. Bot. Gard. Rpt. 10: 48. 1899. Nootka Sound, Vancouver Island, *Haenke.*

Panicum dichotomum var. *pubescens* Munro ex Benth., Pl. Hartw. 341. 1857. Name only. Sacramento, Calif., *Hartweg* 2024 (344).

Panicum brodiei St. John, Fl. Southeast. Wash. and Adj. Idaho 51. 1937. Wawawai, Wash., *Brodie* in 1898.

(96) **Panicum oligosanthes** Schult., Mantissa 2: 256. 1824. Based on *P. pauciflorum* Ell.

Panicum pauciflorum Ell., Bot. S. C. and Ga. 1: 120. 1816. Not *P. pauciflorum* R. Br., 1810. Georgia.

Panicum scoparium var. *angustifolium*

Vasey, U. S. Dept. Agr., Div. Bot. Bul. 8: 32. 1889. South Carolina, *Ravenel*.

Panicum scoparium var. *pauciflorum* Scribn., Tenn. Agr. Expt. Sta. Bul. 7: 48. 1894. Based on *P. pauciflorum* Ell.

(72) **Panicum oricola** Hitchc. and Chase, Rhodora 8: 208. 1906. Lewes, Del., *Hitchcock* 47.

Panicum columbianum var. *oricola* Fernald, Rhodora 36: 79. 1934. Based on *P. oricola* Hitchc. and Chase.

(62) **Panicum ovale** Ell., Bot. S. C. and Ga. 1: 123. 1816. St. Marys, Ga., *Baldwin*.

Panicum ciliiferum Nash, Torrey Bot. Club Bul. 24: 195. 1897. Eustis, Fla., *Nash* 147.

Panicum erythrocarpon Ashe, Elisha Mitchell Sci. Soc. Jour. 16: 90. 1900. New Hanover County, N. C., *Ashe* in 1899.

(22) **Panicum ovinum** Scribn. and Smith, U. S. Dept. Agr., Div. Agrost. Cir. 16: 3. 1899. Waller County, Tex., *Thurow*.

Panicum redivivum Trin. ex Steud., Nom. Bot. ed. 2. 2: 262. 1841. Name only. [Jalapa], Mexico, *Schiede*.

(56) **Panicum pacificum** Hitchc. and Chase, U. S. Natl. Herb. Contrib. 15: 229. f. 241. 1910. Castle Crags, Calif., *Hitchcock* 3070.

(117) **Panicum paludivagum** Hitchc. and Chase, U. S. Natl. Herb. Contrib. 15: 32. f. 13. 1910. Eustis, Fla., *Nash* 746.

Paspalidium paludivagum Parodi, Gram. Bonar. ed. 3. 89. 1939. Based on *Panicum paludivagum* Hitchc. and Chase.

(136) **Panicum pampinosum** Hitchc. and Chase, U. S. Natl. Herb. Contrib. 15: 66. f. 48. 1910. Wilmot, Ariz., *Thornber* 193.

Panicum capillare var. *pampinosum* Gould, Madroño 10: 94. 1949. Based on *P. pampinosum* Hitchc. and Chase.

(91) **Panicum patentifolium** Nash, Torrey Bot. Club Bul. 26: 574. 1899. Eustis, Fla., *Nash* 72.

(89) **Panicum patulum** (Scribn. and Merr.) Hitchc., Rhodora 8: 209. 1906. Based on *P. nashianum* var. *patulum* Scribn. and Merr.

Panicum nashianum var. *patulum* Scribn. and Merr., U. S. Dept. Agr., Div. Agrost. Cir. 27: 9. 1900. "Braidentown" (Bradenton), Fla., *Combs* 1296.

Panicum lancearium var. *patulum* Fernald, Rhodora 36: 80. 1934. Based on *P. nashianum* var. *patulum* Scribn. and Merr.

(100) **Panicum pedicellatum** Vasey, U. S. Dept. Agr., Div. Bot. Bul. 8: 28. 1889. [Kimble County] Tex., *Reverchon*.

(6) **Panicum perlongum** Nash, Torrey Bot. Club Bul. 26: 575. 1899. Creek Nation, Okla., *Carleton* 98.

Panicum pammeli Ashe, N. C. Agr. Expt.

Sta. Bul. 175: 116. 1900. Iowa [*Cratty* in 1881].

Panicum strictum var. *perlongum* Farwell, Amer. Midl. Nat. 11: 44. 1928. Based on *P. perlongum* Nash.

(130) **Panicum philadelphicum** Bernh. ex Trin., Gram. Pan. 216. 1826; Nees, Agrost. Bras. 198. 1829. [Philadelphia, Pa., *Bernhardi*.]

Panicum capillare var. *sylvaticum* Torr., Fl. North. and Mid. U. S. 149. 1824. Not *P. sylvaticum* Lam., 1798. New York City.

Panicum torreyi Fourn. in Hemsl., Biol. Centr. Amer. Bot. 3: 497. 1885. Based on *P. capillare* var. *sylvaticum* Torr.

Panicum capillare var. *minimum* Engelm. ex Gattinger, Tenn. Fl. 94. 1887. [Green Brier, Tenn., *Gattinger*.]

Panicum minus Nash, Torrey Bot. Club Bul. 22: 421. 1895. Based on "*Panicum capillare* var. *minus* Muhl.''

Panicum capillare var. *minus* Muhl. ex Nash, Torrey Bot. Club Bul. 22: 421. 1895, as synonym of *P. minus* Nash. Muhlenberg does not give this a varietal name, noting only "varietas minor occurrit ubique in cultis magis aridis."

Panicum minimum Scribn. and Merr., U. S. Dept. Agr., Div. Agrost. Cir. 27: 4. 1900. Based on *P. capillare* var. *minimum* Engelm.

Panicum pilcomayense Hack., Herb. Boiss. Bul. II. 7: 449. 1907. Pilcomayo, Paraguay, *Rojas* 105.

(19) **Panicum pinetorum** Swallen, Biol. Soc. Wash. Proc. 55: 93. 1942. Bonita Springs, Fla., *Silveus* 6604.

(147) **Panicum plenum** Hitchc. and Chase, U. S. Natl. Herb. Contrib. 15: 80. f. 69. 1910. Mangas Springs, N. Mex., *Metcalfe* 739.

(74) **Panicum polyanthes** Schult., Mantissa 2: 257. 1824. Based on *P. multiflorum* Ell.

Panicum multiflorum Ell., Bot. S. C. and Ga. 1: 122. 1816. Not *P. multiflorum* Poir., 1816. Presumably South Carolina.

Panicum microcarpon Muhl., Descr. Gram. 111. 1817. Not *P. microcarpon* Muhl. ex Ell., 1816. Virginia, "Cherokee" [type] and Delaware.

Panicum firmandum Steud., Syn. Pl. Glum. 1: 418. 1855. North Carolina, *M. A. Curtis*.

Panicum microcarpon var. *isophyllum* Scribn., Tenn. Agr. Expt. Sta. Bul. 7: 51. f. 54. 1894. [Alleghany Springs, Tenn., *Gayle*.]

(12) **Panicum polycaulon** Nash, Torrey Bot. Club Bul. 24: 200. 1897. Tampa, Fla., *Nash* 2420a.

Panicum dichotomum var. *glabrescens* Griseb., Fl. Brit. W. Ind. 553. 1864. Jamaica, *Purdie*.

(87) **Panicum portoricense** Desv. ex

Hamilt., Prodr. Pl. Ind. Occ. 11. 1825. Puerto Rico.

Panicum pauciciliatum Ashe, Elisha Mitchell Sci. Soc. Jour. 16: 87. 1900. Wilmington, N. C., *Ashe* in 1899.

(53) **Panicum praecocius** Hitchc. and Chase, Rhodora 8: 206. 1906. Wady Petra, Ill., *V. H. Chase* 649.

(61) **Panicum pseudopubescens** Nash, Torrey Bot. Club Bul. 26: 577. 1899. Auburn, Ala., *Earle* and *Baker* 1537.

Panicum villosissimum var. *pseudopubescens* Fernald, Rhodora 36: 79. 1934. Based on *P. pseudopubescens* Nash.

Panicum euchlamydeum Shinners, Amer. Midl. Nat. 32: 170. 1944. Adams County, Wis., *Shinners* and *Shaw* 4415.

(118) **Panicum purpurascens** Raddi, Agrost. Bras. 47. 1823. Rio de Janeiro, Brazil, *Raddi*. (*P. purpurascens* Opiz, 1822, is a name only.)

Panicum barbinode Trin., Acad. St. Pétersb. Mém. VI. Sci. Nat. 1: 256. 1834. Bahia, Brazil.

Panicum guadaloupense Steud., Syn. Pl. Glum. 1: 61. 1854. Guadaloupe.

Panicum equinum Salzm. ex Steud., Syn. Pl. Glum. 1: 67. 1854. Bahia, Brazil, *Salzmann*.

Panicum pictigluma Steud., Syn. Pl. Glum. 1: 73. 1854. Brazil.

Brachiaria purpurascens Henr., Blumea 3: 434. 1940. Based on *Panicum purpurascens* Raddi.

This species has been referred to *P. numidianum* Lam. Together with that and *P. barbinode* Trin. it is included under *Brachiaria mutica* (Forsk.) Stapf, in Prain, Fl. Trop. Afr. 9: 526. 1919.

(2) **Panicum ramisetum** Scribn., U. S. Dept. Agr., Div. Agrost. Cir. 27: 9. 1900. Based on *P. subspicatum* Vasey.

Panicum subspicatum Vasey, U. S. Dept. Agr., Div. Bot. Bul. 8: 25. 1889. Not *P. subspicatum* Desv., 1831. Texas, *Nealley*.

Chaetochloa ramiseta Smyth, Kans. Acad. Sci. Trans. 25: 89. 1913. Based on *Panicum ramisetum* Scribn.

Setaria ramiseta Pilger in Engl. and Prantl, Pflanzenfam. ed. 2. 14e: 72. 1940. Presumably based on *Panicum ramisetum* Scribn.

(122) **Panicum ramosum** L., Mant. Pl. 1: 29. 1767. "In Indiis."

(97) **Panicum ravenelii** Scribn. and Merr., U. S. Dept. Agr., Div. Agrost. Bul. 24: 36. 1901. Based on *P. scoparium* as described by Elliott. [South Carolina and Georgia.]

Panicum scoparium var. *majus* Vasey, U. S. Dept. Agr., Div. Bot. Bul. 8: 32. 1889. South Carolina, *Ravenel*.

Panicum scoparium var. *genuinum* Scribn., Tenn. Agr. Expt. Sta. Bul. 7: 48. 1894. Based on *P. scoparium* Lam., as described by Elliott.

(104) **Panicum recognitum** Fernald, Rhodora 40: 331. pl. 497, 498. 1938. Camden County, N. J., *Long* 7671.

(149) **Panicum repens** L., Sp. Pl. ed. 2. 87. 1762. Southern Europe.

Panicum littorale Mohr ex. Vasey, Bot. Gaz. 4: 106. 1879. Mobile, Ala., *Mohr*.

(119) **Panicum reptans** L., Syst. Nat. ed. 10. 2: 870. 1759. [Jamaica, *Browne*.]

Panicum grossarium L., Syst. Nat. ed. 10. 2: 871. 1759. [Jamaica, *Browne*, typonym of *P. reptans* L.]

Panicum prostratum Lam., Tabl. Encycl. 1: 171. 1791. West Indies [type from Dominican Republic].

Panicum caespitosum Swartz, Fl. Ind. Occ. 1: 146. 1797. Jamaica, *Swartz*.

Panicum insularum Steud., Syn. Pl. Glum. 1: 61. 1854. Lesser Antilles [*Hohenacker*].

Brachiaria prostrata Griseb., Abhandl. Gesell. Wiss. Göttingen 7: 263. 1857. Based on *Panicum prostratum* Lam.

Panicum aurelianum Hale in Wood, Classbook ed. 1861. 787. 1861. New Orleans, La., *Hale*.

Panicum prostratum var. *pilosum* Eggers, Fl. St. Croix and Virgin Isl. 104. 1879. St. Croix.

Urochloa reptans Stapf in Prain, Fl. Trop. Afr. 9: 601. 1920. Based on *Panicum reptans* L.

Brachiaria reptans Gard. and C. E. Hubb. in Hook. Icon. Pl. 3363: 3. 1938. Based on *Panicum reptans* L.

(3) **Panicum reverchoni** Vasey, U. S. Dept. Agr. Div. Bot. Bul. 8: 25. 1889. [Dallas] Tex., *Reverchon*.

Chaetochloa reverchoni Smyth, Kans. Acad. Sci. Trans. 25: 88. 1913. Based on *Panicum reverchoni* Vasey.

Setaria reverchoni Pilger in Engl. and Prantl, Pflanzenfam. ed. 2. 14e: 72. 1940. Based on *Panicum reverchoni* Vasey.

(163) **Panicum rhizomatum** Hitchc. and Chase, U. S. Natl. Herb. Contrib. 15: 109. f. 104. 1910. Orangeburg, S. C., *Hitchcock* 450.

Panicum anceps var. *rhizomatum* Fernald, Rhodora 36: 73. 1934. Based on *P. rhizomatum* Hitchc. and Chase.

(36) **Panicum roanokense** Ashe, Elisha Mitchell Sci. Soc. Jour. 15: 44. 1898. Roanoke Island, N. C., *Ashe* in 1898.

Panicum curtivaginum Ashe, Elisha Mitchell Sci. Soc. Jour. 16: 85. 1900. Petit Bois Island, Miss., *Tracy* [4584].

(106) **Panicum scabriusculum** Ell., Bot. S. C. and Ga. 1: 121. 1816. Savannah, Ga., *Baldwin*.

Panicum lanuginosum Bosc ex Spreng., Syst. Veg. 1: 319. 1825. Not *P. lanuginosum* Ell., 1816. Georgia.

Panicum eriophorum Schult., Mantissa 3 (Add. 1): 591. 1827. Based on *P. lanuginosum* Bosc.

Panicum nealleyi Vasey, Torrey Bot. Club Bul. 13: 25. 1886. Texas, *Nealley*.

Panicum dichotomum var. *elatum* Vasey, U. S. Dept. Agr., Div. Bot. Bul. 8: 31. 1889. No locality cited. [Mobile, Ala., *Mohr*.]

Panicum viscidum var. *scabriusculum* Beal, Grasses N. Amer. 2: 143. 1896. Based on "*P. scabriusculum* Chapm. non Ell." Chapman uses Elliott's name correctly.

(63) **Panicum scoparioides** Ashe, Elisha Mitchell Sci. Soc. Jour. 15: 53. 1898. Centreville, Del., *Commons* 283.

Panicum villosissimum var. *scoparioides* Fernald, Rhodora 36: 79. 1934. Based on *P. scoparioides* Ashe.

(102) **Panicum scoparium** Lam., Encycl. 4: 744. 1798. South Carolina, *Michaux*.

Panicum pubescens Lam., Encycl. 4: 748. 1798. South Carolina, *Michaux*.

Panicum viscidum Ell., Bot. S. C. and Ga. 1: 123. pl. 7. f. 3. 1816. Presumably South Carolina.

Panicum nitidum var. *velutinum* Doell in Mart., Fl. Bras. 2²: 247. 1877. Based on *P. viscidum* Ell.

Panicum laxiflorum var. *pubescens* Chapm., Fl. South. U. S. ed. 3. 586. 1897. Not *P. laxiflorum* var. *pubescens* Vasey, 1892. Based on *P. pubescens* Lam., but misapplied to *P. strigosum* Muhl.

Chasea pubescens Nieuwl., Amer. Midl. Nat. 2: 64. 1911. Based on *Panicum pubescens* Lam.

(95) **Panicum scribnerianum** Nash, Torrey Bot. Club Bul. 22: 421. 1895. Based on *P. scoparium* as described by Watson in Gray's Manual. [Type, Pennsylvania, *Carey* in 1836.]

Panicum macrocarpon Torr., Fl. North. and Mid. U. S. 143. 1823. Not *P. macrocarpon* LeConte, 1818. Deerfield, Mass., *Cooley*.

Panicum scoparium S. Wats. ex Nash, Torrey Bot. Club Bul. 22: 421. 1895, as synonym of *P. scribnerianum* Nash.

Panicum oligosanthes var. *scribnerianum* Fernald, Rhodora 36: 80. 1934. Based on *P. scribnerianum* Nash.

(64) **Panicum shastense** Scribn. and Merr., U. S. Dept. Agr., Div. Agrost. Cir. 35: 3. 1901. Castle Crags, Calif., *Greata* in 1899.

Panicum sonorum Beal, Grasses N. Amer. 2: 130. 1896. Based on *P. capillare* var. *miliaceum* Vasey.

Panicum capillare var. *miliaceum* Vasey, U. S. Natl. Herb. Contrib. 1: 28. 1890. Not *P. miliaceum* L. 1753. Lerdo, Mexico, *Palmer* 947 in 1889.

(73) **Panicum sphaerocarpon** Ell., Bot. S. C. and Ga. 1: 125. 1816. Georgia, *Baldwin*.

Panicum kalmii Swartz ex Wikstr., Adnot. Bot. 6. 1829. Pennsylvania, ?*Kalm*.

Panicum heterophyllum Swartz ex Wikstr., Adnot. Bot. 6. 1829. Not *P. heterophyllum* Spreng., 1822. As synonym of *P. kalmii* Swartz.

Panicum nitidum var. *crassifolium* A. Gray, N. Amer. Gram. and Cyp. 1: 30. 1834. Name only for specimen from "Pine barrens of New Jersey;" A. Gray ex Doell, in Mart. Fl. Bras. 2²: 247. 1877 (in obs.).

Panicum dichotomum var. *sphaerocarpon* Wood, Class-book ed. 1861. 786. 1861. Presumably based on *P. sphaerocarpon* Ell.

Panicum microcarpon var. *sphaerocarpon* Vasey, Grasses U. S. 12. 1883. Based on *P. sphaerocarpon* Ell.

Panicum vicarium Fourn., Mex. Pl. 2: 20. 1886. Córdoba, Mexico, *Schaffner* 285.

PANICUM SPHAEROCARPON var. INFLATUM (Scribn. and Smith) Hitchc. and Chase, U. S. Natl. Herb. Contrib. 15: 253. f. 275. 1910. Based on *P. inflatum* Scribn. and Smith. (Published as *P. sphaerocarpon inflatum*.)

Panicum inflatum Scribn. and Smith, U. S. Dept. Agr., Div. Agrost. Cir. 16: 5. 1899. Biloxi, Miss., *Tracy* 4622.

Panicum mississippiense Ashe, Elisha Mitchell Sci. Soc. Jour. 16: 91. 1900. Mississippi River below New Orleans, La., *Ashe*.

(39) **Panicum sphagnicola** Nash, Torrey Bot. Club Bul. 22: 422. 1895. Lake City, Fla., *Nash* 2500.

(40) **Panicum spretum** Schult., Mantissa 2: 248. 1824. Based on Muhlenberg's *Panicum* No. 37. New England.

Panicum nitidum var. *densiflorum* Rand and Redfield, Fl. Mt. Desert 174. 1894. Mount Desert, Maine, *Rand*.

Panicum eatoni Nash, Torrey Bot. Club Bul. 25: 84. 1898. Seabrook, N. H., *Eaton*.

Panicum octonodum Smith, U. S. Dept. Agr., Div. Agrost. Bul. 17: 73. f. 369. 1899. Waller County, Tex., *Thurow* in 1898.

Panicum paucipilum Nash, Torrey Bot. Club Bul. 26: 573. 1899. Wildwood, N. J., *Bicknell* in 1897.

Panicum nitidum octonodum Scribn. and Merr., U. S. Dept. Agr., Div. Agrost. Bul. 24: 34. 1901. Based on *P. octonodum* Smith.

(159) **Panicum stipitatum** Nash in Scribn., U. S. Dept. Agr., Div. Agrost. Bul. 17. (ed. 2): 56. f. 352. 1901. Based on *P. elongatum* Pursh.

Panicum elongatum Pursh, Fl. Amer. Sept. 1: 69. 1814. Not *P. elongatum* Salisb., 1796. New Jersey to Virginia. [Type, Delaware.]

Panicum agrostoides var. *elongatum* Scribn., Tenn. Agr. Expt. Sta. Bul. 7:

42. pl. 9. f. 34. 1894. Based on *P. elongatum* Pursh.

(137) **Panicum stramineum** Hitchc. and Chase, U. S. Natl. Herb. Contrib. 15: 67. f. 50. 1910. Guaymas, Sonora, *Palmer* 206 in 1887.

Panicum capillare var. *stramineum* Gould, Madroño 10: 94. 1949. Based on *P. stramineum* Hitchc. and Chase.

(13) **Panicum strigosum** Muhl. in Ell., Bot. S. C. and Ga. 1: 126. 1816. [South Carolina and Georgia.]

Panicum laxiflorum var. *pubescens* Vasey, U. S. Natl. Herb. Contrib. 3: 30. 1892. No locality cited. [Type, Duval County, Fla., *Curtiss* (No. H).]

Panicum longipedunculatum Scribn., Tenn. Agr. Expt. Sta. Bul. 7: 53. pl. 16. f. 61. 1894. Tennessee, White Cliff Springs, [*Scribner*, type] Tullahoma.

(54) **Panicum subvillosum** Ashe, Elisha Mitchell Sci. Soc. Jour. 16: 86. 1900. Carlton, Minn., *Ashe*.

Panicum unciphyllum forma *pilosum* Scribn. and Merr., Rhodora 3: 124. 1901. Orono, Maine, *Fernald* 501.

(155) **Panicum tenerum** Beyr. in Trin., Acad. St. Pétersb. Mém. VI Sci. Nat. 1: 341. 1834. Georgia, *Beyrich* [62].

Panicum anceps var. *strictum* Chapm., Fl. South. U. S. 573. 1860. Florida, *Chapman*.

This species has been referred to *Panicum stenodes* Griseb., of tropical America.

(49) **Panicum tennesseense** Ashe, Elisha Mitchell Sci. Soc. Jour. 15: 52. 1898. La Vergne County, Tenn., *Biltmore Herbarium* 7087.

Panicum lindheimeri var. *septentrionale* Fernald, Rhodora 23: 227. 1921. Woodstock, New Brunswick, *Fernald* and *Long* 12527.

Panicum lindheimeri var. *tennesseense* Farwell, Amer. Midl. Nat. 11: 45. 1928. Based on *Panicum tennesseense* Ashe.

Panicum lanuginosum var. *septentrionale* Fernald, Rhodora 36: 77. 1934. Based on *P. lindheimeri* var. *septentrionale* Fernald.

(76) **Panicum tenue** Muhl., Descr. Gram. 118. 1817. No locality cited.

Panicum deustum Brickell and Enslin ex Muhl., Descr. |Gram. 119. 1817. Not *P. deustum* Thunb., 1794. As synonym of *P. tenue*.

Panicum liton Schult., Mantissa 2: 250. 1824. Based on *P. tenue* Muhl., that name changed because of *P. tenue* Roxb., name only, 1813, not described until 1820.

Panicum unciphyllum Trin., Gram. Pan. 242. 1826. North America.

Panicum macrum Kunth, Rév. Gram. 1: 40. 1829. Based on *P. tenue* Muhl.

Panicum parvulum Muhl. ex Scribn. and Merr., U. S. Dept. Agr., Div. Agrost.

Cir. 27: 4. 1900. Not *P. parvulum* Trin., 1834. As synonym of *P. tenue* Muhl.

(124) **Panicum texanum** Buckl., Prel. Rpt. Geol. Agr. Survey Tex. App. 3. 1866. Austin, Tex.

(57) **Panicum thermale** Boland., Calif. Acad. Sci. Proc. 2: 181. 1862. Sonoma County, Calif.

Panicum ferventicola Schmoll, Madroño 5: 92. 1939. Yellowstone National Park, *Chase*.

Panicum ferventicola var. *sericeum* Schmoll, Madroño 5: 93. 1939. Yellowstone National Park, *A.* and *E. Nelson* 6037.

Panicum ferventicola var. *papillosum* Schmoll, Madroño 5: 94. 1939. Alberta, *Hitchcock*, Amer. Gr. Natl. Herb. 220.

Panicum lassenianum Schmoll, Madroño 5: 95. 1939. Plumas County, Calif., *Jepson* 4082.

(52) **Panicum thurowii** Scribn. and Smith, U. S. Dept. Agr., Div. Agrost. Cir. 16: 5. 1899. Waller County, Tex., *Thurow* in 1898.

(145) **Panicum trichoides** Swartz, Prodr. Veg. Ind. Occ. 24. 1788. Jamaica; Hispaniola.

(78) **Panicum trifolium** Nash, Torrey Bot. Club Bul. 26: 580. 1899. Macon, Ga., *Small* in 1895.

(70) **Panicum tsugetorum** Nash, Torrey Bot. Club Bul. 25: 86. 1898. Bronx Park, N. Y., *Nash* 287.

Panicum lanuginosum siccanum Hitchc. and Chase, Rhodora 8: 207. 1906. Starved Rock, Ill., *Chase* 1602.

(132) **Panicum tuckermani** Fernald, Rhodora 21: 112. 1919. Lake Memphremagog, Vt., *Tuckerman*.

Panicum soboliferum Tuckerm. ex Scribn. and Merr., Rhodora 3: 106. 1901. Cited as synonym of *P. minimum* Scribn. and Merr., but the *Tuckerman* specimen from Lake Memphremagog, Vt., cited is not the same species as the type of *P. minimum*, which see under *P. philadelphicum*.

Panicum philadelphicum var. *tuckermani* Steyerm. and Schmoll, Rhodora 41: 90. 1939. Based on *Panicum tuckermani* Fernald.

(167) **Panicum urvilleanum** Kunth, Rév. Gram. 2: 403. pl. 115. 1831. [Concepcion] Chile, *Dumont-d' Urville*.

Panicum megastachyum Presl, Rel. Haenk. 1: 305. 1830. Not *P. megastachyum* Nees, 1826. Huánuco, Peru, *Haenke*.

Panicum preslii Kunth, Rév. Gram. 1: Sup. 10. 1830. Based on *P. megastachyum* Presl.

Panicum urvilleanum var. *longiglume* Scribn., U. S. Dept. Agr., Div. Agrost. Bul. 17 (ed. 2): 49. 1901. San Jacinto, Calif., *Parish Brothers* 887.

(82) **Panicum vernale** Hitchc. and Chase, U. S. Natl. Herb. Contrib. 15: 266. f. 293. 1910. Lake City, Fla., *Hitchcock* 1020.

(165) **Panicum verrucosum** Muhl., Descr. Gram. 113. 1817. New Jersey, Delaware, and Georgia.

Panicum debile Ell., Bot. S. C. and Ga. 1: 129. 1816. Not *P. debile* Desf., 1798. Presumably South Carolina.

Panicum umbraculum Bosc ex Spreng., Syst. Veg. 1: 314. 1825, as synonym of *P. verrucosum.*

Panicum rugosum Bosc ex Spreng., Syst. Veg. 1: 314. 1825, as synonym of *P. verrucosum.* [*Bosc.*]

(5)) **Panicum villosissimum** Nash, Torrey Bot. Club Bul. 23: 149. 1896. Macon, Ga., *Small* in 1895.

Panicum tectum Willd. ex Spreng., Syst. Veg. 1: 313. 1825. Name only. North America.

Panicum nitidum var. *villosum* A. Gray, N. Amer. Gram. and Cyp. 2: 111. 1835.

Panicum dichotomum var. *villosum* Vasey, U. S. Dept. Agr., Div. Bot. Bul. 8: 31. 1889. [Type, District of Columbia, *Vasey.*]

Panicum nitidum pubescens Scribn. in Kearney, Torrey Bot. Club Bul. 20: 479. 1893. Name only. Harlan and Bell Counties, Ky., *Kearney 58* and 141 in part.

Panicum atlanticum Nash, Torrey Bot. Club Bul. 24: 346. 1897. Bronx Park, N. Y., *Nash.*

Panicum haemacarpon Ashe, Elisha Mitchell Sci. Soc. Jour. 15: 55. 1898. District of Columbia, *Kearney* in 1897 [type]; North Carolina, *Ashe* in 1898; Iowa, *Carver 258.*

Panicum xanthospermum Scribn. and Mohr, U. S. Natl. Herb. Contrib. 6: 348. 1901. Greenville, Ala., *Mohr.*

(151) **Panicum virgatum** L., Sp. Pl. 59. 1753. Virginia [*Clayton 578*].

Panicum coloratum Walt., Fl. Carol. 73. 1788. Not *P. coloratum* L., 1767. South Carolina.

Eatonia purpurascens Raf., Jour. Phys. Chym. 89: 104. 1819. New York [type, Long Island].

Panicum pruinosum Bernh. ex Trin., Gram. Pan. 191. 1826, as synonym of *P. virgatum.* North America [Delaware], *Bernhardi.*

Panicum giganteum Scheele, Linnaea 22: 340. 1849. Between San Antonio and New Braunfels, Tex., *Lindheimer.*

Panicum glaberrimum Steud., Syn. Pl. Glum. 1: 94. 1854. Grown at Berlin, seed from North America.

Ichnanthus glaber Link ex Steud., Syn. Pl. Glum. 1: 94. 1854, as synonym of *P. glaberrimum* Steud.

Panicum kunthii Fourn. ex Hemsl., Biol.

Centr. Amer. Bot. 3: 490. 1885. Not *P. kunthii* Steud., 1841. Based on *P. coloratum* L. misapplied by Kunth.

Panicum virgatum var. *confertum* Vasey, Torrey Bot. Club Bul. 13: 26. 1886. No locality cited. [Type, Atlantic City, N. J., *Vasey.*]

Panicum virgatum var. *elongatum* Vasey, Torrey Bot. Club Bul. 13: 26. 1886. No locality cited. [Type, White River, S. Dak., *Wilcox 13.*]

Panicum virgatum var. *diffusum* Vasey, Torrey Bot. Club Bul. 13: 26. 1886. "Kansas, Colorado, etc."

Panicum virgatum var. *glaucephylla* Cassidy, Colo. Agr. Expt. Sta. Bul. 12: 29. 1890. Colorado.

Chasea virgata Nieuwl., Amer. Midl. Nat. 2: 64. 1911. Based on *Panicum virgatum* L.

Milium virgatum Lunell, Amer. Midl. Nat. 4: 212. 1915. Based on *Panicum virgatum* L.

Milium virgatum var. *elongatum* Lunell, Amer. Midl. Nat. 4: 212. 1915. Based on *Panicum virgatum* var. *elongatum* Vasey.

PANICUM VIRGATUM var. CUBENSE Griseb., Cat. Pl. Cub. 233. 1866. [Hanábana] Cuba, *Wright* in 1865.

Panicum virgatum var. *obtusum* Wood, Amer. Bot. and Flor. pt. 2: 392. 1871. New Jersey.

Panicum virgatum var. *breviramosum* Nash, Torrey Bot. Club Bul. 23: 150. 1896. Augusta, Ga., *Small* in 1895.

Panicum virgatum var. *thyrsiforme* Linder, Rhodora 24: 14. 1922. Indian River, Fla., *Fredholm 5580.*

PANICUM VIRGATUM var. SPISSUM Linder, Rhodora 24: 15. 1922. Great Pubnico Lake, Nova Scotia, *Fernald, Long,* and *Linder 19766.*

(90) **Panicum webberianum** Nash, Torrey Bot. Club Bul. 23: 149. 1896. Eustis, Fla., *Nash 781.*

Panicum onslowense Ashe, Elisha Mitchell Sci. Soc. Jour. 16: 88. 1900. Wards Mill, Onslow County, N. C., *Ashe.*

(8) **Panicum werneri** Scribn. in Britt. and Brown, Illustr. Fl. 3: 501. f. 268b. 1898. New York and Ohio [type, Painesville, *Werner 60*].

Panicum delawarense Ashe, N. C. Agr. Expt. Sta. Bul. 175: 116. 1900. Centerville, Del., *Commons* [48] in 1878.

Panicum linearifolium var. *werneri* Fernald, Rhodora 23: 194. 1921. Based on *P. werneri* Scribn.

Panicum strictum var. *linearifolium* subvar. *werneri* Farwell, Amer. Midl. Nat. 11: 44. 1928. Based on *P. werneri* Scribn.

(92) **Panicum wilcoxianum** Vasey, U. S. Dept. Agr., Div. Bot. Bul. 8: 32. 1889. Nebraska [Fort Niobrara], *Wilcox* in 1888.

Milium wilcoxianum Lunell, Amer. Midl. Nat. 4: 213   1915. Based on *Panicum wilcoxianum* Vasey.

(69) **Panicum wilmingtonense** Ashe, Elisha Mitchell Sci. Soc. Jour. 16: 86. 1900. Wilmington, N. C., *Ashe* in 1899.

Panicum alabamense Ashe, N. C. Agr. Expt. Sta. Bul. 175: 116. 1900. Not *P. alabamense* Trin., 1854. Auburn, Ala., *Alabama Biological Survey* 1530.

(44) **Panicum wrightianum** Scribn., U. S. Dept. Agr., Div. Agrost. Bul. 11: 44. f. 4. 1898. Vueltabajo, Cuba, *Wright* 3463.

Panicum strictum Bosc ex Roem. and Schult., Syst. Veg. 2: 447. 1817. Not *P. strictum* R. Br., 1810. North America [type, Carolina, *Bosc*].

Panicum minutulum Desv., Opusc. 87. 1833. Not *P. minutulum* Gaud., 1826. Carolina.

Panicum deminutivum Peck, N. Y. State Mus. Bul. 10: 27. 1907. Suffolk County, N. Y., *Peck* in 1906.

(10) **Panicum xalapense** H. B. K., Nov. Gen. et Sp. 1: 103. 1815. Xalapa [Jalapa], Mexico, *Humboldt* and *Bonpland*.

Panicum pumilum Bosc ex Nees, Agrost. Bras. 228. 1829. Not *P. pumilum* Lam., 1798. Name only. North America [*Bosc*].

Panicum rariflorum Rupr., Acad. Sci. Belg. Bul. 9²: 240. 1842. Not *P. rariflorum* Lam., 1798. Name only. Jalapa, Mexico, *Galeotti* 5733.

Panicum ruprechtii Fourn., Mex. Pl. 2: 21. 1886. Not *P. ruprechtii* Fenzl, 1854. Described from type of *P. rariflorum* Rupr.

Panicum caricifolium Scribn. ex Ashe, Elisha Mitchell Sci. Soc. Jour. 15: 57. 1898. Name only. Washington, D. C., *Kearney* in 1897.

This is the species described as *Panicum acuminatum* Swartz by Muhlenberg (Descr. Gram. 125. 1817).

PANICUM XALAPENSE var. STRICTIRAMEUM Hitchc. and Chase, U. S. Natl. Herb. Contrib. 15: 161. f. 148. 1910. Jackson, Miss., *Hitchcock* 1311. (Published as *P. xalapense strictirameum*.)

Panicum laxiflorum var. *strictirameum* Fernald, Rhodora 36: 75. 1934. Based on *P. xalapense strictirameum* Hitchc. and Chase.

(99) **Panicum xanthophysum** A. Gray, N. Amer. Gram. and Cyp. 1: No. 28. 1834. Oneida Lake, N. Y.

Panicum xanthophysum forma *amplifolium* Scribn. in Brainerd, Jones, and Eggleston, Fl. Vt. 104. 1900. Burlington, Vt., *Jones*.

(35) **Panicum yadkinense** Ashe, Elisha Mitchell Sci. Soc. Jour. 16: 85. 1900. Based on *Panicum maculatum* Ashe. ?*Panicum dumus* Desv., Opusc. 88. 1831.

Tropical America (locality erroneous). *Panicum maculatum* Ashe, Elisha Mitchell Sci. Soc. Jour. 15: 44. 1898. Not *P. maculatum* Aubl., 1775. Raleigh, N. C., *Ashe* in 1895.

(39) **PAPPOPHORUM Schreb.**

(2) **Pappophorum bicolor** Fourn., Mex. Pl. 2: 133. 1886. Toluca, Mexico, *Karwinsky* 1483.

(1) **Pappophorum mucronulatum** Nees, Agrost. Bras. 412. 1829. Bahia and Piauhy, Brazil, *Martius*.

Pappophorum vaginatum Buckl., Prel. Rpt. Geol. Agr. Survey Tex. App. 1. 1866. Western Texas [type, *Wright* 803].

Pappophorum apertum Munro ex Scribn., Torrey Bot. Club Bul. 9: 148. 1882. Camp Lowell, Ariz., *Pringle*.

Pappophorum apertum var. *vaginatum* Scribn. ex L. H. Dewey, U. S. Natl. Herb. Contrib. 2: 535. 1894. Based on *P. vaginatum* Buckl.

Pappophorum pappiferum (Lam.) Kuntze var. *mucronulatum* Kuntze, Rev. Gen. Pl. 3³: 365. 1898. Based on *P mucronulatum* Nees.

(52) **PARAPHOLIS C. E. Hubb.**

(1) **Parapholis incurva** (L.) C. E. Hubb., Blumea Sup. 3. (Henrard Jubilee vol.): 14. 1946. Based on *Aegilops incurva* L.

Aegilops incurva L., Sp. Pl. 1051. 1753. Europe.

Aegilops incurvata L., Sp. Pl. ed. 2. 2: 1490. 1763. Europe.

Agrostis incurvata Scop., Fl. Carn. 1: 62. 1772. Based on *Aegilops incurva* L.

Rottboellia incurvata L. f., Sup. Pl. 114. 1781. Based on *Aegilops incurva* L.

Ophiurus incurvatus Beauv., Ess. Agrost. 116, 168, 176. 1812. Based on *Rottboellia incurvata* L. f.

Rottboellia incurva Roem. and Schult., Syst. Veg. 2: 799. 1817. Presumably based on *Aegilops incurva* L.

Lepturus incurvatus Trin., Fund. Agrost. 123. 1820. Based on *Aegilops incurvata* L.

Lepturus filiformis var. *incurvatus* Hook. f., Stud. Fl. 455. 1870. Based on *L. incurvatus* Trin.

Lepturus incurvus Druce, List Brit. Pl. 85. 1908. Presumably based on *Aegilops incurva* L.

Lepturus incurvus subsp. *incurvatus* Briq., Prodr. Fl. Corse 1: 183. 1910. Based on *Lepturus incurvatus* Trin. "sensu stricto."

Pholiurus incurvatus Hitchc., U. S. Dept. Agr. Bul. 772: 106. 1920. Based on *Aegilops incurvata* L.

Pholiurus incurvus Schinz and Thell., Vierteljahrs. Nat. Gesell. Zürich 66: 265. 1921. Based on *Aegilops incurva* L.

Lepidurus incurvus Janchen in Janchen and Neumayer, Wien. Bot. Zeitschr. 93: 85. 1944. Based on *Aegilops incurva* L., but genus not validly published. This species has been included in *Pholiurus* Trin., but in the type of that, *P. pannonicus* (Host) Trin., the rachis is continuous, the spikelets falling entire, free from the rachis joints.

(136) PASPALUM L.

(2) **Paspalum acuminatum** Raddi, Agrost. Bras. 25. 1823. Rio de Janeiro, Brazil, *Raddi*.

(11) **Paspalum almum** Chase, Wash. Acad. Sci. Jour. 23: 137. f. 1. 1933. Beaumont, Tex., *J. F. Combs* in 1932.

(48) **Paspalum bifidum** (Bertol.) Nash, Torrey Bot. Club Bul. 24: 192. 1897. Based on *Panicum bifidum* Bertol.

Panicum floridanum Trin., Acad. St. Pétersb. Mém. VI. Sci. Nat. 1: 248. 1834. Not *Paspalum floridanum* Michx. Florida and Alabama.

Panicum bifidum Bertol., Accad. Sci. Bologna Mem. 2: 598. pl. 41. f. 2. e–h. 1850. Alabama.

Panicum alabamense Trin. ex Steud., Syn. Pl. Glum. 1: 64. 1854. Alabama, locality erroneously cited as North Carolina.

Paspalum racemulosum Nutt. ex Chapm., Fl. South. U. S. 571. 1860. Florida to North Carolina and westward.

Paspalum interruptum Wood, Class-book ed. 1861. 783. 1861. Louisiana and Texas, *Hale*.

Paspalum bifidum var. *projectum* Fernald, Rhodora 40: 388. pl. 509. 1938. Burt, Sussex County, Va., *Fernald* and *Long 7239*.

(25) **Paspalum blodgettii** Chapm., Fl. South. U. S. 571. 1860. Key West, Fla., *Blodgett*.

Paspalum dissectum Swartz ex Roem. and Schult., Syst. Veg. 2: 308. 1817. Not *P. dissectum* L. 1762. Erroneously given as synonym of *P. caespitosum* Flügge. Jamaica, *Swartz*.

Paspalum simpsoni Nash, Torrey Bot. Club Bul. 24: 39. 1897. No-Name Key, Fla., *Simpson 184*.

Paspalum gracillimum Nash in Small, Fl. Southeast. U. S. 73, 1326. 1903. Key West, Fla., *Blodgett*.

Paspalum yucatanum Chase, U. S. Natl. Herb. Contrib. 28: 121. 1929. Mérida, Yucatan, *Schott 597*.

(46) **Paspalum boscianum** Flügge, Monogr. Pasp. 170. 1810. Carolina, *Bosc*.

Paspalum virgatum Walt., Fl. Carol. 75. 1788. Not *P. virgatum* L., 1759. South Carolina.

Paspalum brunneum Bosc ex Flügge, Monogr. Pasp. 171. 1810, as synonym of

P. boscianum. Carolina, *Bosc*.

Paspalum purpurascens Ell., Bot. S. C. and Ga. 1: 108. pl. 6. f. 3. 1816. South Carolina.

Paspalum confertum LeConte, Jour. Phys. Chym. 91: 285. 1820. Georgia [*Le Conte*].

Paspalum virgatum var. *purpurascens* Wood, Class-book ed. 1861. 781. 1861. Based on *P. purpurascens* Ell.

(26) **Paspalum caespitosum** Flügge, Monogr. Pasp. 161. 1810. Hispaniola, *Poiteau* and *Turpin*.

Paspalum gracile Poir. in Lam., Encycl. Sup. 4: 313. 1816. Not *P. gracile* Rudge, 1805. Dominican Republic.

Paspalum heterophyllum Desv. ex Poir. in Lam., Encycl. Sup. 4: 315. 1816. Dominican Republic.

Paspalum poiretii Roem. and Schult., Syst. Veg. 2: 878. 1817. Based on *P. gracile* Poir.

Paspalum lineare Fourn., Mex. Pl. 2: 12. 1886. Not *P. lineare* Trin., 1826. Mexico, *Liebmann 187* [type; the other specimen cited, *Liebmann 192*, is *P. langei*].

Paspalum caespitosum var. *longifolium* Vasey, Torrey Bot. Club Bul. 13: 164. 1886. No locality cited. [Type, *Garber* in 1877.]

(19) **Paspalum ciliatifolium** Michx., Fl. Bor. Amer. 1: 44. 1803. Carolina, *Michaux*.

Paspalum debile Muhl., Cat. Pl. 8. 1813; Descr. Gram. 91. 1817. Not *P. debile* Michx., 1803. Georgia.

Paspalum spathaceum Desv. ex Poir. in Lam., Encycl. Sup. 4: 314. 1816. America.

Paspalum latifolium LeConte, Jour. Phys. Chym. 91: 284. 1820. Columbia, S. C.

Paspalum ciliatifolium var. *brevifolium* Vasey, Acad. Nat. Sci. Phila. Proc. 1886: 285. 1886. Philadelphia, Pa., *Burk*.

Paspalum setaceum var. *ciliatifolium* Vasey, U. S. Natl. Herb. Contrib. 3: 17. 1892. Based on *P. ciliatifolium* Michx.

Paspalum chapmani Nash, N. Y. Bot. Gard. Bul. 1: 290. 1899. Florida, *Chapman*.

Paspalum eggertii Nash, N. Y. Bot. Gard. Bul. 1: 434. 1900. Arkansas [type, Pine Bluff, *Eggert* in 1896].

Paspalum blepharophyllum Nash in Small, Fl. Southeast. U. S. 71, 1326. 1903. Central Florida, *Nash 1426*.

Paspalum epile Nash in Small, Fl. Southeast. U. S. 72, 1326. 1903. Key West, Fla., *Blodgett*.

(36) **Paspalum circulare** Nash in Britton, Man. 73. 1901. New York to North Carolina; Missouri. [Type, Bergen County, N. J., *Nash* in 1889.]

Paspalum praelongum Nash in Small, Fl. Southeast. U. S. 74, 1326. 1903.

Washington, D. C., *Nash* in 1894.
Paspalum laeve var. *circulare* Stone, N. J.
Mus. Ann. Rpt. 1910: 187. 1911. Based
on *P. circulare* Nash.
(31) **Paspalum conjugatum** Bergius, Act.
Helv. Phys. Math. 7: 129. pl. 8. 1762.
Dutch Guiana.
Paspalum tenue Gaertn., Fruct. et Sem.
2: 2. pl. 80. 1791. Apparently based on
P. conjugatum Bergius.
Paspalum ciliatum Lam., Tabl. Encycl. 1:
175. 1791. Tropical America [French
Guiana, *Leblond*].
Paspalum renggeri Steud., Syn. Pl. Glum.
1: 17. 1854. Paraguay, *Rengger.*
Paspalum longissimum Hochst. ex Steud.,
Syn. Pl. Glum. 1: 19. 1854. Dutch
Guiana, *Kappler* 1556.
Paspalum bicrurum Salzm. ex Doell in
Mart., Fl. Bras. 2²: 55. 1877, as syn-
onym of *P. conjugatum.* Bahia, Brazil,
Salzmann.
Paspalum conjugatum var. *parviflorum*
Doell in Mart., Fl. Bras. 2²: 55. 1877.
Brazil, Manáos, *Spruce* 894; Piauhy,
Gardner 3502.
(47) **Paspalum convexum** Humb. and Bon-
pl., in Flugge, Monogr. Pasp. 175.
1810. Mexico, *Humboldt* and *Bonpland.*
Paspalum hemicryptum Wright, An. Acad.
Cienc. Habana 8: 204. 1871; Wright
and Sauv., Fl. Cubana 196. 1873.
Cuba, *Wright* 3847.
Paspalum inops Vasey, U. S. Natl. Herb.
Contrib. 1: 281. 1893. Mexico, *Palmer*
592 in 1866.
Paspalum inops var. *major* Vasey, in
Beal, Grasses N. Amer. 2: 89. 1896.
Mexico, *Pringle* 1875.
(14) **Paspalum debile** Michx., Fl. Bor.
Amer. 1: 44. 1803. Carolina [type]
and Georgia, *Michaux.*
?*Paspalum dissectum* Walt., Fl. Carol. 75.
1788. Not *P. dissectum* L. 1762. South
Carolina.
Paspalum dubium DC., Cat. Hort.
Monsp. 130. 1813. Native country
unknown.
Paspalum infirmum Roem. and Schult.,
Syst. Veg. 2: 307. 1817. Based on
Paspalum debile Michx.
Paspalum villosissimum Nash, Torrey
Bot. Club Bul. 24: 40. 1897. Eustis,
Fla. *Nash* 946.
(39) **Paspalum difforme** LeConte, Jour.
Phys. Chym. 91: 284. 1820. Georgia.
(32) **Paspalum dilatatum** Poir. in Lam.,
Encycl. 5: 35. 1804. Argentina, *Com-
merson.*
Paspalum platense Spreng., Syst. Veg. 1:
247. 1825. Montevideo, Uruguay.
Paspalum ovatum Nees ex Trin., Gram.
Pan. 113. 1826. Brazil, *Besser.*
Paspalum lanatum Spreng., Syst. Veg. 4:
Cur. Post. 30. 1827. Not *P. lanatum*
H. B. K., 1816. Brazil.
Paspalum eriophorum Schult., Mantissa

2: 560. 1827. Based on *P. lanatum*
Spreng.
Paspalum ovatum var. *grandiflorum* Nees,
Agrost. Bras. 43. 1829. Montevideo,
Uruguay, *Sellow.*
Paspalum selloi Spreng. ex Nees, Agrost.
Bras. 43. 1829, as synonym of *P.
ovatum* var. *grandiflorum* Nees.
Paspalum pedunculare Presl, Rel. Haenk.
1: 217. 1830. Habitat unknown.
Paspalum dilatatum var. *decumbens* Vasey,
Torrey Bot. Club Bul. 13: 166. 1886.
No locality cited. [Type, Point-a-la-
Hache, La., *Langlois* 27.]
Paspalum dilatatum var. *sacchariferum*
Arech., An. Mus. Nac. Montevideo 1:
90. 1894. Uruguay.
Panicum platense Kuntze, Rev. Gen. Pl.
3²: 363. 1898. Based on *Paspalum
platense* Spreng.
Digitaria dilatata Coste, Fl. France 3:
553. 1906. Based on *Paspalum di-
latatum* Poir.
(1) **Paspalum dissectum** (L.) L. Sp. Pl. ed.
2. 81. 1762. Based on *Panicum
dissectum* L.
Panicum dissectum L., Sp. Pl. 57. 1753.
Locality erroneously given as "Indiis,"
the type in the Linnaean Herbarium
being from North America, collected by
Kalm.
Paspalum dimidiatum L., Syst. Nat. ed.
10. 2: 855. 1759. Based on *Panicum
dissectum* L.
Paspalum membranaceum Walt., Fl. Carol.
75. 1788. South Carolina.
Paspalum vaginatum Ell., Bot. S. C. and
Ga. 1: 109. 1816. Not *P. vaginatum*
Swartz, 1788. Savannah, Ga., *Baldwin.*
Paspalum walterianum Schult., Mantissa
2: 166. 1824. Based on *P. membrana-
ceum* Walt. In Chapman's Flora (570.
1860.) the name is given as *P. walteri*
Schult.
Paspalum tectum Steud., Syn. Pl. Glum.
1: 29. 1854. Florida, *Chapman.*
Paspalum drummondii C. Muell., Bot.
Ztg. 19: 332. 1861. St. Louis, Mo.,
Drummond 182.
(5) **Paspalum distichum** L., Syst. Nat. ed.
10. 2: 855. 1759. [Jamaica, *Browne*.]
Digitaria paspalodes Michx., Fl. Bor.
Amer. 1: 46. 1803. Charleston, S. C.,
Michaux.
Paspalum digitaria Poir. in Lam., En-
cycl. Sup. 4: 316. 1816. Charleston,
S. C., *Bosc.*
Milium paspalodes Ell., Bot. S. C. and
Ga. 1: 104. 1816. Based on *Digitaria
paspalodes* Michx., but misapplied to
Axonopus furcatus (Flügge) Hitchc.
Milium distichum Muhl., Descr. Gram.
78. 1817. Presumably based on *Pas-
palum distichum* L. Name only, Muhl.,
Cat. Pl. 10. 1813.
Paspalum michauxianum Kunth, Rév.

Gram. 1: 25. 1829. Based on *Digitaria paspalodes* Michx.

Panicum paspaliforme Presl, Rel. Haenk. 1: 296. 1830. Peru, *Haenke.*

Panicum polyrrhizum Presl, Rel. Haenk. 1: 296. 1830. "Monterey, California," but the type from Baja California.

Paspalum fernandezianum Colla, Mem. 1: 296. 1830. "Monterey, California" [but specimens probably collected in Baja California], *Haenke.*

Paspalum fernandezianum Colla, Mem. Accad. Sci. Torino 39: 27. pl. 59. 1836. Juan Fernandez, Chile, *Bertero.*

Paspalum chepica Steud., Syn. Pl. Glum. 1: 21. 1854. Juan Fernandez, Chile, *Bertero* 1223.

Paspalum vaginatum var. *pubescens* Doell in Mart., Fl. Bras. 2²: 75. 1877. Rio de Janeiro, Brazil, *Glaziou* 3612.

Paspalum schaffneri Griseb. in Fourn., Mex. Pl. 2: 6. 1886. Mexico, Chapultepec, *Schaffner* 19a; San Angel, *Schaffner* 19c; Mirador, *Schaffner* 19b.

Paspalum elliottii S. Wats. in A. Gray, Man. ed. 6. 629. 1890. Based on *Milium paspalodes* Ell. but misapplied to *Axonopus furcatus* (Flügge) Hitchc.

Paspalum paspaloides Scribn., Torrey Bot. Club Mem. 5: 29. 1894. Based on *Digitaria paspalodes* Michx. but misapplied to *Axonopus furcatus* (Flügge) Hitchc.

Digitaria disticha Fiori and Paol., Icon. Fl. Ital. Illustra. 1: 16. f. 136. 1895. Based on *Paspalum distichum* L.

Anastrophus paspaloides Nash in Britton, Man. 75. 1901. Based on *Paspalum paspaloides* Scribn. but misapplied to *Axonopus furcatus* (Flügge) Hitchc.

Paspalum distichum var. *digitaria* Hack. in Stuck., An. Mus. Nac. Buenos Aires 13: 424. 1906. Based on *P. digitaria* Poir.

Paspalum distichum subsp. *paspalodes* Thell., Mém. Soc. Sci. Nat. Cherbourg 38: 77. 1912. Based on *Digitaria paspalodes* Michx.

(40) **Paspalum floridanum** Michx., Fl. Bor. Amer. 1: 44. 1803. Florida and Georgia, *Michaux.*

Paspalum macrospermum Flügge, Monogr. Pasp. 172. 1810. Carolina, *Bosc.*

Paspalum glabrum Bosc ex Flügge, Monogr. Pasp. 172. 1810, as synonym of *P. macrospermum* Flügge.

Paspalum laevigatum Bosc ex Poir., Encycl. Sup. 4: 313. 1816, as synonym of *P. floridanum* Michx.

Paspalum laeve var. *floridanum* Wood, Class-book ed. 1861. 782. 1861. Presumably based on *P. floridanum* Michx.

PASPALUM FLORIDANUM var. GLABRATUM Engelm. ex Vasey, Torrey Bot. Club Bul. 13: 166. 1886. No locality cited. [Type, Mobile, Ala., *Mohr* in 1884.]

?*Paspalum altissimum* LeConte, Jour. Phy. Chym. 91: 285. 1820. Salem, N. C.

?*Paspalum laeve* var. *altissimum* Wood, Class-book ed. 1861. 782. 1861. Based on *P. altissimum* LeConte.

Paspalum glabratum Mohr, Torrey Bot. Club Bul. 24: 21. 1897. Based on *P. floridanum* var. *glabratum* Engelm.

(3) **Paspalum fluitans** (Ell.) Kunth, Rév. Gram. 1: 24. 1829. Based on *Ceresia fluitans* Ell.

Paspalum mucronatum Muhl., Cat. Pl. 8. 1813, name only; Georgia; Descr. Gram. 96. 1817. Mississippi and Georgia.

Ceresia fluitans Ell., Bot. S. C. and Ga. 1: 109. pl. 6. f. 4. 1816. Ogechee, Ga.

Paspalum natans LeConte, Jour. Phys. Chym. 91: 285. 1820. Georgia.

Paspalum frankii Steud., Syn. Pl. Glum. 1: 19. 1854. New Orleans, La., *Frank.*

Cymatochloa fluitans Schlecht., Bot. Ztg. 12: 822. 1854. Based on *Ceresia fluitans* Ell.

This is the species described under *P. paniculatum* by Walter (Fl. Carol. 75. 1788); included in *P. repens* in Manual ed. 1.

(41) **Paspalum giganteum** Baldw. ex Vasey, Torrey Bot. Club Bul. 13: 166. 1886. No locality cited. [Type, Pablo Creek, Fla., *Curtiss* in 1875.]

Paspalum longicilium Nash, N. Y. Bot. Gard. Bul. 1: 435. 1900. Eustis, Fla., *Nash* 1359.

(8) **Paspalum hartwegianum** Fourn., Mex. Pl. 2: 12. 1886. León, Mexico, *Hartweg* 245.

Paspalum buckleyanum Vasey, Torrey Bot. Club Bul. 13: 167. 1886. Texas, *Buckley.* In Jacks., Ind. Kew. Sup. 1: 312. 1906, the name is erroneously listed under *Panicum.*

(45) **Paspalum hydrophilum** Henr., Med. Rijks Herb. Leiden 45: 1. pl. 1922. Paraguay, *Balansa* in 1884.

Paspalum intermedium Munro ex Morong, Ann. N. Y. Acad. Sci. 7: 258. 1893. Paraguay, *Morong* 1019.

(34) **Paspalum laeve** Michx., Fl. Bor. Amer. 1: 44. 1803. Georgia, *Michaux.*

Paspalum undulosum LeConte, Jour. Phys. Chym. 91: 284. 1820. Georgia [*LeConte*].

Paspalum angustifolium LeConte, Jour. Phys. Chym. 91: 285. 1820. Carolina and Georgia [*LeConte*].

Paspalum lecomteanum Schult., Mantissa 2: 168. 1824. Based on *P. undulosum* LeConte.

Paspalum punctulatum Bertol., Accad. Sci. Bologna Mem. 2: 599. pl. 42. f. a–e. 1850. Alabama.

Paspalum alternans Steud., Syn. Pl. Glum. 1: 26. 1854. Louisiana, *Hartman* 40.

Paspalum tenue Darby, Bot. South. States 576. 1857. Not *P. tenue* Gaertn., 1791. Georgia and northward.

Paspalum laeve var. *undulosum* Wood, Class-book ed. 1861. 782. 1861. Based on *P. undulosum* LeConte.

Paspalum angustifolium var. *tenue* Wood, Amer. Bot. and Flor. pt. 2: 390. 1871. New Jersey and south.

Paspalum laeve var. *angustifolium* Vasey, Torrey Bot. Club Bul. 13: 165. 1886. Based on *P. angustifolium* LeConte.

Paspalum laeve var. *brevifolium* Vasey, U. S. Natl. Herb. Contrib. 3: 18. 1892. No locality cited. [Type, Texas, *Nealley* in 1886.]

Paspalum australe Nash in Britton, Man. 1039. 1901. Stone Mt., Ga., *Small* in 1895.

Paspalum laeve australe Nash in Hitchc., Rhodora 8: 205. 1906. Based on *P. australe* Nash.

(24) **Paspalum langei** (Fourn.) Nash, N. Amer. Fl. 17: 179. 1912. Based on *Dimorphostachys langei* Fourn.

Panicum senescens Trin. ex Steud., Nom. Bot. ed. 2. 2: 263. 1841, name only. [Mexico, *Schiede*.]

Paspalum abbreviatum Trin. ex Fourn., Mex. Pl. 2: 10. 1886, name only. Mexico, *Schiede* 888.

Dimorphostachys langei Fourn., Mex. Pl. 2: 14. 1886. Mexico, *Liebmann* 186.

Dimorphostachys drummondii Fourn., Mex. Pl. 2: 15. 1886. Not *Paspalum drummondii* C. Muell., 1861. Texas, *Drummond* [350].

Panicum squamatum Fourn., Mex. Pl. 2: 18. 1886. Not *Paspalum squamatum* Steud., 1854. Mexico, *Karwinsky* 982.

Paspalum drummondii Vasey, U. S. Natl. Herb. Contrib. 3: 18. 1892. Not *P. drummondii* C. Muell., 1861. Based on *Dimorphostachys drummondii* Fourn.

Paspalum oricola Millsp. and Chase, Field Mus. Bot. 3: 28. f. 28, 29. 1903. Island of Cozumel, Yucatan, *Millspaugh Pl. Uto.* 1480.

Dimorphostachys ciliifera Nash in Small, Fl. Southeast. U. S. 78, 1327. 1903. Manatee, Fla., *Simpson* 97.

Paspalum ciliiferum Hitchc., U. S. Natl. Herb. Contrib. 12: 201. 1909. Based on *Dimorphostachys ciliifera* Nash.

(27) **Paspalum laxum** Lam., Tabl. Encycl. 1: 176. 1791. Tropical America [probably St. Croix], *Richard*.

Paspalum glabrum Poir. in Lam., Encycl. 5: 30. 1804. Puerto Rico, *Ledru*.

Paspalum miliodeum Desv. ex Poir. in Lam., Encycl. Sup. 4: 315. 1816. Puerto Rico.

Paspalum miliare Spreng., Syst. Veg. 1: 247. 1825. Based on *P. miliodeum* Desv.

Paspalum ischnocaulon Trin., Gram. Icon. 2: pl. 126. 1828. Source erroneously

given as East Indies, doubtless error for West Indies.

Paspalum floribundum Desv., Opusc. 58. 1831. West Indies.

Paspalum rhizomatosum Steud., Syn. Pl. Glum. 1: 17. 1854. Guadeloupe, *Duchaissing*.

Paspalum koleopodum Steud., Syn. Pl. Glum. 1: 18. 1854. Guadeloupe, *Duchaissing*.

Paspalum laxum var. *lamarckianum* Doell in Mart., Fl. Bras. 2²: 86. 1877. Based on *P. laxum* Lam., but misapplied to a Brazilian species.

Paspalum helleri Nash, Torrey Bot. Club Bul. 30: 376. 1903. Santurce, Puerto Rico, *Heller* 10.

Paspalum tenacissimum Mez, Bot. Jahrb. Engler 56: Beibl. 125: 10. 1921. Puerto Rico, *Hioram* 804.

(38) **Paspalum lentiferum** Lam., Tabl. Encycl. 1: 175. 1791. Carolina, *Fraser*.

Paspalum lanuginosum Bosc ex Beauv., Ess. Agrost. 12. 1812. Name only. [Carolina, *Bosc*.]

Paspalum lanuginosum Willd. ex Steud., Nom. Bot. ed. 2. 2: 271. 1841, as synonym of *P. lentiferum* Lam.

Paspalum curtisianum Steud., Syn. Pl. Glum. 1: 26. 1854. Carolina, *M. A. Curtis*.

Paspalum praecox var. *curtisianum* Vasey, Torrey Bot. Club Bul. 13: 165. 1886. Based on *P. curtisianum* Steud.

Paspalum glaberrimum Nash in Small, Fl. Southeast. U. S. 76, 1326. 1903. Central Florida, *Nash* 1619.

Paspalum tardum Nash in Small, Fl. Southeast. U. S. 76, 1326. 1903. Florida, *Nash* 2047.

Paspalum kearneyi Nash in Small, Fl. Southeast. U. S. 77, 1326. 1903. Nicholson, Miss., *Kearney* 357.

Paspalum amplum Nash in Small, Fl. Southeast. U. S. 77, 1326. 1903. Marianna, Fla., *Tracy* 3682.

(7) **Paspalum lividum** Trin. in Schlecht., Linnaea 26: 383. 1854. Mexico, *Schiede*.

(12) **Paspalum longepedunculatum** LeConte, Jour. Phys. Chym. 91: 284. 1820. North Carolina [*LeConte*].

Paspalum setaceum var. *longepedunculatum* Wood, Class-book, ed. 1861. 782. 1861. Based on *P. longepedunculatum* LeConte.

Paspalum kentuckiense Nash in Britton, Man. 1039. 1901. Poor Fork, Ky., *Kearney* in 1893.

(35) **Paspalum longipilum** Nash, N. Y. Bot. Gard. Bul. 1: 435. 1900. Eustis, Fla., *Nash* 1027.

Paspalum laeve var. *pilosum* Scribn., Tenn. Agr. Expt. Sta. Bul. 7: 34. 1894. Tennessee [type, Madisonville, *Scribner*].

Paspalum plenipilum Nash in Britton,

Man. 73. 1901. New Jersey [type, Clifton, *Nash* in 1892].

Paspalum malacophyllum Trin. Gram. Icon. 3: pl. 271. 1831. Brazil.

(10) **Paspalum minus** Fourn., Mex. Pl. 2: 6. 1886. Mexico, *Bourgeau* 2298 [type], *Liebmann* 154.

(23) **Paspalum monostachyum** Vasey in Chapm., Fl. South. U. S. ed. 2. 665. 1883. South Florida, *Garber* [224].

Paspalum rectum var. *longispicatum* Vasey, Bot Gaz. 9: 54, 55. 1884. Miami, Fla., *Garber*.

Paspalum solitarium Nash in Small, Fl. Southeast. U. S. 77, 1326. 1903. Based on *Paspalum monostachyum* "Vasey not Walp." Walper's is a name only.

(9) **Paspalum notatum** Flügge, Monogr. Pasp. 106. 1810. St. Thomas, West Indies.

Paspalum taphrophyllum Steud., Syn. Pl. Glum. 1: 19. 1854. Martinique, *Sieber* 365 [error for 364].

Paspalum distachyon Willd. ex Doell in Mart., Fl. Bras. 2²: 73. 1877. Not *P. distachyon* Poir., 1834. As synonym of *P. notatum*.

Paspalum notatum var. *latiflorum* Doell in Mart., Fl. Bras. 2²: 73. 1877. Brazil and Uruguay, *Sellow*.

Paspalum saltense Arech., An. Mus. Nac. Montevideo 1: 53. 1894. Department del Salto, Uruguay.

PASPALUM NOTATUM var. SAURAE Parodi, Univ. Nac. Buenos Aires Rev. Agron. 15: 55. 1948. Entre Rios, Argentina, *Parodi* 12670.

Paspalum paucispicatum Vasey, U. S. Natl. Herb. Contrib. 1: 281. 1893. Guadalajara, Mexico, *Palmer* 243 in 1886.

(28) **Paspalum pleostachyum** Doell, in Mart., Fl. Bras. 2²: 58. 1877. Bahia, Brazil (*Salzmann*, herb. Bahiense n. 665).

(43) **Paspalum plicatulum** Michx., Fl. Bor. Amer. 1: 45. 1803. Georgia and Florida, *Michaux*.

Paspalum undulatum Poir. in Lam., Encycl. 5: 29. 1804. Puerto Rico, *Ledru*.

Paspalum plicatum Pers., Syn. Pl. 1: 86. 1805, error for *plicatulum*.

Paspalum lenticulare H. B. K., Nov. Gen. et Sp. 1: 92. 1815. Venezuela, *Humboldt* and *Bonpland*.

Paspalum gracile LeConte, Jour. Phys. Chym. 91: 285. 1820. Not *P. gracile* Rudge, 1805. Georgia, *LeConte*.

Paspalum leptos Schult., Mantissa 2: 173. 1824. Based on *P. gracile* LeConte.

Paspalum montevidense Spreng., Syst. Veg. 1: 246. 1825. Montevideo, Uruguay, *Sellow*.

Paspalum tenue Kunth, Rév. Gram. 1: 26. 1829. Not *P. tenue* Gaertn., 1791. Based on *P. gracile* LeConte.

Paspalum multiflorum Desv., Opusc. 58. 1831. Brazil.

Paspalum orthos Schult. ex Kunth, Enum. Pl. 1: 57. 1833. Apparently misprint for *P. leptos*.

Paspalum marginatum Spreng. ex Steud., Nom. Bot. ed. 2. 2: 272. 1841. Not *P. marginatum* Trin., 1826. As synonym of *P. undulatum* Poir. [Puerto Rico.]

Paspalum campestre Schlecht., Linnaea 26: 131. 1853. Not *P. campestre* Trin., 1834. Venezuela, *Wagener* 392.

Paspalum atrocarpum Steud., Syn. Pl. Glum. 1: 25. 1854. Habitat unknown. *Dumont-d'Urville*.

Paspalum virgatum var. *undulatum* Wood, Amer. Bot. and Flor. pt. 2: 390. 1871. Eastern States.

Paspalum antillense Husnot, Soc. Linn. Normand. Bul. II. 5: 260. 1871. Guadeloupe, *Husnot* 76.

Paspalum saxatile Salzm. ex Doell, in Mart., Fl. Bras. 2²: 76. 1877, as synonym of *P. plicatulum* Michx. Brazil, *Salzmann*.

Paspalum decumbens Sagot ex Doell in Mart., Fl. Bras. 2²: 77. 1877. Not *P. decumbens* Swartz, 1788. As synonym of *P. plicatulum* Michx. French Guiana, *Sagot* 1342.

Paspalum plicatulum var. *intumescens* Doell in Mart., Fl. Bras. 2²: 78. 1877. Lagoa Santa, Brazil, *Warming*.

Paspalum pauperculum Fourn., Mex. Pl. 2: 10. 1886. San Luis Potosí, Mexico, *Virlet* 1320.

Paspalum pauperculum var. *altius* Fourn., Mex. Pl. 2: 10. 1886. Orizaba, Mexico, *Bourgeau* 2033 [probably misprint for 2633].

Panicum plicatulum Kuntze, Rev. Gen. Pl. 3²: 363. 1898. Based on *Paspalum plicatulum* Michx.

(37) **Paspalum praecox** Walt., Fl. Carol. 75. 1788. South Carolina.

(20) **Paspalum propinquum** Nash, N. Y. Bot. Gard. Bul. 1: 291. 1899. Eustis, Fla., *Nash* 1427.

(16) **Paspalum psammophilum** Nash in Hitchc., Rhodora 8: 205. 1906. Based on *P. prostratum* Nash.

Paspalum prostratum Nash in Britton, Man. 74. 1901. Not *P. prostratum* Scribn. and Merr., 1901 (earlier than the preceding). New York to Delaware [type, Kingsbridge, N. Y., *Nash* 514].

(18) **Paspalum pubescens** Muhl. in Willd., Enum. Pl. 89. 1809. Carolina.

Paspalum muhlenbergii Nash in Britton, Man. 75. 1901. Massachusetts to Georgia, Missouri, Oklahoma, and Mississippi. [Type, Van Cortlandt Park, N. Y., *Bicknell* in 1896.]

Paspalum pubescens var. *muhlenbergii* House, N. Y. State Mus. Bul. 243–244: 39. 1923. Based on *Paspalum muhlenbergii* Nash.

Paspalum ciliatifolium var. *muhlenbergii*

Fernald, Rhodora 36: 20. 1934. Based on *P. muhlenbergii* Nash.

(6) **Paspalum pubiflorum** Rupr. ex Fourn., Mex. Pl. 2: 11. 1886. Mexico, *Galeotti* 5747.

Paspalum planifolium Fourn., Mex. Pl. 2: 10. 1886. Mexico, San Luis Potosí, *Virlet* [type; the other specimen cited, *Müller* 2062, is *P. lividum*].

Paspalum pubiflorum var. *viride* Fourn., Mex. Pl. 2: 11. 1886. San Luis Potosí, *Virlet* 1328.

Paspalum hallii Vasey and Scribn., Torrey Bot. Club Bul. 13: 165. 1886, as doubtful synonym of *P. remotum* Remy, a Bolivian species. Description drawn from *Hall* 804, Texas.

Paspalum remotum var. *glaucum* Scribn. in Vasey, Torrey Bot. Club Bul. 13: 165. 1886. No locality cited. [Type, Grapevine Canyon, Tex. *Havard* in 1883.]

Paspalum pubiflorum var. *glaucum* Scribn., U. S. Natl. Herb. Contrib. 3: 19. 1892. Southwestern Texas and Mexico [type same as preceding].

PASPALUM PUBIFLORUM var. GLABRUM Vasey ex Scribn., Tenn. Agr. Expt. Sta. Bul. 7: 32. pl. 5. f. 18. 1894. Belle Meade, Tenn., *Scribner* in 1892.

Paspalum remotum var. *glabrum* Vasey, Torrey Bot. Club Bul. 13: 166. 1886. No locality cited. [Type, Plaquemines Parish, La., *Langlois* 26.]

Paspalum geminum Nash, N. Y. Bot. Gard. Bul. 1: 434. 1900. Eustis, Fla., *Nash* 680.

Paspalum laeviglume Scribn. ex Nash in Small, Fl. Southeast. U. S. 75, 1326. 1903. Based on *P. remotum* var. *glabrum* Vasey.

Paspalum racemosum Lam., Tabl. Encycl. 1: 176. 1791. Peru.

(21) **Paspalum rigidifolium** Nash, N. Y. Bot. Gard. Bul. 1: 292. 1899. Eustis, Fla., *Nash* 817.

(30) **Paspalum saugetii** Chase, U. S. Natl. Herb. Contrib. 28: 147. f. 90. 1919. Cuba, *Léon* 8982.

Paspalum scrobiculatum L., Mantissa pl. 1: 29. 1767. India.

(13) **Paspalum setaceum** Michx., Fl. Bor. Amer. 1: 43. 1803. South Carolina, *Michaux*.

Paspalum hirsutum Retz., misapplied by Poir., in Lam., Encycl. 5: 28. 1804. Carolina, *Bosc*.

Paspalum leptostachyum DC., Cat. Hort. Monsp. 130. 1813. Not *P. leptostachyum* Humb. and Bonpl., 1810. No locality cited, type without locality.

Paspalum incertum Roem. and Schult., Syst. Veg. 2: 308. 1817. Based on *P. leptostachyum* DC.

Paspalum eriophorum Willd. ex Nees., Agrost. Bras. 56. 1829. Not *P. eriophorum* Schult., 1827. Native country

unknown.

Paspalum setaceum var. *calvescens* Fernald, Rhodora 49: 121. pl. 1057. 1947. Nansemond County, Va., *Fernald, Long*, and *Clement* 15191.

(17) **Paspalum stramineum** Nash in Britton, Man. 74. 1901. Nebraska [type, Hooker County, *Rydberg* 1582], Kansas, and Indian Territory [Oklahoma].

Paspalum bushii Nash, in Britton, Man. 74. 1901. Missouri [type, Bernie, *Bush* 730].

Paspalum ciliatifolium var. *stramineum* Fernald, Rhodora 36: 20. 1934. Based on *P. stramineum* Nash.

(15) **Paspalum supinum** Bosc ex Poir. in Lam., Encycl. 5: 29. 1804. Carolina, *Bosc*.

Paspalum dasyphyllum Ell., Bot. S. C. and Ga. 1: 105. 1816. South Carolina.

Paspalum setaceum var. *supinum* Trin., Gram. Icon. 2: pl. 130. 1828. Based on *P. supinum* Bosc.

Paspalum ciliatifolium var. *dasyphyllum* Chapm., Fl. South. U. S. ed. 3. 578. 1897. Based on *P. dasyphyllum* Ell.

(44) **Paspalum texanum** Swallen, Biol. Soc. Wash. Proc. 55: 94. 1942. Port Lavaca, Calhoun County, Tex., *Mott* 261.

(22) **Paspalum unispicatum** (Scribn. and Merr.) Nash, N. Amer. Fl. 17: 193. 1912. Based on *Panicum unispicatum* Scribn. and Merr.

Panicum unispicatum Scribn. and Merr., U. S. Dept. Agr., Div. Agrost. Bul. 24: 14. 1901. Oaxaca, Mexico, *Pringle* 6717.

(33) **Paspalum urvillei** Steud., Syn. Pl. Glum. 1: 24. 1854. [Brazil] *Dumont-d'Urville*.

Paspalum ovatum var. *parviflorum* Nees, Agrost. Bras. 43. 1829. Brazil, *Martius*.

Paspalum velutinum Trin. ex Nees, Agrost. Bras. 43. 1829, as synonym of *P. ovatum* var. *parviflorum* Nees.

Paspalum dilatatum var. *parviflorum* Doell in Mart., Fl. Bras. 2²: 64. 1877. Pernambuco, *Forsell*; Lagoa Santa, *Warming* [type].

Paspalum virgatum var. *parviflorum* Doell in Mart., Fl. Bras. 2²: 89. 1877. Brazil, *Raben*.

Paspalum virgatum var. *pubiflorum* Vasey, Torrey Bot. Club Bul. 13: 167. 1886. No locality cited. [Type, Atakopus, La., *Langlois* in 1884.]

Paspalum larranagai Arech., An. Mus. Nac. Montevideo 1: 60. pl. 2. 1894. Uruguay.

Paspalum vaseyanum Scribn., U. S. Dept. Agr., Div. Agrost. Bul. 17: 32. f. 328. 1899. Based on *P. virgatum* var. *pubiflorum* Vasey.

Paspalum griseum Hack. ex Corrêa, Fl. Brazil 128. 1909. Name only. Brazil [*Glaziou* 16559].

(4) **Paspalum vaginatum** Swartz, Prodr. Veg. Ind. Occ. 21. 1788. Jamaica, *Swartz*.
Digitaria foliosa Lag., Gen. et Sp. Nov. 4. 1816. Habana, Cuba.
Paspalum tristachyum LeConte, Jour. Phys. Chym. 91: 285. 1820. Georgia [*LeConte*].
Digitaria tristachya Schult., Mantissa 2: 261. 1824. Based on *Paspalum tristachyum* LeConte.
Paspalum brachiatum Trin. ex Nees, Agrost. Bras. 62. 1829, as synonym of *P. vaginatum*. Martinique, *Sieber*.
Paspalum foliosum Kunth, Rév. Gram. 1: 25. 1829. Based on *Digitaria foliosa* Lag.
Paspalum kleineanum Presl, Rel. Haenk. 1: 209. 1830. Peru, *Haenke*.
Paspalum inflatum A. Rich. in Sagra, Hist. Cuba 11: 298. 1850. Habana, Cuba, *Sagra*.
Paspalum didactylum Salzm. ex Steud., Syn. Pl. Glum. 1: 20. 1854, as synonym of *P. vaginatum* Swartz. Brazil, *Salzmann*.
Panicum vaginatum Gren. and Godr., Fl. France 3: 462. 1855. Not *P. vaginatum* Nees, 1829. Based on *Paspalum vaginatum* Swartz.
Paspalum distichum var. *tristachyum* Wood, Class-book ed. 1861. 783. 1861. Presumably based on *P. tristachyum* LeConte.
Paspalum distichum var. *vaginatum* Swartz ex Griseb., Fl. Brit. W. Ind. 541. 1864. Based on *P. vaginatum* Swartz.
Paspalum reptans Poir. ex Doell, in Mart., Fl. Bras. 2²: 75. 1877, as synonym of *P. vaginatum*.
Paspalum vaginatum var. *nanum* Doell in Mart., Fl. Bras. 2²: 75. 1877. Rio de Janeiro, Brazil, *Glaziou* 4346.
Paspalum reimarioides Chapm., Fl. South. U. S. 665. 1883. Not *P. reimarioides* Brongn., 1830. West Florida [*Chapman*].
Paspalum vaginatum var. *reimarioides* Chapm., Fl. South. U. S. ed. 3. 577. 1897. Presumably based on *P. reimarioides* Chapm.
Paspalum distichum var. *nanum* Stapf in Dyer, Fl. Cap. 7: 371. 1898. Based on *P. vaginatum* var. *nanum* Doell.
Sanguinaria vaginata Bubani, Fl. Pyr. 4: 258. 1901. Based on *Paspalum vaginatum* Swartz.

(42) **Paspalum virgatum** L., Syst. Nat. ed. 10. 2: 855. 1759. Jamaica.
Paspalum virgatum var. *linneanum* Flügge, Monogr. Pasp. 190. 1810. Based on *P. virgatum* L.
Paspalum virgatum var. *jacquinianum* Flügge, Monogr. Pasp. 190. 1810. West Indies, *Jacquin*.
Paspalum virgatum var. *willdenowianum*

Flügge, Monogr. Pasp. 190. 1810. Pará, Brazil.
Paspalum virgatum var. *stramineum* Griseb., Fl. Brit. W. Ind. 543. 1864. Antigua, *Wullschlaegel* [the other specimen cited belongs to *P. arundinaceum* Poir.].
Paspalum leucocheilum Wright, An. Acad. Cienc. Habana 8: 203. 1871; Fl. Cub. 194. 1873. Isla de Pinos, *Blain*.
Paspalum virgatum var. *ciliatum* Doell in Mart., Fl. Bras. 2²: 88. 1877. Based on *P. virgatum* var. *linneanum* Flügge.

(29) **Paspalum virletii** Fourn. Mex. Pl. 2: 12. 1886. San Luis Potosí, Mexico, "Virlet 1329" [error for 1319].

(144) PENNISETUM L. Rich.

Pennisetum alopecuroides (L.) Spreng., Syst. Veg. 1: 303. 1825. Based on *Panicum alopecuroides* L.
Panicum alopecuroides L., Sp. Pl. 55. 1753. China.

Pennisetum ciliare (L.) Link, Hort. Berol. 1: 213. 1827. Based on *Cenchrus ciliaris* L.
Cenchrus ciliaris L., Mant. Pl. 2: 302. 1771. Cape of Good Hope, Africa, *Koenig*.

(6) **Pennisetum clandestinum** Hochst. ex Chiov., Ann. Ist. Bot. Roma 8: 41. 1903. Abyssinia.
Pennisetum longistylum var. *clandestinum* Leeke, Ztschr. Naturwiss. 79: 23. 1907. Based on *P. clandestinum* Hochst.

(1) **Pennisetum glaucum** (L.) R. Br., Prodr. Fl. Nov. Holl. 1: 195. 1810. Based on *Panicum glaucum* L.
Panicum glaucum L., Sp. Pl. 56. 1753.[23] India.
Holcus spicatus L., Syst. Nat. ed. 10. 2: 1305. 1759. India.
Pennisetum typhoideum L. Rich. in Pers., Syn. Pl. 1: 72. 1805. Based on *Holcus spicatus* L.
Penicillaria spicata Willd., Enum. Pl. 1037. 1809. Based on *Holcus spicatus* L.
Setaria glauca Beauv., Ess. Agrost. 51, 178. 1812. Based on *Panicum glaucum* L., but misapplied to *S. lutescens* (Weig.) Hubb.
Pennisetum spicatum Willd. ex Roem. and Schult., Syst. Veg. 2: 499. 1817, as synonym of *Penicillaria spicata* Willd. Koern., in Koern. and Wern., Handb. Getreidebau. 1: 284. 1885. Based on *Holcus spicatus* L.
Panicum spicatum Roxb., Fl. Ind. 1: 286. 1820. Based on *Holcus spicatus* L.
Penicillaria typhoidea Fig. and DeNot., Agrost. Aegypt. Frag. 55. 1853. Based on *Pennisetum typhoideum* "Delile" (same as L. Rich.).
Chamaeraphis glauca Kuntze, Rev. Gen.

[23] For discussion of this name see Chase, Amer. Jour. Bot. 8: 41–49. 1921.

Pl. 2: 767. 1891. Based on *Panicum glaucum* L., but misapplied to *Setaria lutescens* (Weig.) Hubb.

Pennisetum spicatum var. *typhoideum* Dur. and Schinz, Consp. Fl. Afr. 5: 785. 1894. Based on *Penicillaria typhoidea* Fig. and DeNot.

Ixophorus glaucus Nash, Torrey Bot. Club Bul. 22: 423. 1895. Based on *Panicum glaucum* L., but misapplied to *Setaria lutescens* (Weig.) Hubb.

Chaetochloa glauca Scribn., U. S. Dept. Agr., Div. Agrost. Bul. 4: 39. 1897. Based on *Panicum glaucum* L., but misapplied to *Setaria lutescens* (Wieg.) Hubb.

Pennisetum americanum Schum. (in Engl. Pflanzenw. Ost-Afr. 5B: 51. 1895), based on *Panicum americanum* L. (Sp. Pl. 56. 1753) has been used for this species, but the Linnaean name was based on an unidentifiable figure in Clusius (Rar. Pl. Hist. 2: 215. 1601).

Pennisetum latifolium Spreng., Syst. Veg. 1: 302. 1825. Montevideo, *Sello*.

Pennisetum macrostachyum (Brongn.) Trin., Acad. St. Pétersb. Mém. VI. Sci. Nat. 1: 177. 1834. Based on *Gymnothrix macrostachys* Brongn.

Gymnothrix macrostachys Brongn. in Duperrey, Bot. Voy. Coquille 2²: 104. pl. 10. 1830. Moluccas.

(3) **Pennisetum nervosum** (Nees) Trin., Acad. St. Pétersb. Mém. VI. Sci. Nat. 1: 177. 1834. Based on *Gymnothrix nervosa* Nees.

Gymnothrix nervosa Nees, Agrost. Bras. 277. 1829. Bahia, Brazil.

Cenchrus nervosus Kuntze, Rev. Gen. Pl. 3²: 347. 1898. Based on *Gymnothrix nervosa* Nees.

Pennisetum purpureum Schumach., Beskr. Guin. Pl. 64. 1827. Guinea, Africa.

(5) **Pennisetum setaceum** (Forsk.) Chiov., Soc. Bot. Ital. Bul. 1923: 113. 1923. Based on *Phalaris setacea* Forsk.

Phalaris setacea Forsk., Fl. Aegypt. Arab. 17. 1775. Arabia.

Pennisetum ruppelii Steud., Syn. Pl. Glum. 1: 107. 1854. Abyssinia.

(2) **Pennisetum setosum** (Swartz) L. Rich. in Pers., Syn. Pl. 1: 72. 1805. Based on *Cenchrus setosus* Swartz.

Cenchrus setosus Swartz, Prodr. Veg. Ind. Occ. 26. 1788. West Indies.

Panicum cenchroides L. Rich., Act. Soc. Hist. Nat. (Paris) 1: 106. 1792. French Guiana, *Leblond*.

Panicum erubescens Willd., Enum. Hort. Berol. 1031. 1809. St. Thomas.

Setaria erubescens Beauv., Ess. Agrost. 51, 169, 178. 1812. Based on *Panicum erubescens* Willd.

Pennisetum purpurascens H. B. K., Nov. Gen. et Sp. 1: 113. 1815. Jorullo, Mexico, *Humboldt* and *Bonpland*.

Pennisetum uniflorum H. B. K., Nov. Gen.

et Sp. 1: 114. pl. 34. 1815. Venezuela, *Humboldt* and *Bonpland*.

Setaria cenchroides Roem. and Schult., Syst. Veg. 2: 495. 1817. Based on *Panicum cenchroides* L. Rich.

Gymnothrix geniculata Schult., Mantissa 2: 284. 1824. Martinique, *Sieber*.

Pennisetum alopecuroides Desv. ex Hamilt., Prodr. Pl. Ind. Occ. 11. 1825. Not *P. alopecuroides* Spreng., 1825. West Indies.

Pennisetum erubescens Link, Hort. Berol. 1: 215. 1827. Based on *Panicum erubescens* Willd.

Pennisetum hirsutum Nees, Agrost. Bras. 284. 1829. Brazil [*Martius*].

Pennisetum pallidum Nees, Agrost. Bras. 285. 1829. Minas Geraes, Brazil, [*Martius*].

Pennisetum richardi Kunth, Rév. Gram. 1: 49. 1829. Based on *Panicum cenchroides* L. Rich.

Pennisetum sieberi Kunth, Rév. Gram. 1: 50. 1829. Based on *Gymnothrix geniculata* Schult.

Pennisetum flavescens Presl, Rel. Haenk. 1 316. 1830. Mexico, *Haenke*.

Pennisetum hamiltonii Steud., Nom. Bot. ed. 2. 2: 297. 1841. Based on *P. alopecuroides* Desv. ex Hamilt.

Pennisetum nicaraguense Fourn., Soc. Bot. France Bul. II. 27: 293. 1880. Granada, Nicaragua, *Levy* 1304.

Pennisetum indicum var. *purpurascens* Kuntze, Rev. Gen. Pl. 2: 787. 1891. Based on *Pennisetum purpurascens* H. B. K.

(4) **Pennisetum villosum** R. Br. in Salt, Voy. Abyss. App. 62. 1814, name only; in Fres., Mus. Senckenb. Abh. 2: 134. 1837. Abyssinia.

Cenchrus villosus Kuntze, Rev. Gen. Pl. 3²: 347. 1898. Not *C. villosus* Spreng., 1825. Based on *Pennisetum villosum* R. Br.

(118) PHALARIS L.

(6) **Phalaris angusta** Nees ex Trin., Gram. Icon. 1: pl. 78. 1827. Uruguay and southern Brazil.

Phalaris ludoviciana Torr. ex Trin., Acad. St. Pétersb. Mém. VI. Sci. Nat. 3¹: 56. 1839, as synonym of *P. angusta* Nees.

Phalaris laxa Spreng. ex Steud., Nom. Bot. ed. 2. 2: 315. 1841, as synonym of *P. angusta* Nees.

Phalaris intermedia var. *angusta* Chapm., Fl. South. U. S. 569. 1860. Based on *P. angusta* Nees.

Phalaris intermedia var. *angustata* Beal, Grasses N. Amer. 2: 182. 1896. "*P. angustata* Hort." [San Diego] Calif., *Pringle* in 1882.

(9) **Phalaris arundinacea** L., Sp. Pl. 55. 1753. Europe. (*P. arundinacea* Michx., listed in Index Kewensis, is the Linnaean species.)

Arundo colorata Ait., Hort. Kew. 1: 116. 1789. Based on *Phalaris arundinacea* L.

Typhoides arundinacea Moench, Meth. Pl. 202. 1794. Based on *Phalaris arundinacea* L.

Calamagrostis variegata With., Bot. Arr. Veg. Brit. ed. 3. 2: 124. 1796. Based on *Phalaris arundinacea* L.

Arundo riparia Salisb., Prodr. Stirp. 24. 1796. Based on *Phalaris arundinacea* L.

Baldingera colorata Gaertn., Mey., and Scherb., Fl. Wett. 1: 96. 1799. Based on *Phalaris arundinacea* L.

Digraphis arundinacea Trin., Fund. Agrost. 127. 1820. Based on *Phalaris arundinacea* L.

Baldingera arundinacea Dum., Obs. Gram. Belg. 130. pl. 10. f. 40. 1823. Based on *Phalaris arundinacea* L.

Digraphis americana Ell. ex Loud., Hort. Brit. 27. 1830. No description, *Phalaris arundinacea* Michx. cited, Loudon assuming the American form to be distinct from the European and that *Phalaris americana* Ell. was the same as the American *P. arundinacea*.

Endallex arundinacea Raf. ex Jacks., Ind. Kew. 1: 839. 1893, as synonym of *Phalaris arundinacea* L.

PHALARIS ARUNDINACEA var. PICTA L., Sp. Pl. 55. 1753. Europe.

Phalaris americana var. *picta* Eaton and Wright, N. Amer. Bot. ed. 8. 352. 1840. Massachusetts, Connecticut, New York, Ontario.

Phalaris arundinacea var. *variegata* Parnell, Grasses Brit. 188. pl. 82. 1845. Scotland.

Digraphis arundinacea var. *picta* Pacher, Jahrb. Nat. Landesmus. Kärnt. 14: 119. 1880. Presumably based on *Phalaris arundinacea* var. *picta* L.

(3) **Phalaris brachystachys** Link, Neu. Jour. Bot. Schrad. 1³: 134. 1806. Based on *P. canariensis* as described by Brotero (Fl. Lusit. 1: 79. 1804). Portugal.

Phalaris canariensis var. *brachystachys* Fedtsch., Jard. Bot. Prin. U. R. S. S. [Pierre le Grand] Bul. 14 (sup. 2): 47. 1915. Based on *P. brachystachys* Link.

(8) **Phalaris californica** Hook. and Arn., Bot. Beechey Voy. 161. 1841. California. This is the species referred to *P. amethystina* Trin., of Chile, by Thurber and others.

(2) **Phalaris canariensis** L., Sp. Pl. 54. 1753. Southern Europe and the Canary Islands.

Phalaris avicularis Salisb., Prodr. Stirp. 17. 1796. Based on *P. canariensis* L.

(5) **Phalaris caroliniana** Walt., Fl. Carol. 74. 1788. South Carolina.

Phalaris intermedia Bosc. ex Poir. in Lam., Encycl. Sup. 1: 300. 1810. Carolina, *Bosc.*

Phalaris microstachya DC., Cat. Hort.

Monsp. 131. 1813. South Carolina, *Fraser; Bosc.*

Phalaris americana Ell., Bot. S. C. and Ga. 1: 101. pl. 5. f. 4. 1816. South Carolina.

Phalaris occidentalis Nutt., Amer. Phil. Soc. Trans. (n.s.) 5: 144. 1837. Fort Smith, Ark., on the Arkansas to Red River [*Nuttall*].

Phalaris trivialis Trin., Acad. St. Pétersb. Mém. VI. Sci. Nat. 3¹: 55. 1839. Charleston, S. C., *Beyrich.*

Phalaris intermedia var. *microstachya* Vasey ex L. H. Dewey, U. S. Natl. Herb. Contrib. 2: 512. 1894. Based on *P. microstachya* DC.

(7) **Phalaris lemmoni** Vasey, U. S. Natl. Herb. Contrib. 3: 42. 1892. Santa Cruz, Calif., *Lemmon.*

(4) **Phalaris minor** Retz., Obs. Bot. 3: 8. 1783. Orient.

(1) **Phalaris paradoxa** L., Sp. Pl. ed. 2. 2: 1665. 1763. Mediterranean region.

PHALARIS PARADOXA var. PRAEMORSA (Lam.) Coss. and Dur., Expl. Sci. Alger. 2: 25. 1854. Based on *P. praemorsa* Lam.

Phalaris praemorsa Lam., Fl. Franç. 3: 566. 1778. France.

Phalaris tuberosa L., Mant. Pl. 2: 557. 1771. Europe.

PHALARIS TUBEROSA var. STENOPTERA (Hack.) Hitchc., Wash. Acad. Sci. Jour. 24: 292. 1934. Based on *P. stenoptera* Hack.

Phalaris stenoptera Hack., Repert. Sp. Nov. Fedde 5: 333. 1908. Melbourne, Australia, *Ewart*, cultivated. This species has been referred to *P. bulbosa* (see under *Phleum subulatum*).

(125) PHARUS L.

(1) **Pharus parvifolius** Nash, Torrey Bot. Club Bul. 35: 301. 1908. Haiti, *Nash* and *Taylor 1482.*

This is the species described under *Pharus latifolius* L. by Chapman.

(72) PHIPPSIA (Trin.) R. Br.

(1) **Phippsia algida** (Phipps) R. Br., Chlor. Melv. 27. 1823. Based on *Agrostis algida* Phipps.

Agrostis algida Phipps, Voy. 200. 1774. Arctic regions.

Trichodium algidum Roem. and Schult., Syst. Veg. 2: 283. 1817. Based on *Agrostis algida* Wahl. [The same as *A. algida* Phipps.]

Colpodium monandrum Trin. in Spreng., Neu. Entd. 2: 37. 1821. Based on *Agrostis algida* Phipps.

Vilfa algida Trin., Gram. Unifl. 159. 1824. Based on *Agrostis algida* Phipps.

Vilfa monandra Trin., Gram. Unifl. 159. 1824. "Sin. Laurentii" [probably St. Lawrence Island, Alaska], *Chamisso.*

Phippsia monandra Trin., Gram. Unifl.

159. 1824, as synonym of *Vilfa monandra* Trin.; Hook. and Arn., Bot. Beechey Voy. 132. 1841. Based on *Vilfa monandra* Trin.
Catabrosa algida Fries, Nov. Fl. Suec. Mant. 3: Add. 173, 174. 1843. Based on *Agrostis algida* Phipps.
Poa algida Rupr., Fl. Samoj. Cisural. 61. 1845. Not *P. algida* Trin. Based on *Agrostis algida* Phipps.

(79) PHLEUM L.

(2) **Phleum alpinum** L., Sp. Pl. 59. 1753. Europe.
Phleum haenkeanum Presl, Rel. Haenk. 1: 245. 1830. Nootka Sound, Vancouver Island, *Haenke.*
Phleum pratense var. *alpinum* Celak., Prodr. Fl. Böhm. 38. 1869. Based on *P. alpinum* L.
Phleum alpinum var. *americanum* Fourn., Mex. Pl. 2: 90. 1886. Nootka Sound, Vancouver Island, *Haenke.*
Phleum alpinum var. *scribnerianum* Pammel, Davenport Acad. Sci. Proc. 7: 238. 1899. Geranium Park, Wyo., *Pammel 6.*
Plantinia alpina Bubani, Fl. Pyr. 4: 272. 1901. Based on *Phleum alpinum* L.
Phleum arenarium L., Sp. Pl. 60. 1753. Europe.
Phleum paniculatum Huds., Fl. Angl. 23. 1762. England.
Phalaris aspera Retz., Obs. Bot. 4: 14. 1786. Europe.
Phleum asperum Jacq., Col. Bot. 1: 110. 1786. Europe.
Plantinia aspera Bubani, Nuov. Gior. Bot. Ital. 5: 317. 1873. Based on *Phleum asperum* L. (error for Jacq.)
(1) **Phleum pratense** L., Sp. Pl. 59. 1753. Europe.
Phleum nodosum var. *pratense* St. Amans, Fl. Agen. 23. 1821. Based on *P. pratense* L.
Plantinia pratensis Bubani, Fl. Pyr. 4: 270. 1901. Based on *Phleum pratense* Huds. (error for L.).
Stelephuras pratensis Lunell, Amer. Midl. Nat. 4: 216. 1915. Based on *Phleum pratense* L.
Phleum subulatum (Savi) Aschers. and Graebn., Syn. Mitteleur. Fl. 2: 154. 1899. Based on *Phalaris subulata* Savi.
Phalaris bulbosa L., Cent. Pl. 1: 4. 1755; Amoen. Acad. 4: 264. 1759. Not *Phleum bulbosum* Gouan, 1765. "In Oriente."
Phalaris subulata Savi, Fl. Pis. 1: 57. 1798. Italy.
Phalaris bellardi Willd., Ges. Naturf. Freund. Berlin Neue Schrift. 3: 415. 1801. Europe.
Phalaris tenuis Host, Gram. Austr. 2: 27. pl. 36. 1802. Europe.
Phleum tenue Schrad., Fl. Germ. 1: 191. 1806. Based on *Phalaris tenue* Host.
Phleum bellardi Willd., Enum. Pl. 1: 85.

1809. Based on *Phalaris bellardi* Willd. *Phleum bulbosum* Richt., Pl. Eur. 1: 37. 1890. Not *P. bulbosum* Gouan, 1765. Based on *Phalaris bulbosa* L.

(28) PHRAGMITES Trin.

(1) **Phragmites communis** Trin., Fund. Agrost. 134. 1820. Based on *Arundo phragmites* L.
Arundo phragmites L., Sp. Pl. 81. 1753. Europe.
Arundo vulgaris Lam., Fl. Franç. 3: 615. 1778. Based on *A. phragmites* L. The name is untenable.
Arundo palustris Salisb., Prodr. Stirp. 24. 1796. Based on *A. phragmites* L.
Reimaria diffusa Spreng., Neu. Entd. 3: 14. 1822. Martinique, *Sieber* [31].
Cynodon phragmites Raspail, Ann. Sci. Nat., Bot. 5: 302. 1825. Based on *Arundo phragmites* L.
Phragmites vulgaris Crép., Man. Fl. Belg. ed. 2. 345. 1866. Based on *Arundo vulgaris* Lam., an untenable name.
Phragmites berlandieri Fourn., Soc. Bot. France Bul. 24: 178. 1877. Laredo, Tex., *Berlandier 1446.*
Phragmites phragmites Karst., Deut. Fl. 379. 1883. Based on *Arundo phragmites* L.
Trichoon phragmites Rendle, Cat. Afr. Pl. Welw. 2¹: 218. 1899. Based on *Arundo phragmites* L.
Oxyanthe phragmites Nieuwl., Amer. Midl. Nat. 3: 332. 1914. Based on *Arundo phragmites* L.
Phragmites communis var. *berlandieri* Fernald, Rhodora 34: 211. 1925. Based on *P. berlandieri* Fourn.
Phragmites maximus var. *berlandieri* Moldenke, Torreya 36: 93. 1936. Based on *P. berlandieri* Fourn.

PHYLLOSTACHYS Sieb. and Zucc.

Phyllostachys aurea A. and C. Riviere, Soc. Acclim. Bul. III. 5: 716. f. 36. 1878. Hamma, Tunis.
Phyllostachys bambusoides Sieb. and Zucc., Abh. Bayer. Akad. Wiss. 3³: 746. pl. 5. f. 3. 1843. Japan.

(90) PIPTOCHAETIUM Presl

(1) **Piptochaetium fimbriatum** (H. B. K.) Hitchc., Wash. Acad. Sci. Jour. 23: 453. 1933. Based on *Stipa fimbriata* H. B. K.
Stipa fimbriata H. B. K., Nov. Gen. et Sp. 1: 126. 1815. Guanajuato, Mexico, *Humboldt* and *Bonpland.*
Milium mexicanum Spreng., Syst. Veg. 1: 251. 1825. Mexico, *Humboldt.*
Piptatherum mexicanum Schult., Mantissa 3 (Add. 1): 564. 1827. Based on *Milium mexicanum* Spreng.
Avena stipoides Willd. ex Steud., Nom. Bot. ed. 2. 2: 146. 1841, as synonym of *Milium mexicanum* Spreng.

Oryzopsis fimbriata Hemsl., Biol. Centr. Amer. Bot. 3: 538. 1885. Based on *Stipa fimbriata* H. B. K.

Oryzopsis seleri Pilger, Verh. Bot. Ver. Brand. 51: 192. 1909. Guatemala, *Seler* 3238.

Piptochaetium fimbriatum var. *confine* I. M. Johnston, Arnold Arb. Jour. 24: 396. 1943. Coahuila, Mexico, *Johnston* and *Muller* 486.

(10) PLEUROPOGON R. Br.

(1) **Pleuropogon californicus** (Nees) Benth. ex Vasey, Grasses U. S. 40. 1883. Based on *Lophochlaena californica* Nees.

Lophochlaena californica Nees, Ann. Nat. Hist. 1: 283. 1838. California [*Douglas*].

Pleuropogon douglasii Trin. ex Steud., Nom. Bot. ed. 2. 2: 355. 1841. Name only, North America.

Lepitoma brevifolia Torr. ex Steud., Nom. Bot. ed. 2. 2: 355. 1841, as synonym of *Pleuropogon douglasii* Trin.

(4) **Pleuropogon davyi** Benson, Amer. Jour. Bot. 28: 360. 1941. Kelseyville, Calif., *Benson* 3666.

(3) **Pleuropogon hooverianus** (Benson) J. T. Howell, West. Bot. Leaflets 4: 247. 1946. Based on *P. refractus* var. *hooverianus* Benson.

Pleuropogon refractus var. *hooverianus* Benson, Amer. Jour. Bot. 28: 360. 1941. Mendocino County, Calif., *Davy* 6626.

(5) **Pleuropogon oregonus** Chase, Wash. Acad. Sci. Jour. 28: 52. f. 1. 1938. Union, Oreg., *Leckenby* in 1901.

(2) **Pleuropogon refractus** (A. Gray) Benth. ex Vasey, Grasses U. S. 40. 1883. Based on *Lophochlaena refracta* A. Gray.

Lophochlaena refracta A. Gray, Amer. Acad. Sci. Proc. 8: 409. 1872. Oregon [*Hall* 636].

(12) POA L.

(45) **Poa alpina** L., Sp. Pl. 67. 1753. Lapland.

Uralepis mutica Fourn., Mex. Pl. 2: 110. 1886. Not *U. mutica* Fourn. ex Hemsl. 1885. Mexico, *Liebmann* 611.

Poa alpina var. *minor* Scribn. in Beal, Grasses N. Amer. 2: 543. 1896. Not *P. alpina* var. *minor* Koch, 1837. Montana, *Scribner* [388] in 1883.

(25) **Poa alsodes** A. Gray, Man. ed. 2. 562. 1856. New England to Wisconsin. [Type, New Hampshire.]

Poa dinantha Wood, Class-book ed. 1861. 797. 1861. Montgomery, Ala. This species was described as *Poa nemoralis* L., in Torr., Fl. North. and Mid. U. S. 1: 111. 1823.

(69) **Poa ampla** Merr., Rhodora 4: 145. 1902. Steptoe, Wash., *G. R. Vasey* 3009.

Poa laeviculmis Williams, Bot. Gaz. 36: 55. 1903. Several specimens from Washington and Oregon mentioned, the first being Steptoe, Wash., *G. R. Vasey* 3026.

Poa truncata Rydb., Torrey Bot. Club Bul. 32: 607. 1905. Dillon, Colo., *Clements* 373.

Poa confusa Rydb., Torrey Bot. Club Bul. 32: 607. 1905. Medicine Bow Mountains, Wyo., *Nelson* 7787.

(5) **Poa annua** L., Sp. Pl. 68. 1753. Europe.

Aira pumila Pursh, Fl. Amer. Sept. 1: 76. 1814. Pennsylvania.

Poa infirma H. B. K., Nov. Gen. et Sp. 1: 158. 1816. Colombia, *Humboldt* and *Bonpland*.

Megastachya infirma Roem. and Schult., Syst. Veg. 2: 585. 1817. Based on *Poa infirma* H. B. K.

Catabrosa pumila Roem. and Schult., Syst. Veg. 2: 696. 1817. Based on *Aira pumila* Pursh.

Poa aestivalis Presl, Rel. Haenk. 1: 272. 1830. Peru, *Haenke*.

Eragrostis infirma Steud., Nom. Bot. ed. 2. 1: 563. 1840. Based on *Poa infirma* H. B. K.

Poa annua var. *aquatica* Aschers., Fl. Prov. Brandenb. 1: 844. 1864. Germany.

Poa annua var. *rigidiuscula* L. H. Dewey, U. S. Natl. Herb. Contrib. 3: 262. 1895. Nez Perce County, Idaho, *Sandberg* 134.

(7) **Poa arachnifera** Torr. in Marcy, Expl. Red Riv. 301. 1853. Headwaters of the Trinity River [Ark., *Marcy Exped.*].

Poa densiflora Buckl., Acad. Nat. Sci. Phila. Proc. 1862: 96. 1862. Northern Texas.

Poa arachnifera var. *glabrata* Vasey, Grasses U. S. Descr. Cat. 79. 1885, name only, [for staminate plants with glabrous spikelets]; Vasey ex Beal, Grasses N. Amer. 2: 535. 1896. [Texas, *Buckley* in 1881.]

Poa glabrescens Nash in Small, Fl. Southeast. U. S. 154, 1327. 1903. Based on *P. arachnifera* var. *glabrata* Torr. (error for Vasey).

(22) **Poa arctica** R. Br., Sup. App. Parry's Voy. 288 (err. typ. 188). 1823. Melville Island, Arctic America, *Parry*.

Poa grayana Vasey, U. S. Natl. Herb. Contrib. 1: 272. 1893. Grays Peak, Colo., *Patterson* 14 in 1885.

Poa laxa occidentalis Vasey ex Rydb. and Shear, U. S. Dept. Agr., Div. Agrost. Bul. 5: 32. 1897. Name only, for *Shear* 690 and *Rydberg* 2440, Grays Peak, Colo.

Poa longipila Nash in Rydb., N. Y. Bot. Gard. Mem. 1: 46. 1900. Electric Peak, Yellowstone Park, *Rydberg* 3614.

Poa alpicola Nash in Rydb., N. Y. Bot. Gard. Mem. 1: 47. 1900. Based on *P.*

laxa Haenke as misapplied by Thurber (in Watson, Bot. Calif. 2: 312. 1880).
Poa williamsii Nash, N. Y. Bot. Gard. Bul. 2: 156. 1901. White Pass, Alaska, *Williams* in 1899.
Poa aperta Scribn. and Merr., U. S. Dept. Agr., Div. Agrost. Cir. 35: 4. 1901. Telluride, Colo., *Shear* 98.
Poa callichroa Rydb., Torrey Bot. Club Bul. 32: 603. 1905. Dead Lake, near Pikes Peak, Colo., *Clements* 457.
Poa phoenicea Rydb., Torrey Bot. Club Bul. 32: 605. 1905. Pikes Peak Valley, Colo., *Clements* 466.
Poa tricholepis Rydb., Torrey Bot. Club Bul. 32: 606. 1905. Pagosa Peak, Colo., *Baker* 210.
Poa chionogenes Gandog., Soc. Bot. France Bul. 67: 302. 1920. Grays Peak, Colo., *Crandall* [in 1898].
Colorado specimens of this species have been described as *Poa cenisia* All. by American authors.

(20) **Poa arida** Vasey, U. S. Natl. Herb. Contrib. 1: 270. 1893. Socorro, N. Mex., *G. R. Vasey* in 1881.
Poa andina Nutt. ex S. Wats. in King, Geol. Expl. 40th Par. 388. 1871. Not *P. andina* Trin., 1836. "Colorado, East and West Humboldt Mountains and in Clover Mountains, Nevada; also in the Trinity Mountains, *Watson* 1319." The name is given as *P. andina* "Nutt., Ms. in Herb.; (not of Trin.)."
Poa californica Munro ex Coulter, Man. Rocky Mount. 420. 1885. Not *P. californica* Steud., 1854. Based on *P. andina* Nutt.
Poa andina var. *purpurea* Vasey ex Macoun, Can. Pl. Cat. 2⁴: 223. 1888. Name only, for *Macoun* 92, Red Deer Lakes, Alberta.
Poa sheldoni Vasey, U. S. Natl. Herb. Contrib. 1: 276. 1893. Buena Vista, Colo., *Sheldon* 615.
Poa pseudopratensis Scribn. and Rydb., U. S. Natl. Herb. Contrib. 3: 531. pl. 20. 1896. Not *P. pseudopratensis* Beyer, 1819. Hot Springs, S. Dak., *Rydberg* 1151.
Poa pratericola Rydb. and Nash, N. Y. Bot. Gard. Mem. 1: 51. 1900. Based on *P. arida* Vasey.
Poa fendleriana var. *arida* Jones, West. Bot. Contrib. 14: 14. 1912. Based on *P. arida* Vasey, not *P. pratensis* var. *arida* Parnell, 1842.
Poa pratensis var. *pseudopratensis* Jones, West. Bot. Contrib. 14: 15. 1912. Based on *P. pseudopratensis* Scribn. and Rydb.
Paneion aridum Lunell, Amer. Midl. Nat. 4: 222. 1915. Based on *Poa arida* Vasey.
Paneion pratericola Lunell, Amer. Midl. Nat. 4: 223. 1915. Based on *Poa pratericola* Rydb. and Nash.

Poa pratensiformis Rydb., Fl. Rocky Mount. 79. 1917. Based on *P. pseudopratensis* Scribn. and Rydb.
Poa overi Rydb., Brittonia 1: 84. 1931. Custer County, S. Dak., *Over* 18100.

(13) **Poa atropurpurea** Scribn., U. S. Dept. Agr., Div. Agrost. Bul. 11: 53. pl. 10. 1898. Bear Valley, San Bernardino Mountains, Calif., *Parish* 2968.

(44) **Poa autumnalis** Muhl. ex Ell., Bot. S. C. and Ga. 1: 159. 1816. Columbia, S. C., *Herbemont*.
Poa flexuosa Muhl., Descr. Gram. 148. 1817. Not *P. flexuosa* Smith, 1800. Virginia, Carolina, and Cherokee [Tennessee]. Name only, Muhl. Cat. Pl. 11. 1813.
Poa campyle Schult., Mantissa 2: 304. 1824 Based on *P. flexuosa* Muhl.
Poa elliottii Spreng., Syst. Veg. 1: 338. 1825. Based on *P. autumnalis* Muhl.
Poa vestita Bosc. ex Steud., Nom. Bot. ed. 2. 2: 363. 1841. Name only. Carolina.
Poa hexantha Wood, Class-book ed. 1861. 797. 1861. Atlanta, Ga.

(3) **Poa bigelovii** Vasey and Scribn. in Vasey, Grasses U. S. Descr. Cat. 81. 1885. Based on *P. annua* var. *stricta* Vasey.
Poa annua var. *stricta* Vasey, Torrey Bot. Club Bul. 10: 31. 1883. Rillita River, Ariz., *Pringle*.

(1) **Poa bolanderi** Vasey, Bot. Gaz. 7: 32. 1882. [Yosemite National Park] Calif., *Bolander* 6115.
Poa howellii chandleri Davy, Calif. Univ. Pubs., Bot. 1: 60. 1902. Siskiyou County, Calif., *Chandler* 1703.
Poa bolanderi chandleri Piper, U. S. Natl. Herb. Contrib. 11: 132. 1906. Based on *P. howellii chandleri* Davy.
Poa horneri St. John, Fl. Southeast. Wash. and Adj. Idaho 54. 1937. Columbia County, Wash., *Darlington* in 1913.

(36) **Poa bulbosa** L., Sp. Pl. 70. 1753. France.
Poa bulbosa var. *vivipara* Koel., Descr. Gram. 189. 1802. Europe.
Paneion bulbosum var. *viviparum* Lunell, Amer. Midl. Nat. 4: 222. 1915. Based on *Poa bulbosum* var. *vivipara* Koch (same as Koel.).

(65) **Poa canbyi** (Scribn.) Piper, U. S. Natl. Herb. Contrib. 11: 132. 1906. Based on *Glyceria canbyi* Scribn.
Aira brevifolia Pursh, Fl. Amer. Sept. 1: 76, 1814. Not *Poa brevifolia* DC., 1806. Plains of the Missouri, *Lewis*.
Airopsis brevifolia Roem. and Schult., Syst. Veg. 2: 578. 1817. Based on *Aira brevifolia* Pursh.
Poa tenuifolia Nutt. in S. Wats. in King, Geol. Expl. 40th Par. 5: 387. 1871. Not. *P. tenuifolia* L. Rich., 1851. Nevada, *Watson* 1318.
Poa tenuifolia var. *rigida* Vasey in Rothr., in Wheeler, U. S. Survey W. 100th

Merid. Rpt. 6: 290. 1878. Name only. Nevada and Colorado [*Wolf*] 1138, 1140.

Poa tenuifolia var. *elongata* Vasey in Rothr., in Wheeler, U. S. Survey W. 100th Merid. Rpt. 6: 290. 1878. Nevada, Colorado [Twin Lakes, *Wolf*] 1141.

Glyceria canbyi Scribn., Torrey Bot. Club Bul. 10: 77. 1883. Cascade Mountains, Wash., *Tweedy* and *Brandegee* in 1882.

Aira missurica Spreng. ex Jacks., Ind. Kew. 1: 68. 1893, as synonym of *A. brevifolia* Pursh, erroneously credited to "Spreng. Syst. 2: 578." *Aira brevifolia* Pursh is given in Spreng., Syst. 1: 276. 1825.

Poa laevis Vasey, U. S. Natl. Herb. Contrib., 1: 273. 1893. Not *P. laevis* R. Br., 1810. Montana, *Scribner* in 1883.

Poa lucida Vasey, U. S. Natl. Herb. Contrib., 1: 274. 1893. Georgetown, Colo., *Patterson 73*.

Atropis laevis Beal, Grasses N. Amer. 2: 577. 1896. Based on *Poa laevis* Vasey.

Atropis laevis var. *rigida* Beal, Grasses N. Amer. 2: 578. 1896. Utah, *Jones*.

Atropis canbyi Beal, Grasses N. Amer. 2: 580. 1896. Based on *Glyceria canbyi* Scribn.

Poa laevigata Scribn., U. S. Dept. Agr., Div. Agrost. Bul. 5: 31. 1897. Based on *P. laevis* Vasey.

Poa wyomingensis Scribn., Davenport Acad. Sci. Proc. 7: 242. 1899. Big Horn, Sheridan County, Wyo., *Pammel* 192.

Poa leckenbyi Scribn., U. S. Dept. Agr., Div. Agrost. Cir. 9: 2. 1899. Scott, Klickitat County, Wash., *Leckenby* in 1898.

Poa helleri Rydb., Torrey Bot. Club Bul. 36: 534. 1909. Lake Waha, Idaho, *Heller 3274*.

Poa buckleyana var. *elongata* Jones, West. Bot. Contrib. 14: 14. 1912. Based on "*P. andina* var. *elongata* Vasey," error for *P. tenuifolia* var. *elongata* Vasey.

Poa nevadensis var. *laevigata* Jones, West. Bot. Contrib. 14: 14. 1912. Based on *P. laevigata* Scribn.

Poa nevadensis var. *leckenbyi* Jones, West. Bot. Contrib. 14: 14. 1912. Based on *P. leckenbyi* Scribn.

(41) **Poa chaixii** Vill., Fl. Delph. 7. 1785. Dauphiné, France.

(4) **Poa chapmaniana** Scribn., Torrey Bot. Club Bul. 21: 38. 1894. Knoxville, Tenn., *Scribner*. "*P. cristata* Chapm. not Walter" cited as synonym, but what Chapman described as *Poa cristata* Walt. is dubious. Scribner's description is ample, and the type is in the National Herbarium.

(6) **Poa compressa** L., Sp. Pl. 69. 1753. Europe and North America.

Poa compressa var. *sylvestris* Torr., Fl.

North. and Mid. U. S. 1: 110. 1823. New York.

Poa compressa forma *depauperata* Millsp., Fl. W. Va. 472. 1892. Monongalia, along Falling Run, W. Va.

Paneion compressum Lunell, Amer. Midl. Nat. 4: 222. 1915. Based on *Poa compressa* L.

(10) **Poa confinis** Vasey, U. S. Dept. Agr., Div. Bot. Bul. 13²: pl. 75. 1893. Oregon to Alaska [type Tillamook Bay, Oreg., *Howell* 69 in 1882].

(14) **Poa curta** Rydb., Torrey Bot. Club Bul. 36: 534. 1909. Spread Creek [Jackson Hole], Wyo., *Tweedy 13*.

(67) **Poa curtifolia** Scribn., U. S. Dept. Agr., Div. Agrost. Cir. 16: 3. 1899. Mount Stuart, Wash., *Elmer* 1148 [type] and 1150.

(53) **Poa cusickii** Vasey, U. S. Natl. Herb. Contrib., 1: 271. 1893. Oregon, *Cusick* 1219.

Poa filifolia Vasey, U. S. Natl. Herb. Contrib. 1: 271. 1893. Hatwai Creek, Nez Perce County, Idaho, *Sandberg* 138.

Poa idahoensis Beal, Grasses N. Amer. 2: 539. 1896. Based on *P. filifolia* Vasey, not *P. filifolia* Schur, that name, however, published as synonym only.

Poa subaristata Scribn. in Beal, Grasses N. Amer. 2: 533. 1896. Not *P. subaristata* Phil., 1896 [earlier than *P. subaristata* Scribn.]. Yellowstone Park, *Tweedy* 633.

Poa scabrifolia Heller, Torrey Bot. Club Bul. 24: 310. 1897. Based on *P. filifolia* Vasey.

Poa spillmani Piper, Erythea 7: 102. 1899. Douglas County, Wash., *Spillman* in 1896.

Poa capillarifolia Scribn. and Williams, U. S. Dept. Agr., Div. Agrost. Cir. 9: 1. 1899. California, *Hansen* 2614.

Poa cottoni Piper, Biol. Soc. Wash. Proc. 18: 146. 1905. Rattlesnake Mountains, Yakima County, Wash., *Cotton* 557.

Poa nematophylla Rydb., Torrey Bot. Club Bul. 32: 606. 1905. Meeker, Colo., *Osterhout* 2601.

Poa scaberrima Rydb., Torrey Bot. Club Bul. 36: 534. 1909. Beaver Canyon, Idaho, *Rydberg* 2055.

(19) **Poa cuspidata** Nutt. in Barton, Compend. Fl. Phila. 1: 61. 1818. Based on *P. pungens* Nutt.

Aira triflora Ell., Bot. S. C. and Ga. 1: 153. 1816. Not *Poa triflora* Gilib., 1792. Athens, Ga., *Green*.

Poa brevifolia Muhl., Descr. Gram. 138. 1817. Not *P. brevifolia* DC., 1806. Pennsylvania. Name only, Muhl., Cat. Pl. 11. 1813.

Poa trinervata Willd. ex Muhl., Descr. Gram. 138. 1817, as synonym of *Poa brevifolia* Muhl.

Poa pungens Nutt., Gen. Pl. 1: 66. 1818.

Not *P. pungens* Georgi, 1800, nor Bieb., 1808. Near Philadelphia.
Poa brachyphylla Schult., Mantissa 2: 304. 1824. Based on *P. brevifolia* Muhl.
Triodia greenii Spreng., Syst. Veg. 1: 330. 1825. Based on *Aira triflora* Ell.
Graphephorum elliottii Kunth, Rév. Gram. 1: 80. 1829. Based on *Aira triflora* Ell.
Graphephorum melicoides var. *triflorum* Wood, Amer. Bot. and Flor. pt. 2: 398. 1871. Based on *Aira triflora* Ell.
(9) **Poa douglasii** Nees, Ann. Nat. Hist. 1: 284. 1838. California, *Douglas*.
Brizopyrum douglasii Hook. and Arn., Bot. Beechey Voy. Suppl. 404. 1840. Based on *Poa douglasii* Nees.
Poa californica Steud., Syn. Pl. Glum. 1: 261. 1854. California.
(56) **Poa epilis** Scribn., U. S. Dept. Agr., Div. Agrost. Cir. 9: 5. 1899. Buffalo Pass, Colo., *Shear* and *Bessey* 1457.
Poa purpurascens Vasey, Bot. Gaz. 6: 297. 1881. Not *P. purpurascens* Spreng., 1819. Mount Hood, *Howell* [in 1881].
Poa alpina var. *purpurascens* Beal, Grasses N. Amer. 2: 543. 1896. Based on *P. purpurascens* Vasey.
Poa paddensis Williams, U. S. Dept. Agr., Div. Agrost. Bul. 17 (ed. 2): 261 f. 557. 1901. Based on *P. purpurascens* Vasey.
Poa subpurpurea Rydb., Torrey Bot. Club Bul. 32: 606. 1905. Based on *P. purpurascens* Vasey.
Poa purpurascens var. *epilis* Jones, West. Bot. Contrib. 14: 14. 1912. Based on *P. epilis* Scribn.
(42) **Poa fendleriana** (Steud.) Vasey, U. S. Dept. Agr., Div. Bot. Bul. 13²: pl. 74. 1893. Based on *Eragrostis fendleriana* Steud.
Eragrostis fendleriana Steud., Syn. Pl. Glum. 1: 278. 1854. "Mexico" [now New Mexico], *Fendler* 932.
Uralepis poaeoides Buckl., Acad. Nat. Sci. Phila. Proc. 1862: 94. 1862. New Mexico, *Fendler* 932.
Atropis californica Munro ex A. Gray, Acad. Nat. Sci. Phila. Proc. 1862: 336. 1862. California, *Douglas* in 1833.
Poa eatoni S. Wats. in King, Geol. Expl. 40th Par. 5: 386. 1871. Wasatch Mountains, Utah, *Eaton* [in 1869].
Poa andina var. *major* Vasey in Rothr., in Wheeler, U. S. Survey W. 100th Merid. Rpt. 6: 290. 1878. Arizona; Colorado.
Poa andina var. *spicata* Vasey in Rothr., in Wheeler, U. S. Survey W. 100th Merid. Rpt. 6: 290. 1878. Colorado, [*Wolf*] 1135.
Atropis californica Munro ex Thurb. in S. Wats., Bot. Calif. 2: 309. 1880. Near San Francisco, *Bolander;* Monterey, *Hartweg.*
Poa californica Scribn., Torrey Bot. Club

Bul. 10: 31. 1883. Not *P. californica* Steud.; 1854. Based on *Atropis californica* Munro.
Panicularia fendleriana Kuntze, Rev. Gen. Pl. 2: 782. 1891. Based on *Eragrostis fendleriana* Steud.
Atropis fendleriana Beal, Grasses N. Amer. 2: 576. 1896. Based on *Eragrostis fendleriana* Steud.
Poa fendleriana spicata Scribn., U. S. Dept. Agr., Div. Agrost. Bul. 5: 31. 1897. Based on *P. arida* var. *spicata* Vasey, error for *P. andina* var. *spicata* Vasey.
Poa longepedunculata Scribn., U. S. Dept. Agr., Div. Agrost. Bul. 11: 54. pl. 11. 1898. Laramie, Wyo., *Nelson* [3292].
Poa brevipaniculata Scribn. and Williams, U. S. Dept. Agr., Div. Agrost. Cir. 9: 2. 1899. Table Rock, Colo., *Breninger* 554.
Poa scabriuscula Williams, U. S. Dept. Agr., Div. Agrost. Cir. 10: 4. 1899. Glenwood, Utah, *Ward* 136.
Poa longepedunculata viridescens Williams, U. S. Dept. Agr., Div. Agrost. Cir. 10: 4. 1899. Sheep Mountain [near Laramie], Wyo., *Williams* 2302.
Poa brevipaniculata subpallida Williams, U. S. Dept. Agr., Div. Agrost. Cir. 10: 5. 1899. Rocky Mountains, Colo., *Hall* and *Harbour* 674 in part.
Poa fendleriana arizonica Williams, U. S. Dept. Agr., Div. Agrost. Cir. 10: 5. 1899. Yavapai Creek, Ariz., *Rusby* in 1883.
(49) **Poa fernaldiana** Nannf., Symb. Bot. Upsal. 5: 50, 55. f. 6. pl. 4. 1935. Mount Washington, N. H., *Williams* and *Robinson* (Pl. exs. *Grayanae* No. 123).
Poa laxa var. *debilior* Jones, West. Bot. Contrib. 14: 15. 1912. "The eastern plant," no particular locality or specimen cited.
Described as *Poa laxa* Haenke in the Manual, ed. 1. That species is not known from America.
(12) **Poa fibrata** Swallen, Wash. Acad. Sci. Jour. 30: 210. 1940. Grenada, Calif., *Wheeler* 3629.
(47) **Poa glauca** Vahl, Fl. Dan. pl. 964. 1790. Norway.
Poa caesia J. E. Smith, Fl. Brit. 1: 103. 1800. England.
Poa nemoralis var. *glauca* Gaud., Agrost. Helv. 1: 182. 1811. Based on *P. glauca* Vahl.
Poa glauca var. *caesia* Hartm. Handb. Skand. Fl. ed. 1. 57. 1820. Based on *P. caesia* J. E. Smith.
Paneion glaucum Lunell, Amer. Midl. Nat. 4: 222. 1915. Based on *Poa glauca* Vahl.
(48) **Poa glaucantha** Gaudin, Alpina 3: 36. 1808. Switzerland. (Published as *P. glaucanthos*.)
Poa nemoralis var. *glaucantha* Reichenb.,

Fl. Germ. 1: 47. 1830. Based on *P. glaucantha* Gaudin. (Referred to *P. nemoralis* L., as a variety but combination not made.) Reichenb., Icon. 1: pl. 86. f. 1644. 1834.

Poa glauca subsp. *glaucantha* Lindm., Bot. Not. 1926: 275. 1926. Based on *P. glaucantha* Gaudin.

Poa tormentosa Butters and Abbe, Rhodora 49: 14. pl. 1052. f. 7–9. 1947. Minnesota, *Butters, Burns,* and *Hendrickson* 2.

Poa scopulorum Butters and Abbe, Rhodora 49: 16. pl. 1051. f. 1–8. 1947. Cook County, Minn., *Butters* and *Abbe* 97.

(21) **Poa glaucifolia** Scribn. and Williams, U. S. Dept. Agr., Div. Agrost. Cir. 10: 6. 1899. Based on *P. planifolia* Scribn. and Williams.

Poa planifolia Scribn. and Williams, U. S. Dept. Agr., Div. Agrost. Cir. 9: 3. 1899. Not *P. planifolia* Kuntze, 1898. Spring Creek, Big Horn Basin, Wyo., *Williams* 2814.

Poa plattensis Rydb., Brittonia 1: 84. 1931. Lawrence Fork, Nebr., *Rydberg* 461.

(63) **Poa gracillima** Vasey, U. S. Natl. Herb. Contrib. 1: 272. 1893. Mount Adams, Wash., *Suksdorf* 33.

Sporobolus bolanderi Vasey, Bot. Gaz. 11: 337. 1886. Not *Poa bolanderi* Vasey, 1882. Multnomah Falls, Oreg., *Bolander.* [The type an overmature specimen from which all but the lowermost floret had fallen from the spikelets.]

Atropis tenuifolia var. *stenophylla* Vasey ex Beal, Grasses N. Amer. 2: 580. 1896. [Roseburg], Oreg., *Howell* in 1887.

Poa saxatilis Scribn. and Williams, U. S. Dept. Agr., Div. Agrost. Cir. 9: 1. 1899. Mount Rainier, Wash., *Piper* 1964.

Poa tenerrima Scribn., U. S. Dept. Agr., Div. Agrost. Cir. 9: 4. 1899. California.

Poa invaginata Scribn. and Williams, U. S. Dept. Agr., Div. Agrost. Cir. 9: 6. 1899. Summit Camp, Sierra Nevada, Calif.

Poa multnomae Piper, Torrey Bot. Club Bul. 32: 435. 1905. Multnomah Falls, Oreg., *Piper* 6459.

Poa alcea Piper, Torrey Bot. Club Bul. 32: 436. 1905. Portland, Oreg., *Piper* 6463.

Poa buckleyana var. *stenophylla* Jones, West. Bot. Contrib. 14: 14. 1912. Based on *Atropis tenuifolia* var. *stenophylla* Vasey.

Poa gracillima var. *saxatilis* Hack., Allg. Bot. Ztschr. 21: 79. 1915. Based on *P. saxatilis* Scribn. and Williams.

Poa englishii St. John and Hardin, Mazama 11: 64. 1929. Mount Baker National Forest, *Hardin* and *English* 1391.

(2) **Poa howellii** Vasey and Scribn., U. S. Dept. Agr., Div. Bot. Bul. 13²: pl. 78. 1893. California to Oregon. [Portland, *Howell* 25 in 1881, type.]

Poa howellii var. *microsperma* Vasey, U. S. Natl. Herb. Contrib. 1: 273. 1893. Santa Cruz, Calif., *Anderson* 99.

Poa bolanderi var. *howellii* Jones, West. Bot. Contrib. 14: 15. 1912. Based on *P. howellii* Vasey and Scribn.

(40) **Poa interior** Rydb., Torrey Bot. Club Bul. 32: 604. 1905. Headwaters of Clear Creek and Crazy Woman River, Wyo., *Tweedy* 3706.

Poa coloradensis Vasey ex Pammel, U. S. Dept. Agr., Div. Agrost. Bul. 9: 41. 1897. Name only, for a specimen collected by Pammel in Colorado in 1895–96.

Poa subtrivialis Rydb., Torrey Bot. Club Bul. 36: 536. 1909. Big Horn Mountains, Wyo., *Tweedy* 2141.

Paneion interius Lunell, Amer. Midl. Nat. 4: 222. 1915. Based on *Poa interior* Rydb.

Poa nemoralis var. *interior* Butters and Abbe, Rhodora 49: 6. 1947. Based on *P. interior* Rydb.

(52) **Poa involuta** Hitchc., Biol. Soc. Wash. Proc. 41: 159. 1928. Chisos Mountains, Brewster County, Tex., *Ferris* and *Duncan* 2811.

(68) **Poa juncifolia** Scribn., U. S. Dept. Agr., Div. Agrost. Bul. 11: 52. pl. 8. 1898. Point of Rocks, Sweetwater County, Wyo., *Nelson* 3721.

Poa brachyglossa Piper, Biol. Soc. Wash. Proc. 18: 145. 1905. Douglas County, Wash., *Sandberg* and *Leiberg* 267.

Poa fendleriana var. *juncifolia* Jones, West. Bot. Contrib. 14: 14. 1912. Based on *P. juncifolia* Scribn.

(16) **Poa kelloggii** Vasey, U. S. Dept. Agr., Div. Bot. Bul. 13²: pl. 79. 1893. [Mendocino County], Calif., *Bolander* 4705.

Poa bolanderi var. *kelloggii* Jones, West. Bot. Contrib. 14: 15. 1912. Based on *P. kelloggii* Vasey.

(26) **Poa languida** Hitchc., Biol. Soc. Wash. Proc. 41: 158. 1928. Based on *P. debilis* Torr.

Poa debilis Torr., Fl. N. Y. 2: 459. 1843. Not *P. debilis* Thuill., 1799. [Gorham], New York.

(17) **Poa laxiflora** Buckl., Acad. Nat. Sci., Phila. Proc. 1862. 96. 1862. Columbia Woods, Oreg., *Nuttall.*

Poa leptocoma elatior Scribn. and Merr., U. S. Natl. Herb. Contrib. 13: 71. 1910. Cape Fox, Alaska, *Trelease* and *Saunders* 2982.

Poa remissa Hitchc., Biol. Soc. Wash. Proc. 41: 158. 1928. Sol Duc Hot Springs, Olympic Mountains, Wash., *Hitchcock* 23468.

(61) **Poa leibergii** Scribn , U. S. Dept. Agr.,

Div. Agrost. Bul. 8: 6. pl. 2. 1897.
Owyhee-Malheur Divide, Oreg., *Leiberg*
2171.
Poa hanseni Scribn., U. S. Dept. Agr.,
Div. Agrost. Bul. 11: 53. pl. 9. 1898.
Silver Lake, Amador County, Calif.,
Hansen 605.
Poa pringlei var. *hanseni* Smiley, Calif.
Univ. Pubs., Bot. 9: 104. 1921. Based
on *P. hanseni* Scribn.

(34) **Poa leptocoma** Trin., Acad. St. Pét-
ersb. Mém. VI. Math. Phys. Nat. 1:
374. 1830. Sitka, Alaska, *Mertens*.
Poa stenantha var. *leptocoma* Griseb. in
Ledeb., Fl. Ross. 4: 373. 1853. Based
on *P. leptocoma* Trin.
Poa crandallii Gandog., Soc. Bot. France
Bul. 66[7]: 301. 1920. Mountains of
Larimer, Colo., *Crandall* in 1898.

(59) **Poa lettermani** Vasey, U. S. Natl.
Herb. Contrib. 1: 273. 1893. Grays
Peak, Colo., *Letterman 7*.
Poa brandegei Scribn. in Beal, Grasses N.
Amer. 2: 544. 1896. Grays Peak, Colo.
Jones 714.
Atropis lettermani Beal, Grasses N. Amer.
2: 579. 1896. Based on *Poa lettermani*
Vasey.

(43) **Poa longiligula** Scribn. and Williams,
U. S. Dept. Agr., Div. Agrost. Cir. 9:
3. 1899. Silver Reef, Utah, *Jones* 5149.
Poa montana Vasey, U. S. Dept. Agr.,
Monthly Rpt. 155. 1874. Not *P.
montana* All., 1785. Nevada, *Watson*
1312.
Poa longiligula var. *wyomingensis* Wil-
liams, U. S. Dept. Agr., Div. Agrost.
Cir. 10: 3. 1899. Tipton, Wyo., *Nelson*
4799a.
Paneion longiligulum Lunell, Amer. Midl.
Nat. 4: 222. 1915. Based on *Poa
longiligula* Scribn. and Williams.
Poa fendleriana var. *longiligula* Gould,
Madroño 10: 94. 1949. Based on *P.
longiligula* Scribn. and Williams.
This species was referred to *Poa alpina* L.
by Watson, in King, Geol. Expl. 40th Par. 5:
386. 1871.

(8) **Poa macrantha** Vasey, Torrey Bot. Club
Bul. 15: 11. 1888. Mouth of Columbia
River, Oreg., *Howell* [in 1887].
Melica macrantha Beal, Torrey Bot. Club
Bul. 17: 153. 1890. Based on *Poa
macrantha* Vasey.

(38) **Poa macroclada** Rydb., Torrey Bot.
Club Bul. 32: 604. 1905. Rogers,
Gunnison Watershed, Colo., *Baker* 802.

(24) **Poa marcida** Hitchc., Biol. Soc. Wash.
Proc. 41: 158. 1928. Sol Duc Hot
Springs, Olympic Mountains, Wash.,
Hitchcock 23466.

(60) **Poa montevansi** Kelso, Biol. Leaflets
29: 2. 1945. Mount Evans, Colo., *L.
and E. H. Kelso* 427.

(54) **Poa napensis** Beetle, West Bot. Leaflets 4: 289. 1946. Napa County, Calif.,
Beetle 4256.

(37) **Poa nemoralis** L., Sp. Pl. 69. 1753.
Europe.
Paneion nemorale Lunell, Amer. Midl.
Nat. 4: 222. 1915. Based on *Poa
nemoralis* L.

(15) **Poa nervosa** (Hook.) Vasey, U. S.
Dept. Agr., Div. Bot. Bul. 13[2]: pl. 81.
1893. Based on *Festuca nervosa* Hook.
Festuca nervosa Hook., Fl. Bor. Amer. 2:
251. pl. 232. 1840. Nootka Sound,
Vancouver Island, *Scouler*.
Poa columbiensis Steud., Syn. Pl. Glum. 1:
261. 1854. Columbia River, *Douglas*.
Poa wheeleri Vasey in Rothr., Cat. Pl.
Survey W. 100th Merid. 55. 1874.
South Park, Colo. [*Wolf*] 1131 [1131a].
Poa pulchella var. *major* Vasey, U. S.
Dept. Agr., Div. Bot. Bul. 13[2]: pl. 82.
1893. Southern Oregon, no specimen
cited, and none so named by Vasey can
be found.
Poa vaseyana Scribn. in Beal, Grasses N.
Amer. 2: 532. 1896. [Georgetown],
Colo., *Patterson* in 1885.
Poa cuspidata Vasey ex Scribn., U.S. Dept.
Agr., Div. Agrost. Cir. 9: 6. 1899.
Not *P. cuspidata* Nutt., 1818. As syno-
nym of *P. wheeleri* Vasey.
Poa olneyae Piper, Erythea 7: 101. 1899.
Spokane, Wash., *Piper* 2820.
Poa subreflexa Rydb., Torrey Bot. Club
Bul. 36: 535. 1909. Steamboat Springs,
Colo., *State Agricultural College* 3731.
Poa wheeleri vaseyana Will. and Pammel,
Iowa Acad. Sci. Proc. 20: 144. 1915.
Presumably based on *P. vaseyana*
Scribn., "(Scribner.)" being cited.

(66) **Poa nevadensis** Vasey ex Scribn.,
Torrey Bot. Club Bul. 10: 66. 1883.
[Austin, Nev., *Jones* in 1882.]
Atropis pauciflora Thurb. in S. Wats.,
Bot. Calif. 2: 310. 1880. Not *Poa
pauciflora* Roem. and Schult., 1817.
Sierra Valley, Calif., *Lemmon* 1871.
(Though credited to Lemmon the type
specimen appears to have been col-
lected by Bolander, Lemmon's name
not appearing on the label.)
Poa pauciflora Benth. ex Vasey, Grasses
U. S. 42. 1883. Not *P. pauciflora*
Roem. and Schult., 1817. Based on
Atropis pauciflora Thurb.
Poa tenuifolia var. *scabra* Vasey ex Scribn.
Torrey Bot. Club Bul. 10: 66. 1883, as
synonym of *P. nevadensis*. [California,
Lemmon.]
Panicularia thurberiana Kuntze, Rev.
Gen. Pl. 2: 783. 1891. Based on
Atropis pauciflora Thurb.
Poa thurberiana Vasey, U. S. Dept. Agr.,
Div. Bot. Bul. 13[2]: pl. 84. 1893. The
name based on *Panicularia thurberiana*
Kuntze, but the plant described and
figured is *Melica imperfecta* Trin.
Atropis nevadensis Beal, Grasses N.
Amer. 2: 577. 1896. Based on *Poa
nevadensis* Vasey.

(28) **Poa occidentalis** Vasey, U. S. Natl. Herb. Contrib. 1: 274. 1893. Las Vegas, N. Mex., *G. R. Vasey* in 1881.
Poa flexuosa var. *occidentalis* Vasey in Rothr., in Wheeler, U S. Survey W. 100th Merid. Rpt. 6: 290. 1878. Twin Lakes, Colo. [*Wolf*] 1132.
Poa trivialis var. *occidentalis* Vasey, Grasses U. S. Descr. Cat. 85. 1885. Colorado and New Mexico, the type being the specimen later described as *P. occidentalis* Vasey.
Poa flexuosa var. *robusta* Vasey, U. S. Natl. Herb. Contrib. 1: 271. 1893. Rocky Mountains, Colo., *Vasey* 673 [Powell's Expedition].
Poa autumnalis var. *robusta* Beal, Grasses N. Amer. 2: 534. 1896. Based on *P. flexuosa* var. *robusta* Vasey.
Poa occidentalis Rydb., N. Y. Bot. Gard. Mem. 1: 50. 1900. Based on *P. flexuosa* var. *occidentalis* Vasey.
Poa platyphylla Nash and Rydb., Torrey Bot. Club Bul. 28: 266. 1901. Based on *P. occidentalis* Vasey, the name changed because of *P. flexuosa* var. *occidentalis* Vasey, thought to be different.
Poa lacustris Heller, Muhlenbergia 6: 12. 1910. Based on *P. flexuosa* var. *occidentalis* Vasey.

(35) **Poa paludigena** Fern. and Wieg., Rhodora 20: 126. 1918. Wayne County, N. Y., *Metcalf* and *Wiegand* 7572.
Poa sylvestris var. *palustris* Dudley, Cornell Univ. Bul. 2: 128. 1886. Michigan Hollow, N. Y.

(39) **Poa palustris** L., Syst. Nat. ed. 10. 2: 874. 1759. Europe.
Poa serotina Ehrh., Beitr. Naturk. 6: 83. 1791, name only; Schrad. Fl. Germ. 1: 299. 1806. Europe.
Poa triflora Gilib., Exerc. Phyt. 2: 531. 1792. Europe.
Poa crocata Michx., Fl. Bor. Amer. 1: 68. 1803. Lake Mistassini, Quebec, *Michaux*. Misspelled *P. crocea* in Muhl., Cat. Pl. 11. 1813.
Poa glauca var. *crocata* Jones, West. Bot. Contrib. 14: 15. 1912. Based on *P. crocata* Michx.
Paneion triflorum Lunell, Amer. Midl. Nat. 4: 223. 1915. Based on *Poa triflora* Gilib.

(50) **Poa pattersoni** Vasey, U. S. Natl. Herb. Contrib. 1: 275. 1893. Grays Peak, Colo., *Patterson* 154.

(33) **Poa paucispicula** Scribn. and Merr., U. S. Natl. Herb. Contrib. 13: 69. pl. 15. 1910. Yakutat Bay, Alaska, *Coville* and *Kearney* 970.

(18) **Poa pratensis** L., Sp. Pl. 67. 1753. Europe.
Poa angustifolia L. Sp. Pl. 67. 1753. Europe.
Poa pratensis var. *angustifolia* Gaudin, Agrost. Helv. 1: 214. 1811. Based on *P. angustifolia* L. This name has been

credited to Smith, Fl. Brit. 105. 1800, but the combination is not there made, "*β. Poa angustifolia L.*" merely cited under *P. pratensis*.
?*Poa viridis* Schreb. ex Pursh, Fl. Amer. Sept. 1: 79. 1814. North America. Name only, Muhl., Cat. Pl. 11. 1813.
Poa angustifolia Ell., Bot. S. C. and Ga. 1: 160. 1816. South Carolina.
Paneion pratense Lunell, Amer. Midl. Nat. 4: 222. 1915. Based on *Poa pratensis* L.
Poa peckii Chase, Wash. Acad. Sci. Jour. 28: 54. f. 2. 1938. Jefferson County, Oreg., *Peck* 19804.

(58) **Poa pringlei** Scribn., Torrey Bot. Club Bul. 10: 31. 1883. Headwaters of the Sacramento River, Calif., *Pringle* [in 1882].
Poa argentea Howell, Torrey Bot. Club Bul. 15: 11. 1888. [Ashland Butte] Siskiyou Mountains, Oreg., *Howell* [in 1887].
Melica argentea Beal, Torrey Bot. Club Bul. 17: 153. 1890. Based on *Poa argentea* Howell.
Melica nana Beal, Grasses N. Amer. 2: 504. 1896. Based on *Poa argentea* Howell. Name changed because of "*M. argentea* Desv." [error for *M. argentata* Desv.].
Atropis suksdorfii Beal, Grasses N. Amer. 2: 574. 1896. [Mount Adams] Wash., *Suksdorf* 1116. Beal gives as synonym "*Poa suksdorfii* Vasey ined."
Atropis pringlei Beal, Grasses N. Amer. 2: 578. 1896. Based on *Poa pringlei* Scribn.
Poa suksdorfii Vasey ex Piper, U. S. Natl. Herb. Contrib. 11: 135. 1906. Based on *Atropis suksdorfii* Beal.

(31) **Poa reflexa** Vasey and Scribn., U. S. Natl. Herb. Contrib. 1: 276. 1893. Kelso Mountain, near Torrey Peak, Colo., *Letterman* in 1885.
Poa acuminata Scribn. in Beal, Grasses N. Amer. 2: 538. 1896. [Mount Blackmore] Mont., *Tweedy* 639 in 1885, 1027 in 1886.
Poa pudica Rydb., Torrey Bot. Club Bul. 32: 603. 1905. Near Grays Peak, Colo., *Rydberg* 2443.
Poa leptocoma var. *reflexa* Jones, West. Bot. Contrib. 14: 15. 1912. Based on *P. reflexa* Vasey and Scribn.

(11) **Poa rhizomata** Hitchc. in Jepson, Fl. Calif. 1: 155. 1912. Oro Fino, Siskiyou County, Calif., *Butler* 1205.
Poa piperi Hitchc. in Abrams, Illustr. Fl. 1: 201. f. 461. 1923. Waldo, Oreg., *Piper* 6496.

(51) **Poa rupicola** Nash, N. Y. Bot. Gard. Mem. 1: 49. 1900. Based on *P. rupestris* Vasey.
Poa rupestris Vasey, Torrey Bot. Club Bul. 14: 94. 1887. Not *P. rupestris*

With., 1796. Rocky Mountains [*Wolf* 341 in 1873].

(27) **Poa saltuensis** Fern. and Wieg., Rhodora 20: 122. 1918. Gaspé County· Quebec, *Fernald* and *Collins* 357.

Poa debilis var. *acutiflora* Vasey ex Macoun, Cat. Can. Pl. 2⁴: 225. 1888. Name only, for *Macoun* 28 and *Burgess* 12 and 13, Truro, Nova Scotia.

Poa saltuensis var. *microlepis* Fern. and Wieg., Rhodora 20: 124. 1918. Newfoundland, *Fernald* and *Wiegand* 4633.

(62) **Poa scabrella** (Thurb.) Benth. ex Vasey, Grasses U. S. 42. 1883. Based on *Atropis scabrella* Thurb.

Sclerochloa californica Munro ex Benth., Pl. Hartw. 342. 1857. Name only, for *Hartweg* 2035, Sacramento Valley, Calif.

Poa tenuifolia Buckl., Acad. Nat. Sci. Phila. Proc. 1862: 96. 1862. Not *P. tenuifolia* A. Rich., 1851. Columbia River, *Nuttall.*

Atropis scabrella Thurb., in S. Wats., Bot. Calif. 2: 310. 1880. Oakland, Calif., *Bolander.*

Atropis tenuifolia Thurb. in S. Wats., Bot. Calif. 2: 310. 1880. Based on *Poa tenuifolia* Buckl.

Poa orcuttiana Vasey, West Amer. Sci. 3: 165. 1887. San Diego, Calif., *Orcutt* [1070] in 1884.

Panicularia scabrella Kuntze, Rev. Gen. Pl. 2: 783. 1891. Based on *Atropis scabrella* Thurb.

Panicularia nuttalliana Kuntze, Rev. Gen. Pl. 2: 783. 1891. Based on "*Atropis tenuifolia* Thurb., *Poa tenuifolia* Nutt., 1862" (error for Buckl.).

Poa buckleyana Nash, Torrey Bot. Club Bul. 22: 465. 1895. Based on *P. tenuifolia* Buckl.

Poa capillaris Scribn., U. S. Dept. Agr., Div. Agrost. Bul. 11: 51. f. 11. 1898. Not *P. capillaris* L., 1753. Potrero, Calif.

Poa nudata Scribn., U. S. Dept. Agr., Div. Agrost. Cir. 9: 1. 1899. Based on *P. capillaris* Scribn.

Poa acutiglumis Scribn., U. S. Dept. Agr., Div. Agrost. Cir. 9: 4. 1899. Grave Creek, Oreg., *Howell* in 1884.

Poa limosa Scribn. and Williams, U. S. Dept. Agr., Div. Agrost. Cir. 9: 5. 1899. Mono Lake, Calif., *Bolander.*

(64) **Poa secunda** Presl, Rel. Haenk. 1: 271. 1830. Chile, *Haenke.*

Poa sandbergii Vasey, U. S. Natl. Herb. Contrib. 1: 276. 1893. Lewiston, Idaho, *Sandberg* 164.

Poa incurva Scribn. and Williams, U. S. Dept. Agr., Div. Agrost. Cir. 9: 6. 1899. Duckaloose Glacier, Olympic Mountains, Wash., *Piper* 1989.

Poa buckleyana var. *sandbergii* Jones, West. Bot. Contrib. 14: 14. 1912. Based on *P. sandbergii* Vasey.

Paneion sandbergii Lunell, Amer. Midl. Nat. 4: 223. 1915. Based on *Poa sandbergii* Vasey.

(46) **Poa stenantha** Trin., Acad. St. Pétersb. Mém. VI. Math. Phys. Nat. 1: 376. 1830. Kamchatka, Unalaska, Sitka, Karaghinski Island.

(30) **Poa sylvestris** A. Gray, Man. 596. 1848. Ohio and Kentucky, *Short, Sullivant,* Michigan and southwestward [type from Ohio, *Short* in 1842].

(29) **Poa tracyi** Vasey, Torrey Bot. Club Bul. 15: 49. 1888. Raton, N. Mex., *Tracy* in 1887.

Poa nervosa var. *tracyi* Beal, Grasses N. Amer. 2: 538. 1896. Based on *P. tracyi* Vasey.

(23) **Poa trivialis** L., Sp. Pl. 67. 1753. Europe.

Poa stolonifera Hall. ex Muhl., Descr. Gram. 139. 1817. Pennsylvania.

Poa trivialis var. *filiculmis* Scribn. in Beal, Grasses N. Amer. 2: 532. 1896. Vancouver Island, *Macoun* 282.

Poa callida Rydb., Torrey Bot. Club Bul. 36: 533. 1909. Helena, Mont., *Rydberg* 2145.

(55) **Poa unilateralis** Scribn. in Vasey, U. S. Dept. Agr., Div. Bot. Bul. 13²: pl. 85. 1893. San Francisco, Calif. [*Jones* 15 in 1882].

Atropis unilateralis Beal, Grasses N. Amer. 2: 581. 1896. Based on *Poa unilateralis* Scribn.

Poa pachypholis Piper, Biol. Soc. Wash. Proc. 18: 146. 1905. Ilwaco, Wash., *Piper* [4900].

(57) **Poa vaseyochloa** Scribn., U. S. Dept. Agr., Div. Agrost. Cir. 9: 1. 1899. Based on *P. pulchella* Vasey.

Poa pulchella Vasey, Bot. Gaz. 7: 32. 1882. Not *P. pulchella* Salisb., 1796. Columbia River [mountains, Klickitat County, Wash.], *Suksdorf* [in 1881].

Atropis pulchella Beal, Grasses N. Amer. 2: 574. 1896. Based on *Poa pulchella* Vasey.

Poa gracillima var. *vaseyochloa* Jones, West. Bot. Contrib. 14: 14. 1912. Based on *P. vaseyochloa* Scribn.

(32) **Poa wolfii** Scribn., Torrey Bot. Club Bul. 21: 228. 1894. [Canton], Ill., *Wolf* [in 1882].

Poa alsodes var. *wolfii* Vasey ex Scribn., Torrey Bot. Club Bul. 21: 228. 1894, as synonym of *P. wolfii* Scribn.

(77) POLYPOGON Desf.

(4) **Polypogon australis** Brongn. in Duperrey, Bot. Voy. Coquille 2²: 21. 1830. Concepción, Chile.

Polypogon crinitus Trin., Gram. Unifl. 171. 1824. Not *P. crinitus* Nutt., 1818. Chile, *Chamisso.*

Polypogon interruptus var. *crinitus* Hack. in Stuck., An. Mus. Nac. Buenos Aires

13: 473. 1906. Based on *P. crinitus* Trin.
(5) **Polypogon elongatus** H. B. K., Nov. Gen. et Sp. 1: 134. 1815. Chillo, Ecuador, *Humboldt* and *Bonpland.*
(3) **Polypogon interruptus** H. B. K., Nov. Gen. et Sp. 1: 134. pl. 44. 1815. Venezuela, *Humboldt* and *Bonpland.*
Alopecurus interruptus Poir. in Lam., Encycl. Sup. 5: 495. 1817. Based on *Polypogon interruptus* H. B. K.
Polypogon lutosus (Poir.) Hitchc., misapplied to *P. interruptus*, appears to be a rare hybrid of southern Europe; not known from America.
(2) **Polypogon maritimus** Willd., Gesell. Naturf. Freund. Berlin (n.s.) 3: 443. 1801. France.
Alopecurus maritimus Poir. in Lam., Encycl. 8: 779. 1808. Based on *Polypogon maritimus* Willd.
Polypogon monspeliensis var. *maritimus* Coss. and Dur., Expl. Sci. Alger. 2: 70. 1854. Based on *P. maritimus* Willd.
(1) **Polypogon monspeliensis** (L.) Desf., Fl. Atlant. 1: 67. 1798. Based on *Alopecurus monspeliensis* L.
Alopecurus monspeliensis L., Sp. Pl. 61. 1753. Europe.
Phleum crinitum Schreb., Beschr. Gräs. 1: 151. 1769. Based on *Alopecurus monspeliensis* L.
Alopecurus aristatus var. *monspeliensis* Huds., Fl. Angl. 28. 1778. Based on *A. monspeliensis* L.
Agrostis alopecuroides Lam., Tabl. Encycl. 1: 160. 1791. Based on *Alopecurus monspeliensis* L.
Phleum monspeliense Koel., Descr. Gram. 57. 1802. Based on *Alopecurus monspeliensis* L.
Polypogon crinitus Nutt., Gen. Pl. 1: 50. 1818. Based on *Phleum crinitum* Smith (error for Schreb.).
Polypogon flavescens Presl, Rel. Haenk. 1: 234. 1830. Peru, *Haenke.*
Santia monspeliensis Parl., Fl. Palerm. 1: 73. 1845. Based on *Alopecurus monspeliensis* L.

PSEUDOSASA Makino

Pseudosasa japonica (Sieb. and Zucc.) Makino, Jour. Jap. Bot. 2(4): 15. 1920. Based on *Arundinaria japonica* Sieb. and Zucc.
Arundinaria japonica Sieb. and Zucc. ex Steud., Syn. Pl. Glum. 1: 334. 1854. Japan; Java erroneously cited as locality.

(6) PUCCINELLIA Parl.

(8) **Puccinellia airoides** (Nutt.) Wats. and Coult. in A. Gray, Man. ed. 6. 668. 1890. Based on *Poa airoides* Nutt.
Poa airoides Nutt., Gen. Pl. 1: 68. 1818.

Not *P. airoides* Koel., 1802. Mandan, N. Dak., *Nuttall.*
Poa nuttalliana Schult., Mantissa 2: 303. 1824. Based on *P. airoides* Nutt.
Festuca nuttalliana Kunth, Rév. Gram. 1: 129. 1829. Based on *Poa nuttalliana* Schult.
Glyceria airoides Fries, Nov. Fl. Suec. Mant. 3: Add. 176. 1843. Not *G. airoides* Reichenb., 1829. Based on *Poa airoides* Nutt.
Glyceria montana Buckl., Acad. Nat. Sci. Phila. Proc. 1862: 96. 1862. Rocky Mountains, *Nuttall.*
Panicularia distans airoides Scribn., Torrey Bot. Club Mem. 5: 54. 1894. Based on *Poa airoides* Nutt.
Atropis airoides Holm, Bot. Gaz. 46: 427. 1908. Based on *Poa airoides* Nutt.
Puccinellia cusickii Weatherby, Rhodora 18: 182. 1916. Grande Ronde Valley, Oreg., *Cusick 3271.*
Atropis nuttalliana Pilger, Notizbl. Bot. Gart. Berlin 9: 291. 1925. Based on *Poa nuttalliana* Schult.
Wyoming specimens cited by Fernald and Weatherby (Rhodora 18: 16. 1916) under *Puccinellia lucida* (the type from Quebec) are here referred to *P. airoides.*
(7) **Puccinellia distans** (L.) Parl., Fl. Ital. 367. 1848. Based on *Poa distans* L.
Poa distans L., Mant. Pl. 1: 32. 1767. Europe.
Aira aquatica var. *distans* Huds., Fl. Angl. 34. 1778. Based on *Poa distans* L.
Hydrochloa distans Hartm., Gen. Gram. Skand. 8. 1819. Presumably based on *Poa distans* L.
Glyceria distans Wahl., Fl. Upsal. 36. 1820. Based on *Poa distans* L.
Festuca distans Kunth, Rév. Gram. 1: 129. 1829. Based on *Poa distans* L.
Sclerochloa distans Bab., Man. Brit. Bot. 370. 1843. Based on *Poa distans* L.
Catabrosa distans Link ex Heynh., Nom. 2: 126. 1846. Based on *Glyceria distans* Wahl.
Atropis distans Griseb. in Ledeb., Fl. Ross. 4: 388. 1853. Based on *Poa distans* L.
Glyceria distans var. *tenuis* Uechtr. in Crép., Notes Pl. Rar. Belg. 229. 1865. Germany.
Sclerochloa multiculmis subsp. *distans* Syme in Sowerby, English, Bot. ed. 3. 11: 104. 1873. Based on *Poa distans* L.
Panicularia distans Kuntze, Rev. Gen. Pl. 2: 782. 1891. Based on *Poa distans* L.
Atropis distans var. *tenuis* Rouy, Fl. France 14: 195. 1913. Based on *Glyceria distans* var. *tenuis* Uechtr.
Puccinellia distans var. *tenuis* Fern. and Weath., Rhodora 18: 12. 1916. Based on *Glyceria distans* var. *tenuis* Uechtr.
Puccinellia suksdorfii St. John, Wash. State Col. Contrib. Dept. Bot. 2: 80. 1928. Rockland, Wash., *Suksdorf 5089.*

(4) **Puccinellia fasciculata** (Torr.) Bicknell, Torrey Bot. Club Bul. 35: 197. 1908. Based on *Poa fasciculata* Torr.
Poa fasciculata Torr., Fl. North. and Mid. U. S. 1: 107. 1823. New York [*Torrey*].
Poa delawarica Link, Hort. Berol. 1: 174. 1827. Delaware.
Festuca delawarica Kunth, Rév. Gram. 1: 129. 1829. Based on *Poa delawarica* Link.
Festuca borreri Bab., Linn. Soc. Trans. 17: 565. 1837. England.
Glyceria delawarica Heynh., Nom. 1: 360. 1840. Based on *Poa delawarica* Link.
Glyceria borreri Bab. in Smith and Sowerby, English Bot. Sup. 3: pl. 2797. 1843. England.
Sclerochloa borreri Bab., Man. Brit. Bot. 370. 1843. Based on *Glyceria borreri* Bab.
Poa borreri Parnell, Grasses Brit. 220. pl. 98. 1845. Based on *Sclerochloa borreri* Bab.
Sclerochloa arenaria var. *fasciculata* A. Gray, Man. 594. 1848. Based on *Poa fasciculata* Torr.
Sclerochloa multiculmis subsp. *borreri* Syme in Sowerby, English Bot. ed. 3. 11: 105. 1873. Based on *S. borreri* Bab.
Atropis borreri Richt., Pl. Eur. 1: 92. 1890. Based on *Glyceria borreri* Bab.
Puccinellia borreri Hitchc., Rhodora 10: 65. 1908. Based on *Festuca borreri* Bab.

(10) **Puccinellia grandis** Swallen, Wash. Acad. Sci. Jour. 34: 18. 1944. Seattle, Wash., *Piper* 1451. Has been confused with *P. nutkaensis* (Presl) Fern. and Weath., that not found in the United States. Has also been referred to *P. festucaeformis* Parl. of Europe.

(5) **Puccinellia lemmoni** (Vasey) Scribn., U. S. Dept. Agr., Div. Agrost. Bul. 17: 276. f. 572. 1899. Based on *Poa lemmoni* Vasey.
Poa lemmoni Vasey, Bot. Gaz. 3: 13. 1878. Sierra County, Calif., *Lemmon*.
Glyceria lemmoni Vasey, Grasses U. S. Descr. Cat. 88. 1885, name only; Torrey Bot. Club Bul. 13: 119. 1886. Based on *Poa lemmoni* Vasey.
Atropis lemmoni Vasey, U. S. Dept. Agr., Div. Bot. Bul. 13²: pl. 90. 1893. Based on *Poa lemmoni* Vasey.
Puccinellia rubida Elmer, Bot. Gaz. 36: 56. 1903. Prineville, Oreg., *Cusick* 2621.

(6) **Puccinellia maritima** (Huds.) Parl., Fl. Ital. 1: 370. 1848. Based on *Poa maritima* Huds.
Poa maritima Huds., Fl. Angl. 35. 1762. England.
Poa maritima Muhl., Descr. Gram. 148. 1817. New England.
Glyceria maritima Wahlb., Fl. Gothob. 17. 1820. Based on *Poa maritima* Huds.
Festuca distans var. *maritima* Mutel., Fl.

Franç. 4: 116. 1837. Based on *Poa maritima* Huds.
Poa maritima Bigel., Fl. Bost. ed. 3. 36. 1840. Cambridge and Dorchester, Mass.
Diachroa maritima Nutt. ex Steud., Nom. Bot. ed. 2. 1: 497. 1840, as synonym of *Glyceria maritima* Wahlb.
Sclerochloa maritima Lindl. in Bab., Man. Brit. Bot. 370. 1843. Based on *Glyceria maritima* Smith (same as Wahlb.).
Sclerochloa arenaria var. *maritima* A. Gray, Man. 594. 1848. Based on *Poa maritima* Huds.
Atropis maritima Griseb. in Ledeb., Fl. Ross. 4: 389. 1853. Based on *Poa maritima* Huds.
Atropis distans var. *maritima* Coss. and Dur., Expl. Sci. Alger. 2: 141. 1855. Based on *Poa maritima* Huds.
Panicularia maritima Scribn., Torrey Bot. Club Mem. 5: 54. 1894. Based on *Poa maritima* Huds.

(1) **Puccinellia parishii** Hitchc., Biol. Soc. Wash. Proc. 41: 157. 1928. Rabbit Springs, Calif., *Parish* 9799.

(9) **Puccinellia pumila** (Vasey) Hitchc., Amer. Jour. Bot. 21: 129. 1934. Based on *Glyceria pumila* Vasey.
Glyceria pumila Vasey, Torrey Bot. Club Bul. 15: 48. 1888. Vancouver Island, *Macoun* [in 1887].
Puccinellia maritima var. *minor* S. Wats. in A. Gray, Man. ed. 6. 668. 1890. Mount Desert, Maine, *Rand*.
This is the species referred by American authors to *Atropis angustata* Griseb., *Glyceria angustata* Vasey, and *Puccinellia angustata* Nash. The names are based on *Poa angustata* R. Br., a species of Arctic America.

(3) **Puccinellia rupestris** (With.) Fern. and Weath., Rhodora 18: 10. f. 17–22. 1916. Based on *Poa rupestris* With.
Poa rupestris With., Bot. Arr. Veg. Brit. ed. 3. 2: 146. 1796. England.
Poa procumbens Curtis, Fl. Lond. 6: pl. 11. 1798. England.
Sclerochloa procumbens Beauv., Ess. Agrost. 98. 1812. Based on *Poa procumbens* Curtis.
Festuca procumbens Kunth, Rév. Gram. 1: 129. 1829. Not *F. procumbens* Muhl., 1817. Based on *Poa procumbens* Curtis.
Scleropoa procumbens Parl., Fl. Ital. 1: 474. 1848. Based on *Poa procumbens* Curtis.
Atropis procumbens Thurb. in S. Wats., Bot. Calif. 2: 309. 1880. Based on *Poa procumbens* Curtis. [The specimen mentioned by Thurber (*Bolander* 6467) is *Poa unilateralis* Scribn., with a fragment of *Puccinellia rupestris*, which is not known to occur in California.]
Panicularia procumbens Kuntze, Rev. Gen. Pl. 2: 782. 1891. Based on *Poa procumbens* Curtis.

(2) **Puccinellia simplex** Scribn., U. S. Dept.

Agr., Div. Agrost. Cir. 16: 1. f. 1. 1899. Woodland, Calif., *Blankinship*.

(19) REDFIELDIA Vasey

(1) **Redfieldia flexuosa** (Thurb.) Vasey, Torrey Bot. Club Bul. 14: 133. pl. 70. 1887. Based on *Graphephorum flexuosum* Thurb.
Graphephorum flexuosum Thurb. in A. Gray, Acad. Nat. Sci. Phila. Proc. 1863: 78. 1863. "Colorado Territory," latitude 41° [probably Nebraska], *Hall and Harbour* 635.

(135) REIMAROCHLOA Hitchc.

(1) **Reimarochloa oligostachya** (Munro) Hitchc., U. S. Natl. Herb. Contrib. 12: 199. 1909. Based on *Reimaria oligostachya* Munro.
Reimaria oligostachya Munro ex Benth., Linn. Soc. Jour., Bot. 19: 34. 1881. [Jacksonville], Fla., *Curtiss* 3566.

(142) RHYNCHELYTRUM Nees

(1) **Rhynchelytrum roseum** (Nees) Stapf and Hubb. ex Bews, World's Grass. 223. 1929, no basis cited; in Prain, Fl. Trop. Afr. 9: 880. 1930. Based on *Tricholaena rosea* Nees. Has been confused with *R. repens* (Willd.) C. E. Hubb., a pale-flowered annual from West Africa.
Tricholaena rosea Nees, "Cat. Sem. Hort. Vratisl. a. 1836"; Fl. Afr. Austr. 17. 1841. South Africa, *Drège*.
Panicum roseum Steud., Syn. Pl. Glum. 1: 92. 1854. Not *P. roseum* Willd., 1825. Based on *Tricholaena rosea* Nees.
Panicum teneriffae var. *rosea* F. M. Bailey, Queensl. Grass. 22. 1888. Based on *Tricholaena rosea* Nees.
Melinis rosea Hack., Oesterr. Bot. Ztschr. 51: 464. 1901. Based on *Tricholaena rosea* Nees.
Tricholaena repens var. *rosea* Alberts, Imp. Bur. Pastures and Forage Crops Bul. 37: 10. 1947. Presumably based on *Tricholaena rosea* Nees.

(162) ROTTBOELLIA L. f.

(1) **Rottboellia exaltata** L. f., Nov. Gram. Gen. 40. pl. 1. 1779; Sup. Pl. 114. 1781. India.
Manisuris exaltata Kuntze, Rev. Gen. Pl. 2: 779. 1891. Based on *Rottboellia exaltata* L. f.
Stegosia exaltata Nash, N. Amer. Fl. 17: 84. 1909. Based on *Rottboellia exaltata* L. f.

(150) SACCHARUM L.

Saccharum bengalense Retz., Obs. Bot. 5: 16. 1789. India.
Saccharum ciliare Anderss., Öfv. Svensk. Vet. Akad. Förh. 12: 155. 1855. India.

(1) **Saccharum officinarum** L., Sp. Pl. 54. 1753. India.

(139) SACCIOLEPIS Nash

Sacciolepis indica (L.) Chase, Biol. Soc. Wash. Proc. 21: 8. 1908. Based on *Aira indica* L.
Aira spicata L., Sp. Pl. 63. 1753. India.
Aira indica L., Sp. Pl. in Errara. 1753. Based on *Aira spicata* L., page 63, the name changed because of *Aira spicata*, page 64, of the same work, the latter the basis of *Trisetum spicatum*.
Panicum indicum L., Mant. Pl. 2: 184. 1771. Not *P. indicum* Mill., 1768. Based on *Aira indica* L.
Hymenachne indica Buse, in Miquel, Pl. Jungh. 377. 1854. Based on *Panicum indicum* L.
Sacciolepis spicata Honda, Tokyo Univ. Faculty. Sci. Jour. sec. 3. Bot. 3: 261. 1930. Based on *Aira spicata* L.
Panicum.spicatum Farwell, Rhodora 32: 262. 1930. Not *P. spicatum* Roxb., 1820. Based on *Aira spicata* L.

(1) **Sacciolepis striata** (L.) Nash, Torrey Bot. Club Bul. 30: 383. 1903. Based on *Holcus striatus* L.
Holcus striatus L., Sp. Pl. 1048. 1753. Virginia [*Clayton* 590].
Panicum striatum Lam., Tabl. Encycl. 1: 172. 1791. Carolina, *Fraser*.
Sorghum striatum Beauv., Ess. Agrost. 132, 165. 1812. Based on *Holcus striatus* L.
Panicum gibbum Ell., Bot. S. C. and Ga. 1: 116. 1816. Presumably South Carolina.
Panicum aquaticum Muhl., Descr. Gram. 126. 1817. Not *P. aquaticum* Poir., 1816. No locality cited.
Panicum fluitans Brickell ex Muhl., Descr. Gram. 126. 1817, as synonym of *P. aquaticum* Muhl.
Panicum hydrophilum Schult., Mantissa 2: 237. 1824. Based on *P. aquaticum* Muhl.
Panicum elliottianum Schult., Mantissa 2: 256. 1824. Based on *P. gibbum* Ell.
Panicum aquaticum Bosc ex Spreng., Syst. Veg. 1: 319. 1825. Not *P. aquaticum* Poir , 1816. Bermuda.
Hymenachne striata Griseb., Fl. Brit. W. Ind. 554. 1864. Based on *Panicum striatum* Lam.
Sacciolepis gibba Nash in Britton, Man. 89. 1901. Based on *Panicum gibbum* Ell. In a second printing of Britton, Man. 1902 (p. 89), the generic name is spelled *Saccolepis*.
Sacciolepis striata forma *gibba* Fernald, Rhodora 44, 381. 1942. Based on *Panicum gibbum* Ell.

(105) SCHEDONNARDUS Steud.

(1) **Schedonnardus paniculatus** (Nutt.)

Trel., in Branner and Coville, Rpt. Geol. Survey Ark. 1888[4]: 236. 1891. Based on *Lepturus paniculatus* Nutt.
Lepturus paniculatus Nutt., Gen. Pl. 1: 81. 1818. Mandan, N. Dak.
Rottboellia paniculata Spreng., Syst. Veg. 1: 300. 1825. Based on *Lepturus paniculatus* Nutt.
Schedonnardus texanus Steud., Syn. Pl. Glum. 1: 146. 1854. Texas, *Drummond* 360.
Spirochloe paniculata Lunell, Amer. Midl. Nat. 4: 220. 1915. Based on *Lepturus paniculatus* Nutt.

(54) SCHISMUS Beauv.

(2) **Schismus arabicus** Nees, Fl. Afr. Austr. 1: 422. 1841. Arabia.
Schismus barbatus subsp. *arabicus* Maire and Weiller, Soc. Hist. Nat. Afr. Nord. Bul. 30: 310. 1939. Based on *S. arabicus* Nees.

(1) **Schismus barbatus** (L.) Thell., Bul. Herb. Boiss. II. 7: 391. 1907 in obs. Based on *Festuca barbata* L.
Festuca barbata L., Amoen. Acad. 3: 400. 1756. Spain.
Schismus fasciculatus Beauv., Ess. Agrost. 74, 177. 1812, name only; Trin., Fund. Agrost. 148. 1820. No locality cited.
Schismus marginatus Beauv., Ess. Agrost. 177. pl. 15. f. 4. 1812. No locality cited.

(31) SCHIZACHNE Hack.

(1) **Schizachne purpurascens** (Torr.) Swallen, Wash. Acad. Sci. Jour. 18: 204. f. 1. 1928. Based on *Trisetum purpurascens* Torr.
Avena striata Michx., Fl. Bor. Amer. 1: 73. 1803. Not *A. striata* Lam., 1783. Between Hudson Bay and Lake Mistassini, *Michaux.*
Trisetum purpurascens Torr., Fl. North. and Mid. U. S. 1: 127. 1823. Williamstown, Mass., *Dewey;* also Boston, Catskill Mountains, N. Y., and Montreal.
Avena callosa Turcz. in Ledeb., Fl. Ross. 4: 416. 1853. Siberia.
Avena striata forma *albicans* Fernald, Rhodora 7: 244. 1905. Mount Albert, Quebec, *Collins* and *Fernald* 26.
Melica striata Hitchc., Rhodora 8: 211. 1906. Based on *Avena striata* Michx.
Melica striata forma *albicans* Fernald, Rhodora 10: 47. 1908. Based on *Avena striata* forma *albicans* Fernald.
Melica purpurascens Hitchc., U. S. Natl. Herb. Contrib. 12: 156. 1908. Based on *Trisetum purpurascens* Torr.
Schizachne fauriei Hack., Repert. Sp. Nov. Fedde 7: 323. 1909. Sachalin Island, *Faurie.*
Avena torreyi Nash in Britt. and Brown, Illustr. Fl. ed. 2. 1: 219. 1913. Based on *Trisetum purpurascens* Torr., not

Avena purpurascens DC., 1813.
Bromelica striata Farwell, Rhodora 21: 77. 1919. Based on *Avena striata* Michx.
Schizachne striata Hultén, Svensk. Bot. Tidskr. 30: 518. 1936. Based on *Avena striata* Michx.
Schizachne purpurascens forma *albicans* Fernald, Rhodora 44: 139. 1942. Based on *Avena striata* forma *albicans* Fernald.
Schizachne callosa Ohwi, Act. Phytotax. and Geobot. 2: 279. 1933. Based on *Avena callosa* Turcz.

(8) SCLEROCHLOA Beauv.

(1) **Sclerochloa dura** (L.) Beauv., Ess. Agrost. 98, 174, 177. pl. 19. f. 4. 1812. Based on *Poa dura* L. (error for Scop.).
Cynosurus durus L., Sp. Pl. 72. 1753. Southern Europe.
Poa dura Scop., Fl. Carn. ed. 2. 1: 70. 1772. Based on *Cynosurus durus* L.
Eleusine dura Lam., Tabl. Encycl. 1: 203. 1791. Based on *Cynosurus durus* L.
Crassipes annuus Swallen, Amer. Jour. Bot. 18: 684. f. 1–4. 1931. Between Salt Lake City and Ogden, foot of Wasatch Mountains, Utah, *Fallas* in 1928.

(5) SCLEROPOA Griseb.

(1) **Scleropoa rigida** (L.) Griseb., Spic. Fl. Rum. 2: 431. 1844. Based on *Poa rigida* L.
Poa rigida L., Cent. Pl. 1: 5. 1755; Amoen. Acad. 4: 265. 1759. Europe.
Poa cristata Walt., Fl. Carol. 80. 1788. Not *P. cristata* L., 1767. South Carolina.
Sclerochloa rigida Link, Enum. Pl. 1: 90. 1821. Based on *Poa rigida* L.
Glyceria rigida J. E. Smith, English Fl. 1: 119. 1824. Based on *Poa rigida* L.
Festuca rigida Raspail, Ann. Sci. Nat., Bot. 5: 445. 1825. Based on *Poa rigida* L.
Synaphe rigida Dulac, Fl. Haut. Pyr. 90. 1867. Based on *Scleropoa rigida* Griseb.
Diplachne rigida Munro ex Chapm., Fl. South. U. S. ed. 3. 609. 1897. Based on *Poa rigida* L.

(41) SCLEROPOGON Phil.

(1) **Scleropogon brevifolius** Phil., An. Univ. Chile 36: 206. 1870. Mendoza, Argentina.
Festuca macrostachya Torr. and Gray, U. S. Rpt. Expl. Miss. Pacif. 2[4]: 177. 1855. Name only. Pecos, Tex. [Staminate specimen.]
Tricuspis monstra Munro ex Hemsl., Diag. Pl. Mex. 56. 1880, as synonym of *Scleropogon brevifolius* Phil.

Lesourdia karwinskyana Fourn., Soc. Bot. France Bul. 27: 102. pl. 4. f. 12. 1880. Mexico, *Karwinsky* 992.

Lesourdia multiflora Fourn., Soc. Bot. France Bul. 27: 102. pl. 3, 4. 1880. Tampico, Mexico, *Bernier.*

Scleropogon karwinskyanus Benth. ex S. Wats., Amer. Acad. Sci. Proc. 18: 181. 1883. Based on *Lesourdia karwinskyana* Fourn.

(9) SCOLOCHLOA Link

(1) **Scolochloa festucacea** (Willd.) Link, Hort. Berol. 1: 137. 1827. Based on *Arundo festucacea* Willd.

Festuca arundinacea Liljebl., Utk. Svensk Fl. ed. 2. 47. 1798. Not *F. arundinacea* Schreb., 1771. Sweden.

Arundo festucacea Willd., Enum. Pl. 1: 126. 1809. Germany.

Triodia festucacea Roth, Enum. Pl. Phaen. Germ. 1¹: 382. 1827. Based on *Arundo festucacea* Willd.

Graphephorum festucaceum A. Gray, Amer. Acad. Sci. Proc. 5: 191. 1861. Based on *Arundo'festucacea* Willd.

Scolochloa arundinacea MacM., Met. Minn. Vall. 79. 1892. Not *S. arundinacea* Mert. and Koch, 1823. Based on *Festuca arundinacea* Liljebl.

Fluminea festucacea Hitchc., U. S. Dept. Agr. Bul. 772: 38. f. 11. 1920. Based on *Arundo festucacea* Willd.

(53) SCRIBNERIA Hack.

(1) **Scribneria bolanderi** (Thurb.) Hack., Bot. Gaz. 11: 105. pl. 5. 1886. Based on *Lepturus bolanderi* Thurb.

Lepturus bolanderi Thurb., Amer. Acad. Sci. Proc. 7: 401. 1868. Russian River Valley, Calif., *Bolander.*

(45) SECALE L.

(1) **Secale cereale** L., Sp. Pl. 84. 1753. Europe.

Triticum cereale Salisb., Prodr. Stirp. 27. 1796. Based on *Secale cereale* L.

Secale montanum Guss., Fl. Sci. Prod. 1: 145. 1827.

(143) SETARIA Beauv.

Setaria barbata (Lam.) Kunth, Rév. Gram. 1: 47. 1829. Based on *Panicum barbatum* Lam.

Panicum barbatum Lam., Tabl. Encycl. 1: 171. 1791. Mauritius.

Panicum costatum Roxb., Fl. Ind. ed. Carey 1: 314. 1820. Mauritius.

Panicum viaticum Salzm. ex Doell, in Mart., Fl. Bras. 2²: 155. 1877. Bahia, Brazil, *Salzmann* 706.

Chamaeraphis viatica Kuntze, Rev. Gen. Pl. 2: 770. 1891. Based on *Panicum viaticum* Salzm.

Chamaeraphis costata Kuntze, Rev. Gen.

Pl. 2: 771. 1891. Based on *Panicum costatum* Roxb.

Chaetochloa barbata Hitchc. and Chase, U. S. Natl. Herb. Contrib. 18: 348. 1917. Based on *Panicum barbatum* Lam.

Setaria carnei Hitchc., Soc. Linn. N. S. W. Proc. 52: 185. 1927. Western Australia.

(8) **Setaria corrugata** (Ell.) Schult., Mantissa 2: 276. 1824. Based on *Panicum corrugatum* Ell.

Panicum corrugatum Ell., Bot. S. C. and Ga. 1: 113. 1816. Savannah, Ga., *Baldwin.*

Pennisetum corrugatum Nutt., Gen. Pl. 1: 55. 1818. Presumably based on *Panicum corrugatum* Ell.

Setaria glauca var. *corrugata* Schrad., Linnaea 12: 429. 1838. Based on *S. corrugata* Schult.

Chamaeraphis corrugata Kuntze, Rev. Gen. Pl. 2: 770. 1891. Based on *Panicum corrugatum* Ell.

Chaetochloa corrugata Scribn., U. S. Dept. Agr., Div. Agrost. Bul. 4: 39. 1897. Based on *Panicum corrugatum* Ell.

Chaetochloa hispida Scribn. and Merr., U. S. Dept. Agr., Div. Agrost. Bul. 21: 25. f. 13. 1900. Cuba, *Wright.*

Setaria hispida Schum., Just's Bot. Jahresber. 28¹: 417. 1902. Based on *Chaetochloa hispida* Scribn. and Merr.

(13) **Setaria faberii** Herrm., Beitr. Biol. Pflanz. 10: 51. 1910. Prov. Szechwan, China "(*Faber* 582–1182)."

(2) **Setaria geniculata** (Lam.) Beauv., Ess. Agrost. 51, 169, 178. 1812. Based on *Panicum geniculatum* Lam.

Panicum geniculatum Lam., Encycl. 4: 727 (err. typ. 737). 1798. Guadeloupe.

Cenchrus parviflorus Poir. in Lam., Encycl. 6: 52. 1804. Puerto Rico.

Setaria gracilis H. B. K., Nov. Gen. et Sp. 1: 109. 1815. Colombia, *Humboldt* and *Bonpland.*

Setaria purpurascens H. B. K., Nov. Gen. et Sp. 1: 110. 1815. Ecuador, *Humboldt* and *Bonpland.*

Pennisetum geniculatum Jacq., Eclog. Gram. 3: pl. 26. 1815–1820. Based on *Panicum geniculatum* Lam.

Panicum imberbe Poir. in Lam., Encycl. Sup. 4: 272. 1816. North America and Brazil.

Panicum laevigatum Muhl. ex Ell., Bot. S. C. and Ga. 1: 112. 1816. Not *P. laevigatum* Lam. 1778. Eddings Island, S. C. (Published as new in Muhl., Descr. Gram. 100. 1817, for the same species.)

Panicum glaucum var. *purpurascens* Ell., Bot. S. C. and Ga. 1: 113. 1816. Parris Island and Charleston Neck, S. C.

Panicum medium Muhl. ex Fll., Bot. S. C. and Ga. 1: 113. 1816, as synonym of *P. glaucum* var. *purpurascens* Ell.

Setaria imberbis Roem. and Schult., Syst.

Veg. 2: 891.   1817. Based on *Panicum imberbe* Poir.

Pennisetum laevigatum Nutt., Gen. Pl. 1: 55.   1818. Presumably based on *Panicum laevigatum* Muhl.

Setoria laevigata Schult., Mantissa 2: 275. 1824. Based on *Panicum laevigatum* Muhl.

Setaria affinis Schult., Mantissa 2: 276. 1824. Based on Muhlenberg's *Panicum* No. 4. Georgia and Pennsylvania.

Setaria berteroniana Schult., Mantissa 2: 276. 1824. Dominican Republic, *Bertero*.

Setaria glauca var. *purpurascens* Torr., Fl. North. and Mid. U. S. 153.   1824. Based on *Setaria purpurascens* H. B. K. Published as new by Urban (Symb. Antill. 4: 96.   1903), based on the same type.

Panicum flavum Nees, Agrost. Bras. 238. 1829. Brazil.

Panicum dasyurum Nees, Agrost. Bras. 241.   1829. Brazil, *Hoffmansegg;* Montevideo, *Sellow*.

Panicum fuscescens Willd. ex Nees, Agrost. Bras. 241.   1829, as synonym of *P. purpurascens* H. B. K. [South America, *Humboldt*].

Panicum penicillatum Willd. ex Nees, Agrost. Bras. 242.   1829. Not *P. penicillatum* Nees ex Trin. 1826. Brazil,

Panicum tejucense Nees, Agrost. Bras. 243. 1829. Tejuco, Brazil.

Setaria flava Kunth, Rév. Gram. 1: 46. 1829. Based on *Panicum flavum* Nees.

Setaria ventenatii Kunth, Rév. Gram. 1: 251. pl. 37.   1830. Puerto Rico.

Setaria tejucensis Kunth, Rév. Gram. 1: Sup. 11.   1830. Based on *Panicum tejucense* Nees.

Setaria penicillata Presl, Rel. Haenk. 1: 314.   1830. Based on *Panicum penicillatum* Willd.

Panicum ventenatii Steud., Nom. Bot. ed. 2. 2: 265.   1841. Based on *Setaria ventenatii* Kunth.

Panicum berteronianum Steud., Syn. Pl. Glum. 1: 50.   1854. Based on *Setaria berteroniana* Schult.

Setaria glauca var. *laevigata* Chapm., Fl. South. U. S. 578.   1860. Based on *Panicum laevigatum* Muhl.

Setaria stipaeculmis C. Muell., Bot. Ztg. 19: 323.   1861. Rio Brazos, Tex., *Drummond*.

Setaria glauca var. *penicillata* Griseb., Fl. Brit. W. Ind. 554.   1864. Based on *Panicum penicillatum* Willd.

Setaria glauca var. *imberbis* Griseb., Fl. Brit. W. Ind. 554.   1864. Based on *Panicum imberbe* Poir.

Panicum imberbe var. *dasyurum* Doell, in Mart., Fl. Bras. 2²: 157.   1877. Based on *P. dasyurum* Nees.

Panicum imberbe var. *purpurascens* Doell, in Mart., Fl. Bras. 2²: 157.   1877. Based on *P. purpurascens* H. B. K.

Setaria streptobotrys Fourn., Mex. Pl. 2: 47.   1886. Mexico, *Galeotti* 5832, *Liebmann* 358, and several other collections cited.

Chamaeraphis glauca var. *imberbis* Kuntze, Rev. Gen. Pl. 2: 767.   1891. Based on *Panicum imberbe* Poir.

Chamaeraphis glauca var. *penicillata* Kuntze, Rev. Gen. Pl. 2: 767.   1891. Based on *Panicum penicillatum* Willd.

Chamaeraphis glauca var. *geniculata* Kuntze, Rev. Gen. Pl. 2: 767.   1891. Based on *Panicum geniculatum* Lam.

Setaria perennis Hall ex Smyth, Check List Pl. Kans. 26.   1892. [Hutchinson] Kans., *Smyth*.

Setaria gracilis var. *dasyura* Arech., An. Mus. Nac. Montevideo 1: 165.   1894. Based on *Panicum dasyurum* Nees.

Chamaeraphis ventenatii Beal, Grasses N. Amer. 2: 153.   1896. Based on *Setaria ventenatii* Kunth.

Chamaeraphis glauca var. *laevigata* Beal, Grasses N. Amer. 2: 155.   1896. Based on *Panicum laevigatum* Muhl.

Chamaeraphis glauca var. *perennis* Beal, Grasses N. Amer. 2: 156.   1896. Florida, *Curtiss* 3614.*

Chaetochloa imberbis Scribn., U. S. Dept. Agr., Div. Agrost. Bul. 4: 39.   1897. Based on *Panicum imberbe* Poir.

Chaetochloa penicillata Scribn., U. S. Dept. Agr., Div. Agrost. Bul. 4: 39. 1897. Based on *Panicum penicillatum* Willd.

Chaetochloa flava Scribn., U. S. Dept. Agr., Div. Agrost. Bul. 4: 39.   1897. Based on *Panicum flavum* Nees.

Chaetochloa versicolor Bicknell, Torrey Bot. Club Bul. 25: 105. pl. 329.   1898. New York City, *Bicknell*.

Chaetochloa perennis Bicknell, Torrey Bot. Club Bul. 25: 107.   1898. Based on *C. glauca* var. *perennis* Beal.

Chaetochloa laevigata Scribn., U. S. Dept. Agr., Div. Agrost. Bul. 21: 10.   1900. Based on *Panicum laevigatum* Muhl.

Chaetochloa imberbis penicillata Scribn. and Merr., U. S. Dept. Agr., Div. Agrost. Bul. 21: 11. f. 2.   1900. Based on *Panicum penicillatum* Willd.

Chaetochloa imberbis perennis Scribn. and Merr., U. S. Dept. Agr., Div. Agrost. Bul. 21: 12.   1900. Based on *Setaria perennis* Hall.

Chaetochloa imberbis geniculata Scribn. and Merr., U. S. Dept. Agr., Div. Agrost. Bul. 21: 12.   1900. Based on *Panicum geniculatum* Lam.

Chaetochloa imberbis streptobotrys Scribn. and Merr., U. S. Dept. Agr., Div. Agrost. Bul. 21: 13.   1900. Based on *Setaria streptobotrys* Fourn.

Chaetochloa purpurascens Scribn. and Merr., U. S. Dept. Agr., Div. Agrost. Bul. 21: 13.   1900. Based on *Setaria purpurascens* H. B. K.

Chaetochloa gracilis Scribn. and Merr.,
U. S. Dept. Agr., Div. Agrost. Bul. 21:
15. 1900. Based on *Setaria gracilis*
H. B. K.

Chaetochloa corrugata parviflora Scribn.
and Merr., U. S. Dept. Agr., Div.
Agrost. Bul. 21: 24. 1900. Based on
Cenchrus parviflorus Poir.

Ixophorus glaucus-laevigata Chapm. ex
Gattinger, Tenn. Fl. 38. 1901. Presumably based on *Setaria glauca* var.
laevigata Chapm.

Panicum glaberrimum Ell. ex. Scribn. and
Merr., U. S. Dept. Agr., Div. Agrost.
Cir. 29: 3. 1901. Not *P. glaberrimum*
Steud., 1854. As synonym of *Chaeto-
chloa imberbis* Scribn.

Chaetochloa ventenatii Nash in Kearney,
U. S. Natl. Herb. Contrib. 5: 515.
1901. Based on *Setaria ventenatii*
Kunth.

Chaetochloa occidentalis Nash in Britton,
Man. 90. 1901. Kansas [type, Hut-
chinson, *Smyth*] and Oklahoma.

Panicum imberbe var. *gracile* Kneucker,
Allg. Bot. Ztschr. 8: 13. 1902. Based
on *Setaria gracilis* H. B. K.

Setaria glauca var. *geniculata* Urban,
Symb. Antill. 4: 96. 1903. Based on
Panicum geniculatum Lam.

Setaria glauca var. *purpurascens* Urban,
Symb. Antill. 4: 96. 1903. Based on
S. purpurascens H. B. K.

Chaetochloa geniculata Millsp. and Chase,
Field Mus. Bot. 3: 37. 1903. Based
on *Panicum geniculatum* Lam.

Chamaeraphis imberbis Kuntze ex Stuck.,
An. Mus. Nac. Buenos Aires 11: 76.
1904. Based on *Panicum imberbe* Poir.

Chamaeraphis gracilis Kuntze ex Stuck.,
An. Mus. Nac. Buenos Aires 11: 76.
1904. Not *C. gracilis* Hack., 1885.
Based on *Setaria gracilis* H. B. K.

Chamaeraphis penicillata Presl. ex Stuck.,
An. Mus. Nac. Buenos Aires 11: 76.
1904. Based on *Setaria penicillata* Presl.

Setaria imberbis var. *perennis* Hitchc.,
Rhodora 8: 210. 1906. Based on *S.
perennis* Hall.

Setaria imberbis var. *purpurascens* Hack.
in Stuck., An. Mus. Nac. Buenos Aires
13: 442. 1906. Based on *S. purpurascens* H. B. K.

Chaetochloa imberbis versicolor Stone, N. J.
Mus. Ann. Rpt. 1910: 213. 1911.
Based on *C. versicolor* Bicknell.

Panicum versicolor Nieuwl., Amer. Midl.
Nat. 2: 64. 1911. Not *P. versicolor*
Doell, 1877. Based on *Chaetochloa
versicolor* Bicknell.

Panicum occidentale Nieuwl., Amer. Midl.
Nat. 2: 64. 1911. Not *P. occidentale*
Scribn., 1899. Based on *Chaetochloa occi-
dentalis* Nash.

Chaetochloa geniculata var. *perennis* House,
N. Y. State Mus. Bul. 254: 85. 1924.
Based on *Setaria perennis* Hall.

Chaetochloa viridis var. *purpurascens*
Honda, Bot. Mag. Tokyo 38: 197.
1924. Based on *Setaria purpurascens*
H. B. K.

Panicum lutescens var. *flavum* Backer,
Handb. Fl. Java 2: 142. 1928. Based
on *P. flavum* Nees.

Chaetochloa geniculata var. *purpurascens*
Farwell, Mich. Acad. Sci. Papers 26:
5. 1941. Based on *Panicum glaucum*
var. *purpurascens* Ell.

(10) **Setaria grisebachii** Fourn., Mex. Pl. 2:
45. 1886. Orizaba, Mexico, [*Schaffner*
36].

Setaria laevis Fourn., Mex. Pl. 2: 45.
1886. Bernal, Mexico, *Karwinsky* 961.

Chaetochloa grisebachii Scribn., U. S.
Dept. Agr., Div. Agrost. Bul. 4: 39.
1897. Based on *Setaria grisebachii*
Fourn.

Chaetochloa grisebachii var. *ampla* Scribn.
and Merr., U. S. Dept. Agr., Div.
Agrost. Bul. 21: 36. f. 21. 1900.
Federal District, Mexico, *Pringle* 4670
[error for 6470].

Chaetochloa grisebachii var. *mexicana*
Scribn. and Merr., U. S. Dept. Agr.,
Div. Agrost. Bul. 21: 37. 1900. San
Luis Potosí, Mexico, *Schaffner* 1044.

Setaria mexicana Schaffn. ex Scribn. and
Merr., U. S. Dept. Agr., Div. Agrost.
Bul. 21: 37. 1900, as synonym of
Chaetochloa grisebachii var. *mexicana*
Scribn. and Merr.

(14) **Setaria italica** (L.) Beauv., Ess. Agrost.
51, 170, 178. 1812. Based on *Panicum
italicum* L.

Panicum italicum L., Sp. Pl. 56. 1753.
India.

Panicum germanicum Mill., Gard. Dict.
ed. 8. *Panicum* No. 1. 1768. Europe.

Panicum italicum var. *germanicum* Koel.,
Descr. Gram. 17. 1802. Europe.

Pennisetum italicum R. Br., Prodr. Fl.
Nov. Holl. 1: 195. 1810. Based on
Panicum italicum L.

Setaria germanica Beauv., Ess. Agrost.
51, 169, 178. 1812. Based on *Panicum
germanicum* Willd. (same as Mill.
1768).

Pennisetum germanicum Baumg., Enum.
Stirp. Transsilv. 3: 277. 1816. Based
on *Setaria germanica* Beauv.

Setaria italica var. *germanica* Schrad.,
Linnaea 12: 430. 1838. Based on
Panicum germanicum Roth (same as
Mill. 1768).

Setaria californica Kellogg, Calif. Acad.
Sci. Proc. 1 (ed. 2): 26. 1873. Shasta,
Calif., *Dash*.

Panicum italicum var. *californicum* Koern.
and Wern., Handb. Getreidebau. 1:
272, 273. 1885. California.

Chamaeraphis italica Kuntze, Rev. Gen.
Pl. 2: 767. 1891. Based on *Panicum
italicum* L.

Chamaeraphis italica var. *germanica*

Kuntze, Rev. Gen. Pl. 2: 768. 1891. Based on *Panicum germanicum* L. (error for Mill.).

Ixophorus italicus Nash, Torrey Bot. Club Bul. 22: 423. 1895. Based on *Panicum italicum* L.

Chaetochloa italica Scribn., U. S. Dept. Agr., Div. Agrost. Bul. 4: 39. 1897. Based on *Panicum italicum* L.

Chaetochloa italica germanica Scribn., U. S. Dept. Agr., Div. Agrost. Bul. 6: 32. 1897. Based on *Panicum germanicum* Mill.

Chaetochloa germanica Smyth, Kans. Acad. Trans. 25: 89. 1913. Based on *Panicum germanicum* Mill.

Setaria italica subsp. *stramineofructa* subvar. *germanica* Hubb., Amer. Jour. Bot. 2: 189. 1915. Based on *Panicum germanicum* Mill.

Setaria italica subsp. *stramineofructa* var. *brunneoseta* subvar. *densior* Hubb., Amer. Jour. Bot. 2: 192. 1915. Weston, Mass., *Williams* in 1895.

(9) **Setaria liebmanni** Fourn., Mex. Pl. 2: 44. 1886. Mexico, *Liebmann* 389.

Setaria rariflora Presl, Rel. Haenk. 1: 313. 1830. Not *S. rariflora* Mikan, 1821. Acapulco, Mexico, *Haenke*.

Panicum rariflorum Presl ex Steud., Syn. Pl. Glum. 1: 51. 1854. Not *P. rariflorum* Lam., 1798. Based on *Setaria rariflora* Presl.

Chamaeraphis caudata var. *pauciflora* Vasey ex Beal, Grasses N. Amer. 2: 158. 1896. [Baja] California [type, Guaymas, Mexico], *Palmer* 191.

Chaetochloa liebmanni Scribn. and Merr., U. S. Dept. Agr., Div. Agrost. Bul. 21: 31. 1900. Based on *Setaria liebmanni* Fourn.

Chaetochloa liebmanni pauciflora Scribn. and Merr., U. S. Dept. Agr., Div. Agrost. Bul. 21: 33. 1900. Based on *Chamaeraphis caudata* var. *pauciflora* Vasey.

(1) **Setaria lutescens** (Weigel) Hubb., Rhodora 18: 232. 1916. Based on *Panicum lutescens* Weigel.

Panicum lutescens Weigel, Obs. Bot. 20. 1772. Germany.

Panicum glaucum var. *elongatum* Pers., Syn. Pl. 1: 81. 1805. America.

Panicum glaucum var. *flavescens* Ell., Bot. S. C. and Ga. 1: 113. 1816. Presumably South Carolina.

Panicum glaucum var. *laevigatum* Le-Conte ex Torr., in Eaton, Man. Bot. ed. 2. 339. 1818. Northern and Middle States.

Setaria glauca var. *elongata* Raddi, Agrost. Bras. 49. 1823. Based on *Panicum glaucum* var. *elongatum* Pers.

Panicum compressum Balb. ex Steud., Nom. Bot. ed. 2. 2: 254. 1841, erroneously cited as synonym of *P. glaucum* R. Br. [Dominican Republic, *Bertero*.]

Chaetochloa lutescens Stuntz, U. S. Dept. Agr., Bur. Plant Indus. Inventory Seeds 31: 36, 86. 1914. Based on *Panicum lutescens* Weigel.

Chaetochloa glauca var. *purpurea* Farwell, Mich. Acad. Sci. Papers 26: 5. 1941. Detroit, Mich., *Farwell* 5661.

Panicum glaucum L. has been shown to apply to pearl millet (see *Pennisetum glaucum*, p. 727). The name at an early date came to be used for the species here called *Setaria lutescens*. The following names have been misapplied to this species:

Panicum glaucum L., Sp. Pl. 56. 1753.

Setaria glauca Beauv., Ess. Agrost. 51, 178. 1812.

Chamaeraphis glauca Kuntze, Rev. Gen. Pl. 2: 767. 1891.

Ixophorus glaucus Nash, Torrey Bot. Club Bul. 22: 423. 1895.

Chaetochloa glauca Scribn., U. S. Dept. Agr., Div. Agrost. Bul. 4: 39. 1897.

(4) **Setaria macrosperma** (Scribn. and Merr.) Schum., Just's Bot. Jahresber. 28¹: 417. 1902. Based on *Chaetochloa macrosperma* Scribn. and Merr.

Chaetochloa macrosperma Scribn. and Merr., U. S. Dept. Agr., Div. Agrost. Bul. 21: 33. f. 18. 1900. St. Johns River, Fla., *Curtiss* 3617.

(6) **Setaria macrostachya** H. B. K., Nov. Gen. et Sp. 1: 110. 1815. [Guanajuato], Mexico, *Humboldt* and *Bonpland*.

Panicum macrostachyum Nees, Agrost. Bras. 245. 1829. Based on *Setaria macrostachya* H. B. K.

Chamaeraphis setosa var. *macrostachya* Kuntze, Rev. Gen. Pl. 2: 769. 1891. Based on *Setaria macrostachya* H. B. K.

Chaetochloa gibbosa Scribn. and Merr., U. S. Dept. Agr., Div. Agrost. Bul. 21: 24. 1900. Mexico [probably Tamaulipas], *Berlandier* 528.

Chaetochloa leucopila Scribn. and Merr., U. S. Dept. Agr., Div. Agrost. Bul. 21: 26, f. 14. 1900. Parras, Coahuila, *Palmer* 1363 in 1880.

Chaetochloa macrostachya Scribn. and Merr., U. S. Dept. Agr., Div. Agrost. Bul. 21: 29. f. 16. 1900. Based on *Setaria macrostachya* H. B. K.

Chaetochloa rigida Scribn. and Merr., U. S. Dept. Agr., Div. Agrost. Bul. 21: 30. 1900. La Paz, Baja California, *Palmer* 125 in 1890.

Setaria leucopila Schum. in Just's Bot. Jahresber. 28¹: 417. 1902. Based on *Chaetochloa leucopila* Scribn. and Merr.

Setaria gibbosa Schum. in Just's Bot. Jahresber. 28¹: 417. 1902. Based on *Chaetochloa gibbosa* Scribn. and Merr.

Setaria rigida Schum. in Just's Bot. Jahresber. 28¹: 417. 1902. Not *S. rigida* Stapf, 1899. Based on *Chaetochloa rigida* Scribn. and Merr.

Chamaeraphis macrostachya Kuntze ex

Stuck., An. Mus. Nac. Buenos Aires 11: 76. 1904. Based on *Setaria macrostachya* H. B. K.
Setaria commutata Hack. ex Stuck., An. Hist. Nat. Buenos Aires 13: 439. 1906. Based on *Chaetochloa composita* as described and figured by Scribner and Merrill (U. S. Dept. Agr., Div. Agrost. Bul. 21: 27. f. 15. 1900), not *Setaria composita* H. B. K., on which the name *Chaetochloa composita* Scribn. is based. The name is published as "*Setaria commutata* (Scribn.) Hack."
Setaria caudata var. *pauciflora* Jones, West. Bot. Contrib. 16: 13. 1930. Arizona, *Jones* 24697, 24698.

(11) **Setaria magna** Griseb., Fl. Brit. W. Ind. 554. 1864. Jamaica, *Purdie.*
Chamaeraphis magna Beal, Grasses N. Amer. 2: 152. 1896. Based on *Setaria magna* Griseb.
Chaetochloa magna Scribn., U. S. Dept. Agr., Div. Agrost. Bul. 4: 39. 1897. Based on *Setaria magna* Griseb.

Setaria nigrirostris (Nees) Dur. and Schinz, Consp. Fl. Afr. 5: 774. 1894. Based on *Panicum nigrirostris* Nees.
Panicum nigrirostris Nees, Fl. Afr. Austr. 55. 1841. South Africa.
Chaetochloa nigrirostris Skeels, U. S. Dept. Agr., Bur. Plant Indus. Bul. 207: 22. 1911. Based on *Panicum nigrirostris* Nees.

Setaria palmifolia (Koen.) Stapf, Linn. Soc. Jour. Bot. 42: 186. 1914. Based on *Panicum palmifolium* Koen. (Naturforscher 23: 208. 1788, same as *P. palmifolium* Willd.)
Panicum plicatum Willd., Enum. Pl. 1033. 1809. Asia. Not *P. plicatum* Lam., 1791.
Panicum palmifolium Willd. ex Poir., in Lam., Encycl. Sup. 4: 282. 1816. Based on *P. plicatum* Willd.
Chamaeraphis palmifolia Kuntze, Rev. Gen. Pl. 2: 771. 1891. Based on *Panicum palmifolium* Willd.
Chaetochloa palmifolia Hitchc. and Chase, U. S. Natl. Herb. Contrib. 18: 348. 1917. Based on *Panicum palmifolium* Willd.

Setaria poiretiana (Schult.) Kunth, Rév. Gram. 1: 47. 1829. Based on *Panicum poiretianum* Schult.
Panicum elongatum Poir. in Lam., Encycl. Sup. 4: 278. 1816. Not *P. elongatum* Salisb., 1796, nor Pursh, 1814. Brazil.
Panicum poiretianum Schult., Mantissa 2: 229. 1824. Based on *P. elongatum* Poir.
Chaetochloa poiretiana Hitchc., U. S. Natl. Herb. Contrib. 22: 159. 1920. Based on *Panicum poiretianum* Schult.

Setaria rariflora Mikan ex Trin., in Spreng., Neu. Entd. 2: 78. 1821. Brazil.
Chaetochloa rariflora Hitchc. and Chase,

U. S. Natl. Herb. Contrib. 18: 349. 1917. Based on *Setaria rariflora* Mikan.
Panicum rariflorum Makino and Nemoto, Fl. Jap. 1475. 1925. Not *P. rariflora* Lam., 1798. Based on *Setaria rariflora* Mikan.

(7) **Setaria scheelei** (Steud.) Hitchc., Biol. Soc. Wash. Proc. 41: 163. 1928. Based on *Panicum scheelei* Steud.
Setaria polystachya Scheele, Linnaea 22: 339. 1849. Not *S. polystachya* Schrad., 1824. New Braunfels, Tex., *Lindheimer* 564.
Panicum scheelei Steud., Syn. Pl. Glum. 1: 51. 1854. Based on *Setaria polystachya* Scheele.
Chaetochloa polystachya Scribn. and Merr., U. S. Dept. Agr., Div. Agrost. Bul. 21: 37. f. 22. 1900. Based on *Setaria polystachya* Scheele.
Chaetochloa scheelei Hitchc., U. S. Natl. Herb. Contrib. 12: 207. f. 62. 1920. Based on *Panicum scheelei* Steud.

Setaria setosa (Swartz) Beauv., Ess. Agrost. 51, 178. 1812. Based on *Panicum setosum* Swartz.
Panicum setosum Swartz, Prodr. Veg. Ind. Occ. 22. 1788. Jamaica, *Swartz.*
Panicum caudatum Lam., Tabl. Encycl. 1: 171. 1791. Brazil.
Setaria caudata Roem. and Schult., Syst. Veg. 2: 495. 1817. Based on *Panicum caudatum* Lam.
Setaria setosa var. *caudata* Griseb., Fl. Brit. W. Ind. 555. 1864. Based on *Setaria caudata* Roem. and Schult.
Pennisetum swartzii F. Muell., Fragm. Phyt. Austr. 8: 110. 1873. Based on *Panicum setosum* Swartz.
Chamaeraphis setosa Kuntze, Rev. Gen. Pl. 2: 768. 1891. Based on *Panicum setosum* Swartz.
Chamaeraphis caudata Britton, Ann. N. Y. Acad. Sci. 7: 264. 1893. Based on *Panicum caudatum* Lam.
Chaetochloa setosa Scribn., U. S. Dept. Agr., Div. Agrost. Bul. 4: 39. 1897. Based on *Panicum setosum* Swartz.
Chaetochloa caudata Scribn., Mo. Bot. Gard. Rpt. 10: 52. 1899. Based on *Panicum caudatum* Lam.

Setaria sphacelata (Schum.) Stapf and C. E. Hubb., Kew Bul. Misc. Inf. 1929: 184, 195. 1929 (basis not given); in Prain, Fl. Trop. Afr. 9: 795. 1930. Based on *Panicum sphacelatum* Schum.
Panicum sphacelatum Schum., Beskr. Guin. Pl. 78. 1827. Guinea, Africa.
Setaria aurea Hochst., Flora 24: 276. 1841. Abyssinia.

(3) **Setaria verticillata** (L.) Beauv., Ess. Agrost. 51, 178. 1812. Based on *Panicum verticillatum* L.
Panicum verticillatum L., Sp. Pl. ed. 2. 1: 82. 1762. Europe.
Pennisetum verticillatum R. Br., Prodr. Fl.

Nov. Holl. 195. 1810. Based on *Panicum verticillatum* L.

Chamaeraphis italica var. *verticillata* Kuntze, Rev. Gen. Pl. 2: 768. 1891. Based on *Panicum verticillatum* L.

Chamaeraphis verticillata Porter, Torrey Bot. Club Bul. 20: 196. 1893. Based on *Panicum verticillatum* L.

Ixophorus verticillatus Nash, Torrey Bot. Club Bul. 22: 422. 1895. Based on *Panicum verticillatum* L.

Chaetochloa verticillata Scribn., U. S. Dept. Agr., Div. Agrost. Bul. 4: 39. 1897. Based on *Panicum verticillatum* L.

Chaetochloa brevispica Scribn. and Merr., U. S. Dept. Agr., Div. Agrost. Bul. 21: 15. f. 5. 1900. Published as a new name for *Panicum verticillatum* var. *parviflorum* Doell, the identity of which is uncertain. The plants described and figured by Scribner and Merrill are *S. verticillata*.

Setaria brevispica Schum. in Just's Bot. Jahresber. 28¹: 417. 1902. Based on *Chaetochloa brevispica* Scribn. and Merr.

Chaetochloa verticillata var. *breviseta* (Godr.) Farwell, Mich. Acad. Sci. Papers 1: 86. 1921, based on a European type not examined.

SETARIA VERTICILLATA var. AMBIGUA (Guss.) Parl., Fl. Palerm. 1: 36. 1845. Based on *Panicum verticillatum* var. *ambiguum* Guss.

Panicum verticillatum var. *ambiguum* Guss., Fl. Sic. Prodr. 80. 1827. Sicily.

Setaria ambigua Guss., Fl. Sic. Syn. 1: 114. 1842. Not *S. ambigua* Merat, 1836. Based on *Panicum verticillatum* var. *ambiguum* Guss.

Setaria viridis var. *ambigua* Coss. and Dur., Expl. Sci. Alger. 2: 36. 1854. Based on *S. ambigua* Guss.

Panicum ambiguum Hausskn., Oesterr. Bot. Ztschr. 25: 345. 1875. Based on *Setaria ambigua* Guss.

Setaria viridis var. *purpurascens* Peck ex Dudley, Cornell Univ. Bul. 2: 122. 1886. Not *S. viridis* var. *purpurascens* Peterm. 1838. New York, *Peck*.

Chamaeraphis italica var. *ambigua* Kuntze, Rev. Gen. Pl. 2: 768. 1891. Based on *Setaria ambigua* Guss.

Chaetochloa ambigua Scribn. and Merr., U. S. Dept. Agr., Div. Agrost. Bul. 21: 18. f. 7. 1900. Based on *Setaria verticillata* var. *ambigua* Guss.

(5) **Setaria villosissima** (Scribn. and Merr.) Schum., Just's Bot. Jahresber. 28¹: 417. 1902. Based on *Chaetochloa villosissima* Scribn. and Merr.

Chaetochloa villosissima Scribn. and Merr., U. S. Dept. Agr., Div. Agrost. Bul. 21: 34. f. 19. 1900. San Diego, Tex., *J. G. Smith* in 1897.

(12) **Setaria viridis** (L.) Beauv., Ess. Agrost.

51, 178. 1812. Based on *Panicum viride* L.

Panicum viride L., Syst. Nat. ed. 10. 2: 870. 1759. Europe.

Pennisetum viride R. Br., Prodr. Fl. Nov. Holl. 1: 195. 1810. Based on *Panicum viride* L.

Setaria weinmanni Roem. and Schult., Syst. Veg. 2: 490. 1817. Europe.

Panicum viride var. *brevisetum* Doell, Rhein. Fl. 128. 1843. Europe.

Setaria viridis var. *weinmanni* Borbás, Math. Termesz. Közlem. 15: 310. 1878. Based on *Setaria weinmanni* Roem. and Schult.

Panicum italicum var. *viride* Koern., in Koern. and Wern., Handb. Getreidebau. 1: 277. 1885. Based on *Panicum viride* L.

Chamaeraphis italica var. *viridis* Kuntze, Rev. Gen. Pl. 2: 767. 1891. Based on *Panicum viride* L.

Chamaeraphis viridis Millsp., W. Va. Agr. Expt. Sta. Bul. 2: 466. 1892. Based on *Panicum viride* L.

Ixophorus viridis Nash, Torrey Bot. Club Bul. 22: 423. 1895. Based on *Panicum viride* L.

Chaetochloa viridis Scribn., U. S. Dept. Agr., Div. Agrost. Bul. 4: 39. 1897. Based on *Panicum viride* L.

Setaria viridis var. *breviseta* Hitchc., Rhodora 8: 210. 1906. Based on *Panicum viride* var. *brevisetum* Doell.

Setaria italica subsp. *viridis* Thell., Mém. Soc. Sci. Nat. Cherbourg 38: 85. 1912. Based on *Panicum viride* L.

Chaetochloa viridis var. *breviseta* Farwell, Mich. Acad. Sci. Papers 1: 86. 1921. Based on *Panicum viride* var. *brevisetum* Doell.

Chaetochloa viridis var. *weinmanni* House, N. Y. State Mus. Bul. 243–244: 39. 1923. Based on *Setaria weinmanni* Roem. and Schult.

Chaetochloa viridis var. *major* (Gaudin) Farwell, Mich. Acad. Sci. Papers 1: 86. 1921, and *C. viridis* var. *minor* (Koch) Farwell (l.c.) based on European types not examined.

(65) SIEGLINGIA Bernh.

(1) **Sieglingia decumbens** (L.) Bernh., Syst. Verz. Erf. 20: 44. 1800. Based on *Festuca decumbens* L.

Festuca decumbens L., Sp. Pl. 75. 1753. Europe.

Poa decumbens Scop., Fl. Carn. ed. 2. 1: 69. 1772. Based on *Festuca decumbens* L.

Danthonia decumbens Lam. and DC., Fl. Franç. ed. 3, 33. 1805. Based on *Festuca decumbens* L.

Triodia decumbens Beauv., Ess. Agrost. 76, 160. pl. 15. f. 9. 1812. Based on "*Danthonia decumbens* Decand."

SINOCALAMUS McClure

Sinocalamus oldhami (Munro) McClure, Lingnan Univ. Sci. Bul. 9: 67. 1940. Based on *Bambusa oldhami* Munro.
Bambusa oldhami Munro, Trans. Linn. Soc. 26: 109. 1868. Formosa.

(47) SITANION Raf.

(1) **Sitanion hanseni** (Scribn.) J. G. Smith, U. S. Dept. Agr., Div. Agrost. Bul. 18: 20. 1899. Based on *Elymus hanseni* Scribn.
Elymus hanseni Scribn., U. S. Dept. Agr., Div. Agrost. Bul. 11: 56. f. 12. 1898. Amador County, Calif., *Hansen* 1742.
Sitanion planifolium J. G. Smith, U. S. Dept. Agr., Div. Agrost. Bul. 18: 19. 1899. Skamania County, Wash., *Suksdorf* 224.
Sitanion anomalum J. G. Smith, U. S. Dept. Agr., Div. Agrost. Bul. 18: 20. pl. 4. 1899. Pasadena, Calif., *Allen* in 1885.
Sitanion leckenbyi Piper, Erythea 7: 100. 1899. Wawawai, Wash., *Piper* 3003.
Sitanion rubescens Piper, Torrey Bot. Club Bul. 30: 234. 1903. Mount Rainier, Wash., *Piper* 1954.
Elymus leckenbyi Piper, U. S. Natl. Herb. Contrib. 11: 151. 1906. Based on *Sitanion leckenbyi* Piper.
Sitanion hanseni anomalum Hitchc., Biol. Soc. Wash. Proc. 41: 160. 1928. Based on *S. anomalum* J. G. Smith.

(3) **Sitanion hystrix** (Nutt.) J. G. Smith, U. S. Dept. Agr., Div. Agrost. Bul. 18: 15. pl. 2. 1899. Based on *Aegilops hystrix* Nutt.
Aegilops hystrix Nutt., Gen. Pl. 1: 86. 1818. Plains of the Missouri.
Sitanion elymoides Raf., Jour. Phys. Chym. 89: 103. 1819. Missouri [River].
Elymus sitanion Schult., Mantissa 2: 426. 1824. Based on *Sitanion elymoides* Raf.
Polyanthrix hystrix Nees, Ann. Nat. Hist. 1: 284. 1838. Based on *Aegilops hystrix* Nutt., but misapplied to *S. jubatum*.
Elymus elymoides Swezey, Nebr. Pl. Doane Col. 15. 1891. Based on *Sitanion elymoides* Raf.
Sitanion minus J. G. Smith, U. S. Dept. Agr., Div. Agrost. Bul. 18: 12. 1899. Jacumba Hot Springs, Calif., *Schoenefeldt* 3277.
Sitanion rigidum J. G. Smith, U. S. Dept. Agr., Div. Agrost. Bul. 18: 13. 1899. Cascade Mountains, Wash., *Allen* 178.
Sitanion californicum J. G. Smith, U. S. Dept. Agr., Div. Agrost. Bul. 18: 13. 1899. San Bernardino Mountains, Calif., *Parish* 3295.
Sitanion glabrum J. G. Smith, U. S. Dept. Agr., Div. Agrost. Bul. 18: 14. 1899. Coso Mountains, Calif., *Coville* and *Funston* 914.
Sitanion cinereum J. G. Smith, U. S.

Dept. Agr., Div. Agrost. Bul. 18: 14. 1899. Reno, Nev., *Tracy* 222.
Sitanion insulare J. G. Smith, U. S. Dept. Agr., Div. Agrost. Bul. 18: 14. 1899. Carrington Island, Salt Lake, Utah, *Watson* 1338.
Chretomeris trichoides Nutt. ex J. G. Smith, U. S. Dept. Agr., Div. Agrost. Bul. 18: 15. 1899, as synonym of *Sitanion hystrix*.
Elymus difformis Nutt. ex J. G. Smith, U. S. Dept. Agr., Div. Agrost. Bull. 18: 15. 1899, as synonym of *Sitanion hystrix*.
Sitanion montanum J. G. Smith, U. S. Dept. Agr., Div. Agrost. Bul. 18: 16. 1899. Spanish Creek, Mont., *Rydberg* 3091.
Sitanion caespitosum J. G. Smith, U. S. Dept. Agr., Div. Agrost. Bul. 18: 16. 1899. Cliff, N. Mex., *J. G. Smith* in 1897.
Sitanion strigosum J. G. Smith, U. S. Dept. Agr., Div. Agrost. Bul. 18: 17. 1899. Sheep Creek, Mont., *Rydberg* 3298.
Sitanion molle J. G. Smith, U. S. Dept. Agr., Div. Agrost. Bul. 18: 17. 1899. Larimer County, Colo., *Shear* and *Bessey* 1469.
Sitanion brevifolium J. G. Smith, U. S. Dept. Agr., Div. Agrost. Bul. 18: 17. 1899. Tucson, Ariz., *Toumey* 797.
Sitanion longifolium J. G. Smith, U. S. Dept. Agr., Div. Agrost. Bul. 18: 18. 1899. Silverton, Colo., *Shear* 1213.
Sitanion pubiflorum J. G. Smith, U. S. Dept. Agr., Div. Agrost. Bul. 18: 19. 1899. Tucson, Ariz., *Toumey* 795.
Sitanion latifolium Piper, Erythea 7: 99. 1899. Blue Mountains, Walla Walla County, Wash., *Piper* in 1896.
Sitanion marginatum Scribn. and Merr., Torrey Bot. Club Bul. 29: 469. 1902. Leigh Lake, Teton Mountains, Wyo., *Merrill* and *Wilcox* 334.
Elymus glaber Davy, Calif. Univ. Pubs., Bot. 1: 57. 1902. Based on *Sitanion glabrum* J. G. Smith.
Elymus pubiflorus Davy, Calif. Univ. Pubs., Bot. 1: 58. 1902. Based on *Sitanion pubiflorum* J. G. Smith.
Sitanion velutinum Piper, Torrey Bot. Club Bul. 30: 233. 1903. Steptoe, Wash., *G. R. Vasey* in 1901.
Sitanion basalticola Piper, Torrey Bot. Club Bul. 30: 234. 1903. Coulee City, Wash., *Piper* 3924.
Sitanion albescens Elmer, Bot. Gaz. 36: 57. 1903. Ellensburg, Wash., *Whited* 670.
Sitanion ciliatum Elmer, Bot. Gaz. 36: 58. 1903. Wenatchee, Wash., *Whited* in 1901.
Hordeum elymoides Schenck, Bot. Jahrb. Engler 40: 109. 1907. Based on *Sitanion elymoides* Raf.
Elymus brevifolius Jones, West. Bot.

Contrib. 14: 20. 1912. Based on *Sitanion brevifolium* J. G. Smith.
Elymus hystrix Jones, West. Bot. Contrib. 14: 20. 1912. Not *E. hystrix* L. Based on *Aegilops hystrix* Nutt.
Elymus insularis Jones, West. Bot. Contrib. 14: 20. 1912. Based on *Sitanion insulare* J. G. Smith.
Elymus minor Jones, West. Bot. Contrib. 14: 20. 1912. Based on *Sitanion minus* J. G. Smith.
Sitanion rigidum var. *californicum* Smiley, Calif. Univ. Pubs., Bot. 9: 99. 1921. Based on *S. californicum* J. G. Smith.
Sitanion hordeoides Suksdorf, Werdenda 1²: 4. 1923. Spangle, Wash., *Suksdorf* 8705.
(2) **Sitanion jubatum** J. G. Smith, U. S. Dept. Agr., Div. Agrost. Bul. 18: 10. 1899. Waitsburg, Wash., *Horner* 573.
Elymus sitanion jubatum J. G. Smith, U. S. Dept. Agr., Div. Agrost. Bul. 18: 10. 1899, as synonym of *S. jubatum*.
Sitanion villosum J. G. Smith, U. S. Dept. Agr., Div. Agrost. Bul. 18: 11. pl. 1. 1899. Almota, Wash., *Elmer* 266.
Sitanion multisetum J. G. Smith, U. S. Dept. Agr., Div. Agrost. Bul. 18: 11. 1899. Tehachapi Valley, Calif., *Coville* and *Funston* 1121.
Sitanion polyanthrix J. G. Smith, U. S. Dept. Agr., Div. Agrost. Bul. 18: 12. 1899. California, *Douglas.* New name given to the species described by Nees under *Polyanthrix hystrix*, that name being based on *Aegilops hystrix* Nutt.
Sitanion breviaristatum J. G. Smith, U. S. Dept. Agr., Div. Agrost. Bul. 18: 12. 1899. Panamint Mountains, Calif., *Coville* and *Funston* 833.
Sitanion strictum Elmer, Bot. Gaz. 36: 59. 1903. Parker, Wash., *Elmer* in 1898.
Elymus multisetus Jones, West. Bot. Contrib. 14: 20. 1912. Based on *Sitanion multisetum* J. G. Smith.

(157) SORGHASTRUM Nash

(2) **Sorghastrum elliottii** (Mohr) Nash, N. Amer. Fl. 17: 130. 1912. Based on *Chrysopogon elliottii* Mohr.
Chrysopogon elliottii Mohr, Torrey Bot. Club Bul. 24: 21. 1897. Based on *Andropogon nutans* as described by Elliott, not *A. nutans* L.
(1) **Sorghastrum nutans** (L.) Nash in Small, Fl. Southeast. U. S. 66. 1903. Based on *Andropogon nutans* L.
Andropogon nutans L., Sp. Pl. 1045. 1753. "Virginia, Jamaica." [Type eastern America, *Kalm;* cited localities erroneous.]
?*Stipa villosa* Walt., Fl. Carol. 78. 1788. South Carolina.
Andropogon avenaceus Michx., Fl. Bor. Amer. 1: 58. 1803. Illinois, *Michaux.*

Andropogon ciliatus Ell., Bot. S. C. and Ga. 1: 144. 1816. Port Royal, S. C.
Andropogon arenaceus Raf., West. Rev. Misc. Mag. 1: 95. 1819. Error for *A. avenaceus.*
Sorghum nutans A. Gray, Man. 617. 1848. Based on *Andropogon nutans* L.
Sorghum avenaceum Chapm., Fl. South. U. S. 583. 1860. Based on *Andropogon avenaceus* Michx.
Chrysopogon nutans Benth., Linn. Soc. Jour. Bot. 19: 73. 1881. Based on *Andropogon nutans* L.
Chrysopogon avenaceus Benth., Linn. Soc. Jour. Bot. 19: 73. 1881. Based on *Andropogon avenaceus* Michx.
Sorghum nutans subsp. *avenaceum* Hack. in Mart., Fl. Bras. 2³: 274. 1883. Based on *Andropogon avenaceus* Michx.
Sorghum nutans subsp. *linnaeanum* Hack. in Mart., Fl. Bras. 2³: 276. 1883. Based on *Andropogon nutans* L.
Andropogon albescens Fourn., Mex. Pl. 2: 56. 1886. Vera Cruz, Mexico, *Gouin* 53.
Andropogon confertus Trin. ex Fourn., Mex. Pl. 2: 55. 1886. Texas, *Berlandier* 1873.
Andropogon nutans var. *avenaceus* Hack. in DC., Monogr. Phan. 6: 530. 1889. Based on *Andropogon avenaceus* Michx.
Andropogon nutans var. *linnaeanus* Hack. in DC., Monogr. Phan. 6: 531. 1889. Based on *Sorghum nutans* subsp. *linnaeanum* Hack.
Chrysopogon nutans var. *avenaceus* Coville and Branner, Rpt. Geol. Surv. Ark. 4: 234. 1891. Based on *Andropogon avenaceus* Michx.
Poranthera nutans Raf. ex Jacks., Ind. Kew. 2:· 606. 1894, as synonym of *Chrysopogon nutans.*
Poranthera ciliata Raf. ex Jacks., Ind. Kew. 2: 606. 1894, as synonym of *Chrysopogon avenaceus.*
Chrysopogon nutans var. *linnaeanus* Mohr, Torrey Bot. Club Bul. 24: 21. 1897. Based on *Sorghum nutans* subsp. *linnaeanum* Hack.
Sorghastrum avenaceum Nash in Britton, Man. 71. 1901. Based on *Andropogon avenaceus* Michx.
Andropogon linnaeanus Scribn. and Kearn. in Scribn. and Ball., U. S. Dept. Agr., Div. Agrost. Bul. 24: 40. 1901. Based on *Sorghum nutans* subsp. *linnaeanum* Hack.
Sorghastrum linnaeanum Nash in Small, Fl. Southeast. U. S. 66. 1903. Based on *Andropogon nutans* var. *linnaeanus* Hack., but misapplied to *S. elliottii* (Mohr) Nash.
Holcus nutans Kuntze ex Stuck., An. Mus. Nac. Buenos Aires 11: 48. 1904. Presumably based on *Andropogon nutans* L.
Holcus nutans var. *avenaceus* Hack. ex

Stuck., An. Mus. Nac. Buenos Aires 11: 48. 1904. Presumably based on *Andropogon avenaceus* Michx.
Chalcoelytrum nutans Lunell, Amer. Midl. Nat. 4: 212. 1915. Based on *Andropogon nutans* L.

(3) **Sorghastrum secundum** (Ell.) Nash in Small, Fl. Southeast. U. S. 67. 1903. Based on *Andropogon secundus* Ell.
Andropogon secundus Ell., Bot. S. C. and Ga. 1: 580. 1821. Between Flint and Chattahoochee Rivers, Ga.
Sorghum secundum Chapm., Fl. South. U. S. 583. 1860. Based on *Andropogon secundus* Ell.
Chrysopogon secundus Benth. ex Vasey, Grasses U. S. 20. 1883. Based on *Sorghum secundum* Chapm.
Andropogon unilateralis Hack. in DC., Monogr. Phan. 6: 533. 1889. Based on *Sorghum secundum* Chapm.

(156) SORGHUM Moench

(1) **Sorghum halepense** (L.) Pers., Syn. Pl. 1: 101. 1805. Based on *Holcus halepensis* L.
Holcus halepensis L., Sp. Pl. 1047. 1753. Syria.
Blumenbachia halepensis Koel., Descr. Gram. 29. 1802. Based on *Holcus halepensis* L.
Milium halepense Cav., Descr. Pl. 306. 1802. Based on *Holcus halepensis* L.
Andropogon halepensis Brot., Fl. Lusit. 1: 89. 1804. Based on *Holcus halepensis* L.
Andropogon sorghum subsp. *halepensis* Hack. in DC., Monogr. Phan. 6: 501. 1889. Based on *Holcus halepensis* L.
Andropogon halepensis var. *anatherus* Piper, Biol. Soc. Wash. Proc. 28: 28. 1915. Marco, Fla., *Hitchcock* Fla. Pl. 1900. Spikelets awnless.
Sorghum lanceolatum Stapf, in Prain, Fl. Trop. Afr. 9: 112. 1917. Tropical Africa.
Sorghum sudanense (Piper) Stapf in Prain, Fl. Trop. Afr. 9: 113. 1917. Based on *Andropogon sorghum sudanensis* Piper.
Andropogon sorghum sudanensis Piper, Biol. Soc. Wash. Proc. 28: 33. 1915. Grown at Arlington Farm (near Washington, D. C.), seed from Sudan.
Holcus sorghum sudanensis Hitchc., Biol. Soc. Wash. Proc. 29: 128. 1916. Based on *Andropogon sorghum sudanensis* Piper.
Holcus suddnensis Bailey, Gentes Herb. 1: 132. 1923. Based on *Andropogon sorghum sudanensis* Piper.
Sorghum vulgare var. *sudanense* Hitchc., Wash. Acad. Sci. Jour., 17: 147. 1927. Based on *Andropogon sorghum* var, *sudanense* Piper.
Sorghum virgatum (Hack.) Stapf in Prain, Fl. Trop. Afr. 9: 111. 1917. Based on

Andropogon sorghum subsp. *halepensis* var. *virgatus* Hack.
Andropogon sorghum subsp. *halepensis* var. *virgatus* Hack. in DC., Monogr. Phan. 6: 504. 1889. Egypt.
Holcus virgatus Bailey, Gentes Herb. 1: 132. 1923. Based on *Andropogon sorghum* subsp. *halepensis* var. *virgatus* Hack.

(2) **Sorghum vulgare** Pers., Syn. Pl. 1: 101. 1805. Based on *Holcus sorghum* L.
Holcus sorghum L., Sp. Pl. 1047. 1753. India.
Holcus bicolor L., Mant. Pl. 2: 301. 1771. Persia.
Sorghum bicolor Moench, Meth. Pl. 207. 1794. Based on *Holcus bicolor* L.
Andropogon sorghum Brot., Fl. Lusit. 1: 88. 1804. Based on *Holcus sorghum* L.
Holcus cernuus Muhl., Descr. Gram. 276. 1817. Garden plant.
Andropogon vulgaris Raspail, Ann. Sci. Nat., Bot. 5: 307. 1825. Based on *Sorghum vulgare* Pers.
Sorghum vulgare var. *bicolor* Eaton and Wright, N. Amer. Bot. ed. 8. 438. 1840. Not *S. vulgare* var. *bicolor* Schrad., 1838. North America.
Sorghum sorghum Karst., Deut. Fl. 367. f. 189. 1880. Based on *Holcus sorghum* L.
Andropogon sorghum var. *sativus* Hack. in DC., Monogr. Phan. 6: 505. 1889. Group name.
Andropogon sorghum subsp. *sativus* var. *vulgaris* Hack. in DC., Monogr. Phan. 6: 515. 1889. Based on *Sorghum vulgare* Pers.
Andropogon sorghum var. *vulgaris* Hack. ex Hook. f., Fl. Brit. Ind. 7: 184. 1896. Based on *A. sorghum* subsp. *sativus* var. *vulgaris* Hack.
SORGHUM VULGARE var. CAFFRORUM (Retz.) Hubb. and Rehder, Harvard Univ. Bot. Mus. Leaflet 1: 10. 1932. Based on *Holcus caffrorum* Thunb., the same as *Panicum caffrorum* Retz.
Panicum caffrorum Retz., Obs. Bot. 2: 7. 1781. Cape of Good Hope, Africa, grown under the name "Cafferkorn."
Holcus caffrorum Thunb., Prodr. Pl. Cap. 1: 20. 1794. Cape of Good Hope.
Sorghum caffrorum Beauv., Ess. Agrost. 131, 164, 178. 1812. Based on *Holcus caffrorum* Thunb.
Holcus sorghum var. *caffrorum* Bailey, Gentes Herb. 1: 133. 1923. Based on *Holcus caffrorum* Thunb. Retzius' publication not cited.
SORGHUM VULGARE var. DRUMMONDII (Nees) Hack. ex Chiov., Result. Sci. Miss. Stefan.-Paoli Somal. Ital. 1 Coll. Bot. 224. 1916. Based on *Andropogon drummondii* Nees in Steud.
Andropogon drummondii Nees in Steud., Syn. Pl. Glum. 1: 393. 1854. New Orleans, La., *Drummond* 588.

Andropogon sorghum subsp. *sativus* var. *drummcndii* Hack. in DC., Monogr. Phan. 6: 507. 1889. Based on *Andropogon drummondii* Nees.

Sorghum drummondii Nees ex Hack., in DC., Monogr. Phan. 6: 507. 1889, as synonym of *Andropogon sorghum* subsp. *sativus* var. *drummondii* Hack.

Sorghum drummondii Millsp. and Chase, Field Columb. Mus. Publ. Bot. 3: 21. 1903. Based on *Andropogon drummondii* Nees.

Holcus sorghum drummondii Hitchc., Biol. Soc. Wash. Proc. 29: 128. 1916. Based on *Andropogon drummondii* Nees.

SORGHUM VULGARE var. DURRA (Forsk.) Hubb. and Rehder, Harvard Univ. Bot. Mus. Leaflets 1: 10. 1932. Based on *Holcus durra* Forsk.

Holcus durra Forsk., Fl. Aegypt. Arab. 174. 1775. Egypt and Arabia.

Andropogon sorghum subsp. *sativus* var. *durra* Hack., in DC., Monogr. Phan. 6: 516. 1889. Based on *Holcus durra* Forsk.

Holcus sorghum var. *durra* Bailey, Gentes Herb. 1: 132. 1923. Based on *Holcus durra* Forsk.

SORGHUM VULGARE var. ROXBURGHII (Stapf) Haines, Bot. Bihar and Orissa pt. 5: 1034. 1924. Based on *Sorghum roxburghii* Stapf.

Sorghum roxburghii Stapf in Prain, Fl. Trop. Afr. 9: 126. 1917. Africa.

SORGHUM VULGARE var. SACCHARATUM (L.) Boerl., Ann. Jard. Bot. Buitenzorg 8: 69. 1890. Based on *Sorghum saccharatum* Pers.

Holcus saccharatum L., Sp. Pl. 1047. 1753. India.

Sorghum saccharatum Moeñch, Meth. Pl. 207. 1794. Based on *Holcus saccharatus* L. Listed as new Pers., Syn. Pl. 1: 101. 1805, same basis.

Andropogon saccharatus Raspail, Ann. Sci. Nat., Bot. 5: 307. 1825. Based on *Sorghum saccharatum* Pers.

Andropogon sorghum var. *saccharatus* Alefeld, Landw. Fl. 313. 1866. Based on *Holcus saccharatus* L.

Sorghum halepense var. *saccharatum* Goiran, Nuov. Gior. Bot. Ital. n. s. 17: 39. 1910. Based on *Holcus saccharatus* L.

Holcus sorghum var. *saccharatus* Bailey, Gentes Herb. 1: 132. 1923. Based on *Holcus saccharatus* L.

SORGHUM VULGARE var. TECHNICUM (Koern.) Jav. Magyar Fl. 1: 63. 1924. Based on *Andropogon sorghum* var. *technicus* Koern.

Andropogon sorghum var. *technicus* Koern. in Koern. and Wern., Handb. Getreidebau. 1: 308. 1885. Cultivated.

Andropogon sorghum subsp. *sativus* var. *technicus* Koern. ex Hack., in DC.,

Monogr. Phan. 6: 508. 1889. Based on *A. sorghum* var. *technicus* Koern.

Holcus saccharatus var. *technicus* Farwell, Mich. Acad. Sci. Ann. Rpt. 20: 163. 1918. Based on *Andropogon sorghum* var. *technicus* Koern.

Holcus sorghum var. *technicus* Bailey, Gentes Herb. 1: 132. 1923. Based on *Andropogon sorghum* var. *technicus* Koern.

(107) SPARTINA Schreb.

(4) **Spartina alterniflora** Loisel., Fl. Gall. 719. 1807. France.

Dartylis maritima Walt., Fl. Carol. 77. 1788. Not *D. maritima* Curtis, 1787. South Carolina.

Trachynotia alterniflora DC., Fl. Franç. 5: 279. 1815. Based on *Spartina alterniflora* Loisel.

Spartina glabra Muhl. ex Ell., Bot. S. C. and Ga. 1: 95. pl. 4. f. 2. 1816. South Carolina and Georgia. Name only, Muhl., Cat. Pl. 8. 1813.

Limnetis glabra Nutt., Gen. Pl. 1: 38. 1818, name only; Eaton and Wright, N. Amer. Bot. 301. 1840. Presumably based on *Spartina glabra* Muhl.

Spartina laevigata Bosc ex Spreng., Schrad. and Link, Gewächsk. 1³: 92. 1820. North America, *Bosc.*

Trachynotia alternifolia Steud., Nom. Bot. ed. 2. 2: 695. 1841, error for *T. alterniflora.*

Spartina stricta var. *alterniflora* A. Gray, Man. ed. 2. 552. 1856. Based on *S. alterniflora* Loisel.

Spartina stricta var. *glabra* A. Gray, Man. ed. 2. 552. 1856. Based on *S. glabra* Muhl.

Spartina stricta maritima Scribn., Torrey Bot. Club Mem. 5: 45. 1894. Based on *Dactylis maritima* Walt.

Spartina glabra alterniflora Merr., U. S. Dept. Agr., Bur. Plant Indus. Bul. 9: 9. 1902. Based on *Spartina alterniflora* Loisel.

Spartina glabra pilosa Merr., U. S. Dept. Agr., Bur. Plant Indus. Bul. 9: 9. 1902. Atlantic City, N. J., *Scribner* in 1895.

Spartina alterniflora var. *glabra* Fernald, Rhodora 18: 178. 1916. Based on *S. glabra* Muhl.

Spartina alterniflora var. *pilosa* Fernald, Rhodora 18: 179. 1916. Based on *S. glabra pilosa* Merr.

Spartina maritima subsp. *glabra* var. *glabra* St.-Yves, Candollea 5: 24, 49. pl. 1. f. b-2. 1932. Based on *S. glabra* Muhl.

Spartina maritima subsp. *glabra* var. *alterniflora* St.-Yves, Candollea 5: 25, 53. pl. 2. f. a-4. 1932. Based on *S. alterniflora* Loisel.

Spartina maritima subsp. *glabra* subvar.

pilosa St.-Yves, Candollea 5: 51. pl. 1. f. c-3. 1932. Based on *S. glabra pilosa* Merr.
✕*Spartina merrillii* Chevalier, Bul. Soc. France 80: 787. pl. 8. f. 3. 1933. Long Island, N. Y., *Bicknell* 11300.
(6) **Spartina bakeri** Merr., U. S. Dept. Agr., Bur. Plant Indus. Bul. 9: 14. 1902. Lake Ola, Fla., *C. H. Baker* 14.
Spartina juncea var. *bakeri* St.-Yves, Candollea 5: 27, 91. pl. 9. f. c. 1932. Based on *S. bakeri* Merr.
(7) **Spartina caespitosa** A. A. Eaton, Torrey Bot. Club Bul. 25: 338. 1898. Seabrook, N. H., *A. A. Eaton.*
Spartina patens var. *caespitosa* Hitchc., Rhodora 8: 210. 1906. Based on *S. caespitosa* A. A. Eaton.
(2) **Spartina cynosuroides** (L.) Roth, Catal. Bot. 3: 10. 1806. Based on *Dactylis cynosuroides* L.
Dactylis cynosuroides L., Sp. Pl. 71. 1753. Virginia, Canada.
Trachynotia polystachya Michx., Fl. Bor. Amer. 1: 64. 1803. New England to Florida. [Type, South Carolina, *Michaux.*]
Trachynotia cynosuroides Michx., Fl. Bor. Amer. 1: 64. 1803. Based on *Dactylis cynosuroides* L., but misapplied to *S. pectinata.*
Paspalum cynosuroides Brot., Fl. Lusit. 1: 83. 1804. Based on *Dactylis cynosuroides* L.
Limnetis cynosuroides L. Rich. in Pers., Syn. Pl. 1: 72. 1805. Based on *Dactylis cynosuroides* L.
Limnetis polystachia L. Rich. in Pers., Syn. Pl. 1: 72. 1805. Based on *Trachynotia polystachya* Michx.
Spartina polystachya Beauv., Ess. Agrost. 25, 178, 179. 1812. Presumably based on *Trachynotia polystachya* Michx.
Cynodon cynosuroides Raspail, Ann. Sci. Nat., Bot. 5: 303. 1825. Based on *Spartina cynosuroides* Roth.
Spartina cynosuroides var. *polystachya* Beal, Grasses N. Amer. 2: 398. 1896. Based on *Trachynotia polystachya* Michx.
(3) **Spartina foliosa** Trin., Acad. St. Pétersb. Mém. VI. Sci. Nat. 4¹: 114. 1840. California.
Spartina leiantha Benth., Bot. Voy. Sulph. 56. 1844. Bay of Magdalena, Baja California [*Barclay*].
Spartina densiflora var. *typica* subvar. *brongniartii* forma *acuta* St.-Yves, Candollea 5: 76, 81. 1932. Eureka, Calif. [*Heller* 13871.]
(8) **Spartina gracilis** Trin., Acad. St. Pétersb. Mém. VI. Sci. Nat. 4¹: 110. 1840. North America.
(9) **Spartina patens** (Ait.) Muhl., Descr. Gram. 55. 1817. Based on *Dactylis patens* Ait.
Dactylis patens Ait., Hort. Kew. 1: 104.

1789. Grown in England, seed from North America.
Spartina pumila Roth, Catal. Bot. 3: 10. 1806. New York.
Spartina juncea var. *patens* St.-Yves, Candollea 5: 27, 86. 1932. Based on *Dactylis patens* Ait.
SPARTINA PATENS var. MONOGYNA (M. A. Curtis) Fernald, Rhodora 49: 114. 1947. Based on *Limnetis juncea* var. *monogyna* M. A. Curtis.
Trachynotia juncea Michx., Fl. Bor. Amer. 1: 64. 1803. South Carolina and Georgia, *Michaux.*
Limnetis juncea L. Rich. in Pers., Syn. Pl. 1: 72. 1805. Based on *Trachynotia juncea* Michx.
Spartina juncea Willd., Enum. Pl. 81. 1809. Based on *Trachynotia juncea* Michx.
Limnetis juncea var. *monogyna* M. A. Curtis, Boston Jour. Nat. Hist. 1: 136. 1835. Mouth of Cape Fear River, N. C.
Spartina americana Roth ex Trin., Acad. St. Pétersb. Mém. VI. Sci. Nat. 4¹: 109. 1840, as synonym of *S. juncea* Willd.
Spartina patens var. *juncea* Hitchc., Rhodora 8: 210. 1906. Based on *Trachynotia juncea* Michx.
Spartina juncea subvar. *americana* St.-Yves, Candollea 5: 27, 84. pl. 8. f. b-20. 1932. Based on *S. juncea* Willd.
(1) **Spartina pectinata** Link, Jahrb. Gewächsk. 1³: 92. 1820. North America [type collected by Bosc probably at Wilmington, N. C.].
Spartina cynosuroides var. *aureo-marginata* Irving, Gard. Chron. 38: 372. 1905. Grown at Kew Gardens, received from New York Botanical Garden.
Spartina michauxiana Hitchc., U. S. Natl. Herb. Contrib. 12: 153. 1908. Based upon the plant described by Michaux as *Trachynotia cynosuroides* (that name based on *Dactylis cynosuroides* L.). [Near Hudson Bay, *Michaux.*]
Spartina michauxiana var. *suttiei* Farwell, Mich. Acad. Sci. Rpt. 21: 352. 1920. Orchard Lake, Mich., *Suttie.*
Spartina michauxiana var. *tenuior* Farwell, Mich. Acad. Sci. Rpt. 21: 352. 1920. River Rouge, Mich., [*Farwell*] 5138.
Spartina cynosuroides var. *michauxiana* St.-Yves, Candollea 5: pl. 3 f. a-7. 1932. Based on *S. michauxiana* Hitchc.
Spartina cynosuroides var. *michauxiana* forma *major* St.-Yves, Candollea 5: 61, 62. 1932. Canada, *Victorin* 11358; *Victorin* and *Germain* 9055; other specimens cited from Nova Scotia, Newfoundland, Massachusetts, Ohio, Illinois, Minnesota, and Missouri.
Spartina cynosuroides ✕ *gracilis* St.-Yves, Candollea 5: 66. pl. 4. f. b-10. 1932.

* * * "Oregon, Ballards Landing, *Cusick 221* in 1890" [error for 2221 in 1899].

Spartina pectinata var. *suttiei* Fernald, Rhodora 35: 260. 1933. Based on *S. michauxiana* var. *suttiei* Farwell.

(5) **Spartina spartinae** (Trin.) Merr., U. S. Dept. Agr., Bur. Plant Indus. Bul. 9: 11. 1902, as synonym of *S. junciformis* Engelm. and Gray ex Hitchc., U. S. Natl. Herb. Contrib. 17: 329. 1913. Based on *Vilfa spartinae* Trin.

Vilfa spartinae Trin., Acad. St. Pétersb. Mém. VI. Sci. Nat. 4¹: 82. 1840. Texas.

Spartina junciformis Engelm. and Gray, Bost. Soc. Nat. Hist. Jour. 5: 238. 1845. Texas, *Lindheimer* [207].

Spartina gouini Fourn., Mex. Pl. 2: 135. 1886. Vera Cruz, *Gouin* 72.

Spartina multiflora Vasey ex Beal, Grasses N. Amer. 2: 400. 1896, as synonym of *S. junciformis* Engelm. and Gray.

Spartina pittieri Hack., Oesterr. Bot. Ztschr. 52: 237. 1902. Costa Rica, *Pittier* 4209.

Spartina densiflora var. *junciformis* St.-Yves, Candollea 5: 26, 77. pl. 7. f. a-16. 1932. Based on *S. junciformis* Engelm. and Gray.

(56) SPHENOPHOLIS Scribn.

(5) **Sphenopholis filiformis** (Chapm.)Scribn., Rhodora 8: 144. 1906. Based on *Eatonia pennsylvanica* var. *filiformis* Chapm.

Eatonia pennsylvanica var. *filiformis* Chapm., Fl. South. U. S. 560. 1860. Florida [type, *Chapman*] to South Carolina.

Eatonia filiformis Vasey, Bot. Gaz. 11: 117. 1886. Based on *Eatonia pennsylvanica* var. *filiformis* Chapm.

Eatonia hybrida Vasey ex Beal, Grasses N. Amer. 2: 491. 1896. Florida, *Curtiss* in 1886. (The Hunting Creek, Va., specimen referred to is *Trisetum pennsylvanicum* (L.) Beauv., which see.)

Reboulea filiformis Farwell, Mich. Acad. Sci. Rpt. 17: 182. 1916. Based on *Eatonia pennsylvanica* var. *filiformis* Chapm.

(2) **Sphenopholis intermedia** (Rydb.)Rydb., Torrey Bot. Club Bul. 36: 533. 1909. Based on *Eatonia intermedia* Rydb.

Koeleria truncata var. *major* Torr., Fl. North. and Mid. U. S. 1: 117. 1823. Deerfield, Mass., *Cooley*.

Koeleria? pennsylvanica var. *major* Torr., Fl. N. Y. 2: 469. 1843. Based on *Koeleria truncata* var. *major* Torr.

Reboulea pennsylvanica var. *major* A. Gray, Man. 591. 1848. Presumably based on *Koeleria pennsylvanica* var. *major* Torr.

?Aira controversa Steud., Syn. Pl. Glum.

1: 224. 1854. Cincinnati and Miami, Ohio.

?Aira capillacea Frank ex Steud., Syn. Pl. Glum. 1: 224. 1854, as synonym of *A. controversa* Steud.

Eatonia pennsylvanica var. *major* A. Gray, Man. ed. 2. 558. 1856. Presumably based on *Koeleria truncata* var. *major* Torr.

Vilfa alba Buckl., Acad. Nat. Sci. Phila. Proc. 1862: 89. 1862. Not *V. alba* Beauv., 1812. "Oregon, *Spalding*" (locality probably erroneous, the ticket on the type specimen crossed out).

Eatonia intermedia Rydb., Torrey Bot. Club Bul. 32: 602. 1905. East Gallatin Swamps, Mont., *Rydberg* 3174.

Sphenopholis pallens major Scribn., Rhodora 8: 145. 1906. Based on *Koeleria truncata* var. *major* Torr.

Sphenopholis pallens var. *major* Scribn. ex Robinson, Rhodora 10: 65. 1908. Based on *Koeleria truncata* var. *major* Torr.

Reboulea pallens var. *major* Farwell, Mich. Acad. Sci. Rpt. 17: 182. 1916. Based on *Koeleria truncata* var. *major* Torr.

This is the species which has recently been called *Sphenopholis pallens* Scribn., but it is not the same as *Aira pallens* Bieler, on which that name is based.

(3) **Sphenopholis longiflora** (Vasey) Hitchc., Wash. Acad. Sci. Jour. 23: 453. 1933. Based on *Eatonia pennsylvanica* var. *longiflora* Vasey.

Eatonia pennsylvanica var. *longiflora* Vasey ex L. H. Dewey, U. S. Natl. Herb. Contrib. 2: 544. 1894. Houston, Tex., *Nealley* in 1892.

Eatonia longiflora Beal, Grasses N. Amer. 2: 494. 1896. Based on *E. pennsylvanica* var. *longiflora* Vasey.

Sphenopholis pallens longiflora Scribn., Rhodora 8: 145. 1906. Based on *Eatonia pennsylvanica* var. *longiflora* Vasey.

Reboulea pallens var. *longiflora* Farwell, Mich. Acad. Sci. Rpt. 17: 182. 1916. Based on *Eatonia longiflora* Beal.

(4) **Sphenopholis nitida** (Bieler) Scribn., Rhodora 8: 144. 1906. Based on *Aira nitida* Bieler.

Aira nitida Bieler, Pl. Nov. Herb. Spreng. Cent. 8. 1807. Pennsylvania, *Muhlenberg*.

Aira pennsylvanica Spreng., Acad. St. Pétersb. Mém. 2: 299. pl. 7. 1807–08. Pennsylvania.

Koeleria pennsylvanica DC., Cat. Hort. Monsp. 117. 1813. Based on *Aira pennsylvanica* Spreng.

Aira mollis Muhl., Descr. Gram. 82. 1817. Not *A. mollis* Schreb., 1771. Pennsylvania. Name only in Muhl., Cat. Pl. 11. 1813.

Trisetum pennsylvanicum Trin., Acad. St.

Pétersb. Mém. VI. Math. Phys. Nat.
1: 66. 1830. Not *T. pennsylvanicum*
Beauv. Based on *Aira pennsylvanica*
Spreng.
Glyceria pennsylvanica Heynh., Nom. 1:
361. 1840. Based on *Aira pennsyl-vanica* Spreng.
Reboulea pennsylvanica A. Gray, Man.
591. 1848. Based on *Koeleria pennsyl-vanica* DC.
Eatonia pennsylvanica A. Gray, Man. ed.
2. 558. 1856. Based on *Koeleria pennsylvanica* DC.
Eatonia dudleyi Vasey, Bot. Gaz. 11: 116.
1886. Michigan to Long Island and
Pennsylvania to North Carolina. [Type,
Ithaca, N. Y., *Dudley* in 1882.]
Eatonia nitida Nash, Torrey Bot. Club
Bul. 22: 511. 1895. Based on *Aira
nitida* Bieler.
Eatonia glabra Nash, in Britton, Man.
1043. 1901. Madison County, Tenn.,
Bain 507.
Sphenopholis nitida glabra Scribn., Rho-dora 8: 145. 1906. Based on *Eatonia
glabra* Nash.
Sphenopholis nitida var. *glabra* Scribn. ex
Robinson, Rhodora 10: 65. 1908.
Based on *Eatonia glabra* Nash.
Sphenopholis glabra Heller, Muhlenbergia
6: 12. 1910. Based on *Eatonia glabra*
Nash.
Reboulea nitida Farwell, Mich. Acad. Sci.
Rpt. 17: 181. 1916. Based on *Aira
nitida* Bieler.
Reboulea nitida var. *glabra* Farwell, Mich.
Acad. Sci. Rpt. 17: 181. 1916. Based
on *Eatonia glabra* Nash.

(1) **Sphenopholis obtusata** (Michx.) Scribn.,
Rhodora 8: 144. 1906. Based on *Aira
obtusata* Michx.
Aira obtusata Michx., Fl. Bor. Amer. 1:
62. 1803. Carolina to Florida [type],
Michaux.
Airopsis obtusata Desv., Jour. Bot. 1:
200. 1808. Based on *"Agrostis"*
[error for *Aira*] *obtusata* Michx.
Festuca obtusata Michx. ex Beauv., Ess.
Agrost. 163. 1812. Name only, prob-ably error for *Aira obtusata* Michx.
Aira truncata Muhl., Descr. Gram. 83.
1817. Pennsylvania. Name only, Muhl.,
Cat. Pl. 11. 1813.
Koeleria paniculata Nutt., Gen. Pl. 2:
(Add. 2): 1818. East Florida, *T. Say*.
Aira obtusa Raf., Jour. de Phys. 89: 104.
1819. Name only under *Eatonia*. Doubt-less either error for *A. obtusata* Michx.
or change of name.
Koeleria truncata Torr., Fl. North. and
Mid. U. S. 1: 116. 1823. Based on
Aira truncata Muhl.
Poa obtusata Link, Hort. Berol 1: 76.
1827. Based on *Aira obtusata* Michx.
Reboulea gracilis Kunth, Rév. Gram. 2:
341. pl. 84. 1830. New England to
Florida.

Trisetum lobatum Trin., Acad. St. Pétersb.
Mém. VI. Math. Phys. Nat. 1: 66.
1830. North America.
Agrostis obtusata Steud., Nom. Bot. ed. 2.
1: 41. 1840, as synonym of *Airopsis
obtusata* Desv.
Koeleria lobata Trin. ex Steud., Nom.
Bot. ed. 2. 1: 849. 1840. Not *K. lobata*
Roem. and Schult., 1817. As synonym
of *Reboulea gracilis* Kunth.
Koeleria obtusata Trin. ex Steud., Nom.
Bot. ed. 2. 1: 849. 1840, as synonym of
Airopsis obtusata Desv.
Reboulea obtusata A. Gray, Man. 591.
1848. Based on *Aira obtusata* Michx.
Eatonia obtusata A. Gray, Man. ed. 2.
558. 1856. Based on *Aira obtusata*
Michx.
Reboulea truncata Torr. ex Munro, Jour.
Linn. Soc. Bot. 6: 43. 1862, as syn-onym of *R. gracilis* Kunth.
Graphephorum densiflorum Fourn., Soc.
Bot. France Bul. 24: 182. 1877.
Name only. Mexico [Texas], *Berlandier*
1617.
Eatonia densiflora Fourn., Mex. Pl. 2:
111. 1886. Bejar, Tex., *Berlandier*
1617.
Aira mexicana Trin. ex Fourn., Mex. Pl. 2:
111. 1886, as synonym of *Eatonia
densiflora* Fourn.
Eatonia obtusata var. *robusta* Vasey ex
L. H. Dewey, U. S. Natl. Herb. Con-trib. 2: 544. 1894. Western Texas
[Wallisville, *Wallis* in 1881] to Arizona.
Eatonia obtusata var. *robusta* Vasey ex
Rydb., U. S. Natl. Herb. Contrib. 3:
190. 1895. Mullen, Nebr., *Rydberg*
1807.
Eatonia obtusata var. *purpurascens* Vasey
ex Rydb. and Shear, U. S. Dept. Agr.,
Div. Agrost. Bul. 5: 30. 1897. "Vasey
in U. S. Natl. Herb." This, the type,
from False Washita, Okla., *Palmer*
404; Nebraska, *Shear* 252, 252½, *Ryd-berg* 2002, *Kearney* 271, also cited.
Eatonia pubescens Scribn. and Merr.,
U. S. Dept. Agr., Div. Agrost. Cir. 27:
6. 1900. Starkville, Miss., *Tracy*.
Eatonia robusta Rydb., Torrey Bot. Club
Bul. 32: 602. 1905. Based on *E.
obtusata* var. *robusta* Vasey.
Sphenopholis obtusata lobata Scribn., Rho-dora 8: 144. 1906. Based on *Trisetum
lobatum* Trin.
Sphenopholis obtusata pubescens Scribn.,
Rhodora 8: 144. 1906. Based on
Eatonia pubescens Scribn. and Merr.
Eatonia annua Suksdorf, West. Amer.
Sci. 15: 50. 1906. Dalles on Columbia
River, Oreg., *Suksdorf* 1553. [Plants
depauperate, flowering first year.]
Sphenopholis obtusata var. *pubescens*
Scribn. in Robinson, Rhodora 10: 65.
1908. Based on *Eatonia pubescens*
Scribn. and Merr.
Sphenopholis obtusata var. *lobata* Scribn.

in Robinson, Rhodora 10: 65. 1908.
Based on *Trisetum lobatum* Trin.
Sphenopholis annua Heller, Muhlenbergia
6: 12. 1910. Based on *Eatonia annua*
Suksdorf.
Sphenopholis pubescens Heller, Muhlen-
bergia 6: 12. 1910. Based on *Eatonia
pubescens* Scribn. and Merr.
Sphenopholis robusta Heller, Muhlen-
bergia 6: 12. 1910. Based on *Eatonia
obtusata* var. *robusta* Vasey.
Reboulea obtusata var. *lobata* Farwell,
Mich. Acad. Sci. Rpt. 17: 182. 1916.
Based on *Trisetum lobatum* Trin.
Reboulea obtusata var. *pubescens* Farwell,
Mich. Acad. Sci. Rpt. 17: 181. 1916.
Based on *Eatonia pubescens* Scribn. and
Merr.
Sphenopholis obtusata var. *lobata* forma
purpurascens Waterfall, Rhodora 50:
93. 1948. Based on *Eatonia obtusata*
var. *purpurascens* Vasey ex Rydb. and
Shear.
(6) **Sphenopholis pallens** (Bieler) Scribn.,
Rhodora 8: 145. 1906. Based on *Aira
pallens* Bieler.
Aira pallens Bieler, Pl. Nov. Herb.
Spreng. Cent. 8. 1807. Pennsylvania,
Muhlenberg.
Aira pallens Muhl., Desc. Gram. 84.
1817. No locality cited. Name only in
Muhl., Cat. Pl. 11. 1813. *Aira penn-
sylvanica* Spreng., erroneously given as
synonym, Muhlenberg's description
agreeing with that of *A. pallens*
Beiler, not with that of *A. pennsyl-
vanica* Spreng.
Eatonia aristata Scribn. and Merr., U. S.
Dept. Agr., Div. Agrost. Cir. 27: 7.
1900. South Carolina, *Curtiss* in 1875.
Eatonia pallens Scribn. and Merr., U. S.
Dept. Agr., Div. Agrost. Cir. 27: 7.
1900. Based on *Aira pallens* Bieler,
Trisetum aristatum Nash in Small, Fl.
Southeast. U. S. 130. 1903. Presumably
based on *Eatonia aristata* Scribn. and
Merr.
Sphenopholis aristata Heller, Muhlen-
bergia 6: 12. 1910. Based on *Eatonia
aristata* Scribn. and Merr.
Reboulea pallens Farwell, Mich. Acad.
Sci. Rpt. 17: 181. 1916. Based on
Aira pallens Bieler.

(83) SPOROBOLUS R. Br.

(27) **Sporobolus airoides** (Torr.) Torr., U. S.
Rpt. Expl. Miss. Pacif. 7: 21. 1856.
Based on *Agrostis airoides* Torr.
Agrostis airoides Torr., Ann. Lyc. N. Y. 1:
151. 1824. Branches of the Arkansas
River near the Rocky Mountains,
James.
Vilfa airoides Trin. ex Steud., Nom. Bot.
ed. 2. 2: 766. 1841. Based on *Agrostis
airoides* Torr.
Sporobolus diffusissimus Buckl., Acad.

Nat. Sci. Phila. Proc. 1862: 90. 1862.
Western Texas [*Wright* 726].
(5) **Sporobolus asper** (Michx.) Kunth, Rév.
Gram. 1: 68. 1829. Based on *Agrostis
aspera* Michx.
Agrostis aspera Michx., Fl. Bor. Amer. 1:
52. 1803. Illinois, *Michaux.*
Agrostis composita Poir. in Lam., En-
cycl. Sup. 1: 254. 1810. Carolina,
Bosc.
Vilfa aspera Beauv., Ess. Agrost. 16, 147,
181. 1812. Based on *Agrostis aspera*
Michx.
Vilfa composita Beauv., Ess. Agrost. 16,
147, 181. 1812. Based on *Agrostis
composita* Poir.
Agrostis involuta Muhl., Descr. Gram. 72.
1817. Susquehanna, Pa., and New
Jersey. Name only in Muhl., Cat. Pl.
11. 1813.
Agrostis longifolia Torr., Fl. North. and
Mid. U. S. 1: 90. 1823. Kingsbridge,
N. Y.; Hoboken, N. J.; Deerfield,
Mass.; Pennsylvania, *Muhlenberg.*
Muhlenbergia aspera Trin. ex Kunth,
Enum. Pl. 1: 210. 1833, as synonym of
Sporobolus asper Kunth.
Muhlenbergia composita Trin. ex Kunth,
Enum. Pl. 1: 229. 1833, as synonym of
Agrostis composita Poir.
Vilfa longifolia Torr. in A. Gray, N. Amer.
Gram. and Cyp. 1: 4. 1834. Based on
Agrostis longifolia Torr.
Sporobolus longifolius Wood, Class-book
ed. 1861: 775. 1861. Based on *Agrostis
longifolia* Torr.
Sporobolus compositus Merr., U. S. Dept.
Agr., Div. Agrost. Cir. 35: 6. 1901.
Based on *Agrostis composita* Poir.
SPOROBOLUS ASPER var. HOOKERI (Trin.)
Vasey, Descr. Cat. Grasses U. S. 43.
1885. Based on *Vilfa hookeri* Trin.
Vilfa drummondii Trin., Acad. St.
Pétersb. Mém. VI. Sci. Nat. 4¹: 106.
1840. Texas, received from Hooker and
Endlicher [the type *Drummond* II.
306b].
Vilfa hookeri Trin., Acad. St. Pétersb.
Mém. VI. Sci. Nat. 4¹: 106. 1840.
Texas, received from Hooker [type
Drummond II. 306].
Glyceria stricta Buckl.,' Acad. Nat. Sci.
Phila. Proc. 1862: 95. 1862. Middle
Texas, *Buckley.* Inflorescence abnormal,
the spikelets diseased, with 2 or 3
several-nerved lemmas.
Sporobolus drummondii Vasey, Grasses
U. S. Descr. Cat. 44. 1885. Based on
Vilfa drummondii Trin.
Sporobolus asper var. *drummondii* Vasey,
U. S. Natl. Herb. Contrib. 3: 60. 1892.
Based on *Vilfa drummondii* Trin.
Sporobolus attenuatus Nash, in Small,
Fl. Southeast. U. S. 123. 1903. Stark-
ville, Miss., *Kearney* 83.
SPOROBOLUS ASPER var. PILOSUS (Vasey)
Hitchc., Biol. Soc. Wash. Proc. 41: 161.

1928. Based on *S. pilosus* Vasey.
(Published as *S. asper pilosus.*)
Sporobolus pilosus Vasey, Bot. Gaz. 16:
26. 1891. Kansas, *Smyth.*
(26) **Sporobolus buckleyi** Vasey, Torrey
Bot. Club Bul. 10: 128. 1883. Texas,
Buckley.
(7) **Sporobolus clandestinus** (Bieler)
Hitchc., U. S. Natl. Herb. Contrib. 12:
150. 1908. Based on *Agrostis clandestina* Beiler.
Agrostis clandestina Bieler, Pl. Nov. Herb.
Spreng. Cent. 8. 1807. Pennsylvania,
Muhlenberg.
Muhlenbergia clandestina Trin., Gram.
Unifl. 190. 1824. Based on *Agrostis
clandestina* Bieler.
Vilfa clandestina Nees ex Steud., Nom.
Bot. ed. 2. 2: 767. 1841. Based on
Agrostis clandestina Bieler.
Sporobolus canovirens Nash in Britton,
Man. 1042. 1901. Tennessee to Kansas
[type, St. George, *Kellerman* in 1890],
Mississippi, and Texas.
(24) **Sporobolus contractus** Hitchc., Amer.
Jour. Bot. 2: 303. 1915. Based on
Sporobolus strictus Merr.
Sporobolus cryptandrus var. *strictus* Scribn.,
Torrey Bot. Club Bul. 9: 103. 1882.
Camp Lowell, Ariz., *Pringle.*
Sporobolus strictus Merr., U. S. Dept.
Agr., Div. Agrost. Cir. 32: 6. 1901.
Not *S. strictus* Franch., 1893. Based on
Sporobolus cryptandrus var. *strictus*
Scribn.
(21) **Sporobolus cryptandrus** (Torr.) A.
Gray, Man. 576. 1848. Based on
Vilfa cryptandra Torr.
Agrostis cryptandra Torr., Ann. Lyc.
N. Y. 1: 151. 1824. Canadian River
[Texas or Oklahoma], *James.*
Vilfa tenacissima var. *fuscicola* Hook.,
Fl. Bor. Amer. 2: 239. 1839. Menzies
Island, Columbia River, Wash.
Vilfa cryptandra Torr. ex Trin., Acad. St.
Pétersb. Mém. VI. Sci. Nat. 4¹: 69.
1840. Based on *Agrostis cryptandra*
Torr.
Vilfa triniana Steud., Syn. Pl. Glum. 1:
156. 1854. [British] Columbia.
Sporobolus cryptandrus vaginatus Lunell,
Amer. Midl. Nat. 2: 123. 1911.
Benson County, N. Dak., *Lunell* in
1911.
Sporobolus cryptandrus var. *involutus* Farwell, Mich. Acad. Sci. Rpt. 22: 179.
1921. Rochester, Mich., *Farwell* 5393.
(14) **Sporobolus curtissii** (Vasey) Small ex
Scribn., U. S. Dept. Agr., Div. Agrost.
Bul. 7: 142. f. 124. 1897. Based on
"*S. floridanus curtissii* Vasey ex Beal."
Sporobolus curtissii Small ex Kearney,
U. S. Dept. Agr., Div. Agrost. Bul. 1:
24. 1895. Description inadequate.
"*Sporobolus floridanus curtissii* Vasey
in herb." cited. Jacksonville, Fla.,
Curtiss 4053, 5181. The Curtiss speci-

men named *Sporobolus floridanus* var.
curtissii by Vasey is without number.
Sporobolus floridanus var. *curtissii* Vasey
ex Beal, Grasses N. Amer. 2: 290.
1896. "Florida, *Curtiss.*"
(20) **Sporobolus domingensis** (Trin.) Kunth,
Rev. Gram. 1: Sup. 17. 1830. Based
on *Vilfa domingensis* Trin.
Vilfa domingensis Trin., in Spreng., Neu.
Entd. 2: 59. 1821. Dominican Republic.
Agrostis domingensis Schult., Mantissa 3
(Add. 1): 570. 1827. Based on *Vilfa
domingensis* Trin.
Sporobolus inordinatus Mez, Repert. Sp.
Nov. Fedde 17: 294. 1921. Cuba,
Ramon de la Sagra.
(22) **Sporobolus flexuosus** (Thurb.) Rydb.,
Torrey Bot. Club Bul. 32: 601. 1905.
Based on *Sporobolus cryptandrus* var.
flexuosus Thurb.
Vilfa cryptandra var. *flexuosa* Thurb. ex
Vasey, in Rothr., in Wheeler U. S.
Survey W. 100th Merid. Rpt. 6: 282.
1878. Nevada and Arizona, *Wheeler
Exped.*
Sporobolus cryptandrus var. *flexuosus*
Thurb. in S. Wats., Bot. Calif. 2: 269.
1880. Based on *Vilfa cryptandra* var.
flexuosa Thurb.
(16) **Sporobolus floridanus** Chapm., Fl.
South. U. S. 550. 1860. Middle and
west Florida [*Chapman*].
(25) **Sporobolus giganteus** Nash, Torrey
Bot. Club Bul. 25: 88. 1898. Doña
Ana County, N. Mex., *Wooton* 394.
Sporobolus cryptandrus var. *robustus* Vasey, U. S. Natl. Herb. Contrib. 1: 56.
1890. Texas, *Nealley* [746].
Sporobolus cryptandrus var. *giganteus*
Jones, West. Bot. Contrib. 14: 11. 1912.
Based on *S. giganteus* Nash.
(12) **Sporobolus heterolepis** (A. Gray) A.
Gray, Man. 576. 1848. Based on
Vilfa heterolepis A. Gray.
Vilfa heterolepis A. Gray, Ann. Lyc. N. Y.
3: 233. 1835. Watertown, N. Y.,
Crawe.
Agrostis heterolepis Wood, Class-book ed.
2. 598. 1847. Based on *Vilfa heterolepis*
A. Gray.
(9) **Sporobolus indicus** (L.) R. Br., Prodr.
Fl. Nov. Holl. 170. 1810. Based on
Agrostis indica L.
Agrostis indica L., Sp. Pl. 63. 1753.
"India," but the type from Jamaica,
sent by Patrick Browne.
Sporobolus jacquemontii Kunth, Rév.
Gram. 2: 427. pl. 127. 1831. Dominican Republic.
Vilfa jacquemontii Trin., Acad. St.
Pétersb. Mém. VI. Sci. Nat. 4¹: 92.
1840. Based on *Sporobolus jacquemontii*
Kunth.
Vilfa indica Trin. ex Steud., Nom. Bot.
ed. 2. 2: 767. 1841. Based on *Agrostis
indica* L.

(11) **Sporobolus interruptus** Vasey, Torrey Bot. Club Bul. 15: 8. 1888. Arizona, *Coues* and *Palmer* 66 in 1886; San Francisco Forest, *Rusby* 15 in 1883 [the Rusby specimen, distributed as No. 885, the type].

Sporobolus arizonicus Thurb. ex Vasey, Torrey Bot. Club Bul. 15: 8. 1888, as synonym of *Sporobolus interruptus* Vasey.

(17) **Sporobolus junceus** (Michx.) Kunth, Rév. Gram. 1: 68. 1829. Based on *Agrostis juncea* Michx.

Agrostis juncea Michx., Fl. Bor. Amer. 1: 52. 1803. Not *A. juncea* Lam., 1783. Carolina, *Michaux*.

Heleochloa juncea Beauv., Ess. Agrost. 24, 147. 1812. Based on *Agrostis juncea* Michx.

Colpodium junceum Trin. in Spreng., Neu. Entd. 2: 37. 1821. Based on *Agrostis juncea* Michx.

Crypsis juncea Steud., Nom. Bot. 1: 242. 1821. Based on *Agrostis juncea* Michx.

Vilfa juncea Trin., Gram. Unifl. 157. 1824. Based on *Agrostis juncea* Michx.

Vilfa schiedeana Trin., Acad. St. Pétersb. Mém. VI. Sci. Nat. 4¹: 73. 1840. Arkansas, "Schiede." [Type specimen annotated by Ruprecht "Beyrich non Schiede."]

Vilfa gracilis Trin., Acad. St. Pétersb. Mém. VI. Sci. Nat. 4¹: 74. 1840. Carolina.

Vilfa fulvescens Trin., Acad. St. Pétersb. Mém. VI. Sci. Nat. 4¹: 76. 1840. North America, *Bosc*, Willdenow Herb. No. 1750.

Agrostis thyrsoides Bosc ex Trin., Acad. St. Pétersb. Mém. VI. Sci. Nat. 4¹: 76. 1840, as synonym of *Vilfa fulvescens* Trin.

Vilfa subsetacea Trin., Acad. St. Pétersb. Mém. VI. Sci. Nat. 4¹: 133 (in note). 1840. Based on *V. gracilis* Trin., op. cit. (page 74, not op. cit. page 104). (See synonymy under *Muhlenbergia cuspidata*.) Discovering that he had named two distinct species *Vilfa gracilis*, Trinius changed the first to *V. subsetacea*.

Vilfa vinzenti Steud., Syn. Pl. Glum. 1: 155. 1854. [Rusk County] Tex., *Vinzent 62*.

Aira triglumis Steud., Syn. Pl. Glum. 1: 223. 1854. [Rusk County] Tex., *Vincent 62*.

Bennetia juncea Raf. ex Jacks., Ind. Kew. 1: 291. 1893, as synonym of *Sporobolus junceus*. Rafinesque (Bul. Bot. Seringe 1: 220. 1830) cites *Agrostis juncea* Michx., after his description of the new genus *Bennetia*, but does not transfer the specific name.

Sporobolus ejuncidus Nash in Britton, Man. 106. 1901. Based on *Sporobolus junceus* Kunth.

Sporobolus gracilis Merr., Rhodora 4: 48. 1902. Based on *Vilfa gracilis* Trin.

(6) **Sporobolus macer** (Trin.) Hitchc., Amer. Jour. Bot. 2: 303. 1915. Based on *Vilfa macra* Trin.

Vilfa macra Trin., Acad. St. Pétersb. Mém. VI. Sci. Nat. 4¹: 79. 1840. Louisiana.

(23) **Sporobolus nealleyi** Vasey, Torrey Bot. Club Bul. 15: 49. 1888, name only; U. S. Natl. Herb. Contrib. 1: 57. 1890. Brazos Santiago, Tex., *Nealley*.

(4) **Sporobolus neglectus** Nash, Torrey Bot. Club Bul. 22: 464. 1895. Massachusetts to Kentucky, Tennessee, and Kansas. [Type, Woodruff Gap, N. J., *Britton* in 1887.]

Sporobolus vaginiflorus var. *neglectus* Scribn., U. S. Dept. Agr., Div. Agrost. Bul. 17 (ed. 2): 170. f. 466. 1901. Based on *S. neglectus* Nash.

Sporobolus ozarkanus Fernald, Rhodora 35: 109. 1933. Webb City, Mo., *Palmer 3133*.

(2) **Sporobolus patens** Swallen, Wash. Acad. Sci. Jour. 31: 352. f. 5. 1941. Wilcox, Ariz., *Silveus 3504*.

(8) **Sporobolus poiretii** (Roem. and Schult.) Hitchc., Bartonia 14: 32. 1932. Based on *Axonopus poiretii* Roem. and Schult.

Agrostis elongata Lam., Tabl. Encycl. 1: 162. 1791. Not *Sporobolus elongatus* R. Br., 1810. South America.

Agrostis compressa Poir. in Lam., Encycl. Sup. 1: 258. 1810. Not *A. compressa* Willd., 1790, nor Poir. (op. cit.) 1: 259. 1810, nor *Sporobolus compressus* Kunth, 1833. Carolina, *Bosc*.

Milium compressum Poir. in Lam., Encycl. Sup. 1: 258. 1810. Not *M. compressum* Swartz, 1788. As synonym of *Agrostis compressa* Poir.

Vilfa elongata Beauv., Ess. Agrost. 16, 147, 181. 1812. Based on *Agrostis elongata* Lam.

Axonopus poiretii Roem. and Schult., Syst. Veg. 2: 318. 1817. Based on *Agrostis compressa* Poir., "n. 78," not *A. compressa* Willd., 1790, nor Poir. (op. cit.) No. 82, on the following page.

Sporobolus lamarckii Desv. ex Hamilt., Prodr. Pl. Ind. Occ. 4. 1825. Based on *Agrostis elongata* Lam.

Agrostis tenuissima Spreng., Syst. Veg. 1: 258. 1825. West Indies and South America.

Vilfa exilis Trin., Acad. St. Pétersb. Mém. VI. Sci. Nat. 4¹: 89. 1840. Jalapa, Mexico [*Schiede*].

Vilfa berteroana Trin., Acad. St. Pétersb. Mém. VI. Sci. Nat. 4¹: 100. 1840. Dominican Republic, *Bernhardi*.

Sporobolus angustus Buckl., Acad. Nat. Sci. Phila. Proc. 1862: 88. 1863. "Buchanan county" [probably error for Buchanan] Tex., [*Buckley*].

Vilfa tenacissima var. *exilis* Fourn., Mex. Pl. 2: 99. 1886. Based on *Vilfa exilis* Trin.

Sporobolus littoralis var. *elongatus* Dur. and Schinz, Consp. Fl. Afr. 5: 821. 1894. Based on *Vilfa elongata* Beauv.

Sporobolus berteroanus Hitchc. and Chase, U. S. Natl. Herb. Contrib. 18: 370. 1917. Based on *Vilfa berteroana* Trin.

This species has been included in *Sporobolus indicus* in some manuals.

(1) **Sporobolus pulvinatus** Swallen, Wash. Acad. Sci. Jour. 31. 351. f. 4. 1941. Adamana, Ariz., *Griffiths 5107*.

(18) **Sporobolus purpurascens** (Swartz) Hamilt., Prodr. Pl. Ind. Occ. 5. 1825. Based on *Agrostis purpurascens* Swartz.

Agrostis purpurascens Swartz, Prodr. Veg. Ind. Occ. 25. 1788. Jamaica, *Swartz*.

Vilfa purpurascens Beauv., Ess. Agrost. 16, 182. 1812. Based on *Agrostis purpurascens* Swartz.

Vilfa grisebachiana Fourn., Mex. Pl. 2: 98. 1886. Cuba, *Wright 3427a*.

Vilfa liebmanni Fourn., Mex. Pl. 2: 100. 1886. Mexico, *Liebmann 693*.

(19) **Sporobolus pyramidatus** (Lam.) Hitchc., U. S. Dept. Agr., Misc. Pub. 243: 84. 1936. Based on *Agrostis pyramidata* Lam.

Agrostis pyramidata Lam., Tabl. Encycl. 1: 161. 1791. South America.

Vilfa arguta Nees, Agrost. Bras. 395. 1829. Brazil.

Sporobolus argutus Kunth, Rév. Gram. 1: Sup. 17. 1830. Based on *Vilfa arguta* Nees.

Vilfa arkansana Trin., Acad. St. Pétersb. Mém. VI. Sci. Nat. 4¹: 64. 1840. Arkansas, *Beyrich*.

Vilfa subpyramidata Trin., Acad. St. Pétersb. Mém. VI. Sci. Nat. 4¹: 61. 1840. Texas [received from Hooker, the type being *Drummond 377*].

Vilfa richardi Steud., Syn. Pl. Glum. 1: 153. 1854. West Indies.

Agrostis pyramidalis Rich. ex Steud., Syn. Pl. Glum. 1: 153. 1854, as synonym of *Vilfa richardi* Steud.

Vilfa agrostoidea Buckl., Acad. Nat. Sci. Phila. Proc. 1862: 88. 1862. Llano County, Tex.

Vilfa sabeana Buckl., Acad. Nat. Sci. Phila. Proc. 1862: 90. 1862. San Saba County, Tex., *Buckley*. Given as *Vilfa (Sporobolus) sabeana*.

Sporobolus arkansanus Nutt. ex Vasey, U. S. Natl. Herb. Contrib. 3: 61. 1892. as synonym of *S. argutus* Kunth.

Sporobolus sabeanus Buckl. ex Vasey, U. S. Natl. Herb. Contrib. 3: 61. 1892, as synonym of *S. argutus* Kunth.

(13) **Sporobolus silveanus** Swallen, Wash. Acad. Sci. Jour. 31: 350. f. 3. 1941. Orange, Tex., *Silveus 6441*.

(15) **Sporobolus teretifolius** Harper, Torrey Bot. Club Bul. 33: 229. 1906. Near Moultrie, Ga., *Harper 1642*.

(29) **Sporobolus texanus** Vasey, U. S. Natl. Herb. Contrib. 1: 57. 1890. Screw Bean, Presidio County, Tex., *Nealley* [755].

(30) **Sporobolus tharpii** Hitchc., Biol. Soc. Wash. Proc. 41: 161. 1928. Padre Island, Tex., *Tharp 4772*.

(3) **Sporobolus vaginiflorus** (Torr.) Wood, Class-book ed. 1861. 775. 1861. Based on *Vilfa vaginiflora* Torr.

Vilfa vaginiflora Torr. ex Gray, N. Amer. Gram. and Cyp. 1: No. 3. 1834; Trin., Mém. Acad. St. Pétersb. VI. Sci. Nat. 4¹: 56. 1840. New Jersey.

Cryptostachys vaginata Steud., Flora 33: 229. 1850, name only; Syn. Pl. Glum. 1: 181. 1854. North America.

Vilfa riehlii Steud., Syn. Pl. Glum. 1: 154. 1854. North America.

Sporobolus minor Vasey ex A. Gray, Man. ed. 6. 646. 1890. Not *S. minor* Kunth, 1830. Virginia to North Carolina [type, *Boynton*], Tennessee and Texas.

Sporobolus filiculmis L. H. Dewey, U. S. Natl. Herb. Contrib. 2: 519. 1894. Not *S. filiculmis* Vasey, 1885. Based on *S. minor* Vasey.

Sporobolus ovatus Beal, Grasses N. Amer. 2: 300. 1896. Based on *S. minor* Vasey.

Sporobolus vaginatus Scribn., Bot. Gaz. 21: 15. 1896. Based on *Cryptostachys vaginata* Steud.

?Sporobolus vaginiflorus var. *minor* Scribn. ex Chapm., Fl. South. U. S. ed. 3. 598. 1897. North Carolina and Tennessee.

Sporobolus vaginiflorus var. *inaequalis* Fernald, Rhodora 35: 109. 1933. Concord, N. H., *Batchelder in 1901*.

(10) **Sporobolus virginicus** (L.) Kunth, Rév. Gram. 1: 67. 1829. Based on *Agrostis virginica* L.

Agrostis virginica L., Sp. Pl. 63. 1753. Virginia.

Agrostis littoralis Lam., Tabl. Encycl. 1: 161. 1791. South America, *Richard*.

Vilfa littoralis Beauv., Ess. Agrost. 16, 147, 181. 1812. Based on *Agrostis littoralis* Lam.

Vilfa virginica Beauv., Ess. Agrost. 16, 182. 1812. Based on *Agrostis virginica* L.

Agrostis pungens Muhl., Descr. Gram. 72. 1817. Not *A. pungens* Schreb., 1769. Eastern United States. Name only in Muhl., Cat. Pl. 11. 1813.

Crypsis virginica Nutt., Gen. Pl. 1: 49. 1818. Based on *Agrostis virginica* Willd. [error for L.].

Podosaemum virginicum Link, Hort. Berol. 1: 85. 1827. Based on *Agrostis virginica* L.

Sporobolus littoralis Kunth, Rév. Gram. 1: 68. 1829. Based on *Agrostis littoralis* Lam.

(28) **Sporobolus wrightii** Munro ex Scribn., Torrey Bot. Club Bul. 9: 103. 1882. Pantano, Ariz., *Pringle*.

Bauchea karwinskyi Fourn., Mex. Pl. 2: 87. 1886. Mexico, *Karwinsky* 1015, 1015b.

Sporobolus altissimus Vasey, Calif. Acad. Sci. Proc. II. 2: 212. 1889. San Diego, Calif., *Palmer* [in 1888].

Sporobolus altissimus var. *minor* Vasey, Calif. Acad. Sci. Proc. II. 2: 213. 1889. San Enrique, Calif. [*Brandegee*].

Sporobolus airoides var. *wrightii* Gould, Madroño 10: 94. 1949. Based on *S. wrightii* Munro ex Scribn.

(131) STENOTAPHRUM Trin.

(1) **Stenotaphrum secundatum** (Walt.) Kuntze, Rev. Gen. Pl. 2: 794. 1891. Based on *Ischaemum secundatum* Walt. Kuntze misspells the specific name *"secundum."*

Ischaemum secundatum Walt., Fl. Carol. 249. 1788. South Carolina.

Rottboellia stolonifera Poir. in Lam., Encycl. 6: 310. 1804. Puerto Rico, *Ledru*.

Stenotaphrum americanum Schrank, Pl. Rar. Hort. Monac. pl. 98. 1822.

Stenotaphrum sarmentosum Nees, Agrost. Bras. 93. 1829. Based on *Rottboellia stolonifera* Poir.

Stenotaphrum glabrum var. *americanum* Doell in Mart., Fl. Bras. 2²: 300. 1877. Based on *Stenotaphrum americanum* Schrank.

Stenotaphrum dimidiatum var. *americanum* Hack. in Stuck., An. Mus. Nac. Buenos Aires 21: 57. 1911. Based on *Stenotaphrum americanum* Schrank.

Stenotaphrum dimidiatum var. *secundum* [*secundatum*] Domin, Bibl. Bot. 85: 332. 1915. Based on *Ischaemum secundatum* Walt.

Stenotaphrum secundatum var. *variegatum* Hitchc. in Bailey, Stand. Cycl. Hort. 6: 3237. 1917. Greenhouse plant.

(91) STIPA L.

(33) **Stipa arida** Jones, Calif. Acad. Sci. Proc. II. 5: 725. 1895. Marysvale, Utah, *Jones* 5377.

Stipa mormonum Mez, Repert. Sp. Nov. Fedde 17: 209. 1921. Utah, *Jones* [2106].

(8) **Stipa avenacea** L., Sp. Pl. 78. 1753. Virginia.

Stipa barbata Michx., Fl. Bor. Amer. 1: 53. 1803. Not *S. barbata* Desf. 1798. Virginia and Carolina, *Michaux*.

Stipa virginica Pers., Syn. Pl. 1: 99. 1805. Based on *S. barbata* Michx.

Stipa diffusa Willd. ex Steud., Nom. Bot. ed. 2. 2: 643. 1841, as synonym of *Stipa avenacea* L.

Stipa avenacea var. *bicolor* Eaton and Wright, N. Amer. Bot. ed. 8. 444. 1848. Philadelphia and Chester, Pa.; Boston, Mass.; Ontario; Florida.

Podopogon avenaceus Raf. ex Jacks., Ind. Kew. 2: 580. 1894, as synonym of *Stipa avenacea*.

Podopogon barbatus Raf. ex Jacks., Ind. Kew. 2: 580. 1894, as synonym of *Stipa avenacea*.

Piptochaetium avenaceum Parodi, Rev. Mus. La Plata Bot. n. ser. 6: 223, 229. f. 1, B. 1944. Based on *Stipa avenacea* L.

(7) **Stipa avenacioides** Nash, Torrey Bot. Club Bul. 22: 423. 1895. Cassia, Lake County, Fla., *Nash* 2051.

Stipa brachychaeta Godr., Mem. Acad. Monsp. (Sec. Medic.) 1: 450. 1853. Originally described from specimens from unknown source. Native of southern South America.

(22) **Stipa californica** Merr. and Davy, Calif. Univ. Pubs., Bot. 1: 61. 1902. San Jacinto Mountains, Calif., *Hall* 2556.

(12) **Stipa cernua** Stebbins and Love, Madroño 6: 137. f. 1. 2. 1941. Alameda County, Calif., *Stebbins* 2732.

Stipa pulchra var. *cernua* Beetle and Tofsrud, West. Bot. Leaflets 5: 35. 1947. Based on *S. cernua* Stebbins and Love.

(28) **Stipa columbiana** Macoun, Can. Pl. Cat. 2⁴: 191. 1888. Yale, British Columbia, *Macoun* [28,940]; Victoria, Vancouver Island, *Macoun* [28,941].

Stipa viridula var. *minor* Vasey, U. S. Natl. Herb. Contrib. 3: 50. 1892. [Kelso Mountain, Colo., *Letterman* 95.]

Stipa minor Scribn., U. S. Dept. Agr., Div. Agrost. Bul. 11: 46. 1898. Based on *S. viridula* var. *minor* Vasey.

Stipa columbiana var. *nelsoni* (Scribn.) Hitchc., U. S. Natl. Herb. Contrib. 24: 254. 1925. Based on *S. nelsoni* Scribn. (Published as *S. columbiana nelsoni*.)

Stipa columbiana var. *nelsoni* St. John, Fl. Southeast. Wash. and Adj. Idaho 61. 1937. Same basis.

Stipa occidentalis [Thurb.; misapplied by] Boland., Calif. Acad. Sci. Proc. 4: 169. 1872. Larger plant with "awns almost entirely smooth," confused with true *S. occidentalis*.

Stipa nelsoni Scribn., U. S. Dept. Agr., Div. Agrost. Bul. 11: 46. 1898. Albany County, Wyo., *A. Nelson* 3963.

(10) **Stipa comata** Trin. and Rupr., Acad. St. Pétersb. Mém. VI. Sci. Nat. 5¹: 75. 1842. Carlton House Fort, Saskatchewan River, *Drummond;* Columbia River, near Missouri Portage, *Douglas*.

Stipa comata subsp. *intonsa* Piper, U. S. Natl. Herb. Contrib. 11: 109. 1906. Rockland, Klickitat County, Wash., *Suksdorf* 1026.

Stipa comata var. *suksdorfii* St. John, Fl. Southeast. Wash. and Adj. Idaho 61. 1937. Spokane County, Wash., *Suksdorf* 8990.
This is the species described by Pursh (Fl. Amer. Sept. 1: 72. 1814), and Nuttall (Gen. Pl. 1: 58. 1818) under *Stipa juncea* L., and by Hooker (Fl. Bor. Amer. 2: 257. 1840) under *S. capillata* L.
STIPA COMATA var. INTERMEDIA Scribn. and Tweedy, Bot. Gaz. 11: 171. 1886. Junction Butte, Yellowstone Park, *Tweedy* 610.
Stipa tweedyi Scribn., U. S. Dept. Agr., Div. Agrost. Bul. 11: 47. 1898. Based on *S. comata* var. *intermedia* Scribn. and Tweedy.
Stipa spartea var. *tweedyi* Jones, West Bot. Contrib. 14: 11. 1912. Based on *S. tweedyi* Scribn.
(5) **Stipa coronata** Thurb. in S. Wats., Bot. Calif. 2: 287. 1880. California, Julian, *Bolander;* San Bernardino, *Parry* and *Lemmon* 422.
STIPA CORONATA var. DEPAUPERATA (Jones) Hitchc., Jour. Wash. Acad. Sci. 24: 292. 1934. Based on *S. parishii* var. *depauperata* Jones.
Stipa parishii Vasey, Bot. Gaz. 7: 33. 1882. San Bernardino Mountains, Calif., *Parish* 1079.
Stipa parishii var. *depauperata* Jones, West. Bot. Contrib. 14: 11. 1912. Detroit, Utah [*Jones*].
Stipa coronata parishii Hitchc., U. S. Natl. Herb. Contrib. 24: 227. 1925. Based on *S. parishii* Vasey.
(23) **Stipa curvifolia** Swallen, Wash. Acad. Sci. Jour. 23: 456. 1933. Guadalupe Mountains, N. Mex., *Wilkins* 1660.
(31) **Stipa diegoensis** Swallen, Wash. Acad. Sci. Jour. 30: 212. f. 2. 1940. San Diego County, Calif., *Gander* 5778.
Stipa elegantissima Labill., Nov. Holl. Pl. 1: 23. pl. 29. 1804. Australia.
(18) **Stipa elmeri** Piper and Brodie ex Scribn., U. S. Dept. Agr., Div. Agrost. Bul. 11: 46. 1898. Based on *S. viridula* var. *pubescens* Vasey.
Stipa viridula var. *pubescens* Vasey, U. S. Natl. Herb. Contrib. 3: 50. 1892. Not *S. pubescens* R. Br., 1810. Washington, *Suksdorf.*
(14) **Stipa eminens** Cav., Icon. Pl. 5: 42. pl. 467. f. 1. 1799. Chalma, Mexico.
Stipa erecta Fourn., Mex. Pl. 2: 75. 1886. Not *S. erecta* Trin., 1824. Tehuacan, Mexico, *Liebmann* 654.
Stipa flexuosa Vasey, Torrey Bot. Club Bul. 15: 49. 1888. Western Texas [Chenate Mountains], *Nealley.*
(19) **Stipa latiglumis** Swallen, Wash. Acad. Sci. Jour. 23: 198. f. 1. 1933. Camp Lost Arrow, Yosemite Valley, Calif., *Abrams* 4469.
(25) **Stipa lemmoni** (Vasey) Scribn., U. S. Dept. Agr., Div. Agrost. Cir. 30: 3.

1901. Based on *S. pringlei* var. *lemmoni* Vasey.
Stipa pringlei var. *lemmoni* Vasey, U. S. Natl. Herb. Contrib. 3: 55. 1892. Plumas County, Calif., *Lemmon* [5456].
Stipa lemmoni var. *jonesii* Scribn., U. S. Dept. Agr., Div. Agrost. Cir. 30: 4. 1901. Emigrant Gap, Calif., *Jones* 3298.
(15) **Stipa lepida** Hitchc., Amer. Jour. Bot. 2: 302. 1915. Santa Ynez Forest, Calif., *Chase* 5611.
STIPA LEPIDA var. ANDERSONII (Vasey) Hitchc., Amer. Jour. Bot. 2: 303. 1915. Based on *S. eminens* var. *andersonii* Vasey. (Published as *S. lepida andersonii.*)
Stipa eminens var. *andersonii* Vasey, U. S. Natl. Herb. Contrib. 3: 54. 1892. California [Santa Cruz, *Anderson* 58, type]. "Lower California," cited by Vasey is erroneous.
Stipa hassei Vasey, U. S. Natl. Herb. Contrib. 1: 267. 1893. Santa Monica, Calif., *Hasse.* Abnormal specimen, the spikelets distorted by a smut.
(29) **Stipa lettermani** Vasey, Torrey Bot. Club Bul. 13: 53. 1886. Snake River, Idaho, *Letterman* [102].
Stipa viridula var. *lettermani* Vasey, U. S. Natl. Herb. Contrib. 3: 50. 1892. Presumably based on *S. lettermani* Vasey.
(3) **Stipa leucotricha** Trin. and Rupr., Acad. St. Pétersb. Mém., VI. Sci. Nat. 5¹: 54. 1842. Texas, from Hooker.
Stipa ciliata Scheele, Linnaea 22: 342. 1849. New Braunfels, Tex., Römer.
(21) **Stipa lobata** Swallen, Wash. Acad. Sci. Jour. 23: 199. f. 2. 1933. Ranger Station. Queen, Guadalupe Mountains, N. Mex., *Hitchcock* (*Amer. Gr. Natl. Herb.* 819).
Stipa neesiana Trin. and Rupr., Acad. St., Pétersb. Mém. VI. Sci. Nat. 5¹: 27. 1842. Montevideo, *Sellow.*
(1) **Stipa neomexicana** (Thurb.) Scribn., U. S. Dept. Agr., Div. Agrost. Bul. 17: 132. f. 428. 1899. Based on *S. pennata* var. *neo-mexicana* Thurb.
Stipa pennata var. *neo-mexicana* Thurb. in Coulter, Man. Rocky Mount. 408. 1885. New Mexico [type Rio Mimbres, *Thurber* 269], Colorado, and Texas.
(20) **Stipa occidentalis** Thurb. in S. Wats., in King, Geol. Expl. 40th Par. 5: 380. 1871. Yosemite Trail, Calif., *Bolander* 5038.
Stipa stricta Vasey, Torrey Bot. Club Bul. 10: 42. 1883. Not *S. stricta* Lam. 1791. Washington (erroneously cited as Oregon), *Suksdorf.*
Stipa stricta var. *sparsiflora* Vasey, U. S. Natl. Herb. Contrib. 3: 51. 1892. Yosemite Trail, Calif., *Bolander* 5038.
Stipa oregonensis Scribn., U. S. Dept. Agr., Div. Agrost. Bul. 17: 130. f. 426. 1899. Based on *S. stricta* Vasey.
Stipa occidentalis montana Merr. and Davy, Calif. Univ. Pubs., Bot. 1: 62.

1902. Yosemite Trail, *Bolander* 5038.
Stipa pennata L., Sp. Pl. 78. 1753. Europe.
(32) **Stipa pinetorum** Jones, Calif. Acad. Sci. Proc. II. 5: 724. 1895. Panguitch Lake, Utah, *Jones* 6023 p.
(16) **Stipa porteri** Rydb., Torrey Bot. Club Bul. 32: 599. 1905. Based on the plant described as *S. mongolica* Turcz. by Porter and Coulter (Syn. Fl. Colo. 145. 1874). [Rocky Mountains, *Hall* and *Harbour* 648, error for 646.]
This is the species described under the name *Oryzopsis mongolica* (Turcz.) Beal (Bot. Gaz. 15: 111. 1890), but that name is based on *Stipa mongolica* Turcz., an Asiatic species.
(13) **Stipa pringlei** Scribn. in Vasey, U. S. Natl. Herb. Contrib. 3: 54. 1892. Mexico, *Pringle* [1410 type], and Arizona, *Pringle, Lemmon, Tracy*. No reference to *Oryzopsis pringlei* Beal.
Oryzopsis pringlei Beal, Bot. Gaz. 15: 112. 1890. Chihuahua, Mexico, *Pringle* 1410.
Stipa pringlei Scribn. ex Beal, Bot. Gaz. 15: 112. 1890, as synonym of *Oryzopsis pringlei* Beal.
Oryzopsis erecta Beal, Grasses N. Amer. 2: 230. 1896. Apparently based on *O. pringlei* Beal, *Pringle* 1410 being cited, the name changed because of *O. pringlei* Scribn. ex Beal 1896 (page 226 of the same work). The latter is the same as *Stipa virescens* H. B. K. of Mexico, not known from the United States. Beal erroneously gives the authority of *O. erecta* as "(Scribn.) Beal."
Piptochaetium pringlei Parodi, Rev. Mus. La Plata Bot. (n. s.) 6: 223, 230. f. 1, D. 1944. Based on *Oryzopsis pringlei* Beal.
(11) **Stipa pulchra** Hitchc., Amer. Jour. Bot. 2: 301. 1915. Healdsburg, Sonoma County, Calif., *Heller* 5252.
(6) **Stipa richardsoni** Link, Hort. Berol. 2: 245. 1833. Western North America. Grown at Berlin from seed sent by Richardson.
Stipa richardsoni var. *major* Macoun, Can. Pl. Cat. 2⁴: 191. 1888, without description. Columbia Valley, British Columbia, *Macoun*.
Oryzopsis richardsoni Beal, Bot. Gaz. 15: 111. 1890. Based on *Stipa richardsoni* Link, but misapplied to *Oryzopsis canadensis*.
(27) **Stipa robusta** (Vasey) Scribn., U. S. Dept. Agr., Div. Agrost. Bul. 5: 23. 1897. Based on *S. viridula* var. *robusta* Vasey. Not invalidated by *S. robusta* Nutt. ex Trin. and Rupr., published as synonym of *S. spartea*.
Stipa viridula var. *robusta* Vasey, U. S. Natl. Herb. Contrib. 1: 56. 1890. Presidio County, Tex., *Nealley* [714].
Stipa vaseyi Scribn., U. S. Dept. Agr., Div. Agrost. Bul. 11: 46. 1898. Based on *S. viridula* var. *robusta* Vasey.

(24) **Stipa scribneri** Vasey, Torrey Bot. Club Bul. 11: 125. 1884. Santa Fe, N. Mex. [*Vasey*].
(9) **Stipa spartea** Trin., Acad. St. Pétersb. Mém. VI. Math. Phys. Nat. 1: 82. 1830. North America [Rocky Mountains near the Missouri]. By typographical error the name is spelled "sparta."
Stipa robusta Nutt. ex Trin. and Rupr., Acad. St. Pétersb. Mém. VI. Sci. Nat. 5¹: 69. 1842, as synonym of *S. spartea*.
STIPA SPARTEA var. CURTISETA Hitchc., U. S. Natl. Herb. Contrib. 24: 230. 1925. Hound Creek Valley, Mont., *Scribner* 339. (Published as *S. spartea curtiseta*.)
(2) **Stipa speciosa** Trin. and Rupr., Acad. St. Pétersb. Mém. VI. Sci. Nat. 5¹: 45. 1842. Chile, *Cuming*.
Stipa californica Vasey, Amer. Acad. Proc. 24: 80. 1889. Name only for Palmer's No. 505 in 1887 from Los Angeles Bay, Baja California.
Stipa speciosa var. *minor* Vasey, U. S. Natl. Herb. Contrib. 3: 52. 1892. Empire City, Nev., *Jones*.
Stipa humilis var. *jonesiana* Kuntze, Rev. Gen. Pl. 3²: 371. 1898. Empire City, Nev., *Jones* 4111.
Stipa humilis var. *speciosa* Kuntze, Rev. Gen. Pl. 3²: 371. 1898. Based on *S. speciosa* Trin. and Rupr.
Stipa splendens Trin. in Spreng., Neu. Entd. 2: 54. 1821. Siberia.
(4) **Stipa stillmanii** Boland., Calif. Acad. Sci. Proc. 4: 169. 1872. Blue Canyon, Sierra Nevada, Calif., *Bolander*.
Stipa tenacissima L., Cent. Pl. 1: 6. 1755; Amoen. Acad. 4: 266. 1759. Spain.
(34) **Stipa tenuissima** Trin., Acad. St. Pétersb. Mém. VI. Sci. Nat. 2¹: 36. 1836. Mendoza "Chile," [Argentina], *Gillies*.
Stipa cirrosa Fourn., Mex. Pl. 2: 75. 1886. Mexico, *Karwinsky* 1009.
Stipa subulata Fourn., Mex. Pl. 2: 75. 1886. Mexico, *Karwinsky* 1009b.
(17) **Stipa thurberiana** Piper, U. S. Dept. Agr., Div. Agrost. Cir. 27: 10. 1900. Washington, north branch of the Columbia and Okanagan, *Pickering* and *Brackenridge*.
Stipa occidentalis Thurb. in Wilkes, U. S. Expl. Exped. Bot. 17: 483. 1874. Not *S. occidentalis* Thurb. in S. Wats., 1871. North Branch of the Columbia River [Washington, *Pickering* and *Brackenridge*].
(26) **Stipa viridula** Trin., Acad. St. Pétersb. Mém. VI. Sci. Nat. 2¹: 39. 1836. North America [Saskatchewan].
Stipa parviflora [Desf., misapplied by] Nuttall, Gen. Pl. 1: 59. 1818. Plains of the Missouri.
Stipa nuttalliana Steud., Nom. Bot. ed. 2. 2: 643. 1841. Based on *Stipa parviflora* as described by Nuttall.

Stipa sparta Trin. ex Hook., Fl. Bor. Amer. 2: 237. 1840. Name only, *S. parviflora* Nutt., not Desf., cited as synonym.

(30) **Stipa williamsii** Scribn., U. S. Dept. Agr., Div. Agrost. Bul. 11: 45. 1898. Big Horn Mountain, Wyo., *Williams* 2804.

THEMEDA Forsk.

Themeda quadrivalvis (L.) Kuntze, Rev. Gen. Pl. 2: 794. 1891. Based on *Andropogon quadrivalvis* L.
Andropogon quadrivalvis L., Syst. Veg. ed. 13. 758. 1774. India.
Anthistiria ciliata L. f., Sup. 113. 1781. Based on *Andropogon quadrivalvis* L.
Themeda ciliata Hack. in DC., Monogr. Phan. 6: 664. 1889. Based on *Anthistiria ciliata* L. f.

Thysanolaena maxima (Roxb.) Kuntze, Rev. Gen. Pl. 2: 794. 1891. Based on *Agrostis maxima* Roxb.
Agrostis maxima Roxb., Fl. Ind. 1: 319. 1820. India.
Thysanolaena agrostis Nees, Edinburgh New Phil. Jour. 18: 180. 1835. Based on *Agrostis maxima* Roxb.

(160) **TRACHYPOGON Nees**

(1) **Trachypogon secundus** (Presl) Scribn., U. S. Dept. Agr., Div. Agrost. Cir. 32: 1. 1901. Based on *Heteropogon secundus* Presl.
Heteropogon secundus Presl, Rel. Haenk. 1: 355. 1830. Mexico, *Haenke*.
Andropogon secundus Kunth, Rév. Gram. 1: Sup. 39. 1830. Not *A. secundus* Ell., 1821. Based on *Heteropogon secundus* Presl.
Trachypogon .. preslii var. *secundus* Anderss., Öfvers. Svensk. Vetensk. Akad. Förhandl. 14: 50. 1857. Based on *Heteropogon secundus* Presl.
Trachypogon plumosus var. *montufari* subvar. *secundus* Hack. ex Henr., Med. Rijks Herb. Leiden 40: 40. 1921. Based on *Heteropogon secundus* Presl. Included in *Trachypogon montufari* (H. B. K.) Nees in Manual ed. 1. That species has not been found north of Mexico.

(93) **TRAGUS Hall.**

(1) **Tragus berteronianus** Schult., Mantissa 2: 205. 1824. Dominican Republic, *Bertero*.
Tragus occidentalis Nees, Agrost. Bras. 286. 1829. Brazil.
Lappago berteroniana Schult. ex Steud., Syn. Pl. Glum. 1: 112. 1854. erroneously cited as synonym of *L. aliena* Spreng.
Tragus racemosus var. *brevispicula* Doell in Mart., Fl. Bras. 2²: 123. pl. 18. 1877. Brazil.

Nazia occidentalis Scribn., Zoe 4: 386. 1894. Based on *Tragus occidentalis* Nees.
Lappago occidentalis Nees ex Hook. f., Fl. Brit. Ind. 7: 97. 1896. Presumably based on *Tragus occidentalis* Nees; erroneously cited as synonym of *Tragus racemosus* All.

The following two names refer to *Tragus berteronianus*, though they are based on *Lappago alienus* Spreng., which is *Pseudechinolaena polystachya* (H. B. K.) Stapf, of the Tropics.

Nazia racemosa aliena Scribn. and Smith, U. S. Dept. Agr., Div. Agrost. Bul. 4: 12. 1897. Based on *Lappago aliena* Spreng.
Nazia aliena Scribn., U. S. Dept. Agr., Div. Agrost. Bul. 17: 28. f. 324. 1899. Based on *Lappago aliena* Spreng.

(2) **Tragus racemosus** (L.) All., Fl. Pedem. 2: 241. 1785. Based on *Cenchrus racemosus* L.
Cenchrus racemosus L., Sp. Pl. 1049. 1753. Southern Europe.
Lappago racemosa Honck., Syn. Pl. Germ. 1: 440. 1792. Based on *Cenchrus racemosus* L.
Tragus muricatus Moench, Meth. Pl. 53. 1794. Based on *Cenchrus racemosus* L.
Tragus racemosus var. *longispicula* Doell, in Mart., Fl. Bras. 2²: 122. 1877. Based on *T. racemosus* Desf. (Same as *T. racemosus* All.)
Nazia racemosa Kuntze, Rev. Gen. Pl. 2: 780. 1891. Based on *Cenchrus racemosus* L.

(128) **TRICHACHNE Nees**

(2) **Trichachne californica** (Benth.) Chase, Wash. Acad. Sci. Jour. 23: 455. 1933. Based on *Panicum californicum* Benth.
Panicum californicum Benth., Bot. Voy. Sulph. 55. 1840. Bay of Magdalena, Baja California.
Panicum lachnanthum Torr., U. S. Expl. Miss. Pacif. Rpt. 7³: 21. 1858. Not *P. lachnanthum* Hochst., 1855. Burro Mountains, N. Mex.
Panicum saccharatum Buckl., Prel. Rpt. Geol. Agr. Survey Tex. App. 2. 1866. Texas, *Buckley*.
Panicum insulare var. *lachnanthum* Kuntze, Rev. Gen. Pl. 3³: 361. 362. 1898. Based on *P. lachnanthum* Torr.
Trichachne saccharata Nash in Small, Fl. Southeast. U. S. 83. 1903. Based on *Panicum saccharatum* Buckl.
Valota saccharata Chase, Biol. Soc. Wash. Proc. 19: 188. 1906. Based on *Panicum saccharatum* Buckl.
Digitaria californica Henr., Blumea 1: 99. 1934. Based on *Panicum californicum* Benth.

(4) **Trichachne hitchcockii** (Chase) Chase, Wash. Acad. Sci. Jour. 23: 454. 1933. Based on *Valota hitchcockii* Chase.

Valota hitchcockii Chase, Biol. Soc. Wash. Proc. 24: 110. 1911. San Antonio, Tex., *Hitchcock 5329.*
Digitaria hitchcockii Stuck., Ann. Cons. Jard. Genève 17: 287. 1914. Based on *Valota hitchcockii* Chase.
(1) **Trichachne insularis** (L.) Nees, Agrost. Bras. 86. 1829. Based on *Andropogon insularis* L.
Andropogon insularis L., Syst. Nat. ed. 10. 2: 1304. 1759. Jamaica, *Sloane.*
Panicum lanatum Rottb., Act. Lit. Univ. Hafn. 1: 269. 1778. Dutch Guiana.
Milium villosum Swartz, Prodr. Veg. Ind. Occ. 24. 1788. Based on *Andropogon insularis* L.
Milium hirsutum Beauv., Ess. Agrost. 13. pl. 5. f. 5. No locality cited.
Panicum leucophaeum H. B. K., Nov. Gen. et Sp. 1: 97. 1815. Venezuela and Colombia, *Humboldt* and *Bonpland.*
Panicum insulare G. Meyer, Prim. Fl. Esseq. 60. 1818. Based on *Andropogon insularis* L.
Saccharum polystachyum Sieb. ex Kunth, Enum. Pl. 1: 124. 1833. Not *S. polystachyum* Swartz, 1788. As synonym of *Panicum leucophaeum* H. B. K.
Agrostis villosa Poir ex Steud., Nom., Bot. ed. 2. 1: 43. 1840. Not *A. villosa* Poir., 1786. As synonym of *Milium villosum* Swartz.
Panicum saccharoides A. Rich. in Sagra, Hist. Cuba 11: 306. 1850. Not *P. saccharoides* Trin., 1826. Cuba.
Panicum falsum Steud., Syn. Pl. Glum. 1: 67. 1854. Cuba.
Panicum duchaissingii Steud., Syn. Pl. Glum. 1: 93. 1854. Guadeloupe, *Duchaissing.*
Tricholaena insularis Griseb., Abhandl. Gesell. Wiss. Göttingen 7: 265. 1857. Based on *Andropogon insularis* L.
Digitaria leucophaea Stapf in Dyer, Fl. Cap. 7: 382. 1898. Based on *Panicum leucophaeum* Swartz (error for H. B. K.)
Panicum insulare var. *leucophaeum* Kuntze, Rev. Gen. Pl. 3³: 361, 362. 1898. Based on *P. leucophaeum* H. B. K.
Syntherisma insularis Millsp. and Chase, Field Mus. Bot. 1: 473. 1902. Based on *Andropogon insularis* L.
Valota insularis Chase, Biol. Soc. Wash. Proc. 19: 188. 1906. Based on *Andropogon insularis* L.
Digitaria insularis Mez ex Ekman, Arkiv Bot. 13: 22. 1913. Based on *Andropogon insularis* L.
Andropogon fabricii Herzog ex Henr., Med. Rijks Herb. Leiden 40: 44. 1921. Jamaica, *Swartz.* (Sterile specimen with large galls.)
(3) **Trichachne patens** Swallen, Amer. Jour. Bot. 19: 442. f. 5. 1932. Near Lake Mitchell, San Antonio, Tex., *Amer. Gr. Natl. Herb. 294 (Hitchcock 5328).*
Digitaria patens Henr., Blumea 1: 99.

1934. Based on *Trichachne patens* Swallen.

(111) TRICHLORIS Fourn.

(1) **Trichloris crinita** (Lag.) Parodi, Rev. Argentina Agron. 14: 63. 1947. Based on *Chloris crinita* Lag.
Chloris crinita Lag., Var. Cienc. 4: 143. 1805. Erroneously said to come from Philippine Islands (collected by Née), but the type in the Madrid Herbarium and the brief description agree with *Trichloris mendocina.* Née collected grasses in both Mexico and Argentina.
Chloris mendocina R. A. Phil., An. Univ. Chile 36: 208. 1870. Mendoza, Argentina [Philippi].
Trichloris blanchardiana Fourn. ex Scribn., Torrey Bot. Club Bul. 9: 146. 1882. Tucson, Ariz., *Pringle.*
Chloridiopsis [error for *Chloropsis*] *blanchardiana* Gay ex Scribn., Torrey Bot. Club Bul. 9: 146. 1882, as synonym of *Trichloris blanchardiana* Fourn.
Trichloris verticillata, Fourn. ex Vasey, Grasses U. S. Descr. Cat. 61. 1885, name only; U. S. Dept. Agr., Div. Bot. Bul. 12²: pl. 25. 1891. Arizona [Tucson, *Pringle*].
Trichloris fasciculata Fourn., Mex. Pl. 2: 142. 1886. San Luis de Potosí, Mexico, *Virlet 1440.*
Chloropsis fasciculata Kuntze, Rev. Gen. Pl. 2: 771. 1891. Based on *Trichloris fasciculata* Fourn.
Chloropsis blanchardiana Kuntze, Rev. Gen. Pl. 2: 771. 1891. Based on *Trichloris blanchardiana* Hack. (error for Fourn.).
Chloropsis crinita Kuntze, Rev. Gen. Pl. 2: 771. 1891. Based on *Chloris crinita* Lag.
Leptochloris crinita Munro ex Kuntze, Rev. Gen. Pl. 2: 771. 1891. Name in Kew Herbarium.
Trichloris mendocina Kurtz, Mem. Fac. Cienc. Exact. Univ. Córdoba 1896.: 37. 1897. Based on *Chloris mendocina* R. A. Phil.
Chloropsis mendocina Kuntze, Rev. Gen. Pl. 3²: 348. 1898. Based on *Chloris mendocina* R. A. Phil.
Trichloris mendocina forma *blanchardiana* Kurtz, Bol. Acad. Cienc. Córdoba 16: 270. 1900. Based on *T. blanchardiana* Fourn.
Leptochloris greggii Munro ex Merrill, U. S. Dept. Agr., Div. Agrost. Cir. 32: 7. 1901, as synonym of *Chloropsis mendocina* Kuntze.
Chloris trichodes Lag. ex Parodi, Rev. Argentina Agron. 14: 62. 1947, as synonym of *Trichloris crinita* (Lag.) Parodi.
Trichloris crinita var. *typica* Parodi, Rev. Argentina Agron. 14: 63. 1947.
(2) **Trichloris pluriflora** Fourn., Mex. Pl. 2:

142. 1886. Mexico, *Karwinsky;* Texas, between Laredo and Bejar [Bexar], *Berlandier* 1430.

Trichloris latifolia Vasey, U. S. Dept. Agr. Spec. Rpt. 63: 32. 1883. Texas and New Mexico [*Wright* 763]. Name only.

Chloropsis pluriflora Kuntze, Rev. Gen. Pl. 2: 771. 1891. Based on *Trichloris pluriflora* Fourn.

(98) TRICHONEURA Anderss.

(1) Trichoneura elegans Swallen, Amer. Jour. Bot. 19: 439. f. 4. 1932. Devine, Tex., *Silveus* 343.

(33) TRIDENS Roem. and Schult.

(15) Tridens albescens (Vasey) Woot. and Standl., N. Mex. Col. Agr. Bul. 81: 129. 1912. Based on *Triodia albescens* Vasey.

Triodia albescens Vasey, U. S. Dept. Agr., Div. Bot. Bul. 12²: pl. 33. 1891. Texas [type, *Hall* 782] and New Mexico. *Tricuspis albescens* Munro erroneously cited as synonym (see this name under *T. congestus*).

Sieglingia albescens Kuntze, Rev. Gen. Pl. 2: 789. 1891. Based on *Triodia albescens* Vasey.

Rhombolytrum albescens Nash in Britton, Man. 129. 1901. Based on *Triodia albescens* Vasey.

(8) Tridens ambiguus (Ell.) Schult., Mantissa 2: 333. 1824. Based on *Poa ambigua* Ell.

Poa ambigua Ell., Bot. S. C. and Ga. 1: 165. 1816. South Carolina and Georgia. *Windsoria ambigua* Nutt., Gen. Pl. 1: 70. 1818. Based on *Poa ambigua* Ell. *Uralepis ambigua* Kunth, Rév. Gram. 1: 108. 1829. Based on *Poa ambigua* Ell. *Tricuspis ambigua* Chapm., Fl. South. U. S. 559. 1860. Based on *Poa ambigua* Ell. *Triodia ambigua* Benth. ex Vasey, U. S. Dept. Agr. Spec. Rpt. 63: 35. 1883. Not *T. ambigua* R. Br., 1810. Based on *Tricuspis ambigua* Chapm. *Sieglingia ambigua* Kuntze, Rev. Gen. Pl. 2: 789. 1891. Based on *Poa ambigua* Ell.

Tricuspis langloisii Nash, N. Y. Bot. Gard. Bul. 1: 293. 1899. Louisiana, *Langlois*. *Triodia elliottii* Bush, Acad. Sci. St. Louis, Trans. 12: 73. 1902. Based on *Poa ambigua* Ell. *Triodia langloisii* Bush, Acad. Sci. St. Louis, Trans. 12: 72. 1902. Based on *Tricuspis langloisii* Nash. *Tridens langloisii* Nash in Small, Fl. Southeast. U. S. 142. 1903. Based on *Tricuspis langloisii* Nash.

(6) Tridens buckleyanus (L. H. Dewey) Nash in Small, Fl. Southeast. U. S. 143. 1903. Based on *Sieglingia buckleyana* L. H. Dewey.

Sieglingia buckleyana L. H. Dewey, U. S. Natl. Herb. Contrib. 2: 540. 1894. Southern Texas, *Buckley*. *Triodia buckleyana* Vasey, U. S. Natl. Herb. Contrib. 2: 540. 1894, as synonym of *Sieglingia buckleyana* L. H. Dewey. *Triodia buckleyana* Vasey ex Hitchc., Wash. Acad. Sci. Jour. 23: 452. 1933. Based on *Sieglingia buckleyana* L. H. Dewey.

(7) Tridens carolinianus (Steud.) Henr., Blumea 3: 424. 1940. Based on *Festuca caroliniana* Steud.

Festuca caroliniana Steud., Syn. Pl. Glum. 1: 312. 1854. Carolina, *Bosc*. *Triodia drummondii* Scribn. and Kearn., U. S. Dept. Agr., Div. Agrost. Bul. 4: 37. 1897. Jacksonville, "Fla." [Louisiana], *Drummond*. *Tridens drummondii* Nash ex Small, Fl. Southeast. U. S. 143. 1903. Based on *Triodia drummondii* Scribn. and Kearn. *Triodia caroliniana* Chase, Amer. Jour. Bot. 24: 34. 1937. Based on *Festuca caroliniana* Steud.

(12) Tridens chapmani (Small) Chase, new combination. Based on *Sieglingia chapmani* Small.

Sieglingia chapmani Small, Torrey Bot. Club Bul. 22: 365. 1895. Florida, *Chapman*. *Triodia chapmani* Bush, Acad. Sci. St. Louis, Trans. 12: 74. 1902. Based on *Sieglingia chapmani* Small. *Triodia flava* var. *chapmani* Fern. and Grisc., Rhodora 37: 133. 1935. Based on *Sieglingia chapmani* Small.

(5) Tridens congestus (L. H. Dewey) Nash in Small, Fl. Southeast. U. S. 143. 1903. Based on *Sieglingia congesta* L. H. Dewey.

Tricuspis albescens Munro ex A. Gray, Acad. Nat. Sci. Phila. Proc. 1862: 335. 1863. Name only for *Drummond* 314, Texas.

Sieglingia congesta L. H. Dewey, U. S. Natl. Herb. Contrib. 2: 538. 1894. Corpus Christi, Tex., *Nealley* 24. *Tricuspis congesta* Heller, N. Amer. Pl. Cat. ed. 2. 28. 1900. Based on *"Triodia"* [error for *Sieglingia*] *congesta* L. H. Dewey. *Triodia congesta* Bush, Acad. Sci. St. Louis, Trans. 12: 67. pl. 10. 1902. Based on *Sieglingia congesta* L. H. Dewey.

(17) Tridens elongatus (Buckl.) Nash in Small, Fl. Southeast. U. S. 143. 1903. Based on *Uralepis elongata* Buckl.

Uralepis elongata Buckl., Acad. Nat. Sci. Phila. Proc. 1862: 89. 1862. Northern Texas. *Triodia trinerviglumis* Benth. ex Vasey, U. S. Dept. Agr. Spec. Rpt. 63: 35. 1883, name only, with *Tricuspis trinerviglumis* Munro, also name only, as synonym. Texas. Described in Vasey, U. S. Dept. Agr., Div. Bot. Bul. 12²: pl. 40. 1891. Texas to Arizona, northward to Colorado.

Tricuspis trinerviglumis Munro ex Vasey, U. S. Dept. Agr., Spec. Rpt. 63: 35. 1883, as synonym of *Triodia trinerviglumis* Benth.

Sieglingia trinerviglumis Kuntze, Rev. Gen. Pl. 2: 789. 1891. Based on *Tricuspis trinerviglumis* "Buckl." (error for Munro).

Sieglingia elongata Nash in Britt. and Brown, Illustr. Fl. 3: 504. 1898. Based on *Uralepis elongata* Buckl.

Tricuspis elongata Heller, Cat. N. Amer. Pl. ed. 2. 28. 1900. Based on "*Triodia*" [error for *Uralepis*] *elongata* Buckl.

Triodia elongata Scribn., U. S. Dept. Agr., Div. Agrost. Bul. 17 (ed. 2): 210: f. 506. 1901. Based on *Uralepis elongata* Buckl.

(9) **Tridens eragrostoides** (Vasey and Scribn.) Nash in Small, Fl. Southeast. U. S. 142. 1903. Based on *Triodia eragrostoides* Vasey and Scribn.

Triodia eragrostoides Vasey and Scribn., U. S. Natl. Herb. Contrib. 1: 58. 1890. Texas, *Nealley*.

Sieglingia eragrostoides L. H. Dewey, U. S. Natl. Herb. Contrib. 2: 539. 1894. Based on *Triodia eragrostoides* Vasey and Scribn.

Sieglingia eragrostoides var. *scabra* Vasey ex Beal, Grasses N. Amer. 2: 465. 1896. Texas, *Nealley* [probably No. 96].

Triodia eragrostoides var. *scabra* Bush, Acad. Sci. St. Louis, Trans. 12: 71. 1902. Based on *Sieglingia eragrostoides* var. *scabra* Vasey.

(10) **Tridens flavus** (L.) Hitchc., Rhodora 8: 210. 1906. Based on *Poa flava* L.

Poa flava L., Sp. Pl. 68. 1753. Virginia.

Poa sesleroides Michx., Fl. Bor. Amer. 1: 68. 1803. Not *P. seslerioides* All., 1785. Illinois and the mountains of Carolina [type], *Michaux*.

Tricuspis caroliniana Beauv., Ess. Agrost. 179. pl. 3. f. 29, pl. 15. f. 10. 1812. South Carolina.

Tricuspis novae-boracensis Beauv., Ess. Agrost. 77, 179. 1812. Name only. New York, *Delille*.

Poa caerulescens Michx. ex Beauv., Ess. Agrost. 77. 1812, name only; Kunth, Rév. Gram. 1: 108. 1829, as synonym of *Uralepis cuprea* Kunth.

Festuca quadridens Poir. in Lam., Encycl. Sup. 2: 640. 1812. Carolina, *Bosc*.

Triodia cuprea Jacq., Eclog. Gram. 2: 21. pl. 16. 1814. Grown in botanic garden, source unknown.

Poa quinquefida Pursh, Fl. Amer. Sept. 1: 81. 1814. New England to Carolina.

?*Panicum festucoides* Poir. in Lam., Encycl. Sup. 4: 283. 1816. East Indies, Desvaux, but Desvaux later (see

Triodia festucoides below) corrects this to North America.

Poa arundinacea Poir. in Lam., Encycl. Sup. 4: 329. 1816. Based on *P. sesleroides* Michx.

Tridens quinquefidus Roem. and Schult., Syst. Veg. 2: 599. 1817. Based on *Poa quinquefida* Pursh.

Windsoria poaeformis Nutt., Gen. Pl. 1: 70. 1818. Based on *Poa sesleroides* Michx.

Tricuspis sesleroides Torr., Fl. North. and Mid. U. S. 118. 1823. Based on *Poa sesleroides* Michx.

Cynodon carolinianus Raspail, Ann. Sci. Nat., Bot. 5: 302. 1825. Based on *Tricuspis caroliniana* Beauv.

Windsoria seslerioides Eaton, Man. ed. 5. 447. 1829. Based on *Poa sesleroides* Michx.

Uralepis cuprea Kunth, Rév. Gram. 1: 108. 1829. Based on *Triodia cuprea* Jacq.

Eragrostis tricuspis Trin., Acad. St. Pétersb. Mém. VI. Math. Phys. Nat. 1: 414. 1830. Based on *Tricuspis caroliniana* Beauv.

Tricuspis quinquifida Beauv. ex Don, Loud. Hort. Brit. 31. 1830. Based on *Poa caerulescens* Michx.

?*Triodia festucoides* Desv., Opusc. 98. 1831. North America, *Panicum festucoides* Desv., in Poir., cited as synonym.

Triodia caerulescens Desv., Opusc. 99. 1831. Based on *Poa caerulescens* Michx.

Triodia novaeboracensis Desv., Opusc. 99. 1831. Based on *Tricuspis novaeboracensis* Beauv.

Uralepis tricuspis Steud., Nom. Bot. ed. 2. 1: 564. 1840. Based on *Eragrostis tricuspis* Trin.

Festuca purpurea Schreb. ex Steud., Nom. Bot. ed. 2. 1: 632. 1840, as synonym of *Uralepis cuprea* Kunth.

Tricuspis sesleroides var. *flexuosa* Wood, Amer. Bot. and Flor. pt. 2: 398. 1871. Pennsylvania.

Festuca flava F. Muell., Sel. Pl. Indus. Cult. 87. 1876. Based on *Poa flava* "Gronov" [L.].

Triodia sesleroides Benth. ex Vasey, U. S. Dept. Agr. Spec. Rpt. 63: 35. 1883. Based on *Tricuspis sesleroides* Torr.

Sieglingia flava Kuntze, Rev. Gen. Pl. 2: 789. 1891. Based on *Poa flava* L.

Sieglingia cuprea Millsp., Fl. W. Va. 471. 1892. Presumably based on *Triodia cuprea* Jacq.

Sieglingia sesleroides Scribn., Torrey Bot. Club Mem. 5: 48. 1894. Based on *Poa sesleroides* Michx.

Sieglingia sesleroides var. *intermedia* Vasey ex L. H. Dewey, U. S. Natl. Herb. Contrib. 2: 539. 1894. Texas to Oklahoma, *Sheldon* in 1891.

Triodia sesleroides var. *aristata* Scribn.

and Ball, U. S. Dept. Agr., Div. Agrost. Bul. 24: 45. 1901. Clarcona, Fla., *Meislahn* 90.
Tricuspis seslerioides var. *pallida* Holm, Biol. Soc. Wash. Proc. 14: 19. 1901. Marshall Hall, Md., *Holm*.
Tridens seslerioides Nash in Small, Fl. Southeast. U. S. 142. 1903. Based on *Poa sesleroides* Michx.
Tricuspis flava Hubb., Rhodora 14: 186. 1912. Based on *Poa flava* L.
Eragrostis arundinacea Jedw., Bot. Archiv Mez 5: 192. 1924. Texas.
Triodia flava Smyth, Kans. Acad. Trans. 25: 95. 1913. Based on *Poa flava* L.
Triodia flava var. *aristata* Fern. and Grisc., Rhodora 37: 134. 1935. Based on *Triodia sesleroides* var. *aristata* Scribn. and Ball.
Triodia flava forma *flava* Fosberg, Castanea 11: 66. 1946. Based on *Poa flava* L.
Triodia flava forma *cuprea* Fosberg, Castanea 11: 67. 1946. Based on *Triodia cuprea* Jacq.

(2) **Tridens grandiflorus** (Vasey) Woot. and Standl., N. Mex. Col. Agr. Bul. 81: 129. 1912. Based on *Triodia grandiflora* Vasey.
Uralepis avenacea var. *viridiflora* Fourn., Mex. Pl. 2: 110. 1886. San Luis de Potosí, *Virlet* 1379. No description, but specimen cited is *Tridens grandiflorus*.
Triodia grandiflora Vasey, U. S. Natl. Herb. Contrib. 1: 59. 1890. Chenate Mountains, Presidio County, Tex., *Nealley* 823.
Sieglingia avenacea var. *grandiflora* L. H. Dewey, U. S. Natl. Herb. Contrib. 2: 538. 1894. Based on *Triodia grandiflora* Vasey.
Sieglingia grandiflora Beal, Grasses N. Amer. 2: 471. 1896. Based on *Triodia grandiflora* Vasey.

(16) **Tridens muticus** (Torr.) Nash in Small, Fl. Southeast. U. S. 143. 1903. Based on *Tricuspis mutica* Torr.
Tricuspis mutica Torr., U. S. Expl. Miss. Pacif. Rpt. 4: 156. 1857. Laguna Colorado, N. Mex. [*Bigelow*].
Uralepis pilosa Buckl., Acad. Nat. Sci. Phila. Proc. 1862: 95. 1862. Not *U. pilosa* Buckl., op. cit. 94. "Northern Texas" cited, but the type is from western Texas, collected by *Wright*.
Triodia mutica Scribn., Torrey Bot. Club Bul. 10: 30. 1883. Based on *Tricuspis mutica* Torr.
Uralepis mutica Fourn. ex Hemsl., Biol. Centr. Amer. Bot. 3: 569. 1885, as synonym of *Triodia mutica* Benth. (*U. mutica* Fourn., Mex. Pl. 2: 110. 1886, based on *Liebmann* 611, is *Poa alpina*.)
Sieglingia mutica Kuntze, Rev. Gen. Pl. 2: 789. 1891. Based on *Tricuspis mutica* Torr.

(3) **Tridens nealleyi** (Vasey) Woot. and Standl., N. Mex. Col. Agr. Bul. 81: 129. 1912. Based on *Triodia nealleyi* Vasey.
Triodia nealleyi Vasey, Torrey Bot. Club Bul. 15: 49. 1888, name only; U. S. Dept. Agr., Div. Bot. Bul. 12²: pl. 36. 1891. Western Texas, *Nealley*.
Sieglingia nealleyi L. H. Dewey, U. S. Natl. Herb. Contrib. 2: 538. 1894. Based on *Triodia nealleyi* Vasey.
Tricuspis nealleyi Heller, N. Amer. Pl. Cat. ed. 2. 28. 1900. Presumably based on *Triodia nealleyi* Vasey.

(11) **Tridens oklahomensis** (Feath.) Feath., new combination. Based on *Triodia oklahomensis* Feath.
Triodia oklahomensis Feath., Rhodora 40: 243. 1938. Stillwater, Okla., *Wade* in 1937.

(4) **Tridens pilosus** (Buckl.) Hitchc., U. S. Natl. Herb. Contrib. 17: 357. 1913. Based on *Uralepis pilosa* Buckl.
Uralepis pilosa Buckl., Acad. Nat. Sci. Phila. Proc. 1862: 94. 1862. Middle Texas, [*Buckley*].
Tricuspis acuminata Munro ex A. Gray, Acad. Nat. Sci. Phila. Proc. 1862: 335. 1862, as synonym of *Uralepis pilosa* Buckl.
Triodia acuminata Benth. ex Vasey, U. S. Dept. Agr. Spec. Rpt. 63: 35. 1883, name only, with *Tricuspis acuminata* Munro given as synonym; Vasey, U. S. Dept. Agr., Div. Bot. Bul. 12²: pl. 32. 1891. Texas [type, Austin, *Hall* 779] to Arizona and Mexico.
Sieglingia acuminata Kuntze, Rev. Gen. Pl. 2. 789. 1891. Based on *Triodia acuminata* Vasey.
Sieglingia pilosa Nash in Britt. and Brown, Illustr. Fl. 3: 504. 1898. Based on *Uralepis pilosa* Buckl.
Tricuspis pilosa Heller, Cat. N. Amer. Pl. ed. 2: 28. 1900. Presumably based on *Uralepis pilosa* Buckl.
Triodia pilosa (Buckl.) Merr., U. S. Dept. Agr., Div. Agrost. Cir. 32: 9. 1901. Based on *Uralepis pilosa* Buckl.
Erioneuron pilosum Nash in Small, Fl. Southeast. U. S. 144. 1903. Based on *Uralepis pilosa* Buckl.

(1) **Tridens pulchellus** (H. B. K.) Hitchc. in Jepson, Fl. Calif. 1: 141. 1912. Based on *Triodia pulchella* H. B. K.
Triodia pulchella H. B. K., Nov. Gen. et Sp. 1: 155. pl. 47. 1816. Mexico, *Humboldt* and *Bonpland*.
Koeleria pulchella Spreng., Syst. Veg. 1: 332. 1825. Based on *Triodia pulchella* H. B. K.
Uralepis pulchella Kunth, Rév. Gram. 1: 108. 1829. Based on *Triodia pulchella* H. B. K.
Dasyochloa pulchella Willd. ex Steud., Nom. Bot. ed. 2. 1: 484. 1840, as synonym of *Uralepis pulchella* Kunth ex

Rydb., Fl. Rocky Mount. 67. 1917.
Based on *Triodia pulchella* H. B. K.
Tricuspis pulchella Torr., U. S. Expl.
Miss. Pacif. Rpt. 4: 156. 1857. Based
on "*Trichodia*" [error for *Triodia*] pulchella H. B. K.
Trichodiclida prolifera Cervant., Naturaleza 1870: 346. 1870. Near Mexico
City.
Sieglingia pulchella Kuntze, Rev. Gen. Pl.
2: 789. 1891. Based on *Triodia pulchella* H. B. K.
Sieglingia pulchella var. *parviflora* Vasey ex
Beal, Grasses N. Amer. 2: 468. 1896.
Southern California, *Orcutt.*

(14) **Tridens strictus** (Nutt.) Nash in Small,
Fl. Southeast. U. S. 143. 1903. Based
on *Windsoria stricta* Nutt.
Windsoria stricta Nutt., Amer. Phil. Soc.
Trans. (n. s.) 5: 147. 1837. Arkansas
(probably Arkansas Post), *Nuttall.*
Tricuspis stricta Wood, Class-book, ed.
1861. 792. 1861. Based on *Windsoria
stricta* Nutt.
Uralepis densiflora Buckl., Acad. Nat.
Sci. Phila. Proc. 1862: 94. 1862.
Middle Texas, [*Buckley*].
Triodia stricta Benth. ex Vasey, U. S.
Dept. Agr. Spec. Rpt. 63: 35. 1883.
Based on "*Tricuspis*" [error for *Windsoria*] *stricta* Nutt.
Sieglingia stricta Kuntze, Rev. Gen. Pl. 2:
789. 1891. Based on *Windsoria stricta*
Nutt.

(13) **Tridens texanus** (S. Wats.) Nash in
Small, Fl. Southeast. U. S. 142. 1903.
Based on *Triodia texana* Thurb. (error
for S. Wats.).
Triodia texana S. Wats., Amer. Acad. Sci.
Proc. 18: 180. 1883. Coahuila, Mexico;
western Texas and New Mexico, *Wright*
776, 777, and 2045 [error for 2055],
type, from Texas.
Tricuspis texana Thurb. ex S. Wats.,
Amer. Acad. Sci. Proc. 18: 180. 1883,
as synonym of *Triodia texana* S. Wats.
Sieglingia texana Kuntze, Rev. Gen. Pl.
2: 789. 1891. Based on *Triodia texana*
S. Wats.

(34) TRIPLASIS Beauv.

(2) **Triplasis americana** Beauv., Ess. Agrost.
81. pl. 16. f. 10. 1812. United States,
Delille.
Uralepsis cornuta Ell., Bot. S. C. and Ga.
1: 580. 1821. South Carolina and
Georgia.
Tricuspis cornuta A. Gray, Man. 590.
1848. Based on *Uralepsis cornuta* Ell.
Triplasis cornuta Benth. ex Jacks., Ind.
Kew. 2: 1121. 1895, as synonym of
Triplasis americana Beauv.
Sieglingia americana Beal, Grasses N.
Amer. 2: 466. 1896. Based on *Triplasis americana* Beauv.

(1) **Triplasis purpurea** (Walt.) Chapm., Fl.
South. U. S. 560. 1860. Based on *Aira*

purpurea Walt.
Aira purpurea Walt., Fl. Carol. 78. 1788.
South Carolina.
Festuca brevifolia Muhl., Descr. Gram.
167. 1817. Delaware, Georgia, and
New York. Name only, Muhl., Cat.
Pl. 13. 1813.
Diplocea barbata Raf., Amer. Jour. Sci. 1:
252. 1818. Carolina; Long Island.
Uralepsis purpurea Nutt., Gen. Pl. 1: 62.
1818. Based on *Aira purpurea* Walt.
Uralepsis aristulata Nutt., Gen. Pl. 1: 63.
1818. Wilmington, Del., *Baldwin.*
Glyceria? brevifolia Schult., Mantissa 2:
387. 1824. Based on *Festuca brevifolia*
Muhl.
Tricuspis purpurea A. Gray, Man. 589.
1848. Based on *Aira purpurea* Walt.
Merisachne drummondii Steud., Syn. Pl.
Glum. 1: 117. 1854. Texas, *Drummond*
330.
Festuca purpurea F. Muell., Sel. Pl. Indus.
Cult. 88. 1876. Based on *Uralepsis
purpurea* Nutt.
Triplasis sparsiflora Chapm., Bot. Gaz. 3:
19. 1878. Punta Rassa, Fla., [*Chapman*, specimen affected by fungus.]
Sieglingia purpurea Kuntze, Rev. Gen.
Pl. 2: 789. 1891. Based on *Aira
purpurea* Walt.
Panicularia brevifolia Porter, Torrey Bot.
Club Bul. 20: 205. 1893. Based on
Festuca brevifolia Muhl.
Triplasis intermedia Nash, Torrey Bot.
Club Bul. 25: 564. 1898. Tampa, Fla.,
Nash 2426.
Triplasis floridana Gandog., Soc. Bot.
France Bul. 66[7]: 303. 1920. Punta
Rassa, Fla., *Hitchcock* 533.
Triplasis glabra Gandog., Soc. Bot.
France Bul. 66[7]: 303. 1920. Rhode
Island and Florida.
Triodia purpurea Smyth, Kans. Acad.
Sci. Trans. 25: 95. 1913. Based on
Triplasis purpurea Chapm.

(99) TRIPOGON Roth

(1) **Tripogon spicatus** (Nees) Ekman, Arkiv
Bot. 11[4]: 36. 1912. Based on *Bromus
spicatus* Nees.
Bromus spicatus Nees, Agrost. Bras. 471.
1829. Piauhy, Brazil.
Diplachne spicata Doell in Mart., Fl.
Bras. 2[3]: 159. pl. 28. f. 2. 1878. Based
on *Bromus spicatus* Nees.
Triodia schaffneri S. Wats., Amer. Acad.
Sci. Proc. 18: 181. 1883. San Luis
Potosí, Mexico, *Schaffner* 1077.
Diplachne reverchoni Vasey, Torrey Bot.
Club Bul. 13: 118. 1886. Llano
County, Tex., *Reverchon.*
Leptochloa spicata Scribn., Acad. Nat.
Sci. Phila. Proc. 1891: 304. 1891.
Based on *Diplachne spicata* Doell.
Sieglingia schaffneri Kuntze, Rev. Gen.
Pl. 2: 789. 1891. Based on *Triodia
schaffneri* S. Wats.

Rabdochloa spicata Kuntze ex Stuck., An. Mus. Nac. Buenos Aires 11: 121. 1904. Based on *Bromus spicatus* Nees.

Sieglingia spicata Kuntze ex Stuck., An. Mus. Nac. Buenos Aires 11: 128. 1904. Based on *Bromus spicatus* Nees.

(166) TRIPSACUM L.

(1) **Tripsacum dactyloides** (L.) L., Syst. Nat. ed. 10. 2: 1261. 1759. Based on *Coix dactyloides* L.

Coix dactyloides L., Sp. Pl. 972. 1753. America.

Coix angulatis Mill., Gard. Dict. ed. 8. Coix No. 2. 1768. North America.

Ischaemum glabrum Walt., Fl. Carol. 249. 1788. South Carolina.

Tripsacum monostachyum Willd., Sp. Pl. 4: 202. 1805. South Carolina.

Tripsacum dactyloides var. *monostachyon* Eaton and Wright, N. Amer. Bot. ed. 8. 461. 1840. Connecticut. Wood, Classbook 453. 1845. Gray, Man. Bot. 616. 1848. No basis given.

Tripsacum dactyloides var. *monostachyum* Hack. in Mart., Fl. Bras. 2³: 316. 1883. Based on *T. monostachyum* Willd.

Dactylodes angulatum Kuntze, Rev. Gen. Pl. 2: 773. 1891. Based on *Coix angulatis* Mill.

Dactylodes dactylodes Kuntze, Rev. Gen. Pl. 3²: 349. 1898. Based on *Tripsacum dactyloides* L.

Tripsacum dactyloides var. *occidentale* Cutler and Anders., Mo. Bot. Gard. Ann. 28: 258. 1941. Jeff Davis County, Tex., *Moore* and *Steyermark* 3092.

(2) **Tripsacum floridanum** Porter ex Vasey, U. S. Natl. Herb. Contrib. 3: 6. 1892. Florida, *Garber*.

Tripsacum dactyloides var. *floridanum* Beal, Grasses N. Amer. 2: 19. 1896. Based on *T. floridanum* Porter.

(3) **Tripsacum lanceolatum** Rupr. in Fourn., Mex. Pl. 2: 68. 1886. Aguas Calientes, Mexico, *Hartweg* 252.

Tripsacum acutiflorum Fourn., Soc. Bot. Belg. Bul. 15: 466. 1876, name only; Nash, N. Amer. Fl. 17: 81. 1909. Same type as *T. lanceolatum* Rupr.

Tripsacum lemmoni Vasey, U. S. Natl. Herb. Contrib. 3: 6. 1892. Huachuca Mountains, Ariz., *Lemmon* [2932].

Tripsacum dactyloides var. *lemmoni* Beal, Grasses N. Amer. 2: 19. 1896. Based on *T. lemmoni* Vasey.

Tripsacum dactyloides hispidum Hitchc., Bot. Gaz. 41: 295. 1906. Las Canóas, Mexico, *Pringle* 3811.

(57) TRISETUM Pers.

Trisetum aureum (Ten.) Ten., Fl. Napol. 2: 378. 1820. Based on *Koeleria aurea* Ten.

Koeleria aurea Ten., Cors. Bot. Lez. 1: 58. 1806. Europe.

(6) **Trisetum canescens** Buckl., Acad. Nat. Sci. Phila. Proc. 1862: 100. 1862. Columbia Plains, Oreg., *Nuttall*.

Trisetum elatum Nutt. ex A. Gray, Acad. Nat. Sci. Phila. Proc. 1862: 337. 1862, as synonym of *T. canescens* Buckl.

Trisetum cernuum var. *canescens* Beal, Grasses N. Amer. 2: 380. 1896. Based on *T. canescens* Buckl.

Trisetum canescens forma *tonsum* Louis-Marie, Rhodora 30: 216. 1928. Trinity County, Calif., *Yates* 522.

Trisetum canescens forma *velutinum* Louis-Marie, Rhodora 30: 216. 1928. Lassens Peak, Calif., *Austin* in 1879.

Trisetum projectum Louis-Marie, Rhodora 30: 217. 1928. Fresno County, Calif., *Hall* and *Chandler* 359.

Trisetum cernum var. *projectum* Beetle, West. Bot. Leaflets 4: 288. 1946. Based on *T. projectum* Louis-Marie.

(4) **Trisetum cernuum** Trin., Acad. St. Pétersb. Mém. VI. Math. Phys. Nat. 1: 61. 1830. Sitka, Alaska.

Avena nutkaensis Presl, Rel. Haenk. 1: 254. 1830. Nootka Sound, Vancouver Island, *Haenke*.

Avena cernua Kunth, Rév. Gram. 1: Sup. 26. 1830. Based on *Trisetum cernuum* Trin.

Trisetum sandbergii Beal, Grasses N. Amer. 2: 378. 1896. Mount Stuart, Wash., *Sandberg* and *Leiberg* 823.

Trisetum nutkaense Scribn. and Merr. ex Davy, Calif. Univ. Pubs., Bot. 1: 63. 1902. Based on *Avena nutkaensis* Presl.

Tristetum cernuum var. *luxurians* Louis-Marie, Rhodora 30: 213. 1928. Seaside, Oreg., *Shear* and *Scribner* 1705.

Trisetum cernuum var. *luxurians* forma *pubescens* Louis-Marie, Rhodora 30: 213. 1928. Eureka, Calif.

Trisetum cernuum var. *sandbergii* Louis-Marie, Rhodora 30: 214. 1928. Based on *T. sandbergii* Beal.

Trisetum cernuum forma *pubescens* G. N. Jones, Wash. Univ. Pubs. Biol. 5: 108. 1936. Based on *T. cernuum* var. *luxurians* forma *pubescens* Louis-Marie.

(8) **Trisetum flavescens** (L.) Beauv., Ess. Agrost. 88, 153. pl. 18. f. 1. 1812. Based on *Avena flavescens* L.

Avena flavescens L., Sp. Pl. 80. 1753. Europe.

Trisetum pratense Pers., Syn. Pl. 1: 97. 1805. Europe.

Trisetaria flavescens Baumg., Enum. Stirp. Transsilv. 3: 263. 1816. Based on *Avena flavescens* Schreb. (error for L.).

Rebentischia flavescens Opiz, Lotos 4: 104. 1854, as synonym of *Trisetum flavescens* Beauv.

(10) **Trisetum interruptum** Buckl., Acad. Nat. Sci. Phila. Proc. 1862: 100. 1862. Middle Texas [*Buckley*].

?*Calamagrostis longirostris* Buckl., Prel. Rpt. Geol. Agr. Survey Tex. App. 2.

1866. Texas.
Trisetum hallii Scribn., Torrey Bot. Club Bul. 11: 6. 1884. Texas, *Hall* 799 in part.
Sphenopholis interrupta Scribn., Rhodora 8: 145. 1906. Based on *Trisetum interruptum* Buckl.
Sphenopholis hallii Scribn., Rhodora 8: 146. 1906. Based on *Trisetum hallii* Scribn.
Trisetum interruptum hallii Hitchc., Biol. Soc. Wash. Proc. 41: 160. 1928. Based on *T. hallii* Scribn.

(1) **Trisetum melicoides** (Michx.) Scribn., Bot. Gaz. 9: 169. 1884. Based on *Aira melicoides* Michx.
Aira melicoides Michx., Fl. Bor. Amer. 1: 62. 1803. Canada.
?*Arundo airoides* Poir. in Lam., Encycl. 6: 270. 1804. North America, *Michaux.*
Graphephorum melicoideum Desv., Nouv. Bul. Soc. Philom. Paris 2: 189. 1810. Based on *Aira melicoides* Michx.
?*Deyeuxia airoides* Beauv., Ess. Agrost. 44, 152, 160. 1812. Based on *Arundo airoides* Michx. [error for Poir.].
Poa melicoides Nutt., Gen. Pl. 1: 68. 1818. Based on *Aira melicoides* Michx.
Triodia melicoides Spreng., Syst. Veg. 1: 331. 1825. Based on *Aira melicoides* Michx.
?*Agrostis airoides* Raspail, Ann. Sci. Nat., Bot. 5: 449. 1825. Based on *Deyeuxia airoides* Beauv.
?*Calamagrostis airoides* Steud., Nom. Bot. ed. 2. 1: 249. 1840. Based on *Arundo airoides* Poir.
Dupontia cooleyi A. Gray, Man. ed. 2. 556. 1856. Washington, Mich. [*Cooley*].
Graphephorum melicoides var. *major* A. Gray, Amer. Acad. Sci. Proc. 5: 191. 1861. Based on *Dupontia cooleyi* A. Gray.
Graphephorum melicoideum cooleyi Scribn., Torrey Bot. Club Mem. 5: 53. 1894. Based on *Dupontia cooleyi* A. Gray.
Trisetum melicoideum cooleyi Scribn., Rhodora 8: 87. 1906. Based on *Dupontia cooleyi* A. Gray.
Trisetum melicoides var. *majus* Hitchc. in Robinson, Rhodora 10: 65. 1908. Based on *Graphephorum melicoides* var. *major* A. Gray.
Graphephorum cooleyi Farwell, Mich. Acad. Sci. Papers 1: 88. 1921. Based on *Dupontia cooleyi* A. Gray.

(7) **Trisetum montanum** Vasey, Torrey Bot. Club Bul. 13: 118. 1886. No locality cited. [Type, Las Vegas, N. Mex., *G. R. Vasey* in 1881.]
Trisetum argenteum Scribn., U. S. Dept. Agr., Div. Agrost. Bul. 11: 49. f. 8. 1898. Not *T. argenteum* Roem. and Schult., 1817. Silverton, Colo., *Shear* 1214.
Trisetum shearii Scribn., U. S. Dept. Agr.,

Div. Agrost. Cir. 30: 8. 1901. Based on *T. argenteum* Scribn.
Graphephorum shearii Rydb., Torrey Bot. Club Bul. 32: 602. 1905. Based on *Trisetum shearii* Scribn.
Trisetum canescens var. *montanum* Hitchc., Biol. Soc. Wash. Proc. 41: 160. 1928. Based on *T. montanum* Vasey.
Trisetum montanum var. *pilosum* Louis-Marie, Rhodora 30: 212. 1928. Caroles, N. Mex., *Standley* 4536.
Trisetum montanum var. *shearii* Louis-Marie, Rhodora 30: 213. 1928. Based on *Trisetum shearii* Scribn.

(3) **Trisetum orthochaetum** Hitchc., Amer. Jour. Bot. 21: 134. f. 3. 1934. Lolo Hot Springs, Bitterroot Mountains, Mont., *Chase* 5129.

(9) **Trisetum pennsylvanicum** (L.) Beauv. ex Roem. and Schult., Syst. Veg. 2: 658. 1817. Based on *Avena pennsylvanica* L.
Avena pennsylvanica L., Sp. Pl. 79. 1753. Pennsylvania, *Kalm.*
?*Avena caroliniana* Walt., Fl. Carol. 81. 1788. South Carolina.
Avena palustris Michx., Fl. Bor. Amer. 1: 72. 1803. Carolina and Georgia, *Michaux.*
Aira pallens var. *aristata* Muhl. ex Ell., Bot. S. C. and Ga. 1: 151. 1816. South Carolina.
Avena pennsylvanica Muhl., Descr. Gram. 185. 1817. Pennsylvania and North Carolina. No authority cited but the Muhlenberg specimen belongs to the Linnaean species.
Trisetum palustre Torr., Fl. North. and Mid. U. S. 126. 1823. Based on *Avena palustris* Michx.
Arrhenatherum pennsylvanicum Torr., Fl. North. and Mid. U. S. 1: 130. 1823. Based on *Avena pennsylvanica* L.
Arrhenatherum kentuckensis Torr., Fl. North. and Mid. U. S. 1: 131. 1823. Kentucky, sent by Rafinesque.
The name was spelled *"Kentuckenensis"* in Eaton, Man. Bot. N. Amer. ed. 5. 115. 1829, and *A. kentuckiensis* in Eaton and Wright, N. Amer. Bot. ed. 8. 136. 1840, both credited to Torrey.
Trisetum ludovicianum Vasey, Torrey Bot. Club Bul. 12: 6. 1885. Pointe à la Hache, La., *Langlois.*
Sphenopholis palustris Scribn., Rhodora 8: 145. 1906. Based on *Avena palustris* Michx.
Sphenopholis palustris flexuosa Scribn., Rhodora 8: 143, 145. 1906. Wilmington, Del., *Commons* 274.
Sphenopholis palustris var. *flexuosa* Scribn. in Robinson, Rhodora 10: 65. 1908. Based on *S. palustris flexuosa* Scribn.
Sphenopholis pennsylvanica Hitchc., Amer. Jour. Bot. 2: 304. 1915. Based on *Avena pennsylvanica* L.
Sphenopholis pennsylvanica var. *flexuosa* Hubb., Rhodora 18: 234. 1916. Based

on *S. palustris flexuosa* Scribn.
The plant from Hunting Creek, Va., discussed by Vasey (Bot. Gaz. 9: 165. 1884) as a hybrid between *Trisetum palustre* and *Eatonia pennsylvanica*, is an exceptional specimen of *Trisetum pennsylvanicum* (L.) Beauv. with short-awned and awnless spikelets.

(5) **Trisetum spicatum** (L.) Richt., Pl. Eur. 1: 59. 1890. Based on *Aira spicata* L.
Aira spicata L., Sp. Pl. 64. 1753. Lapland.
Aira subspicata L., Syst. Nat. ed. 10. 2: 873. 1759. Based on *A. spicata* L. (Sp. Pl. 64. 1753), the diagnosis copied.
Avena airoides Koel., Descr. Gram. 298. 1802. Based on *Aira subspicata* L.
Avena mollis Michx., Fl. Bor. Amer. 1: 72. 1803. Canada. Not *Avena mollis* Salisb., 1796, nor Koel., 1802.
Avena subspicata Clairv., Man. Herbor. 17. 1811. Based on a phrase name in Haller which refers to *Aira spicata* L.
Trisetum subspicatum Beauv., Ess. Agrost. 88, 149. 1812. Based on *Aira subspicata* L.
Melica triflora Bigel., New England Jour. Med. and Surg. 5: 334. 1816. Mount Washington, N. H., Boott. (In Eaton, Man. ed. 2. 317. 1818, misspelled *Melia triflora* and placed under the genus *Melia*, preceding *Melica*.)
Trisetaria airoides Baumg., Enum. Stirp. Transsilv. 3: 265. 1816. Based on *Avena airoides* Koel.
Trisetum airoides Beauv. ex Roem. and Schult., Syst. Veg. 2: 666. 1817. Based on *Avena airoides* Koel.
Trisetum molle Kunth, Rév. Gram. 1: 101. 1829. Based on *Avena mollis* Michx.
Koeleria subspicata Reichenb., Fl. Germ. 49. 1830. Based on *Aira subspicata* L.
Koeleria canescens Torr. ex. Trin., Acad. St. Pétersb. Mém. VI. Sci. Nat. 2¹: 13. 1836, as synonym of *Trisetum molle* Kunth.
Trisetum subspicatum var. *molle* A. Gray, Man. ed. 2. 572. 1856. Based on *Avena mollis* Michx.
Koeleria spicata Reichenb ex. Willk. and Lange, Prodr. Fl. Hisp. 1: 72. 1861. as synonym of *Trisetum subspicatum* Beauv.
Rupestrina pubescens Provancher, Fl. Canad. 689. 1862. Based on *Avena mollis* Michx.
Trisetum spicatum var. *molle* Beal, Grasses N. Amer. 2: 377. 1896. Based on *Avena mollis* Michx.
Trisetum brittonii Nash, N. Y. Bot. Gard. Bul. 1: 437. 1900. Marquette, Mich., *Britton* in 1883.
Trisetum congdoni Scribn. and Merr., Torrey Bot. Club Bul. 29: 470. 1902. Mariposa County, Calif., *Congdon*.
Trisetum americanum Gandog., Soc. Bot. France Bul. 49: 182. 1902. Colorado;

Idaho.
Trisetum majus Rydb., Colo. Agr. Expt. Sta. Bul. 100: 34. 1906. "*T. subspicatum major* Vasey," an unpublished name, cited as basis. A tall specimen collected by Vasey, Pen Gulch, Colo., in 1884 and marked "var. *major* Vasey" in his script is taken as type. No description by Rydberg except the distinctions given in the key.
Avena spicata Fedtsch., Act. Hort. Petrop. 28: 76. 1908. Not *A. spicata* L. Based on *Aira spicata* L.
Trisetum spicatum var. *pilosiglume* Fernald, Rhodora 18: 195. 1916. Newfoundland, *Fernald*, *Wiegand*, and *Bartram 4593*.
Trisetum spicatum congdoni Hitchc., Biol. Soc. Wash. Proc. 41: 160. 1928. Based on *Trisetum congdoni* Scribn. and Merr.
Trisetum spicatum var. *brittonii* Louis-Marie, Rhodora 30: 239. 1929. Based on *T. brittonii* Nash.
Trisetum spicatum var. *michauxii* St. John, Fl. Southeast. Wash. and Adj. Idaho 62. 1937. Based on *Avena mollis* Michx., not *A. mollis* Salisb., 1796, nor Koel., 1802.

(2) **Trisetum wolfii** Vasey, U. S. Dept. Agr. Monthly Rpt. Feb. Mar. 156. 1874. Twin Lakes, Colo., *Wolf*.
Trisetum subspicatum var. *muticum* Boland. in S. Wats., Bot. Calif. 2: 296. 1880. Upper Tuolumne, Calif., *Bolander 5019*.
Trisetum brandegei Scribn., Torrey Bot. Club Bul. 10: 64. 1883. Cascade Mountains, *Brandegee* and *Tweedy* in 1882.
Graphephorum wolfii Vasey ex Coult., Man. Rocky Mount. 423. 1885. Based on *Trisetum wolfii* Vasey.
Trisetum muticum Scribn., U. S. Dept. Agr., Div. Agrost. Bul. 11: 50. f. 10. 1898. Based on *Trisetum subspicatum* var. *muticum* Boland.
Graphephorum muticum Heller, Cat. N. Amer. Pl. ed. 2. 31. 1900. Presumably based on *Trisetum subspicatum* var. *muticum* Boland.
Trisetum wolfii muticum Scribn., Rhodora 8: 88. 1906. Based on *T. subspicatum* var. *muticum* Thurb. (error for Boland.).
Graphephorum brandegei Rydb., Fl. Rocky Mount. 61. 1917. Based on *Trisetum brandegei* Scribn.
Trisetum wolfii var. *brandegei* Louis-Marie, Rhodora 30: 241. 1929. Based on *T. brandegei* Scribn.
Trisetum wolfii var. *brandegei* forma *muticum* Louis-Marie, Rhodora 30: 241. 1929. Based on *T. wolfii muticum* Scribn.

(43) TRITICUM L.

(1) **Triticum aestivum** L., Sp. Pl. 85. 1753. Cultivated in Europe.

Triticum estivum Raf., Fl. Ludovic. 16. 1817. Error for *T. aestivum*.

Triticum hybernum L., Sp. Pl. 86. 1753. Cultivated in Europe.

Triticum compositum L., Syst. Veg. ed. 13. 108. 1774. Egypt. Form with branched spike.

Triticum sativum Lam., Fl. Franç. 3: 625. 1778. Cultivated in Europe.

Triticum vulgare Vill., Hist. Pl. Dauph. 2: 153. 1787. Cultivated in Europe.

Triticum vulgare var. *aestivum* Spenner, Fl. Friburg. 1: 163. 1825. Based on *T. aestivum* L.

Triticum sativum vulgare Desv., Opusc. 162. 1831. France.

Triticum sativum var. *aestivum* Wood, Class-book ed. 2. 619. 1847. Presumably based on *T. aestivum* L.

Triticum sativum var. *compositum* Wood, Class-book ed. 2. 619. 1847. Presumably based on *T. compositum* L.

Triticum sativum var. *vulgare* Hack. in Engler and Prantl, Nat. Pflanzenfam. 7: 85. 1887. Based on *T. vulgare* Vill.

Triticum sativum var. *vulgare* Vilm., Blumengartn. 1: 1217. 1896. Based on *T. vulgare* Vill.

Triticum aestivum var. *hybernum* Farwell, Mich. Acad. Sci. Rpt. 6: 203. 1904. Based on *T. hybernum* L.

Triticum aestivum subsp. *vulgare* Thell., Mém. Soc. Sci. Nat. Cherbourg 38: 142. 1912. Based on *T. vulgare* Vill.

Zeia vulgaris var. *aestiva* Lunell, Amer. Midl. Nat. 4: 226. 1915. Based on "*Triticum vulgare aestivum* L." error for *T. aestivum*.

Triticum orientale Perciv., Wheat Pl. Monogr. 155, 204, f. 134. 1921. Not *T. orientale* Biebers. 1808. Cultivated race from Persia.

Triticum pyramidale Perciv., Wheat Pl. Monogr. 156, 262, f. 161, 162. 1921. Cultivated race from Egypt.

Triticum persicum Vavilov, in Zhukov., Bul. Appl. Bot. Petrograd 13: 46. 1923. Transcaucasia, *Zhukovsky*. Not *T. persicum* Aitch. and Hemsley 1888, a species of *Aegilops*.

Triticum dicoccum var. *timopheevi* Zhukov., Sci. Papers Appl. Sect. Tiflis Bot. Gard. No. 3: 1. f. 1. 1924. Transcaucasia.

Triticum timopheevi Zhukov., Bul. Appl. Bot. Genet., and Plant Breed. 19²: 64. f. 1–3. 1928. Based on *T. dicoccum* var. *timopheevi* Zhukov.

Triticum compactum Host, Gram. Austr. 4: 4. pl. 7. 1809. Cultivated in Austria.

Triticum dicoccoides Koern., Bericht. Deutsch. Bot. Ges. 26: 309. 1908; Aaronsohn, Verh. Zool. Bot. Ges. Wien 59¹⁰: 485. 1909. Palestine.

Triticum dicoccum Schrank, Baier. Fl. 1: 389. 1789. Cultivated in Europe.

Triticum aestivum subsp. *dicoccum* Thell.,

Mém. Soc. Sci. Nat. Cherbourg 38: 141. 1912. Based on *T. dicoccum* Schrank.

Triticum aestivum var. *dicoccum* Bailey, Gentes Herb. 1: 133. 1923. Based on *T. dicoccum* Schrank.

Triticum durum Desf., Fl. Atlant. 1: 114. 1798. North Africa.

Triticum aestivum subsp. *durum* Thell., Mém. Soc. Sci. Nat. Cherbourg 38: 143. 1912. Based on *T. durum* Desf.

Triticum macha Dekap. and Menab., Bul. Appl. Bot. Genet., and Plant Breed. V. 1: 14, 38. 1932. Transcaspia.

Triticum monococcum L., Sp. Pl. 86. 1753. Cultivated in Europe.

Triticum aestivum var. *monococcum* Bailey, Gentes Herb. 1: 133. 1923. Based on *T. monococcum* L.

Triticum polonicum L., Sp. Pl. ed. 2. 127. 1762. Cultivated in Europe.

Triticum aestivum var. *polonicum* Bailey, Man. Cult. Pl. 116. 1924. Based on *T. polonicum* L.

Triticum spelta L., Sp. Pl. 86. 1753. Cultivated in Europe.

Triticum aestivum subsp. *spelta* Thell., Mitt. Naturw. Ges. Winterthur. 12: 147. 1918. Based on *T. spelta* L.

Triticum aestivum var. *spelta* Bailey, Gentes Herb. 1: 133. 1923. Based on *T. spelta* L.

Triticum sphaerococcum Perciv., Wheat Pl. Monogr. 157, 321. f. 202. 1921. India and Persia.

Triticum turgidum L., Sp. Pl. 86. 1753. Cultivated in Europe.

(22) UNIOLA L.

(2) **Uniola latifolia** Michx., Fl. Bor. Amer. 1: 70. 1803. The locality as published is Allegheny Mountains, but the type specimen is from Illinois.

(6) **Uniola laxa** (L.) B. S. P., Prel. Cat. N. Y. 69. 1888. Based on *Holcus laxus* L.

Holcus laxus L., Sp. Pl. 1048. 1753. Virginia.

Uniola gracilis Michx., Fl. Bor. Amer. 1: 71. 1803. Carolina to Georgia, *Michaux*.

Uniola virgata Bartr. ex Pursh, Fl. Amer. Sept. 1: 82. 1814, as synonym of *Uniola gracilis* Michx.

Chasmanthium gracile Link, Hort. Berol. 1: 159. 1827. Based on *Uniola gracilis* Michx.

Uniola uniflora Benke, Rhodora 31: 148. 1929. Memphis, Tenn., *Benke 4874.*

(3) **Uniola nitida** Baldw. in Ell., Bot. S. C. and Ga. 1: 167. 1816. Camden County, Ga., *Baldwin*.

Uniola intermedia Bosc ex Beauv., Ess. Agrost. 75, 181. 1812. Name only. [A Bosc specimen so named in Padua is *U. nitida*; another in Paris is *U. sessiflora*.]

(4) **Uniola ornithorhyncha** Steud., Syn. Pl. Glum. 1: 280. 1854. Alabama, *Drummond 51.*

Chasmanthium ornithorhynchum Nees ex Steud., Syn. Pl. Glum. 1: 280. 1854, as synonym of *Uniola ornithorhyncha* Steud.

(1) **Uniola paniculata** L., Sp. Pl. 71. 1753. Carolina.

Briza caroliniana Lam., Encycl. 1: 465. 1785. Carolina.

Uniola maritima Michx., Fl. Bor. Amer. 1: 71. 1803. Carolina, *Michaux.*

Trisiola paniculata Raf., Fl. Ludov. 144. 1817. Based on *Uniola paniculata* L.

Nevroctola maritima Raf. ex Jacks., Ind. Kew. 2: 311. 1894, as synonym of *Uniola paniculata* L.

Nevroctola paniculata Raf. ex Jacks., Ind. Kew. 2: 311. 1894, as synonym of *Uniola paniculata* L.

Uniola floridana Gandog., Soc. Bot. France Bul. 66[7]: 304. 1920. Santa Rosa Island, Fla., *Tracy 4545.*

Uniola heterochroa Gandog., Soc. Bot. France Bul. 66[7]: 304. 1920. Punta Rassa, Fla., *Hitchcock 535.*

Uniola macrostachys Gandog., Soc. Bot. France Bul. 66[7]: 304. 1920. Breton Island, La., *Tracy 462.*

(5) **Uniola sessiliflora** Poir. in Lam., Encycl. 8: 185. 1808. Carolina, *Bosc.*

Poa sessiliflora Kunth, Rév. Gram. 1: 111. 1829. Based on *Uniola sessiliflora* Poir.

Uniola longifolia Scribn., Torrey Bot. Club Bul. 21: 229. 1894. Georgia [type, De Kalb County, *Small* in 1893], Florida, Mississippi, Tennessee.

(32) VASEYOCHLOA Hitchc.

(1) **Vaseyochloa multinervosa** (Vasey) Hitchc., Wash. Acad. Sci. Jour. 23: 452. 1933. Based on *Melica multinervosa* Vasey.

Melica multinervosa Vasey, Bot. Gaz. 16: 235. 1891. Brazos Santiago, Tex., *Nealley.*

Distichlis multinervosa Piper, Biol. Soc. Wash. Proc. 18: 147. 1905. Based on *Melica multinervosa* Vasey.

Triodia multinervosa Hitchc., Biol. Soc. Wash. Proc. 41: 159. 1928. Based on *Melica multinervosa* Vasey.

VETIVERIA Bory

Vetiveria zizanioides (L.) Nash in Small, Fl. Southeast. U. S. 67. 1903. Based on *Phalaris zizanioides* L.

Phalaris zizanioides L., Mant. Pl. 2: 183. 1771. India.

Andropogon muricatus Retz., Obs. Bot. 3: 43 [31]. 1783. India.

Agrostis verticillata Lam., Encycl. 1: 59. 1783. Not *Agrostis verticillata* Vill., 1779. India.

Anatherum muricatum Beauv., Ess. Agrost. 150. pl. 22. f. 10. 1812. Based on *Andropogon muricatus* Retz.

Vetiveria odoratissima Bory in Lem., Bul. Soc. Philom. (Paris) 1822: 43. 1822. Ceylon, island of Bourbon.

Vetiveria odorata Virey, Jour. de Pharm. I. 13: 501. 1827. East Indies.

Vetiveria muricata Griseb., Fl. Brit. W. Ind. 560. 1864. Based on *Andropogon muricatus* Retz.

Vetiveria arundinacea Griseb., Fl. Brit. W. Ind. 559. 1864. Jamaica and Trinidad.

Sorghum zizanioides Kuntze, Rev. Gen. Pl. 2: 791. 1891. Based on *Phalaris zizanioides* L.

Andropogon zizanioides Urban, Symb. Antill. 4: 79. 1903. Based on *Phalaris zizanioides* L.

Holcus zizanioides Kuntze ex Stuck., An. Mus. Nac. Buenos Aires 11: 48. 1904. Based on *Phalaris zizanioides* L.

Anatherum zizanioides Hitchc. and Chase, U. S. Natl. Herb. Contrib. 18: 285. 1917. Based on *Phalaris zizanioides* L.

(104) WILLKOMMIA Hack.

(1) **Willkommia texana** Hitchc., Bot. Gaz. 35: 283. f. 1. 1903. Ennis, Tex., *J. G. Smith* in 1897.

Craspedorhachis texana Pilger, Bot. Jahrb. 74: 27. 1945. Based on *Willkommia texana* Hitchc.

(168) ZEA L.

(1) **Zea mays** L., Sp. Pl. 971. 1753. America.

Zea americana Mill., Gard. Dict. ed. 8. Zea No. 1. 1768. West Indies.

Zea vulgaris Mill., Gard. Dict. ed. 8. Zea No. 3. 1768. Northern parts of America.

Mays zea Gaertn., Fruct. et Sem. 1: 6. pl. 1. f. 9. 1788. Based on *Zea mays* L.

Zea segetalis Salisb., Prodr. Stirp. 28. 1796. Based on *Zea mays* L.

Mays americana Baumg., Enum. Stirp. Transsilv. 3: 281. 1816. Based on *Zea mays* L.

Zea mays var. *precox* Torr., in Eaton, Man. Bot. ed. 2. 500. 1818. Northern and Middle States.

Mayzea cerealis Raf., Med. Fl. 2: 241. 1830. Based on *Zea mays* L.

Mayzea cerealis var. *gigantea* Raf., Med. Fl. 2: 241. 1830. Mexico.

Zea hirta Bonaf., Hist. Nat. Mais 29. pl. 4, 39. pl. 4. 1836. Cultivated, seed from California.

Zea mays pensylvanica Bonaf., Hist. Nat. Mais 33. pl. 7. f. 4. 1836. Cultivated.

Zea mays virginica Bonaf., Hist. Nat. Mais 37. pl. 10. f. 15. 1836. Cultivated.

Zea erythrolepis Bonaf., Hist. Nat. Mais

30. pl. 5; 38. pl. 11. f. 17. 1836. Cultivated along Missouri River.
Zea mais hirta Alefeld, Landw. Fl. 309. 1866. Based on Zea hirta Bonaf.
Zea saccharata Sturtev., N. Y. State Agr. Expt. Sta. Rpt. 1884³. 156: 1885. Group name for sweet corn.
Zea canina S. Wats., Amer. Acad. Sci. Proc. 26: 160. 1891. Mexico. Hybrid with Euchlaena mexicana Schrad., fide G. N. Collins.
Zea mays saccharata Bailey, Cycl. Hort. 4: 2006. 1902. Based on Z. saccharata Sturtev.
ZEA MAYS var. EVERTA (Sturtev.) Bailey, Cycl. Hort. 4: 2005. 1902. Based on Z. everta Sturtev.
Zea everta Sturtev., N. Y. State Agr. Expt. Sta. Rpt. 1884³: 183. 1885. Group name for popcorn.
ZEA MAYS var. JAPONICA (Van Houtte) Wood, Amer. Bot. and Flor. pt. 2: 409. 1871. Presumably based on Z. japonica Van Houtte.
Zea japonica Van Houtte, Fl. Serr. Jard. 16: 121. 1865. Japan.
ZEA MAYS var. TUNICATA Larr. ex St. Hil., Ann. Sci. Nat., Bot. 16: 144. 1829. Uruguay.
Zea cryptosperma Bonaf., Hist. Nat. Mais 30, 40. pl. 5 bis. 1836. Based on Z. mais var. tunicata St. Hil.
Zea tunicata Sturtev., Torrey Bot. Club Bul. 21: 335. 1894. Based on Z. mays var. tunicata St. Hil.
Of the many names published for forms of Zea mays only those based on material from the United States are given above, and of these only such as apply to the better known races. See Sturtevant, N. Y. State Agr. Expt. Sta. Rpt., and the following: Montgomery, The Corn Crops, 15, 1913; Tapley, Enzie, and Van Eseltine, N. Y. State Agr. Exp. Sta. Rpt. 1934: 9–13. 1934.

(121) ZIZANIA L.

(1) **Zizania aquatica** L., Sp. Pl. 991. 1753. Virginia. [Jamaica, also cited, is erroneous.]
Zizania clavulosa Michx., Fl. Bor. Amer. 1: 75. 1803. North America, Michaux.
Hydropyrum esculentum Link, Hort. Berol. 1: 252. 1827. North America.
Stipa angulata L. ex Steud., Nom. Bot. ed. 2. 2: 642. 1841, as synonym of Hydropyrum esculentum Link.
Zizania effusa Munro, Linn. Soc. Jour. Proc. 6: 52. 1862, as synonym of Z. aquatica L.
Ceratochaete aquatica Lunell, Amer. Midl. Nat. 4: 214. 1915. Based on Zizania aquatica L.
ZIZANIA AQUATICA var. ANGUSTIFOLIA Hitchc., Rhodora 8: 210. 1906. Belgrade, Maine, Scribner in 1895.
Zizania palustris L., Mant. Pl. 295. 1771. North America.

Melinum palustre Link, Handb. Gewächs. 1: 96. 1829. Based on Zizania palustris L.
ZIZANIA AQUATICA var. INTERIOR Fassett, Rhodora 26: 158. 1924. Armstrong, Iowa, Pammel and Cratty 764.
Zizania interior Rydb., Brittonia 1: 82. 1931. Based on Z. aquatica var. interior Fassett.
(2) **Zizania texana** Hitchc., Wash. Acad. Sci. Jour. 23: 454. 1933. San Marcos, Tex., Silveus.

(122) ZIZANIOPSIS Doell and Aschers.

(1) **Zizaniopsis miliacea** (Michx.) Doell and Aschers. in Doell in Mart., Fl. Bras. 2²: 13. 1871. Presumably based on Zizania miliacea Michx.
Zizania miliacea Michx., Fl. Bor. Amer. 1: 74. 1803. North America, Michaux.

(94) ZOYSIA Willd.

Zoysia japonica Steud., Syn. Pl. Glum. 1: 414. 1854. Japan.
Zoysia pungens var. japonica Hack., Bul. Herb. Boiss. 7: 642. 1899. Based on Z. japonica Steud.
Osterdamia japonica Hitchc., U. S. Dept. Agr. Bul. 772: 166. 1920. Based on Zoysia japonica Steud.
Zoysia matrella (L.) Merr., Philippine Jour. Sci. Bot. 7: 230. 1912. Based on Agrostis matrella L.
Agrostis matrella L., Mant. Pl. 2: 185. 1771. Malabar, India.
Zoysia pungens Willd., Gesell. Naturf. Freund. Berlin Neue Schrift. 3: 441. 1801. Malabar, India.
Osterdamia matrella Kuntze, Rev. Gen. Pl. 2: 781. 1891. Based on Agrostis matrella L.
Osterdamia zoysia Honda, Bot. Mag. [Tokyo] 36: 113. 1922. Based on Zoysia pungens Willd.
Zoysia tenuifolia Willd. ex Trin, Acad. St. Pétersb. Mém. VI. Sci. Nat. 2¹: 96. 1836. Mascarene Islands.
Osterdamia tenuifolia Kuntze, Rev. Gen. Pl. 2: 781. 1891. Based on Zoysia tenuifolia Willd.
Zoysia pungens var. tenuifolia Dur. and Schinz, Consp. Fl. Afr. 5: 734. 1894. Based on Z. tenuifolia Willd.
Osterdamia zoysia var. tenuifolia Honda, Bot. Mag. [Tokyo] 36: 113. 1922. Based on Zoysia tenuifolia Willd.

UNIDENTIFIED NAMES

The following names of grasses, applied to specimens collected in the United States, cannot be identified from the descriptions, and the types have not been located. Several of these names are not effectively published.

Agrestis viridis Raf., Amer. Month. Mag. 3: 356. 1818. Error for Agrostis. Name only. Allegheny Mountains or Ohio.

Agropyron repens var. *nemorale* Anderss. ex Farwell, Mich. Acad. Sci. Rpt. 6: 203. 1904. No basis given, but presumably based on *Triticum repens* var. *nemorale* Anderss., Scandinavia. Specimens so named in the Farwell Herbarium are *Agropyron repens* with awned lemmas. The name was misspelled *"nemorak"* in Bingham, Cranbrook Inst. Sci. Mich. Bul. 22: 93. 1945.

Agrostis affinis Schult., Mantissa 2: 195. 1824. Based on *Agrostis* No. 17 in Muhlenberg's Descriptio Graminum p. 75. *Sporobolus muhlenbergii* Kunth, Rév. Gram. 1: 68. 1829, and *Vilfa muhlenbergii* Steud., Syn. Pl. Glum. 1: 162. 1854, are also based on this. (See Hitchcock, Bartonia 14: 33. 1932.)

Agrostis altissima var. *laxa* Tuckerm., Amer. Jour. Sci. 45: 44. 1843. White Mountains, N. H., *Trichodium altissimum* var. *laxum* Wood, Class-book ed. 2. 600. 1847, presumably based on this.

Agrostis cylindrica Muhl., Amer. Phil. Soc. Trans. 3: 160. 1793. Name only. Pennsylvania.

Agrostis drummondi Torrey ex Hook., Fl. Antarct. 2: 372. 1847. "East side of the Rocky Mountains." Incidental mention as a form of *"A. exarata β."*

Agrostis michauxii Zuccagni, in Roemer, Col. Bot. 123. 1809. Seed received from Thouin, collected in Kentucky by Michaux. Not *A. michauxii* Trin., 1824?

Agrostis pauciflora Pursh, Fl. Amer. Sept. 1: 63. 1814. Not *A. pauciflora* Schrad., 1806. "On high mountains in Virginia and Carolina." In the Kew Herbarium is a specimen of *Muhlenbergia schreberi* marked "N. Amer. Mr. Fred. Pursh, Herb. propr." but with no name on the label. The description does not agree with this specimen, though it suggests some species of *Muhlenbergia*. *A. oligantha* Roem. and Schult., Syst. Veg. 2: 372. 1817, *Polypogon pauciflorus* Spreng., Syst. Veg. 1: 243. 1825, and *Muhlenbergia tenuiflora pauciflora* Scribn., Torrey Bot. Club Mem. 5: 37. 1894, are based on this.

Agrostis viridis Raf., ex Jacks., Ind. Kew. 1: 65. 1893. Correction for *Agrestis viridis* Raf.

Aira compressa Raf., Amer. Monthly Mag. 3: 356. 1818. [Allegheny Mountains] Name only.

Aira navicularis Schreb. ex Muhl., Amer. Phil. Soc. Trans. 3: 161. 1793. Pennsylvania. Name only.

Aira serotina Torr. ex Trin. in Steud., Nom. Bot. ed. 2. 1: 45. 1840. North America. Name only.

Aira speciosa Muhl., Amer. Phil. Soc. Trans. 3: 161. 1793. Pennsylvania. Name only.

Andropogon digitatus Muhl., Amer. Phil. Soc. Trans. 3: 181. 1793. Pennsylvania. Name only.

Andropogon sessiliflorus Raf., Bot. Seringe Bul. 1: 221. 1830. United States. Name only, under section *Dimeiostemon*. In Index Kewensis (1: 760. 1893) the name is listed as *Dimeiostemon sessiliflorus* Raf.

Andropogon tener Muhl. ex Merr. and Hu, Bartonia 25: 42. 1949. Name only, error for *Holcus tener* Muhl.

Apluda scirpoides Walt., Fl. Carol. 250. 1788. South Carolina. Not a grass, apparently a sedge.

Arundo confinis Willd., Enum. Pl. 127. 1809. North America. *Calamagrostis confinis* Beauv., Ess. Agrost. 15, 152. 1812. *Deyeuxia confinis* Kunth, Rév. Gram. 1: 76. 1829, and *C. neglecta* var. *confinis* Beal, Grasses N. Amer. 2: 353. 1896, are based on this.

Arundo glauca Hornem., Hort. Hafn. 1: 74. 1813. Not *A. glauca* Bieb., 1808. North America.

Arundo pallens Muhl. ex Steud., Nom. Bot. ed. 2. 1: 144. 1840. Pennsylvania. Name only, in Schrader Herbarium.

Briza virens Walt., Fl. Carol. 79. 1788. Not *B. virens* L., 1762. See Hitchcock, Mo. Bot. Gard. Rpt. 16: 49. 1905. *Poa virens* Jacq., Eclog. Gram. 54. pl. 36. 1820, is based on this. The figure represents a species of *Poa*.

Bromus poaeformis Beiler. Pl. Nov. Herb. Spreng. Cent. 11. 1807. North America. A glabrous annual, possibly *B. secalinus* L.

Bromus pubescens var. *ciliatus* Eaton and Wright, N. Amer. Bot. ed. 8. 161. 1848. Probably a form of *B. purgans* L., not based on *B. ciliatus* L.

Bromus pubescens var. *canadensis* Eaton and Wright, N. Amer. Bot. ed. 8. 161. 1848. Probably a form of *B. purgans* L. Ontario, the only Canadian locality cited.

Calamagrostis pumilia Nutt. ex A. Gray, Acad. Nat. Sci. Phila. Proc. 1862: 334. 1863. Not *C. pumila* Hook., 1851. Name only for a plant collected in the Rocky Mountains by Nuttall.

Calotheca macrostachya Presl, Rel. Haenk. 1: 268, 351. 1830. In Addenda et Corrigenda (p. 351) the original "in montanis Peruviae. . . " is changed to "ad Monte-Rey Californiae." This locality, as in the case of several other species described by Presl, is erroneous. (See Hitchcock, U. S. Natl. Herb. Contrib. 24: 335. 1927.)

Cenchrus carolinianus Walt., Fl. Carol. 79. 1788. South Carolina. (See Chase, U. S. Natl. Herb. Contrib. 22: 76. 1920.)

Cenchrus gracilis Beauv., Ess. Agrost. 57, 157. 1812. Name only for a specimen sent by Bosc, presumably from the Carolinas.

Chloris longibarba Michx. ex Beauv., Ess. Agrost. 79, 158. 1812. Name only.

Deyeuxia airoides Beauv., Ess. Agrost. 44, 152, 160. 1812. *"Arundo airoides* Mich. ined." is referred to *Deyeuxia. Arundo airoides* Lam. was described from a plant collected in North America by Michaux

and is probably the species Beauvois had
in mind. Lamarck's description suggests
Trisetum melicoides (Michx.) Scribner,
which was collected by Michaux and described by him as *Aira melicoides*.

Deyeuxia halleriana Vasey, Grasses U. S.
Descr. Cat. 50. 1885. Name only for a specimen from Washington Territory.

Digitaria setigera Roth in Roem. and
Schult., Syst. Veg. 2: 474. 1817. "India
orientali, Heyne." Link (Hort. Berol. 1: 225.
1827.) uses this name for *D. horizontalis*
Willd., giving "Brasilia" as locality. The
name is used in the same sense by Grisebach
(Fl. Brit. W. Ind. 544. 1864.) and by
others. So far as known, *S. horizontalis* has
not been found in India.

Dilepyrum angustifolium Raf., Med. Fl.
2: 249. 1830. Name only, *Dilepyrum* is
change of name for "Orizopsis Michx."

Eleusine ciliata Raf., Precis Decour.
Somiol. 45. 1814. Name only.

Eragrostis alba Presl, Rel. Haenk. 1: 279.
1830. "Monte-Rey, California," *Haenke*.
Locality erroneous, the plant probably collected in Peru.

Eragrostis caroliniana Scribn., Torrey
Bot. Club Bul. 5: 49. 1894. Based on *Poa
caroliniana* Bieler.

Eragrostis lugens var. *major* Vasey ex
L. H. Dewey, U. S. Natl. Herb. Contrib. 2:
542. 1894. "Texas to Arizona and east-
ward to Florida."

Eragrostis pilosa var. *caroliniana* Farwell,
Mich. Acad. Sci. Rpt. 17: 182. 1916.
Based on *Poa caroliniana* Bieler.

Festuca duriuscula var. *pubiculmis* Hack.
ex Rohlena, Sitzb. Bohm. Ges. Wiss. Math.
Naturw. Cl. 24: 4. 1899, descr. in Bohe-
mian. "Roztok" [Bohemia]. This name,
published as a new combination by Farwell, Mich. Acad. Sci. Papers 26: 7. 1940,
based on "*Festuca ovina* var. *pubiculmis*
Hackel," error for *F. ovina* var. *pubiculmis*
(Hack.) Aschers. and Graebn., Syn. Mitteleur. Fl. 2: 470. 1900. Specimens so named
by Farwell in his herbarium in Cranbrook
Institute are *F. rubra*.

Festuca glabra Spreng., Syst. Veg. 1: 353.
1825. Not *F. glabra* Lightf., 1777. Long
Island, N. Y. The descripton suggests
Puccinellia distans (L.) Parl.

Flexularia compressa Raf., Jour. Phys.
Chym. 89: 105. 1819. Kentucky and
Ohio.

Holcus tener Schreb. ex Muhl., Amer.
Phil. Soc. Trans. 3: 182. 1793. Name only.

Koeleria airoides Nutt. ex Steud., Nom.
Bot. 456. 1821. Name only. Referred
doubtfully in Index Kewensis to *Arundo
airoides* Lam.

Leptopyrum tenellum Raf., Med. Repos.
N. Y. 5: 351. 1808. [United States.] Name
only.

Lolium canadense Michx. ex Brouss.,
Elench. Pl. Hort. Monsp. 35. 1805, name
only; Roem. and Schult., Syst. Veg. 2: 893.

1817. Grown in Montpellier. The description
rather suggests a tall plant of *L. perenne* L.
Lolium temulentum var. *canadense* Wood,
Amer. Bot. and Flor. pt. 2: 406. 1871,
based on this.

Melica altissima Walt., Fl. Carol. 78.
1788. Not *M. altissima* L., 1753. (See Hitchcock, Mo. Bot. Gard. Rpt. 16: 47. 1905.)

Muhlenbergia anemagrostoides Trin. ex
Steud., Nom. Bot. ed. 2. 2: 164. 1841.
America. Name only.

Muhlenbergia sylvatica var. *vulpina* Wood,
Amer. Bot. and Flor. pt. 2: 86. 1871.
New York, *Lord*.

Paneion buckleyanum var. *maius* Lunell,
Amer. Midl. Nat. 4: 222. 1915. Change of
name for "*Poa tenuifolia* var. *maior*
(Vasey)," but that name was never published, and no specimen so named by Vasey
can be found.

Panicum americanum L., Sp. Pl. 56.
1753. America. This name and *Pennisetum
americanum* Schum., based on it, have been
used for *P. glaucum* (L.) R. Br. The original
description is unidentifiable, probably based
on a confusion of two or more species.
(See Chase, U. S. Natl. Herb. Contrib. 22:
218. 1921; Amer. Jour. Bot. 8: 43. 1921.)

Panicum anomalum Walt., Fl. Carol. 72.
1788. South Carolina. A species of *Setaria*.
(See Hitchcock, Mo. Bot. Gard. Rpt. 16: 35.
1905.)

Panicum barbatum LeConte ex Torr.,
Eaton, Man. Bot. ed. 2. 342. 1818. Not *P.
barbatum* Lam., 1791. New York. The description rather suggests *P. barbulatum*
Michx.

Panicum cartilagineum Muhl., Descr.
Gram. 128. 1817. Georgia. (See Hitchcock,
Bartonia 14: 41. 1932.)

Panicum debile Torr. ex Steud., Nom.
Bot. ed. 2. 2: 255, 262. 1841. Not *P.
debile* Desf., 1798. As synonym of *P.
pubescens* Lam.

Panicum densum Muhl., Descr. Gram.
122. 1817. No locality given. The description suggests one of the Lanuginosa group.

Panicum dichotomum var. *curvatum* Torr.,
Fl. North. and Mid. U. S. 145. 1824. No
locality given.

Panicum dichotomum var. *gracile* Torr.,
Fl. North. and Mid. U. S. 145. 1824.
"Common in swamps, New York."

Panicum dichotomum var. *pubescens* Munro, in Benth., Pl. Hartw. 341. 1857.
Sacramento, Calif., *Hartweg*. Name only.

Panicum dichotomum var. *spathaceum*
Wood, Amer. Bot. and Flor. pt. 2: 393.
1871. No locality mentioned.

Panicum discolor Bieler, Pl. Nov. Herb.
Spreng. Cent. 4. 1807. Pennsylvania. A
species of the subgenus *Dichanthelium*.

Panicum elliottii Spreng. ex Steud.,
Nom. Bot. ed. 2. 2: 256. 1841. Not *P.
elliottii* Trin., 1829. As synonym of *P.
pubescens* [Lam., p. 262].

Panicum fimbriatum Willd. ex Spreng.,

Syst. Veg. 1: 316. 1825, as synonym of *P. viscidum* Ell. [*P. scoparium* Lam.] South Carolina. A specimen in the Willdenow Herbarium so named is *P. albomaculatum* Scribn., from Mexico, collected by Humboldt.

Panicum flexuosum Raf., Precis. Découv. Somiol. 45. 1814; Jour. Bot. Desv. 4: 273. 1814. Not *P. flexuosum* Retz., 1783. New Jersey. *P. rafinesquianum* Schult., Mantissa 2: 257. 1824, is based on this.

Panicum gracilescens Desv. ex Poir., in Lam., Encycl. Sup. 4: 279. 1816. Carolina. Desvaux gives a later description (Opusc. 95. 1831), which disagrees in some respects with that of Poiret.

Panicum hirtellum Bartr., Travels 430. 1791. Not *P. hirtellum* L., 1759. Banks of the Mississippi River in Louisiana. The description suggests a species of *Echinochloa*.

Panicum iowense Ashe, N. C. Agr. Expt. Sta. Bul. 175: 115. 1900. Iowa to Kansas. The description suggests *P. huachucae* or *P. praecocius*. (See Contrib. U. S. Natl. Herb. 15: 330. 1910.)

Panicum muhlenbergianum Schult., Mantissa 2: 230. 1824. Based on *Panicum* No. 27 of Muhlenberg's Descriptio Graminum, the description of which is copied. Muhlenberg gives "Habitat in Georgia."

Panicum nitidum var. *glabrum* Torr., Fl. North. and Mid. U. S. 146. 1824. No locality cited. The description suggests *P. commutatum* Schult.

Panicum nitidum var. *gracile* Torr., Fl. North. and Mid. U. S. 146. 1824. Near New York. The description applies fairly well to the vernal phase of *P. dichotomum* L.

Panicum nitidum var. *majus* Vasey, U. S. Natl. Herb. Contrib. 3: 30. 1892. No locality cited. Vasey says, "Here could be placed several variable forms."

Panicum pensylvanicum Spreng., Nachtr. Bot. Gart. Halle 30. 1801. Pennsylvania.

Panicum pilosum Muhl., Amer. Phil. Soc. Trans. 4: 236. 1799. Pennsylvania. Name only.

Panicum pumilum Raf., Med. Repos. N. Y. 5: 353. 1808. Name only.

Panicum speciosum Walt., Fl. Carol. 73. 1788. South Carolina. The description faintly suggests *Sporobolus junceus* (Michx.) Kunth.

Panicum uniflorum Raf., Amer. Monthly Mag. 2: 120. 1817. Flatbush, N. Y. Some species of subgenus *Dichanthelium*.

Panicum vilfiforme Wood, Class-book ed. 3. 785. 1861. East Tennessee. Appears to be a species of the group Agrostoidia.

Paspalum compressum Raf., Fl. Ludov. 15. 1817. Louisiana prairies, Robin.

Paspalum dasyphyllum var. *floridanum* Wood, Amer. Bot. and Flor. pt. 2: 390. 1871. [Florida.]

Paspalum geniculatum Raf., Fl. Ludov. 15. 1817. Louisiana, Robin.

Paspalum supinum Rich. ex Hornem.,

Hort. Hafn. 1: 77. 1813. Not *P. supinum* Bosc, 1804. Baltimore, introduced in the Royal Botanic Garden in Copenhagen in 1807. Probably *P. pubescens* Muhl.

Paspalum virgatum var. *latifolium* Wood, Amer. Bot. and Flor. pt. 2: 390. 1871. Eastern States. Wood's *P. virgatum* appears to be *Paspalum boscianum* Flügge; the variety may be a luxuriant form of this species.

Pennisetum glaucum var. *purpurascens* Eaton and Wright, N. Amer. Bot. ed. 8. 346. 1840. Virginia and northward.

Poa alata Desv., Opusc. 102. 1831. "Carolina?" Locality erroneous, the type is *Eragrostis maypurensis* (H. B. K.) Steud., of the American tropics.

Poa caesia var. *strictior* A. Gray, Man. ed. 5. 629. 1867. "Lake Superior, *C. G. Loring*, especially Isle Royale, *Prof. Whitney*."

Poa capillaris L., misapplied by Link, Enum. Pl. 1: 88. 1821. "*P. caroliniana* Spreng." cited as synonym.

Poa caroliniana Bieler, Pl. Nov. Herb. Spreng. Cent. 10. 1807. North Carolina. Said to be similar to "*P. cilianensis.*" [*Eragrostis cilianensis* (All.) Lutati.]

Poa glauca var. *strictior* Jones, West. Bot. Contrib. 14: 14. 1912. Based on *P. caesia* var. *strictior* A. Gray.

Poa multicaulis Raf. ex M'Murtrie, Sk. Louisv. 223. 1819. Kentucky. Name only.

Poa nemoralis [L., misapplied by] Pursh, Fl. Amer. Sept. 1: 79. 1814. North America.

Poa nutans Muhl., Amer. Phil. Soc. Trans. 3: 161. 1793. Pennsylvania. Name only.

Poa repens Muhl., Amer. Phil. Soc. Trans. 3: 161. 1793. Pennsylvania. Name only.

Poa rubra Muhl., Amer. Phil. Soc. Trans. 4: 436. 1799. Pennsylvania. Name only.

Poa subaristata (Scribn.) *orendensis* Williams ex Pammel, Iowa Acad. Sci. Proc. 20: 144. 1915. Name only for *Pammel, Johnson, Lummis, Buchanan* 940. Uintah Mountains, Utah.

Poa tenuiflora Raf., Med. Repos. N. Y. 5: 353. 1808. [United States.] Name only.

Saccharifera spontanea Stokes, Bot. Mat. Med. 1: 132. 1812. South Carolina. Probably a species of *Erianthus*.

Sesleria americana Nees ex Steud. "Ins. Staatenland. Am. septr." The type, in the Lindley Herbarium in Cambridge, is labeled "Staten Island. 'Chanticleer.' Webster in 1829." The specimen is one of the group related to *Poa flabellata* (Lam.) Rasp. found in the region of the Straits of Magellan, "Staten Island" obviously referring to the island of that name to the east of Terra del Fuego.

Stipa expansa Poir. in Lam., Encycl. 7: 453. 1806. Carolina, *Bosc.* This has been taken for the basis of *Muhlenbergia expansa* Trin., but Poiret's description does not apply to that, and both DeCandolle and Trinius question Poiret's species.

Stipa spicata Walt., Fl. Carol. 78. 1788. Not *S. spicata* L. f., 1781. South Carolina. Apparently a species of *Andropogon*.

Stipa stricta Lam., Tabl. Encycl. 1: 158. 1791.; Encycl. 7: 453. 1806. South Carolina. *Fraser*. Said to have the aspect of *Andropogon*. Possibly *Sorghastrum nutans* (L.) Nash.

Triodia repens Vasey, Torrey Bot. Club Bul. 15: 49. 1888. Name only for a specimen collected by "Nealley, Western Texas."

Triticum aegilopoides Thurb. ex A. Gray, Acad. Nat. Sci. Phila. Proc. 1863: 79. 1863. Name only. Rocky Mts., *Hall* and *Harbour* 656.

Vilfa varians Buckl., Acad. Nat. Sci. Phila. Proc. 1862: 89. 1863. Rocky Mountains, *Nuttall*. Apparently a species of *Sporobolus*.

The following names, based on Old World types, have been applied to species of the United States. The types have not been examined.

Echinochloa crusgalli forma *longiseta* Farwell, Mich. Acad. Sci. Rpt. 21: 349. 1920. Based on *Panicum crusgalli* var. *longisetum* Trin. This variety, from Astrakhan, U. S. S. R., as represented in Trin., Gram. Icon. 2: pl. 162. 1828, is not known from America. Farwell probably had *Panicum longisetum* Torr., 1822, in mind.

Echinochloa stagnina (Retz.) Beauv., Ess. Agrost. 53, 161, 171. 1812, based on *Panicum stagninum* Retz., a species of the East Indies and the Pacific Islands has a coarsely hairy ligule, while the specimen distributed by Gray Herbarium under this name (*Fernald, Long,* and *Clement* 15182, Princess Anne County, Va.) is entirely without ligule, as in *E. crusgalli* and its allies.

Lolium multiflorum submuticum Mutel, Fl. Franç. 4: 139. 1839. France.

Phleum pratense var. *nodosum* (L.) Huds., Fl. Ang. ed. 2. 26. 1778. Based on *P. nodosum* L. A specimen of *P. pratense* L. from Virginia (*Fernald* and *Long* 12935) with slightly curved base has been recorded under this name. It is not *P. nodosum* L. (upon which the variety is based), which is a much smaller plant, decumbent at base, with few to several swollen nodes and short internodes, the panicle shorter and more slender; not known from America.

Phragmites communis forma *repens* G. F. W. Meyer, Chloris Hanov. 650. 1836. Germany. Applied to a Michigan specimen with long stolons.

Poa annua var. *aquatica* Aschers., Fl. Brand. 1: 844. 1864. Germany. Applied to a specimen from flooded place.

Poa annua var. *reptans* Hausskn., Mitt. Thüring. Bot. Ver. 9: 7. 1891. Germany.

Poa glauca subsp. *conferta* (Blytt) Lindm.

in Holmb., Skand. Fl. 2: 208. 1926. Based on *P. conferta* Blytt. Minnesota specimens distributed under this name are referred to *P. glauca* Vahl.

Poa glauca subsp. *conferta* var. *laxiuscula* (Blytt) Lindm. in Holmb., Skand. Fl. 2: 208. 1926. Based on *P. aspera* var. *laxiuscula* Blytt. Minnesota specimens so named are referred to *P. glauca* Vahl.

Poa nemoralis var. *montana* Gaudin, Alpina 3: 27. 1808. Switzerland. Minnesota specimens distributed under this name do not agree with Gaudin's description, nor that of Ascherson and Graebner. They appear to be rather small specimens of *Poa interior* Rydb.

PERSONS FOR WHOM GRASSES HAVE BEEN NAMED[24]

This list includes names of persons for whom valid genera, species, or varieties of grasses in the Manual have been named.

Addison. See Brown.

Alexander, Annie M. (1867–   ). Botanical and zoological collector, Oakland, Calif.; collections mainly from western North America and Hawaii. *Ectosperma alexandrae.*

Anderson, Charles Lewis (1827–1910). Practicing physician of Carson City, Nev., and Santa Cruz, Calif.; correspondent of Asa Gray. *Stipa lepida* var. *andersoni.*

Arsène, Hermano Gerfroy (1867–1938). Professor in Sacred Heart Training College, Las Vegas, N. Mex.; collected extensively in Mexico. *Muhlenbergia arsenei.*

Ashe, William Willard (1872–1932). Botanist and forester, U. S. Forest Service. *Panicum ashei.*

Baker, Charles Fuller (1872–1927). Botanist and entomologist, teacher and administrator, who collected in Colorado, California, Cuba, and the Philippine Islands. *Agropyron bakeri; Agrostis bakeri.*

Baker, Charles Henry (1848–   ). Horticulturist, collector of fruits and seeds, resident of Oakland, Allegheny County, Pa., and Orange County, Fla. *Spartina bakeri.*

Barrelieri, Jacques (1606–73). French medical botanist, author of a work on the plants of France, Spain, and Italy. *Eragrostis barrelieri.*

Beckmann, Johann (1739–1811). German botanist, author of a botanical lexicon. *Beckmannia.*

Bélanger, Charles Paulus (1805–81). French botanist, who collected extensively in the Old World. When Steudel described *Anthephora belangeri* (*Hilaria belangeri*) the specific name *belangeri* was used, apparently through inadvertence, instead of one for Jean Louis Berlandier (1805–51), who collected the type specimen in Mexico. Bélanger botanized in Martinique, but apparently never in Mexico.

[24] Revised by Joseph A. Ewan, Tulane University.

Berg, Federico Guillermo Carlos (1843–1902). Director, Museo Nacional de Buenos Aires. *Panicum bergii.*

Bertero, Carlo Giuseppe (1789–1831). Italian botanical explorer, resident in Chile 1827–30, lost at sea on his return from Tahiti. *Tragus berteronianus.*

Beyrich, Heinrich Karl (1796–1834). Prussian botanical explorer, visited Brazil 1822–23 and subsequently Virginia, the Carolinas, and Georgia; died at Fort Gibson when exploring Arkansas Territory. *Eragrostis beyrichii.*

Bicknell, Eugene Pintard (1859–1925). New York banker, amateur botanist, and collector of local flora. *Panicum bicknellii.*

Bigelow, John Milton (1804–78). Surgeon-botanist, with Mexican Boundary Survey and Lieutenant Whipple's railroad survey along 35th parallel. *Blepharidachne bigelovii; Poa bigelovii.*

Blasdale, Walter Charles (1871–   ). Professor of chemistry at University of California and amateur botanist. *Agrostis blasdalei.*

Blodgett, John Loomis (1809–53). Physician and druggist of Key West, first important botanical collector among the lower Florida Keys. *Paspalum blodgettii.*

Bloomer, Hiram G. (1821–74). Pioneer botanist of California, active member of the California Academy of Sciences. *Oryzopsis bloomeri.*

Bolander, Henry Nicholas (1831–97). California botanist, teacher, collaborator in State Geological Survey, and special student of cryptogams. *Calamagrostis bolanderi; Poa bolanderi; Scribneria bolanderi.*

Bosc, Louis Augustin Guillaume (1759–1828). French botanist, who visited the Carolinas 1798–1800, author of a treatise on oaks. *Panicum boscii; Paspalum boscianum.*

Boutelou, Claudio (1774–1842) and his brother Estéban (1776–1813). Lagasca named the genus *Bouteloua* for them; Claudio was professor of agriculture in Madrid.

Brewer, William Henry (1828–1910). California botanist, onetime professor at Yale University, whose narrative journal was published under the title "Up and Down California." *Calamagrostis breweri.*

Brown, Addison (1830–1913). New York judge, amateur botanist, patron of New York Botanical Garden. *Panicum addisonii.*

Buckley, Samuel Botsford (1809–83). Southern naturalist and collector, twice State Geologist of Texas, described grasses from Texas and Oregon. *Sporobolus buckleyi; Tridens buckleyanus.*

Cabanis, Jean (1816–1906). German ornithologist, who collected plants in Florida. *Andropogon cabanisii.*

Cain, Stanley Adair (1902–   ). Plant geographer and botanist of Indiana and Tennessee. *Calamagrostis cainii.*

Canby, William Marriott (1831–1904).

Wilmington, Del., merchant and banker, amateur botanist. *Poa canbyi.*

Chaix, Abbé Dominique (1731–1800). French botanist, collaborator with Dominique Villars on treatise on French plants. *Poa chaixii.*

Chapman, Alvan Wentworth (1809–99). Botanist of Apalachicola, Fla., and author of Flora of the Southern United States. *Gymnopogon chapmanianus; Panicum chapmani; Poa chapmaniana; Tridens chapmani.*

Clute, Willard Nelson (1869–   ). Professor of botany, Butler University, Ind., and student of vascular cryptogams. *Panicum clutei.*

Combs, Robert (1872–99). Botanical collector in Florida and Cuba. *Panicum combsii.*

Commons, Albert (1829–1919). Amateur botanist of Delaware, collector of local flora. *Panicum commonsianum.*

Cooke, William Bridge (1908–   ). California botanist, devoted to the flora of the Mount Shasta region. *Glyceria cookei (G. declinata.)*

Cotta, Heinrich (1763–1844). German plant physiologist. *Cottea.*

Curtiss, Allen Hiram (1845–1907). Botanical collector of Jacksonville, Fla. *Aristida curtissii; Calamovilfa curtissii; Sporobolus curtissii.*

Cusick, William Conklin (1842–1922). Oregon botanist, who explored the Wallowa Mountains and eastern Oregon. *Poa cusickii.*

Danthione, Étienne (fl. 1800–15). French botanist, author of an unpublished account of grasses of Marseille region. *Danthonia.*

Davy, Joseph Burtt (1870–1940). English botanist, professor at Oxford University, onetime resident of California and author of a grass flora of central California. *Pleuropogon davyi.*

Deam, Charles Clemon (1865–   ). Veteran Indiana botanist, forester, author of a Flora of Indiana and Grasses of Indiana. *Panicum deamii.*

Deschamps, L. A. (1766–   ). Surgeon-naturalist on *Recherche* sent out by French Government under D'Entrecasteaux in 1791 in search of La Pérouse. *Deschampsia.*

Desmazières, Jean Baptiste Henri Joseph (1796–1862). French botanist, author of a work on grasses of northern France. *Desmazeria.*

Deyeux, Nicholas (1753–1837). French botanist. *Deyeuxia.*

Douglas, David (1799–1834). British botanical explorer, who visited the Pacific Northwest, California, and the Hawaiian Islands. *Poa douglasii.*

Drummond, Thomas (1780–1835). Scotch nurseryman and botanical explorer, curator of Belfast Botanic Garden, member of Second Franklin Expedition, who collected in the Canadian Rockies and in Texas. *Sorghum vulgare var. drummondii.*

Dumont-d'Urville, Jules Sebastien Cesar (1790–1842). French explorer, commander of the expeditions of the *Astrolabe* and the *Zélée* around the world. *Panicum urvilleanum; Paspalum urvillei.*

Eastwood, Alice (1859–   ). California botanist, longtime curator of botany at California Academy of Sciences. *Festuca eastwoodae.*

Ehrhart, Friedrich (1742–95). Swiss-born assistant to an apothecary in Germany, pupil of Linnaeus; especially interested in grasses, rushes, and ferns. *Ehrharta.*

Elliott, Stephen (1771–1830). Pioneer South Carolina botanist and legislator. *Agrostis elliottiana; Andropogon elliottii; Sorghastrum elliottii.*

Elmer, Adolph Daniel Edward (1870–1942). Botanist and collector, first in California, then in Washington State, and a longtime resident of the Philippine Islands, author of an enumeration of Philippine flora. *Agropyron elmeri; Festuca elmeri; Stipa elmeri.*

Emersley, J. D. Botanical collector in the Southwestern States. *Muhlenbergia emersleyi.*

Faber, Ernest (1839–99). Missionary, botanical collector, and student of Chinese botany. *Setaria faberii.*

Fendler, August (1813–83). German-American botanical explorer of New Mexico, Venezuela, Panama, and Trinidad. *Aristida fendleriana; Poa fendleriana.*

Fernald, Merritt Lyndon (1873–   ). Professor of botany, Harvard University, and longtime Director of Gray Herbarium. *Glyceria fernaldii; Poa fernaldiana; Calamagrostis fernaldii.*

Frank, Joseph C. (1782–1835). German botanical collector, who visited Ohio and New Orleans. *Eragrostis frankii.*

Gattinger, Augustin (1825–1903). Pioneer botanist of Tennessee and author of a flora of that State. *Panicum gattingeri.*

Gay, Jacques Étienne (1786–1864). French botanist, onetime Secretary of Chamber of Peers, who visited Africa for plants. *Chloris gayana.*

Geyer, Carl Andreas (1809–53). Botanical explorer, born in Dresden, Germany, who collected first in Illinois, later in the Missouri River country and in Oregon Territory. *Melica geyeri.*

Ghiesbreght, August (1810–93). Belgian botanical collector, who repeatedly visited Mexico for short to long residences. *Panicum ghiesbreghtii.*

Gouin, ——— (fl. 1860–70). French physician, chief of military hospital at Vera Cruz, member of French Scientific Commission to Mexico, 1865–66. *Panicum gouini.*

Gray, Asa (1818–88). Distinguished professor of botany, Harvard University, and best known American botanist of nineteenth century. *Festuca grayi.*

Greene, Edward Lee (1843–1915). First professor of botany, University of California, botanical explorer in New Mexico, Colorado, and California, botanical editor and critic. *Orcuttia greenei.*

Griffiths, David (1867–1935). Botanist, U. S. Department of Agriculture, devoted to xerophytic flora of Southwest. *Agropyron griffithsii.*

Grisebach, August Heinrich Rudolf (1814–79). German botanist, author of Flora of the British West Indian Islands. *Setaria grisebachii.*

Gussone, Giovanni (1787–1866). Italian botanist, professor of botany in Naples. *Bromus rigidus* var. *gussonii.*

Hackel, Eduard (1850–1926). Eminent Austrian agrostologist. *Hackelochloa.*

Hall, Elihu (1822–82). Illinois botanical collector, who visited Texas, Colorado, and Oregon. *Agrostis hallii; Andropogon hallii; Panicum hallii.*

Hall, Harvey Monroe (1874–1932). Professor of botany, University of California, specialist in taxonomy of Compositae and pioneer in use of transplant method. *Bromus orcuttianus* var. *hallii.*

Hansen, George (1863–1908). Resident botanical collector of Amador County, California. *Sitanion hanseni.*

Harford, William George Washington (1825–1911). Pioneer California conchologist, colleague of Bolander and Kellogg in the early California Academy of Sciences. *Melica harfordii.*

Hartweg, Carl Theodor (1812–71). German botanical explorer, sent by Horticultural Society of London to Mexico, California, and Andes to collect plants and seeds. *Paspalum hartwegianum.*

Havard, Valery (1846–1927). Major surgeon, U. S. Army, born in France, who collected in Texas. *Panicum havardii.*

Heller, Amos Arthur (1867–1944). Botanist, founder and editor of journal Muhlenbergia, who collected in the western United States, also Hawaiian Islands, and Puerto Rico. *Panicum helleri.*

Henderson, Louis Fourniquet (1853–1942). Pioneer botanist of Pacific Northwest, longtime curator, University of Oregon Herbarium, Eugene. *Agrostis hendersonii; Oryzopsis hendersoni.*

Hilaire. See St. Hilaire.

Hillman, Frederick Hebard (1863–   ). Botanist, U. S. Department of Agriculture, engaged upon seed morphology. *Panicum hillmani.*

Hitchcock, Albert Spear (1865–1935). Eminent American agrostologist and widely traveled plant explorer. *Trichachne hitchcockii.*

Hooker, William Jackson (1785–1865). Distinguished British botanist, Director, Royal Botanic Gardens, Kew, editor and author of many botanical works. *Helictotrichon*

trichon hookeri; Imperata hookeri; Sporobolus asper var. *hookeri.*

Hoover, Robert Francis (1913–   ). California botanist, devoted to flora of the Great Valley. *Pleuropogon hooverianus; Agrostis hooveri.*

Howell, Thomas Jefferson (1842–1912). Oregon botanist, author of Flora of Northwest America. *Agrostis howellii; Alopecurus howellii; Calamagrostis howellii; Festuca howellii; Poa howellii.*

Imperato, Ferrante (1550–1625). Apothecary in Naples, author of a rare folio work on natural history. *Imperata.*

James, Edwin (1797–1861). Surgeon-botanist with Stephen H. Long's expedition to the Rocky Mountains, and first white man to ascend a 14,000-foot peak in the United States (Pikes Peak). *Hilaria jamesii.*

Jepson, Willis Linn (1867–1946). Longtime professor of botany, University of California, author of a Flora of California. *Elymus glaucus* var. *jepsoni.*

Jones, Marcus Eugene (1852–1934). Onetime teacher, mining engineer, and botanist, who collected widely in the western United States and Mexico. *Muhlenbergia jonesii.*

Joor, Joseph Finley (1849–92). Native of Louisiana, onetime physician of New Orleans and at least three small towns in Texas, professor of botany, Tulane University, 1886–92. *Panicum joori.*

Kalm, Pehr (Peter) (1715–79). Swedish botanist and correspondent of Linnaeus, who collected in southeastern Canada and the northeastern United States. *Bromus kalmii.*

Kellogg, Albert (1813–87). Physician, pioneer botanist of California, one of the founders of California Academy of Sciences. *Poa kelloggii.*

Kennedy, Patrick Beveridge (1874–1930). Agronomist, University of California, Berkeley. *Agrostis kennedyana.*

King, Clarence (1842–1901). Mountaineer, geologist, explorer, in charge of survey of fortieth parallel across the Great Basin. *Blepharidachne kingii; Hesperochloa kingii; Oryzopsis kingii.*

Koeler, George Ludwig (1765–1807). German botanist, professor in Mainz, author of a work on grasses of France and Germany. *Koeleria.*

Lamarck, Jean Baptiste Antoine Pierre Monnet de (1744–1829). Eminent French naturalist, author of works on botany, heredity, and conchology. *Lamarckia.*

Lange, Johan Martin Christian (1818–98). Danish botanist, professor of botany in Copenhagen. *Paspalum langei.*

Leers, Johann Daniel (1727–74). German apothecary, author of a work on local flora. *Leersia.*

Leiberg, John Bernhard (1853–1913). American forest surveyor, who collected plants in Idaho and the Pacific States. *Panicum leibergii; Poa leibergii.*

Lemmon, John Gill (1832–1908). Botanist of California, onetime State Forester, correspondent of Asa Gray, and botanical explorer. *Eriochloa lemmoni; Phalaris lemmoni; Puccinellia lemmoni; Stipa lemmoni.*

Leprieur, F. R. (–1869). French botanical explorer, who traveled in Senegal and French Guiana, 1830–36. *Chloris prieurii.*

Letterman, George Washington (1841–1913). Teacher in public schools of Allenton, Mo., botanical collector chiefly in Missouri and the Southern States. *Poa lettermani; Stipa lettermani.*

Liebmann, Frederik Michael (1813–56). Danish botanist, who collected in Mexico. *Setaria liebmanni.*

Lindheimer, Ferdinand Jakob (1801–79). German-born resident, botanical collector and newspaper editor of New Brunfels, Tex., who sent plants to Asa Gray 1843–52. *Muhlenbergia lindheimeri; Panicum lindheimeri.*

Macoun, James Melville (1862–1920). Canadian botanist, son of John Macoun. *Calamagrostis canadensis* var. *macouniana.*

Macoun, John (1832–1920). Canadian botanist. *Elymus macounii.*

Marsh, Ernest George, Jr. (1915–   ). Wildlife technician of Austin, Tex., who traveled in northern Texas on Farmer Fellowship during 1936–38. *Muhlenbergia marshii.*

Metcalfe, Orrick Baylor (1879–1936). Plant ecologist who botanized in New Mexico between 1902–04. *Muhlenbergia metcalfei.*

Michaux, André (1746–1802). French botanist, who explored eastern United States, author of Flora Boreali-Americana. *Eriochloa michauxii.*

Mohr, Charles Theodore (1824–1901). German-born botanist, who traveled widely, for more than 40 years pharmacist of Mobile, Ala., and author of Plant Life of Alabama. *Andropogon mohrii; Aristida mohrii.*

Molina, Juan Ignazio (later Giovanni Ignazio) (1740–1829). Chilean Jesuit missionary and botanist, author of first comprehensive summary of Chilean plants. *Molinia.*

Morton, Julius Sterling (1832–1902). Agriculturalist and historian, onetime Nebraska magazine editor, Secretary of Agriculture 1893–97. *Helictotrichon mortonianum.*

Muhlenberg, Gotthilf Heinrich Ernst (1753–1815). Pennsylvania born, pastor of a Lutheran church at Lancaster, pioneer botanist, author of Descriptio Uberior Graminum. *Muhlenbergia; Amphicarpum muhlenbergianum.*

Munro, William (1818–80). British botanist, who wrote on grasses. *Munroa.*

Nealley, Greenleaf Cilley (1846–96). Botanical collector, went to Texas in 1882, later commissioned by U. S. Department of Agriculture to explore southwestern Texas for grasses and forage plants. *Leptochloa nealleyi; Sporobolus nealleyi; Tridens nealleyi.*

Nees von Esenbeck, Christian Gottfried Daniel (1776–1858). Eminent German botanist, professor of botany in Breslau, author of Agrostologia Brasiliensis. *Stipa neesiana.*

Nelson, Aven (1859–). Longtime professor of botany, University of Wyoming, author of a manual of Rocky Mountain plants. *Stipa columbiana* var. *nelsoni.*

Nuttall, Thomas (1786–1859). English-American naturalist, onetime professor of botany, Harvard University ("Old Curious" of Dana's Two Years Before the Mast), collector and author. *Puccinellia nuttalliana* (*P. airoides*).

Orcutt, Charles Russell (1864–1929). Resident botanist of San Diego, Calif., who explored northern Baja California. *Orcuttia; Aristida orcuttiana; Bromus orcuttianus; Eragrostis orcuttiana.*

Otis, Ira Clinton (1861–1938). Botanical collector of the State of Washington. *Glyceria otisii.*

Palmer, Edward (1831–1911). Naturalist-explorer, ethnobotanist, and collector in Paraguay, Mexico, and the southwestern United States, first naturalist to visit Guadalupe Island. *Agropyron smithii* var. *palmeri; Eragrostis palmeri.*

Parish, Samuel Bonsall (1838–1928). Resident botanist of San Bernardino, Calif., collector of local flora. *Agropyron parishii; Aristida parishii; Puccinellia parishii.*

Parry, Charles Christopher (1823–90). British-American "veteran botanist and tireless explorer," first with the Mexican Boundary Survey, later in Colorado, Utah, California, and Mexico. *Bouteloua parryi; Danthonia parryi.*

Patterson, Harry Norton (1853–1919). Illinois printer, resident botanist of Oquawka, who collected in Colorado. *Poa pattersoni.*

Phipps, Constantine John (1744–92). Second baron of Mulgrave, British naval commander, explorer, politician, leader of an unsuccessful Arctic expedition to discover a northern passage to India. *Phippsia.*

Pickering, Charles (1805–78). Botanist, ethnologist, historian, who accompanied the U. S. Exploring Expedition under Wilkes. *Calamagrostis pickeringii.*

Poiret, Jean Louis Marie (1755–1834). French botanist, who completed Lamarck's Encyclopédie Méthodique. Botanique. *Sporobolus poiretii; Setaria poiretiana.*

Porter, Thomas Conrad (1822–1901), Classicist, poet, professor of botany, Lafayette College, Pa., author of first Synopsis of Flora of Colorado. *Calamagrostis porteri; Melica porteri; Muhlenbergia porteri; Stipa porteri.*

Prieur. See Leprieur.

Pringle, Cyrus Guernsey (1838–1911). Vermont botanist, pioneer plant breeder,

"prince of botanical collectors," who collected in Arizona and California and repeatedly visited Mexico through 26 years for plants. *Agropyron pringlei; Agrostis hallii* var. *pringlei; Poa pringlei; Stipa pringlei.*

Puccinelli, Benedetto (1808–50). Italian botanist, professor in Lyceum at Lucca. *Puccinellia.*

Pumpelly, Raphael (1837–1923). Geologist, U. S. Geological Survey. *Bromus pumpellianus.*

Pursh, Frederick (1774–1820). German-American botanist, collected in Middle Atlantic States, author of Flora Americae Septentrionalis, which first included discoveries of Lewis and Clark in the Pacific Northwest. *Amphicarpum purshii.*

Ravenel, Henry William (1814–87). Native of South Carolina, planter, agricultural editor, onetime botanist, South Carolina Department of Agriculture, who first issued published series of named specimens of American fungi. *Panicum ravenelii.*

Redfield, John Howard (1815–95). Philadelphia business man, long associated with Academy of Natural Sciences, amateur botanist. *Redfieldia.*

Reimarus, J. A. H. (1729–1814). German botanist, professor of Natural History and Physics at Hamburg. *Reimarochloa.*

Reverchon, Julien (1837–1905). Resident of Dallas, Tex., who came from Lyons, France, in 1856 and collected plants in Texas. *Muhlenbergia reverchoni; Panicum reverchoni.*

Reynaud, J. J. (1773–1842). Surgeon on French exploring vessel *Chevrette,* who collected plants in the Orient. *Neyraudia,* an anagram of *Reynaudia,* a genus of West Indian grasses; *Neyraudia reynaudiana.*

Richardson, Sir John (1787–1865). English naturalist, author of Fauna Boreali-Americana, Arctic explorer, surgeon to three expeditions to the Arctic, the last in search of Sir John Franklin. *Stipa richardsoni; Muhlenbergia richardsonis.*

Roemer, Karl Ferdinand (von) (1818–91). German geologist, who collected plants in Texas, 1845–47. *Aristida roemeriana.*

Ross, Edith A. (fl. 1885–95). Amateur botanical collector of Davenport, Iowa, who visited Yellowstone Park in 1890. *Agrostis rossae.*

Rothrock, Joseph Trimble (1839–1922). Professor of botany, University of Pennsylvania, earlier surgeon-botanist to Wheeler's exploring expedition west of 100th meridian. *Bouteloua rothrockii.*

Rottboell, Christem Friss (1727–97). Danish botanist, professor of botany in Copenhagen. *Rottboellia.*

Roxburgh, William (1751–1815). Scotch botanist, who collected in India, Director of the botanical garden, Calcutta. *Sorghum vulgare* var. *roxburghii.*

Runyon, Robert (1881–). Photographer

and amateur botanist, Brownsville, Tex. *Digitaria runyoni.*

St. Hilaire, Auguste de (1779–1853). French botanist who traveled in Brazil and Paraguay for 6 years, 1816–22. *Hilaria.*

Saunders, William (1822–1900). Scotchborn horticulturist, first botanist and Superintendent of Horticulture, U. S. Department of Agriculture, instrumental in introduction of Bahia orange into California. *Agropyron saundersii.*

Scheele, Adolf (1808–64). German botanist, who described grasses from Texas. *Setaria scheelei.*

Schreber, Johann Christian Daniel (von) (1739–1810). German botanist, professor in Erlangen, who wrote on grasses. *Muhlenbergia schreberi.*

Scribner, Frank Lamson (1851–1938). Agrostologist, U. S. Department of Agriculture. *Scribneria; Agropyron scribneri; Calamagrostis scribneri; Panicum scribnerianum; Stipa scribneri.*

Sello (or Sellow), Friedrich (1789–1831). German botanist, who went to Brazil in 1814 and collected from Bahia on the north to Uruguay on the south. *Cortaderia selloana.*

Siegling, — (fl. ca. 1800). Professor of botany at Erfurt, Germany, associate of Johann Jakob Bernhardi. *Sieglingia.*

Silveus, William Arents (1875–). Agrostologist of San Antonio, Tex.; author of works on Texas grasses. *Eragrostis silveana; Sporobolus silveanus.*

Simpson, Joseph Herman (1841–1918). Resident naturalist of Florida. *Digitaria simpsoni; Eriochloa michauxii var. simpsoni.*

Smith, Charles Eastwick (1820–1900). Engineer, onetime railroad president, amateur botanist. *Melica smithii.*

Smith, Jared Gage (1866–). Onetime botanist, U. S. Department of Agriculture, later resident of Hawaii. *Agropyron smithii.*

Stapf, Otto (1857–1933). Botanist, Royal Botanic Gardens, Kew, England. *Neostapfia.*

Stillman, Jacob Davis Babcock (1819–88). Practicing physician of California, onetime coeditor of California Medical Gazette, amateur botanist. *Stipa stillmanii.*

Suksdorf, Wilhelm Nikolaus (1850–1932). Born in Dransau, Holstein, longtime resident of Bingen, Wash., pioneer collector in Klickitat County region. *Bromus suksdorfii.*

Swallen, Jason Richard (1903–). Agrostologist, U. S. Department of Agriculture; Curator, Division of Grasses, U. S. National Museum. *Eragrostis swalleni; Hilaria swalleni.*

Tharp, Benjamin Carroll (1885–). Professor of botany, University of Texas. *Sporobolus tharpii.*

Thurber, George (1821–90). New York botanist, agricultural editor, who wrote on grasses of California. *Agrostis thurberiana; Festuca thurberi; Muhlenbergia thurberi; Stipa thurberiana.*

Thurow, Friedrich Wilhelm (1852–1930).

German amateur botanist who came to Texas in 1876 and collaborated with Vasey in his study of Texas grasses. *Panicum thurowi.*

Torrey, John (1796–1873). American botanist of distinction and physician of New York City. *Melica torreyana; Muhlenbergia torreyana; Muhlenbergia torreyi.*

Tracy, Joseph Prince (1879–). Business accountant of Eureka, Calif., amateur botanist and collector. *Festuca tracyi.*

Tracy, Samuel Mills (1847–1920). Agronomist and botanical collector of Biloxi, Miss., who collected in the Southern and Western States. *Andropogon tracyi; Eragrostis tracyi; Poa tracyi.*

Trinius, Karl Bernhard (1778–1844). Agrostologist of St. Petersburg, Russia, author of important works on grasses. *Bromus trinii.*

Tuckerman, Edward (1817–86). American lichenologist, professor of botany, Amherst College. *Panicum tuckermani.*

Tweedy, Frank (1854–1937). Topographic engineer, U. S. Geological Survey, who collected in Yellowstone Park and the Pacific Northwest. *Bromus pumpellianus* var. *tweedyi; Calamagrostis tweedyi.*

Urville. See Dumont-d'Urville.

Vasey, George (1822–93). Eminent American agrostologist; botanist, U. S. Department of Agriculture. *Vaseyochloa; Poa vaseyochloa.*

Walter, Thomas (1740–89). South Carolina planter, pioneer botanist, author of Flora Caroliniana. *Echinochloa walteri.*

Webber, David Gould (1809–). Physician, miner, miller, who went to California in 1849, onetime owner of Webber Lake, and friend of J. G. Lemmon. *Oryzopsis webberi.*

Webber, Herbert John (1865–1946). Botanist, U. S. Department of Agriculture, who early collected in Nebraska and Florida, was later devoted to citrus studies. *Panicum webberianum.*

Werner, William C. (1851–1935). Ohio florist and botanical collector. *Panicum werneri.*

Wilcox, Timothy Erastus (1840–1932). Surgeon, U. S. Army, "born naturalist," who collected in the Western States. *Panicum wilcoxianum.*

Williams, Thomas Albert (1865–1900). Agrostologist, U. S. Department of Agriculture. *Stipa williamsii.*

Willkomm, Heinrich Moritz (1821–95). German botanist, professor of botany and Director of gardens in Dorpat. *Willkommia.*

Wolf, John (1820–97). Botanist of Canton, Ill., who collected in Illinois and Colorado. *Poa wolfii; Trisetum wolfii.*

Wright, Charles (1811–85). Botanical explorer in Texas and New Mexico, with Mexican Boundary Survey, and later with North Pacific Ringgold Expedition, also visited Cuba and Santo Domingo. *Andropogon wrightii; Aristida wrightii; Muhlen-*

bergia wrightii; Panicum wrightianum; Sporobolus wrightii.

Zois, Karl von (1756–1800). German botanist. *Zoysia.*

GLOSSARY

Abortive. Imperfectly developed.

Acuminate. Gradually tapering to a sharp point. Compare acute.

Acute. Sharp-pointed, but less tapering than acuminate.

Aggregate. Collected together in tufts, groups, or bunches. Applied especially to inflorescences. The racemes are aggregate in several species of *Andropogon.*

Annual. Within 1 year. Applied to grasses which do not live more than 1 year. *Winter annual.* A plant which germinates in the fall, lives over winter, and produces its seed the following spring, after which it dies.

Anthesis. The period during which a flower is open. In grasses, when the lemma and palea are expanded and the anthers and stigmas are mature.

Antrorse. Directed upwards or forwards. Applied especially to scabrous or pubescent stems, sheaths, awns, and so on. Opposed to retrorse.

Apiculate. Having a minute pointed tip. Applied especially to fertile lemmas in fruit, such as certain species of *Eriochloa.*

Appressed. Lying against an organ. The branches of an inflorescence may be appressed to the main axis or the hairs on a stem may be appressed to the surface.

Aristate. Awned; provided with a bristle at the end or at the back or edge of an organ. In grasses applies especially to the awns at the end of the bracts of the spikelet. Compare awn. *Aristulate.* Bearing a short awn.

Articulate. Jointed. Joined by a line of demarcation between two parts which at maturity separate by a clean-cut scar. Certain spikelets are articulate with the pedicel; certain awns with the lemma. *Articulation.* The point of union of two articulate organs.

Ascending. Sloping upward. Applied to stems which curve upward from the base, to the branches of an inflorescence which slope upward at angle of about 40° to 70°, and to other parts such as blades and hairs. Compare appressed and spreading.

Attenuate. Gradually narrowed to a slender apex or base.

Auricle. An ear. Applied to earlike lobes at the base of blades and to the small lobes at the summit of the sheath in *Hordeae. Auriculate.* Provided with ears.

Awn. A slender bristle at the end or on the back or edge of an organ. In grasses the awn is usually a continuation of the midnerve (sometimes also of the lateral nerves) of the glumes or lemmas, rarely of the palea.

Axil. The angle between an organ and its axis. Applied especially to the angle between a leaf and its stem and between a branch or pedicel and its axis. *Axillary.* Growing in an axil.

Axis. The main stem of an inflorescence, especially of a panicle. Compare *rachis.*

Barbed. Furnished with retrorse projections. Applied to the spines of *Cenchrus.*

Beak. A hard point or projection. Applied to seeds and fruits.

Bearded. Furnished with long stiff hairs, as the nodes of *Andropogon barbinodis,* the callus of *Stipa spartea,* the throat of the sheath of *Sporobolus cryptandrus,* and the main axils of the panicle of *Eragrostis spectabilis.*

Bifid. Two-cleft or two-lobed, applied to the summit of glumes, lemmas, and paleas. The lemmas of *Bromus* are usually bifid at apex.

Blade. The part of a leaf above the sheath.

Bract. The reduced leaves of the inflorescence and upper part of a shoot. Compare scale.

Branch. A lateral stem. Applied to the foliage stems or culms, and to the lateral stems of an inflorescence. *Branchlet.* A branch of the second or higher order. In open much-branched panicles the main branches from the axis are branches of the first order, the branchlets from these are branches of the second order and so on.

Bristle. A stiff slender appendage likened to a hog's bristle. An awn is a kind of bristle. In grasses the term is applied to the modified branchlets at the base of the spikelets in *Setaria* and allied genera, and to the prolongation of the rachis in *Panicum,* sect. *Paurochaetium,* and a few other groups.

Bulb. A subterranean bud with fleshy scales like the onion. The so-called bulbs of grasses are corms (which see). *Bulbous.* Swollen at base like a bulb or corm. Said of the base of the stem of some species of *Melica, Phleum, Phalaris,* and so on. *Bulblets.* Small bulbs or corms. Applied also to the proliferous buds in the inflorescence of certain grasses, as *Poa bulbosa,* proliferous forms of *P. arctica, P. alpina,* and others.

Callus. The indurate downward extension of the mature lemma in *Stipa, Aristida,* and some other genera. Morphologically, such a callus is a part of the rachilla. In *Heteropogon* and other *Andropogoneae* the callus is an oblique part of the rachis which extends downward from the spikelet. In *Chrysopogon* the callus is a part of the peduncle. The term callus is also applied to the thickened lower joint and first glume of *Eriochloa* (callus, a thickened part). *Callus hairs.* The hairs at the base of the floret of *Calamagrostis* and some other genera.

Canescent. Gray-pubescent or hairy.

Capillary. Very slender or hairlike.

Capitate. In a globular cluster or head.

Carinate. Keeled. Said of glumes, lemmas, and other parts when flattened laterally, with a sharp keel.

Cartilaginous. Hard and tough but elastic, like cartilage.

Caryopsis. The grain or fruit of grasses. The seed coat is grown fast to the pericarp as in the grain of wheat or corn. In a few grasses the seed is free within the pericarp, as in *Sporobolus* and *Eleusine.*

Cespitose. Tufted; several or many stems in a close tuft.

Chartaceous. Having the texture of writing paper.

Ciliate. Fringed with hairs on the margin (like an eyelash). *Ciliolate.* Minutely ciliate.

Circinate. Coiled from the top downward.

Clavate. Club-shaped; gradually thickened upward, and more or less circular in cross section.

Cleistogamous. Applied to flowers or florets when fertilized without opening. *Cleistogene.* A cleistogamous flower, such as found in *Triplasis* and *Danthonia.*

Collar. The area on the outer side of a leaf at the junction of sheath and blade.

Column. The lower undivided part of the awns of certain species of *Aristida;* the lower twisted segment of the awn in *Andropogoneae.*

Compact. Said of closely flowered inflorescences. Compare dense.

Compressed. Flattened laterally, as the compressed spikelets of *Uniola latifolia* and the compressed sheaths of *Andropogon virginicus.* If the organ is also sharply keeled, it is said to be compressed-keeled.

Conduplicate. Folded together lengthwise with the upper surface within, as in the blades of many grasses.

Continuous. Said of the rachis or other organ which does not disarticulate. The opposite of articulate or disarticulating.

Contracted. Said of inflorescences that are narrow or dense, the branches short or appressed. The opposite of open or spreading.

Convex. Rounded on the surface. Said especially of glumes and lemmas that are rounded on the back instead of keeled.

Convolute. Rolled longitudinally. Said mostly of blades, one edge being inside and the other outside.

Cordate. Heart-shaped. Said mostly of the base of blades. *Cordate-clasping.* Heart-shaped at base with the lobes overlapping around the stem.

Coriaceous. Leathery in texture.

Corm. The hard swollen base of a stem. In *Melica* the corm is a single enlarged lower internode. In *Panicum bulbosum* several internodes are involved. Compare *bulb.*

Crown. The persistent base of a tufted perennial herbaceous grass. Also the hard ring or zone at the summit of some species of *Stipa.* The "pappuslike crown" of dissected teeth is mentioned under *Pappophorum.*

Culm. The jointed stem of grasses.

Cuneate. Wedge-shaped with the narrow part below.

Cuspidate. Tipped with a sharp short rigid point.

Deciduous. Falling away, as the awn of *Oryzopsis,* the spikelets of some species with articulate pedicels, and the blades of some bamboos. The opposite of persistent.

Decumbent. Curved upward from a horizontal or inclined base. Said of stems or culms.

Decurrent. Extending down an organ below the insertion. Said especially of ligules decurrent on the margins of the sheath.

Dehiscence. Spontaneous opening of an organ, as the opening of anthers to let out the pollen.

Dense. Said of inflorescences in which the spikelets are crowded. The opposite of open or loose. Compare *compact.*

Depauperate. Reduced or undeveloped. Said especially of impoverished or dwarfed plants below the average size.

Diffuse. Open and much-branched. Said of panicles.

Digitate. Several members arising from the summit of a support. Said especially of racemes or spikes from the summit of a peduncle, as in *Digitaria* and *Cynodon.*

Dioecious. Unisexual, the two kinds of flowers on separate plants, as in *Buchloë.*

Disarticulating. Separating at maturity. Compare *articulate.*

Distichous. Conspicuously two-ranked, as the leaves of *Distichlis* and *Zea.*

Divaricate. Widely and stiffly divergent as the branches of certain open panicles (e.g., *Oryzopsis hymenoides*).

Dorsal. Relating to the back of an organ.

Dorsiventral. With a distinct upper and lower surface. Said of shoots bearing broad flat blades in a horizontal position, the blades turned into the same plane.

Drooping. Erect to spreading at base but inclining downward above, as the branches of a panicle.

Ellipsoid. An elliptic solid. Said of the shape of panicles, spikelets, and fruits.

Elliptic. Shaped like an ellipse. Said of blades and other flat surfaces.

Elongate. Narrow, the length many times the width or thickness.

Emarginate. Notched at the apex.

Equitant. Astride. Said of approximate compressed-keeled sheaths or blades at the base of a culm that infold each other like the leaves of *Iris.*

Erose. Irregularly notched at apex as if gnawed. Said of glumes and lemmas.

Excurrent. Running beyond. The midnerve is excurrent from the lemma as an awn in many grasses.

Exserted. Protruding. The awns of some

species of *Calamagrostis* are exserted, protruding beyond the spikelet.

Falcate. Scimiter-shaped, curved sidewise and flat, tapering upward. Said of certain asymmetric blades.

Fascicle. A little bundle or cluster. Said of clustered leaves, branches of a panicle, and spikes or racemes on an axis.

Ferruginous. Rust-colored.

Fertile. Capable of producing fruit, having pistils. A fertile floret may be pistillate or perfect.

Fibrillose. Furnished with fibers. Said especially of the old basal sheaths of some grasses.

Filiform. Threadlike.

Fimbriate. Fringed, the hairs longer or coarser as compared with ciliate.

Flabellate. Fan-shaped. Said of the lemmas of *Neostapfia* and the inflorescence of *Miscanthus sinensis*.

Flexuous. Bent alternately in opposite directions.

Floret. The lemma and palea with included flower (stamens and pistil). Florets may be perfect, staminate, pistillate, neuter, sterile, and so on.

Folded. Conduplicate. Said chiefly of blades.

Fruit. The ripened pistil. In grasses the fruit is usually a caryopsis. The term fruit is also applied to the caryopsis and parts that may enclose it permanently at maturity. In *Panicum* the indurate fertile lemma and palea with the enclosed caryopsis is the fruit. In *Cenchrus* it is the entire bur.

Fuscous. Dusky, brownish gray.

Fusiform. Spindle-shaped. A solid that is terete in the middle and tapering toward each end.

Geniculate. Bent abruptly. Said of awns and of the lower nodes of the culm.

Gibbous. Swollen on one side, as the second glume of *Sacciolepis*.

Glabrous. Without hairs of any sort.

Gland. A protuberance or depression, usually minute, that secretes, or appears to secrete, a fluid. *Glandular.* Supplied with glands. The glands may be depressed as in *Eragrostis cilianensis* and *Heteropogon melanocarpus*.

Glaucous. Covered with a waxy coating that gives a blue-green color as in the leaf of the cabbage, and the bloom of the grape.

Glomerate. Collected in heads.

Glumes. The pair of bracts at the base of a spikelet.

Gregarious. Growing in groups or masses.

Herbaceous. Having the characters of an herb; opposed to woody; thin in texture and green in color, as the herbaceous lemmas of *Poa*.

Hirsute. Pubescent with straight rather stiff hairs. *Hirsutulous, hirtellous.* Minutely hirsute.

Hispid. Pubescent with stiff or rigid hairs. *Hispidulous.* Diminutive of hispid.

Hyaline. Thin and translucent or transparent.

Imbricate. Overlapping, as the lemmas in many spikelets.

Implicate. Tangled, as the branches of the panicle of *Panicum implicatum*.

Indurate. Hard. Compare chartaceous and coriaceous.

Inflated. Puffed up, bladdery.

Inflexed. Turned in at the margins. Said especially of the margin of the glumes or lemmas in some species.

Inflorescence. The flowering part of a plant.

Innovation. The basal shoot of a perennial grass.

Internerves. The spaces between the nerves. Said of glumes and lemmas.

Internode. The part of a stem between two successive nodes.

Interrupted. The continuity broken. Said especially of dense inflorescences whose continuity is broken by gaps.

Involucre. A circle of bracts below a flower or flower cluster. In grasses applied to the cluster of bristles or sterile branchlets below the spikelets in *Pennisetum* and a few other genera, and to the bony bead of *Coix*.

Involute. Rolled inward from the edges, the upper surface within. Said of blades.

Joint. The node of a grass culm. The internode of an articulate rachis.

Keel. The sharp fold at the back of a compressed sheath, blade, glume, or lemma. The palea and sometimes the glumes and lemmas may be two-keeled. Keel is used because of the similarity to the keel of a boat.

Lacerate. Torn at the edge or irregularly cleft, as in some ligules.

Lanate. Woolly, clothed with long tangled hairs.

Lanceolate. Rather narrow (surface), tapering to both ends, the broadest part below the middle.

Laterally (compressed). Flattened from the sides, as certain spikelets, glumes, and lemmas.

Lax. Loose. Said of a soft or open inflorescence and of soft or drooping foliage.

Leaf. The lateral organ of a stem, in grasses consisting of sheath and blade.

Lemma. The bract of a spikelet above the pair of glumes.

Ligule. The thin appendage or ring of hairs on the inside of a leaf at the junction of sheath and blade.

Linear. Long and narrow with parallel sides. Said of surfaces, such as a blade. Said also of spikelets and other organs, having in mind the shape of a longitudinal section.

Lobe. A segment of an organ, usually rounded or obtuse. Applied especially to the divisions of a cleft lemma.

Loose. Open. Said of panicles. The opposite of dense or compact.

Membranaceous. Thin, like a membrane.

Monoecious. Unisexual, the two kinds of flowers on the same plant, as in *Zea* and *Zizania*.

Mucro. A minute awn or excurrent midnerve of an organ. *Mucronate.* Provided with a mucro.

Navicular. Boat-shaped. Shaped like the bow of a canoe. Applied especially to the tip of blades.

Nerve. The vascular veins (mostly longitudinal) of the blades, glumes, and lemmas.

Neuter. Without stamens or pistils. Said of florets or spikelets.

Nodding. Inclined somewhat from the vertical. Said of panicles.

Node. The joint of a culm.

Nodulose. Roughened by minute knots.

Ob-. A prefix meaning inversely, as obovate.

Oblong. Longer than wide, with parallel sides, but not so long as linear. Applied also to panicles and other parts, having in mind a longitudinal section.

Obsolete. Almost wanting. Applied to organs usually present.

Obtuse. Rounded at the apex. Contrasted with acute.

Open. Loose. Said of panicles. Opposite of dense or compact.

Oval. Broadly elliptic.

Ovate. The shape of the longitudinal section of an egg, broadest below the middle.

Ovoid. An egg-shaped solid.

Palea. The inner bract of a floret.

Panicle. An inflorescence with a main axis and subdivided branches. It may be compact and spikelike (*Phleum pratense*) or open (*Avena sativa*).

Papery. See chartaceous.

Papilla. A minute nipple-shaped projection. *Papillose.* Bearing papillae. *Papillose-pilose.* Bearing stiff hairs arising from papillae.

Pappus. In grasses mentioned under *Pappophorum*, referring to the awns as forming a pappuslike crown, similar to the pappus in certain species of *Compositae*.

Pectinate. Comblike. Used especially with some species of *Bouteloua* where the spikelets are set close together, parallel and divergent from the rachis like the teeth of a comb.

Pedicel. The stalk of a spikelet. *Pedicellate.* Having a pedicel. Opposed to sessile.

Peduncle. The stalk or stem of an inflorescence. *Peduncled.* Having a peduncle.

Pendent. Hanging down.

Perennial. Lasting more than 1 year. Applied to grasses in which the underground parts last more than 1 year; and to woody culms to distinguish them from those which die to the ground (herbaceous) even though the underground parts are perennial.

Perfect. Applied to flowers having both stamens and pistil.

Pericarp. The ripened walls of the ovary when it becomes a fruit.

Persistent. Remaining attached, either after other parts have been shed, or for a considerable period. The paleas of certain species of *Eragrostis* persist after the fall of the lemmas. Also used as the opposite of deciduous.

Petiole. The stalk of a leaf blade. Used with the leaves of many bamboos and with some other broadleaved species in which the blade contracts into a petiole. *Petiolate.* Having a petiole.

Pilose. Pubescent with soft straight hairs.

Pistillate. Applied to flowers bearing pistils only and to an inflorescence or a plant with pistillate flowers.

Pitted. Marked with small depressions or pits. Applied to the fruit (fertile lemma) of certain species of *Olyra*. Also applied to the pinhole depression in the glume of certain species of Andropogoneae.

Plicate. Folded in plaits lengthwise as the blades of *Setaria* sect. *Ptycophyllum*.

Plumbeous. Lead-colored, greenish drab, as the spikelets of *Eragrostis cilianensis*.

Plumose. Feathered, having fine hairs on each side. Said chiefly of awns and slender teeth.

Proliferous. Bearing vegetative buds or bulblets in the inflorescence. Compare *bulblets*.

Pruinose. Having a waxy powdery secretion on the surface. Having a more pronounced bloom than when glaucous.

Puberulent. Diminutive of pubescent. Minutely pubescent.

Pubescent. Covered with hairs. Applied especially when the hairs are short and soft. *Pubescence.* A hairy covering.

Pulvinus. The swelling at the base of the branches of some panicles which cause them to spread.

Pustulose. Blistery, furnished with pustules or irregularly raised pimples, as in the spikelets of *Panicum angustifolium.* Not as definitely roughened as papillose.

Pyramidal. Pyramid-shaped. Applied sometimes to panicles that are actually conical.

Pyriform. Pear-shaped. Obovoid with attenuate base. Applied to the shape of spikelets.

Raceme. An inflorescence in which the spikelets are pediceled on a rachis. *Racemose.* In racemes.

Rachilla. A small rachis. Applied especially to the axis of a spikelet.

Rachis. The axis of a spike or raceme.

Reticulate. In a network. Applied especially to the cross-veining on some spikelets, as *Panicum fasciculatum.*

Retrorse. Pointing backward, as the hairs on the sheaths of certain species of *Bromus.*

Revolute. Turned or rolled backward from both edges. Said chiefly of blades.

Rhizome. An underground stem; rootstock. The rhizomes of grasses are usually slender and creeping. They bear scales at the nodes, the scales sometimes remote and inconspicuous (*Poa pratensis*), sometimes imbricate and prominent (*Spartina*).

Rhizomatous. Having rhizomes or appearing like rhizomes, as the base of a decumbent stem.

Rosette. A cluster of spreading or radiating basal leaves, as in the overwintering stage of *Panicum*, sect. *Dichanthelium.*

Rudiment. An imperfectly developed organ or part. *Rudimentary.* Underdeveloped. Applied also to one or more rudimentary florets at the summit of the spikelet of some genera, as *Melica, Bouteloua, Chloris.*

Rugose. Wrinkled. Said especially of the fruit of some species of *Panicum* and allied groups.

Saccate. Bag or sac-shaped, as the second glume of *Sacciolepis.*

Scabrous. Rough to the touch. Covered with minute points, teeth, or very short stiff hairs. *Scaberulous.* Minutely scabrous.

Scale. The reduced leaves at the base of a shoot. Said especially of the reduced or rudimentary leaves on a rhizome.

Scarious. Thin, dry, and membranaceous, not green.

Secondary. Subordinate; below or less than primary. Said of branches arising from primary branches.

Secund. One-sided or arranged along one side.

Self-pollinated. Pollinated in the bud or by pollen from the same flower. The opposite of cross-pollinated.

Serrate. Saw-toothed; having sharp teeth. *Serrulate.* Minutely serrate.

Sessile. Without a pedicel or stalk. The opposite of pediceled. Said of blades, spikelets, and other organs.

Setaceous. Bristlelike. Said especially of slender teeth attenuate to an awn.

Sheath. The lower part of a leaf that encloses the stem.

Sinuous. Wavy.

Smooth. Not rough to the touch. Compare glabrous, without hairs but which may be rough to the touch.

Spathe. A sheathing bract of the inflorescence found especially in the Andropogoneae.

Spike. An unbranched inflorescence in which the spikelets are sessile on a rachis. *Spikelike.* A dense panicle in which the pedicels and branches are short and hidden by the spikelets as in *Phleum.*

Spikelet. The unit of the inflorescence in grasses, consisting of two glumes and one or more florets.

Spreading. Having an outward direction. Said especially of the branches of a panicle when they lie between ascending and the horizontal direction (right angles).

Squarrose. Spreading or recurved at the tip. Said of the tips of lemmas.

Stamen. The part of the flower that bears the pollen. *Staminate.* Containing stamens only. Also applied to an inflorescence or a plant with staminate flowers.

Sterile. Without pistils. A sterile floret may

be staminate or neuter. It may even lack a palea, and consist of nothing but a lemma.

Stipe. A minute stalk to an organ. Applied especially to a pistil. Also sometimes to the prolongation of a rachilla as in *Calamagrostis. Stipitate.* Having a stipe.

Stolon. A modified propagating stem above ground creeping and rooting or curved over and rooting at the tip. *Stoloniferous.* Bearing stolons.

Stramineous. Straw-colored, pale yellow.

Striate. Marked with fine parallel lines or minute ridges

Strict. Stiffly upright.

Strigose. Rough with short stiff hairs; harshly pubescent.

Sub-. A prefix to denote somewhat, slightly, or in a less degree; as subacute, somewhat acute.

Subtend. To be below, as a bract subtends a branch in its axil.

Subulate. Awl-shaped.

Succulent. Fleshy or juicy.

Sulcate. Grooved or furrowed. Said chiefly of stems, sheaths, and slender blades.

Tawny. Pale brown or dirty yellow.

Teeth. Pointed lobes or divisions.

Terete. Cylindric and slender, as the usual unflattened stems or culms of grasses.

Tessellate. The surface marked with square or oblong depressions.

Triad. A group of 3, applied to the central and 2 lateral spikelets in *Hordeum* and to ultimate racemes in *Sorghum.*

Trifid. Divided into three parts as the awns of *Aristida.*

Truncate. Ending abruptly, as if cut off horizontally.

Tuberculate. Furnished with small projections.

Turgid. Swollen, as the pulvini of a panicle during anthesis.

Unilateral. One-sided or turned to one side.

Unisexual. Said of flowers containing only stamens or only pistils.

Verticillate. In verticils or whorls.

Villous. Pubescent with long soft hairs.

Virgate. Straight and erect; wand-shaped.

Web. The cluster of slender soft hairs at the base of the floret in certain species of *Poa.*

Whorl. A cluster of several branches around the axis of an inflorescence.

Wing. A thin projection or border; for example, the thin borders on the rachis of certain species of *Digitaria* and *Paspalum.*

APPENDIX

The following genera are additions to or changes from the genera in the first edition of the Manual, and which are not in Hitchcock's Genera of Grasses of the United States (United States Department of Agriculture Bulletin 772, revised edition, 1936.) The place of publication and the type species are here given, the descriptions being given in the text.

PHYLLOSTACHYS Sieb. and Zucc.

Type species: *Phyllostachys bambusoides* Sieb. and Zucc.

Phyllostachys Sieb. and Zucc., Abh. Bayer. Akad. Wiss. 3³: 745, pl. 5. 1843. A single species included, *Phyllostachys bambusoides* Sieb. and Zucc.

A large genus of Asiatic bamboos.

PSEUDOSASA Makino

Type species: *Pseudosasa japonica* (Sieb. and Zucc.) Makino.

Pseudosasa Makino, Jour. Jap. Bot. 2⁴: 15. 1920. No generic description. "The diagnosis will appear in the forecoming number." Three species are transferred to the genus, the first being *Pseudosasa japonica* (Sieb. and Zucc.) Makino. The diagnosis was published in English by Makino, Jour. Jap. Bot. 5⁴: 15. 1928, *P. japonica* (Sieb. and Zucc.) Makino being one of the seven species included.

(3) BRACHYPODIUM Beauv.

Type species: *Brachypodium pinnatum* (L.) Beauv.

Brachypodium Beauv., Ess. Agrost. 100, pl. 19, f. 3. 1812. Twenty-two names are listed under the genus, several of them not congeneric with *B. pinnatum*, based on *Bromus pinnatus* L., which is taken as the type because it is the only one illustrated.

Trachynia Link, Hort. Berol. 1: 42. 1827. Two species are included, *Trachynia distachya* (L.) Link, based on *Bromus distachyos* L., and *T. rigida* (Roth) Link, based on *Festuca rigida* Roth, differentiated from *Brachypodium* on glumes longer than the lower floret.

Perennials or annuals with racemes of subsessile, many-flowered spikelets. Several species in Eurasia and Africa, one native of Mexico and Central America, and a few introduced in North and South America.

(9) SCOLOCHLOA Link

(*Fluminea* Fries)

Type species: *Scolochloa festucacea* (Willd.) Link.

Scolochloa Link, Hort. Berol. 1: 136. 1827; not *Scolochloa* Mert. and Koch, 1823. A single species included, *Scolochloa festucacea* (Willd.) Link, based on *Arundo festucacea* Willd.

Fluminia Fries, Summa Veg. Scand. 247. 1846. Based on *Festuca borealis* Mert. and Koch. A single species included, its name being given as "*Festuca borealis* or *Fluminea arundinacea.*" This is the same as *Scolochloa festucacea*.

The genus consists of a single marsh grass of Eurasia, Canada, and northern United States.

(11) HESPEROCHLOA (Piper) Rydb.

Type species: *Festuca confinis* Vasey (*F. kingii* Cassidy).

Festuca subgenus *Hesperochloa* Piper, U. S. Nat. Herb. Contrib. 10: 40, pl. 15. 1906. A single species is included, *F. confinis* Vasey (*Hesperochloa kingii* (S. Wats.) Rydb.).

Hesperochloa (Piper) Rydb., Torrey Bot. Club Bul. 39: 106. 1912. Based on *Festuca* subgenus *Hesperochloa* Piper.

Wasatchia Jones, West. Bot. Contrib. 14: 16. 1912. A single species is included, *W. kingii* (S. Wats.) Jones.

The genus consists of a single dioecious species of the Western States.

(32 A.) ECTOSPERMA Swallen

Spikelets several-flowered, glumes and lemmas persistent on the continuous short-jointed rachilla, the caryopsis falling free; glumes subequal, about reaching the summit of the spikelet, broad, spreading, 7- to 11-nerved; lemmas rounded on the back, closely imbricate, thin, 5- to 7-nerved, densely long villous on the lower half to two-thirds; palea as long as the lemma or slightly exceeding it, the broad margins densely long villous nearly to the summit, the apex more or less erose or lacerate; caryopsis readily falling from the floret, broadly elliptic, the embryo broad, about two-thirds the length of the grain; stamens 3. Rigid perennial with firm pungent blades and narrow, simple panicles of broad spikelets. Only known from the type species, *Ectosperma alexandrae*. Name from *ectos*, free from, and *sperma*, seed.

1. **Ectosperma alexándrae** Swallen. (Fig. 1200.) Rigid perennial, branching at base from an erect or creeping thick scaly rhizome with woolly nodes; flowering culms erect or ascending, 30 to 35 cm. tall, sulcate-ridged, puberulent at the summit, otherwise glabrous; leaves 2 or 3 above the base, distant, the sheaths much shorter than the internodes, the uppermost about reaching the base of the panicle, villous on the margin toward the summit; ligule a ring of hairs about 1 mm. long; blades rigid, 5 to 9 cm. long, 3 to 5 mm. wide, tapering to a pungent apex, the upper 2 blades subulate; leaves of the few to several stout erect sterile branches at base numerous, the sheaths much overlapping, long-villous on the margin and densely so at the summit; blades conspicuously distichous, rigid, 4 to 14, mostly 5 to 7 cm., long (the lower shorter), 3 to 6 mm. wide, tapering to a pungent apex; panicle erect or nearly so, simple, 6 to 10 cm. long, the axis and few, short, 2- to 3-flowered branches compressed, sparsely pubescent; spikelets on short, pubescent pedicels, scarcely imbricate, palea, 1 to 1.5 cm. long, nearly as wide; rachilla compressed, bearded at the nodes; glumes 9 to

FIGURE 1200.—*Ectosperma alexandrae*. Plant × ½; glumes, dorsal view of lemma, ventral view of palea with lodicules, stamens, and pistil, and two views of caryopsis, × 5. (Type.)

14 mm. long, acuminate, glabrous; |the lemmas 7 to 9 mm. long, apiculate, the margins conspicuously white-villous except the apex; palea deeply sulcate between the arched keels, the margins conspicuously white-villous; caryopsis readily falling, brown, 4 mm. long, 2 mm. wide, the embryo conspicuous; anthers 3.5 mm. long, pale. 21 —Sand hill, altitude 3,050 feet, Eureka Valley, Inyo County, Calif.

(33) TRIDENS Roem. and Schult.

Type species: *Tridens quinquifidus* (Pursh) Roem. and Schult. (*Tridens flavus* (L.) Hitchc.)
Tricuspis Beauv., Ess. Agrost. 77, pl. 15, f. 10. 1812. Not *Tricuspis* Pers., 1807. *Tricuspis caroliniana* Beauv., the species illustrated, is taken as the type. This is *Tridens flavus* (L.) Hitchc. Two other names mentioned are nomina nuda.
Tridens Roem. and Schult., Syst. Veg. 2: 34, 599. 1817; *Tricuspis* Beauv., pl. 15. f. 10 is cited on page 34, and *Tridens quinquifida* Roem. and Schult., based on *Poa quinquifida* Pursh, is published on page 599.
Windsoria Nutt., Gen. Pl. 1: 70. 1818. Two species are described, *W. poaeformis* Nutt., which is *Tridens flavus*, and *W. ambiguus* (Ell.) Nutt. The first is selected as the type.
Erioneuron Nash in Small, Fl. Southeast. U. S. 143. 1903. The type, *Uralepis pilosa* Buckl. (*Tridens pilosus* (Buckl.) Hitchc.), is indicated on page 1327. Only one species is included.
Dasyochloa Willd. ex Rydb., Colo. Agr. Expt. Sta. Bul. 100: 18, 37. 1906. There is no description except in the key. *Dasyochloa pulchella* Willd. is listed in Steud. Nom. Bot. ed. 2. 1: 484. 1840, as synonym of *Uralepis* ("*Uralepsis*") *pulchella* Kunth (*Tridens pulchellus* (H. B. K.) Hitchc.) and is the only species included in the genus by Rydberg.
Perennials, diverse in habit and spikelets, which have been included in *Triodia* R. Br., of Australia and New Zealand, which also consists of species with somewhat diverse spikelets. *Tridens* differs from that in the strictly 3-nerved lemmas, the lateral nerves marginal or nearly so, the lemmas of *Triodia* being mostly in 3 groups of 2 or 3 nerves each (sometimes indistinct), the lateral nerves not marginal.
In habit the species of *Triodia* are very different from those of *Tridens*, being tussock grasses with rigid pungently pointed blades. *Tridens* is confined to the western hemisphere. Two new combinations are necessary, see pages 971, 973.

(35) NEOSTAPFIA Davy

Type species: *Stapfia colusana* Davy.
Stapfia Davy, Erythea 6: 110, pl. 3. 1898. Not *Stapfia* Chodat, 1897. A single species included, *Stapfia colusana* Davy.

Neostapfia Davy, Erythea 7: 43. 1899. Change of name for *Stapfia*, the species renamed *Neostapfia colusana* (Davy) Davy.
Davyella Hack., Oesterr. Bot. Zeitschr. 49: 133. 1899. Change of name for *Stapfia* Davy, the species renamed *Davyella colusana* (Davy) Hack.
The one species, confined to California, was included in the related South American genus, *Anthochloa* Nees, by Scribner (U. S. Dept. Agr., Div. Agrost. Bul. 17: 221, f. 517. 1899). In *Anthochloa* the leaves are differentiated into sheath and blade as in other grasses, the axis of the inflorescence is not extended and foliaceous as in *Neostapfia*, and the glumes are developed and persistent. (See R. F. Hoover, West. Bot. Leaflets 11: 274. 1940.)

(40) ENNEAPOGON Desv.

Type species: *Enneapogon desvauxii* Beauv.
Enneapogon Desv. ex Beauv., Ess. Agrost. 81, pl. 16, f. 11. 1812. *Enneapogon desvauxii* Beauv. and four species described by Robert Brown under *Pappophorum* are included, *E. desvauxii* being the only one illustrated. No locality is here given, but in a later paper by Desvaux. (Jour. de Bot. 1: 70. 1813.) "iles Manilles" was erroneously given for the locality. What is undoubtedly part of the type collection was recently found in the British Museum and proves to be the American species known as *Pappophorum wrightii* S. Wats., with which Beauvois' illustration agrees. The collection was probably made by Née near Mendoza, Argentina, where the species is still found. *See* Burbidge, N. T., Linn. Soc. London Proc. 153 Sess. (1940–41): 52–91, f. 1–5. 1941; *also* Chase, A., Madroño 8: 187–189. 1946.
Tufted perennials of subarid regions of Asia, Africa and Australia, one species in America.

(51) MONERMA Beauv.

Type species: *Monerma monandra* Beauv.
Monerma Beauv., Ess. Agrost. 116, pl. 20, f. 10. 1812. Three names are listed. *Monerma monandra* Beauv., the species illustrated, is taken as the type. This is same as *M. cylindrica* (Willd.) Coss. and Dur. (See p. 898.) This species has commonly been included in *Lepturus* R. Br., the type of which is *L. repens* (G. Forst.) R. Br. of Australia and the Pacific islands. *Monerma* consists of a single species of the Mediterranean region, introduced in America.

(52) PARAPHOLIS C. E. Hubbard

Type species: *Parapholis incurva* (L.) C. E. Hubbard, based on *Aegilops incurva* L.
Parapholis C. E. Hubb., Blumea Sup. 3 (Henrard Jubilee vol.): 14. 1946. Differentiated from *Pholiurus*, to which the four

species included had been referred. *Pholiurus* Trin. is based on a single species, *P. pannonicus* (Host) Trin., in which the rachis is continuous, the spikelets falling alone at maturity (as in *Scribneria*).

Low annuals with slender cylindric spikes. Species 4, in the Eastern Hemisphere, one introduced in the United States.

(60) CORYNEPHORUS Beauv.

Type species: *Corynephorus canescens* (L.) Beauv.

Weingaertneria Bernh., Syst. Verz. Pflanz. 23: 51. 1800. A single species, *W. canescens* (L.) Bernh., based on *Aira canescens* L., is included.

Corynephorus Beauv., Ess. Agrost. 190. 1812. Two species are included, *C. articulatus* (Desf.) Beauv., based on *Aira articulata* "Lin." (error for Desf.) and *C. canescens* (L.) Beauv., based on *Aira canescens* L. The latter species, being illustrated, is taken as the type.

(62) HELICTOTRICHON Besser

Type species: *Avena sempervirens* Host.

Elictotrichon Bess. ex Andrzej., Rys. Bot. 9. 1823. Undescribed; *E. sempervirens* Bess., presumably based on *Avena sempervirens* Host, is included in a list of plants.

Helictotrichon Bess. in Schult., Mantissa 3. (Add. 1): 526 (error 326). 1827. A generic description is given and five species listed, "*Av. sempervirens* Host, *versicolor* Vill., *pratensis* L., *pubescens* L., *planiculmis* Schrad." None are here transferred to *Helictotrichon*, but all have been transferred in recent years. In Schur, Enum. Pl. Transsilv. 762. 1866, the name is misspelled "*Heliotrichum.*"

Avena sect. *Avenastrum* Koch, Syn. Fl. Germ. Helv. 795. 1837. Six species are included, *A. planiculmis* Schrad., *A. pubescens* Huds., *A. alpina* J. E. Smith, *A. pratensis* L., *A. versicolor* Vill., and *A. sempervirens* Vill., all European and all later transferred to *Helictotrichon*.

Avenastrum Jessen, Deutschl. Gräser 214. 1863. Presumably based on *Avena*, sect. *Avenastrum*, but Koch is not mentioned except in the list of authors (p. 297). Besides 2 species included by Koch, *A. pubescens* and *A. pratense*, Jessen transferred one species of *Trisetum*, one of *Arrhenatherum*, and two of *Aira* to his *Avenastrum*.

Heuffelia Schur, Enum. Pl. Transsilv. 760. 1866. "*Avena* sect. II. *Avenastrum* Koch" is cited and 12 species listed.

Numerous perennials of Eurasia and Africa, one introduced and two native in western North America.

(65) SIEGLINGIA Bernh.

Type species: *Sieglingia decumbens* (L.) Bernh.

Sieglingia Bernh., Syst. Verz. Erf. 20, 44. 1800. A single species is included, *Sieglingia decumbens* (L.) Bernh., based on *Festuca decumbens* L.

The genus consists of a single tufted perennial of Europe and British America, recently found in northern Washington.

(70) APERA Adans.

Type species: *Agrostis spica-venti* L.

Apera Adans., Fam. Pl. 2: 495. 1763. "*Agrostis* 1. Lin. Sp. 61" is cited. The first species of *Agrostis* in Linnaeus, Species Plantarum, is "*A. spica venti.*"

Anemagrostis Trin., Fund. Agrost. 128. 1820. Two species, *Agrostis spica-venti* L. and *A. interrupta* L., are included.

The genus consists of two annuals of Eurasia, both introduced in the United States. Previously included in *Agrostis*.

(86) HELEOCHLOA Host ex Roemer

Type species: *Heleochloa alopecuroides* (Pill. and Mitterp.) Host.

Heleochloa Host ex Roemer, Collect. Rem. Bot. 233. 1809. Generic description given for *Heleochloa* Host, Icon. Gram. Austr. 1: 23. pl. 29. 1801, including *H. alopecuroides* (Pill. and Mitterp.) Host and *H. schoenoides* (L.) Host, described and figured by Host but without generic description. Roemer includes the same species.

(102) MICROCHLOA R. Br.

Type species: *Microchloa setacea* (Roxb.) R. Br.

Microchloa R. Br., Prodr. Fl. Nov. Holl. 208. 1810. A single species is included, *Microchloa setacea* (Roxb.) R. Br., based on *Rottboellia setacea* Roxb.

Wiry annuals or perennials with slender curved spikes. Several species in Africa and Australia, one introduced and one native in America.

EHRHARTA Thunb.

Type species: *Ehrharta capensis* Thunb.

Ehrharta Thunb., Svensk. Vet. Akad. Handl. 40: 217, pl. 8. 1779. A single species is included, *Ehrharta capensis* Thunb., of South Africa.

Trochera L. C. Rich., Jour. de Phys. (Obs. Phys.) 13: 225, pl. 3. 1779. A single species is included, *Trochera striata* L. C. Rich., a garden plant from unknown source, referred by Stapf to *Ehrharta bulbosa* Smith. Though the title-page date is the same, Kuntze (Rev. Gen. Pl. 2: 795. 1891) gives March for month of publication for *Trochera* and July-September for *Ehrharta*.

Placed in *Phalarideae*, having a pair of sterile lemmas below the single fertile floret, the sterile lemmas exceeding the glumes and usually the fertile floret. Species numerous in South Africa, a few introduced elsewhere, two in California.

(142) RHYNCHELYTHRUM Nees

Type species: *Rhynchelythrum dregeanum* Nees (*R. repens* (Willd.) C. E. Hubbard).

Rhynchelytrum Nees in Lindl., Nat. Syst. Bot. 446. 1836. A single species, *R. dregeanum* from Cape of Good Hope, is included. (Nees corrected the spelling to *Rhynchelytrum* in Errata, following page 490 in Nees, Agrostographia, 1841.) There are some 40 species in Africa, southern Europe, and southern Asia, one species, *R. repens* (Willd.) C. E. Hubbard, widely introduced in the warmer parts of America and commonly known as Natal grass. Most of the species were formerly included in *Tricholaena* Schrad., the type species of which is *T. micrantha* Schrad. (*T. teneriffae* (L.f.) Link, based on *Saccharum teneriffae* L.f.) In his 1836 publication Nees gives "C. b. Sp." (Cape bonae Spei) as the locality for *Rhynchelytrum dregeanum*, but in his Agrostographia Capensis 64. 1841, Nees, repeating the earlier generic description almost verbatim and describing the species in detail, gives as locality "In loco depresso humido ad Port Natal vix 100' [pedales] alt., (Drège.)." Port Natal is now known as Durban, but Nees seems to have included the south African regions explored by Drège under the general name of colonia Capensis, or Capstadt.

In the Agrostographia Capensis, pages 16–20, Nees includes *Tricholaena* Schrad., with four species, *T. tonsa* Nees and *T. rosea* Nees, which have the characters of *Rhynchelytrum*, and *T. capensis* (Licht.) Nees and *T. arenaria* Nees, with the characters now restricted to *Tricholaena*.

Perennials or annuals, the panicles with capillary branchlets and pedicels and silky, often reddish, spikelets.

(152) MICROSTEGIUM Nees

Type species: *Microstegium willdenovianum* Nees.

Microstegium Nees in Lindl., Nat. Syst. 447. 1836. A single species is included, *M. willdenovianum* Nees, which is the same as *M. vimineum* (Trin.) A. Camus, based on *Andropogon vimineus* Trin.

Leptatherum Nees, Proc. Linn. Soc. 1: 92. 1841. A single species is included, *L. royleanum* Nees, which is the same as *Microstegium nudum* (Trin.) A. Camus, based on *Pollinia nuda* Trin.

Nemastachys Steud., Syn. Pl. Glum. 1: 357. 1854. A single species is included, *N. taitensis* Steud., which is the same as *Microstegium glabratum* (Brongn.) A. Camus, based on *Eulalia glabrata* Brongn.

Mostly decumbent species with lanceolate blades and digitate racemes, numerous in southern Asia and East Indies, several species in the Pacific Islands, and a few in Africa, one introduced in eastern United States.

(158) CHRYSOPOGON Trin.

Type species: *Chrysopogon gryllus* (L.) Trin.

Rhaphis Lour., Fl. Cochinch. 552. 1790. A single species, *Rhaphis trivialis* Lour., which is the same as *Andropogon aciculatus* Retz. (*Rhaphis aciculatus* (Retz.) Desv.) and *Chrysopogon aciculatus* (Retz.) Trin. is included.

Pollinia Spreng., Pugill. 2: 10. 1815. Not *Pollinia* Trin., 1832. Several species are described, but the generic characters are given under the first, *P. gryllus* Spreng., based on *Andropogon gryllus* L.

Centrophorum Trin., Fund. Agrost. 106, pl. 5. 1820. A single species, *C. chinense*, is included. This is the same as *Chrysopogon aciculatus* (Retz.) Trin.

Chrysopogon Trin., Fund. Agrost. 187. 1820. Two species are included, *C. gryllus* (L.) Trin. and *C. aciculatus* (Retz.) Trin. An illustration of the first is cited and that species is taken as the type.

Chrysopogon sect. *Rhaphis* (Lour.) Ohwi, Acta Phytotax. and Geobot. 11: 163. 1942. Based on *Rhaphis* Lour.

Bentham (Linn. Soc. Jour., Bot. 19: 73. 1881) transferred the American species, *Andropogon nutans*, *A. avenaceus*, and allied species, to *Chrysopogon* Trin., and that name was adopted by Vasey and others until *Sorghastrum* Nash was described for these species.

Awned perennials of Eurasia, Africa, and the Pacific islands. Only one annual species, *C. pauciflorus* (Chapm.) Benth., known from America.

VETIVERIA Bory

Type species: *Vetiveria odoratissima* Bory. (*V. zizanioides* (L.) Nash.)

Vetiveria Bory in Lem., Bul. Soc. Philom. (Paris) 1822: 43. 1822. A single species, *V. odoratissima* Bory, mentioned in an account of the rhizome, and *Agrostis verticillata* Lam. are cited, both the same as *V. zizanioides* (L.) Nash, based on *Phalaris zizanioides* L. The species is described under the name *Vetiveria odorata* Virey in Dupetit-Thouars ex Virey, Jour. de Pharm. 13: 501. 1827, the preceding paper cited.

Mandelorna Steud., Syn. Pl. Glum. 1: 359. 1854. A single species included, *M. insignis* Steud., the same as *Vetiveria nigritana* (Benth.) Stapf, a species closely allied to *V. zizanioides*.

Tall perennials of the Old World, one species introduced in America.

THEMEDA Forsk.

Type species: *Themeda triandra* Forsk. *Themeda* Forsk., Fl. Aegypt. Arab. 178. 1775. A single species included, *Themeda triandra* Forsk.

Anthistiria L. f., Nov. Gram. Gen. 38, pl. 1. 1779. The genus is described and figured, but no species is mentioned; L. f., Suppl. Pl. 113. 1781. A single species described, *Anthistiria ciliata* L. f., "*Andropogon quadrivalvis* [L.] Syst. Veg. ed. 13. p. 758" cited.

Androscepia Brongn. in Duperrey, Bot. Voy. Coquille 77. 1831. A single species included, *A. gigantea* (Cav.) Brongn., based on *Anthistiria gigantea* Cav. (*Themeda gigantea* (Cav.) Hack.

Annuals or robust perennials of the Old World, one species introduced in the United States and two in the West Indies.

Names published by Johann Friedrich Theodor Bieler in his doctor's thesis entitled, "Plantarum Novarum ex Herbario Sprengelii Centuriam," issued May 30, 1807, were entirely overlooked until discovered by Prof. M. L. Fernald a few years ago. The names that are based on specimens from North America, mostly sent to Sprengel by Muhlenberg, were listed by Fernald in Rhodora 47: 198–204 (1945). Among them are seven names of grasses. Bieler's paper was republished unchanged by Sprengel under the title, "Novarum Plantarum ex Herbario Meo Centuria," in the second part of his Mantiss Prima, pages 27–28 (1807), evidently later than Bieler's paper. Bieler's name is not mentioned, and the species have been credited to Sprengel and are so listed in the Index Kewensis.

In this edition of the Manual of Grasses, Bieler is given as author of these names, valid and synonyms. They are found in *Festuca, Sphenopholis, Sporobolus,* and *Panicum.* One name listed by Fernald, "*Panicum pensylvanicum* Bieler, Plant. Nov. Herb. Spreng. Cent. 4. 1807," was published by Sprengel in Natchrag. Bot. Gart. Halle 30 (1801), the description reworded in Bieler's paper, but agreeing with the earlier description. This and *Panicum discolor* Bieler, apparently belonging in subgenus *Dichanthelium,* are given in Unidentified Names (p. 982).

It is possible that Sprengel, who was working on grasses, may have written the descriptions of the grasses in Bieler's paper. The century includes many genera in different families, from many parts of the world, as well as garden plants and a good many ferns, mosses, and lichens. But since Bieler's paper was published before Sprengel's Novarum Plantarum, he is here accepted as author.

ADDENDA

Page 821: Transfer "*Aristida pallens* [Cav. misapplied by] Nutt., Gen. Pl. 1: 57. 1818. Fort Mandan, N. Dak. [Nuttall]" to page 821 and insert after (26) *Aristida longiseta* Steud. 1855. A Pursh specimen from the Lambert Herbarium, in the herbarium of the Academy of Sciences, Philadelphia, named "*Aristida pallens* Cav. ic. 468" and bearing on the back of the sheet in Lambert's script "Louisiana. Bradbury.," was recently examined through the kindness of Prof. J. A. Ewan. It consists of a single flowering culm of *Aristida longiseta* Steud. Many of Bradbury's collections on the Lewis and Clark expedition are marked "Louisiana. Bradbury." only, "Louisiana" at that time being applied to the trans-Mississippi region. It had been assumed that the collection was made by Nuttall. Nuttall had access to the Pursh Herbarium, later purchased by Lambert, Pursh's patron.

Page 850: **Danthonia purpurea** (Thunb.) Beauv. ex Roem. and Schult., Syst. Veg. 2: 690. 1817. Based on *Avena purpurea* Thunb.

Avena purpurea Thunb., Cat. 23. 1794. Cape Good Hope, Africa.

Page 871: *Erianthus giganteus* var. *compactus* Fernald, Rhodora 52: 71. 1950. Based on *E. compactus* Nash.

Page 872: *Eriochloa lemmoni* var. *gracilis* Gould, West. Bot. Leaflets 6: 51. 1950. Based on *Helopus gracilis* Fourn.

Page 931: **Paspalum nicorae** Parodi, Nat. Mus. La Plata (Bot.) 8: 82. 1943. Based on *P. plicatulum* var. *arenarium* Arech., not *P. arenarium* Schrad., 1824.

Paspalum plicatulum var. *arenarium* Arech., Anal. Mus. Nac. Montevideo 1: 58. 1894. Uruguay.

Page 982: *Elymus pilosus* Muhl., Amer. Phil. Soc. Trans. 3: 161. 1793. Pennsylvania.

Page 983: *Panion buckleyanum* var. *maius* Lunell, Amer. Midl. Nat. 4: 222. 1915. Change of name for "*Poa tenuifolia* var. *maior* (Vasey)," but that name was never published, and no specimen so named by Vasey can be found.

Poa viridis Raf., Med. Repos. 5: 353. 1808. Not *Gilib* 1792.

INDEX

[This comprehensive index covers both volumes of the work; volume I contains pages 1 to 570 and volume II contains pages 570 through 1000. Synonyms are in *italic* type. The page numbers of the principal entries are in **heavy-face** type.]

A CATALOGUE OF SELECTED DOVER BOOKS
IN ALL FIELDS OF INTEREST

A CATALOGUE OF SELECTED DOVER
BOOKS IN ALL FIELDS OF INTEREST

CONDITIONED REFLEXES, Ivan P. Pavlov. Full translation of most complete statement of Pavlov's work; cerebral damage, conditioned reflex, experiments with dogs, sleep, similar topics of great importance. 430pp. 5⅜ x 8½.   60614-7 Pa. $4.50

NOTES ON NURSING: WHAT IT IS, AND WHAT IT IS NOT, Florence Nightingale. Outspoken writings by founder of modern nursing. When first published (1860) it played an important role in much needed revolution in nursing. Still stimulating. 140pp. 5⅜ x 8½.   22340-X Pa. $3.00

HARTER'S PICTURE ARCHIVE FOR COLLAGE AND ILLUSTRA-TION, Jim Harter. Over 300 authentic, rare 19th-century engravings selected by noted collagist for artists, designers, decoupeurs, etc. Machines, people, animals, etc., printed one side of page. 25 scene plates for backgrounds. 6 collages by Harter, Satty, Singer, Evans. Introduction. 192pp. 8⅞ x 11¾.   23659-5 Pa. $5.00

MANUAL OF TRADITIONAL WOOD CARVING, edited by Paul N. Hasluck. Possibly the best book in English on the craft of wood carving. Practical instructions, along with 1,146 working drawings and photographic illustrations. Formerly titled Cassell's Wood Carving. 576pp. 6½ x 9¼.
                                                                 23489-4 Pa. $7.95

THE PRINCIPLES AND PRACTICE OF HAND OR SIMPLE TURN-ING, John Jacob Holtzapffel. Full coverage of basic lathe techniques—history and development, special apparatus, softwood turning, hardwood turning, metal turning. Many projects—billiard ball, works formed within a sphere, egg cups, ash trays, vases, jardiniers, others—included. 1881 edition. 800 illustrations. 592pp. 6⅛ x 9¼.   23365-0 Clothbd. $15.00

THE JOY OF HANDWEAVING, Osma Tod. Only book you need for hand weaving. Fundamentals, threads, weaves, plus numerous projects for small board-loom, two-harness, tapestry, laid-in, four-harness weaving and more. Over 160 illustrations. 2nd revised edition. 352pp. 6½ x 9¼.
                                                                 23458-4 Pa. $6.00

THE BOOK OF WOOD CARVING, Charles Marshall Sayers. Still finest book for beginning student in wood sculpture. Noted teacher, craftsman discusses fundamentals, technique; gives 34 designs, over 34 projects for panels, bookends, mirrors, etc. "Absolutely first-rate"—E. J. Tangerman. 33 photos. 118pp. 7¾ x 10⅝.   23654-4 Pa. $3.50

AN AUTOBIOGRAPHY, Margaret Sanger. Exciting personal account of hard-fought battle for woman's right to birth control, against prejudice, church, law. Foremost feminist document. 504pp. 5⅜ x 8½.
20470-7 Pa. $5.50

MY BONDAGE AND MY FREEDOM, Frederick Douglass. Born as a slave, Douglass became outspoken force in antislavery movement. The best of Douglass's autobiographies. Graphic description of slave life. Introduction by P. Foner. 464pp. 5⅜ x 8½. 22457-0 Pa. $5.50

LIVING MY LIFE, Emma Goldman. Candid, no holds barred account by foremost American anarchist: her own life, anarchist movement, famous contemporaries, ideas and their impact. Struggles and confrontations in America, plus deportation to U.S.S.R. Shocking inside account of persecution of anarchists under Lenin. 13 plates. Total of 944pp. 5⅜ x 8½.
22543-7, 22544-5 Pa., Two-vol. set $12.00

LETTERS AND NOTES ON THE MANNERS, CUSTOMS AND CONDITIONS OF THE NORTH AMERICAN INDIANS, George Catlin. Classic account of life among Plains Indians: ceremonies, hunt, warfare, etc. Dover edition reproduces for first time all original paintings. 312 plates. 572pp. of text. 6⅛ x 9¼. 22118-0, 22119-9 Pa.. Two-vol. set $12.00

THE MAYA AND THEIR NEIGHBORS, edited by Clarence L. Hay, others. Synoptic view of Maya civilization in broadest sense, together with Northern, Southern neighbors. Integrates much background, valuable detail not elsewhere. Prepared by greatest scholars: Kroeber, Morley, Thompson, Spinden, Vaillant, many others. Sometimes called Tozzer Memorial Volume. 60 illustrations, linguistic map. 634pp. 5⅜ x 8½.
23510-6 Pa. $10.00

HANDBOOK OF THE INDIANS OF CALIFORNIA, A. L. Kroeber. Foremost American anthropologist offers complete ethnographic study of each group. Monumental classic. 459 illustrations, maps. 995pp. 5⅜ x 8½.
23368-5 Pa. $13.00

SHAKTI AND SHAKTA, Arthur Avalon. First book to give clear, cohesive analysis of Shakta doctrine, Shakta ritual and Kundalini Shakti (yoga). Important work by one of world's foremost students of Shaktic and Tantric thought. 732pp. 5⅜ x 8½. (Available in U.S. only)
23645-5 Pa. $7.95

AN INTRODUCTION TO THE STUDY OF THE MAYA HIEROGLYPHS, Syvanus Griswold Morley. Classic study by one of the truly great figures in hieroglyph research. Still the best introduction for the student for reading Maya hieroglyphs. New introduction by J. Eric S. Thompson. 117 illustrations. 284pp. 5⅜ x 8½. 23108-9 Pa. $4.00

A STUDY OF MAYA ART, Herbert J. Spinden. Landmark classic interprets Maya symbolism, estimates styles, covers ceramics, architecture, murals, stone carvings as artforms. Still a basic book in area. New introduction by J. Eric Thompson. Over 750 illustrations. 341pp. 8⅜ x 11¼.
21235-1 Pa. $6.95

HISTORY OF BACTERIOLOGY, William Bulloch. The only comprehensive history of bacteriology from the beginnings through the 19th century. Special emphasis is given to biography-Leeuwenhoek, etc. Brief accounts of 350 bacteriologists form a separate section. No clearer, fuller study, suitable to scientists and general readers, has yet been written. 52 illustrations. 448pp. 5⅝ x 8¼. 23761-3 Pa. $6.50

THE COMPLETE NONSENSE OF EDWARD LEAR, Edward Lear. All nonsense limericks, zany alphabets, Owl and Pussycat, songs, nonsense botany, etc., illustrated by Lear. Total of 321pp. 5⅜ x 8½. (Available in U.S. only) 20167-8 Pa. $3.95

INGENIOUS MATHEMATICAL PROBLEMS AND METHODS, Louis A. Graham. Sophisticated material from Graham *Dial*, applied and pure; stresses solution methods. Logic, number theory, networks, inversions, etc. 237pp. 5⅜ x 8½. 20545-2 Pa. $4.50

BEST MATHEMATICAL PUZZLES OF SAM LOYD, edited by Martin Gardner. Bizarre, original, whimsical puzzles by America's greatest puzzler. From fabulously rare *Cyclopedia*, including famous 14-15 puzzles, the Horse of a Different Color, 115 more. Elementary math. 150 illustrations. 167pp. 5⅜ x 8½. 20498-7 Pa. $2.75

THE BASIS OF COMBINATION IN CHESS, J. du Mont. Easy-to-follow, instructive book on elements of combination play, with chapters on each piece and every powerful combination team—two knights, bishop and knight, rook and bishop, etc. 250 diagrams. 218pp. 5⅜ x 8½. (Available in U.S. only) 23644-7 Pa. $3.50

MODERN CHESS STRATEGY, Ludek Pachman. The use of the queen, the active king, exchanges, pawn play, the center, weak squares, etc. Section on rook alone worth price of the book. Stress on the moderns. Often considered the most important book on strategy. 314pp. 5⅜ x 8½. 20290-9 Pa. $4.50

LASKER'S MANUAL OF CHESS, Dr. Emanuel Lasker. Great world champion offers very thorough coverage of all aspects of chess. Combinations, position play, openings, end game, aesthetics of chess, philosophy of struggle, much more. Filled with analyzed games. 390pp. 5⅜ x 8½. 20640-8 Pa. $5.00

500 MASTER GAMES OF CHESS, S. Tartakower, J. du Mont. Vast collection of great chess games from 1798-1938, with much material nowhere else readily available. Fully annotated, arranged by opening for easier study. 664pp. 5⅜ x 8½. 23208-5 Pa. $7.50

A GUIDE TO CHESS ENDINGS, Dr. Max Euwe, David Hooper. One of the finest modern works on chess endings. Thorough analysis of the most frequently encountered endings by former world champion. 331 examples, each with diagram. 248pp. 5⅜ x 8½. 23332-4 Pa. $3.75

SECOND PIATIGORSKY CUP, edited by Isaac Kashdan. One of the greatest tournament books ever produced in the English language. All 90 games of the 1966 tournament, annotated by players, most annotated by both players. Features Petrosian, Spassky, Fischer, Larsen, six others. 228pp. 5⅜ x 8½. 23572-6 Pa. $3.50

ENCYCLOPEDIA OF CARD TRICKS, revised and edited by Jean Hugard. How to perform over 600 card tricks, devised by the world's greatest magicians: impromptus, spelling tricks, key cards, using special packs, much, much more. Additional chapter on card technique. 66 illustrations. 402pp. 5⅜ x 8½. (Available in U.S. only) 21252-1 Pa. $4.95

MAGIC: STAGE ILLUSIONS, SPECIAL EFFECTS AND TRICK PHOTOGRAPHY, Albert A. Hopkins, Henry R. Evans. One of the great classics; fullest, most authorative explanation of vanishing lady, levitations, scores of other great stage effects. Also small magic, automata, stunts. 446 illustrations. 556pp. 5⅜ x 8½. 23344-8 Pa. $6.95

THE SECRETS OF HOUDINI, J. C. Cannell. Classic study of Houdini's incredible magic, exposing closely-kept professional secrets and revealing, in general terms, the whole art of stage magic. 67 illustrations. 279pp. 5⅜ x 8½. 22913-0 Pa. $4.00

HOFFMANN'S MODERN MAGIC, Professor Hoffmann. One of the best, and best-known, magicians' manuals of the past century. Hundreds of tricks from card tricks and simple sleight of hand to elaborate illusions involving construction of complicated machinery. 332 illustrations. 563pp. 5⅜ x 8½. 23623-4 Pa. $6.00

MADAME PRUNIER'S FISH COOKERY BOOK, Mme. S. B. Prunier. More than 1000 recipes from world famous Prunier's of Paris and London, specially adapted here for American kitchen. Grilled tournedos with anchovy butter, Lobster a la Bordelaise, Prunier's prized desserts, more. Glossary. 340pp. 5⅜ x 8½. (Available in U.S. only) 22679-4 Pa. $3.00

FRENCH COUNTRY COOKING FOR AMERICANS, Louis Diat. 500 easy-to-make, authentic provincial recipes compiled by former head chef at New York's Fitz-Carlton Hotel: onion soup, lamb stew, potato pie, more. 309pp. 5⅜ x 8½. 23665-X Pa. $3.95

SAUCES, FRENCH AND FAMOUS, Louis Diat. Complete book gives over 200 specific recipes: bechamel, Bordelaise, hollandaise, Cumberland, apricot, etc. Author was one of this century's finest chefs, originator of vichyssoise and many other dishes. Index. 156pp. 5⅜ x 8. 23663-3 Pa. $2.75

TOLL HOUSE TRIED AND TRUE RECIPES, Ruth Graves Wakefield. Authentic recipes from the famous Mass. restaurant: popovers, veal and ham loaf, Toll House baked beans, chocolate cake crumb pudding, much more. Many helpful hints. Nearly 700 recipes. Index. 376pp. 5⅜ x 8½. 23560-2 Pa. $4.50

A MAYA GRAMMAR, Alfred M. Tozzer. Practical, useful English-language grammar by the Harvard anthropologist who was one of the three greatest American scholars in the area of Maya culture. Phonetics, grammatical processes, syntax, more. 301pp. 5⅜ x 8½. 23465-7 Pa. $4.00

THE JOURNAL OF HENRY D. THOREAU, edited by Bradford Torrey, F. H. Allen. Complete reprinting of 14 volumes, 1837-61, over two million words; the sourcebooks for *Walden*, etc. Definitive. All original sketches, plus 75 photographs. Introduction by Walter Harding. Total of 1804pp. 8½ x 12¼. 20312-3, 20313-1 Clothbd., Two-vol. set $70.00

CLASSIC GHOST STORIES, Charles Dickens and others. 18 wonderful stories you've wanted to reread: "The Monkey's Paw," "The House and the Brain," "The Upper Berth," "The Signalman," "Dracula's Guest," "The Tapestried Chamber," etc. Dickens, Scott, Mary Shelley, Stoker, etc. 330pp. 5⅜ x 8½. 20735-8 Pa. $4.50

SEVEN SCIENCE FICTION NOVELS, H. G. Wells. Full novels. *First Men in the Moon, Island of Dr. Moreau, War of the Worlds, Food of the Gods, Invisible Man, Time Machine, In the Days of the Comet.* A basic science-fiction library. 1015pp. 5⅜ x 8½. (Available in U.S. only) 20264-X Clothbd. $8.95

ARMADALE, Wilkie Collins. Third great mystery novel by the author of *The Woman in White* and *The Moonstone*. Ingeniously plotted narrative shows an exceptional command of character, incident and mood. Original magazine version with 40 illustrations. 597pp. 5⅜ x 8½. 23429-0 Pa. $6.00

MASTERS OF MYSTERY, H. Douglas Thomson. The first book in English (1931) devoted to history and aesthetics of detective story. Poe, Doyle, LeFanu, Dickens, many others, up to 1930. New introduction and notes by E. F. Bleiler. 288pp. 5⅜ x 8½. (Available in U.S. only) 23606-4 Pa. $4.00

FLATLAND, E. A. Abbott. Science-fiction classic explores life of 2-D being in 3-D world. Read also as introduction to thought about hyperspace. Introduction by Banesh Hoffmann. 16 illustrations. 103pp. 5⅜ x 8½. 20001-9 Pa. $2.00

THREE SUPERNATURAL NOVELS OF THE VICTORIAN PERIOD, edited, with an introduction, by E. F. Bleiler. Reprinted complete and unabridged, three great classics of the supernatural: *The Haunted Hotel* by Wilkie Collins, *The Haunted House at Latchford* by Mrs. J. H. Riddell, and *The Lost Stradivarius* by J. Meade Falkner. 325pp. 5⅜ x 8½. 22571-2 Pa. $4.00

AYESHA: THE RETURN OF "SHE," H. Rider Haggard. Virtuoso sequel featuring the great mythic creation, Ayesha, in an adventure that is fully as good as the first book, *She*. Original magazine version, with 47 original illustrations by Maurice Greiffenhagen. 189pp. 6½ x 9¼. 23649-8 Pa. $3.50

CATALOGUE OF DOVER BOOKS

PRINCIPLES OF ORCHESTRATION, Nikolay Rimsky-Korsakov. Great classical orchestrator provides fundamentals of tonal resonance, progression of parts, voice and orchestra, tutti effects, much else in major document. 330pp. of musical excerpts. 489pp. 6½ x 9¼.　21266-1 Pa. $7.50

TRISTAN UND ISOLDE, Richard Wagner. Full orchestral score with complete instrumentation. Do not confuse with piano reduction. Commentary by Felix Mottl, great Wagnerian conductor and scholar. Study score. 655pp. 8⅛ x 11.　22915-7 Pa. $13.95

REQUIEM IN FULL SCORE, Giuseppe Verdi. Immensely popular with choral groups and music lovers. Republication of edition published by C. F. Peters, Leipzig, n. d. German frontmaker in English translation. Glossary. Text in Latin. Study score. 204pp. 9⅜ x 12¼.
23682-X Pa. $6.00

COMPLETE CHAMBER MUSIC FOR STRINGS, Felix Mendelssohn. All of Mendelssohn's chamber music: Octet, 2 Quintets, 6 Quartets, and Four Pieces for String Quartet. (Nothing with piano is included). Complete works edition (1874-7). Study score. 283 pp. 9⅜ x 12¼.
23679-X Pa. $7.50

POPULAR SONGS OF NINETEENTH-CENTURY AMERICA, edited by Richard Jackson. 64 most important songs: "Old Oaken Bucket," "Arkansas Traveler," "Yellow Rose of Texas," etc. Authentic original sheet music, full introduction and commentaries. 290pp. 9 x 12.　23270-0 Pa. $7.95

COLLECTED PIANO WORKS, Scott Joplin. Edited by Vera Brodsky Lawrence. Practically all of Joplin's piano works—rags, two-steps, marches, waltzes, etc., 51 works in all. Extensive introduction by Rudi Blesh. Total of 345pp. 9 x 12.　23106-2 Pa. $14.95

BASIC PRINCIPLES OF CLASSICAL BALLET, Agrippina Vaganova. Great Russian theoretician, teacher explains methods for teaching classical ballet; incorporates best from French, Italian, Russian schools. 118 illustrations. 175pp. 5⅜ x 8½.　22036-2 Pa. $2.50

CHINESE CHARACTERS, L. Wieger. Rich analysis of 2300 characters according to traditional systems into primitives. Historical-semantic analysis to phonetics (Classical Mandarin) and radicals. 820pp. 6⅛ x 9¼.
21321-8 Pa. $10.00

EGYPTIAN LANGUAGE: EASY LESSONS IN EGYPTIAN HIERO-GLYPHICS, E. A. Wallis Budge. Foremost Egyptologist offers Egyptian grammar, explanation of hieroglyphics, many reading texts, dictionary of symbols. 246pp. 5 x 7½. (Available in U.S. only)
21394-3 Clothbd. $7.50

AN ETYMOLOGICAL DICTIONARY OF MODERN ENGLISH, Ernest Weekley. Richest, fullest work, by foremost British lexicographer. Detailed word histories. Inexhaustible. Do not confuse this with *Concise Etymological Dictionary*, which is abridged. Total of 856pp. 6½ x 9¼.
21873-2, 21874-0 Pa., Two-vol. set $12.00

YUCATAN BEFORE AND AFTER THE CONQUEST, Diego de Landa. First English translation of basic book in Maya studies, the only significant account of Yucatan written in the early post-Conquest era. Translated by distinguished Maya scholar William Gates. Appendices, introduction, 4 maps and over 120 illustrations added by translator. 162pp. 5⅜ x 8½.
23622-6 Pa. $3.00

THE MALAY ARCHIPELAGO, Alfred R. Wallace. Spirited travel account by one of founders of modern biology. Touches on zoology, botany, ethnography, geography, and geology. 62 illustrations, maps. 515pp. 5⅜ x 8½.
20187-2 Pa. $6.95

THE DISCOVERY OF THE TOMB OF TUTANKHAMEN, Howard Carter, A. C. Mace. Accompany Carter in the thrill of discovery, as ruined passage suddenly reveals unique, untouched, fabulously rich tomb. Fascinating account, with 106 illustrations. New introduction by J. M. White. Total of 382pp. 5⅜ x 8½. (Available in U.S. only)   23500-9 Pa. $4.00

THE WORLD'S GREATEST SPEECHES, edited by Lewis Copeland and Lawrence W. Lamm. Vast collection of 278 speeches from Greeks up to present. Powerful and effective models; unique look at history. Revised to 1970. Indices. 842pp. 5⅜ x 8½.   20468-5 Pa. $8.95

THE 100 GREATEST ADVERTISEMENTS, Julian Watkins. The priceless ingredient; His master's voice; 99 44/100% pure; over 100 others. How they were written, their impact, etc. Remarkable record. 130 illustrations. 233pp. 7⅞ x 10 3/5.   20540-1 Pa. $5.95

CRUICKSHANK PRINTS FOR HAND COLORING, George Cruickshank. 18 illustrations, one side of a page, on fine-quality paper suitable for watercolors. Caricatures of people in society (c. 1820) full of trenchant wit. Very large format. 32pp. 11 x 16.   23684-6 Pa. $5.00

THIRTY-TWO COLOR POSTCARDS OF TWENTIETH-CENTURY AMERICAN ART, Whitney Museum of American Art. Reproduced in full color in postcard form are 31 art works and one shot of the museum. Calder, Hopper, Rauschenberg, others. Detachable. 16pp. 8¼ x 11.
23629-3 Pa. $3.00

MUSIC OF THE SPHERES: THE MATERIAL UNIVERSE FROM ATOM TO QUASAR SIMPLY EXPLAINED, Guy Murchie. Planets, stars, geology, atoms, radiation, relativity, quantum theory, light, antimatter, similar topics. 319 figures. 664pp. 5⅜ x 8½.
21809-0, 21810-4 Pa., Two-vol. set $11.00

EINSTEIN'S THEORY OF RELATIVITY, Max Born. Finest semi-technical account; covers Einstein, Lorentz, Minkowski, and others, with much detail, much explanation of ideas and math not readily available elsewhere on this level. For student, non-specialist. 376pp. 5⅜ x 8½.
60769-0 Pa. $4.50

THE EARLY WORK OF AUBREY BEARDSLEY, Aubrey Beardsley. 157 plates, 2 in color: *Manon Lescaut, Madame Bovary, Morte Darthur, Salome,* other. Introduction by H. Marillier. 182pp. 8⅛ x 11. 21816-3 Pa. $4.50

THE LATER WORK OF AUBREY BEARDSLEY, Aubrey Beardsley. Exotic masterpieces of full maturity: *Venus and Tannhauser, Lysistrata, Rape of the Lock, Volpone,* Savoy material, etc. 174 plates, 2 in color. 186pp. 8⅛ x 11. 21817-1 Pa. $5.95

THOMAS NAST'S CHRISTMAS DRAWINGS, Thomas Nast. Almost all Christmas drawings by creator of image of Santa Claus as we know it, and one of America's foremost illustrators and political cartoonists. 66 illustrations. 3 illustrations in color on covers. 96pp. 8⅜ x 11¼. 23660-9 Pa. $3.50

THE DORÉ ILLUSTRATIONS FOR DANTE'S DIVINE COMEDY, Gustave Doré. All 135 plates from Inferno, Purgatory, Paradise; fantastic tortures, infernal landscapes, celestial wonders. Each plate with appropriate (translated) verses. 141pp. 9 x 12. 23231-X Pa. $4.50

DORÉ'S ILLUSTRATIONS FOR RABELAIS, Gustave Doré. 252 striking illustrations of *Gargantua and Pantagruel* books by foremost 19th-century illustrator. Including 60 plates, 192 delightful smaller illustrations. 153pp. 9 x 12. 23656-0 Pa. $5.00

LONDON: A PILGRIMAGE, Gustave Doré, Blanchard Jerrold. Squalor, riches, misery, beauty of mid-Victorian metropolis; 55 wonderful plates, 125 other illustrations, full social, cultural text by Jerrold. 191pp. of text. 9⅜ x 12¼. 22306-X Pa. $7.00

THE RIME OF THE ANCIENT MARINER, Gustave Doré, S. T. Coleridge. Dore's finest work, 34 plates capture moods, subtleties of poem. Full text. Introduction by Millicent Rose. 77pp. 9¼ x 12. 22305-1 Pa. $3.50

THE DORE BIBLE ILLUSTRATIONS, Gustave Doré. All wonderful, detailed plates: Adam and Eve, Flood, Babylon, Life of Jesus, etc. Brief King James text with each plate. Introduction by Millicent Rose. 241 plates. 241pp. 9 x 12. 23004-X Pa. $6.00

THE COMPLETE ENGRAVINGS, ETCHINGS AND DRYPOINTS OF ALBRECHT DURER. "Knight, Death and Devil"; "Melencolia," and more—all Dürer's known works in all three media, including 6 works formerly attributed to him. 120 plates. 235pp. 8⅜ x 11¼. 22851-7 Pa. $6.50

MECHANICK EXERCISES ON THE WHOLE ART OF PRINTING, Joseph Moxon. First complete book (1683-4) ever written about typography, a compendium of everything known about printing at the latter part of 17th century. Reprint of 2nd (1962) Oxford Univ. Press edition. 74 illustrations. Total of 550pp. 6⅛ x 9¼. 23617-X Pa. $7.95

THE ANATOMY OF THE HORSE, George Stubbs. Often considered the great masterpiece of animal anatomy. Full reproduction of 1766 edition, plus prospectus; original text and modernized text. 36 plates. Introduction by Eleanor Garvey. 121pp. 11 x 14¾. 23402-9 Pa. $6.00

BRIDGMAN'S LIFE DRAWING, George B. Bridgman. More than 500 illustrative drawings and text teach you to abstract the body into its major masses, use light and shade, proportion; as well as specific areas of anatomy, of which Bridgman is master. 192pp. 6½ x 9¼. (Available in U.S. only) 22710-3 Pa. $3.50

ART NOUVEAU DESIGNS IN COLOR, Alphonse Mucha, Maurice Verneuil, Georges Auriol. Full-color reproduction of *Combinaisons ornementales* (c. 1900) by Art Nouveau masters. Floral, animal, geometric, interlacings, swashes—borders, frames, spots—all incredibly beautiful. 60 plates, hundreds of designs. 9⅜ x 8-1/16. 22885-1 Pa. $4.00

FULL-COLOR FLORAL DESIGNS IN THE ART NOUVEAU STYLE, E. A. Seguy. 166 motifs, on 40 plates, from *Les fleurs et leurs applications decoratives* (1902): borders, circular designs, repeats, allovers, "spots." All in authentic Art Nouveau colors. 48pp. 9⅜ x 12¼.
23439-8 Pa. $5.00

A DIDEROT PICTORIAL ENCYCLOPEDIA OF TRADES AND IN-DUSTRY, edited by Charles C. Gillispie. 485 most interesting plates from the great French Encyclopedia of the 18th century show hundreds of working figures, artifacts, process, land and cityscapes; glassmaking, paper-making, metal extraction, construction, weaving, making furniture, clothing, wigs, dozens of other activities. Plates fully explained. 920pp. 9 x 12.
22284-5, 22285-3 Clothbd., Two-vol. set $40.00

HANDBOOK OF EARLY ADVERTISING ART, Clarence P. Hornung. Largest collection of copyright-free early and antique advertising art ever compiled. Over 6,000 illustrations, from Franklin's time to the 1890's for special effects, novelty. Valuable source, almost inexhaustible.
Pictorial Volume. Agriculture, the zodiac, animals, autos, birds, Christmas, fire engines, flowers, trees, musical instruments, ships, games and sports, much more. Arranged by subject matter and use. 237 plates. 288pp. 9 x 12.
20122-8 Clothbd. $14.50

Typographical Volume. Roman and Gothic faces ranging from 10 point to 300 point, "Barnum," German and Old English faces, script, logotypes, scrolls and flourishes, 1115 ornamental initials, 67 complete alphabets, more. 310 plates. 320pp. 9 x 12. 20123-6 Clothbd. $15.00

CALLIGRAPHY (CALLIGRAPHIA LATINA), J. G. Schwandner. High point of 18th-century ornamental calligraphy. Very ornate initials, scrolls, borders, cherubs, birds, lettered examples. 172pp. 9 x 13.
20475-8 Pa. $7.00

THE COMPLETE WOODCUTS OF ALBRECHT DURER, edited by Dr. W. Kurth. 346 in all: "Old Testament," "St. Jerome," "Passion," "Life of Virgin," Apocalypse," many others. Introduction by Campbell Dodgson. 285pp. 8½ x 12¼. 21097-9 Pa. $7.50

DRAWINGS OF ALBRECHT DURER, edited by Heinrich Wölfflin. 81 plates show development from youth to full style. Many favorites; many new. Introduction by Alfred Werner. 96pp. 8⅛ x 11. 22352-3 Pa. $5.00

THE HUMAN FIGURE, Albrecht Dürer. Experiments in various techniques—stereometric, progressive proportional, and others. Also life studies that rank among finest ever done. Complete reprinting of *Dresden Sketchbook*. 170 plates. 355pp. 8⅜ x 11¼. 21042-1 Pa. $7.95

OF THE JUST SHAPING OF LETTERS, Albrecht Dürer. Renaissance artist explains design of Roman majuscules by geometry, also Gothic lower and capitals. Grolier Club edition. 43pp. 7⅞ x 10¾ 21306-4 Pa. $3.00

TEN BOOKS ON ARCHITECTURE, Vitruvius. The most important book ever written on architecture. Early Roman aesthetics, technology, classical orders, site selection, all other aspects. Stands behind everything since. Morgan translation. 331pp. 5⅜ x 8½. 20645-9 Pa. $4.50

THE FOUR BOOKS OF ARCHITECTURE, Andrea Palladio. 16th-century classic responsible for Palladian movement and style. Covers classical architectural remains, Renaissance revivals, classical orders, etc. 1738 Ware English edition. Introduction by A. Placzek. 216 plates. 110pp. of text. 9½ x 12¾. 21308-0 Pa. $10.00

HORIZONS, Norman Bel Geddes. Great industrialist stage designer, "father of streamlining," on application of aesthetics to transportation, amusement, architecture, etc. 1932 prophetic account; function, theory, specific projects. 222 illustrations. 312pp. 7⅞ x 10¾. 23514-9 Pa. $6.95

FRANK LLOYD WRIGHT'S FALLINGWATER, Donald Hoffmann. Full, illustrated story of conception and building of Wright's masterwork at Bear Run, Pa. 100 photographs of site, construction, and details of completed structure. 112pp. 9¼ x 10. 23671-4 Pa. $5.50

THE ELEMENTS OF DRAWING, John Ruskin. Timeless classic by great Viltorian; starts with basic ideas, works through more difficult. Many practical exercises. 48 illustrations. Introduction by Lawrence Campbell. 228pp. 5⅜ x 8½. 22730-8 Pa. $3.75

GIST OF ART, John Sloan. Greatest modern American teacher, Art Students League, offers innumerable hints, instructions, guided comments to help you in painting. Not a formal course. 46 illustrations. Introduction by Helen Sloan. 200pp. 5⅜ x 8½. 23435-5 Pa. $4.00

THE DEPRESSION YEARS AS PHOTOGRAPHED BY ARTHUR ROTH-STEIN, Arthur Rothstein. First collection devoted entirely to the work of outstanding 1930s photographer: famous dust storm photo, ragged children, unemployed, etc. 120 photographs. Captions. 119pp. 9¼ x 10¾.
23590-4 Pa. $5.00

CAMERA WORK: A PICTORIAL GUIDE, Alfred Stieglitz. All 559 illustrations and plates from the most important periodical in the history of art photography, Camera Work (1903-17). Presented four to a page, reduced in size but still clear, in strict chronological order, with complete captions. Three indexes. Glossary. Bibliography. 176pp. 8⅜ x 11¼.
23591-2 Pa. $6.95

ALVIN LANGDON COBURN, PHOTOGRAPHER, Alvin L. Coburn. Revealing autobiography by one of greatest photographers of 20th century gives insider's version of Photo-Secession, plus comments on his own work. 77 photographs by Coburn. Edited by Helmut and Alison Gernsheim. 160pp. 8⅛ x 11.
23685-4 Pa. $6.00

NEW YORK IN THE FORTIES, Andreas Feininger. 162 brilliant photographs by the well-known photographer, formerly with Life magazine, show commuters, shoppers, Times Square at night, Harlem nightclub, Lower East Side, etc. Introduction and full captions by John von Hartz. 181pp. 9¼ x 10¾.
23585-8 Pa. $6.95

GREAT NEWS PHOTOS AND THE STORIES BEHIND THEM, John Faber. Dramatic volume of 140 great news photos, 1855 through 1976, and revealing stories behind them, with both historical and technical information. Hindenburg disaster, shooting of Oswald, nomination of Jimmy Carter, etc. 160pp. 8¼ x 11.
23667-6 Pa. $5.00

THE ART OF THE CINEMATOGRAPHER, Leonard Maltin. Survey of American cinematography history and anecdotal interviews with 5 masters—Arthur Miller, Hal Mohr, Hal Rosson, Lucien Ballard, and Conrad Hall. Very large selection of behind-the-scenes production photos. 105 photographs. Filmographies. Index. Originally Behind the Camera. 144pp. 8¼ x 11.
23686-2 Pa. $5.00

DESIGNS FOR THE THREE-CORNERED HAT (LE TRICORNE), Pablo Picasso. 32 fabulously rare drawings—including 31 color illustrations of costumes and accessories—for 1919 production of famous ballet. Edited by Parmenia Migel, who has written new introduction. 48pp. 9⅜ x 12¼. (Available in U.S. only)
23709-5 Pa. $5.00

NOTES OF A FILM DIRECTOR, Sergei Eisenstein. Greatest Russian filmmaker explains montage, making of Alexander Nevsky, aesthetics; comments on self, associates, great rivals (Chaplin), similar material. 78 illustrations. 240pp. 5⅜ x 8½.
22392-2 Pa. $4.50

HOLLYWOOD GLAMOUR PORTRAITS, edited by John Kobal. 145 photos capture the stars from 1926-49, the high point in portrait photography. Gable, Harlow, Bogart, Bacall, Hedy Lamarr, Marlene Dietrich, Robert Montgomery, Marlon Brando, Veronica Lake; 94 stars in all. Full background on photographers, technical aspects, much more. Total of 160pp. 8⅜ x 11¼. 23352-9 Pa. $6.00

THE NEW YORK STAGE: FAMOUS PRODUCTIONS IN PHOTO-GRAPHS, edited by Stanley Appelbaum. 148 photographs from Museum of City of New York show 142 plays, 1883-1939. *Peter Pan, The Front Page, Dead End, Our Town,* O'Neill, hundreds of actors and actresses, etc. Full indexes. 154pp. 9½ x 10. 23241-7 Pa. $6.00

DIALOGUES CONCERNING TWO NEW SCIENCES, Galileo Galilei. Encompassing 30 years of experiment and thought, these dialogues deal with geometric demonstrations of fracture of solid bodies, cohesion, leverage, speed of light and sound, pendulums, falling bodies, accelerated motion, etc. 300pp. 5⅜ x 8½. 60099-8 Pa. $4.00

THE GREAT OPERA STARS IN HISTORIC PHOTOGRAPHS, edited by James Camner. 343 portraits from the 1850s to the 1940s: Tamburini, Mario, Caliapin, Jeritza, Melchior, Melba, Patti, Pinza, Schipa, Caruso, Farrar, Steber, Gobbi, and many more—270 performers in all. Index. 199pp. 8⅜ x 11¼. 23575-0 Pa. $7.50

J. S. BACH, Albert Schweitzer. Great full-length study of Bach, life, background to music, music, by foremost modern scholar. Ernest Newman translation. 650 musical examples. Total of 928pp. 5⅜ x 8½. (Available in U.S. only) 21631-4, 21632-2 Pa., Two-vol. set $11.00

COMPLETE PIANO SONATAS, Ludwig van Beethoven. All sonatas in the fine Schenker edition, with fingering, analytical material. One of best modern editions. Total of 615pp. 9 x 12. (Available in U.S. only) 23134-8, 23135-6 Pa., Two-vol. set $15.50

KEYBOARD MUSIC, J. S. Bach. Bach-Gesellschaft edition. For harpsichord, piano, other keyboard instruments. English Suites, French Suites, Six Partitas, Goldberg Variations, Two-Part Inventions, Three-Part Sinfonias. 312pp. 8⅛ x 11. (Available in U.S. only) 22360-4 Pa. $6.95

FOUR SYMPHONIES IN FULL SCORE, Franz Schubert. Schubert's four most popular symphonies: No. 4 in C Minor ("Tragic"); No. 5 in B-flat Major; No. 8 in B Minor ("Unfinished"); No. 9 in C Major ("Great"). Breitkopf & Hartel edition. Study score. 261pp. 9⅜ x 12¼. 23681-1 Pa. $6.50

THE AUTHENTIC GILBERT & SULLIVAN SONGBOOK, W. S. Gilbert, A. S. Sullivan. Largest selection available; 92 songs, uncut, original keys, in piano rendering approved by Sullivan. Favorites and lesser-known fine numbers. Edited with plot synopses by James Spero. 3 illustrations. 399pp. 9 x 12. 23482-7 Pa. $9.95

AMERICAN BIRD ENGRAVINGS, Alexander Wilson et al. All 76 plates. from Wilson's *American Ornithology* (1808-14), most important ornithological work before Audubon, plus 27 plates from the supplement (1825-33) by Charles Bonaparte. Over 250 birds portrayed. 8 plates also reproduced in full color. 111pp. 9⅜ x 12½. 23195-X Pa. $6.00

CRUICKSHANK'S PHOTOGRAPHS OF BIRDS OF AMERICA, Allan D. Cruickshank. Great ornithologist, photographer presents 177 closeups, groupings, panoramas, flightings, etc., of about 150 different birds. Expanded *Wings in the Wilderness*. Introduction by Helen G. Cruickshank. 191pp. 8¼ x 11. 23497-5 Pa. $6.00

AMERICAN WILDLIFE AND PLANTS, A. C. Martin, et al. Describes food habits of more than 1000 species of mammals, birds, fish. Special treatment of important food plants. Over 300 illustrations. 500pp. 5⅜ x 8½. 20793-5 Pa. $4.95

THE PEOPLE CALLED SHAKERS, Edward D. Andrews. Lifetime of research, definitive study of Shakers: origins, beliefs, practices, dances, social organization, furniture and crafts, impact on 19th-century USA, present heritage. Indispensable to student of American history, collector. 33 illustrations. 351pp. 5⅜ x 8½. 21081-2 Pa. $4.50

OLD NEW YORK IN EARLY PHOTOGRAPHS, Mary Black. New York City as it was in 1853-1901, through 196 wonderful photographs from N.-Y. Historical Society. Great Blizzard, Lincoln's funeral procession, great buildings. 228pp. 9 x 12. 22907-6 Pa. $8.95

MR. LINCOLN'S CAMERA MAN: MATHEW BRADY, Roy Meredith. Over 300 Brady photos reproduced directly from original negatives, photos. Jackson, Webster, Grant, Lee, Carnegie, Barnum; Lincoln; Battle Smoke, Death of Rebel Sniper, Atlanta Just After Capture. Lively commentary. 368pp. 8⅜ x 11¼. 23021-X Pa. $8.95

TRAVELS OF WILLIAM BARTRAM, William Bartram. From 1773-8, Bartram explored Northern Florida, Georgia, Carolinas, and reported on wild life, plants, Indians, early settlers. Basic account for period, entertaining reading. Edited by Mark Van Doren. 13 illustrations. 141pp. 5⅜ x 8½. 20013-2 Pa. $5.00

THE GENTLEMAN AND CABINET MAKER'S DIRECTOR, Thomas Chippendale. Full reprint, 1762 style book, most influential of all time; chairs, tables, sofas, mirrors, cabinets, etc. 200 plates, plus 24 photographs of surviving pieces. 249pp. 9⅞ x 12¾. 21601-2 Pa. $7.95

AMERICAN CARRIAGES, SLEIGHS, SULKIES AND CARTS, edited by Don H. Berkebile. 168 Victorian illustrations from catalogues, trade journals, fully captioned. Useful for artists. Author is Assoc. Curator, Div. of Transportation of Smithsonian Institution. 168pp. 8½ x 9½.
23328-6 Pa. $5.00

"OSCAR" OF THE WALDORF'S COOKBOOK, Oscar Tschirky. Famous American chef reveals 3455 recipes that made Waldorf great; cream of French, German, American cooking, in all categories. Full instructions, easy home use. 1896 edition. 907pp. 6⅝ x 9⅜.   20790-0 Clothbd. $15.00

COOKING WITH BEER, Carole Fahy. Beer has as superb an effect on food as wine, and at fraction of cost. Over 250 recipes for appetizers, soups, main dishes, desserts, breads, etc. Index. 144pp. 5⅜ x 8½. (Available in U.S. only)   23661-7 Pa. $2.50

STEWS AND RAGOUTS, Kay Shaw Nelson. This international cookbook offers wide range of 108 recipes perfect for everyday, special occasions, meals-in-themselves, main dishes. Economical, nutritious, easy-to-prepare: goulash, Irish stew, boeuf bourguignon, etc. Index. 134pp. 5⅜ x 8½.
23662-5 Pa. $2.50

DELICIOUS MAIN COURSE DISHES, Marian Tracy. Main courses are the most important part of any meal. These 200 nutritious, economical recipes from around the world make every meal a delight. "I . . . have found it so useful in my own household,"—N.Y. Times. Index. 219pp. 5⅜ x 3½.   23664-1 Pa. $3.00

FIVE ACRES AND INDEPENDENCE, Maurice G. Kains. Great back-to-the-land classic explains basics of self-sufficient farming: economics, plants, crops, animals, orchards, soils, land selection, host of other necessary things. Do not confuse with skimpy faddist literature; Kains was one of America's greatest agriculturalists. 95 illustrations. 397pp. 5⅜ x 8½.
20974-1 Pa. $3.95

A PRACTICAL GUIDE FOR THE BEGINNING FARMER, Herbert Jacobs. Basic, extremely useful first book for anyone thinking about moving to the country and starting a farm. Simpler than Kains, with greater emphasis on country living in general. 246pp. 5⅜ x 8½.
23675-7 Pa. $3.50

PAPERMAKING, Dard Hunter. Definitive book on the subject by the foremost authority in the field. Chapters dealing with every aspect of history of craft in every part of the world. Over 320 illustrations. 2nd, revised and enlarged (1947) edition. 672pp. 5⅜ x 8½.   23619-6 Pa. $7.95

THE ART DECO STYLE, edited by Theodore Menten. Furniture, jewelry, metalwork, ceramics, fabrics, lighting fixtures, interior decors, exteriors, graphics from pure French sources. Best sampling around. Over 400 photographs. 183pp. 8⅜ x 11¼.   22824-X Pa. $6.00

ACKERMANN'S COSTUME PLATES, Rudolph Ackermann. Selection of 96 plates from the Repository of Arts, best published source of costume for English fashion during the early 19th century. 12 plates also in color. Captions, glossary and introduction by editor Stella Blum. Total of 120pp. 8⅜ x 11¼.   23690-0 Pa. $4.50

UNCLE SILAS, J. Sheridan LeFanu. Victorian Gothic mystery novel, considered by many best of period, even better than Collins or Dickens. Wonderful psychological terror. Introduction by Frederick Shroyer. 436pp. 5⅜ x 8½.   21715-9 Pa. $6.00

JURGEN, James Branch Cabell. The great erotic fantasy of the 1920's that delighted thousands, shocked thousands more. Full final text, Lane edition with 13 plates by Frank Pape. 346pp. 5⅜ x 8½.
                                                                              23507-6 Pa. $4.50

THE CLAVERINGS, Anthony Trollope. Major novel, chronicling aspects of British Victorian society, personalities. Reprint of Cornhill serialization, 16 plates by M. Edwards; first reprint of full text. Introduction by Norman Donaldson. 412pp. 5⅜ x 8½.   23464-9 Pa. $5.00

KEPT IN THE DARK, Anthony Trollope. Unusual short novel about Victorian morality and abnormal psychology by the great English author. Probably the first American publication. Frontispiece by Sir John Millais. 92pp. 6½ x 9¼.   23609-9 Pa. $2.50

RALPH THE HEIR, Anthony Trollope. Forgotten tale of illegitimacy, inheritance. Master novel of Trollope's later years. Victorian country estates, clubs, Parliament, fox hunting, world of fully realized characters. Reprint of 1871 edition. 12 illustrations by F. A. Faser. 434pp. of text. 5⅜ x 8½.   23642-0 Pa. $5.00

YEKL and THE IMPORTED BRIDEGROOM AND OTHER STORIES OF THE NEW YORK GHETTO, Abraham Cahan. Film *Hester Street* based on *Yekl* (1896). Novel, other stories among first about Jewish immigrants of N.Y.'s East Side. Highly praised by W. D. Howells—Cahan "a new star of realism." New introduction by Bernard G. Richards. 240pp. 5⅜ x 8½.   22427-9 Pa. $3.50

THE HIGH PLACE, James Branch Cabell. Great fantasy writer's enchanting comedy of disenchantment set in 18th-century France. Considered by some critics to be even better than his famous *Jurgen*. 10 illustrations and numerous vignettes by noted fantasy artist Frank C. Pape. 320pp. 5⅜ x 8½.   23670-6 Pa. $4.00

ALICE'S ADVENTURES UNDER GROUND, Lewis Carroll. Facsimile of ms. Carroll gave Alice Liddell in 1864. Different in many ways from final Alice. Handlettered, illustrated by Carroll. Introduction by Martin Gardner. 128pp. 5⅜ x 8½.   21482-6 Pa. $2.50

FAVORITE ANDREW LANG FAIRY TALE BOOKS IN MANY COLORS, Andrew Lang. The four Lang favorites in a boxed set—the complete *Red, Green, Yellow* and *Blue* Fairy Books. 164 stories; 439 illustrations by Lancelot Speed, Henry Ford and G. P. Jacomb Hood. Total of about 1500pp. 5⅜ x 8½.   23407-X Boxed set, Pa. $15.95

HOUSEHOLD STORIES BY THE BROTHERS GRIMM. All the great Grimm stories: "Rumpelstiltskin," "Snow White," "Hansel and Gretel," etc., with 114 illustrations by Walter Crane. 269pp. 5⅜ x 8½.
21080-4 Pa. $3.50

SLEEPING BEAUTY, illustrated by Arthur Rackham. Perhaps the fullest, most delightful version ever, told by C. S. Evans. Rackham's best work. 49 illustrations. 110pp. 7⅞ x 10¾. 22756-1 Pa. $2.50

AMERICAN FAIRY TALES, L. Frank Baum. Young cowboy lassoes Father Time; dummy in Mr. Floman's department store window comes to life; and 10 other fairy tales. 41 illustrations by N. P. Hall, Harry Kennedy, Ike Morgan, and Ralph Gardner. 209pp. 5⅜ x 8½. 23643-9 Pa. $3.00

THE WONDERFUL WIZARD OF OZ, L. Frank Baum. Facsimile in full color of America's finest children's classic. Introduction by Martin Gardner. 143 illustrations by W. W. Denslow. 267pp. 5⅜ x 8½.
20691-2 Pa. $3.50

THE TALE OF PETER RABBIT, Beatrix Potter. The inimitable Peter's terrifying adventure in Mr. McGregor's garden, with all 27 wonderful, full-color Potter illustrations. 55pp. 4¼ x 5½. (Available in U.S. only)
22827-4 Pa. $1.25

THE STORY OF KING ARTHUR AND HIS KNIGHTS, Howard Pyle. Finest children's version of life of King Arthur. 48 illustrations by Pyle. 131pp. 6⅛ x 9¼. 21445-1 Pa. $4.95

CARUSO'S CARICATURES, Enrico Caruso. Great tenor's remarkable caricatures of self, fellow musicians, composers, others. Toscanini, Puccini, Farrar, etc. Impish, cutting, insightful. 473 illustrations. Preface by M. Sisca. 217pp. 8⅜ x 11¼. 23528-9 Pa. $6.95

PERSONAL NARRATIVE OF A PILGRIMAGE TO ALMADINAH AND MECCAH, Richard Burton. Great travel classic by remarkably colorful personality. Burton, disguised as a Moroccan, visited sacred shrines of Islam, narrowly escaping death. Wonderful observations of Islamic life, customs, personalities. 47 illustrations. Total of 959pp. 5⅜ x 8½.
21217-3, 21218-1 Pa., Two-vol. set $12.00

INCIDENTS OF TRAVEL IN YUCATAN, John L. Stephens. Classic (1843) exploration of jungles of Yucatan, looking for evidences of Maya civilization. Travel adventures, Mexican and Indian culture, etc. Total of 669pp. 5⅜ x 8½. 20926-1, 20927-X Pa., Two-vol. set $7.90

AMERICAN LITERARY AUTOGRAPHS FROM WASHINGTON IRVING TO HENRY JAMES, Herbert Cahoon, et al. Letters, poems, manuscripts of Hawthorne, Thoreau, Twain, Alcott, Whitman, 67 other prominent American authors. Reproductions, full transcripts and commentary. Plus checklist of all American Literary Autographs in The Pierpont Morgan Library. Printed on exceptionally high-quality paper. 136 illustrations. 212pp. 9⅛ x 12¼. 23548-3 Pa. $12.50

THE AMERICAN SENATOR, Anthony Trollope. Little known, long un-available Trollope novel on a grand scale. Here are humorous comment on American vs. English culture, and stunning portrayal of a heroine/villainess. Superb evocation of Victorian village life. 561pp. 5⅜ x 8½.
23801-6 Pa. $6.00

WAS IT MURDER? James Hilton. The author of Lost Horizon and Good-bye, Mr. Chips wrote one detective novel (under a pen-name) which was quickly forgotten and virtually lost, even at the height of Hilton's fame. This edition brings it back—a finely crafted public school puzzle resplen-dent with Hilton's stylish atmosphere. A thoroughly English thriller by the creator of Shangri-la. 252pp. 5⅜ x 8. (Available in U.S. only)
23774-5 Pa. $3.00

CENTRAL PARK: A PHOTOGRAPHIC GUIDE, Victor Laredo and Henry Hope Reed. 121 superb photographs show dramatic views of Central Park: Bethesda Fountain, Cleopatra's Needle, Sheep Meadow, the Blockhouse, plus people engaged in many park activities: ice skating, bike riding, etc. Captions by former Curator of Central Park, Henry Hope Reed, provide historical view, changes, etc. Also photos of N.Y. landmarks on park's periphery. 96pp. 8½ x 11. 23750-8 Pa. $4.50

NANTUCKET IN THE NINETEENTH CENTURY, Clay Lancaster. 180 rare photographs, stereographs, maps, drawings and floor plans recreate unique American island society. Authentic scenes of shipwreck, light-houses, streets, homes are arranged in geographic sequence to provide walking-tour guide to old Nantucket existing today. Introduction, captions. 160pp. 8⅞ x 11¾. 23747-8 Pa. $6.95

STONE AND MAN: A PHOTOGRAPHIC EXPLORATION, Andreas Feininger. 106 photographs by Life photographer Feininger portray man's deep passion for stone through the ages. Stonehenge-like megaliths, forti-fied towns, sculpted marble and crumbling tenements show textures, beau-ties, fascination. 128pp. 9¼ x 10¾. 23756-7 Pa. $5.95

CIRCLES, A MATHEMATICAL VIEW, D. Pedoe. Fundamental aspects of college geometry, non-Euclidean geometry, and other branches of mathe-matics: representing circle by point. Poincare model, isoperimetric prop-erty, etc. Stimulating recreational reading. 66 figures. 96pp. 5⅝ x 8¼.
63698-4 Pa. $2.75

THE DISCOVERY OF NEPTUNE, Morton Grosser. Dramatic scientific history of the investigations leading up to the actual discovery of the eighth planet of our solar system. Lucid, well-researched book by well-known historian of science. 172pp. 5⅜ x 8½. 23726-5 Pa. $3.50

THE DEVIL'S DICTIONARY. Ambrose Bierce. Barbed, bitter, brilliant witticisms in the form of a dictionary. Best, most ferocious satire America has produced. 145pp. 5⅜ x 8½. 20487-1 Pa. $2.25

CATALOGUE OF DOVER BOOKS

AMERICAN ANTIQUE FURNITURE, Edgar G. Miller, Jr. The basic coverage of all American furniture before 1840: chapters per item chronologically cover all types of furniture, with more than 2100 photos. Total of 1106pp. 7⅞ x 10¾.   21599-7, 21600-4 Pa., Two-vol. set $17.90

ILLUSTRATED GUIDE TO SHAKER FURNITURE, Robert Meader. Director, Shaker Museum, Old Chatham, presents up-to-date coverage of all furniture and appurtenances, with much on local styles not available elsewhere. 235 photos. 146pp. 9 x 12.   22819-3 Pa. $6.00

ORIENTAL RUGS, ANTIQUE AND MODERN, Walter A. Hawley. Persia, Turkey, Caucasus, Central Asia, China, other traditions. Best general survey of all aspects: styles and periods, manufacture, uses, symbols and their interpretation, and identification. 96 illustrations, 11 in color. 320pp. 6⅛ x 9¼.   22366-3 Pa. $6.95

CHINESE POTTERY AND PORCELAIN, R. L. Hobson. Detailed descriptions and analyses by former Keeper of the Department of Oriental Antiquities and Ethnography at the British Museum. Covers hundreds of pieces from primitive times to 1915. Still the standard text for most periods. 136 plates, 40 in full color. Total of 750pp. 5⅝ x 8½.
23253-0 Pa. $10.00

THE WARES OF THE MING DYNASTY, R. L. Hobson. Foremost scholar examines and illustrates many varieties of Ming (1368-1644). Famous blue and white, polychrome, lesser-known styles and shapes. 117 illustrations, 9 full color, of outstanding pieces. Total of 263pp. 6⅛ x 9¼. (Available in U.S. only)   23652-8 Pa. $6.00

Prices subject to change without notice.

Available at your book dealer or write for free catalogue to Dept. GI, Dover Publications, Inc., 180 Varick St., N.Y., N.Y. 10014. Dover publishes more than 175 books each year on science, elementary and advanced mathematics, biology, music, art, literary history, social sciences and other areas.